INHERENTLY PARALLEL ALGORITHMS IN FEASIBILITY AND OPTIMIZATION AND THEIR APPLICATIONS

STUDIES IN COMPUTATIONAL MATHEMATICS 8

ELSEVIER
Amsterdam – London – New York – Oxford – Paris – Shannon – Tokyo

INHERENTLY PARALLEL ALGORITHMS IN FEASIBILITY AND OPTIMIZATION AND THEIR APPLICATIONS

Edited by

Dan BUTNARIU
Department of Mathematics
University of Haifa, Israel

Yair CENSOR
Department of Mathematics
University of Haifa, Israel

Simeon REICH
Department of Mathematics
The Technion – Israel Institute of Technology, Haifa, Israel

2001

ELSEVIER
Amsterdam – London – New York – Oxford – Paris – Shannon – Tokyo

ELSEVIER SCIENCE B.V.
Sara Burgerhartstraat 25
P.O. Box 211, 1000 AE Amsterdam, The Netherlands

First edition 2001

Library of Congress Cataloging in Publication Data
A catalog record from the Library of Congress has been applied for.

ISBN: 0-444-50595-4

∞ The paper used in this publication meets the requirements of ANSI/NISO Z39.48-1992 (Permanence of Paper).

Transferred to digital printing 2005

In memory of our colleague and friend

YAKI STERNFELD
1944 – 2001

D.B., Y.C. and S.R.

PREFACE

This book brings together twenty seven state-of-the-art, carefully refereed and subsequently revised, research and review papers in the field of parallel feasibility and optimization algorithms and their applications – with emphasis on *inherently parallel* algorithms. By this term we mean algorithms which are logically (i.e., in their mathematical formulations) parallel, not just parallelizable under some conditions, such as when the underlying problem is decomposable in a certain manner. As this volume shows, pure mathematical work in this field goes hand-in-hand with real-world applications and the mutual "technology transfer" between them leads to further progress.

The Israel Science Foundation, founded by the Israel Academy of Sciences and Humanities, recognizing the importance of the field and the need for interaction between theoreticians and practitioners, provided us with a special grant to organize a Research Workshop on *Inherently Parallel Algorithms in Feasibility and Optimization and Their Applications.* This Research Workshop was held jointly at the University of Haifa and the Technion–Israel Institute of Technology in Haifa, Israel, on March 13–16, 2000, with sessions taking place in both campuses.

Thirty five experts from around the world were invited and participated. They came from Argentina, Belgium, Brazil, Canada, Cyprus, France, Germany, Hungary, Israel, Japan, Norway, Poland, Russia, USA, and Venezuela. Most of the papers in this volume originated from the lectures presented at this Workshop while others were written in the wake of discussions held during the Workshop.

We thank the other members of the Scientific Committee of the Research Workshop, Lev Bregman (Beer-Sheba, Israel), Tommy Elfving (Linköping, Sweden), Gabor T. Herman (Philadelphia, PA, USA) and Stavros A. Zenios (Nicosia, Cyprus) for their cooperation. Many thanks are due to the referees whose high-quality work greatly enhanced the final versions of the papers which appear here. Last but not least, we thank the participants of the Research Workshop and the authors who contributed their work to this volume. We gratefully acknowledge the help of Production Editor Erik Oosterwijk from the Elsevier editorial office in Amsterdam.

Additional financial support was provided by the Institute of Advanced Studies in Mathematics at the Technion, the Research Authority and the Faculty of Social and Mathematical Sciences of the University of Haifa, and the Israel Mathematical Union. We appreciate very much this help as well as the organizational assistance of the staff members of the Technion and the University of Haifa, without which the Workshop could have never been the interesting and enjoyable event it was.

Dan Butnariu, Yair Censor and Simeon Reich
Haifa, Israel, March, 2001

CONTENTS

Inherently Parallel Algorithms in Feasibility and Optimization and their Applications
D. Butnariu, Y. Censor and S. Reich (Editors)

A LOG-QUADRATIC PROJECTION METHOD FOR CONVEX FEASIBILITY PROBLEMS

A. Auslender[a*] and Marc Teboulle[b†]

[a]Laboratoire d'Econometrie de L'Ecole Polytechnique,
1 Rue Descartes, Paris 75005, France

[b]School of Mathematical Sciences
Tel-Aviv University, Ramat-Aviv 69978, Israel

The convex feasibility problem which consists of finding a point in the intersection of convex sets is considered. We suggest a barycenter type projection algorithm, where the usual squared Euclidean distance is replaced by a logarithmic-quadratic distance-like functional. This allows in particular for handling efficiently feasibility problems arising in the nonnegative orthant. The proposed method includes approximate projections and is proven globally convergent under the sole assumption that the given intersection is nonempty and the errors are controllable. Furthermore, we consider the important special case involving the intersection of hyperplanes with the nonnegative orthant and show how in this case the projections can be efficiently computed via Newton's method.

1. INTRODUCTION

The convex feasibility (CF) problem consists of finding a point in the intersection of convex sets. It occurs in a wide variety of contexts within mathematics as well as in many applied sciences problems and particularly in image reconstruction problems. For nice surveys on CF problems, projections algorithms toward their solutions and an extensive bibliography we refer for example to [4] and [8]. Let C_i, $i = 1, \ldots, m$ be given closed convex set of $\mathbb{R}^p$ with a nonempty intersection $C := C_1 \cap \ldots \cap C_m$. The CF problem then consists of finding some point $x \in C$.
One the basic approach to solve the CF problem is simply to use projections onto each set C_i and to generate a sequence of points converging a solution of the CF problem. There are many ways of using projections to generate a solution to the CF problem, (see e.g., [4]. In this paper we will focus on the simple idea of averaging projections (barycenter method) which goes back at least to Cimmino [7] who has considered the special case when each C_i is a half-space. Later this method has been extended by Auslender [1], to general convex sets. Denoting by P_{C_i} the projection onto the set C_i the barycenter projection method consists of two main steps.

---

*Partially supported by the French-Israeli Scientific Program Arc-en-Ciel.
†Partially supported by the French-Israeli Scientific Program Arc-en-Ciel and The Israeli Ministry of Science under Grant No. 9636-1-96.

Barycenter Projection Method. Start with $x_0 \in \mathbb{R}^n$ and generate the sequence $\{x_k\}$ via:
(i) Project on $C_i : p_k^i = P_{C_i}(x_k), \ \forall i = 1, \ldots, m,$
(ii) Average Projections: $x_{k+1} = m^{-1} \sum_{i=1}^m p_k^i$.

More generally, in the averaging step, one can assign nonnegative weights such that $w_k^i \geq w > 0$ and replace step (ii) above by

$$x_{k+1} = (\sum_{i=1}^m (w_k^i)^{-1} p_k^i)/(\sum_{i=1}^m (w_k^i)^{-1}).$$

Clearly step (ii) above is then recovered with the choice $w_k^i = m^{-1}, \ \forall k$.

This basic algorithm produces a sequence $\{x_k\}$ which converges to a solution of the CF problem, provided the intersection C is assumed nonempty, see e.g., [1]. As already mentioned, there exists several other variants ([4], [8]) of projection types algorithms for solving the CF problem, but in order to make this paper self-contained and transparent we will focus only on the above idea of averaging projections. Yet, we would like to emphasize that the algorithm and results we developed below can be extended as well to many of these variants involving projections-based methods.

In many applications, the convex feasibility problem occurs in the nonnegative orthant, namely we search for a point $x \in C \cap \mathbb{R}_+^n$. In such cases, even when the C_i are simple hyperplanes we do not have an explicit formula for the projection $P_{C_i \cap \mathbb{R}_+^n}$, i.e., when intersecting the hyperplanes with the nonnegative orthant. In fact, in order to take into account the geometry of $C \cap \mathbb{R}_+^n$, it appears more natural to use projections not necessarily based on Euclidean quadratic squared distances, but rather on some other type of projections-like operators that which will be able to ensure automatically the positiveness of the required point solving CF in these cases. This has lead several authors to consider non-quadratic distance-like functions. Most common has been the use of Bregman type distances, see for example the work of Censor and Elfving [6] and references therein. Yet, when using non quadratic distance-like functions, the projections cannot be computed exactly or analytically. It is therefore required to build a nonorthogonal (nonlinear) projection algorithm which first allows to control the errors with inexact computations, and secondly allows for computing the projections efficiently, e.g., via Newton's method. Previous works using non-quadratic projections do not address these issues. Motivated by these recent approaches based on nonorthogonal Bregman type projections, see e.g., [5], [6], in this paper we also suggest a nonorthogonal projection algorithm for CF, which is based on a Logarithmic-Quadratic (LQ) distance-like functional and is not of the Bregman-type. As we shall see, the LQ functional and its associated conjugate functional which also plays a key role in nonlinear projections, enjoy remarkable and very important properties not shared by any other non-quadratic distance-like functions previously used and proposed in the literature. We refer the reader to [2], [3], where the LQ-distance like functional has been introduced, for further details explaining its advantages over Bregman type distances and Csizar's φ-divergence ([12]) when used for solving variational inequalities and convex programming problems. Further properties of the LQ functional are also given in Section 3. We derive a convergent projection algorithm, under very mild assumptions on the problems data and which address the two issues alluded above. We believe and hope that our contribution is a first step toward the development of various (other than barycentric type) efficient non-orthogonal projections methods for the convex feasibility problem arising in the non-negative orthant.

2. THE LOG-QUADRATIC PROJECTION

We begin by introducing the LQ distance-like functional used to generate the nonorthogonal projections. Let $\nu > \mu > 0$ be given fixed parameters, and define

$$\varphi(t) = \begin{cases} \frac{\nu}{2}(t-1)^2 + \mu(t - \log t - 1) & \text{if } t > 0 \\ +\infty & \text{otherwise} \end{cases} \tag{1}$$

Given φ as defined above , we define d_φ for $x, y \in \mathbb{R}^p_{++}$ by

$$d_\varphi(x, y) = \sum_{j=1}^{p} y_j^2 \varphi(x_j/y_j). \tag{2}$$

The functional d_φ with φ as defined in (1) was first introduced in [2] and enjoys the following basic properties:

- d_φ is an homogeneous function of order 2, i.e.,

  $$d_\varphi(\alpha x, \alpha y) = \alpha^2 d_\varphi(x, y), \ \forall \alpha > 0.$$

- $\forall (x, y) \in \mathbb{R}^p_{++} \times \mathbb{R}^p_{++}$ we have:

$$\begin{aligned} d_\varphi(x, y) &\geq 0 \text{ and,} \\ d_\varphi(x, y) &= 0 \text{ iff } x = y. \end{aligned} \tag{3}$$

The first property is obvious from the definition (1), while the second property follows from the strict convexity of φ and noting that $\varphi(1) = \varphi'(1) = 0$, which implies,

$\varphi(t) \geq 0$, and $\varphi(t) = 0$ if and only if $t = 1$.

We define the *Log-Quadratic* (LQ for short) projection onto $C \cap \mathbb{R}^n_+$, based on d_φ via: for each $y \in \mathbb{R}^p_{++}$,

$E_C(y) := \operatorname{argmin}\{d_\varphi(x, y) : \ x \in C \cap \mathbb{R}^p_+\}$.

Throughout this paper we make the following assumption: $C \cap \mathbb{R}^p_{++} \neq \emptyset$.
Note that this assumption is very minimal (in fact a standard constraint qualification) and is necessary to make the LQ projection meaningful. It will be useful to use the notation

$$\Phi'(a, b) := (a_1 \varphi'(b_1/a_1), \ldots, a_p \varphi'(b_p/a_p))^T \ \forall a, b, \in \mathbb{R}^p_{++} \tag{4}$$

where $\varphi'(t) = \nu(t-1) + \mu(1 - t^{-1}), \ t > 0$.

We begin with the following key lemma proven in [3, Lemma 3.4].

Lemma 2.1 *Let φ be given in (1). Then, for any a , $b \in \mathbb{R}^p_{++}$ and $c \in \mathbb{R}^p_+$, we have*

$$2\langle c-b, \Phi'(a,b)\rangle \leq (\nu+\mu)(||c-a||^2 - ||c-b||^2) - (\nu-\mu)||b-a||^2.$$

The next result can in fact be obtained as a consequence of the more general results established in Auslender-Teboulle-Bentiba, [3], but for completeness we include a direct and short proof.

Proposition 2.1 *Let C be a closed convex set of $\mathbb{R}^p$ such that $C \cap \mathbb{R}^p_{++} \neq \emptyset$. Then,*
(i) For each $y \in \mathbb{R}^p_{++}$ the projection $E_C(y)$ exists, is unique and one has $E_C(y) \in C \cap \mathbb{R}^p_{++}$.
(ii) For any $x \in C \cap \mathbb{R}^p_+$ it holds

$$\alpha||E_C(y) - y||^2 \leq ||x-y||^2 - ||x - E_C(y)||^2,$$

where we set $\alpha := (\nu-\mu)(\nu+\mu)^{-1}$.

Proof. For any $y \in \mathbb{R}^p_{++}$ the function $x \to d_\varphi(x,y)$ is strictly convex on $\mathbb{R}^p_{++}$ and thus if the minimum exists, it is unique. Since here one has φ cofinite, i.e., $\varphi_\infty(d) = +\infty$, $\forall d \neq 0$, (here φ_∞ denotes the recession function of φ, see [11]) it follows that the LQ projection $E_C(y)$ exists. Furthermore, since $\lim_{t\to 0^+} \varphi'(t) = -\infty$, and we assume the nonemptyness of $C \cap \mathbb{R}^p_{++}$, then it follows that $E_C(y) \in C \cap \mathbb{R}^p_{++}$ and is characterized by :

$$E_C(y) \in C \cap \mathbb{R}^p_{++},\ 0 \in \partial\delta(E_C(y)|C) + \Phi'(y, E_C(y)),$$

where $\delta(\cdot|C)$ denotes the indicator of the set C. But, since $\delta(\cdot|C) = N_C(\cdot)$, (the normal cone of the closed convex set C (see [11]), we then obtain

$$\langle \Phi'(y, E_C(y)), x - E_C(y)\rangle \geq 0,\ \ \forall x \in C \cap \mathbb{R}^p_+.$$

Invoking Lemma 1 with $a = y, b = E_C(y)$ and $c = x$ together with the above inequality we obtain the desired result (ii). □

It is interesting to remark that the above result essentially shows that LQ projections share properties of orthogonal projections yet eliminate the difficulty associated with the nonnegative constraint as it produces automatically projections in the positive orthant. Moreover, formally, by looking at the extreme case $\mu = 0$, the function φ defined by (1), (but now defined on the whole line $\mathbb{R}$) reduces to $\varphi(t) = \frac{\nu}{2}(t-1)^2$ and with this particular choice the corresponding d_φ is nothing else but

$$d_\varphi(x,y) = \frac{\nu}{2}||x-y||^2,$$

i.e., the usual squared Euclidean distance. Thus we also formally recover the well known orthogonal projection on C.

We are now considering the CF problem, namely, given a collection of closed convex sets C_i, we let $C = \cap_{i=1}^m C_i$ assumed nonempty and look for a point $x \in C \cap \mathbb{R}^p_+$. We suggest the following "approximate" barycentric method based on LQ projections.

The LQ Projection Method Let φ be defined in (1).

Let $\{\epsilon_k\}$ be a nonnegative sequence such that $\sum_{k=1}^{\infty} \epsilon_k < +\infty$. Start with $x_0 \in \mathbb{R}^p_{++}$ and generate iteratively the sequence $\{x_k\}$ by :

Step 1. *Compute approximate LQ projections.* For each $i = 1, \ldots, m$ compute x_k^i such that

$$x_k^i \in \mathbb{R}^p_{++}, \quad \epsilon_k^i = x_k^i - E_{C_i}(x_k), \quad \text{with } ||\epsilon_k^i|| \leq \epsilon_k.$$

Step 2. *Average:* $x_{k+1} = m^{-1} \sum_{i=1}^{m} x_k^i$.

Before proving the convergence of the LQ projection method, we recall the following result on nonnegative sequences (see Polyak [10] for a proof).

Lemma 2.2 *Let $a_k \geq 0$ such that $\sum_{k=1}^{\infty} a_k < \infty$. Then any nonnegative sequence $\{v_k\}$ satisfying $v_{k+1} \leq v_k + a_k$ is a convergent sequence.*

Theorem 2.1 *The sequence $\{x_k\}$ produced by the LQ projection method converges to a point $x \in C \cap \mathbb{R}^p_+$.*

Proof. Set $z_k^i = E_{C_i}(x_k)$ the exact LQ projection of $x_k \in \mathbb{R}^p_{++}$ on the set C_i. Using Proposition 2.1, (ii) we thus obtain for all $i = 1, \ldots, m$ and any $k \geq 0$,

$$\alpha ||z_k^i - x_k||^2 \leq ||x_k - x||^2 - ||x - z_k^i||^2, \quad \forall x \in C \cap \mathbb{R}^p_+,$$

which in turns implies that

$$||x - z_k^i||^2 \leq ||x_k - x||^2. \tag{5}$$

Now from step 2 of the algorithm, $x_{k+1} = m^{-1} \sum_{i=1}^{m} x_k^i$, then since $|| \cdot ||^2$ is convex we obtain,

$$\begin{aligned} ||x_{k+1} - x||^2 &= ||m^{-1} \sum_{i=1}^{m} (x_k^i - x)||^2 \\ &\leq m^{-1} \sum_{i=1}^{m} ||x_k^i - x||^2 \\ &= m^{-1} \sum_{i=1}^{m} ||z_k^i - x + \epsilon_k^i||^2 \\ &\leq m^{-1} \sum_{i=1}^{m} (||z_k^i - x||^2 + 2\epsilon_k ||z_k^i - x||) + \epsilon_k^2 &\quad (6) \\ &\leq (||x - x_k|| + \epsilon_k)^2, &\quad (7) \end{aligned}$$

where for each $i = 1, \ldots, m$, in the second equality we use $x_k^i = z_k^i + \epsilon_k^i$, in the second inequality the Cauchy-Schwartz inequality and $||\epsilon_k^i|| \leq \epsilon_k$ and in the last one we use the estimates (5). We have thus proved that $||x - x_{k+1}|| \leq ||x - x_k|| + \epsilon_k$ and therefore from Lemma 2.2, it follows that the sequence $\{||x - x_k||\}$ converges. Now, summing over $i = 1, \ldots, m$ the inequalities given at the beginning of the proof, we obtain $\forall x \in C \cap \mathbb{R}^p_+$:

$$m^{-1} \alpha \sum_{i=1}^{m} ||z_k^i - x_k||^2 \leq ||x_k - x||^2 - m^{-1} \sum_{i=1}^{m} ||x - z_k^i||^2,$$

and using (5) and (6) it follows that

$$m^{-1}\alpha\sum_{i=1}^{m}||z_k^i - x_k||^2 \le ||x_k - x||^2 - ||x_{k+1} - x||^2 + \epsilon_k^2 + 2\epsilon_k||x_k - x||^2.$$

But since we have $\epsilon_k \to 0$ and we proved that $\{||x_k - x||\}$ is a convergent sequence, we thus have:

$$\lim_{k\to+\infty} ||x_k - z_k^i|| = 0, \quad \forall i = 1, \dots, m.$$

To finish the proof, note that $\{x_k\}$ is also bounded and hence admits a limit point x_∞. Moreover, since by the exact LQ projection, $z_k^i \in C_i \cap \mathbb{R}^p_{++}$, it follows that $x_\infty \in C \cap \mathbb{R}^p_+$. Passing to subsequences if necessary and using a standard argument based on the properties of norms, since the sequence $\{||x_k - x||\}$ is convergent for each $x \in C \cap \mathbb{R}^p_+$ it follows that the whole sequence x_k converges to a point $x \in C \cap \mathbb{R}^p_+$. □

It is important to notice that the convergence result above indicates that while the algorithm is generating strictly positive iterates x_k, it globally converges to a point $x \in C \cap \mathbb{R}^p_+$ which can thus admits some of its components to be zeros.

3. AN IMPORTANT SPECIAL CASE: FURTHER ANALYSIS

In several applications we are often interested to solve the CF problem whenever the closed convex sets C_i are described for each i by hyperplanes $H_i = \{x \in \mathbb{R}^p : \langle a_i, x\rangle = b_i\}$, where $0 \ne a_i \in \mathbb{R}^p, b_i \in \mathbb{R}$. In this case, applying the LQ projection algorithm (for simplicity in exact form) to find a point in $\mathbb{R}^p_+ \cap (\cap_{i=1}^m H_i)$ thus leads to solve for each i the following problem: Given $y \in \mathbb{R}^p_{++}$ find

$$E_{H_i}(y) = \text{argmin}\{d_\varphi(x, y) : \langle a_i, x\rangle = b_i,\ x \ge 0\}.$$

Projections on an hyperplane which intersect with the nonnegative orthant are characterized in the following result.

Proposition 3.1 *Let φ be given in (1) and let H be the hyperplane $H = \{x \in \mathbb{R}^p : \langle a, x\rangle = b\}$ for some $0 \ne a \in \mathbb{R}^p$ and $b \in \mathbb{R}$. Assume that $H \cap \mathbb{R}^p_{++} \ne \emptyset$ and $y \in \mathbb{R}^p_{++}$. Then, $x = E_H(y)$ if and only if*

$$x \in \mathbb{R}^p_{++},\ \langle a, x\rangle = b,\ \ x_j = y_j(\varphi')^{-1}(\frac{\eta a_j}{y_j}),\ j = 1, \dots, p.$$

for some unique $\eta \in \mathbb{R}$.

Proof. Since dom$\varphi = \mathbb{R}^p_{++}$ and $\varphi'(t) \to -\infty$ as $t \to 0^+$, the result follows immediately by writing the Khun-Tucker optimality conditions or simply from Proposition 2.1. □

Note that the basic condition $H \cap \mathbb{R}^p_{++} \ne \emptyset$ is satisfied in most applications, e.g., in computerized tomography, since in these type of problems we have in fact $0 \ne a \in \mathbb{R}^p_+$ and $b > 0$, so that the nonemptyness condition on $H \cap \mathbb{R}^p_{++}$ holds.

To compute the projection $x = E_H(y)$, which by Proposition 2.1 is guaranteed to be in the positive orthant, we need to solve the one dimensional equation in η:

$$\langle a, x(\eta)\rangle = b, \quad (x(\eta) > 0), \tag{8}$$

where for each $j = 1, \ldots, p$,

$$x_j(\eta) = y_j(\varphi')^{-1}(\frac{\eta a_j}{y_j}) = y_j(\varphi^*)'(\frac{\eta a_j}{y_j}) > 0, \tag{9}$$

and φ^* denotes the conjugate of φ. A direct computation (see [3]) shows that :

$$\varphi^*(s) \quad = \quad \frac{\nu}{2}t^2(s) + \mu \log t(s) - \frac{\nu}{2}, \text{ where} \tag{10}$$

$$t(s) \quad := \quad (2\nu)^{-1}\{(\nu - \mu) + s + \sqrt{((\nu - \mu) + s)^2 + 4\mu\nu}\} = (\varphi^*)'(s) > 0. \tag{11}$$

The function φ^* possesses remarkable properties, rendering the solution of the one dimensional equation (8) by Newton's method an easy task. Indeed, the conjugate function enjoys the following properties. For simplicity we now set $\nu = 2, \mu = 1$.

Proposition 3.2 *Let φ be given by (1) with associated conjugate φ^* given in (10). Then,*
(i) dom$\varphi^* = \mathbb{R}$ and $\varphi^* \in C^\infty(\mathbb{R})$.
(ii) $(\varphi^*)'(s) = (\varphi')^{-1}(s)$ is Lipschitz for all $s \in \mathbb{R}$, with constant 2^{-1}.
(iii) $(\varphi^*)''(s) \leq 2^{-1}, \quad \forall s \in \mathbb{R}$.
(iv) φ^* and $(\varphi^*)'$ are strictly convex functions on $\mathbb{R}$.

Proof. The proofs of (i)-(iii) can be found in [3, Proposition 7.1]. To show (iv) let $\theta(s) := \varphi^*(s)$. We will verify that $\theta''(s) > 0, \ \theta'''(s) > 0, \ \forall s \in \mathbb{R}$. Using (10)-(11) (with $\nu = 2, \mu = 1$) we have for any $s \in \mathbb{R}$:

$$\theta(s) = \theta'^2(s) + \log \theta'(s) - 1.$$

Deriving the above identity with respect to s one obtains

$$\theta''(s) = (1 + 2\theta'^2(s))^{-1}\theta'^2(s) > 0, \quad \forall s \in \mathbb{R}. \tag{12}$$

Deriving one more time the later equation we obtain

$$\theta'''(s) = \frac{2\theta'(s)\theta''(s)}{(1 + 2\theta'^2(s))^2}, \tag{13}$$

showing that for all $s \in \mathbb{R}$, $\theta'''(s) > 0$ since $\theta'(s) > 0$. □

Since $(\varphi^*)'$ is strictly convex and monotone, we can thus apply efficiently Newton's method to solve the one dimensional equation in η (cf. (8):

$$\sum_{j=1}^{p} a_j y_j (\varphi^*)'(\frac{\eta a_j}{y_j}) = b. \tag{14}$$

As an alternative to Proposition 3.1, to compute the projection $E_H(y)$ we can use also the dual formulation. Thus, instead of solving the one dimensional equation (14) in η, we will have to solve a one dimensional optimization problem. It is straightforward to verify that the dual problem to compute the LQ projection on $H \cap \mathbb{R}^p_+$ is simply given by the strictly convex problem :

$$(D) \quad \min\{\sum_{j=1}^{p} y_j^2 \varphi^*(\frac{\eta a_j}{y_j}) - b\eta : \ \eta \in \mathbb{R}\}.$$

Given an optimal solution of the dual, which is unique, one then recovers the LQ projection through the formula (9).

The objective function of the dual optimization problem (D) has the remarkable property to be self-concordant, a key property needed to develop an efficient Newton type algorithm, see [9].

Lemma 3.1 *Let φ be given by (1). Then, the conjugate φ^* is self-concordant with parameter 2.*

Proof. As in the proof of Proposition 3.2 we set $\theta(s) = \varphi^*(s)$. By the definition of self-concordance (see [9]), one has to show that

$\theta'''(s) \le 2(\theta'')^{3/2}(s), \ \forall s \in \mathbb{R}$.

Using (13) and (12)we obtain for all $s \in \mathbb{R}$:

$$\begin{aligned}\frac{\theta'''(s)}{(\theta''(s))^{3/2}} &= \frac{2\theta'(s)\theta''(s)}{(1+2\theta'^2(s))^2}\frac{(1+2\theta'^2(s))^{3/2}}{\theta'^3(s)} \\ &= \frac{2\theta''(s)}{\theta'^2(s)}\frac{1}{(1+2\theta'^2(s))^{1/2}} = 2(\frac{\theta''(s)}{\theta'^2(s)})^{3/2}.\end{aligned}$$

But from (12) we also deduce that

$$\frac{\theta''(s)}{\theta'^2(s)} = 1 - 2\theta''(s),$$

and it follows from the last equation above that

$$\frac{\theta'''(s)}{(\theta''(s))^{3/2}} = 2(1-2\theta''(s))^{3/2} \le 2,$$

thus proving the desired inequality. □

Therefore, the objective function of the dual problem

$$F(\eta) := \sum_{j=1}^{p} y_j^2 \varphi^*(\frac{\eta a_j}{y_j}) - b\eta,$$

is a self-concordant $C^\infty(\mathbb{R})$ function, strictly convex and from Proposition 3.2 we have in addition that $F'(\eta)$ is Lipschitz with constant 2^{-1} and $F''(\eta) \le 2^{-1}$, $\forall\eta$. We thus have the best possible ingredients to solve the one dimensional optimization dual problem in a fast and most efficient way via Newton's method. This is important regarding the overall efficiency of the algorithm whenever m is very large, since the main step of the LQ projection algorithm, i.e., solving (D), has to be performed m times.

REFERENCES

1. A. Auslender, *Pour la Resolution des Problemes d'Optimisation avec contraintes* These de Doctorat, Faculty des Sciences de Grenoble (1969).
2. A. Auslender, M. Teboulle and S. Ben-Tiba, A Logarithmic-Quadratic Proximal Method for Variational Inequalities, *Computational Optimization and Applications* **12** (1999) 31–40.
3. A. Auslender, M. Teboulle and S. Ben-Tiba, Interior Proximal and Multiplier Methods based on Second Order Homogeneous Kernels, *Mathematics of Operations Research* **24** (1999) 645–668.
4. H. H. Bauschke and J. M. Borwein, On projection algorithms for solving convex feasibility problems, *SIAM Review* **38** (1996) 367–426.
5. H. H. Bauschke and J. M. Borwein, Legendre functions and the method of random Bregman projections, *Journal of Convex Analysis* **4** (1997) 27–67.
6. Y. Censor and T. Elfving, A multiprojection algorithm using Bregman projections in a product space, *Numerical Algorithms* **8** (1994) 221-239.
7. G. Cimmino, Calcolo appprssimato per le soluzioni dei sistemi di equazioni lineari, *La Ricerca Scientifica Roma* **1** (1938) 326–333.
8. P. L. Combettes, Hilbertian convex feasibility problem: convergence and projection methods, *Applied Mathematics and Optimization* **35** (1997) 311–330.
9. Y. Nesterov, A. Nemirovski, *Interior point polynomial algorithms in convex programming* (SIAM Publications, Philadelphia, PA, 1994).
10. B. T. Polyak, *Introduction to Optimization* (Optimization Software Inc., New York, 1987).
11. R. T. Rockafellar, *Convex Analysis* (Princeton University Press, Princeton, NJ, 1970).
12. M. Teboulle, Convergence of Proximal-like Algorithms, *SIAM J. of Optimization* **7** (1997), 1069-1083.

Inherently Parallel Algorithms in Feasibility and Optimization and their Applications
D. Butnariu, Y. Censor and S. Reich (Editors)

PROJECTION ALGORITHMS: RESULTS AND OPEN PROBLEMS

Heinz H. Bauschke[a]*

[a]Department of Mathematics and Statistics, Okanagan University College, Kelowna, British Columbia V1V 1V7, Canada.

In this note, I review basic results and open problems in the area of projection algorithms. My aim is to generate interest in this fascinating field, and to highlight the fundamental importance of bounded linear regularity.
Keywords: acceleration, alternating projections, bounded linear regularity, convex feasibility problem, cyclic projections, Fejér monotone sequence, metric regularity, orthogonal projection, projection algorithms, random projections.

1. INTRODUCTION

We assume throughout that

X is a real Hilbert space with inner product $\langle\cdot,\cdot\rangle$ and induced norm $\|\cdot\|$.

The first general projection algorithm — the *method of alternating projections* — was studied by John von Neumann in 1933:

Fact 1.1 (von Neumann). [38] Suppose C_1, C_2 are two closed subspaces in X with corresponding projections P_1, P_2. Let $C := C_1 \cap C_2$ and fix a starting point $x_0 \in X$. Then the *sequence of alternating projections* generated by

$$x_1 := P_1 x_0, x_2 := P_2 x_1, x_3 := P_1 x_2, \ldots$$

converges in norm to the projection of x_0 onto C.

In view of its conceptual simplicity and elegance, it is not surprising that Fact 1.1 has been generalized and rediscovered many times. (See the Deutsch's [26,25] for further information. Other algorithmic approaches are possible via generalized inverses [2].)

In this note, I consider some of the many generalizations of von Neumann's result, and discuss the intriguing open problems these generalizations spawned. My aim is to demonstrate that bounded linear regularity, *a quantitative geometric property of a collection of sets, is immensely useful and plays a crucial role in several results related to the open problems.*

*Research supported by NSERC.

The material is organized as follows.

In Section 2, we review basic convergence results by Halperin, by Bregman, and by Gubin, Polyak, and Raik. After recalling helpful properties of Fejér monotone sequences and of boundedly linearly regular collections of sets, we show how these notions work together in the proof a prototypical convergence result (Theorem 2.10).

Bounded linear regularity is reviewed in Section 3. Metric regularity, a notion ubiquitous in optimization, is shown to be genuinely stronger than bounded linear regularity. We also mention the beautiful relationship to conical open mapping theorems.

In the remaining sections, we discuss old and new open problems related to the inconsistent case (Section 4), to random projections (Section 5), and to acceleration (Section 6).

2. WEAK VS. NORM VS. LINEAR CONVERGENCE

It is very natural to try to generalize Fact 1.1 from two to finitely many subspaces. Israel Halperin achieved this, with a proof very different from von Neumann's, in 1962.

Fact 2.1 (Halperin). [31] Suppose $C_1, \ldots, C_N$ are finitely many closed subspaces in X with corresponding projections $P_1, \ldots, P_N$. If $C := \bigcap_{i=1}^N C_i$ and $x_0 \in X$, then the *sequence of cyclic projections*

$$x_1 := P_1 x_0, x_2 := P_2 x_1, \ldots, x_N := P_N x_{N-1}, x_{N+1} := P_1 x_N, \ldots$$

converges *in norm* to the projection of x_0 onto C.

See also Bruck and Reich's [18] and Baillon, Bruck, and Reich's [4] for various extensions of Halperin's result to uniformly convex Banach spaces and more general (not necessarily linear) mappings.

We assume from now on that

$$\boxed{C_1, \ldots, C_N \text{ are finitely many } (N \geq 2) \text{ closed convex sets with projections } P_1, \ldots, P_N,}$$

and

$$\boxed{C := \textstyle\bigcap_{i=1}^N C_i.}$$

For the reader's convenience, we recall some basic properties of projections.

Fact 2.2. Suppose S is a closed convex nonempty set in X, and $x \in X$. Then there exists a unique point in S, denoted $P_S x$ and called the *projection of x onto S*, with $\|x - P_S x\| = \min_{s \in S} \|x - s\| =: d(x, S)$. This point is characterized by

$$P_S x \in S \text{ and } \langle S - P_S x, x - P_S x \rangle \leq 0.$$

Moreover, P_S is *firmly nonexpansive*:

$$\|P_S x - P_S y\|^2 + \|(x - P_S x) - (y - P_S y)\|^2 \leq \|x - y\|^2;$$

in particular, P_S is *nonexpansive*:

$$\|P_S x - P_S y\| \leq \|x - y\|, \quad \forall x \in X, \forall y \in X.$$

What happens if we drop the assumption on linearity of the sets in Fact 2.1? In 1965, Lev Bregman proved the following fundamental result on a projection algorithm with general convex sets:

Fact 2.3 (Bregman). [17] If $C \neq \emptyset$, then the sequence of cyclic projections converges *weakly* to some point in C.

Remark 2.4. Several comments are in order.

(i) Many optimization problems can be recast as a *Convex Feasibility Problem*:

find $x \in C$.

For instance, *Linear Programming* has such a reformulation (in terms of primal feasibility, dual feasibility, and complementary slackness). Thus Bregman's result opened the door to solve important classes of convex optimization problems by using projection algorithms; see Censor and Zenios's recent monograph [19].

(ii) In contrast to Halperin's result, Bregman obtained weak convergence only.

(iii) Halperin's result identifies the limit of the sequence of cyclic projections as the point in C that is nearest to the starting point. Simple examples (consider two distinct but intersecting halfspaces in $\mathbb{R}^2$) show that cyclic projections may fail to converge to the nearest point in C.

Because of Remark 2.4.(ii), the following problem is of outstanding importance:

Open Problem 1. In Bregman's result (Fact 2.3), can the convergence be only weak?

Remark 2.5. During the Haifa workshop on inherently parallel algorithms, Hein Hundal announced the existence of two closed convex intersecting sets in ℓ_2 and a particular starting point where the method of alternating projections converges weakly but not in norm. (Full details were not available at the time of the writing of this note.)

However, it will always be interesting to know when the convergence is *linear*. Two disks in $\mathbb{R}^2$ that touch in exactly one point demonstrate that *something* is needed in order to guarantee linear convergence. This is best achieved by imposing bounded linear regularity:

Definition 2.6. [10] Suppose $C \neq \emptyset$. Then the collection $\{C_1, \ldots, C_N\}$ is *boundedly linearly regular*, if for every bounded set S in X, there exists $\kappa > 0$ such that

$$d(s, C) \leq \kappa \max\{d(s, C_1), \ldots, d(s, C_N)\}, \quad \forall s \in S.$$

If we can find $\kappa > 0$ so that the last inequality is true for every $s \in X$, then $\{C_1, \ldots, C_N\}$ is *linearly regular*. Finally, $\{C_1, \ldots, C_N\}$ is *boundedly regular*, if $d(y_n, C) \to 0$, for every bounded sequence (y_n) with $\max_{1 \leq i \leq N} d(y_n, C_i) \to 0$.

Clearly, linear regularity $\Rightarrow$ bounded linear regularity $\Rightarrow$ bounded regularity.

Gubin, Polyak, and Raik implicitly proved the next result in their fundamental paper from 1965:

Fact 2.7 (Gubin, Polyak, and Raik). [30] If $C \neq \emptyset$ and $\{C_1, \ldots, C_N\}$ is boundedly linearly regular, then the sequence of cyclic projections converges *linearly* to some point in C.

We now recall the definition of Fejér monotonicity.

Definition 2.8. Suppose S is a closed convex nonempty set in X, and $(y_n)_{n \geq 0}$ is a sequence in X. Then (y_n) is *Fejér monotone with respect to* S, if

$$\|y_{n+1} - s\| \leq \|y_n - s\|, \quad \forall n \geq 0, \forall s \in S.$$

Fejér monotone sequence have various pleasant properties; see [22,23], [10], and [6]. Here, we focus on characterizations of convergence, which will come handy when studying algorithms:

Fact 2.9. Suppose $(y_n)_{n \geq 0}$ is Fejér monotone with respect to a closed convex nonempty set S in X. Then:

(i) (y_n) is bounded, and $(d(y_n, S))$ is decreasing.

(ii) $P_S(y_n)$ converges in norm to some point $\bar{s} \in S$.

(iii) (y_n) converges weakly to $\bar{s} \Leftrightarrow$ all weak cluster points of (y_n) lie in S.

(iv) (y_n) converges in norm to $\bar{s} \Leftrightarrow d(y_n, S) \to 0$.

(v) (y_n) converges linearly to $\bar{s} \Leftrightarrow \exists\, \theta \in [0,1)$ with $d(y_{n+1}, S) \leq \theta d(y_n, S)$, $\forall n \geq 0$.

It is highly instructive to see the general structure of the proofs of Fact 2.3 and Fact 2.7. For clarity, we consider only alternating projections.

Theorem 2.10. (Prototypical Convergence Result) Suppose $N = 2$, $C = C_1 \cap C_2 \neq \emptyset$, and $(x_n)_{n \geq 0}$ is a sequence of alternating projections. Then (x_n) is Fejér monotone with respect to C, and

$$\max\{d^2(x_n, C_1), d^2(x_n, C_2)\} \leq d^2(x_n, C) - d^2(x_{n+1}, C), \quad \forall n \geq 0. \tag{$*$}$$

Let $\bar{c} = \lim_n P_C(x_n)$. Then:

(i) (x_n) *always* converges *weakly* to $\bar{c}$.

(ii) If $\{C_1, C_2\}$ is *boundedly regular*, then (x_n) converges *in norm* to $\bar{c}$.

(iii) If $\{C_1, C_2\}$ is *boundedly linearly regular*, then (x_n) converges *linearly* to $\bar{c}$.

(iv) If $\{C_1, C_2\}$ is *linearly regular*, then (x_n) converges *linearly* to $\bar{c}$ with a *rate independent of the starting point.*

Proof. Using (firm) nonexpansiveness of projections (Fact 2.2), we obtain easily $(*)$ and Fejér monotonicity of (x_n). Now the right-hand side of $(*)$ tends to 0 (use Fact 2.9.(i)); hence, so does the left-hand side of $(*)$, and its square root:

$$\max\{d(x_n, C_1), d(x_n, C_2)\} \to 0. \qquad (**)$$

(i): Suppose x is an arbitrary weak cluster point of (x_n). Use $(**)$ and the weak lower semicontinuity of $d(\cdot, C_1), d(\cdot, C_2)$ to conclude that $d(x, C_1) = d(x, C_2) = 0$. Thus $x \in C = C_1 \cap C_2$, and we are done by Fact 2.9.(iii).

(ii): By $(**)$ and bounded regularity, $d(x_n, C) \to 0$. Apply Fact 2.9.(iv).

(iii): The set $\{x_n : n \geq 0\}$ is bounded (Fact 2.9.(i)). Bounded linear regularity yields $\kappa > 0$ such that

$$d(x_n, C) \leq \kappa \max\{d(x_n, C_1), d(x_n, C_2)\}, \quad \forall n \geq 0.$$

Square this, and combine with "κ^2 times $(*)$" to get

$$d^2(x_n, C) \leq \kappa^2 \big(d^2(x_n, C) - d^2(x_{n+1}, C)\big);$$

equivalently, after re-arranging, $d(x_{n+1}, C) \leq \sqrt{1 - 1/\kappa^2} d(x_n, C)$. (Note that κ must be greater than or equal to 1 — because C is a subset of both C_1 and C_2 — and so the radical is always nonnegative.) The result follows now from Fact 2.9.(v).

(iv): is analogous to (iii). □

Remark 2.11. A second look at the proof of Theorem 2.10 reveals that only Fejér monotonicity and $(*)$ is required to arrive at the conclusions.

We conclude this section by noting that all results have analogues for much broader classes of algorithms, including various *parallel schemes*. As a starting point to the vast literature, we refer the reader to [21,20], [34], [35], [36], [37], and [6].

3. BOUNDED LINEAR REGULARITY

We now collect the facts on bounded linear regularity that are most relevant to us.

Fact 3.1. Suppose $N = 2$ and $C = C_1 \cap C_2 \neq \emptyset$. Then $\{C_1, C_2\}$ is boundedly linearly regular whenever one of the following conditions holds.

(i) $C_1 \cap \operatorname{int} C_2 \neq \emptyset$;

(ii) $0 \in \operatorname{int}(C_2 - C_1)$;

(iii) C_1, C_2 are subspaces and $C_1 + C_2$ is closed;

(iv) $\{r(c_2 - c_1) : r \geq 0, c_1 \in C_1, c_2 \in C_2\}$ is a closed subspace.

Proof. (i): [30]. (ii), (iii), and (iv): [8]. □

Condition (iv) of Fact 3.1 subsumes, in fact, conditions (i)—(iii). We now turn to the general case.

Fact 3.2. Suppose $N \geq 2$ and $C = \bigcap_{i=1}^{N} C_i \neq \emptyset$. Then $\{C_1, \ldots, C_N\}$ is boundedly linearly regular whenever one of the following conditions holds.

(i) *reduction to two sets*: each $\{C_1 \cap \cdots C_i, C_{i+1}\}$ is boundedly linearly regular;

(ii) *standard constraint qualification*: $X = \mathbb{R}^M$, $\bigcap_{i=1}^{r} \mathrm{ri}(C_i) \cap \bigcap_{i=r+1}^{N} C_i \neq \emptyset$, and the sets $C_{r+1}, \ldots, C_N$ are polyhedral, for some $0 \leq r \leq N$;

(iii) *for subspaces*: each C_i is a subspace, and $C_1^{\perp} + \cdots + C_N^{\perp}$ is closed;

(iv) *Hoffman's inequality*: each C_i is a polyhedron.

Proof. (i): [10, Theorem 5.11]. (ii): [6, Theorem 5.6.2] or [12, Corollary 5]. (iii): [10, Theorem 5.19]. (iv): essentially [32]; see also [6, Corollary 5.7.2]. □

Remark 3.3. • Combining Fact 3.1 and Fact 3.2 with Fact 2.7 yields a number of classical results on cyclic projections. (See [6, Section 9.5].) • However, this approach is incapable of recovering the basic results by von Neumann (Fact 1.1) and by Halperin (Fact 2.1). (Those results are best derived via *Dykstra's algorithm*; see [16] for this beautiful and powerful method. In contrast to the method of cyclic projections (Remark 2.4.(iii)), Dykstra's algorithm yields the projection of the starting point onto C and thus a well-recognized and most useful limit.)

Fact 3.4. [8, Theorem 5.3.(iv)] In $X := \ell_2$, let C_1 be the cone of nonnegative sequences and C_2 be an arbitrary hyperplane with $C = C_1 \cap C_2 \neq \emptyset$. Then the sequence of alternating projections always converges *in norm*.

The proof of Fact 3.4 considers two cases: presence and absence of bounded (linear) regularity. In the latter case, an order-theoretic argument yields norm, but not linear, convergence. It would be interesting to find out how to tackle the following, closely related problem:

Open Problem 2. What happens in Fact 3.4 if replace C_2 by an affine subspace of finite codimension?

Bounded linear regularity vs metric regularity

The notion of metric regularity of a set-valued map is well-known. How does it compare to bounded linear regularity? Metric regularity turns out to be a much stronger property.

Definition 3.5. Suppose $C = \bigcap_{i=1}^{N} C_i \neq \emptyset$. Then $\{C_1, \ldots, C_N\}$ is *metrically regular* if for every $c \in C$, there exists $K > 0$ such that for every $x \in X$ sufficiently close to c and for all $(b_1, \ldots, b_N) \in X^N$ sufficiently close to $0 \in X^N$, we have

$$d\big(x, (C_1 - b_1) \cap \cdots \cap (C_N - b_N)\big) \leq K\big(d(x + b_1, C_1) + \cdots + d(x + b_N, C_N)\big).$$

Remark 3.6. In the language of set-valued analysis, Definition 3.5 precisely states that the set-valued map $\Omega(x) := (C_1 \times \cdots \times C_N) - (x, \ldots, x)$ is metrically regular on $C \times (0, \ldots, 0) \subseteq X \times X^N$. See Ioffe's [33] for further information.

Theorem 3.7. If $C \neq \emptyset$ and $\{C_1, \ldots, C_N\}$ is metrically regular, then $\{C_1, \ldots, C_N\}$ is boundedly linearly regular.

Proof. We give details for $N = 2$ and the general case; the former is somewhat simpler.

Special case $N = 2$: The inequality of Definition 3.5 holds for all (b_1, b_2) sufficiently close to $(0, 0)$; in particular, the left-hand side is finite for such (b_1, b_2). Hence there exists $\delta > 0$ such that

$$(C_1 - b_1) \cap (C_2 - b_2) \neq \emptyset, \quad \text{whenever } \max\{\|b_1\|, \|b_2\|\} \leq \delta.$$

Now fix $b \in X$ with $\|b\| \leq \delta$ and set $b_1 := b$ and $b_2 := 0$. By the above, there exist $c_1 \in C_1$ and $c_2 \in C_2$ such that $c_1 - b_1 = c_2 - b_2$, or $b = c_1 - c_2 \in C_1 - C_2$. Denote the unit ball $\{x \in X : \|x\| \leq 1\}$ by B_X. Since b has been chosen arbitrarily in δB_X, it follows that $\delta B_X \subseteq C_1 - C_2$. Thus $0 \in \operatorname{int}(C_1 - C_2)$ and therefore $\{C_1, C_2\}$ is boundedly linearly regular by Fact 3.1.(ii).

General case $N \geq 2$: We work in $\mathbf{X} := X^N$ with $\mathbf{C} := C_1 \times \cdots \times C_N$ and $\boldsymbol{\Delta} := \{\mathbf{x} = (x_i) \in \mathcal{X} : x_1 = \cdots = x_N\}$.

Claim: $\exists \rho > 0$ such that $\rho B_{\mathbf{X}} \subseteq \mathbf{C} - \boldsymbol{\Delta}$.

(This follows from a general result on metric regularity; see [33, Proposition 5.2]. We repeat the argument here for the reader's convenience.)

Fix an arbitrary $c \in C$. The set-valued map Ω from Remark 3.6 is metrically regular at $(c, \mathbf{0})$. Hence there is $K > 0$ and $\delta > 0$ such that

$$d\big(c, \Omega^{-1}(\mathbf{b})\big) \leq K d(\mathbf{b}, \Omega(c)), \quad \text{for all } \|\mathbf{b}\| \leq \delta.$$

Let $0 < \rho < \min\{\delta, \delta/K\}$ and fix an arbitrary $b \in \rho B_{\mathbf{X}}$. Since $\|\mathbf{b}\| \leq \delta$, we have

$$d\big(c, \Omega^{-1}(\mathbf{b})\big) \leq K d(\mathbf{b}, \Omega(c)) \leq K\|\mathbf{b} - \mathbf{0}\| \leq K\rho < \delta.$$

Hence there exists $x \in \Omega^{-1}(\mathbf{b})$ such that $\|c - x\| < \delta$. In particular, $\mathbf{b} \in \Omega(x) \subseteq \operatorname{ran} \Omega$. The Claim is thus proven.

The Claim and Fact 3.1.(ii) yield bounded linear regularity of $\{\mathbf{C}, \boldsymbol{\Delta}\}$ in $\mathbf{X}$. We now argue as in the proof of [10, Theorem 5.19.(i)$\Rightarrow$(ii)] to conclude that $\{C_1, \ldots, C_N\}$ is boundedly linearly regular. ☐

Remark 3.8. A finite collection of zero subspaces in $X := \mathbb{R}$ shows that the converse of Theorem 3.7 is false in general. We thus observe that metric regularity is a genuinely more restrictive property than bounded linear regularity. In passing, we mention that Deutsch et al.'s property *strong CHIP* is genuinely less restrictive than bounded linear regularity. See [13] for details and further information.

Bounded linear regularity and the conical Open Mapping Theorem

In [12], bounded linear regularity was put in the broader context of convex optimization. Perhaps surprisingly, bounded linear regularity also relates to the classical Open Mapping Theorem. Since the latter is of general appeal, we now describe the connection.

Definition 3.9. [7] Suppose K is a closed convex cone in X and $T : X \to Y$ is a bounded linear operator to another real Hilbert space Y. Then T is *open relative to* K, if there exists $\delta > 0$ such that $\delta B_Y \cap T(K) \subseteq T(B_X \cap K)$.

Fact 3.10. (Conical Open Mapping Theorem) [7] Suppose K is a closed convex cone in X and $T : X \to Y$ is a bounded linear operator to another real Hilbert space Y. In $X \times Y$, set

$$K_1 := X \times 0, K_2 := \text{gra}(T|_K),$$

i.e., K_2 is the graph of the restriction of T to K, and then take negative polar cones:

$$C_1 = K_1^{\ominus}, C_2 = K_2^{\ominus}.$$

Then T is open relative to $K \Leftrightarrow \{C_1, C_2\}$ is boundedly linearly regular.

Remark 3.11. Suppose $T : X \to Y$ is also onto. We obtain the classical Open Mapping Theorem through the following equivalences: T is an open mapping $\Leftrightarrow$ T is open relative to $X \Leftrightarrow \{(X \times 0)^{\ominus}, (\text{gra}\, T)^{\ominus}\}$ is boundedly linearly regular (Fact 3.10) $\Leftrightarrow \{(X \times 0)^{\perp}, (\text{gra}\, T)^{\perp}\}$ is boundedly linearly regular $\Leftrightarrow (X \times 0) + \text{gra}\, T$ is closed (Fact 3.2.(iv)) $\Leftrightarrow X \times Y$ is closed (since T is onto), which is clearly true!

4. THE INCONSISTENT CASE

What happens if we apply the method of alternating projections to two (possibly non-intersecting) sets? This question is of great importance in applications (for instance, in Medical Imaging), where measurement errors may render the corresponding convex feasibility problem inconsistent.

The answer to this question is — up to the weak vs norm convergence problem (see Open Problem 1) — known:

Fact 4.1. (Dichotomy) [9] Suppose $N := 2$ and let $v := P_{\text{cl}(C_2 - C_1)}(0)$. Then $x_{2n} - x_{2n-1} \to v$ and $x_{2n} - x_{2n+1} \to v$. Either $v \notin C_2 - C_1$ and $\|x_n\| \to +\infty$, or $v \in C_2 - C_1$ and

$$(x_{2n+1}) \text{ converges weakly to } e_1, \quad \text{and} \quad (x_{2n}) \text{ converges weakly to } e_2 = e_1 + v,$$

for some $e_1 \in E_1 = \{c_1 \in C_1 : d(c_1, C_2) = \|v\|\}$, and $e_2 \in E_2 = \{c_2 \in C_2 : d(c_2, C_1) = \|v\|\}$, where E_1, E_2 are closed convex sets with $E_1 + v = E_2$.

It would be highly satisfying to settle the following:

Open Problem 3. What is the behavior of the sequence of cyclic projections for three or more sets with empty intersection?

Some positive results are available, but the geometry of the problem is far from being understood. The crux of the problem is this: the vector v in Fact 4.1 is well-defined and important for the formulation of the behavior of the sequence of alternating projections even when the distance $d(C_1, C_2) = \inf\{\|c_1 - c_2\| : c_1 \in C_1, c_2 \in C_2\}$ is not attained, i.e., $v \notin C_2 - C_1$; however, it is not known how to define the vector(s) analogous to v in the N-set case. See [11] for an in-depth survey on the state of this problem.

5. RANDOM PROJECTIONS

In this section, we assume that

$$\boxed{C = \bigcap_{i=1}^{N} C_i \neq \emptyset} \text{ and } \boxed{r : \{1, 2, \ldots\} \to \{1, 2, \ldots, N\} \quad \text{is a } \textit{random map},}$$

i.e., it is onto and assumes every value infinitely often. We are interested in the *method of random projections*, i.e., the behavior of the sequence (z_n) defined by

$$\boxed{z_0 \in X, \qquad z_{n+1} := P_{r(n+1)} z_n, \qquad \text{for } n \geq 0.}$$

If we let r be the "mod N" function (with remainders in $\{1, \ldots, N\}$), then the sequence (z_n) is precisely a sequence of cyclic projections. In general, we just "roll a die" — this "N-die" could be unfair, but not to an extent where one set would be ignored eventually.

In 1965, Amemiya and Ando proved the following fundamental result.

Fact 5.1 (Amemiya and Ando). [1] Suppose each C_i is a subspace. Then the sequence of random projections converges *weakly* to a point in C.

For extensions to more general Banach spaces and mappings; see Dye, Khamsi, and Reich's [27] and Dye and Reich's [28].

In contrast to Fact 2.1, only weak convergence is asserted in Fact 5.1 — thus we are forced to ask the obvious question:

Open Problem 4. In Fact 5.1, can the convergence be only weak?

The situation is even less clear for general closed convex sets:

Open Problem 5. Does the sequence of random projections converge weakly to some point in C?

Remark 5.2. We point out that the case $N = 2$, which shaped our intuition earlier, is not helpful at all for these problems: indeed, since projections are idempotents, a sequence of random projections is essentially a sequence of alternating projections. Thus the answer to Open Problem 4 is a resounding "No" because of von Neumann's Fact 1.1, whereas Bregman's Fact 2.3 yields an affirmative answer to Open Problem 5.

In 1992, Dye and Reich showed that — for *three* sets — Open Problem 5 does have an affirmative answer:

Fact 5.3 (Dye and Reich). [28] If $N = 3$, then the sequence of random projections converges weakly to some point in C.

The reader is referred to Baillon and Bruck's [3] for further references on the random projection problem as well as the following stunning

Conjecture (Baillon and Bruck). [3] Suppose each C_i is a subspace. No matter what the random map r and the starting point z_0 is, there is a constant $K > 0$, depending only on N, such that

$$\|z_n - z_{n+l}\|^2 \leq K\big(\|z_n\|^2 - \|z_{n+l}\|^2\big), \quad \forall n \geq 0, \forall l \geq 0.$$

In fact, $K \leq \binom{N}{N/2}$.

If this conjecture is true, then every sequence of random projections is Cauchy in the subspace case; consequently, Open Problem 4 would be resolved.

Here is a positive result for the general case:

Fact 5.4. [5] Suppose $\{C_i : i \in I\}$ is boundedly regular, for all $\emptyset \neq I \subseteq \{1, \ldots, N\}$. Then each sequence of random projections converges *in norm* to a point in C.

The combination of Fact 5.4 with Fact 3.2.(iii) leads to a quite flexible norm convergence result in the subspace case.

6. ACCELERATION

In this last section, we assume that

$$\boxed{\text{each } C_i \text{ is a subspace, and } T = P_N P_{N-1} \cdots P_1.}$$

We now turn to an acceleration scheme that was first explicitly suggested by Gearhart and Koshy [29] in 1989. (It is, however, already implicit in the classical paper by Gubin, Polyak, and Raik [30], and closely related to work by Dax [24].)

Define

$$\boxed{A : X \to X : x \mapsto (1 - t_x)x + t_x Tx,}$$

where $t_x \in \mathbb{R}$ is chosen so that $\|A(x)\|$ is minimal. (There is a simple closed form for t_x.)

Then consider the sequence with

$$\boxed{\text{starting point } z_0 \in X, \text{ and } z^n := A^{n-1}(Tz_0), \text{ for all } n \geq 1.}$$

Fact 6.1. [15,14]

(i) (z_n) *always* converges *weakly* to $P_C z_0$.

(ii) (z_n) converges *in norm*, if $N = 2$.

(iii) (z_n) converges *linearly*, if $\{C_1, \ldots, C_N\}$ is boundedly linearly regular.

Once again, bounded linear regularity played a crucial role! We conclude with one last question:

Open Problem 6. Can (z_n) fail to converge in norm when $N \geq 3$ or when $\{C_1, \ldots, C_N\}$ is not boundedly linearly regular?

ACKNOWLEDGMENT

I wish to thank Adi Ben-Israel, Achiya Dax, Simeon Reich, and an anonymous referee for helpful comments and pointers to various articles.

REFERENCES

1. I. Amemiya and T. Andô. Convergence of random products of contractions in Hilbert space. *Acta Sci. Math. (Szeged)*, **26**:239–244, 1965.
2. W. N. Anderson, Jr. and R. J. Duffin. Series and parallel addition of matrices. *J. Math. Anal. Appl.*, **26**:576–594, 1969.
3. J.-B. Baillon and R. E. Bruck. On the random product of orthogonal projections in Hilbert space. 1998. Available at `http://mathlab.usc.edu/~bruck/research/papers/dvi/naca98.dvi`.
4. J. B. Baillon, R. E. Bruck, and S. Reich. On the asymptotic behavior of nonexpansive mappings and semigroups in Banach spaces. *Houston J. Math.*, **4**:1–9, 1978.
5. H. H. Bauschke. A norm convergence result on random products of relaxed projections in Hilbert space. *Trans. Amer. Math. Soc.*, **347**:1365–1373, 1995.
6. H. H. Bauschke. *Projection algorithms and monotone operators.* PhD thesis, 1996. Available at `http://www.cecm.sfu.ca/preprints/1996pp.html`.
7. H. H. Bauschke and J. M. Borwein. Conical open mapping theorems and regularity. Proceedings of the Centre for Mathematics and its Applications (Australian National University). National Symposium on Functional Analysis, Optimization and Applications, March 1998.
8. H. H. Bauschke and J. M. Borwein. On the convergence of von Neumann's alternating projection algorithm for two sets. *Set-Valued Anal.*, **1**:185–212, 1993.
9. H. H. Bauschke and J. M. Borwein. Dykstra's alternating projection algorithm for two sets. *J. Approx. Theory*, **79**:418–443, 1994.
10. H. H. Bauschke and J. M. Borwein. On projection algorithms for solving convex feasibility problems. *SIAM Rev.*, **38**:367–426, 1996.
11. H. H. Bauschke, J. M. Borwein, and A. S. Lewis. The method of cyclic projections for closed convex sets in Hilbert space. In *Recent developments in optimization theory and nonlinear analysis (Jerusalem, 1995)*, pages 1–38. Amer. Math. Soc., Providence, RI, 1997.
12. H. H. Bauschke, J. M. Borwein, and W. Li. Strong conical hull intersection property, bounded linear regularity, Jameson's property (G), and error bounds in convex optimization. *Math. Program. (Ser. A)*, **86**:135–160, 1999.
13. H. H. Bauschke, J. M. Borwein, and P. Tseng. Bounded linear regularity, strong CHIP, and CHIP are distinct properties. To appear in *J. Convex Anal.*
14. H. H. Bauschke, F. Deutsch, H. Hundal, and S.-H. Park. Accelerating the convergence of the method of alternating projections. 1999. Submitted. Available at `http://www.cecm.sfu.ca/preprints/1999pp.html`.
15. H. H. Bauschke, F. Deutsch, H. Hundal, and S.-H. Park. Fejér monotonicity and weak convergence of an accelerated method of projections. In *Constructive, Experimental, and Nonlinear Analysis (Limoges, 1999)*, pages 1–6. Canadian Math. Soc. Conference Proceedings Volume **27**, 2000.
16. J. P. Boyle and R. L. Dykstra. A method for finding projections onto the intersection of convex sets in Hilbert spaces. In *Advances in order restricted statistical inference (Iowa City, Iowa, 1985)*, pages 28–47. Springer, Berlin, 1986.
17. L. M. Bregman. The method of successive projection for finding a common point of

convex sets. *Soviet Math. Dokl.*, **6**:688–692, 1965.
18. R. E. Bruck and S. Reich. Nonexpansive projections and resolvents of accretive operators in Banach spaces. *Houston J. Math.*, **3**:459–470, 1977.
19. Y. Censor and S. A. Zenios. *Parallel optimization.* Oxford University Press, New York, 1997.
20. P. L. Combettes. *The Convex Feasibility Problem in Image Recovery*, volume 95 of *Advances in Imaging and Electron Physics*, pages 155–270. Academic Press, 1996.
21. P. L. Combettes. Hilbertian convex feasibility problem: convergence of projection methods. *Appl. Math. Optim.*, **35**:311–330, 1997.
22. P. L. Combettes. Fejér-monotonicity in convex optimization. In C. A. Floudas and P. M. Pardalos, editors, *Encyclopedia of Optimization.* Kluwer, 2000.
23. P. L. Combettes. Quasi-Fejérian analysis of some optimization algorithms. In D. Butnariu, Y. Censor, and S. Reich, editors, *Inherently Parallel Algorithms in Feasibility and Optimization and their Applications (Haifa, 2000).* To appear.
24. A. Dax. Line search acceleration of iterative methods. *Linear Algebra Appl.*, **130**:43–63, 1990. Linear algebra in image reconstruction from projections.
25. F. Deutsch. *Best approximation in inner product spaces.* Monograph. To appear.
26. F. Deutsch. The method of alternating orthogonal projections. In *Approximation theory, spline functions and applications (Maratea, 1991)*, pages 105–121. Kluwer Acad. Publ., Dordrecht, 1992.
27. J. Dye, M. A. Khamsi, and S. Reich. Random products of contractions in Banach spaces. *Trans. Amer. Math. Soc.*, **325**:87–99, 1991.
28. J. M. Dye and S. Reich. Unrestricted iterations of nonexpansive mappings in Hilbert space. *Nonlinear Anal.*, **18**:199–207, 1992.
29. W. B. Gearhart and M. Koshy. Acceleration schemes for the method of alternating projections. *J. Comput. Appl. Math.*, **26**:235–249, 1989.
30. L. G. Gubin, B. T. Polyak, and E. V. Raik. The method of projections for finding the common point of convex sets. *Comput. Math. Math. Phys.*, **7**:1–24, 1967.
31. I. Halperin. The product of projection operators. *Acta Sci. Math. (Szeged)*, **23**:96–99, 1962.
32. A. J. Hoffman. On approximate solutions of systems of linear inequalities. *J. Research Nat. Bur. Standards*, **49**:263–265, 1952.
33. A. D. Ioffe. Codirectional compactness, metric regularity and subdifferential calculus. In *Constructive, Experimental, and Nonlinear Analysis (Limoges, 1999)*, pages 123–163. Canadian Math. Soc. Conference Proceedings Volume **27**, 2000.
34. K. C. Kiwiel. Block-iterative surrogate projection methods for convex feasibility problems. *Linear Algebra Appl.*, **215**:225–259, 1995.
35. K. C. Kiwiel and B. Łopuch. Surrogate projection methods for finding fixed points of firmly nonexpansive mappings. *SIAM J. Optim.*, **7**:1084–1102, 1997.
36. Y. I. Merzlyakov. On a relaxation method of solving systems of linear inequalities. *Comput. Math. Math. Phys.*, **2**:504–510, 1963.
37. S. Reich. A limit theorem for projections. *Linear and Multilinear Algebra*, **13**:281–290, 1983.
38. J. von Neumann. *Functional Operators. II. The Geometry of Orthogonal Spaces.* Princeton University Press, Princeton, N. J., 1950. Annals of Math. Studies, no. **22**.

Inherently Parallel Algorithms in Feasibility and Optimization and their Applications
D. Butnariu, Y. Censor and S. Reich (Editors)

JOINT AND SEPARATE CONVEXITY OF THE BREGMAN DISTANCE

Heinz H. Bauschke[a*] and Jonathan M. Borwein[b†]

[a]Department of Mathematics and Statistics, Okanagan University College, Kelowna, British Columbia V1V 1V7, Canada.

[b]Centre for Experimental and Constructive Mathematics, Simon Fraser University, Burnaby, British Columbia V5A 1S6, Canada.

Algorithms involving Bregman projections for solving optimization problems have been receiving much attention lately. Several of these methods rely crucially on the joint convexity of the Bregman distance. In this note, we study joint and separate convexity of Bregman distances.

To bring out the main ideas more clearly, we consider first functions defined on an open interval. Our main result states that the Bregman distance of a given function is jointly convex if and only if the reciprocal of its second derivative is concave. We observe that Bregman distances induced by the two most popular choices — the *energy* and the *Boltzmann-Shannon entropy* — are limiting cases in a profound sense. This result is generalized by weakening assumptions on differentiability and strict convexity.

We then consider general, not necessarily separable, convex functions. The characterization of joint convexity has a natural and beautiful analog. Finally, we discuss spectral functions, where the situation is less clear. Throughout, we provide numerous examples to illustrate our results.
Keywords: Bregman distance, convex function, joint convexity, separate convexity.

1. INTRODUCTION

Unless stated otherwise, we assume throughout that

$$\boxed{I \text{ is a nonempty open interval in } \mathbb{R}, \text{ and that } f \in C^3(I) \text{ with } f'' > 0 \text{ on } I.}$$

Clearly, f is strictly convex. Associated with f is the *Bregman "distance"* D_f:

$$\boxed{D_f : I \times I \to [0, +\infty) : (x, y) \mapsto f(x) - f(y) - f'(y)(x - y).}$$

For further information on Bregman distances and their applications, see [1], [6], [14], [22–24], and [34, Section 2]. (See also [15] for pointers to related software.)

Since f is convex, it is clear that the Bregman distance is convex in the first variable:

*Research supported by NSERC.
†Research supported by NSERC.

- $x \mapsto D_f(x, y)$ is convex, for every $y \in I$.

In this note, we are interested in the following two stronger properties:

- D_f is *jointly convex*: $(x, y) \mapsto D_f(x, y)$ is convex on $I \times I$;
- D_f is *separately convex*: $y \mapsto D_f(x, y)$ is convex, for every $x \in I$.

Clearly, if D_f is jointly convex, then it is separately convex. Joint convexity lies at the heart of the analysis in many recent papers. It was used explicitly by Butnariu, Censor, and Reich [9, Section 1.5], by Butnariu, Iusem, and Burachik [12, Section 6], by Butnariu and Iusem [11, Section 2.3], by Butnariu, Reich, and Zaslavski [13], by Byrne and Censor [7,8], by Csiszár and Tusnády [17], by Eggermont and LaRiccia [18], by Iusem [20]. Separate convexity is a sufficient condition for results on the convergence of certain algorithms; see Butnariu and Iusem's [10, Theorem 1], and Bauschke et al.'s [2].

Despite the usefulness of joint and separate convexity of D_f, we are not aware of any work that studies these concepts in their own right — except for an unpublished manuscript [4] on which the present note is based.

The objective of this note is to systematically study separate and joint convexity of the Bregman distance.

The material is organized as follows.

In Section 2, we collect some preliminary results. Joint and separate convexity of D_f for a one-dimensional convex function f are given in Section 3. Our main result states that D_f is jointly convex if and only if $1/f''$ is concave. The well-known examples of functions inducing jointly convex Bregman distances — the energy and the Boltzmann-Shannon entropy — are revealed as limiting cases in a profound sense. Section 4 discusses asymptotic behavior whereas in Section 5 we relax some of our initial assumptions. We turn to the general discussion of (not necessarily separable) convex function in the final Section 6. Our main result has a beautiful analog: D_f is jointly convex if and only if the inverse of the Hessian of f is (Loewner) concave. Finally, we discuss spectral functions where the situation appears to be less clear.

2. PRELIMINARIES

The results in this section are part of the folklore and listed only for the reader's convenience.

Fact 2.1. [31, Theorem 13.C] Suppose ψ is a convex function, and ϕ is an increasing and convex function. Then the composition $\phi \circ \psi$ is convex.

Corollary 2.2. Suppose g is a positive function. Consider the following three properties:

(i) $1/g$ is concave.

(ii) g is *log-convex*: $\ln \circ g$ is convex.

(iii) g is convex.

Then: (i)$\Rightarrow$(ii)$\Rightarrow$(iii).

Proof. "(i)$\Rightarrow$(ii)": Let $\psi = -1/g$ and $\phi : (-\infty, 0) \to \mathbb{R} : x \mapsto -\ln(-x)$. Then ψ is convex, and ϕ is convex and increasing. By Fact 2.1, $\ln \circ g = \phi \circ \psi$ is convex.
"(ii)$\Rightarrow$(iii)": Let $\psi = \ln \circ g$ and $\phi = \exp$. Then $g = \phi \circ \psi$ is convex, again by Fact 2.1. □

Remark 2.3. It is well-known that the implications in Corollary 2.2 are not reversible: • exp is log-convex on $\mathbb{R}$, but $1/\exp$ is not concave; • $x \mapsto x$ is convex on $(0 + \infty)$, but not log-convex.

Fact 2.4. Suppose g is a differentiable function on I. Then:

(i) g is convex $\Leftrightarrow D_g(x,y) \geq 0$, for all $x, y \in I$.

(ii) g is affine $\Leftrightarrow D_g(x,y) = 0$, for all $x, y \in I$.

Proof. (i): [31, Theorem 42.A]. (ii): use (i) with g and $-g$, and $D_{-g} = -D_g$. □

Proposition 2.5. If g is convex and proper, then $\lim_{z\to+\infty} g(z)/z$ exists in $(-\infty, +\infty]$.

Proof. Fix x_0 in $\operatorname{dom} g$ and check that $\Psi : x \mapsto \big(g(x_0+x) - g(x_0)\big)/x$ is increasing. Hence $\lim_{x\to+\infty} \Psi(x)$ exists in $(-\infty, +\infty]$. The result follows from

$$\frac{g(x_0+x)}{x_0+x} = \frac{g(x_0+x) - g(x_0)}{x} \frac{x}{x_0+x} + \frac{g(x_0)}{x_0+x}$$

and the change of variables $z = x_0 + x$. □

The next two results characterize convexity of a function.

Theorem 2.6. Suppose U is a convex nonempty open set in $\mathbb{R}^N$ and $g : U \to \mathbb{R}$ is continuous. Let $A := \{x \in U : \nabla g(x) \text{ exists}\}$. Then the following are equivalent.

(i) g is convex.

(ii) $U \setminus A$ is a set of measure zero, and $\nabla g(x)(y - x) \leq g(y) - g(x)$, for all $x \in A$ and $y \in U$.

(iii) A is dense in U, and $\nabla g(x)(y - x) \leq g(y) - g(x)$, for all $x \in A$ and $y \in U$.

Proof. "(i)$\Rightarrow$(ii)": g is almost everywhere differentiable, either by [31, Theorem 44.D] or by local Lipschitzness and Rademacher's Theorem ([16, Theorem 3.4.19]). On the other hand, the subgradient inequality holds on U. Altogether, (ii) follows.

"(ii)$\Rightarrow$(iii)": trivial.

"(iii)$\Rightarrow$(i)": Fix u, v in U, $u \neq v$, and $t \in (0,1)$. It is not hard to see that there exist sequences (u_n) in U, (v_n) in U, (t_n) in $(0,1)$ with $u_n \to u$, $v_n \to v$, $t_n \to t$, and $x_n := t_n u_n + (1-t_n)v_n \in A$, for every n. By assumption, $\nabla g(x_n)(u_n - x_n) \leq g(u_n) - g(x_n)$ and $\nabla g(x_n)(v_n - x_n) \leq g(v_n) - g(x_n)$. Equivalently, $(1-t_n)\nabla g(x_n)(v_n - u_n) \leq g(u_n) - g(x_n)$ and $t_n \nabla g(x_n)(u_n - v_n) \leq g(v_n) - g(x_n)$. Multiply the former inequality by t_n, the latter by $1 - t_n$, and add. We obtain that $g(x_n) \leq t_n g(u_n) + (1 - t_n) g(v_n)$. Let n tend to $+\infty$ to deduce that $g\big(tu + (1-t)v\big) \leq tg(u) + (1-t)g(v)$. The convexity of g follows and the proof is complete. □

Theorem 2.7. Suppose U is a convex nonempty open set in $\mathbb{R}^N$ and $g : U \to \mathbb{R}$ is continuously differentiable on U. Let $A := \{x \in U : g''(x) \text{ exists}\}$. Then g is convex $\Leftrightarrow$ $U \setminus A$ is a set of measure zero, and g'' is positive semidefinite on A.

Proof. "$\Rightarrow$": Aleksandrov's Theorem (see [33, Theorem 13.51]) states that the set $U \setminus A$ is of measure zero. Fix $x \in U$ and $y \in \mathbb{R}^N$ arbitrarily. The function $t \mapsto g(x+ty)$ is convex in a neighborhood of 0; consequently, its second derivative $\langle y, \big(g''(x+ty)\big)(y)\rangle$ is nonnegative whenever it exists. Since $g''(x)$ does exist, it must be that $g''(x)$ is positive semidefinite, for every $x \in A$.

"$\Leftarrow$": Fix x and y in U. By assumption and a Fubini argument, we obtain two sequences (x_n) and (y_n) in U with $x_n \to x$, $y_n \to y$, and $t \mapsto g''(x_n + t(y_n - x_n))$ exists almost everywhere on $[0,1]$. Integrating $t \mapsto \langle y_n - x_n, \big(g''(x_n + t(y_n - x_n))\big)(y_n - x_n)\rangle \geq 0$ from 0 to 1, we deduce that $\big(g'(y_n) - g'(x_n)\big)(y_n - x_n) \geq 0$. Taking limits and recalling that g' is continuous, we see that g' is monotone and so g is convex [31, Theorem 42.B]. □

3. JOINT AND SEPARATE CONVEXITY ON $\mathbb{R}$

BASIC RESULTS

Definition 3.1. We say that D_f is:

(i) *separately convex*, if $y \mapsto D_f(x,y)$ is convex, for every $x \in I$.

(ii) *jointly convex*, if $(x,y) \mapsto D_f(x,y)$ is convex on $I \times I$.

The following result will turn out to be useful later.

Lemma 3.2. Suppose $h : I \to (0,+\infty)$ is a differentiable function. Then:

(i) $1/h$ is concave $\Leftrightarrow h(y) + h'(y)(y-x) \geq \big(h(y)\big)^2/h(x)$, for all x, y in I.

(ii) $1/h$ is affine $\Leftrightarrow h(y) + h'(y)(y-x) = \big(h(y)\big)^2/h(x)$, for all x, y in I.

(iii) $1/h$ is concave $\Rightarrow h$ is log-convex $\Rightarrow h$ is convex.

(iv) If h is twice differentiable, then: $1/h$ is concave $\Leftrightarrow hh'' \geq 2(h')^2$.

Proof. Let $g := -1/h$ so that $g' = h'/h^2$.

"(i)": $1/h$ is concave $\Leftrightarrow g$ is convex $\Leftrightarrow D_g$ is nonnegative (Lemma 2.4) $\Leftrightarrow 0 \leq -1/h(x) + 1/h(y) - \big(h'(y)/h^2(y)\big)(x-y)$, $\forall x, y \in I \Leftrightarrow 0 \leq h(x)h(y) - h^2(y) - h(x)h'(y)(x-y)$, $\forall x, y \in I$. "(ii)": is similar to (i). "(iii)": restates Corollary 2.2. "(iv)": $1/h$ is concave $\Leftrightarrow g$ is convex $\Leftrightarrow g'' = \big(h^2h'' - 2h(h')^2\big)/h^4 \geq 0 \Leftrightarrow hh'' \geq 2(h')^2$. □

Theorem 3.3. Let $h := f''$. Then:

(i) D_f is jointly convex $\Leftrightarrow 1/h$ is concave $\Leftrightarrow$

$$h(y) + h'(y)(y-x) \geq \big(h(y)\big)^2/h(x), \quad \text{for all } x, y \text{ in } I. \tag{j}$$

In particular, if f'''' exists, then: D_f is jointly convex $\Leftrightarrow hh'' \geq 2(h')^2$.

(ii) D_f is separately convex $\Leftrightarrow$

$$h(y) + h'(y)(y - x) \geq 0, \quad \text{for all } x, y \text{ in } I. \tag{s}$$

Proof. "(i)": $\nabla^2 D_f(x, y)$, the Hessian of D_f at $(x, y) \in I \times I$, is equal to

$$\begin{pmatrix} f''(x) & -f''(y) \\ -f''(y) & f''(y) + f'''(y)(y - x) \end{pmatrix} = \begin{pmatrix} h(x) & -h(y) \\ -h(y) & h(y) + h'(y)(y - x) \end{pmatrix}.$$

Using [31, Theorem 42.C], we have the following equivalences: D_f is jointly convex $\Leftrightarrow$ $\nabla^2 D_f(x, y)$ is positive semidefinite, $\forall x, y \in I \Leftrightarrow h(x) \geq 0$, $h(y) + h'(y)(y - x) \geq 0$, and $\det \nabla^2 D_f(x, y) \geq 0$, $\forall x, y \in I \Leftrightarrow 1/h$ is concave (using $h > 0$ and Lemma 3.2.(i)). The "In particular" part is now clear from Lemma 3.2.(iv).

"(ii)": for fixed x, the second derivative of $y \mapsto D_f(x, y)$ equals $h(y) + h'(y)(y - x)$. Hence the result follows. □

Separate convexity is genuinely less restrictive than joint convexity:

Example 3.4. Let $f(x) = \exp(-x)$. Then $f''(x) = \exp(-x)$, and $1/f''(x) = \exp(x)$ is nowhere concave. Set $I = (0, 1)$. By Theorem 3.3.(i), *D_f is not jointly convex on I.* On the other hand, fix $x, y \in I$ arbitrarily. Then $y - x \leq |y - x| \leq 1$ and hence

$$f''(y) + f'''(y)(y - x) = \exp(-y)\big(1 - (y - x)\big) > 0 > -\exp(-y) = f'''(y).$$

The first inequality and Theorem 3.3.(ii) imply that *D_f is separately convex.*

Example 3.5. (Fermi-Dirac entropy) Let $f(x) = x\ln(x) + (1 - x)\ln(1 - x)$ on $I = (0, 1)$. Then $f''(x) = 1/\big(x(1 - x)\big)$. Thus $1/f''(x) = x(1 - x)$ is concave (but not affine). By Theorem 3.3.(i), D_f is jointly convex.

LIMITING CASES

Remark 3.6. A limiting case occurs when f as in Theorem 3.3 "barely" produces a jointly convex Bregman distance, by which we mean that the inequality (j) is always an equality. By the proof of Theorem 3.3 and Lemma 3.2.(ii), this occurs precisely when $1/h = 1/f''$ is affine, i.e., there exist real α, β such that

$$f''(x) = \frac{1}{\alpha x + \beta} > 0, \quad \text{for every } x \in I.$$

This condition determines f *up to additive affine perturbations.* We discover that either

$$f(x) = \frac{x^2}{2\beta}, \quad \text{if } \alpha = 0 \text{ and } \beta > 0;$$

or

$$f(x) = \frac{(\alpha x + \beta)\big(-1 + \ln(\alpha x + \beta)\big)}{\alpha^2}, \quad \text{if } \alpha \neq 0.$$

Hence f must be either a "general energy" ($\alpha = 0$ and $\beta = 1$ yields the *energy* $f(x) = \frac{1}{2}x^2$ on the real line) or a "general entropy" ($\alpha = 1$ and $\beta = 0$ results in the *Boltzmann-Shannon entropy* $f(x) = x\ln x - x$ on $(0, +\infty)$).

We saw in Remark 3.6 how limiting the requirement of joint convexity of D_f on the entire real line is — only quadratic functions have this property. Is this different for the weaker notion of separate convexity? The answer is negative:

Corollary 3.7. Suppose $I = \mathbb{R}$. Then D_f is separately convex if and only if f is essentially the *energy*: there exist $a, b, c \in \mathbb{R}$, $a > 0$, such that $f(x) = a\frac{1}{2}x^2 + bx + c$.

Proof. Consider inequality (s) of Theorem 3.3.(ii) and let x tend to $\pm\infty$. We conclude that $h'(y) = f'''(y) = 0$, for all $y \in \mathbb{R}$, and the result follows. □

Remark 3.8. If $I = \mathbb{R}$, then the above yields following characterization:

D_f is jointly convex $\Leftrightarrow$ D_f is separately convex $\Leftrightarrow$ f is a quadratic.

Hence an example such as Example 3.4 requires I to be a proper subset of the real line.

It is tempting to guarantee separate convexity of D_f by imposing *symmetry*; however, this sufficent condition is too restrictive:

Lemma 3.9. (Iusem; [21]) If D_f is symmetric, then f is a quadratic.

Proof. Differentiate both sides of $D_f(x, y) = D_f(y, x)$ with respect to x to learn that the gradient map $y \mapsto f'(y)$ is affine. It follows that f is a quadratic. □

4. ASYMPTOTIC RESULTS

THE CASE WHEN $I = (0, +\infty)$

We now provide a more usable characterization of separate convexity for the important case when $I = (0, +\infty)$.

Corollary 4.1. If $I = (0, +\infty)$, then: D_f is separately convex $\Leftrightarrow$ $f''(x) + xf'''(x) \geq 0 \geq f'''(x)$, for every $x > 0$.

Proof. By Theorem 3.3.(ii), D_f is separately convex if and only if

$$f''(y) + f'''(y)(y - x) \geq 0, \quad \text{for all } x, y > 0. \tag{s}$$

"$\Rightarrow$": Consider (s). Let x tend to $+\infty$ to learn that $f'''(y) \leq 0$. Then let x tend to 0 from the right to deduce that $f''(y) + yf'''(y) \geq 0$. Altogether, $f''(y) + yf'''(y) \geq 0 \geq f'''(y)$, for every $y > 0$. "$\Leftarrow$": straight-forward. □

Discussing limiting cases leads to no new class of functions:

Remark 4.2. Suppose D_f is separately convex and $I = (0, +\infty)$; equivalently, by Corollary 4.1,

$$f''(x) + xf'''(x) \geq 0 \geq f'''(x), \quad \text{for every } x > 0. \tag{$*$}$$

When considering limiting solutions of ($*$), we have two choices: Either we require the right inequality in ($*$) to be an equality throughout — this results in $f''' = 0$, and we obtain (as in Corollary 3.7) essentially the *energy*. Or we impose equality in the left inequality of $(*)$: $f''(x) + xf'''(x) = 0$, for all $x > 0$. But this differential equation readily leads to (essentially) the *Boltzmann-Shannon entropy*:

$$f(x) = a\big(x\ln(x) - x\big) + bx + c,$$

where $a, b, c \in \mathbb{R}$ and $a > 0$.

Remark 3.8 and Remark 4.2 may make the reader wonder whether separate convexity differs from joint convexity when $I = (0, +\infty)$. Indeed, they do differ, and a counter-example will be constructed (see Example 4.5) with the help of the following result.

Theorem 4.3. Suppose $I = (0, +\infty)$ and f'''' exists. Let $\psi := \ln \circ (f'')$. Then:

(i) D_f is jointly convex $\Leftrightarrow$ $\psi''(x) \geq \big(\psi'(x)\big)^2$, $\forall x > 0$.

(ii) D_f is separately convex $\Leftrightarrow$ $0 \geq \psi'(x) \geq -1/x$, $\forall x > 0$.

(iii) f'' is log-convex $\Leftrightarrow$ $\psi''(x) \geq 0$, $\forall x > 0$.

(iv) f'' is convex $\Leftrightarrow$ $\psi''(x) \geq -\big(\psi'(x)\big)^2$, $\forall x > 0$.

Proof. "(i)": clear from Theorem 3.3.(i). "(ii)": use Corollary 4.1. "(iii)": f'' is log-convex $\Leftrightarrow$ ψ is convex $\Leftrightarrow$ $\psi'' \geq 0$. "(iv)": use f'' is convex $\Leftrightarrow$ $h'' \geq 0$. □

Remark 4.4. *Joint (or separate) convexity of D_f is* not preserved *under Fenchel conjugation*: indeed, if $f(x) = x\ln(x) - x$ is the Boltzmann-Shannon entropy, then $f^* = \exp$ on $\mathbb{R}$. Now D_f is jointly convex, whereas D_{f^*} is not separately convex on $(0, +\infty)$ (by Theorem 4.3.(ii)).

Example 4.5. On $I = (0, +\infty)$, let

$$\psi(x) := \frac{-\ln(x) - \mathrm{Si}(x)}{2}, \quad \text{where} \quad \mathrm{Si}(x) := \int_0^x \frac{\sin(t)}{t}\,dt$$

denotes the sine integral function. Let f be a second anti-derivative of $\exp \circ \psi$. Then $\psi'(x) = -\big(1 + \sin(x)\big)/(2x)$; therefore, by Theorem 4.3.(ii), *D_f is separately convex.* However, since condition (iv) of Theorem 4.3 fails at $x = 2\pi$, f'' cannot be convex and *D_f is not jointly convex.* (It appears there is no elementary closed form for f.)

Example 4.6. (Burg entropy) Suppose $f(x) = -\ln(x)$ on $(0, +\infty)$. Then $f'(x) = -1/x$ and $f''(x) = 1/x^2$. Hence f'' is convex. Let $\psi(x) := \ln\big(f''(x)\big) = -2\ln(x)$. Since $\psi'(x) = -2/x < -1/x$, Theorem 4.3.(ii) implies that D_f is not separately convex. (In fact, there is no $x > 0$ such that $y \mapsto D_f(x, y)$ is convex on $(0, +\infty)$.) However, $\psi''(x) = 2/x^2 > 0$, so f'' is log-convex by Theorem 4.3.(iii).

Remark 4.7. Example 4.5 and Example 4.6 show that *separate convexity of D_f and (log-) convexity of f'' are independent properties.*

5. ASYMPTOTIC RESULTS

This subsection once again shows the importance of the energy and the Boltzmann-Shannon entropy — they appear naturally when studying asymptotic behavior.

Lemma 5.1. Suppose $\sup I = +\infty$ and D_f is separately convex. Then f *does not grow faster than the energy*: $0 \leq \lim_{x\to+\infty} f(x)/x^2 < +\infty$.

Proof. Let $L := \lim_{x\to+\infty} f(x)/x^2$. We must show that L exists as a nonnegative finite real number. Since D_f is separately convex, Theorem 3.3.(ii) yields $f''(y) + f'''(y)(y-x) \geq 0$. Letting tend x to $+\infty$ shows that $f'''(y) \leq 0$. Hence f' is concave and f'' is decreasing. Also, $f'' > 0$. It follows that

$$f''(+\infty) := \lim_{x\to+\infty} f''(x) \in [0, +\infty).$$

We will employ similar notation when there is no cause for confusion. Since f is convex, $f(+\infty)$ exists in $[-\infty, +\infty]$. If $f(+\infty)$ is finite, then $L = 0$ and we are done. Thus assume $f(+\infty) = \pm\infty$. Consider the quotient

$$q(x) := \frac{f'(x)}{2x}$$

for large x. Since f' is concave, $q(+\infty)$ exists by Proposition 2.5. L'Hospital's Rule (see [35, Theorem 4.22]) yields $q(+\infty) = L$. Since f' is increasing, $f'(+\infty)$ exists in $(-\infty, +\infty]$. Thus if $f'(+\infty)$ is finite, then $L = 0$. So we assume $f'(+\infty) = +\infty$. Now $f''(+\infty)/2 \in [0, +\infty)$; hence, again by L'Hospital Rule, so is $q(+\infty) = L$. □

Example 5.2. (p-norms) For $1 < p$, let $f(x) = \frac{1}{p}x^p$ on $I = (0, +\infty)$. Then:

D_f is jointly convex $\Leftrightarrow$ D_f is separately convex $\Leftrightarrow$ $1 < p \leq 2$.

Proof. *Case 1:* $1 < p \leq 2$. Then $f''(x) = (p-1)x^{p-2}$, hence $1/f''(x) = x^{2-p}/(p-1)$ is concave. Thus D_f is jointly convex by Theorem 3.3.(i).

Case 2: $2 < p < +\infty$. Then $f(z)/z^2 = z^{p-2}/p$ tends to $+\infty$ as z does. Consequently, by Lemma 5.1, D_f is not separately convex. □

Lemma 5.3. Suppose $\inf I = 0$, $\lim_{x\to 0^+} f(x) = 0$, and D_f is separately convex. Then $0 \leq \lim_{x\to 0^+} f(x)/(x\ln(x) - x) < +\infty$.

Proof. Let $L := \lim_{x\to 0^+} f(x)/(x\ln(x) - x)$. We must show that L exists and is a nonnegative finite real number. As in the proof of Lemma 5.1, we employ space-saving notation such as $f(0) := \lim_{x\to 0^+} f(x) = 0$. Consider

$$q(x) := \frac{f'(x)}{\ln(x)}$$

for small $x > 0$.

It suffice to show that $q(0)$ exists in $[0, \infty)$: indeed, this implies $q(0) = L$ by L'Hospital's Rule, which is what we want.

Since f is convex, f' is increasing. Hence $f'(0)$ exists in $[-\infty, +\infty)$. If $f'(0) > -\infty$, then $q(0) = 0$ and we are done. Hence assume $f'(0) = -\infty$. Since D_f is separately convex, Theorem 3.3.(ii) yields again $f''(y) + f'''(y)(y - x) \geq 0$. Letting tend x to 0 from the right shows that $f''(y) + yf'''(y) \geq 0$, for all $y \in I$. Hence $y \mapsto yf''(y)$ is increasing. It follows that

$$\lim_{x \to 0^+} xf''(x) \in [0, +\infty).$$

By L'Hospital Rule, $q(0) \in [0, +\infty)$ as is L. □

Example 5.4. Suppose $f(x) = -\frac{1}{p}x^p$ on $I = (0, +\infty)$ for $p \in (-\infty, 1) \setminus \{0\}$. L'Hospital's Rule applied twice easily yields $\lim_{x \to 0^+} f(x)/(x \ln(x) - x) = +\infty$; thus, by Lemma 5.3, D_f is not separately convex.

6. RELAXING THE ASSUMPTIONS

RELAXING $f'' > 0$

The function $x \mapsto \frac{1}{4}x^4$ is strictly convex, but its second derivative has a zero; consequently, none of the above results is applicable. The following technique allows to handle such function within our framework. For $\epsilon > 0$, let

$$f_\epsilon := f + \epsilon\tfrac{1}{2}|\cdot|^2.$$

Then we obtain readily the following.

Observation 6.1. D_f is jointly (resp. separately) convex if and only if each D_{f_ϵ} is.

We only give one example to show how this Observation can be used rather than writing down several slightly more general results.

Example 6.2. Suppose $f(x) = \frac{1}{4}x^4$ on $I = (-1, 1)$. Since $f''(0) = 0$, we cannot use our previous results. Now $f_\epsilon''(x) = f''(x) + \epsilon = 3x^2 + \epsilon$ and $f_\epsilon'''(x) = 6x$. Pick $0 < \epsilon < \frac{1}{2}$, let $x = 2\sqrt{\epsilon}$ and $y = \frac{1}{3}x$. With this assignment, $f_\epsilon''(y) + f_\epsilon'''(y)(y - x) = -3\epsilon < 0$. Thus, by Theorem 3.3.(ii), D_{f_ϵ} is not separately convex. Hence, by Observation 6.1, D_f is not separately convex.

RELAXING $f \in C^3(I)$

The assumption on differentiability on f can be weakened and several slightly more general result could be obtained — we illustrate this through the following variant of Theorem 3.3.(i).

Theorem 6.3. Suppose $f \in C^2(I)$, $f'' > 0$ on I, and f''' exists almost everywhere in I. Then D_f is jointly convex if and only if $1/f''$ is concave.

Proof. D_f is jointly convex $\Leftrightarrow$ $\nabla^2 D_f$ exists almost everywhere in $I \times I$, and whenever it exists,

$$\nabla^2 D_f(x, y) = \begin{pmatrix} f''(x) & -f''(y) \\ -f''(y) & f''(y) + f'''(y)(y - x) \end{pmatrix}$$

is positive semidefinite (by Theorem 2.7) $\Leftrightarrow$ for almost every $y \in I$, $f'''(y)$ exists and $f''(y) + f'''(y)(y - x) \geq \left(f''(y)\right)^2$, for all $x \in I$ (by using a Fubini-type argument such as [35, Lemma 6.120]) $\Leftrightarrow$ $1/f''$ is concave (by applying Theorem 2.6 to $g := -1/f''$). □

Example 6.4. Let $I := (-\infty, 1)$ and $f(x) := \frac{1}{2}x^2$, if $x \leq 0$; $x + (1-x)\ln(1-x)$, if $0 \leq x < 1$. Then $f \in C^2(I)\backslash C^3(I)$ so that Theorem 3.3 does not apply directly. However: since $1/f''(x) = \min\{1, 1-x\}$ is concave, Theorem 6.3 yields the joint convexity of D_f.

7. JOINT AND SEPARATE CONVEXITY IN GENERAL

The results above concern a function defined on an interval. This allows us to handle functions on $\mathbb{R}^N$ that are separable. The general case is a little more involved; however, the main patterns observed so far generalize quite beautifully.

Throughout this section, we assume that

$$\boxed{U \text{ is a convex nonempty open set in } \mathbb{R}^N}$$

and that

$$\boxed{f \in C^3(U) \text{ with } f'' \text{ positive definite on } U.}$$

Clearly, f is strictly convex. Let

$$\boxed{H := H_f := \nabla^2 f = f''}$$

denote the Hessian of f. $H(x)$ is a real symmetric positive semidefinite matrix, for every $x \in U$. Recall that the real symmetric matrices can be partially ordered by the *Loewner ordering*

$$\boxed{H_1 \succeq H_2 \quad :\Leftrightarrow \quad H_1 - H_2 \text{ is positive semidefinite,}}$$

and they form a Euclidean space with the inner product

$$\boxed{\langle H_1, H_2 \rangle := \operatorname{trace}(H_1 H_2).}$$

Further information can be found in [19, Section 7.7], [30, Section 16.E], and in the very recent [5]. We will be using results from these sources throughout.

We start by generalizing Theorem 3.3.

Theorem 7.1. (i) D_f is jointly convex if and only if

$$H(y) + \big(\nabla H(y)\big)(y - x) \succeq H(y)H^{-1}(x)H(y), \quad \text{for all } x, y \in U. \tag{J}$$

(ii) D_f is separately convex if and only if

$$H(y) + \big(\nabla H(y)\big)(y - x) \succeq 0, \quad \text{for all } x, y \in U. \tag{S}$$

Proof. "(i)": The Hessian of the function $U \times U \to [0, +\infty) : (x, y) \mapsto D_f(x, y)$ is (compare to the proof of Theorem 3.3) the block matrix

$$\nabla^2 D_f(x, y) = \begin{pmatrix} H(x) & -H(y) \\ -H(y) & H(y) + \big(\nabla H(y)\big)(y - x) \end{pmatrix}.$$

Using standard criteria for positive semidefiniteness of block matrices (see [19, Section 7.7]) and remembering that H is positive definite, we obtain that $\nabla^2 D_f(x, y)$ is positive semidefinite for all x, y if and only if (J) holds. "(ii)": For fixed $x \in U$, similarly discuss positive semidefiniteness of the Hessian of the map $y \mapsto D_f(x, y)$. $\square$

Corollary 7.2. The following are equivalent:

(i) D_f is jointly convex.

(ii) $\nabla H^{-1}(y)(x-y) \succeq H^{-1}(x) - H^{-1}(y)$, for all $x, y \in U$.

(iii) H^{-1} is *(matrix) concave*, i.e.,

$$H^{-1}(\lambda x + \mu y) \succeq \lambda H^{-1}(x) + \mu H^{-1}(y),$$

for all $x, y \in U$ and $\lambda, \mu \in [0,1]$ with $\lambda + \mu = 1$.

(iv) $x \mapsto \langle P, H^{-1}(x)\rangle$ is concave, for every $P \succeq 0$.

Proof. Consider the mapping $U \to \mathbb{R}^{N\times N} : y \mapsto H(y)H^{-1}(y)$. It is constant, namely the identity matrix. Take the derivative with respect to y. Using an appropriate product rule (see, for instance, [25, page 469f in Section 17.3]) yields $0 = H(y)\big((\nabla H^{-1}(y))(z)\big) + \big((\nabla H(y))(z)\big)H^{-1}(y)$, for every $z \in \mathbb{R}^N$. In particular, after setting $z = x - y$, multiplying by $H^{-1}(y)$ from the left, and re-arranging, we obtain

$$H^{-1}(y)\big((\nabla H(y))(y-x)\big)H^{-1}(y) = \big(\nabla H^{-1}(y)\big)(x-y).$$

"(i)⇔(ii)": The equivalence follows readily from the last displayed equation and Theorem 7.1.(i). "(ii)⇔(iii)": The proof of [31, Theorem 42.A] works without change in the present positive semidefinite setting. "(iii)⇔(iv)": is clear, since the cone of positive semidefinite matrices is self-dual. □

Example 7.3. Suppose Q is an N-by-N real symmetric and positive definite matrix. Let $U = \mathbb{R}^N$ and $f(x) = \frac{1}{2}\langle x, Qx\rangle$. We assume that Q is not a diagonal matrix. Then f is not separable, hence none of the results in previous chapters are applicable. Now $f''(x) = Q = H(x)$, for all $x \in \mathbb{R}^N$. Hence $H^{-1}(x) = Q^{-1}$ is constant, and thus trivially matrix-concave. By Corollary 7.2, the Bregman distance D_f is jointly convex.

SPECTRAL FUNCTIONS

In this last subsection, we discuss convex functions defined on the N-by-N real symmetric matrices, denoted by $\mathcal{S}$ and equipped with the inner product

$$\boxed{\langle S_1, S_2\rangle := \operatorname{trace}(S_1 S_2).}$$

Suppose further

$$\boxed{U \subseteq \mathbb{R}^N \text{ is convex, nonempty, and open,}}$$

and

$$\boxed{f \in C^3(U) \text{ is convex, symmetric, with } f'' > 0.}$$

Then f induces a *spectral function*

$$\boxed{F := f \circ \lambda,}$$

where $\lambda : \mathcal{S} \to \mathbb{R}^N$ is the *eigenvalue map* (ordered decreasingly). Then the function F is *orthogonally invariant*: $F(X) = F(U^T X U)$, for all $X \in \mathcal{S}$ and every orthogonal N-by-N matrix U, and $F \circ \Lambda = f \circ \lambda$. For further information, we refer the reader to [26] and [5]; in [1, Section 7.2], we discussed the Bregman distance D_F in detail.

Our one-dimensional results showed the special status of the separable energy and the separable Boltzmann-Shannon entropy: if $f(x) = \sum_{n=1}^N \frac{1}{2}x_n^2$ or $f(x) = \sum_{n=1}^N x_n \ln(x_n) - x_n$, then D_f is jointly convex. It is natural to inquire about D_F, the Bregman distance induced by the corresponding spectral function $F = f \circ \lambda$. Now if f is the separable energy, then F is the energy on $\mathcal{S}$ and thus D_F is jointly convex. This is rather easy.

The same is true for the Boltzmann-Shannon entropy — this is essentially known as *Lindblad's Theorem* [29]:

Theorem 7.4. If $f(x) = \sum_{n=1}^N x_n \ln(x_n) - x_n$, then D_F is jointly convex.

Proof. For $X \in \mathcal{S}$ positive definite with eigenvalues $\lambda(X)$ ordered decreasingly, denote the diagonal matrix with $\lambda(X)$ on the diagonal by $\Lambda(X)$ so that $X = U\Lambda(X)U^T$ for some orthogonal matrix U. Recall that (see [3]) if g is a function from real interval to $\mathbb{R}$, then g acting on a diagonal matrix is defined as the diagonal matrix obtained by applying g to each diagonal entry; moreover, this is used to define $g(X) := Ug(\Lambda(X))U^T$. Also, we diagonalize a positive definite Y by $Y = V\Lambda(Y)V^T$, for some orthogonal matrix V. Then

$$\begin{aligned} D_F(X,Y) &= F(X) - F(Y) - \langle \nabla F(Y), X - Y\rangle \\ &= F(U\Lambda(X)U^T) - F(V\Lambda(Y)V^T) - \langle \nabla F(Y), X - Y\rangle \\ &= f(\lambda(X)) - f(\lambda(Y)) - \langle \nabla F(Y), X - Y\rangle. \end{aligned}$$

On the other hand, using [26, Corollary 3.3],

$$\begin{aligned} \langle \nabla F(Y), X\rangle &= \langle V(\nabla f)(\Lambda(Y))V^T, X\rangle = \langle V\ln(\Lambda(Y))V^T, X\rangle = \langle \ln(Y), X\rangle \\ &= \operatorname{trace}(X\ln(Y)) \end{aligned}$$

and similarly

$$\begin{aligned} \langle \nabla F(Y), Y\rangle &= \langle V\ln(\Lambda(Y))V^T, Y\rangle = \operatorname{trace}(V\ln(\Lambda(Y))V^T Y) \\ &= \operatorname{trace}(\ln(\Lambda(Y))V^T Y V) = \operatorname{trace}(\ln(\Lambda(Y))\Lambda(Y)) \\ &= \textstyle\sum_{n=1}^N \lambda_n(Y)\ln(\lambda_n(Y)) = f(\lambda(Y)) + \operatorname{trace}(Y). \end{aligned}$$

Altogether,

$$\begin{aligned} D_F(X,Y) &= f(\lambda(X)) - f(\lambda(Y)) - \langle \nabla F(Y), X\rangle + \langle \nabla F(Y), Y\rangle \\ &= f(\lambda(X)) - \operatorname{trace}(X\ln(Y)) + \operatorname{trace}(Y) \\ &= \textstyle\sum_{n=1}^N \big(\lambda_n(X)\ln(\lambda_n(X)) - \lambda_n(X)\big) - \operatorname{trace}(X\ln(Y)) + \operatorname{trace}(Y) \\ &= \operatorname{trace}(\Lambda(X)\ln(\Lambda(X))) - \operatorname{trace}(X) - \operatorname{trace}(X\ln(Y)) + \operatorname{trace}(Y) \\ &= \operatorname{trace}(U\Lambda(X)U^T U\ln(\Lambda(X))U^T) + \operatorname{trace}(Y - X) - \operatorname{trace}(X\ln(Y)) \\ &= \operatorname{trace}(X\ln(X)) + \operatorname{trace}(Y - X) - \operatorname{trace}(X\ln(Y)) \\ &= \operatorname{trace}(Y - X) + \operatorname{trace}(X(\ln(X) - \ln(Y))). \end{aligned}$$

Clearly, trace$(Y-X)$ is jointly convex. Finally, Lindblad's Theorem [29] (see also [3, Theorem IX.6.5]) precisely states that trace$(X(\ln(X) - \ln(Y)))$ is jointly convex. Therefore, $D_F(X,Y)$ is jointly convex. □

Remark 7.5. Let f be the (separable) Boltzmann-Shannon entropy as in Theorem 7.4, and let F be its corresponding spectral function. Then Corollary 7.2.(iii) implies that $(\nabla^2 F)^{-1}$ is Loewner concave. (See [27,28] for further information on computing second derivatives of spectral functions.)

It would be interesting to find out about the general case. We thus conclude with a question.

Open Problem 7.6. *Is D_f jointly convex if and only if D_F is?*

We believe that Lewis and Sendov's recent work on the second derivative of a spectral function F [27,28] in conjunction with our second derivative characterization of joint convexity of D_F (Corollary 7.2.(iii)) will prove useful in deciding this problem.

ACKNOWLEDGMENT

We wish to thank Hristo Sendov for his pertinent comments on an earlier version of this manuscript.

REFERENCES

1. H. H. Bauschke and J. M. Borwein. Legendre functions and the method of random Bregman projections. *Journal of Convex Analysis* **4**:27–67, 1997.
2. H. H. Bauschke, D. Noll, A. Celler, and J. M. Borwein. An EM-algorithm for dynamic SPECT tomography. *IEEE Transactions on Medical Imaging*, **18**:252–261, 1999.
3. R. Bhatia. *Matrix Analysis.* Springer-Verlag, 1996.
4. J. M. Borwein and L. C. Hsu. On the Joint Convexity of the Bregman Distance. Unpublished manuscript, 1993.
5. J. M. Borwein and A. S. Lewis. *Convex Analysis and Nonlinear Optimization.* Springer-Verlag, 2000.
6. L. M. Bregman. The relaxation method of finding the common point of convex sets and its application to the solution of problems in convex programming. *U.S.S.R. Computational Mathematics and Mathematical Physics* **7**:200–217, 1967.
7. C. Byrne and Y. Censor. Proximity function minimization and the convex feasibility problem for jointly convex Bregman distances. Technical Report, 1998.
8. C. Byrne and Y. Censor. Proximity Function Minimization Using Multiple Bregman Projections, with Applications to Split Feasibility and Kullback-Leibler Distance Minimization. Technical Report: June 1999, Revised: April 2000, *Annals of Operations Research*, to appear.
9. D. Butnariu, Y. Censor, and S. Reich. Iterative Averaging of Entropic Projections for Solving Stochastic Convex Feasibility Problems. *Computational Optimization and Applications* **8**:21–39, 1997.
10. D. Butnariu and A. N. Iusem. On a proximal point method for convex optimization in Banach spaces. *Numerical Functional Analysis and Optimization* **18**:723–744, 1997.

11. D. Butnariu and A. N. Iusem. *Totally Convex Functions for Fixed Points Computation and Infinite Dimensional Optimization.* Kluwer, 2000.
12. D. Butnariu, A. N. Iusem, and R. S. Burachik. Iterative methods for solving stochastic convex feasibility problems. To appear in *Computational Optimization and Applications.*
13. D. Butnariu, S. Reich, and A. J. Zaslavski. Asymptotic behaviour of quasi-nonexpansive mappings. Preprint, 2000.
14. Y. Censor and S. A. Zenios. *Parallel Optimization.* Oxford University Press, 1997.
15. Centre for Experimental and Constructive Mathematics. *Computational Convex Analysis* project at `www.cecm.sfu.ca/projects/CCA`.
16. F. H. Clarke, Y. S. Ledyaev, R. J. Stern, and P. R. Wolenski. *Nonsmooth Analysis and Control Theory.* Springer-Verlag, 1998.
17. I. Csiszár and G. Tusnády. Information geometry and alternating minimization procedures. *Statistics and Decisions* (Supplement **1**), 205–237, 1984.
18. P. P. B. Eggermont and V. N. LaRiccia. On EM-like algorithms for minimum distance estimation. Preprint, 2000.
19. R. A. Horn and C. R. Johnson. *Matrix Analysis.* Cambridge University Press, 1985.
20. A. N. Iusem. A short convergence proof of the EM algorithm for a specific Poisson model. *Revista Brasileira de Probabilidade e Estatistica* **6**:57–67, 1992.
21. A. N. Iusem. Personal communication.
22. K. C. Kiwiel. Free-steering relaxation methods for problems with strictly convex costs and linear constraints. *Mathematics of Operations Research* **22**:326-349, 1997.
23. K. C. Kiwiel. Proximal minimization methods with generalized Bregman functions. *SIAM Journal on Control and Optimization* **35**:1142–1168, 1997.
24. K. C. Kiwiel. Generalized Bregman projections in convex feasibility problems. *Journal of Optimization Theory and Applications* **96**:139–157, 1998.
25. S. Lang. *Undergraduate Analysis (Second Edition).* Springer-Verlag, 1997.
26. A. S. Lewis. Convex analysis on the Hermitian matrices. *SIAM Journal on Optimization* **6**:164–177, 1996.
27. A. S. Lewis and H. S. Sendov. Characterization of Twice Differentiable and Twice Continuously Differentiable Spectral Functions. Preprint, 2000.
28. A. S. Lewis and H. S. Sendov. Quadratic Expansions of Spectral Functions. Preprint, 2000.
29. G. Lindblad. Entropy, information and quantum measurements. *Communications in Mathematical Physics* **33**:305–322, 1973.
30. A. W. Marshall and I. Olkin. *Inequalities: Theory of Majorization and Its Applications.* Academic Press, 1979.
31. A. W. Roberts and D. E. Varberg. *Convex Functions.* Academic Press, 1973.
32. R. T. Rockafellar. *Convex Analysis.* Princeton University Press, 1970.
33. R. T. Rockafellar and R. J.-B. Wets. *Variational Analysis.* Springer-Verlag, 1998.
34. M. V. Solodov and B. F. Svaiter. An inexact hybrid generalized proximal point method algorithm and some new results on the theory of Bregman functions. To appear in *Mathematics of Operations Research.*
35. K. R. Stromberg. *An Introduction to Classical Real Analysis.* Wadsworth, 1981.

Inherently Parallel Algorithms in Feasibility and Optimization and their Applications
D. Butnariu, Y. Censor and S. Reich (Editors)

A PARALLEL ALGORITHM FOR NON-COOPERATIVE RESOURCE ALLOCATION GAMES

L. M. Bregman[a] and I. N. Fokin[b*]

[a]Institute for Industrial Mathematics,
4 Yehuda Hanakhtom, Beer-Sheva 84311, Israel

[b]Institute for Economics and Mathematics, Russian Academy of Sciences,
38 Serpuchovskaya str., St. Petersburg 198013, Russia

Non-cooperative resource allocation games which generalize the Blotto game are considered. These games can be reduced to matrix games, but the size of the obtained matrices is huge. We establish special properties of these matrices which make it possible to develop algorithms operating with relatively small arrays for the construction of the Nash equilibrium in the considered games. The computations in the algorithms can be performed in parallel.

1. INTRODUCTION

Let us consider a non-cooperative game (game Γ) with a finite set I of players. Player $i \in I$ owns integer capacity $K_i > 0$ of some resource. The players allocate their resources among a finite set N of terrains. A pure strategy of player i is defined as a vector

$$x_i := (x_{i\nu}),\ \nu \in N,\ x_{i\nu} \text{ are non-negative integers, and } \sum_{\nu \in N} x_{i\nu} = K_i.$$

The component $x_{i\nu}$ is interpreted as the number of resource units which player i allocates to terrain ν.

The payoff of each player is assumed to be additive, that is, if all players use pure strategies (player i uses a strategy x_i), and $x := (x_i),\ i \in I$, is a strategy profile, then player i obtains payoff

$$\pi_i(x) := \sum_{j \in I} \pi_{ij}(x_i, x_j),$$

and his payoff against rival j is assumed to be additive subject to the terrains:

$$\pi_{ij}(x_i, x_j) := \sum_{\nu \in N} \pi_{ij\nu}(x_{i\nu}, x_{j\nu}). \tag{1}$$

*We thank the anonymous referee for many constructive comments and suggestions which helped to improve the first version of the manuscript.

Here $\pi_{ii}(x_i, x_i)$ is assumed to be 0, for all x_i. Moreover, any pairwise interaction is assumed to be zero-sum, i.e.

$$\pi_{ij}(x_i, x_j) + \pi_{ji}(x_j, x_i) = 0. \tag{2}$$

Evidently, (2) implies that $\sum_{i \in I} \pi_i(x) = 0$, for all profiles of strategies x, so Γ is a zero-sum game.

Note that player i has

$$m_i := \binom{K_i + |N| - 1}{|N| - 1}$$

pure strategies, it is the number of ordered partitions of integer $K_i + |N|$ in $|N|$ parts (see [1]).

The game Γ is a generalization of the well-known Blotto game (see, for example, [2,3]).

We are interested to construct a method for finding a Nash equilibrium in Γ. As shown in section 2, the Nash equilibrium can be obtained from the solution of some fair matrix game, that is, from the solution of some feasibility problem. The order of the game matrix is huge, it is $\prod_{i \in I} m_i$. However, the rank of the matrix is much smaller than its order. Because of this a method can be constructed which operates with arrays and matrices of relatively small size.

The resource allocation game considered in the present paper is a particular case of the separable non-cooperative game introduced in [4]. The method for the construction of the Nash equilibrium described there operates with vectors and matrices with size equal to the rank of the game matrix.

Special properties of the resource allocation game make it possible to reduce the magnitude of arrays required by the method and to organize the computations so that they can be performed in parallel.

2. NASH EQUILIBRIUM AND EQUIVALENT MATRIX GAME

Let X_i be the set of pure strategies of player i in game Γ. We denote $\prod_{i \in I} X_i$ by X. For the resource allocation game,

$$|X_i| = m_i = \binom{K_i + |N| - 1}{|N| - 1}.$$

Let s_i be a mixed strategy for player i, i.e. s_i is a probability distribution on X^i:

$$s_i(x_i) \geq 0, \sum_{x_i \in X_i} s_i(x_i) = 1.$$

Let S_i be the set of mixed strategies for player i, and $S := \prod_{i \in I} S_i$ the set of randomized-strategy profiles. For $s \in S$, the expected value of the payoff function for player i is

$$\pi_i(s) = \sum_{x \in X} \pi_i(x) \prod_{j \in I} s_j(x_j). \tag{3}$$

However, taking into account that player interactions are pairwise, we can represent $\pi_i(s)$ in a simpler form:

$$\pi_i(s) = \sum_{j \in I} \pi_{ij}(s_i, s_j),$$

where

$$\pi_{ij}(s_i, s_j) := \sum_{x_i \in X_i, x_j \in X_j} \pi_{ij}(x_i, x_j) s_i(x_i) s_j(x_j).$$

For any $t_i \in S_i$, we let (s_{-i}, t_i) denote the randomized-strategy profile in which the i-th component is t_i and all other components are as in s. Thus,

$$\pi(s_{-i}, t_i) = \sum_{x \in X} \left(\prod_{j \in I \setminus i} s_j(x_j) \right) t_i(x_i) \pi_i(x),$$

or

$$\pi(s_{-i}, t_i) = \sum_{j \in I} \sum_{x_i \in X_i, x_j \in X_j} \pi_{ij}(x_i, x_j) t_i(x_i) s_j(x_j). \tag{4}$$

Definition 2.1 *We say that $s \in S$ is a Nash equilibrium if*

$\pi_i(s) \geq \pi_i(s_{-i}, t_i)$, *for each* $i \in I, t_i \in S_i$.

Now we construct a fair matrix game Δ which is equivalent to game Γ, that is, each randomized-strategy profile in Γ corresponds to some mixed strategy for a player in Δ and vice versa. The strategy corresponding to the Nash equilibrium is an optimal mixed strategy in Δ. Since game Δ is fair (i.e. its value is 0 and the sets of optimal strategies for the first and second player are the same), an optimal mixed strategy can be found by the solution of some linear feasibility problem.

Construct $|X_i| \times |X_j|$-matrices H_{ij}: an element in the row corresponding to the strategy $x_i \in X_i$ and in the column corresponding to the strategy $x_j \in X_j$ is defined as $H_{ij}[x_i, x_j] := \pi_{ij}(x_i, x_j)$. H_{ii} is defined as $|X_i| \times |X_i|$-matrix of zeros. It is clear that $H_{ij} = -H_{ji}^T$.

Then we consider a matrix H consisting of blocks H_{ij}:

$$H := \begin{pmatrix} H_{11} & H_{12} & \dots & H_{1|I|} \\ \cdot & \cdot & \dots & \cdot \\ \cdot & \cdot & \dots & \cdot \\ H_{|I|1} & H_{|I|2} & \dots & H_{|I||I|}. \end{pmatrix}$$

Matrix H is a square matrix of order $\sum_{i \in I} |X_i|$. It is a skew-symmetric matrix, that is, $H = -H^T$.

Consider a matrix B consisting of zeros and ones. Each row of B corresponds to some $x \in X$, that is, B has $m := \prod_{i \in I} |X_i|$ rows. The columns of B are divided into $|I|$ blocks. The i-th block B_i corresponds to set X_i and consists of $|X_i|$ columns, each column corresponding to some $x_i \in X_i$. Matrix B has $|I|$ ones in each row: one 1 in each block, namely, if some row corresponds to $x = (x_i),\ i \in I$, then it has 1 in the i-th block in the column corresponding to strategy x_i.

Let us consider now a matrix game Δ with $(m \times m)$-matrix

$$A = BHB^T. \tag{5}$$

It is clear that matrix A is skew-symmetric. It means that game Δ is fair, its value is 0 and the sets of optimal strategies for the first and second player are the same.

Lemma 2.2 *If a vector y is a mixed strategy for the second player in Δ, then $s = B^T y$ is a randomized-strategy profile in Γ. The reverse: if s is a randomized-strategy profile in Γ, then there exists a vector y satisfying $B^T y = s$ which is a mixed strategy for the second player in Δ.*

Proof. Let y be a mixed strategy for the second player in Δ, and $s = B^T y$. Then $s = (s_i),\ i \in I$, where $s_i = B_i^T y$. Since B_i has one unity in each row, and $\sum_{x \in X} y(x) = 1$, we have $\sum_{x_i \in X_i} s_i(x_i) = 1$, for all $i \in I$. It is clear that $s_i \geq 0$, for all i and x^i, because B_i and y are non-negative. So s_i is a mixed strategy for player i in Γ, and s is a randomized-strategy profile.

Let $s = (s_i),\ i \in I$, be a randomized-strategy profile in Γ.

Consider a vector y with components $y(x),\ x \in X$, defined as

$$y(x) := \prod_{i \in I} s_i(x_i).$$

It is clear that y is a probability vector, that is, y is a mixed strategy for the second player in Δ. Show that $B_i^T y = s_i$, for all $i \in I$.

Let t be a component of vector $B_i^T y$ corresponding to some $z \in X_i$. Then

$$t = \sum_{\{x \in X:\ x_i = z\}} \prod_{j \in I} s_j(x_j) = s_i(z) \prod_{j \neq i} \sum_{x_j \in X_j} s_j(x_j).$$

Since $\sum_{x_j \in X_j} s_j(x_j) = 1$, we have $B_i^T y = s_i$, that is, $B^T y = s$. ∎

Obviously, the lemma's statements also relate to the first player strategies.

Theorem 2.3 *Game Δ is equivalent to Γ, i.e. if y is an optimal strategy for the second player in Δ, then $s = B^T y$ is a Nash equilibrium in Γ, and if $s \in S$ is a Nash equilibrium in Γ, then each y satisfying $s = B^T y$ is an optimal strategy for the second player in Δ.*

Proof. Let y be an optimal strategy for the second player in Δ. Since Δ is fair, this means that

$$\begin{aligned} &Ay \leq 0, \\ &\textstyle\sum_{x \in X} y(x) = 1, \\ &y \geq 0. \end{aligned} \tag{6}$$

Let $s = B^T y,\ s = (s_i),\ i \in I,\ s_i$ is a mixed strategy for the i-th player. Let t_i be another mixed strategy for player i, and $u = (s_{-i}, t_i)$.

By lemma 2.2, there exists a mixed strategy for the first player z such that $zB = u$. By (6), we have

$$zAy \leq 0,$$

or taking into account representation (5),

$$uHs \leq 0.$$

The latter inequality is equivalent to

$$\sum_{k \in I, j \in I} u_k H_{kj} s_j \leq 0.$$

By definition of H_{kj},

$$u_k H_{kj} s_j = \sum_{x_k \in X_k, x_j \in X_j} \pi(x_k, x_j) u_k(x_k) s_j(x_j)$$

For $k \neq i$,

$$\sum_{j \in I} u_k H_{kj} s_j = \sum_{j \in I} s_k H_{kj} s_j = \sum_{j \in I} \sum_{x_k \in X_k, x_j \in X_j} \pi(x_k, x_j) s_k(x_k) s_j(x_j) = \pi_k(s),$$

and for $k = i$, by (4),

$$\sum_{j \in I} u_i H_{ij} s_j = \sum_{j \in I} t_i H_{ij} s_j = \sum_{j \in I} \sum_{x_i \in X_i, x_j \in X_j} \pi(x_i, x_j) t_i(x_i) s_j(x_j) = \pi_i(s_{-i}, t_i).$$

These relations imply that

$$\sum_{k \neq i} \pi_k(s) + \pi_i(s_{-i}, t_i) \leq 0,$$

and since Γ is a zero sum game, i.e. $\sum_{k \in I} \pi_k(s) = 0$, we have

$$\pi_i(s_{-i}, t_i) \leq \pi_i(s).$$

Since $i \in I$ and $t_i \in S_i$ was chosen arbitrarily, this means that s is a Nash equilibrium in Γ.

Let now s be a Nash equilibrium in Γ. Take $x \in X$: $x = (x_i)$, $i \in I$. Let u_i be the mixed strategy for the i-th player which chooses the strategy x_i with the probability 1, and $u := (u_i)$, $i \in I$. Then $u = b_x$, where b_x is the row of matrix B corresponding to the strategy profile x. For each $i \in I$, we have

$$\pi_i(s_{-i}, u_i) \leq \pi_i(s).$$

Summing these inequalities, we obtain

$$\sum_{i \in I} \pi_i(s_{-i}, u_i) \leq \sum_{i \in I} \pi_i(s) = 0.$$

Using (4), we have

$$uHs \leq 0.$$

Let y be a mixed strategy for the second player in Δ such that $s = B^T y$ (such y exists by lemma 2.2). Then $b_x H B^T y \leq 0$. Since x was chosen arbitrarily, we have

$$BHB^T y \leq 0,$$

that is, y is an optimal strategy for the second player in Δ. ■

Thus, the problem of finding a Nash equilibrium in Γ has been reduced to the solution of a system of linear inequalities (6). However, the number of variables and constraints in (6) is huge because the order of matrix A is

$$\prod_{i \in I} |X_i| = \prod_{i \in I} \binom{K_i + |N| - 1}{|N| - 1}.$$

We need to exploit specific properties of matrix A to construct an efficient method for the solution of system (6). The basic property is that the rank of matrix A is significantly smaller than its order, and we can represent this matrix as a product of matrices with relatively small size.

Representation (5) gives some factorization for matrix A, and we can see that the rank of A is not greater than the number of rows in H which is equal to the sum of the numbers of pure strategies of all the players. This is significantly smaller than the number of rows (or columns) in A which is equal to the product of these numbers, but it is still a large number. We construct a factorization for matrix H which shows that in fact the rank of matrix A is much smaller than that of H. This factorization makes it possible to construct an efficient method for the solution of (6) which operates with arrays of relatively small size.

We begin with a factorization of matrices H_{ij}. The factorization reflects the additive form of the elements of these matrices expressed by (1).

Let us introduce $(K_i + 1) \times (K_j + 1)$- matrices $p_{ij\nu}$, $\nu \in N$, whose elements are payoffs for player i from player j on terrain ν, i.e. $p_{ij\nu}[k_i, k_j] := \pi_{ij\nu}(k_i, k_j)$, $0 \leq k_i \leq K_i$, $0 \leq k_j \leq K_j$. Construct a block-angular matrix G_{ij}, which is composed of blocks $p_{ij\nu}$:

$$G_{ij} := \begin{pmatrix} p_{ij1} & & & \\ & p_{ij2} & & \\ & & \cdots & \\ & & & p_{ij|N|} \end{pmatrix}.$$

G_{ij} is an $(|N|(K_i + 1) \times |N|(K_j + 1))$-matrix.

Let us consider zero-one matrices $R_{i\nu}$, $i \in I, \nu \in N$. Each row of $R_{i\nu}$ corresponds to some pure strategy of player i. The number of columns in this matrix is equal to $K_i + 1$. The element in the row corresponding to strategy x_i and in column k $(0 \leq k \leq K_i)$ is equal to one, if player i allocates k units of the resource to terrain ν in his strategy x^i. $R_{i\nu}$ are $(|X_i| \times (K_i + 1))$-matrices. Compose a matrix R_i from matrices $R_{i\nu}$ merging their columns for all $\nu \in N$: $R_i := (R_{i1}, R_{i2}, ..., R_{i|N|})$, R_i is an $(|X_i| \times |N|(K_i + 1))$-matrix.

Then matrix H_{ij} can be represented in the following form:

$$H_{ij} = R_i G_{ij} R_j^T,$$

and matrix H can be represented in the form $H = RGR^T$, where matrices R and G are composed of R_i and G_{ij}:

$$R := \begin{pmatrix} R_1 & & & \\ & R_2 & & \\ & & \dots & \\ & & & R_{|I|} \end{pmatrix},$$

$$G := \begin{pmatrix} G_{11} & G_{12} & \dots & G_{1\,|I|} \\ . & . & \dots & . \\ . & . & \dots & . \\ G_{|I|\,1} & G_{|I|\,2} & \dots & G_{|I|\,|I|} \end{pmatrix}.$$

R is a $\left(\sum_{i\in I}|X_i| \times |N|(\sum_{i\in I} K_i + |I|)\right)$-matrix, and G is a square matrix of the order $|N|(\sum_{i\in I} K_i + |I|)$.

Hence matrix A can be represented as a product of five matrices: $A = BRGR^T B^T$. The rank of A is not greater than the minimal rank of these matrices. Therefore the rank of A is significantly lower than its order. The order of A is

$$\prod_{i=1}^{n} \begin{pmatrix} K_i + |N| - 1 \\ |N| - 1 \end{pmatrix},$$

and the rank of A is less than or equal to the rank of matrix G, that is,

$$\text{rank } A \le |N|(\sum_{i\in I} K_i + |I|).$$

3. METHODS FOR FINDING A NASH EQUILIBRIUM

We should find a solution of a matrix game with an $(m \times m)$-matrix A, where

$$m = \prod_{i=1}^{n} \begin{pmatrix} K_i + |N| - 1 \\ |N| - 1 \end{pmatrix},$$

and $A = BRGR^T B^T$.

The value of the game is 0, therefore the game is equivalent to the following feasibility problem:

$$\begin{aligned} & Ay \le 0, \\ & y \in S_m, \end{aligned} \tag{7}$$

where S_m is a simplex

$$S_m = \{y \in E^m : \sum_{j=1}^{m} y_j = 1, \ y_j \ge 0\}.$$

The solution of problem (7) is an optimal strategy for the second player in Δ. By theorem 2.3, $s = B^T y = (s_i)$, $i \in I$, is a Nash equilibrium in game Γ. Therefore the problem of finding a Nash equilibrium can be reformulated as follows:

$$\begin{aligned} &BRGR^T s \leq 0, \\ &\textstyle\sum_{x_i \in X_i} s_i(x^i) = 1, \;\; s_i \geq 0, \;\text{for } i \in I. \end{aligned} \tag{8}$$

We develop a method for the solution of (8) which allows parallel computations. The method operates with arrays and matrices of the size which is not greater than

$$\mu := |N|(\sum_{i \in I} K_i + |I|),$$

the order of matrix G. We suppose that μ is a relatively small number, that is, we can store vectors and matrices of order μ.

Define sets Ω_i, for $i \in I$, as

$$\Omega_i = \{z_i \in E^{|N|(K_i+1)} : \; z_i = R_i^T s_i, \; \sum_{x_i \in X_i} s_i(x_i) = 1, \;\; s_i \geq 0\}.$$

Then, taking into consideration the structure of matrices B, G and R, problem (8) can be rewritten as

$$\begin{aligned} &\textstyle\sum_{i \in I} B_i R_i \sum_{j \in I} G_{ij} z_j \leq 0, \\ &z_i \in \Omega_i, \text{ for } i \in I. \end{aligned} \tag{9}$$

Since Ω_i are polyhedra, they can be represented as a finite intersection of half-spaces. This means that there exist matrices Q_i and vectors q_i such that

$$\Omega_i = \{z_i \in E^{|N|(K_i+1)} : \; Q_i z_i \leq q_i\}.$$

Denoting $\sum_{i \in I} B_i R_i G_{ij}$ by P_j, we rewrite (9) in the following form:

$$\begin{aligned} &\textstyle\sum_{j \in I} P_j z_j \leq 0, \\ &Q_i z_i \leq q_i, \text{ for } i \in I. \end{aligned} \tag{10}$$

Thus, the process of obtaining a Nash equilibrium consists of the following stages:

1. Calculation of z_i, $i \in I$, from system (10).
2. Calculation of a Nash equilibrium $s = (s_i)$ from systems

$$z_i = R_i^T s_i, \; \sum_{x_i \in X_i} s_i(x_i) = 1, \;\; s_i \geq 0.$$

But first of all we fix the representation of Ω_i as the intersection of half-spaces because such representation is not unique.

3.1. Representation of Ω_i

Consider a polyhedron in E^{p+1}, where $p := |N|(K_i + 1)$, consisting of vectors (π, ρ) (π is a p-vector, and ρ is a number) satisfying the following conditions:

$$\begin{aligned} &R_i\pi + \rho e \leq 0 \\ &-1 \leq \pi_j \leq 1 \ (j = 1, 2, \ldots, p). \end{aligned} \tag{11}$$

Here e is an $|X_i|$-vector of ones.

Let $\{(\pi_k, \rho_k),\ k \in K\}$ be a (finite) set of all extreme points of this polyhedron. Define Q_i as a $(|K| \times p)$-matrix whose k-th row is π_k, and q_i as a $|K|$-vector with $q_i[k] := -\rho_k$. Show that Ω_i can be represented by the system of inequalities $Q_i z \leq q_i$.

Theorem 3.1 $\Omega_i = \{z \in E^{|N|(K_i+1)} : \ Q_i z \leq q_i\}$.

Proof. Take $z \in \Omega_i$ and show that $Q_i z \leq q_i$.

Consider the linear programming problem:

$$\max\ (\langle \pi, z \rangle + \rho) \tag{12}$$

subject to conditions (11).

The dual problem to (12) is

$$\begin{aligned} &\min\ \textstyle\sum_{j=1}^{p}(\omega_j^+ + \omega_j^-) \\ &R_i^T s_i + \omega^+ - \omega^- = z \\ &\textstyle\sum_{x_i \in X_i} s_i(x_i) = 1, \\ &s_i, \omega^+, \omega^- \geq 0. \end{aligned} \tag{13}$$

Since $z \in \Omega_i$, the minimal value of the target function in (13) is zero. Therefore the maximal value in (12) is also zero, that is,

$$\langle \pi_k, z \rangle + \rho_k \leq 0, \text{ for all } k \in K.$$

This means that $Q_i z \leq q_i$.

Take now z satisfying $Q_i z \leq q_i$ and show that $z \in \Omega_i$.

Suppose $z \bar{\in} \Omega_i$. Then the minimal value of the target function in (13) is positive. Hence the maximal value in (12) is also positive, that is, there exists an extreme point (π_k, ρ_k) such that

$$\langle \pi_k, z \rangle + \rho_k > 0.$$

This means that z does not satisfy conditions $Q_i z \leq q_i$. ■

3.2. Solution of system (10)

System (10) is a system of linear inequalities with a huge number of constraints: the first group of them $\sum_{j \in I} P_j z_j \leq 0$ has $m = |X|$ inequalities, and the number of inequalities in other groups is equal to the number of extreme points in the polyhedra described in section 3.1. However, the number of variables in (10) is relatively small, it is equal to $\mu = |N|(\sum_{i \in I} K_i + |I|)$.

We consider two approaches to the solution of (10). The first is the simplex method with the column generation technique (see [5,6]). In this method, we have a current basic solution z_i, $i \in I$, together with the current basis, i.e. a set of no more than μ constraints of (10) which are satisfied as equalities for the current z_i. The next steps are to find the most violated restriction in (10), introduce it into the basis, define the basic constraint to be removed from the basis and recalculate z_i according to the new basis. The process stops if for the current z_i all conditions (10) are satisfied.

The other approach is the row action method (see [7]). In this method, we have a current approximation z_i^k for the solution of (10). Then we find the most violated restriction in (10) for z_i^k and define the next approximation z_i^{k+1} as the projection (orthogonal or non-orthogonal) of z_i^k onto the corresponding hyperplane. The sequence z_i^k converges to the solution of (10) (see [8]).

The common part of both approaches is the following problem: for given z_i, $i \in I$, find the most violated restriction in (10). We concentrate on this problem because other actions (projections onto a hyperplane and reconstructions of a basis in the simplex method) are performed as usual and operate with vectors and matrices with the order no more than μ.

Let us consider a method for determining the most violated restriction for the first group:

$$\sum_{j \in I} P_j z_j \leq 0. \tag{14}$$

Denote $v_i := \sum_{j \in I} G_{ij} z_j$, v_i can be calculated easily because z_j are known vectors. Since $P_j = \sum_{i \in I} B_i R_i G_{ij}$, we need to find the maximal element in the vector

$$\sum_{i \in I} B_i R_i v_i.$$

We may search for the maximal element in $B_i R_i v_i$ separately for each i. Indeed, let x_i be a pure strategy for the i-th player in Γ which corresponds to the maximal element of $B_i R_i v_i$. Then the row of B corresponding to $x \in X$, where $x = (x_i)$, $i \in I$, is the row with the maximal element of $\sum_{i \in I} B_i R_i v_i$.

Matrix B_i contains one unity in each row, so the product of a row of B_i by R_i is a row of R_i, that is, an $(|N|(K_i + 1))$-vector containing $|N|$ blocks with one unity in each block. The row corresponds to some pure strategy for the i-th player, namely, to the strategy $x_i = (x_{i\nu})$ such that a unity in block $\nu \in N$ is on the position $x_{i\nu}$ (positions are numbered from 0 to K_i).

Vector v_i is also an $(|N|(K_i + 1))$-vector, its components also regarded as divided into $|N|$ blocks with $(K_i + 1)$ elements in each block. Let the element in block ν at position x_ν be denoted by $v_{i\nu}[x_\nu]$.

Hence the maximal element in $B_i R_i v_i$ can be calculated from the following problem:

$$\max \sum_{\nu \in N} v_{i\nu}[x_\nu] \tag{15}$$

subject to conditions

$$\begin{array}{l} \sum_{\nu \in N} x_\nu = K_i, \\ x_\nu \text{ are integer and non-negative.} \end{array} \tag{16}$$

This problem can be solved by the dynamic programming technique. The solution of this problem defines x_i a pure strategy for the i-th player. The set (x_i), $i \in I$, corresponds to the most violated constraint in (14).

Problems (15)-(16) can be solved in parallel, for each i.

Consider now how to define the most violated restriction for the system

$$Q_i z_i \leq q_i, \tag{17}$$

where z_i is a given $(|N|(K_i+1))$-vector, and Q_i and q_i are defined as in section 3.1. According to the representation of Ω_i, we need to find a p-vector π and a number ρ which form the solution of the linear programming problem:

$$\max\ (\langle \pi, z \rangle + \rho) \tag{18}$$

subject to conditions (11).

If the maximal value in (18) is positive, then the solution of this problem is an extreme point of polyhedron (11) corresponding to the most violated restriction in (17), otherwise conditions (17) are satisfied.

We solve (18) together with its dual problem (13) by the simplex method with the column generation technique. The number of constraints in (13) is $p+1$, it is less than μ, so we can store and operate vectors and matrices of this order.

Let us have a feasible basic solution for (13). Corresponding dual variables π_j, $1 \leq j \leq p$, and ρ are defined from the following conditions:

$\langle R_i[x_i], \pi \rangle + \rho = 0$, if $s_i(x_i)$ is a basic variable,
$\pi_j = 1$, if ω^+ is a basic variable,
$\pi_j = -1$, if ω^- is a basic variable.

Here $R_i[x_i]$ is a row of matrix R_i corresponding to $x_i \in X_i$.

If these π and ρ satisfy (11), then they are a solution of (18). Otherwise, we need to change the current basis and basic solution in (13).

If $\pi_j > 1$, for some j, introduce variable ω^+ into the basis. If $\pi_j < -1$, for some j, introduce variable ω^- into the basis.

Let all π_j lie between -1 and 1. Verify the conditions

$$R_i \pi + \rho e \leq 0. \tag{19}$$

Find

$$\tau := \max_{x_i \in X_i} \langle R_i[x_i], \pi \rangle \tag{20}$$

and the row $R_i[x_i^*]$ on which this maximum is attained. If $\tau + \rho \leq 0$, then conditions (19) are satisfied. Otherwise, we introduce variable $s_i(x_i^*)$ into the basis in (13).

We define the variable to be removed from the basis and reconstruct the basis as in the usual simplex method. It requires processing vectors and matrices of order no more than $p+1$.

The process of changing the basis is repeated till we get the solution of problem (18). At that moment, the solution of (18) (π, ρ) defines the most violated restriction in (17).

It remains to describe how to find τ and x_i^* in (20). Row $R_i[x_i]$ consists of $|N|$ blocks with a unique unity in each block: block ν contains 1 at position $x_{i\nu}$ (positions are numbered from 0 to K_i). The components of vector π are also regarded as divided into $|N|$ blocks with $K_i + 1$ components in each block. Let $\pi_\nu[x_\nu]$ be the component in block ν at position x_ν. Then (20) can be rewritten as

$$\tau = \max \sum_{\nu \in N} \pi_\nu[x_\nu] \tag{21}$$

subject to conditions

$$\begin{array}{l} \sum_{\nu \in N} x_\nu = K_i, \\ x_\nu \text{ are integer and non-negative.} \end{array} \tag{22}$$

Problem (21)-(22) is similar to problem (15)-(16), and can also be solved by the dynamic programming technique.

Note that problems (17) may be solved in parallel, for each i.

3.3. Calculation of a Nash equilibrium

After we have solved the inequality system (10) and obtained its solution z_i, $i \in I$, we can calculate mixed strategies s_i forming a Nash equilibrium. We can obtain them as a solution of linear programming problems (13) with $z = z_i$, for all $i \in I$. A method for the solution of these problems is described in the previous section. Note that although s_i has $|X_i|$ components, the basic solution of (13) has no more than $|N|(K_i + 1) + 1$ non-zero components, so we obtain a Nash equilibrium with a relative small number of non-zero components by the described method.

In fact, we need not solve problems (13) as a separate stage because we have already obtained their solution while we verified inequalities (17) for the last time.

REFERENCES

1. M. Hall, *Combinatorial Theory*. (Blaisdell Publishing Company, Waltham (Massachusetts), Toronto, London, 1967).
2. J. W. Tukey, A Problem of Strategy, *Econometrica* **17** (1949) 73.
3. D. W. Blackett, Pure Strategy Solutions to Blotto Games, *Naval Research Logistics Quarterly* **5** (1958) 107–109.
4. L. M. Bregman and I. N. Fokin, On Separable Non-cooperative Zero-sum Games, *Optimization* **44** (1998) 69–84.
5. L. V. Kantorovich and J. V. Romanovskij, The Column Generation in the Simplex-method, *Economics and Mathematical Methods* **21** (1985) 128–138 (Russian).
6. L. S. Lasdon, *Optimization Theory for Large Systems* (MacMillan, New York, 1970).
7. Y. Censor and S. A. Zenios, *Parallel Optimization: Theory, Algorithms and Applications*. (Oxford University Press, New York, 1997).
8. L. M. Bregman, The relaxation method of finding the common point of convex sets and its application to the solution of problems in convex programming, *USSR Computational Mathematics and Mathematical Physics* **7** (1967) 200–217.

Inherently Parallel Algorithms in Feasibility and Optimization and their Applications
D. Butnariu, Y. Censor and S. Reich (Editors)

ASYMPTOTIC BEHAVIOR OF QUASI-NONEXPANSIVE MAPPINGS

Dan Butnariu[a]*, Simeon Reich[b] and Alexander J. Zaslavski[c]

[a]Department of Mathematics, University of Haifa,
31905 Haifa, Israel

[b]Department of Mathematics, The Technion-Israel Institute of Technology,
32000 Haifa, Israel

[c]Department of Mathematics, The Technion-Israel Institute of Technology,
32000 Haifa, Israel

Let K be a closed convex subset of a Banach space X and let F be a nonempty closed convex subset of K. We consider complete metric spaces of self-mappings of K which fix all the points of F and are quasi-nonexpansive with respect to a given convex function f on X. We prove (under certain assumptions on f) that the iterates of a generic mapping in these spaces converge strongly to a retraction on F.

1. INTRODUCTION

Assume that $(X, ||\cdot||)$ is a Banach space, $f: X \to R^1 \cup \{\infty\}$ is convex, the algebraic interior $\mathcal{D}^0$ of

$$\mathcal{D} = \mathrm{dom}(f) = \{x \in X : f(x) < \infty\}$$

is nonempty, and K is a nonempty closed convex subset of $\mathcal{D}^0$.

For each $x \in \mathcal{D}^0$ and $y \in X$ set

$$D_f(y,x) = f(y) - f(x) + f^0(x, x-y),$$

where

$$f^0(x,v) = \lim_{t \to 0^+} t^{-1}(f(x+tv) - f(x)).$$

The basic properties of the Bregman distance D_f are presented in detail in [9,14]. It is not a distance in the usual sense of the term because, although nonnegative, it is not, in general, symmetric nor does it satisfy the triangle inequality. Denote by $\mathcal{M}$ the set of all

*The work of the first two authors was partially supported by the Israel Science Foundation founded by the Israel Academy of Sciences and Humanities (Grants 293/97 and 592/00). The second author was also partially supported by the Fund for the Promotion of Research at the Technion, and by the Technion VPR Fund. All three authors thank the referee for several useful comments.

mappings $T: K \to K$ which are bounded on bounded subsets of K. We endow the set $\mathcal{M}$ with the uniformity determined by the following base:

$$E(N,\epsilon) = \{(T_1,T_2) \in \mathcal{M} \times \mathcal{M} : \ ||T_1x - T_2x|| \le \epsilon \\ \text{for all } x \in K \text{ satisfying } ||x|| \le N\},$$

where $N, \epsilon > 0$. Clearly this uniform space is metrizable and complete [21]. We also endow the space $\mathcal{M}$ with the topology induced by this uniformity. Denote by $\mathcal{M}_c$ the set of all continuous $T \in \mathcal{M}$. Clearly $\mathcal{M}_c$ is a closed subset of $\mathcal{M}$. We consider the topological subspace $\mathcal{M}_c \subset \mathcal{M}$ with the relative topology.

Denote by $\mathcal{M}_u$ the set of all those mappings $T: K \to K$ which are uniformly continuous on bounded subsets of K. Clearly any $T \in \mathcal{M}_u$ is bounded on bounded subsets of K and $\mathcal{M}_u$ is a closed subset of the complete uniform space $\mathcal{M}$. We consider the topological subspace $\mathcal{M}_u \subset \mathcal{M}$ with the relative topology.

Let $F \subset K$ be a nonempty closed convex subset of K. Denote by $\mathcal{M}^{(F)}$ the set of all $T \in \mathcal{M}$ such that

$$Tx = x \text{ for all } x \in F \tag{1}$$

and

$$D_f(z,Tx) \le D_f(z,x) \text{ for all } z \in F \text{ and all } x \in K. \tag{2}$$

Set

$$\mathcal{M}_c^{(F)} = \mathcal{M}^{(F)} \cap \mathcal{M}_c \text{ and } \mathcal{M}_u^{(F)} = \mathcal{M}^{(F)} \cap \mathcal{M}_u.$$

We assume that

$$D_f(z,\cdot): K \to R^1 \text{ is convex and continuous for all } z \in F. \tag{3}$$

This assumption implies that $\mathcal{M}^{(F)}$, $\mathcal{M}_c^{(F)}$ and $\mathcal{M}_u^{(F)}$ are closed subsets of $\mathcal{M}$. We consider the topological subspaces $\mathcal{M}^{(F)}, \mathcal{M}_c^{(F)}, \mathcal{M}_u^{(F)} \subset \mathcal{M}$ with the relative topologies.

We also assume that for each $x \in K$, each $y, z \in F$ and each $\gamma \in (0,1)$,

$$D_f((1-\gamma)z + \gamma y, (1-\gamma)x + \gamma y) \le (1-\gamma)D_f(z,x). \tag{4}$$

Clearly (3) and (4) are true if the function $D_f(\cdot,\cdot): F \times K \to R^1$ is convex as a function of two variables.

Although these two assumptions are rather restrictive we note that there are quite a few examples of convex functions f for which the corresponding Bregman distance D_f is convex as a function of two variables. These examples include the square of the norm in Hilbert space, the negentropy in R^n [7] and the examples discussed in [3].

For each $x \in K$ and each nonempty $G \subset K$ set

$$\rho_f(x,G) = \inf\{D_f(z,x): \ z \in G\}. \tag{5}$$

Finally, we assume the following uniform continuity condition:

For any $\epsilon > 0$ there exists $\delta > 0$ such that if $x \in K$, $z \in F$ and

$$D_f(z,x) \le \delta, \text{ then } ||z - x|| \le \epsilon. \tag{6}$$

If K is bounded, then condition (6) is a weakened version of the so-called sequential compatibility condition studied in [9].

The study of the asymptotic behavior of various classes of operators of contractive type is a central topic in applied nonlinear analysis and optimization. In recent years Bregman functions and distances [5,12] have been playing an ever-increasing role in the literature devoted to this topic. See, for instance, [6–10,13,14,24] and the references mentioned therein.

In this paper we study the asymptotic behavior of quasi-nonexpansive (see (2)) self-mappings of K. For appropriate complete metric spaces of such mappings we establish the existence of an everywhere dense G_δ subset such that the iterates of any of its elements converge strongly to a retraction onto F. Results of this kind for powers of a nonexpansive operator were already established by De Blasi and Myjak [15]. This approach, when a certain property is investigated not for a single point of a complete metric space, but for the whole space, has also been successfully applied in the theory of dynamical systems [10,16,23,25], approximation theory [17], optimization [18–20], variational analysis [1,4, 22], the calculus of variations [2,26], as well as in optimal control [27,28].

Our paper is organized as follows. In Section 2 we state a theorem on the convergence of powers for the space $\mathcal{M}_u^{(F)}$. This theorem is proved in Section 3. In Section 4 we state two convergence theorems for the space $\mathcal{M}^{(F)}$ when the Bregman distance D_f is uniformly continuous. These two theorems are established in Section 5. In Section 6 we consider the asymptotic behavior of continuous mappings without assuming uniform continuity of either the mappings or the Bregman distance. We obtain three theorems which are proved in Section 7.

2. CONVERGENCE OF POWERS FOR A CLASS OF UNIFORMLY CONTINUOUS MAPPINGS

In this section we make the following assumption:

$$\text{For each bounded set } K_0 \subset K \text{ and each } \epsilon > 0 \text{ there is } \delta > 0$$
$$\text{such that if } x \in K_0,\ z \in F \text{ and } ||z - x|| \leq \delta, \text{ then } D_f(z, x) \leq \epsilon. \tag{7}$$

Remark. Note that (7) is true if the function f is Lipschitzian on each bounded subset of K.

We also assume that for each bounded set $K_0 \subset K$,

$$\sup\{\rho_f(x, F) :\ x \in K_0\} < \infty. \tag{8}$$

Theorem 2.1 *Assume that $P \in \mathcal{M}_u^{(F)}$ and $P(K) = F$. Then there exists a set $\mathcal{F} \subset \mathcal{M}_u^{(F)}$ which is a countable intersection of open everywhere dense subsets of $\mathcal{M}_u^{(F)}$ such that for each $B \in \mathcal{F}$ the following assertions hold:*

(i) There exists $P_B \in \mathcal{M}_u^{(F)}$ such that $P_B(K) = F$ and $B^n x \to P_B x$ as $n \to \infty$, uniformly on bounded subsets of K;

(ii) for each $\epsilon > 0$ and each bounded set $C \subset K$ there exist a neighborhood U of B in $\mathcal{M}_u^{(F)}$ and an integer $N \geq 1$ such that for each $S \in U$, each $x \in C$ and each integer $n \geq N$, we have $||S^n x - P_B x|| \leq \epsilon$

In this theorem, as well as in our subsequent theorems, we assume that certain retractions belong to $\mathcal{M}^{(F)}$. Sufficient conditions for the existence of such retractions can be found in [11].

3. PROOF OF THEOREM 2.1

We first recall the following result established in [10].

Proposition 3.1 *Let K_0 be a bounded subset of K, $T \in \mathcal{M}_u$, $\epsilon > 0$, and let $n \geq 1$ be an integer. Then there exists a neighborhood U of T in $\mathcal{M}_u$ such that for each $S \in U$ and each $x \in K_0$, the inequality $||T^n x - S^n x|| \leq \epsilon$ holds.*

Now let $\gamma \in (0,1)$ and $T \in \mathcal{M}_u^{(F)}$. Define a mapping $T_\gamma : K \to K$ by

$$T_\gamma x = \gamma P x + (1-\gamma)Tx, \; x \in K. \tag{9}$$

Clearly, $T_\gamma \in \mathcal{M}_u$ and $T_\gamma x = x$ for all $x \in F$. By (9), (3) and (2), for each $z \in F$ and each $x \in K$,

$$D_f(z, T_\gamma x) = D_f(z, \gamma P x + (1-\gamma)Tx) \leq \gamma D_f(z, Px) +$$

$$(1-\gamma) D_f(z, Tx) \leq D_f(z, x). \tag{10}$$

Clearly for each $T \in \mathcal{M}_u^{(F)}$ we have

$$T_\gamma \to T \text{ as } \gamma \to 0 \text{ in } \mathcal{M}. \tag{11}$$

Lemma 3.1 *Let $T \in \mathcal{M}_u^{(F)}$ and $\gamma \in (0,1)$. Then for each $x \in K$,*

$$\rho_f(T_\gamma x, F) \leq (1-\gamma)\rho_f(x, F). \tag{12}$$

Proof. Let $x \in K$ and $\epsilon > 0$. There exists $z \in F$ such that (see (5))

$$D_f(z, x) \leq \rho_f(x, F) + \epsilon. \tag{13}$$

Then $\gamma P x + (1-\gamma) z \in F$ and by (9), (5), (4), (2) and (13),

$$\rho_f(T_\gamma x, F) \leq D_f(\gamma P x + (1-\gamma)z, (1-\gamma)Tx + \gamma P x) \leq$$

$$D_f(z, Tx)(1-\gamma) \leq (1-\gamma) D_f(z, x) \leq (1-\gamma)\rho_f(x, F) + \epsilon(1-\gamma).$$

Since ϵ is an arbitrary positive number we conclude that (12) is true. Lemma 3.1 is proved.

Lemma 3.2 *Let $T \in \mathcal{M}_u^{(F)}$, $\gamma \in (0,1)$ and let K_0 be a nonempty bounded subset of K. Then the set $\{T_\gamma^i x : \; x \in K_0, \; i \geq 1\}$ is bounded.*

Proof. By (6) there exists $\delta \in (0,1)$ such that

$$\text{if } x \in K,\ z \in F \text{ and } D_f(z,x) \leq 2\delta, \text{ then } ||z-x|| \leq 2^{-1}. \tag{14}$$

By (8) there is a number $c_0 > 0$ such that

$$\rho_f(x,F) \leq c_0 \text{ for all } x \in K_0. \tag{15}$$

Choose a natural number N such that

$$(1-\gamma)^N(c_0+2) \leq 4^{-1}\delta. \tag{16}$$

It follows from Lemma 3.1, (15) and (16) that for each $x \in K_0$,

$$\rho_f(T_\gamma^N x, F) = (1-\gamma)^N \rho_f(x,F) \leq (1-\gamma)^N c_0 \leq 4^{-1}\delta.$$

Thus (see (5)) for each $x \in K_0$ there is $Qx \in F$ such that

$$D_f(Qx, T_\gamma^N x) \leq 2^{-1}\delta.$$

Combined with (10) this inequality implies that

$$D_f(Qx, T_\gamma^n x) \leq 2^{-1}\delta$$

for each $x \in K_0$ and each integer $n \geq N$. It follows from the latter inequality and (14) that

$$||Qx - T_\gamma^n x|| \leq 2^{-1} \text{ for each } x \in K_0 \text{ and each integer } n \geq N. \tag{17}$$

Since the operator T_γ is bounded on bounded subsets of K it follows from (17) that the sets $\{T_\gamma^i x : x \in K_0\}$, $i = 1, 2, \ldots, N$, $\{Qx : x \in K_0\}$ and $\{T_\gamma^n x : x \in K_0,\ n \geq N\}$ are all bounded. Lemma 3.2 is proved.

Lemma 3.3 *Let $T \in \mathcal{M}_u^{(F)}$, $\gamma \in (0,1)$, $\epsilon > 0$, and let K_0 be a nonempty bounded subset of K. Then there exist a neighborhood U of T_γ in $\mathcal{M}_u^{(F)}$ and a natural number N such that for each $x \in K_0$ there exists $Qx \in F$ such that for each integer $n \geq N$ and each $S \in U$, we have*

$$||S^n x - Qx|| \leq \epsilon. \tag{18}$$

Proof. Set

$$K_1 = K_0 \cup \{T_\gamma^i x : x \in K_0,\ i \geq 1\}$$

and

$$K_2 = \{y \in K : \inf\{||y-x|| : x \in K_1\} \leq 2\}. \tag{19}$$

By Lemma 3.2 the sets K_1 and K_2 are bounded. By (8) there is $c_0 > 0$ such that

$$\rho_f(x,F) \leq c_0 \text{ for all } x \in K_2. \tag{20}$$

By (6) there exists $\delta \in (0,\epsilon)$ such that

$$\text{if } x \in K,\ z \in F \text{ and } D_f(z,x) \leq 2\delta, \text{ then } ||z-x|| \leq 2^{-1}\epsilon. \tag{21}$$

By (7) there is $\epsilon_0 \in (0, 2^{-1}\delta)$ such that

$$\text{if } x \in K_2,\ z \in F \text{ and } ||x - z|| \leq 2\epsilon_0, \text{ then } D_f(z, x) \leq 2^{-1}\delta. \tag{22}$$

By (6) there is $\delta_0 \in (0, 2^{-1}\epsilon)$ such that

$$\text{if } x \in K,\ z \in F,\ D_f(z, x) \leq 2\delta_0, \text{ then } ||x - z|| \leq 2^{-1}\epsilon_0. \tag{23}$$

Choose a natural number N such that

$$(1 - \gamma)^N (c_0 + 2) \leq 4^{-1}\delta_0.$$

It follows from this inequality, Lemma 3.1, (20) and (19) that for each $x \in K_0$,

$$\rho_f(T_\gamma^N x, F) \leq (1 - \gamma)^N \rho_f(x, F) \leq (1 - \gamma)^N c_0 \leq 4^{-1}\delta_0.$$

Thus for each $x \in K_0$ there is $Qx \in F$ such that $D_f(Qx, T_\gamma^N x) \leq 2^{-1}\delta_0$. Combined with (23) this implies that

$$||T_\gamma^N x - Qx|| \leq 2^{-1}\epsilon_0 \text{ for all } x \in K_0. \tag{24}$$

By Proposition 3.1 there exists a neighborhood U of T_γ in $\mathcal{M}_u^{(F)}$ such that for each $x \in K_0$ and each $S \in U$,

$$||S^N x - T_\gamma^N x|| \leq 4^{-1}\epsilon_0. \tag{25}$$

Assume that $x \in K_0$ and $S \in U$. Then (25) is true and by (19) we have

$$S^N x \in K_2. \tag{26}$$

By (25) and (24), $||S^N x - Qx|| \leq 3 \cdot 4^{-1}\epsilon_0$. It follows from this inequality, (26) and (22) that $D_f(Qx, S^N x) \leq 2^{-1}\delta$. Since $S \in \mathcal{M}_u^{(F)}$ it follows from the latter inequality that $D_f(Qx, S^n x) \leq 2^{-1}\delta$ for all integers $n \geq N$. Combined with (21) this implies that $||Qx - S^n x|| \leq \epsilon$ for all integers $n \geq N$. Lemma 3.3 is proved.

Completion of the proof of Theorem 2.1. By (11), the set $\{T_\gamma : T \in \mathcal{M}_u^{(F)},\ \gamma \in (0,1)\}$ is an everywhere dense subset of $\mathcal{M}_u^{(F)}$. Fix $\theta \in K$. For each natural number i set

$$K_i = \{x \in K : ||x - \theta|| \leq i\}. \tag{27}$$

By Lemma 3.3, for each $T \in \mathcal{M}_u^{(F)}$, each $\gamma \in (0,1)$ and each integer $i \geq 1$, there exist an open neighborhood $\mathcal{U}(T, \gamma, i)$ of T_γ in $\mathcal{M}_u^{(F)}$ and a natural number $N(T, \gamma, i)$ such that the following property holds:

P(i) For each $x \in K_{2^i}$ there is $Qx \in F$ such that

$$||S^n x - Qx|| \leq 2^{-i} \text{ for all integers } n \geq N(T, \gamma, i) \text{ and all } S \in \mathcal{U}(T, \gamma, i).$$

Define

$$\mathcal{F} = \cap_{q=1}^{\infty} \cup \{\mathcal{U}(T, \gamma, i) : T \in \mathcal{M}_u^{(F)},\ \gamma \in (0,1),\ i \geq q\}.$$

Clearly $\mathcal{F}$ is a countable intersection of open everywhere dense sets in $\mathcal{M}_u^{(F)}$.

Let $B \in \mathcal{F}$, $\epsilon > 0$ and let C be a bounded subset of K. There exists an integer $q \geq 1$ such that

$$C \subset K_{2^q} \text{ and } 2^{-q} < 4^{-1}\epsilon. \tag{28}$$

There exist $T \in \mathcal{M}_u^{(F)}$, $\gamma \in (0,1)$ and an integer $i \geq q$ such that

$$B \in \mathcal{U}(T, \gamma, i). \tag{29}$$

It follows from Property P(i), (28) and (29) that the following property also holds:

P(ii) For each $x \in C$ there is $Qx \in F$ such that

$$||S^n x - Qx|| \leq 4^{-1}\epsilon$$

for each integer $n \geq N(T, \gamma, i)$ and each $S \in \mathcal{U}(T, \gamma, i)$.

Property P(ii) and (29) now imply that for each $x \in C$ and each integer $n \geq N(T, \gamma, i)$,

$$||B^n x - Qx|| \leq 4^{-1}\epsilon. \tag{30}$$

Since ϵ is an arbitrary positive number and C is an arbitrary bounded subset of K we conclude that for each $x \in K$, $\{B^n x\}_{n=1}^{\infty}$ is a Cauchy sequence. Therefore for each $x \in K$ there exists

$$P_B x = \lim_{n \to \infty} B^n x. \tag{31}$$

By (30) and (31), for each $x \in C$ we have

$$||P_B x - Qx|| \leq 4^{-1}\epsilon. \tag{32}$$

Once again, since ϵ is an arbitrary positive number and C is an arbitrary bounded subset of K, we conclude that

$$P_B(K) = F. \tag{33}$$

It follows from Property (Pii) and (32) that for each $x \in C$, each $S \in \mathcal{U}(T, \gamma, i)$ and each integer $n \geq N(T, \gamma, i)$,

$$||S^n x - P_B x|| \leq 2^{-1}\epsilon.$$

This completes the proof of Theorem 2.1.

4. CONVERGENCE OF POWERS FOR A CLASS OF MAPPINGS WITH A UNIFORMLY CONTINUOUS BREGMAN DISTANCE

In this section we continue to assume that for each bounded set $K_0 \subset K$,

$$\sup\{\rho_f(x, F) : x \in K_0\} < \infty. \tag{34}$$

We also assume that there exists $P \in \mathcal{M}^{(F)}$ for which $P(K) = F$ and that the function $D_f(\cdot,\cdot) : F \times K \to R^1$ is uniformly continuous on bounded subsets of $F \times K$. In the next section we will establish the following two theorems.

Theorem 4.1 *There exists a set $\mathcal{F} \subset \mathcal{M}^{(F)}$ which is a countable intersection of open everywhere dense subsets of $\mathcal{M}^{(F)}$ such that for each $B \in \mathcal{F}$ the following assertions hold:*

1. There exists $P_B \in \mathcal{M}^{(F)}$ such that $P_B(K) = F$ and $B^n x \to P_B x$ as $n \to \infty$, uniformy on bounded subses of K; if $B \in \mathcal{M}_c^{(F)}$, then $P_B \in \mathcal{M}_c^{(F)}$.

2. For each $\epsilon > 0$ and each nonempty bounded set $C \subset K$, there exists a neighborhood U of B in $\mathcal{M}^{(F)}$ and a natural number N such that for each $S \in U$ and each $x \in C$, there is $z(S,x) \in F$ such that $||S^n x - z(S,x)|| \le \epsilon$ for all integers $n \ge N$.

Moreover, if $P \in \mathcal{M}_c^{(F)}$, then there exists a set $\mathcal{F}_c \subset \mathcal{F} \cap \mathcal{M}_c^{(F)}$ which is a countable intersection of open everywhere dense subsets of $\mathcal{M}_c^{(F)}$.

Theorem 4.2 *Let the set $\mathcal{F} \subset \mathcal{M}^{(F)}$ be as guaranteed by Theorem 4.1, $B \in \mathcal{F} \cap \mathcal{M}_c^{(F)}$, $P_B z = \lim_{n\to\infty} B^n z$, $z \in K$, and let $x \in K$, $\epsilon > 0$. Then there exist a neighborhood U of B in $\mathcal{M}^{(F)}$, a number $\delta > 0$ and a natural number N such that for each $y \in K$ satisfying $||x - y|| \le \delta$, each $S \in U$ and each integer $n \ge N$, we have $||S^n y - P_B x|| \le \epsilon$.*

5. PROOFS OF THEOREMS 4.1 AND 4.2

We begin with several lemmas.

Lemma 5.1 *Let F_0 be a nonempty bounded subset of F and let β be a positive number. Then the set $\{(z,y) \in F_0 \times K : D_f(z,y) \le \beta\}$ is bounded.*

Proof. Assume the contrary. Then there exists a sequence $\{(z_i, x_i)\}_{i=1}^{\infty} \subset F_0 \times K$ such that

$$D_f(z_i, x_i) \le \beta,\ i = 1, 2, \ldots, \text{ and } ||x_i|| \to \infty \text{ as } i \to \infty. \tag{35}$$

We may assume that $||z_i - x_i|| \ge 16$, $i = 1, 2, \ldots$. For each integer $i \ge 1$ there exists $\alpha_i > 0$ such that

$$||[(1-\alpha_i)z_i + \alpha_i x_i] - z_i|| = 1. \tag{36}$$

Clearly $\alpha_i \to 0$ as $i \to \infty$. By (3) and (35), for each integer $i \ge 1$ we have

$$D_f(z_i, (1-\alpha_i)z_i + \alpha_i x_i) \le \alpha_i D_f(z_i, x_i) \le \alpha_i \beta \to 0 \text{ as } i \to \infty.$$

Combined with (6) this implies that

$$||z_i - [(1-\alpha_i)z_i + \alpha_i x_i]|| \to 0 \text{ as } i \to \infty,$$

which contradicts (36). This proves Lemma 5.1.

In our next lemma we use the convention that the empty set is bounded.

Lemma 5.2 *Let K_0 be a nonempty bounded subset of K and let β be a positive number. Then the set $\{(z,y) \in F \times K_0 : D_f(z,y) \le \beta\}$ is bounded.*

Proof. Let us suppose the contrary. Then there exists a sequence $\{(z_i, x_i)\}_{i=1}^{\infty} \subset F \times K_0$ such that

$$D_f(z_i, x_i) \le \beta,\ i = 1, 2, \ldots, \text{ and } ||z_i|| \to \infty \text{ as } i \to \infty. \tag{37}$$

We may assume that $||z_i - x_i|| \geq 16$, $i = 1, 2, \ldots$. For each integer $i \geq 1$ there exists $\alpha_i > 0$ such that

$$||[(1-\alpha_i)z_i + \alpha_i x_i] - z_i|| = 1. \tag{38}$$

Clearly $\alpha_i \to 0$ as $i \to \infty$. By (3), for each integer $i \geq 1$,

$$D_f(z_i, (1-\alpha_i)z_i + \alpha_i x_i) \leq \alpha_i D_f(z_i, x_i) \leq \alpha_i \beta \to 0 \text{ as } i \to \infty.$$

Combined with (6) this implies that $||z_i - [(1-\alpha_i)z_i + \alpha_i x_i]|| \to 0$ as $i \to \infty$. This contradicts (38). Thus Lemma 5.2 is proved.

Let $T \in \mathcal{M}^{(F)}$ and $\gamma \in (0,1)$. Define a mapping $T_\gamma : K \to K$ by

$$T_\gamma x = (1-\gamma)Tx + \gamma Px, \; x \in K. \tag{39}$$

Clearly $T_\gamma \in \mathcal{M}$, $T_\gamma x = x$, $x \in F$, and if $T, P \in \mathcal{M}_c$, then $T_\gamma \in \mathcal{M}_c$. By (39) and (3), for each $z \in F$ and each $x \in K$,

$$D_f(z, T_\gamma x) = D_f(z, (1-\gamma)Tx + \gamma Px) \leq (1-\gamma)D_f(z, Tx) + \gamma D_f(z, Px) \leq D_f(z, x).$$

Therefore,

$$T_\gamma \in \mathcal{M}^{(F)} \text{ and if } T, P \in \mathcal{M}_c, \text{ then } T_\gamma \in \mathcal{M}_c^{(F)}. \tag{40}$$

Clearly, for each $T \in \mathcal{M}^{(F)}$ we have

$$T_\gamma \to T \text{ as } \gamma \to 0 \text{ in } \mathcal{M}^{(F)}. \tag{41}$$

Lemma 5.3 *Let $T \in \mathcal{M}^{(F)}$, $\gamma, \epsilon \in (0,1)$ and let K_0 be a nonempty bounded subset of K. Then there exists a neighborhood U of T_γ in $\mathcal{M}^{(F)}$ such that the set $\{Sx : \; S \in U, \; x \in K_0\}$ is bounded and for each $S \in U$ and each $x \in K_0$ satisfying $\rho_f(x, F) > \epsilon$, the following inequality holds:*

$$\rho_f(Sx, F) \leq \rho_f(x, F) - \epsilon\gamma/4. \tag{42}$$

Proof. Set

$$U_0 = \{S \in \mathcal{M}^{(F)} : \; ||Sx - T_\gamma x|| \leq 1 \text{ for all } x \in K_0\}. \tag{43}$$

Clearly U_0 is a neighborhood of T_γ in $\mathcal{M}^{(F)}$ and

$$\sup\{||Sx|| : \; S \in U_0, \; x \in K_0\} < \infty. \tag{44}$$

By (34) there is $c_0 > 0$ such that

$$4 + \sup\{\rho_f(x, F) : \; x \in K_0\} < c_0. \tag{45}$$

By Lemma 5.2 there exists a number $c_1 > 0$ such that

$$\text{if } (z, x) \in F \times K_0 \text{ and } D_f(z, x) \leq c_0 + 2, \text{ then } ||z|| \leq c_1. \tag{46}$$

We may assume without loss of generality that

$$c_1 > \sup\{||Px|| : \; x \in K_0\}. \tag{47}$$

Since $D_f(\cdot,\cdot)$ is assumed to be uniformly continuous on bounded subsets of $F \times K$, there exists a number $\delta \in (0, 2^{-1})$ such that for each

$$(z, x_1),\ (z, x_2) \in \{\xi \in F :\ ||\xi|| \le c_1\} \times \cup\{S(K_0) :\ S \in U_0\}$$

satisfying $||x_1 - x_2|| \le \delta$, the following inequality holds:

$$|D_f(z, x_1) - D_f(z, x_2)| \le 4^{-1}\epsilon\gamma. \tag{48}$$

Set

$$U = \{S \in \mathcal{M}^{(F)} :\ ||Sx - T_\gamma x|| \le \delta \text{ for all } x \in K_0\}. \tag{49}$$

Clearly U is a neighborhood of T_γ in $\mathcal{M}^{(F)}$ and the set $\{Sx :\ S \in U,\ x \in K_0\}$ is bounded.

Assume that

$$S \in U,\ x \in K_0 \text{ and } \rho_f(x, F) > \epsilon. \tag{50}$$

We will show that (42) is valid. Let

$$\Delta \in (0, 4^{-1}\gamma\epsilon). \tag{51}$$

There is

$$z \in F \text{ such that } D_f(z, x) \le \rho_f(x, F) + \Delta. \tag{52}$$

By (45), (46), (50) and (52),

$$D_f(z, x) \le c_0 \text{ and } ||z|| \le c_1. \tag{53}$$

By (49) and (4),

$$\begin{aligned} D_f((1-\gamma)z + \gamma Px, T_\gamma x) &= D_f((1-\gamma)z + \gamma Px, (1-\gamma)Tx + \gamma Px) \\ &\le (1-\gamma)D_f(z, Tx) = (1-\gamma)D_f(z, x). \end{aligned} \tag{54}$$

By (50) and (49) we have

$$||T_\gamma x - Sx|| \le \delta. \tag{55}$$

By (53) and (47),

$$\begin{aligned} &((1-\gamma)z + \gamma Px, T_\gamma x), ((1-\gamma)z + \gamma Px, Sx) \\ &\in \{\xi \in F :\ ||\xi|| \le c_1\} \times \cup\{H(K_0) :\ H \in U_0\}. \end{aligned} \tag{56}$$

By (55), (56) and the definition of δ (see (48)),

$$|D_f((1-\gamma)z + \gamma Px, T_\gamma x) - D_f((1-\gamma)z + \gamma Px, Sx)| \le 4^{-1}\epsilon\gamma.$$

Combined with (54), (52) and (51) this inequality implies that

$$\begin{aligned} \rho_f(Sx, F) \le D_f((1-\gamma)z + \gamma Px, Sx) \le 4^{-1}\epsilon\gamma + D_f((1-\gamma)z + \gamma Px, T_\gamma x) \le \\ 4^{-1}\epsilon\gamma + (1-\gamma)D_f(z, x) \le 4^{-1}\epsilon\gamma + (1-\gamma)\rho_f(x, F) + \\ (1-\gamma)\Delta \le (1-\gamma)\rho_f(x, F) + 2^{-1}\epsilon\gamma. \end{aligned}$$

Since (42) follows from the latter inequality and (50), Lemma 5.3 is proved.

We use the convention that $S^0 = I$, the identity operator, for each $S \in \mathcal{M}^{(F)}$.

Lemma 5.4 *Let $T \in \mathcal{M}^{(F)}$, $\gamma, \epsilon \in (0,1)$ and let K_0 be a nonempty bounded subset of K. Then there exists a neighborhood U of T_γ in $\mathcal{M}^{(F)}$ and a natural number N such that for each $S \in U$ and each $x \in K_0$,*

$$\rho_f(S^N x, F) \le \epsilon. \tag{57}$$

Proof. By (34) there is

$$c_0 > \sup\{\rho_f(x,F) : \ x \in K_0\}. \tag{58}$$

Choose a natural number $N \ge 2$ for which

$$8^{-1}\epsilon\gamma N > c_0 + 1. \tag{59}$$

Using Lemma 5.3 and induction we construct a sequence of nonempty bounded sets $K_i \subset K$, $i = 0, \dots, N$, and a sequence of neighborhoods U_i, $i = 0, \dots N-1$, of T_γ in $\mathcal{M}^{(F)}$ such that for all $i = 0, \dots, N-1$ the following properties hold:

(Pi) $K_{i+1} = \{Sx : \ S \in U_i, \ x \in K_i\}$;

(Pii) for each $S \in U_i$ and each $x \in K_i$ satisfying $\rho_f(x,F) > \epsilon$ the following inequality holds:

$$\rho_f(Sx, F) \le \rho_f(x,F) - \epsilon\gamma/4.$$

Set $U = \cap_{i=0}^{N-1} U_i$.

Assume that $S \in U$ and $x \in K_0$. We will show that (57) is valid. Assume the contrary. Then $\rho(S^i x, F) > \epsilon$, $i = 0, \dots, N$. Combined with Property (Pii) this implies that for all $i = 0, \dots, N-1$ we have

$$\rho_f(S^{i+1}x, F) \le \rho_f(S^i x, F) - \epsilon\gamma/4.$$

Therefore

$$\rho_f(S^N x, F) \le \rho_f(x,F) - \epsilon\gamma N/8.$$

Using this inequality, (58) and (59) we obtain

$$0 \le \rho_f(S^n x, F) \le c_0 - 8^{-1}\epsilon\gamma N \le -1,$$

a contradiction. Thus (57) is indeed true and Lemma 5.4 is proved.

Lemma 5.5 *Let $T \in \mathcal{M}^{(F)}$, $\gamma, \epsilon \in (0,1)$ and let K_0 be a nonempty bounded subset of K. Then there exists a neighborhood U of T_γ in $\mathcal{M}^{(F)}$ and a natural number N such that for each $S \in U$ and each $x \in K_0$ there is $z(S,x) \in F$ for which*

$$||S^i x - z(S,x)|| \le \epsilon \text{ for all integers } i \ge N. \tag{60}$$

Proof. By (41) there is $\delta \in (0,1)$ such that

$$\text{if } x \in K, \ z \in F \text{ and } D_f(z,x) \le \delta, \text{ then } ||x - z|| \le 2^{-1}\epsilon. \tag{61}$$

By Lemma 5.4 there exist a neighborhood U of T_γ in $\mathcal{M}^{(F)}$ and a natural number N such that

$$\rho_f(S^N x, F) \le \delta/2 \text{ for each } S \in U \text{ and } x \in K_0.$$

This implies that for each $x \in K_0$ and each $S \in U$, there exists $z(S,x) \in F$ for which $D_f(z(S,x), S^N x) < \delta$. When combined with (51) this inequality implies that for each $x \in K_0$, each $S \in U$, and each integer $i \ge N$ we have

$$D_f(z(S,x), S^i x) < \delta \text{ and } ||S^i x - z(S,x)|| \le 2^{-1}\epsilon.$$

Lemma 5.5 is proved.

Completion of the proof of Theorem 4.1. By (41), the set $\{T_\gamma : T \in \mathcal{M}^{(F)}, \gamma \in (0,1)\}$ is an everywhere dense subset of $\mathcal{M}^{(F)}$ and if $P \in \mathcal{M}_c^{(F)}$, then $\{T_\gamma : T \in \mathcal{M}_c^{(F)}, \gamma \in (0,1)\}$ is an everywhere dense subset of $\mathcal{M}_c^{(F)}$. Fix $\theta \in K$. For each natural number i set

$$K_i = \{x \in K : ||x - \theta|| \le i\}. \tag{62}$$

By Lemma 5.5, for each $T \in \mathcal{M}^{(F)}$, each $\gamma \in (0,1)$ and each integer $i \ge 1$ there exist an open neighborhood $\mathcal{U}(T,\gamma,i)$ of T_γ in $\mathcal{M}^{(F)}$ and a natural number $N(T,\gamma,i)$ such that the following property holds:

P(iii) For each $x \in K_{2^i}$ and each $S \in \mathcal{U}(T,\gamma,i)$, there is $z(S,x) \in F$ for which

$$||S^n x - z(S,x)|| \le 2^{-i} \text{ for all integers } n \ge N(T,\gamma,i).$$

Define

$$\mathcal{F} = \cap_{q=1}^{\infty} \cup \{\mathcal{U}(T,\gamma,i) : T \in \mathcal{M}^{(F)},\ \gamma \in (0,1),\ i \ge q\}.$$

Clearly $\mathcal{F}$ is a countable intersection of open everywhere dense subsets of $\mathcal{M}^{(F)}$. If $P \in \mathcal{M}_c^{(F)}$, then define

$$\mathcal{F}_c = [\cap_{q=1}^{\infty} \cup \{\mathcal{U}(T,\gamma,i) : T \in \mathcal{M}_c^{(F)},\ \gamma \in (0,1),\ i \ge q\}] \cap \mathcal{M}_c^{(F)}.$$

In this case $\mathcal{F}_c \subset \mathcal{F}$ and $\mathcal{F}_c$ is a countable intersection of open everywhere dense subsets of $\mathcal{M}_c^{(F)}$.

Let $B \in \mathcal{F}$, $\epsilon > 0$ and let C be a bounded subset of K. There exists an integer $q \ge 1$ such that

$$C \subset K_{2^q} \text{ and } 2^{-q} < 4^{-1}\epsilon. \tag{63}$$

There also exist $T \in \mathcal{M}^{(F)}$, $\gamma \in (0,1)$ and an integer $i \ge q$ such that

$$B \in \mathcal{U}(T,\gamma,i). \tag{64}$$

Note that if $P \in \mathcal{M}_c^{(F)}$ and $B \in \mathcal{F}_c$, then $T \in \mathcal{M}_c^{(F)}$.

It follows from Property P(iii), (63) and (64) that the following property holds:

(Piv) for each $S \in \mathcal{U}(T,\gamma,i)$ and each $x \in C$, there is $z(S,x) \in F$ such that $||S^n x - z(S,x)|| \le 4^{-1}\epsilon$ for each integer $n \ge N(T,\gamma,i)$.

Property P(iv) and (64) imply that for each $x \in C$ and each integer $n \ge N(T,\gamma,i)$,

$$||B^n x - z(B,x)|| \le 4^{-1}\epsilon. \tag{65}$$

Since ϵ is an arbitrary positive number and C is an arbitrary bounded subset of K, we conclude that for each $x \in K$, $\{B^n x\}_{n=1}^{\infty}$ is a Cauchy sequence. Therefore for each $x \in K$ there exists the limit

$$P_B x = \lim_{n \to \infty} B^n x.$$

In view of (65) we see that, for each $x \in C$,

$$||P_B x - z(B, x)|| \leq 4^{-1}\epsilon. \tag{66}$$

Once again, since ϵ is an arbitrary positive number and C is an arbitrary bounded subset of K, we conclude that $P_B(K) = F$. It follows from (65) and (66) that for each $x \in C$ and each integer $n \geq N(T, \gamma, i)$,

$$||B^n x - P_B x|| \leq 2^{-1}\epsilon.$$

This implies that $P_B \in \mathcal{M}^{(F)}$ and if $B \in \mathcal{M}_c^{(F)}$, then $P_B \in \mathcal{M}_c^{(F)}$. This completes the proof of Theorem 4.1.

In the proof of Theorem 4.2 we will use the following lemma which can be obtained by induction on n.

Lemma 5.6 *Let $B \in \mathcal{M}_c^{(F)}$, $x \in K$, $\epsilon \in (0, 1)$ and let $n \geq 1$ be an integer. Then there exists a neighborhood U of B in $\mathcal{M}^{(F)}$ and a number $\delta > 0$ such that for each $S \in U$ and each $y \in K$ satisfying $||y - x|| \leq \delta$, the following inequality holds:*

$$||S^n y - B^n x|| \leq \epsilon.$$

Proof of Theorem 4.2. By Theorem 4.1 there exists a natural number N and a neighborhood U_0 of B in $\mathcal{M}^{(F)}$ such that

$$||P_B y - B^n y|| \leq 8^{-1}\epsilon \text{ for each } y \in K \text{ satisfying } ||y - x|| \leq 1; \tag{67}$$

for each $S \in U_0$ and each $y \in K$ satisfying $||y - x|| \leq 1$, there is $z(S, y) \in F$ such that

$$||S^n y - z(S, y)|| \leq 8^{-1}\epsilon \text{ for all integers } n \geq N. \tag{68}$$

By Lemma 5.6 there exist a number $\delta \in (0, 1)$ and a neighborhood U of B in $\mathcal{M}^{(F)}$ such that $U \subset U_0$ and

$$||S^N y - B^N x|| \leq 8^{-1}\epsilon \text{ for each } S \in U \text{ and each } y \in K \text{ for which } ||y - x|| \leq \delta. \tag{69}$$

Assume that

$$y \in K, \ ||x - y|| \leq \delta \text{ and } S \in U. \tag{70}$$

By (70), (69) and (67),

$$||S^N y - B^N x|| \leq 8^{-1}\epsilon, \ ||S^N y - z(S, y)|| \leq 8^{-1}\epsilon \text{ and } ||P_B x - B^N x|| \leq 8^{-1}\epsilon.$$

These three inequalities imply that

$$||z(S, y) - P_B x|| \leq 3 \cdot 8^{-1}\epsilon.$$

Combined with (68) the latter inequality implies that

$$||S^n y - P_B x|| \leq 2^{-1}\epsilon \text{ for all integers } n \geq N.$$

Theorem 4.2 is proved.

6. CONVERGENCE OF POWERS FOR CONTINUOUS MAPPINGS

In this section we assume that there is $P \in \mathcal{M}_c^{(F)}$ such that

$$P(K) = F. \tag{71}$$

The following three theorems will be established in Section 7.

Theorem 6.1 *Let $x \in K$. Then there exists a set $\mathcal{F} \subset \mathcal{M}_c^{(F)}$ which is a countable intersection of open everywhere dense subsets of $\mathcal{M}_c^{(F)}$ such that for each $B \in \mathcal{F}$ the following assertions hold:*

1. There exists $\lim_{n\to\infty} B^n x \in F$.

2. For each $\epsilon > 0$ there exist a neighborhood U of B in $\mathcal{M}_c^{(F)}$, a natural number N and a number $\delta > 0$ such that for each $S \in U$, each $y \in K$ satisfying $||y - x|| \le \delta$ and each integer $n \ge N$, $||S^n y - \lim_{i\to\infty} B^i x|| \le \epsilon$.

In the next two theorems we consider the space $K \times \mathcal{M}_c^{(F)}$ with the product topology.

Theorem 6.2 *There exists a set $\mathcal{F} \subset K \times \mathcal{M}_c^{(F)}$ which is a countable intersection of open everywhere dense subsets of $K \times \mathcal{M}_c^{(F)}$ such that for each $(z, B) \in \mathcal{F}$ the following two assertions hold:*

1. There exists $\lim_{n\to\infty} B^n z \in F$.

2. For each $\epsilon > 0$ there exist a neighborhood U of (z, B) in $K \times \mathcal{M}_c^{(F)}$ and a natural number N such that for each $(y, S) \in U$ and each integer $n \ge N$,

$$||S^n y - \lim_{i\to\infty} B^i z|| \le \epsilon.$$

Theorem 6.3 *Assume that the set K_0 is a nonempty separable closed subset of K. Then there exists a set $\mathcal{F} \subset \mathcal{M}_c^{(F)}$ which is a countable intersection of open everywhere dense subsets of $\mathcal{M}_c^{(F)}$ such that for each $T \in \mathcal{F}$ there exists a set $\mathcal{K}_T \subset K_0$ which is a countable intersection of open everywhere dense subsets of K_0 with the relative topology such that the following two assertions hold:*

1. For each $x \in \mathcal{K}_T$ there exists $\lim_{n\to\infty} T^n x \in F$.

2. For each $x \in \mathcal{K}_T$ and each $\epsilon > 0$ there exist an integer $N \ge 1$ and a neighborhood U of (x, T) in $\mathcal{K} \times \mathcal{M}_c^{(F)}$ such that for each $(y, S) \in U$ and each integer $i \ge N$, $||S^i y - \lim_{n\to\infty} T^n x|| \le \epsilon$.

7. PROOFS OF THEOREMS 6.1-6.3

Let $T \in \mathcal{M}_c^{(F)}$ and $\gamma \in (0, 1)$. Define a mapping $T_\gamma : K \to K$ by

$$T_\gamma x = (1 - \gamma)Tx + \gamma Px, \; x \in K. \tag{72}$$

Clearly, $T_\gamma \in \mathcal{M}_c$, $T_\gamma x = x$, $x \in F$, and for each $z \in F$ and each $x \in K$,

$$D_f(z, T_\gamma x) = D_f(z, (1 - \gamma)Tx + \gamma Px) \le$$
$$(1 - \gamma)D_f(z, Tx) + \gamma D_f(z, Px) \le D_f(z, x).$$

Therefore $T_\gamma \in \mathcal{M}_c^{(F)}$. Clearly, for each $T \in \mathcal{M}_c^{(F)}$ we have

$$T_\gamma \to T \text{ as } \gamma \to 0 \text{ in } \mathcal{M}_c^{(F)}. \tag{73}$$

In the proofs of Theorems 6.1-6.3 we will use the followings two lemmas. The proof of the first one is identical with that of Lemma 3.1 and therefore we omit it.

Lemma 7.1 *Let $T \in \mathcal{M}_c^{(F)}$ and $\gamma \in (0,1)$. Then for each $x \in K$,*

$$\rho_f(T_\gamma x, F) \leq (1-\gamma)\rho_f(x, F).$$

Lemma 7.2 *Let $T \in \mathcal{M}_c^{(F)}$, $\gamma, \epsilon \in (0,1)$ and let $x \in K$. Then there exist a neighborhood U of T_γ in $\mathcal{M}_c^{(F)}$, a natural number N, $\hat{z} \in F$ and a number $\delta > 0$ such that for each $S \in U$, each $y \in K$ satisfying $||y - x|| \leq \delta$ and each integer $n \geq N$,*

$$||S^n y - \hat{z}|| \leq \epsilon. \tag{74}$$

Proof. By (6) there is $\epsilon_0 \in (0, \epsilon/2)$ such that

$$\text{if } z \in F,\ y \in K \text{ and } D_f(z, y) \leq 2\epsilon_0, \text{ then } ||z - y|| \leq \epsilon/2. \tag{75}$$

Choose a natural number N for which

$$(1-\gamma)^N(\rho_f(x, F) + 1) < \epsilon_0/8. \tag{76}$$

When combined with Lemma 7.1 this inequality implies that

$$\rho_f(T_\gamma^N x, F) < \epsilon_0/8.$$

Hence there exists $\hat{z} \in F$ for which

$$D_f(\hat{z}, T_\gamma^N x) < \epsilon_0/8. \tag{77}$$

Since the function $D_f(\hat{z}, \cdot) : K \to R^1$ is continuous (see (3)), there exists $\epsilon_1 \in (0, \epsilon_0/2)$ such that

$$D_f(\hat{z}, \xi) < \epsilon_0/8 \text{ for all } \xi \in K \text{ satisfying } ||\xi - T_\gamma^N x|| \leq \epsilon_1. \tag{78}$$

It follows from the continuity of T_γ that there exist a neighborhood U of T_γ in $\mathcal{M}_c^{(F)}$ and a number $\delta > 0$ such that for each $S \in U$ and each $y \in K$ satisfying $||y - x|| \leq \delta$ we have

$$||S^N y - T_\gamma^N x|| \leq \epsilon_1. \tag{79}$$

Assume that

$$S \in U,\ y \in K \text{ and } ||y - x|| \leq \delta.$$

By the definition of U and δ, the inequality (79) is valid. By (79) and (78), $D_f(\hat{z}, S^N y) < \epsilon_0/8$. This implies that $D_f(\hat{z}, S^n y) < \epsilon_0/8$ for all integers $n \geq N$. Combined with (75) the latter inequality implies that $||\hat{z} - S^n y|| \leq \epsilon$ for all integers $n \geq N$. This completes the proof of Lemma 7.2.

Proof of Theorem 6.1. Let $x \in K$. By Lemma 7.2, for each $T \in \mathcal{M}_c^{(F)}$, each $\gamma \in (0,1)$ and each integer $i \geq 1$, there exist an open neighborhood $\mathcal{U}(T,\gamma,i)$ of T_γ in $\mathcal{M}_c^{(F)}$, a natural number $N(T,\gamma,i)$, a point $z(T,\gamma,i) \in F$ and a number $\delta(T,\gamma,i) > 0$ such that the following property holds:

(Pi) For each $S \in \mathcal{U}(T,\gamma,i)$, each $y \in K$ satisfying $||x-y|| \leq \delta(T,\gamma,i)$ and each integer $n \geq N(T,\gamma,i)$,

$$||S^n y - z(T,\gamma,i)|| \leq 2^{-i}.$$

Define

$$\mathcal{F} = \cap_{q=1}^{\infty} \cup \{\mathcal{U}(T,\gamma,i) : \ T \in \mathcal{M}_c^{(F)}, \ \gamma \in (0,1), \ i \geq q\}.$$

Clearly $\mathcal{F}$ is a countable intersection of open everywhere dense subsets of $\mathcal{M}_c^{(F)}$.

Let $B \in \mathcal{F}$ and $\epsilon > 0$. There exists an integer $q \geq 1$ such that

$$2^{-q} < 4^{-1}\epsilon. \tag{80}$$

There also exist $T \in \mathcal{M}_c^{(F)}$, $\gamma \in (0,1)$ and an integer $i \geq q$ such that

$$B \in \mathcal{U}(T,\gamma,i). \tag{81}$$

It follows from Property (Pi) and (80) that the following property also holds:

(Pii) For each $S \in \mathcal{U}(T,\gamma,i)$, each $y \in K$ satisfying $||y-x|| \leq \delta(T,\gamma,i)$ and each integer $n \geq N(T,\gamma,i)$,

$$||S^n y - z(T,\gamma,i)|| \leq 4^{-1}\epsilon. \tag{82}$$

Since ϵ is an arbitrary positive number we conclude that $\{B^n x\}_{n=1}^{\infty}$ is a Cauchy sequence and there exists $\lim_{n\to\infty} B^n x$. The inequality (82) implies that

$$||\lim_{n\to\infty} B^n x - z(T,\gamma,i)|| \leq 4^{-1}\epsilon.$$

Since ϵ is an arbitrary positive number we conclude that $\lim_{n\to\infty} B^n x \in F$. It follows from this relation and Property (Pi) that for each $S \in \mathcal{U}(T,\gamma,i)$, each $y \in K$ satisfying $||y-x|| \leq \delta(T,\gamma,i)$ and each integer $n \geq N(T,\gamma,i)$,

$$||S^n y - \lim_{i\to\infty} B^i x|| \leq 2^{-1}\epsilon.$$

Theorem 6.1 is established.

Proof of Theorem 6.2. By Lemma 7.2, for each $(x,T) \in K \times \mathcal{M}_c^{(F)}$, each $\gamma \in (0,1)$ and each integer $i \geq 1$, there exist an open neighborhood $\mathcal{U}(x,T,\gamma,i)$ of (x,T_γ) in $K \times \mathcal{M}_c^{(F)}$, a natural number $N(x,T,\gamma,i)$ and a point $z(x,T,\gamma,i) \in F$ such that the following property holds:

(Piii) For each $(y,S) \in \mathcal{U}(x,T,\gamma,i)$ and each integer $n \geq N(x,T,\gamma,i)$,

$$||S^n y - z(x,T,\gamma,i)|| \leq 2^{-i}.$$

Define

$$\mathcal{F} = \cap_{q=1}^{\infty} \cup \{\mathcal{U}(x,T,\gamma,i) : \ (x,T) \in K \times \mathcal{M}_c^{(F)}, \ \gamma \in (0,1), \ i \geq q\}.$$

Clearly $\mathcal{F}$ is a countable intersection of open everywhere dense subsets of $K \times \mathcal{M}_c^{(F)}$.

Let $(z, B) \in \mathcal{F}$ and $\epsilon > 0$. There exists an integer $q \geq 1$ such that

$$2^{-q} < 4^{-1}\epsilon. \tag{83}$$

There also exist $x \in K$, $T \in \mathcal{M}_c^{(F)}$, $\gamma \in (0, 1)$ and an integer $i \geq q$ such that

$$(z, B) \in \mathcal{U}(x, T, \gamma, i). \tag{84}$$

By (83) and property (Piii) the following property holds:

(Piv) For each $(y, S) \in \mathcal{U}(x, T, \gamma, i)$ and each integer $n \geq N(x, T, \gamma, i)$,

$$||S^n y - z(x, T, \gamma, i)|| \leq 4^{-1}\epsilon. \tag{85}$$

Since ϵ is an arbitrary positive number we conclude that $\{B^n z\}_{n=1}^{\infty}$ is a Cauchy sequence and there exists $\lim_{n\to\infty} B^n z$. Property (Piv) and (84) imply that

$$|| \lim_{n\to\infty} B^n z - z(x, T, \gamma, i)|| \leq 4^{-1}\epsilon. \tag{86}$$

Since ϵ is an arbitrary positive number we conclude that $\lim_{n\to\infty} B^n z \in F$. It follows from (86) and Property (Piv) that for each $(y, S) \in \mathcal{U}(x, T, \gamma, i)$ and each integer $n \geq N(x, T, \gamma, i)$,

$$||S^n y - \lim_{i\to\infty} B^i z|| \leq 2^{-1}\epsilon.$$

Theorem 6.2 is proved.

Proof of Theorem 6.3. Assume that K_0 is a nonempty closed separable subset of K. Let the sequence $\{x_j\}_{j=1}^{\infty}$ be dense in K_0. By Theorem 6.1, for each integer $p \geq 1$ there exists a set $\mathcal{F}_p \subset \mathcal{M}_c^{(F)}$ which is a countable intersection of open everywhere dense subsets of $\mathcal{M}_c^{(F)}$ such that for each $T \in \mathcal{F}_p$ the following two properties hold:

C(i) There exists $\lim_{n\to\infty} T^n x_p \in F$.

C(ii) For each $\epsilon > 0$, there exist a neighborhood U of T in $\mathcal{M}_c^{(F)}$, a number $\delta > 0$ and a natural number N such that for each $S \in U$, each $y \in K$ satisfying $||y - x_p|| \leq \delta$ and each integer $m \geq N$,

$$||S^m y - \lim_{n\to\infty} T^n x_p|| \leq \epsilon.$$

Set

$$\mathcal{F} = \cap_{p=1}^{\infty} \mathcal{F}_p. \tag{87}$$

Clearly $\mathcal{F}$ is a countable intersection of open everywhere dense subsets of $\mathcal{M}_c^{(F)}$.

Assume that $T \in \mathcal{F}$. Then for each $p \geq 1$ there exists $\lim_{n\to\infty} T^n x_p \in F$.

Now we will construct the set $\mathcal{K}_T \subset K_0$. By property C(ii), for each pair of natural numbers q, i there exist a neighborhood $\mathcal{U}(q, i)$ of T in $\mathcal{M}_c^{(F)}$, a number $\delta(q, i) > 0$ and a natural number $N(q, i)$ such that the following property holds:

C(iii) For each $S \in \mathcal{U}(q, i)$, each $y \in K$ satisfying $||y - x_q|| \leq \delta(q, i)$ and each integer $m \geq N(q, i)$,

$$||S^m y - \lim_{n\to\infty} T^n x_q|| \leq 2^{-i}.$$

Define

$$\mathcal{K}_T = \cap_{n=1}^{\infty} \cup \{\{y \in K_0 : \ ||y - x_q|| < \delta(q,i)\} : \ q \geq 1, \ i \geq n\}. \tag{88}$$

Clearly $\mathcal{K}_T$ is a countable intersection of open everywhere dense subsets of K_0.

Assume that $x \in \mathcal{K}_T$ and $\epsilon > 0$. There exists an integer $n \geq 1$ such that

$$2^{-n} < 4^{-1}\epsilon. \tag{89}$$

By (88) there exist a natural number q and an integer $i \geq n$ such that

$$||x - x_q|| < \delta(q,i). \tag{90}$$

It follows from (89) and C(iii) that the following property also holds:

C(iv) For each $S \in \mathcal{U}(q,i)$, each $y \in K$ satisfying $||y - x_q|| \leq \delta(q,i)$ and each integer $m \geq N(q,i)$,

$$||S^m y - \lim_{j\to\infty} T^j x_q|| \leq 4^{-1}\epsilon.$$

By property C(iv) and (90) we have

$$||T^m x - \lim_{j\to\infty} T^j x_q|| \leq 4^{-1}\epsilon$$

for all integers $m \geq N(q,i)$. Since ϵ is an arbitrary positive number we conclude that $\{T^m x\}_{m=1}^{\infty}$ is a Cauchy sequence, there exists $\lim_{m\to\infty} T^m x$, and

$$||\lim_{m\to\infty} T^m x - \lim_{m\to\infty} T^m x_q|| \leq 4^{-1}\epsilon.$$

Since ϵ is an arbitrary positive number we conclude that $\lim_{m\to\infty} T^m x \in F$. By this relation and property (Civ), for each $S \in \mathcal{U}(q,i)$, each $y \in K$ satisfying $||y - x|| < \delta(q,i) - ||x - x_q||$, and each integer $m \geq N(q,i)$ we have

$$||S^m y - \lim_{j\to\infty} T^j x|| \leq 2^{-1}\epsilon.$$

Theorem 6.3 is established.

REFERENCES

1. E. Asplund, Fréchet differentiability of convex functions, *Acta Math.* **121** (1968) 31–47.
2. J.M. Ball and N.S. Nadirashvili, Universal singular sets for one-dimensional variational problems, *Calc. Var.* **1** (1993) 429–438.
3. H.H. Bauschke and J.M. Borwein, Joint and separate convexity of the Bregman distance, *Inherently Parallel Algorithms in Feasibility and Optimization and their Applications* (2001), this volume.
4. J.M. Borwein and D. Preiss, A smooth variational principle with applications to subdifferentiability and to differentiability of convex functions, *Trans. Amer. Math. Soc.* **303** (1987) 517–527.

5. L.M. Bregman, The relaxation method for finding the common point of convex sets and its application to the solution of problems in convex programming, *USSR Computational Math. and Math. Phys.* **7** (1967) 200–217.
6. L.M. Bregman, Y. Censor and S. Reich, Dykstra's algorithm as the nonlinear extension of Bregman's optimization method, *J. Convex Analysis* **6** (1999) 319–333.
7. D. Butnariu, Y. Censor and S. Reich, Iterative averaging of entropic projections for solving stochastic convex feasibility problems, *Computational Optimization and Applications* **8** (1997) 21–39.
8. D. Butnariu and A.N. Iusem, Local moduli of convexity and their application to finding almost common fixed points of measurable families of operators, *Recent Developments in Optimization Theory and Nonlinear Analysis, Contemporary Mathematics* **204** (1997) 61–91.
9. D. Butnariu and A.N. Iusem, *Totally Convex Functions for Fixed Points Computation and Infinite Dimensional Optimization* (Kluwer, Dordrecht, 2000).
10. D. Butnariu, S. Reich and A.J. Zaslavski, Generic power convergence of operators in Banach spaces, *Numerical Functional Analysis and Optimization* **20** (1999) 629–650.
11. D. Butnariu, S. Reich and A.J. Zaslavski, Asymptotic behavior of relatively nonexpansive operators in Banach spaces, *Preprint* (2000).
12. Y. Censor and A. Lent, An iterative row-action method for interval convex programming, *J. Optim. Theory Appl.* **34** (1981) 321–353.
13. Y. Censor and S. Reich, Iterations of paracontractions and firmly nonexpansive operators with applications to feasibility and optimization, *Optimization* **37** (1996) 323–339.
14. Y. Censor and S.A. Zenios, *Parallel Optimization* (Oxford, New York, 1997).
15. F.S. De Blasi and J. Myjak, Sur la convergence des approximations successives pour les contractions non linéaires dans un espace de Banach, *C. R. Acad. Sci. Paris* **283** (1976) 185–187.
16. F.S. De Blasi and J. Myjak, Generic flows generated by continuous vector fields in Banach spaces, *Adv. in Math.* **50** (1983) 266–280.
17. F.S. De Blasi and J. Myjak, On a generalized best approximation problem, *J. Approximation Theory* **94** (1998) 54–72.
18. R. Deville and J. Revalski, Porosity of ill-posed problems, *Proc. Amer. Math. Soc.* **128** (2000) 1117–1124.
19. A.L. Dontchev and T. Zolezzi, *Well-Posed Optimization Problems. Lecture Notes in Math.* **1543** (Springer, Berlin, 1993).
20. A.D. Ioffe and A.J. Zaslavski, Variational principles and well-posedness in optimization and calculus of variations, *SIAM J. Control Optim.* **38** (2000) 566–581.
21. J.L. Kelley, *General Topology* (D. Van Nostrand, New York, 1955).
22. G. Lebourg, Generic differentiability of Lipschitzian functions, *Trans. Amer. Math. Soc.* **256** (1979) 125–144.
23. J. Myjak, Orlicz type category theorems for functional and differential equations, *Dissertationes Math. (Rozprawy Mat.)* **206** (1983) 1–81.
24. S. Reich, A weak convergence theorem for the alternating method with Bregman distances, *Theory and Applications of Nonlinear Operators of Accretive and Monotone Type (A.G. Kartsatos, Editor)* (1996) 313–318.

25. S. Reich and A.J. Zaslavski, Convergence of generic infinite products of nonexpansive and uniformly continuous operators, *Nonlinear Analysis* **36** (1999) 1049–1065.
26. A.J. Zaslavski, Turnpike property for extremals of variational problems with vector-valued functions, *Trans. Amer. Math. Soc.* **351** (1999) 211–231.
27. A.J. Zaslavski, Existence of solutions of optimal control problems for a generic integrand without convexity assumptions, *Nonlinear Analysis* **43** (2001) 339–361.
28. A.J. Zaslavski, Generic well-posedness of optimal control problems without convexity assumptions, *SIAM J. Control Optim.*, in press.

Inherently Parallel Algorithms in Feasibility and Optimization and their Applications
D. Butnariu, Y. Censor and S. Reich (Editors)

THE OUTER BREGMAN PROJECTION METHOD FOR STOCHASTIC FEASIBILITY PROBLEMS IN BANACH SPACES

Dan Butnariu[a] and Elena Resmerita[b*]

[a]Department of Mathematics, University of Haifa,
31905 Haifa, Israel

[b]Department of Mathematics, University of Haifa,
31905 Haifa, Israel

We present an iterative method for solving, eventually infinite systems of inequalities $g(\omega, x) \leq 0$, a.e.(Ω), in which the functions $g(\omega, \cdot)$ are defined, convex, lower semicontinuous and essentially uniformly bounded on a Banach space. We show that the proposed method produces sequences which accumulate weakly to solutions of the system, provided that such solutions exist. In the most usual Banach spaces, the size of the constraint violations along the sequences generated by the iterative procedure converges in mean to zero. We prove that this method can be implemented for solving consistent operator equations (like first kind Fredholm or Volterra equations, some potential equations, etc.) when they have solutions in Hilbert, Lebesgue or Sobolev spaces.

1. INTRODUCTION

In this paper we consider the following *stochastic convex feasibility problem*: Given a complete probability space $(\Omega, \mathcal{A}, \mu)$, a uniformly convex, separable and smooth Banach space X and a function $g : \Omega \times X \to \mathbb{R}$ satisfying the following conditions:

$(C1)$ For each $x \in X$, the function $g(\cdot, x)$ is measurable;

$(C2)$ For each $\omega \in \Omega$, the function $g(\omega, \cdot)$ is convex and lower semicontinuous;

$(C3)$ The family of functions $g(\omega, \cdot)$, $\omega \in \Omega$, is essentially uniformly bounded on bounded sets, that is, for any bounded set $E \subseteq X$, there exists a real number $M_E > 0$ such that, for all $x \in E$, $|g(\omega, x)| \leq M_E$, a.e.;

Find a point $x^* \in X$ such that

$$g(\omega, x^*) \leq 0, \text{ a.e.}, \tag{1}$$

provided that such a point exists. One should note that this problem is equivalent to that of finding an almost common point (see [5]) of the sets

$$Q_\omega := \{x \in X : g(\omega, x) \leq 0\}, \ \omega \in \Omega, \tag{2}$$

*The authors gratefully acknowledge the financial support of the Israel Science Foundation. Also, they are grateful to Alfredo N. Iusem and Simeon Reich for comments and suggestions which led to improvements of earlier versions of this paper.

where, according to $(C1)$, $(C2)$ and Theorem 8.2.9 in [3], the point-to-set mapping $\omega \to Q_\omega : \Omega \to X$ is closed, convex valued and measurable.

Our purpose is to show that, in the specified context, the algorithm described below, which we call the *Outer-Bregman-Projection Method (OBPM)*, is well defined and produces bounded sequences whose weak accumulation points are solutions of the given stochastic convex feasibility problem. Moreover, under some additional conditions which, as we will show, are satisfied in the most useful spaces like the Hilbert spaces, the Lebesgue spaces $\mathcal{L}^p$ and the Sobolev spaces $\mathcal{W}^{m,p}$ with $p \in (1, \infty)$, the sequences $\{x^k\}_{k \in \mathbb{N}}$ produced by the OBPM have the property that the size of the constraint violation at each iterative step, $\max\left[g(\omega, x^k), 0\right]$, converges in mean to zero.

The OBPM is the following procedure: Fix a number $r \in (1, \infty)$, choose an *initial point* $x^0 \in X$ and, for any nonnegative integer k, choose a measurable selector Γ_k of the point-to-set mapping $\omega \to \partial g(\omega, \cdot)(x^k)$; then, let

$$x^{k+1} = \int_\Omega J_r^* \left(s_k(\omega)\Gamma_k(\omega) + J_r x^k\right) d\mu(\omega), \tag{3}$$

where $J_r : X \to X^*$ and $J_r^* : X^* \to X$ are the duality mappings (see [13]) of weights $\psi_r(t) = rt^{r-1}$ and $\psi_r^{-1}(t) = \left(\frac{t}{r}\right)^{1/(r-1)}$, respectively, and $s_k(\omega)$ is a (necessarily existing) negative solution of the equation

$$\langle \Gamma_k(\omega), J_r^*(s_k(\omega)\Gamma_k(\omega) + J_r x^k) - x^k \rangle = -g(\omega, x^k), \tag{4}$$

when $g(\omega, x^k) > 0$ and $s_k(\omega) = 0$, otherwise.

The geometric idea underlying the OBPM consists of obtaining the new iterate x^{k+1} by averaging the Bregman projections of the current iterate x^k with respect to the function $f(x) = \|x\|^r$ onto the closed half space

$$H_k(\omega) = \{y \in X : \langle \Gamma_k(\omega), y - x^k \rangle + g(\omega, x^k) \le 0\}, \tag{5}$$

which contains the closed convex set Q_ω. Recall that, according to [12], the Bregman projection of $x \in X$ with respect to $f(x) = \|x\|^r$ onto a closed, convex, nonempty set $C \subseteq X$ is the (necessarily unique) point

$$\Pi_C^f(x) = \arg\min \{D_f(y, x) : y \in C\},$$

where $D_f : X \times X \to [0, \infty)$, usually called *Bregman distance,* is given by

$$D_f(y, x) = \|y\|^r - \|x\|^r - \langle f'(x), y - x \rangle, \tag{6}$$

with $f' : X \to X^*$ being the (Gâteaux) derivative of f.

Compared with the methods of solving the stochastic convex feasibility problem (1) which are based on averaging Bregman projections onto the sets Q_ω (see [5], [10], [9], [6], [7], [8] and the references therein) the OBPM presents the advantage that it does not require determining difficult to compute minimizers of nonlinear functions over the sets Q_ω. For the most usual smooth, separable and uniformly convex Banach spaces there are already known formulae for calculating the duality mappings J_r and J_r^* involved in the previous procedure (see [13, pp. 72-73]). Also, a careful analysis of the proof of Theorem 2 in [11], which guarantees the existence of the numbers $s_k(\omega)$ required by the

OBPM, shows that finding the negative solutions $s_k(\omega)$ of (4) amounts to finding zeros of continuous and increasing real functions and this can be done by standard methods (e.g., by Newton's method). Therefore, the implementation of OBPM is quite easy.

The idea of solving convex feasibility problems by using Bregman projections onto half spaces containing the sets Q_ω goes back to [12] and was repeatedly used since then - see, for instance, [15], [17], [18], [19], [20]. Expanding this idea to (possibly infinite) stochastic feasibility problems in infinite dimensional Banach spaces is now possible due to recent results concerning the integrability of measurable families of totally nonexpansive operators (see [9, Chapter 2]) and due to the results established in [11] concerning the total convexity of the powers of the norm in uniformly convex Banach spaces and the relatively simple formulae of computing Bregman projections with respect to them. Guaranteeing that the OBPM works in this more general setting is of interest because of the practically significant applied mathematical problems whose straightforward formulation and natural environment is that of an infinite system of convex inequalities in a uniformly convex Banach space. It was shown in [9, Section 2.4] that finding minima of convex continuous functions, finding Nash equilibria in n-person noncooperative games, solving significant classes of variational inequalities and solving linear operator equations $Tx = b$ in some function spaces is equivalent to solving infinite systems of inequalities (1). Such problems are, usually, unstable in the sense that their "discretization" to finite feasibility problems may lead to "approximate solutions" which are far from the solution set of the problem itself (this phenomenon can be observed, for instance, when one reduce Fredholm integral equations to finite systems of equalities). The use of OBPM for solving the problems in their intrinsic setting may avoid such difficulties.

It is of practical interest to observe that the OBPM described above can be used for solving stochastic convex feasibility problems which are not directly given in the form (1). That is the case of the problem of finding an almost common point of a family of closed convex sets R_ω, $\omega \in \Omega$, where the mapping $\omega \to R_\omega : \Omega \to X$ is measurable. Such problems appear, for instance, from operatorial equations in the form $T(\omega, x) = u(\omega)$ a.e., where $T(\omega, \cdot)$, $\omega \in \Omega$, are continuous linear operators from X to a Banach space Y and the function $u : \Omega \to Y$ as well as the functions $T(\cdot, x)$ whenever $x \in X$, are measurable. Solving this operatorial equation amounts to finding an almost common point of the closed convex sets

$$R_\omega = \{x \in X : T(\omega, x) = u(\omega)\}. \tag{7}$$

In cases like that one can re-write the problem in the form of a system of inequalities (1) with

$$g(\omega, x) := d^2(x, R_\omega), \tag{8}$$

where $d^2(x, A)$ stands for the square of the distance from $x \in X$ to the set $A \subseteq X$. According to [3, Theorem 8.2.11] and [14, Proposition 2.4.1], this function g satisfies the conditions $(C1)$, $(C2)$ and $(C3)$ required above. Moreover, since X is uniformly convex and smooth, the functions $g(\omega, \cdot)$ given by (8) are differentiable and

$$\partial g(\omega, \cdot)(x) = \{J_2(x - P_\omega x)\},$$

where $P_\omega : X \to R_\omega$ is the metric projection operator onto the set R_ω. Thus, in this situation, the OBPM described by (3) reduces to

$$x^{k+1} = \int_\Omega J_r^* \left(s_k(\omega)J_2(x^k - P_\omega x^k) + J_r x^k\right) d\mu(\omega),$$

and it is effectively calculable whenever the metric projections P_ω are so.

A computational feature of the OBPM is its implementability via parallel processing. At each step k of the iterative procedure defined by (3) one can compute in parallel the vector $J_r x^k$ and the subgradients $\Gamma_k(\omega)$ at various points $\omega \in \Omega$ and, then, one can proceed by parallel processing to solve the corresponding equation (4) in order to determine the numbers $s_k(\omega)$.

2. CONVERGENCE ANALYSIS OF THE OBPM

2.1. In this section we consider the stochastic feasibility problem (1) under the conditions $(C1)$, $(C2)$ and $(C3)$ given in Section 1. Our objective is to prove well definedness of the OBPM and to establish useful convergence properties for the sequences it generates. To this end, recall that the *modulus of total convexity of the function f at the point $x \in X$* (see [9, Section 1.2]) is defined by

$$\nu_f(x,t) = \inf\{D_f(y,x) : \|y - x\| = t\}. \tag{9}$$

We associate with any nonempty set $E \subseteq X$ the function $\nu_f(E,\cdot) : [0,\infty) \to [0,\infty)$ given by

$$\nu_f(E,t) = \inf\{\nu_f(x,t) : x \in E\}. \tag{10}$$

Clearly, $\nu_f(E,0) = 0$. It was shown in [11] that for the function $f(x) = \|x\|^r$ with $r \in (1,\infty)$, the function $\nu_f(E,\cdot)$ is strictly positive on $(0,\infty)$ when E is bounded. With these facts in mind we state the following result.

Theorem. *For each $r \in (1,\infty)$ the sequence $\{x^k\}_{k\in\mathbb{N}}$ generated by the OBPM in a uniformly convex, separable and smooth Banach space X is well defined no matter how the initial point $x^0 \in X$ is chosen. Moreover, the sequence $\{x^k\}_{k\in\mathbb{N}}$ has the following properties:*

(I) If the stochastic feasibility problem (1) has a solution $z \in X$ such that the function $D_f(z,\cdot)$ is convex on X, then $\{x^k\}_{k\in\mathbb{N}}$ is bounded, has weak accumulation points, any weak accumulation point x^* of it is a solution of the stochastic convex feasibility problem (1), the following limit exists and we have

$$\lim_{k\to\infty} \|x^{k+1} - x^k\| = 0. \tag{11}$$

(II) If, in addition to the requirements of (I), the function $f(x) = \|x\|^r$ has the property that for each nonempty bounded set $E \subseteq X$ there exists a function $\eta_E : [0,+\infty) \times [0,+\infty) \to [0,+\infty)$ such that, for any $a \in [0,+\infty)$, the function $\eta_E(a,\cdot)$ is continuous, convex, strictly increasing and

$$\nu_f(E,t) \geq \eta_E(a,t), \tag{12}$$

whenever $t \in [0, a]$, *then, for any OBPM generated sequence* $\{x^k\}_{k\in\mathbb{N}}$, *the size of constraint violations converges in mean to zero, i.e.,*

$$\lim_{k\to\infty} \int_{\Omega} \max\left[0, g(\omega, x^k)\right] d\mu(\omega) = 0. \tag{13}$$

The proof of this result consists of a sequence of lemmas which is given below.

2.2. We start our proof by establishing the following technical facts.

Lemma. *For any* $x \in X$ *the following statements hold:*
(i) The point-to-set mapping $G : \Omega \to X^*$ *given by*

$$G(\omega) = \partial g(\omega, \cdot)(x) \tag{14}$$

is nonempty, convex, closed valued, measurable, and it has a measurable selector;
(ii) If $\omega \in \Omega$ *and* $\gamma \in \partial g(\omega, \cdot)(x)$, *then the set* Q_ω, *given by* (2), *is contained in the closed half space*

$$N(\omega, \gamma, x) = \{y \in X : \langle \gamma, y - x\rangle + g(\omega, x) \leq 0\}. \tag{15}$$

(iii) If $\Gamma : \Omega \to X^*$ *is a measurable selector of the mapping* G *given by* (14), *then the point-to-set mapping* $H^\Gamma : \Omega \times X \to X$ *defined by*

$$H^\Gamma(\omega, x) = N(\omega, \Gamma(\omega), x), \tag{16}$$

has the property that $H^\Gamma(\cdot, x)$ *is measurable.*

Proof. (*i*) Note that, for each $\omega \in \Omega$, the set $G(\omega)$ is convex and weakly* closed in X^* (cf. Proposition 1.11 in [23]). Consequently, $G(\omega)$ is closed in X^*. Let $\{y^k\}_{k\in\mathbb{N}}$ be a dense countable subset of X (such a subset exists because X is separable). For each $k \in \mathbb{N}$, define the function $F_k : \Omega \times X^* \to \mathbb{R}$ by

$$F_k(\omega, \zeta) := \langle \zeta, y^k - x\rangle - g(\omega, y^k) + g(\omega, x)$$

and put

$$G_k(\omega) := \left[F_k(\omega, \cdot)\right]^{-1}(-\infty, 0].$$

It is easy to verify that

$$G(\omega) = \bigcap_{k=1}^{\infty} G_k(\omega). \tag{17}$$

Since X is reflexive and separable, Theorem 1.14 in [1] applies and it shows that X^* is separable. Therefore, one can apply Theorem 8.2.9 in [3] to the Carathéodory mapping F_k and deduce that the point-to-set mappings $\omega \to G_k(\omega)$ are measurable for all $k \in \mathbb{N}$. Consequently, Theorem 8.2.4 in [3] combined with (17) show that the point-to-set mapping G defined by (14) is measurable, too. According to [3, Theorem 8.1.3], the measurable mapping G has a measurable selector.

(ii) For any $y \in Q_\omega$, we have $g(\omega, y) \le 0$ and, therefore,

$$-g(\omega, x) \ge g(\omega, y) - g(\omega, x) \ge \langle \gamma, y - x \rangle .$$

(iii) Consider the function $\Phi_x : \Omega \times X \to \mathbb{R}$ given by

$$\Phi_x(\omega, y) = \langle \Gamma(\omega), y - x \rangle + g(\omega, x).$$

For any $y \in X$, the function $\omega \to \langle \Gamma(\omega), y - x \rangle$ is measurable and, therefore, so is the function $\Phi_x(\cdot, y)$. For each $\omega \in \Omega$, the function $\Phi_x(\omega, \cdot)$ is continuous because $g(\omega, \cdot)$ is continuous (it is convex and lower semicontinuous on X). Hence, Φ_x is a Carathéodory function to which Theorem 8.2.9 in [3] applies and it shows that the point-to-set mapping $\omega \to [\Phi_x(\omega, \cdot)]^{-1}((-\infty, 0])$ is measurable. Since we have

$$H^\Gamma(\omega, x) = [\Phi_x(\omega, \cdot)]^{-1}((-\infty, 0]),$$

for any $\omega \in \Omega$, it results that the point-to-set mapping $H^\Gamma(\cdot, x)$ is measurable. □

2.3. A consequence of Lemma 2.2 is that the OBPM generated sequences are well defined as shown by the following result.

Lemma. *For any initial point $x^0 \in X$ and for each integer $k \ge 0$, there exists a measurable selector Γ_k of the point-to-set mapping $\omega \to \partial g(\omega, \cdot)(x^k)$ and the vector x^{k+1} given by* (3) *is well defined.*

Proof. We proceed by induction upon k. Let $k = 0$ and observe that, according to Lemma 2.2(i), there exists a measurable selector Γ_k of the point-to-set mapping $\omega \to \partial g(\omega, \cdot)(x^k)$. According to [11, Theorem 2], when $g(\omega, x^k) > 0$, the Bregman projection of x^k with respect to the function $f(x) = \|x\|^r$ onto the half space $H_k(\omega) := H^{\Gamma_k}(\omega, x^k)$, given by (5), exists and it is exactly

$$\Pi^f_{H_k(\omega)}(x^k) = J_r^* \left[s_k(\omega)\Gamma_k(\omega) + J_r x^k \right] . \tag{18}$$

Applying Proposition 2.1.5 in [9] one deduces that the family of operators $x \to \Pi^f_{H_k(\omega)}(x) : X \to X$, $\omega \in \Omega$, is totally nonexpansive with respect to the function f. This family of operators is also measurable because of [3, Theorem 8.2.11]. Consequently, one can apply Corollary 2.2.5 in [9] and it shows that the function $\omega \to \Pi^f_{H_k(\omega)}(x^k)$ is integrable, that is, the integral in (3) exists and x^{k+1} is well defined.

Now, assume that for some $k > 0$ the vector x^k is well defined. Then, repeating the reasoning above with x^k instead of x^0, we deduce that the measurable selector Γ_k exists and the vector x^{k+1} is well defined. □

2.4. Our further argumentation is based on several properties of the Bregman distance which are summarized in the next lemma. Note that, according to (6), for any $x \in X$, the function $D_f(\cdot, x)$ is continuous and convex. Continuity of the function $D_f(y, \cdot)$ for any $y \in X$ is established below. Convexity of this function occurs only in special circumstances as we will show later. It was noted in Subsection 2.1 that the function $f(x) = \|x\|^r$ with $r > 1$ has the property that the function $\nu_f(E, \cdot)$ is positive on $(0, \infty)$, whenever $E \subset X$ is nonempty and bounded. Strict monotonicity of this function, an essential feature in our analysis of OBPM, is established here.

Lemma. *(i) For any $y \in X$ the function $D_f(y,\cdot)$ defined by (6) is continuous on X and we have*

$$D_f(y,x) = \|y\|^r + (r-1)\|x\|^r - \langle J_r(x), y\rangle; \tag{19}$$

(ii) If the set $E \subset X$ is nonempty and $c \in [1,\infty)$, then $\nu_f(E,ct) \geq c\nu_f(E,t)$, for all $t \in [0,\infty)$;

(iii) If the set $E \subset X$ is nonempty and bounded, then the function $\nu_f(E,\cdot)$ is strictly increasing on $[0,\infty)$.

Proof. (i) In order to prove continuity of $D_f(y,\cdot)$, let $\{u^k\}_{k\in\mathbb{N}}$ be a convergent sequence in X whose limit is u. Then, we have

$$\begin{aligned} 0 \leq |D_f(y,u^k) - D_f(y,u)| &\leq |f(u^k) - f(u)| + |\langle f'(u^k) - f'(u), y\rangle| \\ &+ |\langle f'(u^k), u^k\rangle - \langle f'(u), u\rangle|, \end{aligned}$$

where, obviously, the first and the second term converge to zero as $k \to \infty$ because f is continuous and f' is norm-to-weak* continuous (see [23, Proposition 2.8]), respectively. Taking into account that

$$\begin{aligned} 0 \leq |\langle f'(u^k), u^k\rangle - \langle f'(u), u\rangle| &\leq |\langle f'(u^k), u^k - u\rangle| + |\langle f'(u^k) - f'(u), u\rangle| \\ &\leq \|f'(u^k)\|_* \|u^k - u\| + |\langle f'(u^k) - f'(u), u\rangle| \end{aligned}$$

and using the boundedness of the weakly* convergent sequence $\{f'(u^k)\}_{k\in\mathbb{N}}$, we obtain that the sum on the right hand-side converges to zero as $k \to \infty$. Therefore,

$$\lim_{k\to+\infty} |D_f(y,u^k) - D_f(y,u)| = 0,$$

and the continuity of $D_f(y,\cdot)$ is proved.

Now, for proving (19), we apply Asplund's Theorem (see, for instance, [13, p. 18]) which implies that, in the current setting, $f' = J_r$. Hence, we have

$$\begin{aligned} D_f(y,x) &= \|y\|^r - \|x\|^r - \langle J_r(x), y - x\rangle \\ &= \|y\|^r - \|x\|^r + \langle J_r(x), x\rangle - \langle J_r(x), y\rangle \\ &= \|y\|^r + (r-1)\|x\|^r - \langle J_r(x), y\rangle, \end{aligned}$$

because

$$\langle J_r(x), x\rangle = \psi_r(\|x\|)\|x\| = r\|x\|^r.$$

(ii) results from [9, Proposition 1.2.2] which shows that, for any $x \in X$, we have $\nu_f(x,ct) \geq c\nu_f(x,t)$, whenever $c \geq 1$.

(iii) If $0 < t_1 < t_2 < \infty$, then, applying (ii), we obtain

$$\nu_f(E,t_2) = \nu_f(E, \frac{t_2}{t_1}t_1) \geq \frac{t_2}{t_1}\nu_f(E,t_1).$$

Since, as noted in Subsection 2.1, $\nu_f(E,t_1) > 0$ because E is nonempty and bounded, we deduce

$$\nu_f(E,t_2) \geq \frac{t_2}{t_1}\nu_f(E,t_1) > \nu_f(E,t_1)$$

and the proof is complete. □

2.5. Lemma 2.3 guarantees that, no matter how the initial point x^0 is chosen in X, the OBPM generated sequence $\{x^k\}_{k\in\mathbb{N}}$ with the initial point x^0 is well defined. From now on we assume that for some solution z of (1) the function $D_f(z,\cdot)$ is convex. The next result shows that, under these circumstances, any sequence $\{x^k\}_{k\in\mathbb{N}}$ generated by the OBPM has some of the properties listed at point (I) of Theorem 2.1.

Lemma. *Any sequence $\{x^k\}_{k\in\mathbb{N}}$ generated by the OBPM satisfies*

$$D_f(z,x^{k+1}) + D_f(x^{k+1},x^k) \le D_f(z,x^k). \tag{20}$$

Also, $\{x^k\}_{k\in\mathbb{N}}$ is bounded, has weak accumulations points, the following limits exist and

$$\lim_{k\to\infty} \left\|x^{k+1}-x^k\right\| = \lim_{k\to\infty} D_f(x^{k+1},x^k) = \lim_{k\to\infty} \int_\Omega D_f(\Pi^f_{H_k(\omega)}(x^k),x^k) = 0. \tag{21}$$

Proof. Let $z \in X$ be a solution of the stochastic convex feasibility problem such that $D_f(z,\cdot)$ is convex. It was noted in the proof of Lemma 2.3 that the Bregman projections $\Pi^f_{H_k(\omega)}(x^k)$ exist and are given by (18). According to [9, Proposition 2.1.5] we also have that, for each $k \in \mathbb{N}$,

$$D_f(z,\Pi^f_{H_k(\omega)}(x^k)) + D_f(\Pi^f_{H_k(\omega)}(x^k),x^k) \le D_f(z,x^k), \tag{22}$$

for all $\omega \in \Omega$. The functions $D_f(\cdot,x^k)$ and $D_f(z,\cdot)$ are convex and continuous (cf. Lemma 2.4(i)). Therefore, by integrating the inequality (22) with respect to $\omega \in \Omega$ and applying Jensen's inequality we obtain (20). From (20) we see that the nonnegative sequence $\{D_f(z,x^k)\}_{k\in\mathbb{N}}$ is nonincreasing and, thus, convergent. Now, using Jensen's inequality again in (22) we deduce that

$$0 \le \int_\Omega D_f(\Pi^f_{H_k(\omega)}(x^k),x^k)d\mu(\omega) \le D_f(z,x^k) - D_f(z,x^{k+1}),$$

where, according to (20), the right hand side converges to zero as $k\to\infty$. Consequently,

$$\lim_{k\to\infty} \int_\Omega D_f(\Pi^f_{H_k(\omega)}(x^k),x^k)d\mu(\omega) = 0.$$

Since

$$0 \le D_f(x^{k+1},x^k) \le \int_\Omega D_f(\Pi^f_{H_k(\omega)}(x^k),x^k)d\mu(\omega),$$

we also have that

$$\lim_{k\to\infty} D_f(x^{k+1},x^k) = 0. \tag{23}$$

The monotonicity of the sequence $\{D_f(z,x^k)\}_{k\in\mathbb{N}}$ which results from (20) shows that the entire sequence $\{x^k\}_{k\in\mathbb{N}}$ is contained in the set

$$R^f_0(z) := \{y \in X : D_f(z,y) \le D_f(z,x^0)\}. \tag{24}$$

According to Corollary 1(i) in [11], this set is bounded. Hence, the sequence $\{x^k\}_{k\in\mathbb{N}}$ is bounded and, therefore, it has weak accumulation points (since the space X is reflexive). Let E be the bounded set of all terms of the sequence $\{x^k\}_{k\in\mathbb{N}}$. From (10) and (9) we get that

$$0 \le \nu_f(E, \|x^{k+1} - x^k\|) \le \nu_f(x^k, \|x^{k+1} - x^k\|) \le D_f(x^{k+1}, x^k),$$

for all $k \in \mathbb{N}$. This and (23) implies that

$$\lim_{k\to\infty} \nu_f(E, \|x^{k+1} - x^k\|) = 0.$$

According to Lemma 2.4(iii), this can not happen unless $\lim_{k\to\infty} \|x^{k+1} - x^k\| = 0$. □

2.6. Now we are in position to complete the proof of point (I) of the theorem.

Lemma. *If x^* is a weak accumulation point of a OBPM generated sequence $\{x^k\}_{k\in\mathbb{N}}$, then x^* is a solution of the stochastic convex feasibility problem* (1).

Proof. Denote by

$$\theta_k(\omega) = \|\Pi^f_{H_k(\omega)}(x^k) - x^k\|. \tag{25}$$

Let E be the set of all terms of the sequence $\{x^k\}_{k\in\mathbb{N}}$. This set is bounded (cf. Lemma 2.5). Note that $\omega \to \nu_f(E, \theta_k(\omega))$ is measurable because it is the composition of the monotone function $\nu_f(E, \cdot)$ (see Lemma 2.4(iii)) with the function θ_k which, according to [3, Theorem 8.2.11], is measurable. From (9) and (10) we have

$$0 \le \int_\Omega \nu_f(E, \theta_k(\omega)) d\mu(\omega) \le \int_\Omega D_f(\Pi^f_{H_k(\omega)}(x^k), x^k) d\mu(\omega),$$

for all $k \in \mathbb{N}$. This, together with Lemma 2.5, implies that

$$\lim_{k\to\infty} \int_\Omega \nu_f(E, \theta_k(\omega)) d\mu(\omega) = 0. \tag{26}$$

From (4) and (18) we have that

$$\begin{aligned} g(\omega, x^k) &\le |\langle \Gamma_k(\omega), J_r^*(s_k(\omega)\Gamma_k(\omega) + J_r x^k) - x^k\rangle| \\ &\le \|\Gamma_k(\omega)\|_* \left\|J_r^*(s_k(\omega)\Gamma_k(\omega) + J_r x^k) - x^k\right\| \\ &= \|\Gamma_k(\omega)\|_* \left\|\Pi^f_{H_k(\omega)}(x^k) - x^k\right\| = \|\Gamma_k(\omega)\|_* \theta_k(\omega). \end{aligned} \tag{27}$$

Let

$$E' = \bigcup_{k=0}^{\infty} \{y \in X : \|y - x^k\| \le 2\}.$$

Since E is bounded the set E' is bounded too and, obviously, $E \subseteq E'$. For each $k \in \mathbb{N}$ and $\omega \in \Omega$ for which $\Gamma_k(\omega) \ne 0$ denote

$$x^k_\omega = x^k + \frac{1}{\|\Gamma_k(\omega)\|_*} J_2^*(\Gamma_k(\omega)).$$

Clearly, $x_\omega^k \in E'$ because $\|J_2^*(\Gamma_k(\omega))\| = 2\|\Gamma_k(\omega)\|_*$. Let $\overline{M} = M_{E'}$ be a positive real number such that

$$|g(\omega, x)| \leq \overline{M}, \text{ a.e.}$$

for each $x \in E'$. Such a number exists because of condition $(C3)$. Then, for each $k \in \mathbb{N}$ and for almost any $\omega \in \Omega$, we have

$$\begin{aligned} 2\overline{M} &\geq |g(\omega, x_\omega^k)| + |g(\omega, x^k)| \\ &\geq g(\omega, x_\omega^k) - g(\omega, x^k) \geq \langle \Gamma_k(\omega), x_\omega^k - x^k \rangle \\ &= \frac{1}{\|\Gamma_k(\omega)\|_*} \langle \Gamma_k(\omega), J_2^*(\Gamma_k(\omega)) \rangle = 2\|\Gamma_k(\omega)\|_*, \end{aligned}$$

provided that $\Gamma_k(\omega) \neq 0$. This shows that, for each $k \in \mathbb{N}$, we have $\|\Gamma_k(\omega)\|_* \leq \overline{M}$ for almost all $\omega \in \Omega$. Combining this and (27) we deduce that

$$g(\omega, x^k) \leq \overline{M}\theta_k(\omega), \text{ a.e.,} \tag{28}$$

for all $k \in \mathbb{N}$.

Let $\{x^{i_k}\}_{k\in\mathbb{N}}$ be a subsequence of $\{x^k\}_{k\in\mathbb{N}}$ whose weak limit is x^*. Note that, according to (26), the sequence of functions $\{\nu_f(E, \theta_{i_k}(\omega))\}_{k\in\mathbb{N}}$ converges in mean to zero. Therefore, $\{\nu_f(E, \theta_{i_k}(\omega))\}_{k\in\mathbb{N}}$ is fundamental in measure (cf. [21, Theorem 38.7]). Hence, there exists a set $\Omega_0 \subseteq \Omega$ and a subsequence $\{\nu_f(E, \theta_{j_k}(\omega))\}_{k\in\mathbb{N}}$ of $\{\nu_f(E, \theta_{i_k}(\omega))\}_{k\in\mathbb{N}}$ such that $\mu(\Omega_0) = 1$ and

$$\lim_{k\to\infty} \nu_f(E, \theta_{j_k}(\omega)) = 0, \tag{29}$$

for all $\omega \in \Omega_0$ (see Proposition 24.18 in [22]). Suppose, by contradiction, that there exists $\omega_0 \in \Omega_0$ such that

$$\overline{\lim}_{k\to\infty} \theta_{j_k}(\omega_0) > 0.$$

Then, for some positive real number ε_0 and for some subsequence $\left\{\theta_{j_{h_k}}(\omega_0)\right\}_{k\in\mathbb{N}}$ of $\{\theta_{j_k}(\omega_0)\}_{k\in\mathbb{N}}$ we have $\theta_{j_{h_k}}(\omega_0) \geq \varepsilon_0 > 0$ for all $k \in \mathbb{N}$. Applying Lemma 2.4(iii) we deduce that

$$\nu_f(E, \theta_{j_{h_k}}(\omega_0)) \geq \nu_f(E, \varepsilon_0) > 0,$$

for all $k \in \mathbb{N}$ and this contradicts (29). Hence, $\lim_{k\to\infty} \theta_{j_k}(\omega) = 0$, for almost all $\omega \in \Omega$. This, (28) and condition $(C2)$ taken together imply that, for almost all $\omega \in \Omega$, we have

$$g(\omega, x^*) \leq \underline{\lim}_{k\to\infty} g(\omega, x^{j_k}) \leq \overline{M} \lim_{k\to\infty} \theta_{j_k}(\omega) = 0,$$

that is, x^* is a solution of (1). □

2.7. We proceed to prove the statement (II) of our theorem.

Lemma. *Under the assumptions of (II) any weak accumulation point x^* of an OBPM generated sequence $\{x^k\}_{k\in\mathbb{N}}$ is a solution of problem* (1) *and the sequence of functions $\{\max[0, g(\cdot, x^k)]\}_{k\in\mathbb{N}}$ converges in mean to zero.*

Proof. Observe that (26) still holds with the functions $\theta_k(\omega)$ given by (25). Note that, according to (22), for almost all $\omega \in \Omega$ and for all $k \in \mathbb{N}$, the vectors $\Pi^f_{H_k(\omega)}(x^k)$ are contained in the bounded set $R^f_0(z)$ defined at (24). Since the sequence $\{x^k\}_{k\in\mathbb{N}}$ is bounded (see Lemma 2.5), it results that there exists a real number $a > 0$ such that, for all $k \in \mathbb{N}$,

$$\theta_k(\omega) \leq \|\Pi^f_{H_k(\omega)}(x^k)\| + \|x^k\| \leq a, \text{ a.e.}$$

Hence, for each $k \in \mathbb{N}$,

$$\nu_f(E, \theta_k(\omega)) \geq \eta_E(a, \theta_k(\omega)), \text{ a.e.}$$

Combining this inequality with (26) and with Jensen's inequality, we deduce that

$$0 \leq \lim_{k\to\infty} \eta_E\left(a, \int_\Omega \theta_k(\omega) d\mu(\omega)\right) \leq \lim_{k\to\infty} \int_\Omega \eta_E(a, \theta_k(\omega)) d\mu(\omega) = 0.$$

Since $\eta_E(a, \cdot)$ is strictly increasing, this cannot happen unless

$$\lim_{k\to\infty} \int_\Omega \theta_k(\omega) = 0.$$

Taking into account (28), which still holds in the current circumstances, we obtain that the sequence of functions $\{\max[0, g(\cdot, x^k)]\}_{k\in\mathbb{N}}$ converges in mean to zero. □

3. APPLICABILITY IN PARTICULAR BANACH SPACES

3.1. One of the most practically interesting features of the OBPM is that, under the assumption of Theorem 2.1(II), it produces sequences $\{x^k\}_{k\in\mathbb{N}}$ which not only accumulate weakly to solutions of the given stochastic convex feasibility problem, but also have the property that, along them, the size of the constraint violations converges in mean to zero. This section is aimed at showing that, in the most usual Banach spaces, i. e., in Hilbert spaces, Lebesgue spaces and Sobolev spaces, the main requirement of Theorem 2.1(II) is satisfied in the sense that there are numbers $r \in (1, \infty)$ such that the function $f(x) = \|x\|^r$ has the following property:

(*) *For any nonempty, bounded set $E \subset X$, there exists a function $\eta_E : [0,\infty)\times[0,\infty) \to [0,\infty)$ such that, for each real number $a > 0$, the function $\eta_E(a, \cdot)$ is strictly increasing, continuous, convex, and $\nu_f(E, t) \geq \eta_E(a, t)$ whenever $t \in [0, a)$.*

Combining this fact with the easy implementability of OBPM pointed out in Section 1, we observe that this algorithm is a potentially useful tool for solving stochastic convex feasibility problems and, in particular, optimization and equilibrium problems, variational inequalities, operator equations, etc. (see [9]).

Our argumentation below is based on the fact, established in [11, Formula (12)], that if X is an uniformly convex Banach space, then there exists a real number $K > 0$ such that, for all $z \in X$,

$$\nu_f(z, t) \geq rK\left(\frac{t}{2}\right)^r \int_0^1 \tau^{r-1} \delta_X\left(\frac{\tau t}{2(\|z\| + \tau t)}\right) d\tau, \tag{30}$$

where $\delta_X : [0,2] \to [0,1]$ stands for the modulus of uniform convexity of X.

3.2. We first show that in Hilbert spaces condition (*) holds.

Proposition. *If X is a Hilbert space and $r \in (1,\infty)$, then the function $f(x) = \|x\|^r$ has the property (*).*

Proof. Recall (see, for instance, [2]) that the modulus of uniform convexity of the Hilbert space X has the property that $\delta_X(t) \geq ct^2$ for some constant $c > 0$. Let E be a bounded set in X and let $M > 0$ be of upper bound of E. From this and from (30) we deduce that, for $a > 0$ and $t \in (0, a]$

$$\nu_f(z,t) \geq rK\left(\frac{t}{2}\right)^r \int_0^1 \tau^{r-1}\delta_X\left(\frac{\tau t}{2\left(\|z\| + \tau t\right)}\right) d\tau$$

$$\geq crK\left(\frac{t}{2}\right)^{r+2} \int_0^1 \frac{\tau^{r+1}}{\left(M + \tau a\right)^2} d\tau,$$

for all $z \in E$. Hence, $\nu_f(E,t) \geq \beta(a,r)t^{r+2}$, where $\beta(a,r)$ is a positive real number. □

3.3. The next result shows that condition (*) holds in Lebesgue spaces no matter how the number $r > 1$ is chosen.

Proposition. *Suppose that $p, r \in (1, +\infty)$. In the Lebesgue space $\mathcal{L}^p$ the function $f : \mathcal{L}^p \to \mathbb{R}$, defined by $f(x) = \|x\|_p^r$, satisfies the condition (*).*

Proof. Let t be a positive real number, E a nonempty and bounded subset of $\mathcal{L}^p$, and denote by M a upper bound of E. It follows from (30) that, for any $z \in E$,

$$\nu_f(z,t) \geq rK\left(\frac{t}{2}\right)^r \cdot \int_0^1 \tau^{r-1}\delta_p\left(\frac{\tau t}{2\left(M + \tau t\right)}\right) d\tau. \tag{31}$$

because the modulus of uniform convexity δ_p of $\mathcal{L}^p$ is increasing on $[0,2]$. Recall (see, for instance, [16, Theorem 1, pp. 69]) that

$$\delta_p(t) \geq \begin{cases} c_p t^2 & \text{if } p \in (1,2), \\ c_p t^p & \text{if } p \geq 2, \end{cases} \tag{32}$$

where c_p is some positive constant (depending on p). Define the function $\eta_E : [0,+\infty) \times [0,+\infty) \to [0,+\infty)$ by

$$\eta_E(a,t) = \begin{cases} \beta_1(r,p,a)t^{r+2} & \text{if } p \in (1,2), \\ \beta_2(r,p,a)t^{r+p} & \text{if } p \geq 2, \end{cases}$$

with

$$\beta_1(r,p,a) := \frac{rc_pK}{2^{r+2}} \int_0^1 \frac{\tau^{r+1}}{\left(M + \tau a\right)^2} d\tau \tag{33}$$

and

$$\beta_2(r,p,a) := \frac{rc_pK}{2^{r+p}} \cdot \int_0^1 \frac{\tau^{r+p-1}}{(M+\tau a)^p} d\tau. \tag{34}$$

This function η_E has the properties required by condition (*) when $p \in (1,2)$. Indeed, combining (31) and (32) we obtain

$$\nu_f(z,t) \geq \frac{rc_pK}{2^{r+2}} t^{r+2} \cdot \int_0^1 \frac{\tau^{r+1}}{(M+\tau t)^2} d\tau,$$

for any $z \in E$. Let a be a positive real number and suppose that $t \in (0,a]$. Then, taking the infimum for $z \in E$ in the last inequality, we get that

$$\nu_f(E,t) \geq \beta_1(r,p,a)t^{r+2}, \tag{35}$$

for all $t \in (0,a]$, where $\beta_1(r,p,a)$ is given at (33). A similar argument shows that the function η_E defined above satisfies condition (*) in the case $p \geq 2$. □

3.4. Now we are going to show that, in Sobolev spaces $\mathcal{W}^{m,p}$, the p power of the norm has property (*). To make notations precise, recall (see [1]) that, for an open, bounded, convex, nonempty subset $S \subset \mathbb{R}^n$, the Sobolev space $\mathcal{W}^{m,p} := \mathcal{W}^{m,p}(S)$ consists of all $u := u(t_1,...,t_n) \in \mathcal{L}^p(S)$ such that, for each multi-index $\alpha = (\alpha_1,...,\alpha_n) \in \mathbb{N}^n$ for which $|\alpha| := \sum_{i=1}^n \alpha_i \leq m$, the (generalized) derivative

$$\mathfrak{D}^\alpha u := \frac{\partial^{|\alpha|}u}{\partial t_1^{\alpha_1}...\partial t_n^{\alpha_n}}$$

exists and belongs to $\mathcal{L}^p(S)$. Provided with the norm

$$\|u\|_{m,p} = \left(\sum_{0\leq|\alpha|\leq m} \|\mathfrak{D}^\alpha u\|_p^p \right)^{1/p}, \tag{36}$$

the space $\mathcal{W}^{m,p}$ is uniformly convex and smooth (cf. [1, Theorem 3.5] and [13, Corollary 4.5, pp. 27 and Theorem 3.5, pp. 22]).

Proposition. *Suppose that $p \in (1,+\infty)$ and m is a positive integer. In the Sobolev space $\mathcal{W}^{m,p}$ the function $f(x) = \|x\|_{m,p}^p$ has the property (*).*

Proof. Let t be a positive real number and let E a nonempty and bounded subset of $\mathcal{W}^{m,p}$. Fix $u \in E$ and choose $w \in \mathcal{W}^{m,p}$ be such that $\|u - w\|_{m,p} = t$. There exists a multi-index $\alpha_0 \in \mathbb{N}^n$ with $0 \leq |\alpha_0| \leq m$ such that

$$\|\mathfrak{D}^{\alpha_0}u - \mathfrak{D}^{\alpha_0}w\|_p \geq \frac{t}{\gamma^{1/p}}, \tag{37}$$

where γ is the cardinality of the set of multi-indexes α with $0 \leq |\alpha| \leq m$. Indeed, if we assume, by contradiction, that for any such multi-index α, we have

$$\|\mathfrak{D}^\alpha u - \mathfrak{D}^\alpha w\|_p < \frac{t}{\gamma^{1/p}},$$

then it follows from (36) that $\|u-w\|_{m,p} < t$, and this contradicts the equality $\|u-w\|_{m,p} = t$. Observe that, according to (36), we have

$$\begin{aligned} D_f(w,u) &= \|w\|_{m,p}^p - \|u\|_{m,p}^p - \langle J_p(u), w-u\rangle & (38)\\ &= \sum_{0\le|\alpha|\le m} \int_S [|\mathfrak{D}^\alpha w|^p - |\mathfrak{D}^\alpha u|^p]\,dx - \langle J_p(u), w-u\rangle, & (39) \end{aligned}$$

where J_p is the duality mapping of weight pt^{p-1} in $\mathcal{W}^{m,p}$. Applying Asplund's theorem we deduce that

$$\begin{aligned} \langle J_p(u), v\rangle &= \frac{d}{dt}\left(\|u+tv\|_{m,p}^p\right)\Big|_{t=0} \\ &= \frac{d}{dt}\left(\sum_{0\le|\alpha|\le m}\int_S |\mathfrak{D}^\alpha u + t\mathfrak{D}^\alpha v|^p\,dx\right)\Bigg|_{t=0} \\ &= p\sum_{0\le|\alpha|\le m}\int_S \left[|\mathfrak{D}^\alpha u|^{p-1}\,sign\,(\mathfrak{D}^\alpha u)\,\mathfrak{D}^\alpha v\right]dx. \end{aligned}$$

This and (38) imply

$$\begin{aligned} D_f(w,u) &= \sum_{0\le|\alpha|\le m}\int_S [|\mathfrak{D}^\alpha w|^p - |\mathfrak{D}^\alpha u|^p \\ &\qquad -p\,|\mathfrak{D}^\alpha u|^{p-1}(\mathfrak{D}^\alpha w - \mathfrak{D}^\alpha u)\,sign\,(\mathfrak{D}^\alpha u)]\,dx, \end{aligned}$$

that is,

$$D_f(w,u) = \sum_{0\le|\alpha|\le m} D_{\|\cdot\|_p^p}(\mathfrak{D}^\alpha w, \mathfrak{D}^\alpha u).$$

Consequently, for the multi-index α_0 satisfying (37), we have

$$\begin{aligned} D_f(w,u) &\ge D_{\|\cdot\|_p^p}(\mathfrak{D}^{\alpha_0} w, \mathfrak{D}^{\alpha_0} u) \\ &\ge \nu_{\|\cdot\|_p^p}\left(\mathfrak{D}^{\alpha_0}u, \|\mathfrak{D}^{\alpha_0}w - \mathfrak{D}^{\alpha_0}u\|_p\right) \\ &\ge \nu_{\|\cdot\|_p^p}\left(\mathfrak{D}^{\alpha_0}u, \frac{t}{\gamma^{1/p}}\right), \end{aligned}$$

where $\nu_{\|\cdot\|_p^p}$ denotes the modulus of total convexity of the function $\|\cdot\|_p^p$ on $\mathcal{L}^p(S)$ and the last inequality holds because the function $\nu_{\|\cdot\|_p^p}(\mathfrak{D}^{\alpha_0}u,\cdot)$ is nondecreasing (cf. [9, Proposition 1.2.2]). According to (36), and taking into account that E is bounded by some real number $M>0$, we have that $\|\mathfrak{D}^\alpha u\|_p \le M$, whenever α is a multi-index with $0\le|\alpha|\le m$. Take F to be the closed ball of center zero and radius M in $\mathcal{L}^p(S)$. Then, $\mathfrak{D}^{\alpha_0}u \in F$ and we have

$$D_f(w,u) \ge \nu_{\|\cdot\|_p^p}\left(\mathfrak{D}^{\alpha_0}u, \frac{t}{\gamma^{1/p}}\right) \ge \nu_{\|\cdot\|_p^p}\left(F, \frac{t}{\gamma^{1/p}}\right). \qquad (40)$$

Fix a positive real number a and define

$$\eta_E(a,t) = \begin{cases} \beta_{1,p}(a)\left(\frac{t}{\gamma^{1/p}}\right)^{p+2} & \text{if } p \in (1,2), \\ \\ \beta_{2,p}(a)\left(\frac{t}{\gamma^{1/p}}\right)^{2p} & \text{if } p \geq 2. \end{cases}$$

where $\beta_{i,p}(a) := \beta_i(p,p,a)$, for $i = 1,2$, are given by (33) and (34), respectively. Then, using (40), one deduces that this function η_E satisfies the condition (*) because, as shown in the proof of Proposition 2.2, we have that

$$\nu_{\|\cdot\|_p^p}\left(F, \frac{t}{\gamma^{1/p}}\right) \geq \begin{cases} \beta_1(p,p,a)\left(\frac{t}{\gamma^{1/p}}\right)^{p+2} & \text{if } p \in (1,2), \\ \\ \beta_2(p,p,a)\left(\frac{t}{\gamma^{1/p}}\right)^{2p} & \text{if } p \geq 2, \end{cases}$$

for all $t \in (0,a]$. □

3.5. A restrictive condition involved in Theorem 2.1 concerns the existence of a solution z of the given stochastic convex feasibility problem (1) such that the function $D_f(z,\cdot)$ is convex when $f(x) = \|x\|^r$ for some real number $r > 1$. Instances in which this may happen in $\mathbb{R}^n$ were discussed in [4]. If X is a Hilbert space and if $r = 2$, then this condition is satisfied by any solution z of the problem (recall that we presume that the problem is consistent). That is so because, in this case, $D_f(z,x) = \|z-x\|^2$. In general, i.e., if X is not a Hilbert space or $r \neq 2$, then this condition holds whenever $z = 0$ is a solution of the given problem (note that $D_f(0,x) = (r-1)\|x\|^r$). In such situations and depending on how r is chosen, it may happen that $z = 0$ is the only element of X for which $D_f(z,\cdot)$ is convex. This is the case if $X = \ell^p$ and $r = p > 2$. Indeed, in this specific case, for the twice differentiable function $D_f(z,\cdot)$ with $f(x) = \|x\|_p^p$, $p > 2$, to be convex in ℓ^p it is necessarily and sufficient (cf. [24, Theorem 2.1.7]) that

$$\langle [D_f(z,\cdot)]''(x)h, h\rangle \geq 0,$$

whenever $x, h \in \ell^p$. Also, we have

$$\langle [D_f(z,\cdot)]''(x)h, h\rangle = p(p-1)\sum_{n=0}^{\infty} |x_n|^{p-4}\left[(p-1)x_n^2 - (p-2)x_n z_n\right] h_n^2.$$

If the last sum is nonnegative for all $x, h \in \ell^p$, then it is also nonnegative for all $h = e^k \in \ell^p$, where $e_n^k = 1$, if $n = k$, and $e_n^k = 0$, otherwise. This implies that, for each $n \in \mathbb{N}$, we have

$$|x_n|^{p-4}\left[(p-1)x_n^2 - (p-2)x_n z_n\right] \geq 0,$$

for all $x_n \in \mathbb{R}\setminus\{0\}$. This cannot happen unless $z_n = 0$, for all $n \in \mathbb{N}$. Hence, $z = 0$ is the only element of ℓ^p such that $D_f(z,\cdot)$, with $f(x) = \|x\|_p^p$, is convex when $p > 2$.

3.6. In order to illustrate de behavior of *OBPM*, we consider the following stochastic feasibility problem: Find $x \in \mathcal{L}^p$ such that

$$\int_\Omega K(\omega, t, x(t))dt \leq c(\omega), \text{ a.e.}(\Omega),$$

where $\Omega = [a,b]$ and c, $K(\cdot,t,u) : \Omega \to \mathbb{R}$ are measurable and, for each $\omega \in \Omega$ and $t \in [a,b]$, the function $K(\omega,t,\cdot)$ is convex on $\mathbb{R}$. We apply the *OBPM* to the particular case where $p = 4/3$, $a = 0$, $b = 1$, $c(\omega) = 3e^{\omega}$ and

$$K(\omega,t,x(t)) = e^{\omega^2 t} x^{4/3}(t).$$

In this situation, we define the function g as follows:

$$g(\omega,x) = \int_0^1 e^{\omega^2 t} x^{4/3}(t)dt - 3e^{\omega}.$$

Note that this function g satisfies conditions $(C1)$-$(C3)$ and that $z = 0$ is a solution of the problem.

We start the *OBPM* procedure with $x^0 \equiv 5^{3/4}$ and $r = 1.5$. We have

$$g(\omega,x^0) = (5/\omega^2)\cdot(e^{\omega^2} - 1) - 3e^{\omega} > 0,$$

for all $\omega \in (0,1]$ and, then,

$$\int_\Omega \max\left[0, g(\omega,x^0)\right] d\mu(\omega) = 0.80612. \tag{41}$$

We observe that

$$J_r x = \begin{cases} r\|x\|_p^{r-p}|x|^{p-2}x & \text{if } x \neq 0, \\ 0 & \text{if } x = 0, \end{cases}$$

and

$$J_r^* \xi = \begin{cases} r^{\frac{1}{1-r}}\|\xi\|_q^{\frac{r}{r-1}-q}|\xi|^{q-2}\xi & \text{if } \xi \neq 0, \\ 0 & \text{if } \xi = 0, \end{cases}$$

where $q = p/(p-1) = 4$. Since $g(\omega,\cdot)$ is differentiable, we necessarily have $\Gamma_k(\omega) = g'(\omega,\cdot)(x^k)$, for all k . Thus, we obtain that

$$x^{k+1} = r^{\frac{1}{1-r}} \int_\Omega \|y^k\|_q^{\frac{r}{r-1}-q}|y^k|^{q-2}y^k d\mu(\omega),$$

where

$$y^k := s_k(\omega)g'(\omega,\cdot)(x^k) + r\|x^k\|_p^{r-p}|x^k|^{p-2}x^k$$

and $s_k(\omega)$ is a solution of the equation

$$\int_\Omega g'(\omega,\cdot)(x^k)\left[r^{\frac{1}{1-r}}\|y^k\|_q^{\frac{r}{r-1}-q}|y^k|^{q-2}y^k - x^k\right] d\mu(\omega) = -g(\omega,x^k).$$

Computing x^1 according to the formula above for the given x^0, we obtain the corresponding averaged constraint violation

$$\int_\Omega \max\left[0, g(\omega,x^1)\right] d\mu(\omega) = 0.235225.$$

Comparing that to (41), we note that, after just a single step, we have a significant reduction of the averaged constraint violation.

REFERENCES

1. R.A. Adams, *Sobolev Spaces*, Academic Press, New York, 1975.
2. Y. Alber and A.I. Notik, Geometric properties of Banach spaces and approximate methods for solving nonlinear operator equations, *Soviet Mathematiky Doklady* **29** (1984) 615–625.
3. J.-P. Aubin and H. Frankowska, *Set-Valued Analysis*, Birkhäuser, 1990.
4. H. Bauschke and J. Borwein, Joint and separate convexity of the Bregman distance, paper contained in this volume.
5. D. Butnariu, The expected projection method: Its behavior and applications to linear operator equations and convex optimization, *Journal of Applied Analysis* **1** (1995) 93–108.
6. D. Butnariu, Y. Censor and S. Reich, Iterative averaging of entropic projections for solving stochastic convex feasibility problems, *Computational Optimization and Applications* **8** (1997) 21–39.
7. D. Butnariu and S. Flam, Strong convergence of expected-projection methods in Hilbert spaces, *Numerical Functional Analysis and Optimization* **16** (1995) 601–637.
8. D. Butnariu and A.N. Iusem, Local moduli of convexity and their application to finding almost common fixed points of measurable families of operators, in: *Recent Developments in Optimization Theory and Nonlinear Analysis*, Y. Censor and S. Reich, eds., Contemporary Mathematics, **204**, (1997), 61–91.
9. D. Butnariu and A.N. Iusem, *Totally Convex Functions for Fixed Points Computation and Infinite Dimensional Optimization*, Kluwer Academic Publishers, Dordrecht, 2000.
10. D. Butnariu, A.N. Iusem and R.S. Burachik, Iterative methods for solving stochastic convex feasibility problems and applications, *Computational Optimization and Applications* **15** (2000) 269–307.
11. D. Butnariu, A.N. Iusem and E. Resmerita, Total convexity for powers of the norm in uniformly convex Banach spaces, *Journal of Convex Analysis* **7** (2000) 319–334.
12. Y. Censor and A. Lent, Cyclic subgradient projections, *Mathematical Programming* **24** (1982) 223–235.
13. I. Ciorănescu, *Geometry of Banach Spaces, Duality mappings, and Nonlinear Problems*, Kluwer Academic Publishers, Dordrecht, 1990.
14. F.H. Clarke, *Optimization and Nonsmooth Analysis*, John Wiley and Sons, New York, 1983.
15. P.L. Combettes, Convex set theoretic image recovery by extrapolated iterations of parallel subgradient projections, *IEEE Transactions on Image Processing* **6** (1997) 493–506.
16. J. Diestel, *Geometry of Banach Spaces: Selected Topics*, Springer Verlag, Berlin, 1975.
17. S.D. Flåm and J. Zowe, Relaxed outer projections, weighted averages and convex feasibility, *BIT* **30** (1990) 289–300.
18. A.N. Iusem and L. Moledo, A finitely convergent method of simultaneous subgradient projections for the convex feasibility problem, *Computational and Applied Mathematics* **5** (1986) 169–184.
19. A.N. Iusem and L. Moledo, On finitely convergent iterative methods for the convex

feasibility problem, *Bulletin of the Brazilian Mathematical Society* **18** (1987) 11–18.
20. K.C. Kiwiel and B. Łopuch, Surrogate projection methods for finding fixed points of firmly nonexpansive mappings, *SIAM Journal on Optimization* **7** (1997) 1084–1102.
21. M.E. Munroe, *Measure and Integration*, Addison-Wesley Publishing Company, 1970.
22. K.R. Parthasarathy, *Probability Measures on Metric Spaces*, Academic Press, New York, 1967.
23. R.R. Phelps, *Convex Functions, Monotone Operators and Differentiability, 2-nd Edition*, Springer Verlag, Berlin, 1993.
24. C. Zălinescu, *Mathematical Programming in Infinite Dimensional Normed Linear Spaces* (Romanian), Editura Academiei Române, Bucharest, 1998.

Inherently Parallel Algorithms in Feasibility and Optimization and their Applications
D. Butnariu, Y. Censor and S. Reich (Editors)

BREGMAN-LEGENDRE MULTIDISTANCE PROJECTION ALGORITHMS FOR CONVEX FEASIBILITY AND OPTIMIZATION

Charles Byrne[a]

[a]Department of Mathematical Sciences,
University of Massachusetts Lowell,
1 University Ave., Lowell, MA 01854, USA

The *convex feasibility problem* (CFP) is to find a member of the nonempty set $C = \bigcap_{i=1}^{I} C_i$, where the C_i are closed convex subsets of R^J. The *multidistance successive generalized projection* (MSGP) algorithm extends Bregman's *successive generalized projection* (SGP) algorithm for solving the CFP to permit the use of generalized projections onto the C_i associated with Bregman-Legendre functions f_i that may vary with the index i. The MSGP method depends on the selection of a super-coercive Bregman-Legendre function h whose Bregman distance D_h satisfies the inequality $D_h(x,z) \geq D_{f_i}(x,z)$ for all $x \in \operatorname{dom} h \subseteq \bigcap_{i=1}^{I} \operatorname{dom} f_i$ and all $z \in \operatorname{int} \operatorname{dom} h$, where $\operatorname{dom} h = \{x | h(x) < +\infty\}$.

The MSGP method is used to obtain an iterative solution procedure for the *split feasibility problem* (SFP): given the M by N matrix A and closed convex sets K and Q in R^N and R^M, respectively, find x in K with Ax in Q.

If $I = 1$ and $f := f_1$ has a unique minimizer $\hat{x}$ in $\operatorname{int} \operatorname{dom} h$, then the MSGP iteration using $C_1 = \{\hat{x}\}$ is

$$\nabla h(x^{k+1}) = \nabla h(x^k) - \nabla f(x^k).$$

This suggests an interior point algorithm that could be applied more broadly to minimize a convex function f over the closure of $\operatorname{dom} h$.

1. INTRODUCTION

The *convex feasibility problem* (CFP) is to find a member of the nonempty set $C = \bigcap_{i=1}^{I} C_i$, where the C_i are closed convex subsets of R^J. In most applications the sets C_i are more easily described than the set C and algorithms are sought whereby a member of C is obtained as the limit of an iterative procedure involving (exact or approximate) orthogonal or generalized projections onto the individual sets C_i. Such algorithms are the topic of this paper.

In his often cited paper [3] Bregman introduced a class of functions that have come to be called *Bregman functions* and used the associated *Bregman distances* to define generalized projections onto closed convex sets (see the book by Censor and Zenios [9] for details concerning Bregman functions).

In [2] Bauschke and Borwein introduce the related class of *Bregman-Legendre functions* and show that these functions provide an appropriate setting in which to study Bregman distances and generalized projections associated with such distances.

Bregman's *successive generalized projection* (SGP) method uses projections with respect to Bregman distances to solve the convex feasibility problem. Let $f : R^J \to (-\infty, +\infty]$ be a closed, proper convex function, with essential domain $D = \operatorname{dom} f = \{x | f(x) < +\infty\}$ and $\emptyset \neq \operatorname{int} D$. Denote by $D_f(\cdot, \cdot) : D \times \operatorname{int} D \to [0, +\infty)$ the Bregman distance, given by

$$D_f(x, z) = f(x) - f(z) - \langle \nabla f(z), x - z \rangle \tag{1}$$

and by $P^f_{C_i}$ the Bregman projection operator associated with the convex function f and the convex set C_i; that is

$$P^f_{C_i} z = \arg\min_{x \in C_i \cap D} D_f(x, z). \tag{2}$$

Bregman considers the following iterative procedure:

Algorithm 1.1 Bregman's method of Successive Generalized Projections (SGP): *Beginning with* $x^0 \in \operatorname{int} \operatorname{dom} f$, *for* $k = 0, 1, ...,$ *let* $i = i(k) := k(\operatorname{mod} \mathrm{I}) + 1$ *and*

$$x^{k+1} = P^f_{C_{i(k)}}(x^k). \tag{3}$$

He proves that the sequence $\{x^k\}$ given by (3) converges to a member of $C \cap \operatorname{dom} f$, whenever this set is nonempty and the function f is what came to be called a Bregman function ([3]).

Investigations in [4] into several well known iterative algorithms, including the 'expectation maximization maximum likelihood' (EMML) method, the 'multiplicative algebraic reconstruction technique' (MART) as well as block-iterative and simultaneous versions of MART revealed that the iterative step of each algorithm involved weighted arithmetic or geometric means of Bregman projections onto hyperplanes; interestingly, the projections involved were associated with Bregman distances that differed from one hyperplane to the next. This representation of the EMML algorithm as a weighted arithmetic mean of Bregman projections provided the key step in obtaining block-iterative and row-action versions of EMML. Because it is well known that convergence is not guaranteed if one simply extends Bregman's algorithm to multiple distances by replacing the single distance D_f in (3) with multiple distances D_{f_i}, the appearance of distinct distances in these algorithms suggested that a somewhat more sophisticated algorithm employing multiple Bregman distances might be possible.

In [5] such an iterative multiprojection method for solving the CFP, called the *multi-distance successive generalized projection* (MSGP) method, was presented in the context of Bregman functions. In this paper we prove convergence of the MSGP method in the framework of Bregman-Legendre functions. The MSGP extends Bregman's SGP method by allowing the Bregman projection onto each set C_i to be performed with respect to a Bregman distance D_{f_i} derived from a Bregman-Legendre function f_i. The MSGP method depends on the selection of a super-coercive Bregman-Legendre function h whose Bregman distance D_h satisfies the inequality $D_h(x, z) \geq D_{f_i}(x, z)$ for all $x \in \operatorname{dom} h \subseteq \bigcap_{i=1}^I \operatorname{dom} f_i$ and all $z \in \operatorname{int} \operatorname{dom} h$, where $\operatorname{dom} h = \{x | h(x) < +\infty\}$. By using different Bregman distances for different convex sets, we found that we can sometimes calculate the desired Bregman projections in closed form, thereby obtaining computationally tractable iterative algorithms (see [4]).

To illustrate the MSGP method we use it to obtain an iterative solution procedure for the *split feasibility problem* (SFP): given the M by N matrix A and closed convex sets K and Q in R^N and R^M, respectively, find x in K with Ax in Q.

Consideration of a special case of the MSGP, involving only a single convex set C_1, leads us to an interior point optimization method. If $I = 1$ and $f := f_1$ has a unique minimizer $\hat{x}$ in $\operatorname{int}\operatorname{dom} h$, then the MSGP iteration using $C_1 = \{\hat{x}\}$ is

$$\nabla h(x^{k+1}) = \nabla h(x^k) - \nabla f(x^k). \tag{4}$$

This suggests an *interior point algorithm* (IPA) that could be applied more broadly to minimize a convex function f over the closure of $\operatorname{dom} h$.

We begin with a brief discussion of Bregman-Legendre functions, closely following the treatment in Bauschke and Borwein [2]. Then we present the MSGP method and prove convergence, in the context of Bregman-Legendre functions. In the section following we apply the MSGP to the SFP. Finally, we investigate the IPA suggested by the MSGP algorithm.

2. BREGMAN-LEGENDRE FUNCTIONS

In [2] Bauschke and Borwein show convincingly that the Bregman-Legendre functions provide the proper context for the discussion of Bregman projections onto closed convex sets. The summary here follows closely the discussion given in [2].

A convex function $f : R^J \rightarrow [-\infty, +\infty]$ is *proper* if there is no x with $f(x) = -\infty$ and some x with $f(x) < +\infty$. The *essential domain* of f is $\operatorname{dom} f = D = \{x | f(x) < +\infty\}$. A proper convex function f is *closed* if it is lower semi-continuous. The *conjugate function* associated with f is the function f^*, with $f^*(x^*) = \sup_z(\langle x^*, z\rangle - f(z))$.

Following [14] we say that a closed proper convex function f is *essentially smooth* if $\operatorname{int} D$ is not empty, f is differentiable on $\operatorname{int} D$ and $x^n \in \operatorname{int} D$, with $x^n \rightarrow x \in \operatorname{bdry} D$, implies that $||\nabla f(x^n)|| \rightarrow +\infty$. Here $\operatorname{int} D$ and $\operatorname{bdry} D$ denote the interior and boundary of the set D.

An essentially smooth function f is a *Legendre function* if f is strictly convex on the set $\operatorname{int} D$; the gradient operator ∇f is then a topological isomorphism with ∇f^* as its inverse. The gradient operator ∇f maps $\operatorname{int}\operatorname{dom} f$ onto $\operatorname{int}\operatorname{dom} f^*$. If $\operatorname{int}\operatorname{dom} f^* = R^J$ then the range of ∇f is R^J and the equation $\nabla f(x) = y$ can be solved for every $y \in R^J$. In order for $\operatorname{int}\operatorname{dom} f^* = R^J$ it is necessary and sufficient that the Legendre function f be *super-coercive*, that is,

$$\lim_{||x|| \rightarrow +\infty} \frac{f(x)}{||x||} = +\infty.$$

If f is Legendre and the essential domain of f is bounded, then f is necessarily super-coercive and its gradient operator is a mapping onto the space R^J.

Let K be a nonempty closed convex set with $K \cap \operatorname{int} D \neq \emptyset$. If f is Legendre, then the Bregman projection $P_K^f(z)$ exists, for all $z \in \operatorname{int} D$, is uniquely defined and is in $\operatorname{int} D$;

$P_K^f(z)$ is the unique member of $K \cap \text{int}D$ satisfying the inequality

$$\langle \nabla f(P_K^f(z)) - \nabla f(z), P_K^f(z) - c \rangle \geq 0, \tag{5}$$

for all $c \in K$. From this one obtains the *Bregman inequality*:

$$D_f(c, z) \geq D_f(c, P_K^f(z)) + D_f(P_K^f(z), z), \tag{6}$$

for all $c \in K \cap D$.

Following [2], we say that a Legendre function f is a *Bregman-Legendre* function if the following properties hold; the specific citations below refer to [2].

B1: For x in D and any $a > 0$ the set $\{z | D_f(x, z) \leq a\}$ is bounded (BL0 and BL1 of Def. 5.2).
B2: If x is in D but not in $\text{int}\, D$, for each positive integer n, y^n is in $\text{int}\, D$ with $y^n \to y \in \text{bdry}\, D$ and if $\{D_f(x, y^n)\}$ remains bounded, then $D_f(y, y^n) \to 0$, so that $y \in D$ (BL2 of Def. 5.2).
B3: If x^n and y^n are in $\text{int}\, D$, with $x^n \to x$ and $y^n \to y$, where x and y are in D but not in $\text{int}\, D$, and if $D_f(x^n, y^n) \to 0$ then $x = y$ (BL3 of Def. 5.2) .

The following results are proved in somewhat more generality in [2]; all references below are to that paper. We assume throughout that f is a Bregman-Legendre function with essential domain D.
R1: If $y^n \in \text{int}\, D$ and $y^n \to y \in \text{int}\, D$, then $D_f(y, y^n) \to 0$. If, in addition, $x^n \in \text{int}\, D$, $x^n \to x \in \text{int}\, D$ and $D_f(x^n, y^n) \to 0$, then $x = y$.
Proof: Both f and ∇f are continuous on $\text{int}\, D$.
R2: If x and $y^n \in \text{int}\, D$ and $y^n \to y \in \text{bdry}\, D$, then $D_f(x, y^n) \to +\infty$ (Theorem 3.8 (i)) .
R3: If $x^n \in D$, $x^n \to x \in D$, $y^n \in \text{int}\, D$, $y^n \to y \in D$, $\{x, y\} \cap \text{int}\, D \neq \emptyset$ and $D_f(x^n, y^n) \to 0$, then $x = y$ and $y \in \text{int}\, D$ (Theorem 3.9(iii)).
R4: If x and y are in D, but are not in $\text{int}\, D$, $y^n \in \text{int}\, D$, $y^n \to y$ and $D_f(x, y^n) \to 0$, then $x = y$ (Proposition 5.5).
As a consequence of these results we have the following.
R5: If $\{D_f(x, y^n)\} \to 0$, for $y^n \in \text{int}\, D$ and $x \in D$, then $\{y^n\} \to x$.
Proof of R5: By Property B1 above it follows that the sequence $\{y^n\}$ is bounded; without loss of generality, we assume that $\{y^n\} \to y$, for some $y \in \overline{D}$.

If y is in $\text{int}\, D$, then, by the continuity of both f and ∇f on $\text{int}\, D$, we have $\{D_f(x, y^n)\} \to D_f(x, y)$. Therefore, $D_f(x, y) = 0$ and $x = y$.

If y is in $\overline{D}$ but not in $\text{int}\, D$, then, by R2, x is not in $\text{int}\, D$. Then, using B2, we have that y is in D; by R4, it follows that $x = y$. This completes the proof of R5. ∎

We turn now to the MSGP algorithm.

3. THE MSGP ALGORITHM

We begin by setting out the assumptions we shall make and the notation we shall use in this section.

Assumptions and notation:

We make the following assumptions throughout this section. Let $C = \cap_{i=1}^{I} C_i$ be the nonempty intersection of closed convex sets C_i. The function h is super-coercive and Bregman-Legendre with essential domain $D = \operatorname{dom} h$ and $C \cap \operatorname{dom} h \neq \emptyset$. For $i = 1, 2, ..., I$ the function f_i is also Bregman-Legendre, with $D \subseteq \operatorname{dom} f_i$, so that $\operatorname{int} D \subseteq \operatorname{int} \operatorname{dom} f_i$; also $C_i \cap \operatorname{int} \operatorname{dom} f_i \neq \emptyset$. For all $x \in \operatorname{dom} h$ and $z \in \operatorname{int} \operatorname{dom} h$ we have $D_h(x, z) \geq D_{f_i}(x, z)$, for each i.

The MSGP algorithm:

Algorithm 3.1 The MSGP algorithm: *Let* $x^0 \in \operatorname{int} \operatorname{dom} h$ *be arbitrary. For* $k = 0, 1, ...$ *and* $i(k) := k (\operatorname{mod} I) + 1$ *let*

$$x^{k+1} = \nabla h^* \big(\nabla h(x^k) - \nabla f_{i(k)}(x^k) + \nabla f_{i(k)}(P_{C_{i(k)}}^{f_{i(k)}}(x^k)) \big). \tag{7}$$

A preliminary result:

For each $k = 0, 1, ...$ define the function $G^k(\cdot) : \operatorname{dom} h \to [0, +\infty)$ by

$$G^k(x) = D_h(x, x^k) - D_{f_{i(k)}}(x, x^k) + D_{f_{i(k)}}(x, P_{C_{i(k)}}^{f_{i(k)}}(x^k)). \tag{8}$$

The next proposition provides a useful identity, which can be viewed as an analogue of Pythagoras' theorem. The proof is not difficult and we omit it.

Proposition 3.1 *For each* $x \in \operatorname{dom} h$, *each* $k = 0, 1, ...$, *and* x^{k+1} *given by (7) we have*

$$G^k(x) = G^k(x^{k+1}) + D_h(x, x^{k+1}). \tag{9}$$

Consequently, x^{k+1} *is the unique minimizer of the function* $G^k(\cdot)$.

This identity (9) is the key ingredient in the convergence proof for the MSGP algorithm.

The MSGP convergence theorem:

We shall prove the following convergence theorem:

Theorem 3.1 *Let* $x^0 \in \operatorname{int} \operatorname{dom} h$ *be arbitrary. Any sequence* x^k *obtained from the iterative scheme given by Algorithm 3.1 converges to* $x^\infty \in C \cap \operatorname{dom} h$. *If the sets* C_i *are hyperplanes, then* x^∞ *minimizes the function* $D_h(x, x^0)$ *over all* $x \in C \cap \operatorname{dom} h$; *if, in addition,* x^0 *is the global minimizer of* h, *then* x^∞ *minimizes* $h(x)$ *over all* $x \in C \cap \operatorname{dom} h$.

Proof: Let c be a member of $C \cap \operatorname{dom} h$. From the Pythagorean identity (9) it follows that

$$G^k(c) = G^k(x^{k+1}) + D_h(c, x^{k+1}). \tag{10}$$

Using the definition of $G^k(\cdot)$, we write

$$G^k(c) = D_h(c, x^k) - D_{f_{i(k)}}(c, x^k) + D_{f_{i(k)}}(c, P_{C_{i(k)}}^{f_{i(k)}}(x^k)). \tag{11}$$

From Bregman's inequality (6) we have that

$$D_{f_{i(k)}}(c, x^k) - D_{f_{i(k)}}(c, P_{C_{i(k)}}^{f_{i(k)}}(x^k)) \geq D_{f_{i(k)}}(P_{C_{i(k)}}^{f_{i(k)}}(x^k), x^k). \tag{12}$$

Consequently, we know that

$$D_h(c,x^k) - D_h(c,x^{k+1}) \geq G^k(x^{k+1}) + D_{f_{i(k)}}(P^{f_{i(k)}}_{C_{i(k)}}(x^k), x^k) \geq 0. \tag{13}$$

It follows that $\{D_h(c,x^k)\}$ is decreasing and finite and the sequence $\{x^k\}$ is bounded. Therefore, $\{D_{f_{i(k)}}(P^{f_{i(k)}}_{C_{i(k)}}(x^k), x^k)\} \to 0$ and $\{G^k(x^{k+1})\} \to 0$; from the definition of $G^k(x)$ it follows that $\{D_{f_{i(k)}}(x^{k+1}, P^{f_{i(k)}}_{C_{i(k)}}(x^k))\} \to 0$ as well. Using the Bregman inequality we obtain the inequality

$$D_h(c,x^k) \geq D_{f_{i(k)}}(c,x^k) \geq D_{f_{i(k)}}(c, P^{f_{i(k)}}_{C_{i(k)}}(x^k)), \tag{14}$$

which tells us that the sequence $\{P^{f_{i(k)}}_{C_{i(k)}}(x^k)\}$ is also bounded. Let x^* be an arbitrary cluster point of the sequence $\{x^k\}$ and let $\{x^{k_n}\}$ be a subsequence of the sequence $\{x^k\}$ converging to x^*.

We first show that $x^* \in \operatorname{dom} h$ and $\{D_h(x^*,x^k)\} \to 0$. If x^* is in $\operatorname{int}\operatorname{dom} h$ then our claim is verified, so suppose that x^* is in $\operatorname{bdry}\operatorname{dom} h$. If c is in $\operatorname{dom} h$ but not in $\operatorname{int}\operatorname{dom} h$, then, applying B2 above, we conclude that $x^* \in \operatorname{dom} h$ and $\{D_h(x^*,x^k)\} \to 0$. If, on the other hand, c is in $\operatorname{int}\operatorname{dom} h$ then by R2 x^* would have to be in $\operatorname{int}\operatorname{dom} h$ also. It follows that $x^* \in \operatorname{dom} h$ and $\{D_h(x^*,x^k)\} \to 0$. Now we show that x^* is in C.

Label $x^* = x^*_0$. Since there must be at least one index i that occurs infinitely often as $i(k)$, we assume, without loss of generality, that the subsequence $\{x^{k_n}\}$ has been selected so that $i(k) = 1$ for all $n = 1, 2, \ldots$. Passing to subsequences as needed, we assume that, for each $m = 0, 1, 2, \ldots, I-1$, the subsequence $\{x^{k_n+m}\}$ converges to a cluster point x^*_m, which is in $\operatorname{dom} h$, according to the same argument we used in the previous paragraph. For each m the sequence $\{D_{f_m}(c, P^{f_m}_{C_m}(x^{k_n+m-1}))\}$ is bounded, so, again, by passing to subsequences as needed, we assume that the subsequence $\{P^{f_m}_{C_m}(x^{k_n+m-1})\}$ converges to $c^*_m \in C_m \cap \overline{\operatorname{dom} f_m}$.

Since the sequence $\{D_{f_m}(c, P^{f_m}_{C_m}(x^{k_n+m-1})\}$ is bounded and $c \in \operatorname{dom} f_m$, it follows, from either B2 or R2, that $c^*_m \in \operatorname{dom} f_m$. We know that

$$\{D_{f_m}(P^{f_m}_{C_m}(x^{k_n+m-1}), x^{k_n+m-1})\} \to 0 \tag{15}$$

and both $P^{f_m}_{C_m}(x^{k_n+m-1})$ and x^{k_n+m-1} are in $\operatorname{int}\operatorname{dom} f_m$. Applying R1, B3 or R3, depending on the assumed locations of c^*_m and x^*_{m-1}, we conclude that $c^*_m = x^*_{m-1}$.

We also know that

$$\{D_{f_m}(x^{k_n+m}, P^{f_m}_{C_m}(x^{k_n+m-1}))\} \to 0, \tag{16}$$

from which it follows, using the same arguments, that $x^*_m = c^*_m$. Therefore, we have $x^* = x^*_m = c^*_m$ for all m; so $x^* \in C$.

Since $x^* \in C \cap \operatorname{dom} h$, we may now use x^* in place of the generic c, to obtain that the sequence $\{D_h(x^*,x^k)\}$ is decreasing. However, we also know that the sequence $\{D_h(x^*,x^{k_n})\} \to 0$. So we have $\{D_h(x^*,x^k)\} \to 0$. Applying R5, we conclude that $\{x^k\} \to x^*$.

If the sets C_i are hyperplanes, then we get equality in Bregman's inequality (6)and so

$$D_h(c,x^k) - D_h(c,x^{k+1}) = G^k(x^{k+1}) + D_{f_{i(k)}}(P^{f_{i(k)}}_{C_{i(k)}}(x^k), x^k). \tag{17}$$

Since the right side of this equation is independent of which c we have chosen in the set $C\cap$ dom h, the left side is also independent of this choice. This implies that

$$D_h(c,x^0) - D_h(c,x^M) = D_h(x^*,x^0) - D_h(x^*,x^M), \tag{18}$$

for any positive integer M and any $c \in C \cap \operatorname{dom} h$. Therefore

$$D_h(c,x^0) - D_h(x^*,x^0) = D_h(c,x^M) - D_h(x^*,x^M). \tag{19}$$

Since $\{D_h(x^*,x^M)\} \to 0$ as $M \to +\infty$ and $\{D_h(c,x^M)\} \to \alpha \geq 0$, we have that $D_h(c,x^0) - D_h(x^*,x^0) \geq 0$. This completes the proof. ▮

4. AN INTERIOR POINT ALGORITHM FOR ITERATIVE OPTIMIZATION

We consider now an interior point algorithm (IPA) for iterative optimization. This algorithm was first presented in [6] and applied to transmission tomography in [13]. The IPA is suggested by a special case of the MSGP, involving functions h and $f := f_1$.

Assumptions:
We assume, for the remainder of this section, that h is a super-coercive Legendre function with essential domain $D = \operatorname{dom} h$. We also assume that f is continuous on the set $\overline{D}$, takes the value $+\infty$ outside this set and is differentiable in $\operatorname{int}\operatorname{dom} D$. Thus, f is a closed, proper convex function on R^J. We assume also that $\hat{x} = \operatorname{argmin}_{x\in\overline{D}} f(x)$ exists, but not that it is unique. As in the previous section, we assume that $D_h(x,z) \geq D_f(x,z)$ for all $x \in \operatorname{dom} h$ and $z \in \operatorname{int}\operatorname{dom} h$. As before, we denote by h^* the function conjugate to h.

The IPA:
The IPA is an iterative procedure that, under conditions to be described shortly, minimizes the function f over the closure of the essential domain of h, provided that such a minimizer exists.

Algorithm 4.1 *Let x^0 be chosen arbitrarily in* $\operatorname{int} D$. *For $k = 0, 1, \ldots$ let x^{k+1} be the unique solution of the equation*

$$\nabla h(x^{k+1}) = \nabla h(x^k) - \nabla f(x^k). \tag{20}$$

Note that equation (20) can also be written as

$$x^{k+1} = \nabla h^*(\nabla h(x^k) - \nabla f(x^k)). \tag{21}$$

Motivating the IPA:
As already noted, the IPA was originally suggested by consideration of a special case of the MSGP. Suppose that $\overline{x} \in \operatorname{dom} h$ is the unique global minimizer of the function f, and that $\nabla f(\overline{x}) = 0$. Take $I = 1$ and $C = C_1 = \{\overline{x}\}$. Then $P^f_{C_1}(x^k) = \overline{x}$ always and the iterative MSGP step becomes that of the IPA. Since we are assuming that $\overline{x}$ is in dom h, the convergence theorem for the MSGP tells us that the iterative sequence $\{x^k\}$ converges to $\overline{x}$.

In most cases, the global minimizer of f will not lie within the essential domain of the function h and we are interested in the minimum value of f on the set $\overline{D}$, where $D = \operatorname{dom} h$; that is, we want $\hat{x} = \operatorname{argmin}_{x \in \overline{D}} f(x)$, whenever such a minimum exists. As we shall see, the IPA can be used to advantage even when the specific conditions of the MSGP do not hold.

Preliminary results for the IPA:

Two aspects of the IPA suggest strongly that it may converge under more general conditions than those required for convergence of the MSGP. The sequence $\{x^k\}$ defined by (20) is entirely within the interior of $\operatorname{dom} h$. In addition, as we now show, the sequence $\{f(x^k)\}$ is decreasing. Adding both sides of the inequalities $D_h(x^{k+1}, x^k) - D_f(x^{k+1}, x^k) \geq 0$ and $D_h(x^k, x^{k+1}) - D_f(x^k, x^{k+1}) \geq 0$ gives

$$\langle \nabla h(x^k) - \nabla h(x^{k+1}) - \nabla f(x^k) + \nabla f(x^{k+1}), x^k - x^{k+1} \rangle \geq 0. \tag{22}$$

Substituting according to equation (20) and using the convexity of the function f, we obtain

$$f(x^k) - f(x^{k+1}) \geq \langle \nabla f(x^{k+1}), x^k - x^{k+1} \rangle \geq 0. \tag{23}$$

Therefore, the sequence $\{f(x^k)\}$ is decreasing; since it is bounded below by $f(\hat{x})$, it has a limit, $\hat{f} \geq f(\hat{x})$. We have the following result (see [6], Prop. 3.1).

Lemma 4.1 $\hat{f} = f(\hat{x})$.

Proof: Suppose, to the contrary, that $0 < \delta = \hat{f} - f(\hat{x})$. Select $z \in D$ with $f(z) \leq f(\hat{x}) + \delta/2$. Then $f(x^k) - f(z) \geq \delta/2$ for all k. Writing $H_k = D_h(z, x^k) - D_f(z, x^k)$ for each k, we have

$$H_k - H_{k+1} = D_h(x^{k+1}, x^k) - D_f(x^{k+1}, x^k) + \langle \nabla f(x^{k+1}), x^{k+1} - z \rangle . \tag{24}$$

Since $\langle \nabla f(x^{k+1}), x^{k+1} - z \rangle \geq f(x^{k+1}) - f(z) \geq \delta/2 > 0$ and $D_h(x^{k+1}, x^k) - D_f(x^{k+1}, x^k) \geq 0$, it follows that $\{H_k\}$ is a decreasing sequence of positive numbers, so that the successive differences converge to zero. This is a contradiction; we conclude that $\hat{f} = f(\hat{x})$. ∎

Convergence of the IPA:

We prove the following convergence result for the IPA (see also [6]).

Theorem 4.1 *If $\hat{x} = \operatorname{argmin}_{x \in \overline{D}} f(x)$ is unique, then the sequence $\{x^k\}$ generated by the IPA according to equation (20) converges to $\hat{x}$. If $\hat{x}$ is not unique, but can be chosen in D, then the sequence $\{D_h(\hat{x}, x^k)\}$ is decreasing. If, in addition, the function $D_h(\hat{x}, \cdot)$ has bounded level sets, then the sequence $\{x^k\}$ is bounded and so has cluster points $x^* \in \overline{D}$ with $f(x^*) = f(\hat{x})$. Finally, if h is a Bregman-Legendre function, then $x^* \in D$ and the sequence $\{x^k\}$ converges to x^*.*

Proof: According to Corollary 8.7.1 of [14], if G is a closed, proper convex function on R^J and if the level set $L_\alpha = \{x | G(x) \leq \alpha\}$ is nonempty and bounded for at least one value of α, then L_α is bounded for all values of α. If the constrained minimizer $\hat{x}$ is unique, then, by the continuity of f on $\overline{D}$ and Rockafellar's corollary, we can conclude that the sequence $\{x^k\}$ converges to $\hat{x}$. If $\hat{x}$ is not unique, but can be chosen in D, then, with additional assumptions, convergence can still be established.

Suppose now that $\hat{x}$ is not necessarily unique, but can be chosen in D. Assuming $\hat{x} \in D$, we show that the sequence $\{D_h(\hat{x}, x^k)\}$ is decreasing. Using equation (20) we have

$$D_h(\hat{x}, x^k) - D_h(\hat{x}, x^{k+1}) = D_h(x^{k+1}, x^k) + \langle \nabla h(x^{k+1}) - \nabla h(x^k), \hat{x} - x^{k+1} \rangle$$

$$= D_h(x^{k+1}, x^k) - D_f(x^{k+1}, x^k) + D_f(x^{k+1}, x^k) + \langle \nabla f(x^k), x^{k+1} - \hat{x} \rangle$$

$$= D_h(x^{k+1}, x^k) - D_f(x^{k+1}, x^k) + f(x^{k+1}) - f(x^k) - \langle \nabla f(x^k), \hat{x} - x^k \rangle$$

$$\geq D_h(x^{k+1}, x^k) - D_f(x^{k+1}, x^k) + f(x^{k+1}) - f(x^k) + f(x^k) - f(\hat{x});$$

the final inequality follows from the convexity of f. Since $D_h(x^{k+1}, x^k) - D_f(x^{k+1}, x^k) \geq 0$ and $f(x^{k+1}) - f(\hat{x}) \geq 0$, it follows that the sequence $\{D_h(\hat{x}, x^k)\}$ is decreasing.

If h has bounded level sets, then the sequence $\{x^k\}$ is bounded and we can extract a subsequence $\{x^{k_n}\}$ converging to some x^* in the closure of D.

Finally, assume that h is a Bregman-Legendre function. If $\hat{x}$ is in D but not in $\operatorname{int} D$, then, by B2, $x^* \in \operatorname{bdry} D$ implies that x^* is in D and $\{D_h(x^*, x^{k_n})\} \to 0$. If $\hat{x}$ is in $\operatorname{int} D$, then we conclude, from R2, that x^* is also in $\operatorname{int} D$. Then, by R1, we have $\{D_h(x^*, x^{k_n})\} \to 0$. We can then replace the generic $\hat{x}$ with x^*, to conclude that $\{D_h(x^*, x^k)\}$ is decreasing. But, $\{D_h(x^*, x^{k_n})\}$ converges to zero; therefore, the entire sequence $\{D_h(x^*, x^k)\}$ converges to zero. Applying R5, we conclude that $\{x^k\}$ converges to x^*, completing the proof. ∎

5. APPLYING THE MSGP TO THE SPLIT FEASIBILITY PROBLEM

We close by illustrating use of the MSGP algorithm to solve the split feasibility problem. Other examples are to be found in [6] and [13].

Overview of the SFP:

Given closed convex sets K in R^N and Q in R^M and M by N full-rank matrix A, the *split feasibility problem* (SFP) is to find x in K such that Ax is in Q, if such x exist, which we shall assume throughout this section. With $A^{-1}(Q) = \{x | Ax \in Q\}$, the SFP is to find a member of the intersection of K and $A^{-1}(Q)$, if there are any members. Formulated this way, the SFP becomes a special case of the convex feasibility problem (CFP). Iterative algorithms for solving the CFP, involving orthogonal or generalized projection onto the individual convex sets, can then be applied to solve the SFP, as is done by Censor and Elfving [8]. For examples in Hilbert space see Youla [15], Bauschke and Borwein [1] and Kotzer *et al* [10]. In typical applications, the sets K and Q are easy to describe and the orthogonal projections onto K and Q, denoted P_K and P_Q, respectively, are relatively simple to implement. In contrast, the orthogonal projections onto $A(K) = \{Ax | x \in K\}$ and $A^{-1}(Q)$ are not easily computed. We seek iterative methods that require only the orthogonal projections P_K and P_Q.

Let G be an M by M positive-definite matrix with associated squared norm $||z||_G^2 = z^T G z$. For fixed x in R^N the G-projection of Ax onto the convex set $A(K)$ is the vector $P_{A(K)}^G(Ax) = A\hat{c}$ that minimizes the function $||Ax - Ac||_G^2 = (x - c)^T A^T G A (x - c)$ over all c in K. If there is G for which $A^T G A = I$, then it follows that $\hat{c}$ is the orthogonal projection of x onto K; therefore, we have $P_{A(K)}^G(Ax) = A P_K(x)$.

If $M = N$ then $G = (AA^T)^{-1}$ is a suitable choice; since A is invertible, we then have $P^G_{A(K)}(z) = AP_K(A^{-1}(z))$ for all z in R^M. Our iterative algorithms then involve P_Q and $P^G_{A(K)}$.

If $M < N$ there is no such matrix G. Moreover, the convex set $A(K)$ need not be closed. To obtain iterative algorithms for this case we augment A using a full-rank $N - M$ by N matrix B to obtain the N by N invertible matrix $R = [A^T\, B^T]^T$. Although it is not necessary, we shall assume that B has been chosen so that $AB^T = 0$.

If $M > N$ the choice of $G = A(A^TA)^{-2}A^T$ gives $A^TGA = I$. With $A^+ := (A^TA)^{-1}A^T$ and $A(R^N)$ denoting the range of A, we have $A(K) = (A^+)^{-1}(K) \cap A(R^N)$, so the set $A(K)$ is closed. Proceeding as above, we have $P^G_{A(K)}(Ax) = AP_K(x)$; for those z in R^M that are not in the range of A, we have $P^G_{A(K)}(z) = AP_K(A^TA)^{-1}A^Tz$. Because the function $g(z) = z^TGz$ is not a Legendre function on R^M, we do not consider $P^g_{A(K)}(z)$, a Bregman projection onto the convex set $A(K)$. We replace G with $H = G + U$, where U is a nonnegative-definite symmetric matrix such that H is positive-definite and $A^TU = 0$. The function $f(z) = z^THz$ is a Legendre function on R^M and the associated Bregman projection of z onto $A(K)$ is $P^f_{A(K)}(z) = P^H_{A(K)}(z) = AP_K(A^TA)^{-1}A^Tz$ once again.

When we employ P_Q and $P^H_{A(K)}$ in the simultaneous algorithm of Censor and Elfving [8], we find that the algorithm involves the choice of the matrix U. Here we illustrate the use of the MSGP algorithm and find that we obtain an iterative solution to the SFP, which, for the case in which $M > N$, does not depend on the choice of the matrix U.

Case 1: $M = N$

Now A is an N by N matrix with rank equal to N. In order to apply the MSGP algorithm we define $C_1 = A^{-1}(Q) = \{x|\, Ax \in Q\}$, $C_2 = K$, $f_1(x) = \gamma x^TA^TAx$, and $h(x) = f_2(x) = x^Tx$; here γ is a positive constant chosen so that

$$D_h(x,z) - D_{f_1}(x,z) = (x-z)^T(I - \gamma A^TA)(x-z) \geq 0, \tag{25}$$

for all x and z. Therefore, $\gamma \leq 1/\lambda_{max}(A^TA)$, where $\lambda_{max}(A^TA)$ denotes the largest eigenvalue of the matrix A^TA. Let us normalize A so that each of its rows has Euclidean norm equal to one. Then, since trace (A^TA) =trace$(AA^T) = M$, it follows that we can choose $\gamma \leq 1/M$. The corresponding Bregman projections are

$$P^{f_1}_{C_1}(x) = (A^TA)^{-1}A^TP_Q(Ax) \tag{26}$$

and

$$P^{f_2}_{C_2}(x) = P_K(x), \tag{27}$$

where P_Q and P_K denote the orthogonal projections onto Q and K, respectively.

Applying the MSGP algorithm, we obtain the following iteration step. For $k = 0, 1, ...,$ having obtained x^k, let

$$x^{k+1} = P_K(I + \gamma A^T(P_Q - I)A)x^k. \tag{28}$$

Then the sequence $\{x^k\}$ given by (28) converges to an $x \in K$ for which $Ax \in Q$.

Case 2: $M < N$

In this case we augment the matrix A, as discussed above, to obtain the invertible square matrix R. This puts us in Case 1, where the iterative algorithm for that case gives the following. For $k = 0, 1, ...$, having obtained x^k, we have

$$x^{k+1} = P_K(I + \gamma R^T(P_V - I)R)x^k, \tag{29}$$

where γ is now selected so that the matrix $I - \gamma R^T R$ is nonnegative definite. Recalling that $AB^T = 0$, we rewrite (29) in terms of the matrices A and B:

$$x^{k+1} = P_K(x^k + \gamma(A^T(P_Q - I)A + B^T(I - I)B)x^k) \tag{30}$$

so that

$$x^{k+1} = P_K(x^k + \gamma(A^T(P_Q - I)A)x^k). \tag{31}$$

This is the same algorithm as in the previous case, except that here the γ must be chosen so that $I - \gamma(A^T A + B^T B)$ is nonnegative definite. However, since there is no lower limit on the maximum eigenvalue of $B^T B$, we may conclude that the iteration in (28) applies for all $M \leq N$. This is reasonable, since there is no inversion involved. Notice that we have not needed that $A(K)$ be closed.

For example, suppose we wish to solve the consistent system of linear equations represented by the matrix equation $Ax = b$, with $M \leq N$. If we have no constraints that we wish to impose on the solution, we let $K = R^N$; let $Q = \{b\}$. Then (28) becomes

$$x^{k+1} = x^k + \gamma A^T(b - Ax^k), \tag{32}$$

which is sometimes called the Landweber algorithm [11], known to converge to a solution of $Ax = b$ whenever γ is chosen so that $I - \gamma A^T A$ is nonnegative definite. If we wish to impose the constraint that the solution x be a nonnegative vector, then we take K to be the nonnegative orthant in R^N. The resulting algorithm is a nonnegatively clipped version of the iteration in (32).

Case 3: $M > N$

For simplicity, we formulate the SFP as a CFP involving the three closed convex sets $C_1 = A(R^N)$, $C_2 = A(K)$ and $C_3 = Q$ in R^M. We let $h(x) = f_1(x) = f_3(x) = x^T x/2$ and $f_2(x) = \gamma x^T Hx/2$, where, as above, H is the positive-definite matrix $H = G + U$, built from $G = A(A^T A)^{-2} A^T$ and the nonnegative-definite matrix U and γ chosen so that $I - \gamma H$ is positive-definite. Recall that $A^T U = 0$. Then we have $P_1 := P_{C_1}^{f_1}$ the orthogonal projection onto the range of A in R^M, $P_2 := P_{C_2}^{f_2}$ with $P_2(Ax) = P_{A(K)}^H(Ax) = AP_K(x)$ and $P_3 := P_{C_3}^{f_3} = P_Q$. We now apply the MSGP algorithm.

Beginning with an arbitrary z^0 in R^M, we take

$$z^1 = P_1(z^0) = A(A^T A)^{-1} A^T z^0 = Au^1, \tag{33}$$

where

$$u^1 = (A^T A)^{-1} A^T z^0. \tag{34}$$

The next step gives z^2, which minimizes the function

$$(h - f_2)(z - z^1) + f_2(z - P_2(z^1)). \tag{35}$$

Therefore, we have

$$0 = (I - \gamma H)(z - Au^1) + \gamma H(z - AP_K(u^1)), \tag{36}$$

so that

$$z^2 = A((I - \gamma(A^TA)^{-1}(I - P_K))((A^TA)^{-1}A^Tz^0). \tag{37}$$

Finally,

$$z^3 = P_Q(z^2). \tag{38}$$

Writing the iterative algorithm in terms of completed cycles, we have $w^0 = z^0$ and

$$w^{k+1} = P_QA(I + \gamma(A^TA)^{-1}(P_K - I))(A^TA)^{-1}A^Tw^k. \tag{39}$$

The iterative step for $x^k := (A^TA)^{-1}A^Tw^k$ is then

$$x^{k+1} = (A^TA)^{-1}A^TP_QA(I + \gamma(A^TA)^{-1}(P_K - I))x^k. \tag{40}$$

We must select γ so that $I - \gamma H$ is positive-definite. Because there is no lower limit to the maximum eigenvalue of U, it follows that γ must be chosen so that $I - \gamma G$ is positive-definite. Since $G = A(A^TA)^{-2}A^T$ we have $Gz = \lambda z$ implies that $\lambda A^Tz = A^TGz = (A^TA)^{-1}A^Tz$, so that the nonzero eigenvalues of G are those of $(A^TA)^{-1}$. It follows that we must select γ not greater than the smallest eigenvalue of A^TA.

SUMMARY

In this paper we have considered the iterative method of successive generalized projections onto convex sets for solving the convex feasibility problem. The generalized projections are derived from Bregman-Legendre functions. In particular, we have extended Bregman's method to permit the generalized projections used at each step to be taken with respect generalized distances that vary with the convex set. Merely replacing the distance $D_f(x,z)$ in Bregman's method with distances $D_{f_i}(x,z)$ is not enough; counterexamples show that such a simple extension may not converge. We show that a convergent algorithm, the MSGP, can be obtained through the use of a dominating Bregman-Legendre distance, that is $D_h(x,z) \geq D_{f_i}(x,z)$, for all i, and a form of relaxation based on the notion of generalized convex combination.

Particular problems are solved through the selection of appropriate functions h and f_i. The MSGP algorithm can be used to solve the split feasibility problem. Iterative interior point optimization algorithms can also be based on the MSGP approach.

REFERENCES

1. H. H. Bauschke and J. M. Borwein, On projection algorithms for solving convex feasibility problems, *SIAM Review* **38** (1996) 367–426.
2. H. H. Bauschke and J. M. Borwein, Legendre functions and the method of random Bregman projections, *J. of Convex Analysis* **4** (1997) 27–67.
3. L. M. Bregman, The relaxation method of finding the common point of convex sets and its application to the solution of problems in convex programming, *USSR Computational Mathematics and Mathematical Physics* **7** (1967) 200–217.
4. C. L. Byrne, Block-iterative methods for image reconstruction from projections, *IEEE Transactions on Image Processing* **IP-5** (1996) 792–794.
5. C. L. Byrne, Iterative projection onto convex sets using multiple Bregman distances, *Inverse Problems* **15** (1999) 1295–1313.
6. C. L. Byrne, Block-iterative Interior point optimization methods for image reconstruction from limited data, *Inverse Problems,* **16** (2000) 1405–1419.
7. C. L. Byrne and Y. Censor, Proximity function minimization using multiple Bregman projections, with applications to split feasibility and Kullback-Leibler distance minimization, *Annals of Operations Research* accepted for publication.
8. Y. Censor and T. Elfving, A multiprojection algorithm using Bregman projections in a product space, *Numerical Algorithms,* **8** (1994) 221–239.
9. Y. Censor and S. A. Zenios, *Parallel Optimization: Theory, Algorithms and Applications* (Oxford University Press, New York, 1997).
10. T. Kotzer, N. Cohen and J. Shamir, A projection-based algorithm for consistent and inconsistent constraints, *SIAM J. Optim.* **7(2)** (1997) pp. 527–546.
11. L. Landweber, An iterative formula for Fredholm integral equations of the first kind, *Amer. J. of Math.* **73** (1951) pp. 615–624.
12. K. Lange, M. Bahn and R. Little, A theoretical study of some maximum likelihood algorithms for emission and transmission tomography, *IEEE Trans. Med. Imag.* **6** (1987) 106–114.
13. M. Narayanan, C. L. Byrne and M. King, An interior point iterative reconstruction algorithm incorporating upper and lower bounds, with application to SPECT transmission imaging, *IEEE Trans. Medical Imaging,* submitted for publication.
14. R. T. Rockafellar, *Convex Analysis* (Princeton University Press, Princeton, New Jersey, 1970).
15. D.C. Youla, Mathematical theory of image restoration by the method of convex projections, *in* Stark, H. (Editor) *Image Recovery: Theory and Applications,* (Academic Press, Orlando, Florida, USA, 1987), pp. 29–78.

Inherently Parallel Algorithms in Feasibility and Optimization and their Applications
D. Butnariu, Y. Censor and S. Reich (Editors)

AVERAGING STRINGS OF SEQUENTIAL ITERATIONS FOR CONVEX FEASIBILITY PROBLEMS

Y. Censor[a*], T. Elfving[b†] and G.T. Herman[c‡]

[a]Department of Mathematics, University of Haifa,
Mt. Carmel, Haifa 31905, Israel

[b] Department of Mathematics, Linköping University,
SE-581 83 Linköping, Sweden

[c] Department of Computer and Information Sciences,
Temple University, 1805 North Broad Street, Philadelphia, PA. 19122-6094, USA

An algorithmic scheme for the solution of convex feasibility problems is proposed in which the end-points of strings of sequential projections onto the constraints are averaged. The scheme, employing Bregman projections, is analyzed with the aid of an extended product space formalism. For the case of orthogonal projections we give also a relaxed version. Along with the well-known purely sequential and fully simultaneous cases, the new scheme includes many other inherently parallel algorithmic options depending on the choice of strings. Convergence in the consistent case is proven and an application to optimization over linear inequalities is given.

1. INTRODUCTION

In this paper we present and study a new algorithmic scheme for solving the *convex feasibility problem* of finding a point x^* in the nonempty intersection $C = \cap_{i=1}^m C_i$ of finitely many closed and convex sets C_i in the Euclidean space $\mathbf{R}^n$. Algorithmic schemes for this problem are, in general, either *sequential* or *simultaneous* or can also be *block-iterative* (see, e.g., Censor and Zenios [15, Section 1.3] for a classification of projection algorithms into such classes, and the review paper of Bauschke and Borwein [3] for a variety of specific algorithms of these kinds).

We now explain these terms in the framework of the algorithmic scheme proposed in this paper. For $t = 1, 2, \ldots, M$, let the *string* I_t be an ordered subset of $\{1, 2, \ldots, m\}$ of the form

$$I_t = (i_1^t, i_2^t, \ldots, i_{m(t)}^t), \tag{1}$$

*Work supported by grants 293/97 and 592/00 of the Israel Science Foundation, founded by the Israel Academy of Sciences and Humanities, and by NIH grant HL-28438.
†Work supported by the Swedish Natural Science Research Council under Project M650-19981853/2000.
‡Work supported by NIH grant HL-28438.

with $m(t)$ the number of elements in I_t. We will assume that, for any t, the elements of I_t are distinct from each other; however, the extension of all that we say below to the case without this assumption is trivial (it only complicates the notation). Suppose that there is a set $S \subseteq \mathbf{R}^n$ such that there are operators $R_1, R_2, \ldots, R_m$ mapping S into S and an operator R which maps S^M into S.

Algorithmic Scheme

Initialization: $x^{(0)} \in S$ is arbitrary.

Iterative Step: given the current iterate $x^{(k)}$,
(i) calculate, for all $t = 1, 2, \ldots, M$,

$$T_t x^{(k)} = R_{i^t_{m(t)}} \cdots R_{i^t_2} R_{i^t_1} x^{(k)}, \tag{2}$$

(ii) and then calculate

$$x^{(k+1)} = R(T_1 x^{(k)}, T_2 x^{(k)}, \ldots, T_M x^{(k)}). \tag{3}$$

For every $t = 1, 2, \ldots, M$, this algorithmic scheme applies to $x^{(k)}$ successively the operators whose indices belong to the tth string. This can be done in parallel for all strings and then the operator R maps all end-points onto the next iterate $x^{(k+1)}$. This is indeed an algorithm provided that the operators $\{R_i\}_{i=1}^m$ and R all have algorithmic implementations. In this framework we get a *sequential* algorithm by the choice $M = 1$ and $I_1 = (1, 2, \ldots, m)$ and a *simultaneous* algorithm by the choice $M = m$ and $I_t = (t),\ t = 1, 2, \ldots, M$.

We demonstrate the underlying idea of our algorithmic scheme with the aid of Figure 1. For simplicity, we take the convex sets to be hyperplanes, denoted by H_1, H_2, H_3, H_4, H_5, and H_6, and assume all operators $\{R_i\}$ to be orthogonal projections onto the hyperplanes. The operator R is taken as a convex combination

$$R(x^1, x^2, \ldots, x^M) = \sum_{t=1}^{M} \omega_t x^t, \tag{4}$$

with $\omega_t > 0$, for all $t = 1, 2, \ldots, M$, and $\sum_{t=1}^{M} \omega_t = 1$.

Figure 1(a) depicts a purely sequential algorithm. This is the so-called POCS (Projections Onto Convex Sets) algorithm which coincides, for the case of hyperplanes, with the Kaczmarz algorithm, see, e.g., Algorithms 5.2.1 and 5.4.3, respectively, in [15] and Gubin, Polyak and Raik [26].

The fully simultaneous algorithm appears in Figure 1(b). With orthogonal reflections instead of orthogonal projections it was first proposed, by Cimmino [16], for solving linear equations. Here the current iterate $x^{(k)}$ is projected on all sets simultaneously and the next iterate $x^{(k+1)}$ is a convex combination of the projected points.

In Figure 1(c) we show how a simple averaging of *successive* projections (as opposed to averaging of parallel projections in Figure 1(b)) works. In this case $M = m$ and $I_t = (1, 2, \ldots, t)$, for $t = 1, 2, \ldots, M$. This scheme, appearing in Bauschke and Borwein [3], inspired our proposed Algorithmic Scheme whose action is demonstrated in Figure

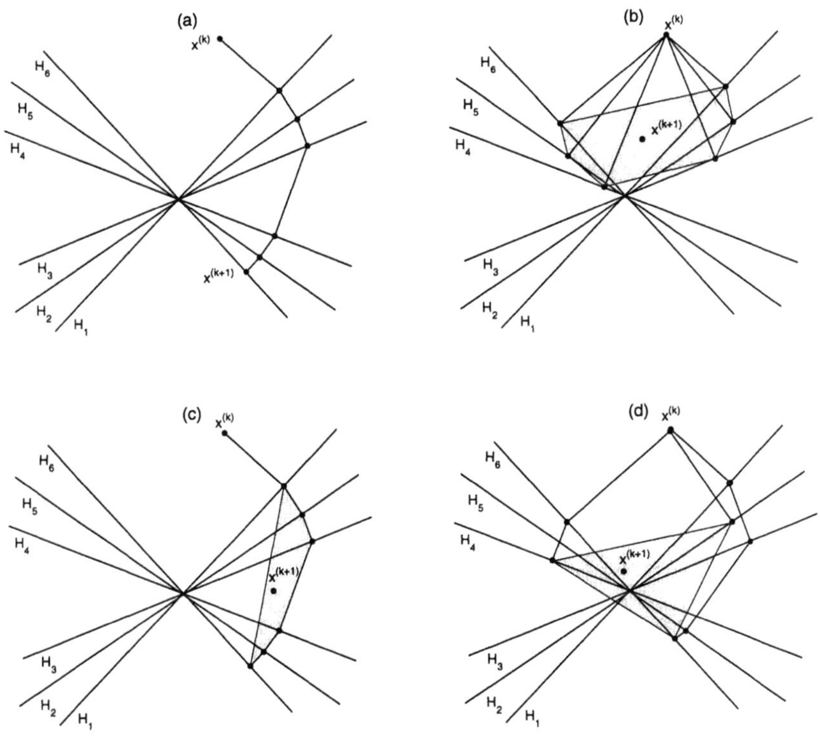

Figure 1. (a) Sequential projections. (b) Fully simultaneous projections. (c) Averaging sequential projections. (d) The new scheme: combining end-points of sequential strings.

1(d). It averages, via convex combinations, the end-points obtained from strings of sequential projections. This proposed scheme offers a variety of options for steering the iterates towards a solution of the convex feasibility problem. It is an inherently parallel scheme in that its mathematical formulation is parallel (like the fully simultaneous method mentioned above). We use this term to contrast such algorithms with others which are sequential in their mathematical formulation but can, sometimes, be implemented in a parallel fashion based on appropriate model decomposition (i.e., depending on the structure of the underlying problem). Being inherently parallel, our algorithmic scheme enables flexibility in the actual manner of implementation on a parallel machine.

We have been able to prove convergence of the Algorithmic Scheme for two special cases. In both cases it is assumed that (i) $C \cap \overline{S} \neq \emptyset$ (where $C = \cap_{i=1}^{m} C_i$ and $\overline{S}$ is the closure of S), (ii) every element of $\{1, 2, \ldots, m\}$ appears in at least one of the strings I_t, and (iii) all weights ω_t associated with the operator R are positive real numbers which sum up to one.

Case I. Each R_i is the Bregman projection onto C_i with respect to a Bregman function f with zone S and the operator R of (3) is a generalized convex combination, with weights ω_t, to be defined in Section 2.1.

Case II. $S = \mathbf{R}^n$ and, for $i = 1, 2, \ldots, m$, $R_i x = x + \theta_i(P_{C_i}x - x)$, with $0 < \theta_i < 2$, where P_{C_i} is the orthogonal projection onto C_i and R is defined by (4).

A generalization of this operator R was used by Censor and Elfving [12] and Censor and Reich [14] in fully simultaneous algorithms which employ Bregman projections. Our proof of convergence for Case I is based on adopting a product space formalism which is motivated by, but is somewhat different from, the product space formalism of Pierra [31]. For the proof of Case II we use results of Elsner, Koltracht and Neumann [25] and Censor and Reich [14].

The details and proofs of convergence are given in Section 2. In Section 3 we describe an application to optimization of a Bregman function over linear equalities. We conclude with a discussion, including some open problems in Section 4. The Appendix in Section 5 describes the role of Bregman projections in convex feasibility problems.

2. PROOFS OF CONVERGENCE

We consider the convex feasibility problem of finding $x^* \in C = \cap_{i=1}^{m} C_i$ where, $C_i \subseteq \mathbf{R}^n$, for all $i = 1, 2, \ldots, m$, are closed convex sets and $C \neq \emptyset$. The two Cases I and II, mentioned in the introduction, are presented in detail and their convergence is proven. For both cases we make the following assumptions.

Assumption 1. $C \cap \overline{S} \neq \emptyset$ where $\overline{S}$ is the closure of S, the domain of the algorithmic operators $R_1, R_2, \ldots, R_m$.

Assumption 2. Every element of $\{1, 2, \ldots, m\}$ appears in at least one of the strings I_t, constructed as in (1).

Assumption 3. The weights $\{\omega_t\}_{t=1}^{M}$ associated with the operator R are positive real numbers and $\sum_{t=1}^{M} \omega_t = 1$.

2.1. Case I: An Algorithm for Bregman Projections

Let $\mathcal{B}(S)$ denote the family of Bregman functions with zone $S \subseteq \mathbf{R}^n$ (see, e.g., Censor and Elfving [12], Censor and Reich [14], or Censor and Zenios [15] for definitions, basic properties and relevant references). For a discussion of the role of Bregman projections in algorithms for convex feasibility problems we refer the reader to the Appendix at the end of the paper.

In Case I we define, for $i = 1, 2, \ldots, m$, the algorithmic operator $R_i x$ to be the Bregman projection, denoted by $P^f_{C_i} x$, of x onto the set C_i with respect to a Bregman function f. Recall that the generalized distance $D_f : \overline{S} \times S \subseteq \mathbf{R}^{2n} \to \mathbf{R}$ is

$$D_f(y, x) = f(y) - f(x) - \langle \nabla f(x), y - x \rangle, \tag{5}$$

where $\langle \cdot, \cdot \rangle$ is the standard inner product in $\mathbf{R}^n$. The Bregman projection $P^f_Q x$ onto a closed convex set Q is then defined by

$$P^f_Q x = \arg\min\{D_f(y, x) \mid y \in Q \cap \overline{S}\}. \tag{6}$$

Such a projection exists and is unique, if $Q \cap \overline{S} \neq \emptyset$, see [15, Lemma 2.1.2].

Following Censor and Reich [14] let us call an x which satisfies, for $(x^1, x^2, \ldots, x^M) \in S^M$,

$$\nabla f(x) = \sum_{t=1}^{M} \omega_t \nabla f(x^t), \tag{7}$$

a *generalized convex combination* of $(x^1, x^2, \ldots, x^M)$ with respect to f. We further assume:

Assumption 4. For any $\boldsymbol{x} = (x^1, x^2, \ldots, x^M) \in S^M$ and any set of weights $\{\omega_t\}_{t=1}^{M}$, as in Assumption 3, there is a unique x in S which satisfies (7).

The operator R is defined by letting $R\boldsymbol{x}$ be the x whose existence and uniqueness is guaranteed by Assumption 4. The applicability of the algorithm depends (similarly to the applicability of its predecessors in [12] and [14]) on the ability to invert the gradient ∇f explicitly. If the Bregman function f is essentially smooth, then ∇f is a one-to-one mapping with continuous inverse $(\nabla f)^{-1}$, see, e.g., Rockafellar [33, Corollary 26.3.1].

We now prove convergence of the Algorithmic Scheme in Case I.

Theorem 2.1 *Let $f \in \mathcal{B}(S)$ be a Bregman function and let $C_i \subseteq \mathbf{R}^n$ be given closed convex sets, for $i = 1, 2, \ldots, m$, and define $C = \cap_{i=1}^m C_i$. If $P^f_{C_i} x \in S$ for any $x \in S$ and Assumptions 1–4 hold, then any sequence $\{x^{(k)}\}_{k \geq 0}$, generated by the Algorithmic Scheme for Case I, converges to a point $x^* \in C \cap \overline{S}$.*

Proof. Let $V = \mathbf{R}^n$ and consider the product space $\boldsymbol{V} = V^M = V \times V \times \cdots \times V$ in which, for any $\boldsymbol{x} \in \boldsymbol{V}$, $\boldsymbol{x} = (x^1, x^2, \ldots, x^M)$ with $x^t \in V$, for $t = 1, 2, \ldots, M$. The scalar product in $\boldsymbol{V}$ is denoted and defined by

$$\langle\langle \boldsymbol{x}, \boldsymbol{y} \rangle\rangle = \sum_{t=1}^{M} \langle x^t, y^t \rangle, \tag{8}$$

and we define in $\boldsymbol{V}$, for $j = 1, 2, \ldots, m$, the product sets

$$\boldsymbol{C}_j = \prod_{t=1}^{M} C_{j,t}, \tag{9}$$

with $C_{j,t}$ depending on the strings I_t as follows:

$$C_{j,t} = \begin{cases} C_{i_j^t}, & \text{if } j = 1, 2, \ldots, m(t), \\ V, & \text{if } j = m(t)+1, m(t)+2, \ldots, m. \end{cases} \tag{10}$$

Let

$$\boldsymbol{\Delta} = \{\boldsymbol{x} \mid \boldsymbol{x} = (x, x, \ldots, x),\ x \in V\}, \tag{11}$$

and

$$\delta : V \to \boldsymbol{\Delta}, \quad \delta(x) = (x, x, \ldots, x). \tag{12}$$

The set $\boldsymbol{\Delta}$ is called the *diagonal set* and the mapping δ is the *diagonal mapping.* In view of Assumption 2, the following equivalence between the convex feasibility problems in V and $\boldsymbol{V}$ is obvious:

$$x^* \in C \ \text{ if and only if } \ \delta(x^*) \in (\cap_{j=1}^m \boldsymbol{C}_j) \cap \boldsymbol{\Delta}. \tag{13}$$

The proof is based on examining Bregman's sequential projections algorithm (see Bregman [7, Theorem 1] or Censor and Zenios [15, Algorithm 5.8.1]) applied to the convex feasibility problem on the right-hand side of (13) in the product space $\boldsymbol{V}$. This is done as follows. With weights $\{\omega_t\}_{t=1}^M$, satisfying Assumption 3, we construct the function

$$\boldsymbol{F}(\boldsymbol{x}) = \sum_{t=1}^{M} \omega_t f(x^t). \tag{14}$$

By [12, Lemma 3.1], $\boldsymbol{F}$ is a Bregman function with zone $\boldsymbol{S}$ in the product space, i.e., $\boldsymbol{F} \in \mathcal{B}(\boldsymbol{S})$, where $\boldsymbol{S} = S^M$. Further, denoting by $\boldsymbol{P}_{\boldsymbol{Q}}^{\boldsymbol{F}}\boldsymbol{x}$ the Bregman projection of a point $\boldsymbol{x} \in \boldsymbol{V}$ onto a closed convex set $\boldsymbol{Q} = Q_1 \times Q_2 \times \cdots \times Q_M \subseteq \boldsymbol{V}$, with respect to $\boldsymbol{F}$, we can express it, by [12, Lemma 4.1], as

$$\boldsymbol{P}_{\boldsymbol{Q}}^{\boldsymbol{F}}\boldsymbol{x} = (P_{Q_1}^f x^1, P_{Q_2}^f x^2, \cdots, P_{Q_M}^f x^M). \tag{15}$$

From (2), (9), (10) and (15) we obtain

$$\boldsymbol{P}_{\boldsymbol{C}_m}^{\boldsymbol{F}} \cdots \boldsymbol{P}_{\boldsymbol{C}_2}^{\boldsymbol{F}} \boldsymbol{P}_{\boldsymbol{C}_1}^{\boldsymbol{F}} \boldsymbol{x} = (T_1 x^1, T_2 x^2, \cdots, T_M x^M). \tag{16}$$

Next we show that, for every $\boldsymbol{x} \in \boldsymbol{V}$,

$$\boldsymbol{P}_{\boldsymbol{\Delta}}^{\boldsymbol{F}} \boldsymbol{x} = \delta(x), \tag{17}$$

with $x = R(\boldsymbol{x})$. By (6), (11) and (12), the x which satisfies (17) is

$$x = \arg\min\{\boldsymbol{D}_{\boldsymbol{F}}(\delta(y), \boldsymbol{x}) \mid \delta(y) \in \overline{\boldsymbol{S}}\}, \tag{18}$$

where $\boldsymbol{D_F}(\delta(y), \boldsymbol{x})$ is the Bregman distance in $\boldsymbol{V}$ with respect to $\boldsymbol{F}$. Noting that

$$\nabla \boldsymbol{F}(\boldsymbol{x}) = (\omega_1 \nabla f(x^1), \omega_2 \nabla f(x^2), \ldots, \omega_M \nabla f(x^M)), \tag{19}$$

we have, by (5), (8) and (14), that

$$\boldsymbol{D_F}(\delta(y), \boldsymbol{x}) = \sum_{t=1}^{M} \omega_t (f(y) - f(x^t) - \langle \nabla f(x^t), y - x^t \rangle). \tag{20}$$

Since a Bregman distance is convex with respect to its first (vector) variable (see, e.g., [15, Chapter 2]), at the point x where (20) achieves its minimum, the gradient (with respect to y) must be zero. Thus, differentiating the right-hand side of (20), we get that this x must satisfy (7) and, therefore, by Assumption 4, it is in fact $R(\boldsymbol{x})$.

The convergence ([7, Theorem 1] or [15, Algorithm 5.8.1]) of Bregman's sequential algorithm guarantees, by taking $\boldsymbol{x}^{(0)} = \delta(x^{(0)})$ with $x^{(0)} \in S$ and, for $k \geq 0$, iterating

$$\boldsymbol{x}^{(k+1)} = \boldsymbol{P}^{\boldsymbol{F}}_{\boldsymbol{\Delta}} \boldsymbol{P}^{\boldsymbol{F}}_{\boldsymbol{C}_m} \cdots \boldsymbol{P}^{\boldsymbol{F}}_{\boldsymbol{C}_2} \boldsymbol{P}^{\boldsymbol{F}}_{\boldsymbol{C}_1} \boldsymbol{x}^{(k)}, \tag{21}$$

that $\lim_{k \to 0} \boldsymbol{x}^{(k)} = \boldsymbol{x}^* \in (\cap_{j=1}^{m} \boldsymbol{C}_j) \cap \boldsymbol{\Delta}$. Observing (3), (16), and the fact that the x of (17) is $R(\boldsymbol{x})$, we get by induction that, for all $k \geq 0$, $\boldsymbol{x}^{(k)} = \delta(x^{(k)})$. By (13), this implies that $\lim_{k \to 0} x^{(k)} = x^* \in C$. ∎

2.2. Case II: An Algorithm for Relaxed Orthogonal Projections

The framework and method of proof used in the previous subsection do not let us introduce relaxation parameters into the algorithm. However, drawing on findings of Elsner, Koltracht and Neumann [25] and of Censor and Reich [14] we do so for the special case of orthogonal projections.

In Case II we define, for $i = 1, 2, \ldots, m$, the algorithmic operators

$$R_i x = x + \theta_i (P_{C_i} x - x), \tag{22}$$

where $P_{C_i} x$ is the orthogonal projection of x onto the set C_i and θ_i are *periodic relaxation parameters*. By this we mean that the θ_i are fixed for each set C_i as in Eggermont, Herman and Lent [23, Theorem 1.2]. The algorithmic operator R is defined by (4) with weights ω_t as in Assumption 3. Equation (4) can be obtained from (7) by choosing the Bregman function $f(x) = ||x||_2^2$ with zone $S = \mathbf{R}^n$. In this case $P^f_{C_i} = P_{C_i}$ is the orthogonal projection and the Bregman distance is $D_f(y, x) = ||y - x||_2^2$, see, e.g., [15, Example 2.1.1]. The convergence theorem for the Algorithmic Scheme in Case II now follows.

Theorem 2.2 *If Assumptions 1–3 hold and if, for all* $i = 1, 2, \ldots, m$, *we have* $0 < \theta_i < 2$, *then any sequence* $\{x^{(k)}\}_{k \geq 0}$, *generated by the Algorithmic Scheme for Case II, converges to a point* $x^* \in C$.

Proof. By [25, Example 2] a relaxed projection operator of the form (22) is *strictly nonexpansive* with respect to the Euclidean norm, for any $0 < \theta_i < 2$. By this we mean that [25, Definition 2], for any pair $x, y \in \mathbf{R}^n$,

$$\text{either} \quad ||R_i x - R_i y||_2 < ||x - y||_2 \quad \text{or} \quad R_i x - R_i y = x - y. \tag{23}$$

Further, since every finite composition of strictly nonexpansive operators is a strictly nonexpansive operator [25, p. 307], any finite composition of relaxed projections operators of the form (2) is strictly nonexpansive. Consequently, each such T_t is also a *paracontracting* operator in the sense of [25, Definition 1]), namely, $T_t : \mathbf{R}^n \to \mathbf{R}^n$ is continuous and for any *fixed point* $y \in \mathbf{R}^n$ of T_t, i.e., $T_t y = y$, and any $x \in \mathbf{R}^n$

$$||T_t x - y||_2 < ||x - y||_2 \quad \text{or} \quad T_t x = x. \tag{24}$$

From Censor and Reich [14, Section 4] we then conclude the convergence of any sequence generated by

$$x^{(k+1)} = \sum_{t=1}^{M} \omega_t T_t x^{(k)}, \tag{25}$$

to a common fixed point x^* of the family $\{T_t\}_{t=1}^M$, which in our case means convergence to a feasible point in C. This is so because, for each $t = 1, 2, \ldots, M$, T_t is a product of the paracontractions R_i, given by (2.18), for all $i \in I_t$, and [25, Corollary 1] then implies that x^* is a fixed point of each R_i, thus of each P_{C_i}. The periodic relaxation and the fixed strings guarantee the finite number of paracontractions, thus enabling the use of the convergence results of [14]. ■

3. APPLICATION TO OPTIMIZATION OVER LINEAR EQUALITIES

In this application, we use the fact that the Algorithmic Scheme for Case I solves the convex feasibility problem to prove its nature as an optimization problem solver. Let f be a Bregman function with zone $S \subseteq \mathbf{R}^n$, let A be a matrix and let $d \in \mathcal{R}(A)$ be a vector in the range of A. Consider the following optimization problem

$$\min\{f(x) \mid x \in \overline{S},\ Ax = d\}. \tag{26}$$

We will show that the Algorithmic Scheme for Case I can be used to solve this problem.

Let the matrix A be of order $\overline{m} \times n$, with $\overline{m} = \sum_{i=1}^m \nu_i$, and partition it into m row-blocks of sizes ν_i as follows,

$$A^T = (A_1^T, A_2^T, \ldots, A_m^T), \tag{27}$$

where we denote vector and matrix transposition by T, and let

$$C_i = \{x \in \mathbf{R}^n \mid A_i x = d_i,\ d_i \in \mathbf{R}^{\nu_i}\}, \quad i = 1, 2, \ldots, m, \tag{28}$$

where $d^T = (d_1^T, d_2^T, \ldots, d_m^T)$. Partitioning a system of linear equations in this way has been shown to be useful in real-world problems, particularly for very large and sparse matrices, see, e.g., Eggermont, Herman and Lent [23].

We prove the following optimization result.

Theorem 3.1 *If $f \in \mathcal{B}(S)$ satisfies the assumptions in Theorem 2.1 and*

$$\nabla f(x^{(0)}) \in \mathcal{R}(A^T), \tag{29}$$

then the Algorithmic Scheme for Case I, applied to the sets C_i of (28), generates a sequence which converges to a solution x^* of (26).

Proof. Applying the Algorithmic Scheme for Case I to the convex feasibility problem with the sets (28), convergence towards a point $x^* \in C \cap \overline{S}$ is guaranteed by Theorem 2.1. Defining

$$Z = \{x \in S \mid \exists z \in \mathbf{R}^{\overline{m}} \text{ such that } \nabla f(x) = A^T z\}, \tag{30}$$

and

$$U = \{x \in \mathbf{R}^n \mid Ax = d, \ x \in \overline{S}\}, \tag{31}$$

we will use the result in [7, Lemma 3], which says that if $x^* \in U \cap \overline{Z}$ then x^* is a solution of (26). Therefore, we show now that $x^{(k)} \in Z$, for all $k \geq 0$, from which $x^* \in \overline{Z}$ follows.

For any $f \in \mathcal{B}(S)$ and $C = \{x \in \mathbf{R}^n \mid Ax = d\}$ such that $P_C^f x$ belongs to S, for any $x \in S$, it is the case that $\nabla f(P_C^f x) - \nabla f(x)$ is in the range of A^T. This follows from [12, Lemma 6.1] (which extends [15, Lemma 2.2.1]). Using this and the fact that, for all $j = 1, 2, \ldots, m(t)$,

$$\mathcal{R}(A_{i_j^t}^T) \subseteq \mathcal{R}(A^T), \tag{32}$$

we deduce that

$$\nabla f(T_t x^{(k)}) - \nabla f(x^{(k)}) \tag{33}$$

is in the range of A^T. Multiplying (33) by ω_t and summing over t we obtain, using (7) and $\sum_{t=1}^{M} \omega_t = 1$, that

$$\nabla f(x^{(k+1)}) - \nabla f(x^{(k)}) \in \mathcal{R}(A^T). \tag{34}$$

Using the initialization (29), we do induction on k with (34) and obtain that $x^{(k)} \in Z$, for all $k \geq 0$. ■

4. DISCUSSION AND SOME OPEN PROBLEMS

All algorithms and results presented here apply, in particular to orthogonal unrelaxed projections, because those are a special case of Bregman projections (see the comments made before Theorem 2.2) as well as of the operators in (22). Thus our Algorithmic Scheme generalizes the method described by Bauschke and Borwein [3, Examples 2.14 and 2.20] where they define an operator $T = \frac{1}{m}(P_1 + P_2 P_1 + \cdots + P_m \cdots P_2 P_1)$, with P_i orthogonal projections onto given sets, for $i = 1, 2, \ldots, m$, and show weak convergence in Hilbert space of $\{T^k x^{(0)}\}_{k \geq 0}$ to some fixed point of T, for every $x^{(0)}$.

Earlier work concerning the convergence of (random) products of averaged mappings is due to Reich and coworkers; see, e.g., Dye and Reich [21], Dye and Reich [20, Theorem 5] and Dye et al. [22, Theorem 5]. In the infinite-dimensional case they require some conditions on the fixed point sets of the mappings which are not needed in the finite-dimensional case. The above-mentioned method of Bauschke and Borwein can also be understood by using the results of Baillon, Bruck and Reich [6, Theorems 1.2 and 2.1], Bruck and Reich [9, Corollary 1.3], and Reich [32, Proposition 2.4]. A more recent study is Bauschke [2].

At the extremes of the "spectrum of algorithms," derivable from our Algorithmic Scheme, are the generically sequential method, which uses one set at a time, and the fully simultaneous algorithm, which employs all sets at each iteration. The "block-iterative projections" (BIP) scheme of Aharoni and Censor [1] (see also Butnariu and Censor [10], Bauschke and Borwein [3], Bauschke, Borwein and Lewis [5] and Elfving [24]) also has the sequential and the fully simultaneous methods as its extremes in terms of block structures. The question whether there are any other relationships between the BIP scheme of [1] and the Algorithmic Scheme of this paper is of theoretical interest. However, the current lack of an answer to it does not diminish the value of the proposed Algorithmic Scheme, because its new algorithmic structure gives users a tool to design algorithms that will average sequential strings of projections.

We have not as yet investigated the behavior of the Algorithmic Scheme, or special instances of it, in the inconsistent case when the intersection $C = \cap_{i=1}^{m} C_i$ is empty. For results on the behavior of the fully simultaneous algorithm with orthogonal projections in the inconsistent case see, e.g., Combettes [18] or Iusem and De Pierro [27]. Another way to treat possible inconsistencies is to reformulate the constraints as $c \leq Ax \leq d$ or $||Ax - d||_2 \leq \epsilon$, see e.g. [15]. Also, variable iteration-dependent relaxation parameters and variable iteration-dependent string constructions could be interesting future extensions.

The practical performance of specific algorithms derived from the Algorithmic Scheme needs still to be evaluated in applications and on parallel machines.

5. APPENDIX: THE ROLE OF BREGMAN PROJECTIONS

Bregman generalized distances and generalized projections are instrumental in several areas of mathematical optimization theory. Their introduction by Bregman [7] was initially followed by the works of Censor and Lent [13] and De Pierro and Iusem [19] and, subsequently, lead to their use in special-purpose minimization methods, in the proximal point minimization method, and for stochastic feasibility problems. These generalized distances and projections were also defined in non-Hilbertian Banach spaces, where, in the absence of orthogonal projections, they can lead to simpler formulas for projections.

In the Euclidean space, where our present results are formulated, Bregman's method for minimizing a convex function (with certain properties) subject to linear inequality constraints employs Bregman projections onto the half-spaces represented by the constraints, see, e.g., [13,19]. Recently the extension of this minimization method to nonlinear convex constraints has been identified with the Han-Dykstra projection algorithm for finding the projection of a point onto an intersection of closed convex sets, see Bregman, Censor and Reich [8].

It looks as if there might be no point in using non-orthogonal projections for solving the convex feasibility problem in $\mathbf{R}^n$ since they are generally not easier to compute. But this is not always the case. In [29,30] Shamir and co-workers have used the multiprojection method of Censor and Elfving [12] to solve filter design problems in image restoration and image recovery posed as convex feasibility problems. They took advantage of that algorithm's flexibility to employ Bregman projections with respect to *different* Bregman functions within the same algorithmic run.

Another example is the seminal paper by Csiszár and Tusnády [17], where the central procedure uses alternating entropy projections onto convex sets. In their "alternating minimization procedure," they alternate between minimizing over the first and second argument of the Bregman distance (Kullback-Leibler divergence, in fact). These divergences are nothing but the generalized Bregman distances obtained by using the negative of Shannon's entropy as the underlying Bregman function.

Recent studies about Bregman projections (Kiwiel [28]), Bregman/Legendre projections (Bauschke and Borwein [4]), and averaged entropic projections (Butnariu, Censor and Reich [11]) – and their uses for convex feasibility problems in $\mathbf{R}^n$ discussed therein – attest to the continued (theoretical and practical) interest in employing Bregman projections in projection methods for convex feasibility problems. This is why we formulated and studied Case I of our Algorithmic Scheme within the framework of such projections.

Acknowledgements. We are grateful to Charles Byrne for pointing out an error in an earlier draft and to the anonymous referees for their constructive comments which helped to improve the paper. We thank Fredrik Berntsson for help with drawing the figures. Part of the work was done during visits of Yair Censor at the Department of Mathematics of the University of Linköping. The support and hospitality of Professor Åke Björck, head of the Numerical Analysis Group there, are gratefully acknowledged.

REFERENCES

1. R. Aharoni and Y. Censor, Block-iterative methods for parallel computation of solutions to convex feasibility problems, *Linear Algebra and Its Applications* **120** (1989) 165–175.
2. H.H. Bauschke, A norm convergence result on random products of relaxed projections in Hilbert space, *Transactions of the American Mathematical Society* **347** (1995) 1365–1373.
3. H.H. Bauschke and J.M. Borwein, On projection algorithms for solving convex feasibility problems, *SIAM Review* **38** (1996) 367–426.
4. H.H. Bauschke and J.M. Borwein, Legendre functions and the method of random Bregman projections, *Journal of Convex Analysis* **4** (1997) 27–67.
5. H.H. Bauschke, J.M. Borwein and A.S. Lewis, The method of cyclic projections for closed convex sets in Hilbert space *Contemporary Mathematics* **204** (1997) 1–38.
6. J.M. Baillon, R.E. Bruck and S. Reich, On the asymptotic behavior of nonexpansive mappings and semigroups in Banach spaces, *Houston Journal of Mathematics* **4** (1978) 1–9.
7. L.M. Bregman, The relaxation method of finding the common point of convex sets and its application to the solution of problems in convex programming, *USSR Com-*

putational Mathematics and Mathematical Physics **7** (1967) 200–217.
8. L.M. Bregman, Y. Censor and S. Reich, Dykstra's algorithm as the nonlinear extension of Bregman's optimization method, *Journal of Convex Analysis* **6** (1999) 319–334.
9. R.E. Bruck and S. Reich, Nonexpansive projections and resolvents of accretive operators in Banach spaces, *Houston Journal of Mathematics* **3** (1977) 459–470.
10. D. Butnariu and Y. Censor, Strong convergence of almost simultaneous block-iterative projection methods in Hilbert spaces, *Journal of Computational and Applied Mathematics* **53** (1994) 33–42.
11. D. Butnariu, Y. Censor and S. Reich, Iterative averaging of entropic projection for solving stochastic convex feasibility problems, *Computational Optimization and Applications* **8** (1997) 21–39.
12. Y. Censor and T. Elfving, A multiprojection algorithm using Bregman projections in a product space, *Numerical Algorithms* **8** (1994) 221–239.
13. Y. Censor and A. Lent, An iterative row-action method for interval convex programming, *Journal of Optimization Theory and Applications* **34** (1981) 321–353.
14. Y. Censor and S. Reich, Iterations of paracontractions and firmly nonexpansive operators with applications to feasibility and optimization, *Optimization* **37** (1996) 323–339.
15. Y. Censor and S.A. Zenios, *Parallel Optimization: Theory, Algorithms, and Applications,* (Oxford University Press, New York, 1997).
16. G. Cimmino, Calcolo approssimato per le soluzioni dei sistemi di equazioni lineari, *La Ricerca Scientifica* **XVI** Series II, Anno IX, 1 (1938), 326–333.
17. I. Csiszár and G. Tusnády, Information geometry and alternating minimization procedures, *Statistics and Decisions* Supplement Issue No. 1 (1984) 205–237.
18. P.L. Combettes, Inconsistent signal feasibility problems: least-squares solutions in a product space, *IEEE Transactions on Signal Processing* **SP-42** (1994), 2955–2966.
19. A.R. De Pierro and A.N. Iusem, A relaxed version of Bregman's method for convex programming, *Journal of Optimization Theory and Applications* **51** (1986) 421–440.
20. J.M. Dye and S. Reich, Unrestricted iteration of projections in Hilbert space, *Journal of Mathematical Analysis and Applications* **156** (1991) 101–119.
21. J.M. Dye and S. Reich, On the unrestricted iterations of nonexpansive mappings in Hilbert space, *Nonlinear Analysis, Theory, Methods and Applications* **18** (1992) 199–207.
22. J.M. Dye, T. Kuczumow, P.-K. Lin and S. Reich, Convergence of unrestricted products of nonexpansive mappings in spaces with the Opial property, *Nonlinear Analysis, Theory, Methods and Applications* **26** (1996) 767–773.
23. P.P.B. Eggermont, G.T. Herman and A. Lent, Iterative algorithms for large partitioned linear systems, with applications to image reconstruction, *Linear Algebra and Its Applications* **40** (1981) 37–67.
24. T. Elfving, Block-iterative methods for consistent and inconsistent linear equations, *Numerische Mathematik* **35** (1980) 1–12.
25. L. Elsner, I. Koltracht and M. Neumann, Convergence of sequential and asynchronous nonlinear paracontractions, *Numerische Mathematik* **62** (1992) 305–319.
26. L.G. Gubin, B.T. Polyak and E.V. Raik, The method of projections for finding the common point of convex sets, *USSR Computational Mathematics and Mathematical Physics* **7** (1967) 1–24.

27. A.N. Iusem and A.R. De Pierro, Convergence results for an accelerated nonlinear Cimmino algorithm, *Numerische Mathematik* **49** (1986) 367–378.
28. K.C. Kiwiel, Generalized Bregman projections in convex feasibility problems, *Journal of Optimization Theory and Applications* **96** (1998) 139–157.
29. T. Kotzer, N. Cohen and J. Shamir, A projection-based algorithm for consistent and inconsistent constraints, *SIAM Journal on Optimization* **7** (1997) 527–546.
30. D. Lyszyk and J. Shamir, Signal processing under uncertain conditions by parallel projections onto fuzzy sets, *Journal of the Optical Society of America A* **16** (1999) 1602–1611.
31. G. Pierra, Decomposition through formalization in a product space, *Mathematical Programming* **28** (1984) 96–115.
32. S. Reich, A limit theorem for projections, *Linear and Multilinear Algebra* **13** (1983) 281–290.
33. R.T. Rockafellar, *Convex Analysis,* (Princeton University Press, Princeton, New Jersey, 1970).

Inherently Parallel Algorithms in Feasibility and Optimization and their Applications
D. Butnariu, Y. Censor and S. Reich (Editors)

QUASI-FEJÉRIAN ANALYSIS OF SOME OPTIMIZATION ALGORITHMS

Patrick L. Combettes[a*]

[a]Laboratoire d'Analyse Numérique, Université Pierre et Marie Curie – Paris 6, 4 Place Jussieu, 75005 Paris, France
and
City College and Graduate Center, City University of New York, New York, NY 10031, USA

A quasi-Fejér sequence is a sequence which satisfies the standard Fejér monotonicity property to within an additional error term. This notion is studied in detail in a Hilbert space setting and shown to provide a powerful framework to analyze the convergence of a wide range of optimization algorithms in a systematic fashion. A number of convergence theorems covering and extending existing results are thus established. Special emphasis is placed on the design and the analysis of parallel algorithms.

1. INTRODUCTION

The convergence analyses of convex optimization algorithms often follow standard patterns. This observation suggests the existence of broad structures within which these algorithms could be recast and then studied in a simplified and unified manner.

One such structure relies on the concept of Fejér monotonicity: a sequence $(x_n)_{n\geq 0}$ in a Hilbert space $\mathcal{H}$ is said to be a *Fejér (monotone) sequence* relative to a *target set* $S \subset \mathcal{H}$ if

$$(\forall x \in S)(\forall n \in \mathbb{N}) \quad \|x_{n+1} - x\| \leq \|x_n - x\|. \tag{1}$$

In convex optimization, this basic property has proven to be an efficient tool to analyze various optimization algorithms in a unified framework, e.g., [8], [9], [10], [13], [20], [22], [29], [30], [31], [45], [54], [63], [64], [69]; see also [24] for additional references and an historical perspective. In this context, the target set S represents the set of solutions to the problem under consideration and (1) states that each iterate generated by the underlying solution algorithm cannot be further from any solution point than its predecessor.

In order to derive unifying convergence principles for a broader class of optimization algorithms, the notion of Fejér monotonicity can be extended in various directions. In this paper, the focus will be placed on three variants of (1).

Definition 1.1 Relative to a nonempty *target set* $S \subset \mathcal{H}$, a sequence $(x_n)_{n\geq 0}$ in $\mathcal{H}$ is

*This work was partially supported by the National Science Foundation under grant MIP-9705504.

- *Quasi-Fejér of Type I* if

$$(\exists\,(\varepsilon_n)_{n\geq 0} \in \ell_+ \cap \ell^1)(\forall x \in S)(\forall n \in \mathbb{N}) \ \ \|x_{n+1} - x\| \leq \|x_n - x\| + \varepsilon_n. \tag{2}$$

- *Quasi-Fejér of Type II* if

$$(\exists\,(\varepsilon_n)_{n\geq 0} \in \ell_+ \cap \ell^1)(\forall x \in S)(\forall n \in \mathbb{N}) \ \ \|x_{n+1} - x\|^2 \leq \|x_n - x\|^2 + \varepsilon_n. \tag{3}$$

- *Quasi-Fejér of Type III* if

$$(\forall x \in S)(\exists\,(\varepsilon_n)_{n\geq 0} \in \ell_+ \cap \ell^1)(\forall n \in \mathbb{N}) \ \ \|x_{n+1} - x\|^2 \leq \|x_n - x\|^2 + \varepsilon_n. \tag{4}$$

The concept of quasi-Fejér sequences goes back to [34], where it was introduced through a property akin to (4) for sequences of $\mathbb{R}^N$-valued random vectors (see also [33] for more recent developments in that direction). Instances of quasi-Fejér sequences of the above three types also appear explicitly in [2], [47], and [49].

The goal of this paper is to study the properties of the above types of quasi-Fejér sequences and to exploit them to derive convergence results for numerous optimization algorithms. Known results will thus be recast in a common framework and new extensions will be obtained in a straightforward fashion. In Section 2, it is shown that most common types of nonlinear operators arising in convex optimization belong to a so-called $\mathfrak{T}$ class whose properties are investigated. The asymptotic properties of quasi-Fejér sequences are discussed in Section 3. In particular, necessary and sufficient conditions for weak and strong convergence are established and convergence estimates are derived. In Section 4, a generic quasi-Fejér algorithm is constructed by iterating $\mathfrak{T}$-class operator with errors and introducing relaxation parameters. Convergence results for this algorithm are obtained through the analysis developed in Section 3 and applied to specific optimization problems in Section 5. In Section 6, a general inexact, parallel, block-iterative algorithm for countable convex feasibility problems is derived from the generic algorithm constructed in Section 4 and analyzed. All the algorithms discussed up to this point are essentially perturbed Fejér monotone algorithms. Section 7 concerns the projected subgradient method and constitutes a different field of applications of the analysis of Section 3.

Notation. Throughout $\mathcal{H}$ is a real Hilbert space with scalar product $\langle \cdot \mid \cdot \rangle$, norm $\|\cdot\|$, and distance d. Given $x \in \mathcal{H}$ and $\rho \in \,]0, +\infty[$, $B(x, \rho)$ is the closed ball of center x and radius ρ. The expressions $x_n \rightharpoonup x$ and $x_n \to x$ denote respectively the weak and strong convergence to x of a sequence $(x_n)_{n\geq 0}$ in $\mathcal{H}$, $\mathfrak{W}(x_n)_{n\geq 0}$ its set of weak cluster points, and $\mathfrak{S}(x_n)_{n\geq 0}$ its set of strong cluster points. $\overline{\operatorname{aff}}\, S$, $\overline{\operatorname{conv}}\, S$, and $\operatorname{conv} S$ are respectively the closed affine hull, the closed convex hull, and the convex hull of a set S. d_S is the distance function to the set S and, if S is closed and convex, P_S is the projector onto S. The sets $\operatorname{dom} A = \{x \in \mathcal{H} \mid Ax \neq \varnothing\}$, $\operatorname{ran} A = \{u \in \mathcal{H} \mid (\exists\, x \in \mathcal{H})\ u \in Ax\}$, and $\operatorname{gr} A = \{(x, u) \in \mathcal{H}^2 \mid u \in Ax\}$ are respectively the domain, the range, and the graph of a set-valued operator $A\colon \mathcal{H} \to 2^{\mathcal{H}}$; the inverse A^{-1} of A is the set-valued operator with graph $\{(u, x) \in \mathcal{H}^2 \mid u \in Ax\}$. The *subdifferential* of a function $f\colon \mathcal{H} \to \mathbb{R}$ is the set-valued operator

$$\partial f\colon \mathcal{H} \to 2^{\mathcal{H}}\colon x \mapsto \big\{u \in \mathcal{H} \mid (\forall y \in \mathcal{H})\ \ \langle y - x \mid u\rangle + f(x) \leq f(y)\big\} \tag{5}$$

and the elements of $\partial f(x)$ are the *subgradients* of f at x. The *lower level set* of f at height $\eta \in \mathbb{R}$ is $\operatorname{lev}_{\leq \eta} f = \{x \in \mathcal{H} \mid f(x) \leq \eta\}$. $\operatorname{Fix} T = \{x \in \mathcal{H} \mid Tx = x\}$ denotes the set of fixed points of an operator $T\colon \mathcal{H} \to \mathcal{H}$. Finally, ℓ_+ denotes the set of all sequences in $[0, +\infty[$ and ℓ^1 [resp. ℓ^2] the space of all absolutely [resp. square] summable sequences in $\mathbb{R}$.

2. NONLINEAR OPERATORS

Convex optimization algorithms involve a variety of (not necessarily linear) operators. In this respect, recall that an operator $T\colon \mathcal{H} \to \mathcal{H}$ with $\operatorname{dom} T = \mathcal{H}$ is *firmly nonexpansive* if

$$(\forall (x,y) \in \mathcal{H}^2) \;\; \|Tx - Ty\|^2 \leq \|x - y\|^2 - \|(T - \operatorname{Id})x - (T - \operatorname{Id})y\|^2; \tag{6}$$

nonexpansive if

$$(\forall (x,y) \in \mathcal{H}^2) \;\; \|Tx - Ty\| \leq \|x - y\|; \tag{7}$$

and *quasi-nonexpansive* if

$$(\forall (x,y) \in \mathcal{H} \times \operatorname{Fix} T) \;\; \|Tx - y\| \leq \|x - y\|. \tag{8}$$

Clearly, (6) $\Rightarrow$ (7) $\Rightarrow$ (8). Now let $A\colon \mathcal{H} \to 2^{\mathcal{H}}$ be a monotone operator. Then the *resolvent* of index $\gamma \in \,]0, +\infty[$ of A is (the single-valued operator) $(\operatorname{Id} + \gamma A)^{-1}$. Next, let $f\colon \mathcal{H} \to \mathbb{R}$ be a continuous convex function such that $\operatorname{lev}_{\leq 0} f \neq \varnothing$ and let g be a selection of ∂f. Then the operator

$$G_f^g : x \mapsto \begin{cases} x - \dfrac{f(x)}{\|g(x)\|^2} g(x), & \text{if } f(x) > 0 \\ x & \text{if } f(x) \leq 0 \end{cases} \tag{9}$$

is a *subgradient projector* onto $\operatorname{lev}_{\leq 0} f$.

The above operators are closely related to the so-called class $\mathfrak{T}$ of [9]. Given $(x, y) \in \mathcal{H}^2$, we shall use the notation

$$H(x,y) = \big\{u \in \mathcal{H} \mid \langle u - y \mid x - y\rangle \leq 0\big\}. \tag{10}$$

Definition 2.1 [9, Def. 2.2] $\mathfrak{T} = \big\{T\colon \mathcal{H} \to \mathcal{H} \,|\, \operatorname{dom} T = \mathcal{H}, (\forall x \in \mathcal{H}) \operatorname{Fix} T \subset H(x, Tx)\big\}$.

Proposition 2.2 [9, Prop. 2.3] *Consider the following statements:*

(i) *T is the projector onto a nonempty closed and convex subset of $\mathcal{H}$.*

(ii) *T is the resolvent of a maximal monotone operator $A\colon \mathcal{H} \to 2^{\mathcal{H}}$.*

(iii) $\operatorname{dom} T = \mathcal{H}$ *and T is firmly nonexpansive.*

(iv) *T is a subgradient projector onto $\operatorname{lev}_{\leq 0} f$, where $f\colon \mathcal{H} \to \mathbb{R}$ is a continuous convex function such that $\operatorname{lev}_{\leq 0} f \neq \varnothing$.*

(v) $\operatorname{dom} T = \mathcal{H}$ *and* $2T - \mathrm{Id}$ *is quasi-nonexpansive.*

(vi) $T \in \mathfrak{T}$.

Then:

$$\begin{array}{ccccc} \text{(i)} & \Rightarrow & \text{(ii)} & \Leftrightarrow & \text{(iii)} \\ \Downarrow & & & & \Downarrow \\ \text{(iv)} & \Rightarrow & \text{(v)} & \Leftrightarrow & \text{(vi).} \end{array}$$

Some properties of $\mathfrak{T}$-class operators are described below.

Proposition 2.3 *Every* T *in* $\mathfrak{T}$ *satisfies the following properties.*

(i) $(\forall (x,y) \in \mathcal{H} \times \operatorname{Fix} T)\ \ \|Tx - x\|^2 \leq \langle y - x \mid Tx - x\rangle$.

(ii) *Put* $T' = \mathrm{Id} + \lambda(T - \mathrm{Id})$, *where* $\lambda \in [0,2]$. *Then* $(\forall (x,y) \in \mathcal{H} \times \operatorname{Fix} T)\ \ \|T'x - y\|^2 \leq \|x - y\|^2 - \lambda(2-\lambda)\|Tx - x\|^2$.

(iii) $(\forall x \in \mathcal{H})\ \ \|Tx - x\| \leq d_{\operatorname{Fix} T}(x)$.

(iv) $\operatorname{Fix} T = \bigcap_{x \in \mathcal{H}} H(x, Tx)$.

(v) $\operatorname{Fix} T$ *is closed and convex.*

(vi) $(\forall \lambda \in [0,1])\ \ \mathrm{Id} + \lambda(T - \mathrm{Id}) \in \mathfrak{T}$.

Proof. From Definition 2.1, we get

$$(\forall (x,y) \in \mathcal{H} \times \operatorname{Fix} T)\ \ \langle y - Tx \mid x - Tx\rangle \leq 0. \tag{11}$$

and (i) ensues. (ii): Fix $(x,y) \in \mathcal{H} \times \operatorname{Fix} T$. It follows from (i) that

$$\begin{aligned} \|T'x - y\|^2 &= \|x - y\|^2 - 2\lambda\langle y - x \mid Tx - x\rangle + \lambda^2\|Tx - x\|^2 \\ &\leq \|x - y\|^2 - \lambda(2-\lambda)\|Tx - x\|^2. \end{aligned} \tag{12}$$

(iii): Fix $x \in \mathcal{H}$. Then (ii) with $\lambda = 1$ implies

$$(\forall y \in \operatorname{Fix} T)\ \ \|Tx - x\| \leq \|y - x\|. \tag{13}$$

Now take the infimum over all $y \in \operatorname{Fix} T$ (with the usual convention $\inf \varnothing = +\infty$). (iv)-(vi): See [9, Prop. 2.6]. □

The next proposition provides a generalization of the operator averaging process described in item (vi) that will be an essential component in the design of block-iterative parallel algorithms in Section 6.

Proposition 2.4 *Let I be a countable index set, $(T_i)_{i\in I}$ a family of operators in $\mathfrak{T}$, and $(\omega_i)_{i\in I}$ strictly positive real numbers such that $\sum_{i\in I}\omega_i = 1$. Suppose that $\bigcap_{i\in I}\operatorname{Fix} T_i \neq \varnothing$ and let*

$$T\colon \mathcal{H} \to \mathcal{H}\colon x \mapsto x + \lambda(x) L\big(x, (T_i)_{i\in I}, (\omega_i)_{i\in I}\big)\left(\sum_{i\in I}\omega_i T_i x - x\right), \tag{14}$$

where, for every $x \in \mathcal{H}$, $\lambda(x) \in\,]0,1]$ and

$$L\big(x, (T_i)_{i\in I}, (\omega_i)_{i\in I}\big) = \begin{cases} 1 & \text{if } x \in \bigcap_{i\in I}\operatorname{Fix} T_i \\ \dfrac{\sum_{i\in I}\omega_i \|T_i x - x\|^2}{\|\sum_{i\in I}\omega_i T_i x - x\|^2} & \text{otherwise.} \end{cases} \tag{15}$$

Then $\operatorname{Fix} T = \bigcap_{i\in I}\operatorname{Fix} T_i$ and $T \in \mathfrak{T}$.

Proof. Fix $(x,y) \in \mathcal{H} \times \bigcap_{i\in I}\operatorname{Fix} T_i$. We first observe that the series $\sum_{i\in I}\omega_i(T_i x - x)$ converges absolutely since, by (13), $\sum_{i\in I}\omega_i\|T_i x - x\| \leq \|y - x\|$. Moreover, by Proposition 2.3(i),

$$(\forall i \in I)\ \ \|T_i x - x\|^2 \leq \langle y - x \mid T_i x - x\rangle. \tag{16}$$

Hence, since the function $\|\cdot\|^2$ is convex and continuous, we arrive at the chain of inequalities

$$\begin{aligned} \left\|\sum_{i\in I}\omega_i T_i x - x\right\|^2 &\leq \sum_{i\in I}\omega_i\|T_i x - x\|^2 \leq \left\langle y - x \,\middle|\, \sum_{i\in I}\omega_i T_i x - x\right\rangle \\ &\leq \|y - x\| \cdot \left\|\sum_{i\in I}\omega_i T_i x - x\right\|, \end{aligned} \tag{17}$$

from which we deduce that

$$\begin{aligned} x \in \operatorname{Fix}\sum_{i\in I}\omega_i T_i &\Leftrightarrow \left\|\sum_{i\in I}\omega_i T_i x - x\right\| = 0 \Leftrightarrow \sum_{i\in I}\omega_i\|T_i x - x\|^2 = 0 \\ &\Leftrightarrow x \in \bigcap_{i\in I}\operatorname{Fix} T_i. \end{aligned} \tag{18}$$

Hence, $L(x)$ in (15) is a well-defined number in $[1,+\infty[$ and

$$\operatorname{Fix} T = \operatorname{Fix}\sum_{i\in I}\omega_i T_i = \bigcap_{i\in I}\operatorname{Fix} T_i. \tag{19}$$

Since $\operatorname{dom} T = \mathcal{H}$, it remains to show that $\operatorname{Fix} T \subset H(x, Tx)$ to establish that $T \in \mathfrak{T}$. To this end, we derive from (16) that

$$\begin{aligned}
\langle y - x \mid Tx - x\rangle &= \lambda(x)L(x)\sum_{i\in I}\omega_i\langle y - x \mid T_i x - x\rangle \\
&\geq \lambda(x)L(x)\sum_{i\in I}\omega_i\|T_i x - x\|^2 \\
&= \frac{1}{\lambda(x)}\left(\lambda(x)L(x)\left\|\sum_{i\in I}\omega_i T_i x - x\right\|\right)^2 \\
&= \|Tx - x\|^2/\lambda(x) \\
&\geq \|Tx - x\|^2. \qquad (20)
\end{aligned}$$

It follows that $\langle y - Tx \mid x - Tx\rangle \leq 0$ and, in turn, that $y \in H(x, Tx)$. Since by virtue of (19) y is an arbitrary fixed point of T, we conclude that $\operatorname{Fix} T \subset H(x, Tx)$. □

Remark 2.5 Taking $\lambda(x) = 1/L(x)$ in (14) yields $\sum_{i\in I}\omega_i T_i \in \mathfrak{T}$.

A couple of additional definitions will be required. An operator $T\colon \mathcal{H} \to \mathcal{H}$ is said to be *demiclosed* at $y \in \mathcal{H}$ if for every $x \in \mathcal{H}$ and every sequence $(x_n)_{n\geq 0}$ in $\mathcal{H}$ such that $x_n \rightharpoonup x$ and $Tx_n \to y$, we have $Tx = y$ [13]; *demicompact* [resp. demicompact at $y \in \mathcal{H}$] if, for every bounded sequence $(x_n)_{n\geq 0}$ such that $(Tx_n - x_n)_{n\geq 0}$ converges strongly [resp. converges strongly to y], we have $\mathfrak{S}(x_n)_{n\geq 0} \neq \emptyset$ [57].

Remark 2.6 Take an operator $T\colon \mathcal{H} \to \mathcal{H}$. Then:

- If T is nonexpansive, then $T - \mathrm{Id}$ is demiclosed on $\mathcal{H}$ [13].
- One will easily check that T is demicompact if its range is boundedly compact (its intersection with any closed ball is compact), e.g., T is the projector onto a boundedly compact convex set. Other examples will be found in [57].

3. PROPERTIES OF QUASI-FEJÉR SEQUENCES

As we shall find in this section, most of the asymptotic properties of Fejér sequences remain valid for quasi-Fejér sequences.

3.1. Basic properties

First, we need

Lemma 3.1 *Let* $\chi \in \left]0, 1\right]$, $(\alpha_n)_{n\geq 0} \in \ell_+$, $(\beta_n)_{n\geq 0} \in \ell_+$, *and* $(\varepsilon_n)_{n\geq 0} \in \ell_+ \cap \ell^1$ *be such that*

$$(\forall n \in \mathbb{N})\quad \alpha_{n+1} \leq \chi\alpha_n - \beta_n + \varepsilon_n. \qquad (21)$$

Then

(i) $(\alpha_n)_{n\geq 0}$ *is bounded.*

(ii) $(\alpha_n)_{n\geq 0}$ *converges.*

(iii) $(\beta_n)_{n\geq 0} \in \ell^1$.

(iv) *If* $\chi \neq 1$, $(\alpha_n)_{n\geq 0} \in \ell^1$.

Proof. Put $\varepsilon = \sum_{n\geq 0} \varepsilon_n$ and $\alpha = \underline{\lim}\, \alpha_n$. (i): We derive from (21) that

$$(\forall n \in \mathbb{N}) \;\; 0 \leq \alpha_{n+1} \leq \chi^{n+1}\alpha_0 + \gamma_n, \;\; \text{where } \gamma_n = \sum_{k=0}^{n} \chi^{n-k}\varepsilon_k. \tag{22}$$

Hence, $(\alpha_n)_{n\geq 0}$ lies in $[0, \alpha_0 + \varepsilon]$. (ii): It follows from the previous inclusion that $\alpha \in [0, \alpha_0 + \varepsilon]$. Now extract from $(\alpha_n)_{n\geq 0}$ a subsequence $(\alpha_{k_n})_{n\geq 0}$ such that $\alpha = \lim \alpha_{k_n}$ and fix $\delta \in \,]0, +\infty[$. Then we can find $n_0 \in \mathbb{N}$ such that $\alpha_{k_{n_0}} - \alpha \leq \delta/2$ and $\sum_{m\geq k_{n_0}} \varepsilon_m \leq \delta/2$. However, by (21),

$$(\forall n \geq k_{n_0}) \;\; 0 \leq \alpha_n \leq \alpha_{k_{n_0}} + \sum_{m\geq k_{n_0}} \varepsilon_m \leq \delta/2 + \alpha + \delta/2 = \alpha + \delta. \tag{23}$$

Hence $\overline{\lim}\, \alpha_n \leq \underline{\lim}\, \alpha_n + \delta$ and, since δ can be arbitrarily small in $]0, +\infty[$, the whole sequence $(\alpha_n)_{n\geq 0}$ converges to α. (iii): It follows from (21) that, for every $N \in \mathbb{N}$, $\beta_N \leq \alpha_N - \alpha_{N+1} + \varepsilon_N$ and, in turn, $\sum_{n=0}^{N} \beta_n \leq \alpha_0 - \alpha_{N+1} + \sum_{n=0}^{N} \varepsilon_n \leq \alpha_0 + \varepsilon$. Hence $\sum_{n\geq 0} \beta_n \leq \alpha_0 + \varepsilon$. (iv): Suppose $\chi \in \,]0, 1[$. Then the sequence $(\gamma_n)_{n\geq 0}$ of (22) is the convolution of the two ℓ^1-sequences $(\chi^n)_{n\geq 0}$ and $(\varepsilon_n)_{n\geq 0}$. As such, it is therefore in ℓ^1 and the inequalities in (22) force $(\alpha_n)_{n\geq 0}$ in ℓ^1 as well. □

Let us start with some basic relationships between (2), (3), and (4).

Proposition 3.2 *Let* $(x_n)_{n\geq 0}$ *be a sequence in* $\mathcal{H}$ *and let* S *be a nonempty subset of* $\mathcal{H}$. *Then the three types of quasi-Fejér monotonicity of Definition 1.1 are related as follows:*

(i) *Type I* $\Rightarrow$ *Type III* $\Leftarrow$ *Type II.*

(ii) *If* S *is bounded, Type I* $\Rightarrow$ *Type II.*

Proof. It is clear that Type II $\Rightarrow$ Type III. Now suppose that $(x_n)_{n\geq 0}$ satisfies (2). Then

$$\begin{aligned}(\forall x \in S)(\forall n \in \mathbb{N}) \;\; \|x_{n+1} - x\|^2 &\leq \big(\|x_n - x\| + \varepsilon_n\big)^2 \\ &\leq \|x_n - x\|^2 + 2\varepsilon_n \sup_{l\geq 0} \|x_l - x\| + \varepsilon_n^2. \end{aligned} \tag{24}$$

Hence, since $(\forall x \in S)\; \sup_{l\geq 0} \|x_l - x\| < +\infty$ by Lemma 3.1(i) and $(\varepsilon_n)_{n\geq 0} \in \ell^1 \subset \ell^2$, (4) holds. To show (ii), observe that (24) yields

$$(\forall x \in S)(\forall n \in \mathbb{N}) \;\; \|x_{n+1} - x\|^2 \leq \|x_n - x\|^2 + 2\varepsilon_n \sup_{z\in S} \sup_{l\geq 0} \|x_l - z\| + \varepsilon_n^2. \tag{25}$$

Therefore, if S is bounded, $\sup_{z\in S}\sup_{l\geq 0}\|x_l - z\| < +\infty$ and (3) ensues. □

Our next proposition collects some basic properties of quasi-Fejér sequences of Type III.

Proposition 3.3 *Let $(x_n)_{n\geq 0}$ be a quasi-Fejér sequence of Type III relative to a nonempty set S in $\mathcal{H}$. Then*

(i) *$(x_n)_{n\geq 0}$ is bounded.*

(ii) *$(x_n)_{n\geq 0}$ is quasi-Fejér of Type III relative to* $\operatorname{conv} S$.

(iii) *For every $x \in \overline{\operatorname{conv}}\, S$, $(\|x_n - x\|)_{n\geq 0}$ converges.*

(iv) *For every $(x, x') \in (\overline{\operatorname{conv}}\, S)^2$, $(\langle x_n \mid x - x'\rangle)_{n\geq 0}$ converges.*

Proof. Suppose that $(x_n)_{n\geq 0}$ satisfies (4). (i) is a direct consequence of Lemma 3.1(i). (ii): Take $x \in \operatorname{conv} S$, say $x = \alpha y_1 + (1-\alpha)y_2$, where $(y_1, y_2) \in S^2$ and $\alpha \in [0,1]$. Then there exist two sequences $(\varepsilon_{1,n})_{n\geq 0}$ and $(\varepsilon_{2,n})_{n\geq 0}$ in $\ell_+ \cap \ell^1$ such that

$$(\forall n \in \mathbb{N}) \quad \begin{cases} \|x_{n+1} - y_1\|^2 \leq \|x_n - y_1\|^2 + \varepsilon_{1,n} \\ \|x_{n+1} - y_2\|^2 \leq \|x_n - y_2\|^2 + \varepsilon_{2,n}. \end{cases} \tag{26}$$

Now put $(\forall n \in \mathbb{N})$ $\varepsilon_n = \max\{\varepsilon_{1,n}, \varepsilon_{2,n}\}$. Then $(\varepsilon_n)_{n\geq 0} \in \ell_+ \cap \ell^1$ and

$$\begin{aligned}(\forall n \in \mathbb{N}) \ \|x_{n+1} - x\|^2 &= \|\alpha(x_{n+1} - y_1) + (1-\alpha)(x_{n+1} - y_2)\|^2 \\ &= \alpha\|x_{n+1} - y_1\|^2 + (1-\alpha)\|x_{n+1} - y_2\|^2 - \alpha(1-\alpha)\|y_1 - y_2\|^2 \\ &\leq \alpha\|x_n - y_1\|^2 + (1-\alpha)\|x_n - y_2\|^2 - \alpha(1-\alpha)\|y_1 - y_2\|^2 + \varepsilon_n \\ &= \|\alpha(x_n - y_1) + (1-\alpha)(x_n - y_2)\|^2 + \varepsilon_n \\ &= \|x_n - x\|^2 + \varepsilon_n. \end{aligned} \tag{27}$$

(iii): Start with $y \in \operatorname{conv} S$, say $y = \alpha y_1 + (1-\alpha)y_2$, where $(y_1, y_2) \in S^2$ and $\alpha \in [0,1]$. Then $(\|x_n - y_1\|)_{n\geq 0}$ and $(\|x_n - y_2\|)_{n\geq 0}$ converge by Lemma 3.1(ii) and so does $(\|x_n - y\|)_{n\geq 0}$ since

$$(\forall n \in \mathbb{N}) \ \|x_n - y\|^2 = \alpha\|x_n - y_1\|^2 + (1-\alpha)\|x_n - y_2\|^2 - \alpha(1-\alpha)\|y_1 - y_2\|^2. \tag{28}$$

Next, take $x \in \overline{\operatorname{conv}}\, S$, say $y_k \to x$ where $(y_k)_{k\geq 0}$ lies in $\operatorname{conv} S$. It remains to show that $(\|x_n - x\|)_{n\geq 0}$ converges. As just shown, for every $k \in \mathbb{N}$, $(\|x_n - y_k\|)_{n\geq 0}$ converges. Moreover,

$$\begin{aligned}(\forall k \in \mathbb{N}) \ -\|y_k - x\| &\leq \underline{\lim}\,\|x_n - x\| - \lim{}_n\|x_n - y_k\| \\ &\leq \overline{\lim}\,\|x_n - x\| - \lim{}_n\|x_n - y_k\| \\ &\leq \|y_k - x\|. \end{aligned} \tag{29}$$

Taking the limit as $k \to +\infty$, we conclude that $\lim_n \|x_n - x\| = \lim_k \lim_n \|x_n - y_k\|$. (iv): Fix $(x, x') \in (\overline{\operatorname{conv}}\, S)^2$. Then

$$(\forall n \in \mathbb{N}) \ \langle x_n \mid x - x'\rangle = \big(\|x_n - x'\|^2 - \|x_n - x\|^2 - \|x - x'\|^2\big)/2 + \langle x \mid x - x'\rangle. \tag{30}$$

However, as the right-hand side converges by (iii), we obtain the claim. □

Not unexpectedly, sharper statements can be formulated for quasi-Fejér sequences of Types I and II.

Proposition 3.4 *Let $(x_n)_{n\geq 0}$ be a quasi-Fejér sequence of Type II relative to a nonempty set S in $\mathcal{H}$. Then $(x_n)_{n\geq 0}$ quasi-Fejér of Type II relative to $\overline{\operatorname{conv}} S$.*

Proof. Suppose that $(x_n)_{n\geq 0}$ satisfies (3). By arguing as in the proof of Proposition 3.3(ii), we obtain that $(x_n)_{n\geq 0}$ is quasi-Fejér of Type II relative to $\operatorname{conv} S$ with the same error sequence $(\varepsilon_n)_{n\geq 0}$. Now take $x \in \overline{\operatorname{conv}} S$, say $y_k \to x$ where $(y_k)_{n\geq 0}$ lies in $\operatorname{conv} S$. Then, for every $n \in \mathbb{N}$, we obtain

$$(\forall k \in \mathbb{N}) \quad \|x_{n+1} - y_k\|^2 \leq \|x_n - y_k\|^2 + \varepsilon_n \tag{31}$$

and, upon taking the limit as $k \to +\infty$, $\|x_{n+1} - x\|^2 \leq \|x_n - x\|^2 + \varepsilon_n$. □

A Fejér monotone sequence $(x_n)_{n\geq 0}$ relative to a nonempty set S may not converge, even weakly: a straightforward example is the sequence $((-1)^n x)_{n\geq 0}$ which is Fejér monotone with respect to $S = \{0\}$ and which does not converge for any $x \notin S$. Nonetheless, if S is closed and convex, the projected sequence $(P_S x_n)_{n\geq 0}$ always converges strongly [8, Thm. 2.16(iv)], [24, Prop. 3] (see also [65, Rem. 1], where this result appears in connection with a fixed point problem). We now show that quasi-Fejér sequence of Types I and II also enjoy this remarkable property.

Proposition 3.5 *Let $(x_n)_{n\geq 0}$ be a quasi-Fejér sequence of Type I relative to a nonempty set S in $\mathcal{H}$ with error sequence $(\varepsilon_n)_{n\geq 0}$. Then the following properties hold.*

(i) $(\forall n \in \mathbb{N}) \;\; d_S(x_{n+1}) \leq d_S(x_n) + \varepsilon_n$.

(ii) $(d_S(x_n))_{n\geq 0}$ *converges.*

(iii) *If* $(\exists\, \chi \in \,]0,1[)(\forall n \in \mathbb{N}) \;\; d_S(x_{n+1}) \leq \chi d_S(x_n) + \varepsilon_n$, *then* $(d_S(x_n))_{n\geq 0} \in \ell^1$.

(iv) *If S is closed and convex, then*

  (a) $(x_n)_{n\geq 0}$ *is quasi-Fejér of Type II relative to the set* $\{P_S x_n\}_{n\geq 0}$ *with error sequence* $(\varepsilon'_n)_{n\geq 0}$, *where*

$$(\forall n \in \mathbb{N}) \quad \varepsilon'_n = 2\varepsilon_n \sup_{(l,k)\in\mathbb{N}^2} \|x_l - P_S x_k\| + \varepsilon_n^2. \tag{32}$$

  (b) $(P_S x_n)_{n\geq 0}$ *converges strongly.*

Proof. (i): Take the infimum over $x \in S$ in (2). (i) $\Rightarrow$ (ii): Use Lemma 3.1(ii). (iii): Use Lemma 3.1(iv). (iv): (a): Since $(x_n)_{n\geq 0}$ is bounded by Proposition 3.3(i) and P_S is (firmly) nonexpansive by Proposition 2.2, $\{P_S x_n\}_{n\geq 0}$ is bounded. The claim therefore follows from Proposition 3.2(ii) and (25). (b): By Proposition 2.2, $P_S \in \mathfrak{T}$. Therefore Proposition 2.3(ii) with $\lambda = 1$ yields

$$(\forall (m,n) \in \mathbb{N}^2) \ \ \|P_S x_{n+m} - P_S x_n\|^2 \leq \|x_{n+m} - P_S x_n\|^2 - d_S(x_{n+m})^2. \tag{33}$$

On the other hand, we derive from (a) that

$$(\forall (m,n) \in \mathbb{N}^2) \ \ \|x_{n+m} - P_S x_n\|^2 \leq \|x_n - P_S x_n\|^2 + \sum_{k=n}^{n+m-1} \varepsilon'_k. \tag{34}$$

Upon combining (33) and (34), we obtain

$$(\forall (m,n) \in \mathbb{N}^2) \ \ \|P_S x_{n+m} - P_S x_n\|^2 \leq d_S(x_n)^2 - d_S(x_{n+m})^2 + \sum_{k \geq n} \varepsilon'_k. \tag{35}$$

However, since $(\varepsilon'_n)_{n\geq 0} \in \ell_+ \cap \ell^1$, $\lim \sum_{k\geq n} \varepsilon'_k = 0$. It therefore follows from (ii) that $\lim_{m,n} \|P_S x_{n+m} - P_S x_n\| = 0$, i.e, $(P_S x_n)_{n\geq 0}$ is a Cauchy sequence. □

Proposition 3.6 *Let $(x_n)_{n\geq 0}$ be a quasi-Fejér sequence of Type II relative to a nonempty set S in $\mathcal{H}$ with error sequence $(\varepsilon_n)_{n\geq 0}$. Then the following properties hold.*

(i) $(\forall n \in \mathbb{N}) \ \ d_S(x_{n+1})^2 \leq d_S(x_n)^2 + \varepsilon_n$.

(ii) $(d_S(x_n))_{n\geq 0}$ *converges.*

(iii) *If* $(\exists\, \chi \in \,]0,1[)(\forall n \in \mathbb{N}) \ \ d_S(x_{n+1})^2 \leq \chi d_S(x_n)^2 + \varepsilon_n$, *then* $(d_S(x_n))_{n\geq 0} \in \ell^2$.

(iv) *If S is closed and convex, then $(P_S x_n)_{n\geq 0}$ converges strongly.*

Proof. Analogous to that of Proposition 3.5, except that $(\varepsilon'_n)_{n\geq 0} = (\varepsilon_n)_{n\geq 0}$ in (iv). □

3.2. Weak convergence

The following proposition records some elementary weak topology properties of quasi-Fejér sequences.

Proposition 3.7 *Let $(x_n)_{n\geq 0}$ be a quasi-Fejér sequence of Type III relative to a nonempty set S in $\mathcal{H}$. Then*

(i) $\mathfrak{W}(x_n)_{n\geq 0} \neq \varnothing$.

(ii) $\big(\forall (x,x') \in (\mathfrak{W}(x_n)_{n\geq 0})^2\big)(\exists\, \alpha \in \mathbb{R}) \ \ S \subset \big\{y \in \mathcal{H} \mid \langle y \mid x - x' \rangle = \alpha\big\}$.

(iii) *If* $\overline{\operatorname{aff}}\, S = \mathcal{H}$ *(for instance* $\operatorname{int} S \neq \varnothing$*), then $(x_n)_{n\geq 0}$ converges weakly.*

(iv) *If* $x_n \rightharpoonup x \in \overline{\operatorname{conv}}\, S$, *then $(\|x_n - y\|)_{n\geq 0}$ converges for every $y \in \mathcal{H}$.*

Proof. (i) follows from Proposition 3.3(i). (ii): Take two points x and x' in $\mathfrak{W}(x_n)_{n\geq 0}$, say $x_{k_n} \rightharpoonup x$ and $x_{l_n} \rightharpoonup x'$, and $y \in S$. Since

$$(\forall n \in \mathbb{N}) \quad \|x_n - y\|^2 - \|y\|^2 = \|x_n\|^2 - 2\langle y \mid x_n \rangle, \tag{36}$$

it follows from Proposition 3.3(iii) that $\beta = \lim \|x_n - y\|^2 - \|y\|^2$ is well defined. Therefore

$$\beta = \lim \|x_{k_n}\|^2 - 2\langle y \mid x \rangle = \lim \|x_{l_n}\|^2 - 2\langle y \mid x' \rangle \tag{37}$$

and we obtain the desired inclusion with $\alpha = (\lim \|x_{k_n}\|^2 - \lim \|x_{l_n}\|^2)/2$. (iii): In view of (ii), if $\overline{\operatorname{aff}} S = \mathcal{H}$ then

$$\big(\forall (x, x') \in (\mathfrak{W}(x_n)_{n\geq 0})^2\big)(\exists\, \alpha \in \mathbb{R})(\forall y \in \mathcal{H}) \quad \langle y \mid x - x' \rangle = \alpha. \tag{38}$$

Consequently $\mathfrak{W}(x_n)_{n\geq 0}$ reduces to a singleton. Since $(x_n)_{n\geq 0}$ lies in a weakly compact set by virtue of Proposition 3.3(i), it therefore converges weakly. (iv): Take $y \in \mathcal{H}$. Then the identities

$$(\forall n \in \mathbb{N}) \quad \|x_n - y\|^2 = \|x_n - x\|^2 + 2\langle x_n - x \mid x - y \rangle + \|x - y\|^2 \tag{39}$$

together with Proposition 3.3(iii) imply that $(\|x_n - y\|)_{n\geq 0}$ converges. □

The following fundamental result has been known for Fejér monotone sequences for some time [13, Lem. 6]. In the present context, it appears in [2, Prop. 1.3].

Theorem 3.8 *Let $(x_n)_{n\geq 0}$ be a quasi-Fejér sequence of Type III relative to a nonempty set S in $\mathcal{H}$. Then $(x_n)_{n\geq 0}$ converges weakly to a point in S if and only if $\mathfrak{W}(x_n)_{n\geq 0} \subset S$.*

Proof. Necessity is straightforward. To show sufficiency, suppose $\mathfrak{W}(x_n)_{n\geq 0} \subset S$ and take x and x' in $\mathfrak{W}(x_n)_{n\geq 0}$. Since $(x, x') \in S^2$, Proposition 3.7(ii) asserts that $\langle x \mid x - x' \rangle = \langle x' \mid x - x' \rangle$ (this identity could also be derived from Proposition 3.3(iv)), whence $x = x'$. In view of Proposition 3.3(i), the proof is complete. □

3.3. Strong convergence

There are known instances of Fejér monotone sequences which converge weakly but not strongly to a point in the target set [7], [9], [38], [42]. A simple example is the following: any orthonormal sequence $(x_n)_{n\geq 0}$ in $\mathcal{H}$ is Fejér monotone relative to $\{0\}$ and, by Bessel's inequality, satisfies $x_n \rightharpoonup 0$; however, $1 \equiv \|x_n\| \not\to 0$. The strong convergence properties of quasi-Fejér sequences must therefore be investigated in their own rights. We begin this investigation with some facts regarding the strong cluster points of quasi-Fejér sequences of Type III. The first two of these facts were essentially known to Ermol'ev [32].

Proposition 3.9 *Let $(x_n)_{n\geq 0}$ be a quasi-Fejér sequence of Type III relative to a nonempty set S in $\mathcal{H}$. Then*

(i) $\big(\forall (x, x') \in (\mathfrak{S}(x_n)_{n\geq 0})^2\big) \;\; S \subset \big\{y \in \mathcal{H} \mid \langle y - (x + x')/2 \mid x - x' \rangle = 0\big\}$.

(ii) *If $\overline{\operatorname{aff}} S = \mathcal{H}$ (for instance $\operatorname{int} S \neq \varnothing$), then $\mathfrak{S}(x_n)_{n\geq 0}$ contains at most one point.*

(iii) $(x_n)_{n\geq 0}$ *converges strongly if there exist* $x \in S$, $(\varepsilon_n)_{n\geq 0} \in \ell_+ \cap \ell^1$, *and* $\rho \in]0, +\infty[$ *such that*

$$(\forall n \in \mathbb{N}) \quad \|x_{n+1} - x\|^2 \leq \|x_n - x\|^2 - \rho\|x_{n+1} - x_n\| + \varepsilon_n. \tag{40}$$

Proof. (i): Take x and x' in $\mathfrak{S}(x_n)_{n\geq 0}$, say $x_{k_n} \to x$ and $x_{l_n} \to x'$, and $y \in S$. Then $\lim \|x_{k_n} - y\| = \|x - y\|$ and $\lim \|x_{l_n} - y\| = \|x' - y\|$. Hence, by Proposition 3.3(iii), $\|x - y\| = \|x' - y\|$ or, equivalently, $\langle y - (x + x')/2 \mid x - x'\rangle = 0$. Since $\mathfrak{S}(x_n)_{n\geq 0} \subset \mathfrak{W}(x_n)_{n\geq 0}$, this identity could also be obtained through Proposition 3.7(ii) where $\alpha = (\|x\|^2 - \|x'\|^2)/2$. (ii) follows from (i) or, alternatively, from Proposition 3.7(iii). (iii): By virtue of Lemma 3.1(iii), $(\|x_{n+1} - x_n\|)_{n\geq 0} \in \ell^1$ and $(x_n)_{n\geq 0}$ is therefore a Cauchy sequence. □

We now extend to quasi-Fejér sequences of Types I and II a strong convergence property that was first identified in the case of Fejér sequences in [64] (see also [8, Thm. 2.16(iii)] and the special cases appearing in [53] and [55, Sec. 6]).

Proposition 3.10 *Let* $(x_n)_{n\geq 0}$ *be a quasi-Fejér sequence of Type I or II relative to a set* S *in* $\mathcal{H}$ *such that* $\operatorname{int} S \neq \varnothing$. *Then* $(x_n)_{n\geq 0}$ *converges strongly.*

Proof. Take $x \in S$ and $\rho \in]0, +\infty[$ such that $B(x, \rho) \subset S$. Proposition 3.2(ii) asserts that $(x_n)_{n\geq 0}$ is quasi-Fejér of Type II relative to the bounded set $B(x, \rho)$. Hence,

$$(\exists\, (\varepsilon_n)_{n\geq 0} \in \ell_+ \cap \ell^1)(\forall z \in B(x, \rho))(\forall n \in \mathbb{N}) \quad \|x_{n+1} - z\|^2 \leq \|x_n - z\|^2 + \varepsilon_n. \tag{41}$$

Now define a sequence $(z_n)_{n\geq 0}$ in $B(x, \rho)$ by

$$(\forall n \in \mathbb{N}) \quad z_n = \begin{cases} x & \text{if } x_{n+1} = x_n \\ x - \rho \dfrac{x_{n+1} - x_n}{\|x_{n+1} - x_n\|} & \text{otherwise.} \end{cases} \tag{42}$$

Then (41) yields $(\forall n \in \mathbb{N})$ $\|x_{n+1} - z_n\|^2 \leq \|x_n - z_n\|^2 + \varepsilon_n$ and, after expanding, we obtain

$$(\forall n \in \mathbb{N}) \quad \|x_{n+1} - x\|^2 \leq \|x_n - x\|^2 - 2\rho\|x_{n+1} - x_n\| + \varepsilon_n. \tag{43}$$

The strong convergence of $(x_n)_{n\geq 0}$ then follows from Proposition 3.9(iii). □

For quasi-Fejér sequences, a number of properties are equivalent to strong convergence to a point in the target set. Such equivalences were already implicitly established in [41] for Fejér monotone sequences relative to closed convex sets (see also [8] and [24]).

Theorem 3.11 *Let* $(x_n)_{n\geq 0}$ *be a quasi-Fejér sequence of Type III relative to a nonempty set* S *in* $\mathcal{H}$. *Then the following statements are equivalent:*

(i) $(x_n)_{n\geq 0}$ *converges strongly to a point in* S.

(ii) $\mathfrak{W}(x_n)_{n\geq 0} \subset S$ *and* $\mathfrak{S}(x_n)_{n\geq 0} \neq \varnothing$.

(iii) $\mathfrak{S}(x_n)_{n\geq 0} \cap S \neq \varnothing$.

If S is closed and $(x_n)_{n\geq 0}$ is quasi-Fejér of Type I or II relative to S, each of the above statements is equivalent to

(iv) $\underline{\lim}\, d_S(x_n) = 0$.

Proof. (i) $\Rightarrow$ (ii): Clearly, $x_n \to x \in S \Rightarrow \mathfrak{W}(x_n)_{n\geq 0} = \mathfrak{S}(x_n)_{n\geq 0} = \{x\} \subset S$. (ii) $\Rightarrow$ (iii): Indeed, $\mathfrak{S}(x_n)_{n\geq 0} \subset \mathfrak{W}(x_n)_{n\geq 0}$. (iii) $\Rightarrow$ (i): Fix $x \in \mathfrak{S}(x_n)_{n\geq 0} \cap S$. Then $x \in \mathfrak{S}(x_n)_{n\geq 0}$ $\Rightarrow \underline{\lim}\, \|x_n - x\| = 0$. On the other hand, $x \in S$, and it follows from Proposition 3.3(iii) that $(\|x_n - x\|)_{n\geq 0}$ converges. Thus, $x_n \to x$.
Now assume that S is closed. (iv) $\Rightarrow$ (i): If $(x_n)_{n\geq 0}$ is quasi-Fejér of Type I with error sequence $(\varepsilon_n)_{n\geq 0}$ then

$$(\forall x \in S)(\forall (m,n) \in \mathbb{N}^2)\ \|x_n - x_{n+m}\| \leq \|x_n - x\| + \|x_{n+m} - x\| \leq 2\|x_n - x\| + \sum_{k=n}^{n+m-1} \varepsilon_k \tag{44}$$

and therefore

$$(\forall (m,n) \in \mathbb{N}^2)\ \|x_n - x_{n+m}\| \leq 2d_S(x_n) + \sum_{k\geq n} \varepsilon_k. \tag{45}$$

Likewise, if $(x_n)_{n\geq 0}$ is quasi-Fejér of Type II with error sequence $(\varepsilon_n)_{n\geq 0}$ then

$$(\forall x \in S)(\forall (m,n) \in \mathbb{N}^2)\ \|x_n - x_{n+m}\|^2 \leq 2\big(\|x_n - x\|^2 + \|x_{n+m} - x\|^2\big) \leq 4\|x_n - x\|^2 + 2\sum_{k=n}^{n+m-1} \varepsilon_k \tag{46}$$

and therefore

$$(\forall (m,n) \in \mathbb{N}^2)\ \|x_n - x_{n+m}\|^2 \leq 4d_S(x_n)^2 + 2\sum_{k\geq n} \varepsilon_k. \tag{47}$$

Now suppose $\underline{\lim}\, d_S(x_n) = 0$. Then Propositions 3.5(ii) and 3.6(ii) yield $\lim d_S(x_n) = 0$ in both cases. On the other hand, by summability, $\lim \sum_{k\geq n} \varepsilon_k = 0$ and we derive from (45) and (47) that $(x_n)_{n\geq 0}$ is a Cauchy sequence in both cases. It therefore converges strongly to some point $x \in \mathcal{H}$. By continuity of d_S, we deduce that $d_S(x) = 0$, i.e., $x \in \overline{S} = S$. (i) $\Rightarrow$ (iv): Indeed, $(\forall x \in S)(\forall n \in \mathbb{N})\ d_S(x_n) \leq \|x_n - x\|$. □

Remark 3.12 With the additional assumption that S is convex, the implication (iv) $\Rightarrow$ (i) can be established more directly. Indeed, Propositions 3.5(ii) and 3.6(ii) yield $x_n - P_S x_n \to 0$ while Propositions 3.5(iv)(b) and 3.6(iv) guarantee the existence of a point $x \in S$ such that $P_S x_n \to x$. Altogether, $x_n \to x$.

3.4. Convergence estimates

In order to compare algorithms or devise stopping criteria for them, it is convenient to have estimates of their speed of convergence. For quasi-Fejér sequences of Type I or II it is possible to derive such estimates.

Theorem 3.13 *Let $(x_n)_{n\geq 0}$ be a quasi-Fejér sequence of Type I [resp. Type II] relative to a nonempty set S in $\mathcal{H}$ with error sequence $(\varepsilon_n)_{n\geq 0}$. Then*

(i) *If $(x_n)_{n\geq 0}$ converges strongly to a point $x \in S$ then*

$$\begin{aligned}&(\forall n \in \mathbb{N}) \quad \|x_n - x\| \leq 2d_S(x_n) + \sum_{k\geq n} \varepsilon_k \\ &\big[\mathit{resp.}\ (\forall n \in \mathbb{N}) \quad \|x_n - x\|^2 \leq 4d_S(x_n)^2 + 2\sum_{k\geq n} \varepsilon_k\big].\end{aligned} \tag{48}$$

(ii) *If S is closed and*

$$\begin{aligned}&(\exists \chi \in \,]0,1[)(\forall n \in \mathbb{N}) \quad d_S(x_{n+1}) \leq \chi d_S(x_n) + \varepsilon_n, \\ &\big[\mathit{resp.}\ (\exists \chi \in \,]0,1[)(\forall n \in \mathbb{N}) \quad d_S(x_{n+1})^2 \leq \chi d_S(x_n)^2 + \varepsilon_n\big],\end{aligned} \tag{49}$$

then $(x_n)_{n\geq 0}$ converges strongly to a point $x \in S$ and

$$\begin{aligned}&(\forall n \in \mathbb{N} \smallsetminus \{0\}) \quad \|x_n - x\| \leq 2\chi^n d_S(x_0) + 2\sum_{k=0}^{n-1} \chi^{n-k-1}\varepsilon_k + \sum_{k\geq n} \varepsilon_k \\ &\big[\mathit{resp.}\ (\forall n \in \mathbb{N} \smallsetminus \{0\}) \quad \|x_n - x\|^2 \leq 4\chi^n d_S(x_0)^2 + 4\sum_{k=0}^{n-1} \chi^{n-k-1}\varepsilon_k + 2\sum_{k\geq n} \varepsilon_k\big].\end{aligned} \tag{50}$$

Proof. (i): Take the limit as $m \to +\infty$ in (45) [resp. in (47)]. (ii): It follows from (49) and Proposition 3.5(iii) [resp. Proposition 3.6(iii)] that $\lim d_S(x_n) = 0$. Therefore, by Theorem 3.11, there exists a point $x \in S$ such that $x_n \to x$. For every $n \in \mathbb{N}$, we obtain from (22) the estimate $d_S(x_{n+1}) \leq \chi^{n+1} d_S(x_0) + \sum_{k=0}^{n} \chi^{n-k}\varepsilon_k$ [resp. $d_S(x_{n+1})^2 \leq \chi^{n+1} d_S(x_0)^2 + \sum_{k=0}^{n} \chi^{n-k}\varepsilon_k$] which, together with (i), provides (50). □

For Type I sequences, item (i) appears in [47, Thm. 5.3]. Item (ii) owes its relevance to the fact that the right-hand side of (50) converges to zero as $n \to +\infty$ since $(\varepsilon_n)_{n\geq 0} \in \ell_+ \cap \ell^1$ and $\chi \in \,]0,1[$ (as seen in the proof of Lemma 3.1(iv), its two first terms belong to ℓ^1). Sharper estimates require additional assumptions on $(\varepsilon_n)_{n\geq 0}$.

Corollary 3.14 *Let $(x_n)_{n\geq 0}$ be a quasi-Fejér sequence of Type I relative to a nonempty closed and convex set S in $\mathcal{H}$ with error sequence $(\varepsilon_n)_{n\geq 0}$, and let $(\varepsilon'_n)_{n\geq 0}$ be as in (32). Then*

(i) *If $(x_n)_{n\geq 0}$ converges strongly to a point $x \in S$ then*

$$(\forall n \in \mathbb{N}) \quad \|x_n - x\|^2 \leq 4d_S(x_n)^2 + 2\sum_{k\geq n} \varepsilon'_k. \tag{51}$$

(ii) *If* $(\exists\,\chi \in \,]0,1[)(\forall n \in \mathbb{N})\ d_S(x_{n+1})^2 \leq \chi d_S(x_n)^2 + \varepsilon'_n$, *then* $(x_n)_{n\geq 0}$ *converges strongly to a point* $x \in S$ *and*

$$(\forall n \in \mathbb{N} \smallsetminus \{0\})\ \ \|x_n - x\|^2 \leq 4\chi^n d_S(x_0)^2 + 4\sum_{k=0}^{n-1} \chi^{n-k-1}\varepsilon'_k + 2\sum_{k\geq n} \varepsilon'_k. \tag{52}$$

Proof. The claim follows from Proposition 3.5(iv)(a) and Theorem 3.13 since $(\forall n \in \mathbb{N})$ $d_S(x_n) = d_{\{P_S x_k\}_{k\geq 0}}(x_n)$. □

In the case of Fejér monotone sequences, Corollary 3.14 captures well-known results that originate in [41] (see also [8] and [24]). Thus, (i) furnishes the estimate $(\forall n \in \mathbb{N})$ $\|x_n - x\| \leq 2d_S(x_n)$ while (ii) states that if

$$(\exists\,\chi \in \,]0,1[)(\forall n \in \mathbb{N})\ \ d_S(x_{n+1}) \leq \chi d_S(x_n), \tag{53}$$

then $(x_n)_{n\geq 0}$ converges linearly to a point in S: $(\forall n \in \mathbb{N})\ \ \|x_n - x\| \leq 2\chi^n d_S(x_0)$.

4. ANALYSIS OF AN INEXACT $\mathfrak{T}$-CLASS ALGORITHM

Let $S \subset \mathcal{H}$ be the set of solutions to a given problem and let T_n be an operator in $\mathfrak{T}$ such that $\operatorname{Fix} T_n \supset S$. Then, for every point x_n in $\mathcal{H}$ and every relaxation parameter $\lambda_n \in [0,2]$, Proposition 2.3(ii) guarantees that $x_n + \lambda_n(T_n x_n - x_n)$ is not further from any solution point than x_n is. This remark suggests that a point in S can be constructed via the iterative scheme $x_{n+1} = x_n + \lambda_n(T_n x_n - x_n)$. Since in some problems one may not want – or be able – to evaluate $T_n x_n$ exactly, a more realistic algorithmic model is obtained by replacing $T_n x_n$ by $T_n x_n + e_n$, where e_n accounts for some numerical error.

Algorithm 4.1 At iteration $n \in \mathbb{N}$, suppose that $x_n \in \mathcal{H}$ is given. Then select $T_n \in \mathfrak{T}$, $\lambda_n \in [0,2]$, and set $x_{n+1} = x_n + \lambda_n(T_n x_n + e_n - x_n)$, where $e_n \in \mathcal{H}$.

The convergence analysis of Algorithm 4.1 will be greatly simplified by the following result, which states that its orbits are quasi-Fejér relative to the set of common fixed points of the operators $(T_n)_{n\geq 0}$.

Proposition 4.2 *Suppose that* $F = \bigcap_{n\geq 0} \operatorname{Fix} T_n \neq \varnothing$ *and let* $(x_n)_{n\geq 0}$ *be an arbitrary orbit of Algorithm 4.1 such that* $(\lambda_n\|e_n\|)_{n\geq 0} \in \ell^1$. *Then*

(i) $(x_n)_{n\geq 0}$ *is quasi-Fejér of Type I relative to* F *with error sequence* $(\lambda_n\|e_n\|)_{n\geq 0}$.

(ii) $\big(\lambda_n(2-\lambda_n)\|T_n x_n - x_n\|^2\big)_{n\geq 0} \in \ell^1$.

(iii) *If* $\overline{\lim}\,\lambda_n < 2$, *then* $(\|x_{n+1} - x_n\|)_{n\geq 0} \in \ell^2$.

Proof. Fix $x \in F$ and put, for every $n \in \mathbb{N}$, $z_n = x_n + \lambda_n(T_n x_n - x_n)$. (i): For every $n \in \mathbb{N}$, $x \in \operatorname{Fix} T_n$ and, since $T_n \in \mathfrak{T}$, Proposition 2.3(ii) yields

$$\|z_n - x\|^2 \leq \|x_n - x\|^2 - \lambda_n(2-\lambda_n)\|T_n x_n - x_n\|^2. \tag{54}$$

Whence,

$$(\forall n \in \mathbb{N}) \ \ \|z_n - x\| \leq \|x_n - x\| \tag{55}$$

and therefore

$$(\forall n \in \mathbb{N}) \ \ \|x_{n+1} - x\| \leq \|z_n - x\| + \lambda_n\|e_n\| \leq \|x_n - x\| + \lambda_n\|e_n\|, \tag{56}$$

which shows that $(x_n)_{n\geq 0}$ satisfies (2). (ii): Set

$$(\forall n \in \mathbb{N}) \ \ \varepsilon'_n(x) = 2\lambda_n\|e_n\| \sup_{l\geq 0} \|x_l - x\| + \lambda_n^2\|e_n\|^2. \tag{57}$$

Then it follows from the assumption $(\lambda_n\|e_n\|)_{n\geq 0} \in \ell^1$ and from Proposition 3.3(i) that $(\varepsilon'_n(x))_{n\geq 0} \in \ell^1$. Using (56), (54), and (55), we obtain

$$\begin{aligned}(\forall n \in \mathbb{N}) \ \ \|x_{n+1} - x\|^2 &\leq \big(\|z_n - x\| + \lambda_n\|e_n\|\big)^2 \\ &\leq \|x_n - x\|^2 - \lambda_n(2-\lambda_n)\|T_n x_n - x_n\|^2 + \varepsilon'_n(x)\end{aligned} \tag{58}$$

and Lemma 3.1(iii) allows us to conclude $(\lambda_n(2-\lambda_n)\|T_n x_n - x_n\|^2)_{n\geq 0} \in \ell^1$. (iii): By assumption, there exist $\delta \in \,]0,1[$ and $N \in \mathbb{N}$ such that $(\lambda_n)_{n\geq N}$ lies in $[0, 2-\delta]$. Hence, for every $n \geq N$, $\lambda_n \leq (2-\delta)(2-\lambda_n)/\delta$ and, in turn,

$$\begin{aligned}\|x_{n+1} - x_n\|^2 &\leq \big(\lambda_n\|T_n x_n - x_n\| + \lambda_n\|e_n\|\big)^2 \\ &\leq 2\lambda_n^2\|T_n x_n - x_n\|^2 + 2\lambda_n^2\|e_n\|^2 \\ &\leq \frac{2(2-\delta)}{\delta}\lambda_n(2-\lambda_n)\|T_n x_n - x_n\|^2 + 2\lambda_n^2\|e_n\|^2.\end{aligned} \tag{59}$$

In view of (ii) and the fact that $(\lambda_n\|e_n\|)_{n\geq 0} \in \ell^2$, the proof is complete. □

We are now ready to prove

Theorem 4.3 *Suppose that* $\varnothing \neq S \subset \bigcap_{n\geq 0} \operatorname{Fix} T_n$ *and let* $(x_n)_{n\geq 0}$ *be an arbitrary orbit of Algorithm 4.1. Then* $(x_n)_{n\geq 0}$ *converges weakly to a point* x *in* S *if*

(i) $(\lambda_n\|e_n\|)_{n\geq 0} \in \ell^1$ *and* $\mathfrak{W}(x_n)_{n\geq 0} \subset S$.

The convergence is strong if any of the following assumptions is added:

(ii) S *is closed and* $\underline{\lim}\, d_S(x_n) = 0$.

(iii) $\operatorname{int} S \neq \varnothing$.

(iv) *There exist a strictly increasing sequence* $(k_n)_{n\geq 0}$ *in* $\mathbb{N}$ *and an operator* $T\colon \mathcal{H} \to \mathcal{H}$ *which is demicompact at* 0 *such that* $(\forall n \in \mathbb{N})$ $T_{k_n} = T$ *and* $\sum_{n\geq 0} \lambda_{k_n}(2-\lambda_{k_n}) = +\infty$.

(v) *S is closed and convex, $(\lambda_n)_{n\geq 0}$ lies in $[\delta, 2-\delta]$, where $\delta \in \,]0,1[$, and*

$$(\exists \chi \in \,]0,1])(\forall n \in \mathbb{N}) \;\; \|T_n x_n - x_n\| \geq \chi d_S(x_n). \tag{60}$$

In this case, for every integer $n \geq 1$, we have

$$\|x_n - x\|^2 \leq 4(1-\delta^2\chi^2)^n d_S(x_0)^2 + 4\sum_{k=0}^{n-1}(1-\delta^2\chi^2)^{n-k-1}\varepsilon'_k + 2\sum_{k\geq n}\varepsilon'_k, \tag{61}$$

where $\varepsilon'_k = 2\lambda_k\|e_k\| \sup_{(l,m)\in\mathbb{N}^2} \|x_l - P_S x_m\| + \lambda_k^2\|e_k\|^2$.

Proof. First, we recall from Propositions 4.2(i) and 3.2(i) that $(x_n)_{n\geq 0}$ is quasi-Fejér of Types I and III relative to S. Hence, (i) is a direct consequence of Theorem 3.8. We now turn to strong convergence. (ii) follows from Theorem 3.11. (iii) is supplied by Proposition 3.10. (iv): Proposition 4.2(ii) yields

$$\sum_{n\geq 0} \lambda_{k_n}(2-\lambda_{k_n})\|Tx_{k_n} - x_{k_n}\|^2 < +\infty. \tag{62}$$

Since $\sum_{n\geq 0}\lambda_{k_n}(2-\lambda_{k_n}) = +\infty$, it therefore follows that $\underline{\lim}\,\|Tx_{k_n} - x_{k_n}\| = 0$. Hence, the demicompactness of T at 0 gives $\mathfrak{S}(x_n)_{n\geq 0} \neq \emptyset$, and the conclusion follows from Theorem 3.11. (v): The assumptions imply

$$(\forall n \in \mathbb{N}) \;\; \lambda_n(2-\lambda_n)\|T_n x_n - x_n\|^2 \geq \delta^2\chi^2 d_S(x_n)^2. \tag{63}$$

Strong convergence therefore follows from Proposition 4.2(ii) and (ii). On the other hand, (58) yields

$$\begin{aligned}(\forall (k,n) \in \mathbb{N}^2) \;\; \|x_{n+1} - P_S x_k\|^2 \;\leq\; & \|x_n - P_S x_k\|^2 \\ & - \lambda_n(2-\lambda_n)\|T_n x_n - x_n\|^2 + \varepsilon'_n\end{aligned} \tag{64}$$

where, just as in Proposition 3.5(iv)(a), $(\varepsilon'_n)_{n\geq 0} \in \ell^1$. Hence, we derive from (63) and (64) that

$$\begin{aligned}(\forall n \in \mathbb{N}) \;\; d_S(x_{n+1})^2 \;&\leq\; \|x_{n+1} - P_S x_n\|^2 \\ &\leq\; \|x_n - P_S x_n\|^2 - \lambda_n(2-\lambda_n)\|T_n x_n - x_n\|^2 + \varepsilon'_n \\ &\leq\; (1-\delta^2\chi^2)d_S(x_n)^2 + \varepsilon'_n.\end{aligned} \tag{65}$$

Corollary 3.14(ii) can now be invoked to complete the proof. □

Remark 4.4 The condition $(\lambda_n\|e_n\|)_{n\geq 0} \in \ell^1$ cannot be removed in Theorem 4.3. Indeed take $(\forall n \in \mathbb{N})$ $T_n = \mathrm{Id}$, $\lambda_n = 1$, and $e_n = x_0/(n+1)$ in Algorithm 4.1. Then $\|x_n\| \to +\infty$.

Remark 4.5 Suppose that $e_n \equiv 0$. Then Algorithm 4.1 describes all Fejér monotone methods [9, Prop. 2.7]. As seen in Theorem 4.3, stringent conditions must be imposed to achieve strong convergence to a point in the target set with such methods. In [9], a slight modification of Algorithm 4.1 was proposed that renders them strongly convergent without requiring any additional restrictions on the constituents of the problem.

5. APPLICATIONS TO PERTURBED OPTIMIZATION ALGORITHMS

Let S be a nonempty set in $\mathcal{H}$ representing the set of solutions to an optimization problem. One of the assumptions in Theorem 4.3 is that $S \subset \bigcap_{n\geq 0} \operatorname{Fix} T_n$. This assumption is certainly satisfied if

$$(\forall n \in \mathbb{N}) \ \ \operatorname{Fix} T_n = S. \tag{66}$$

In this section, we shall consider applications of Algorithm 4.1 in which (66) is fulfilled. In view of Proposition 2.3(v), S is therefore closed and convex.

5.1. Krasnosel'skiĭ-Mann iterates

Let T be an operator in $\mathfrak{T}$ with at least one fixed point. To find such a point, we shall use a perturbed version of the Krasnosel'skiĭ-Mann iterates.

Algorithm 5.1 At iteration $n \in \mathbb{N}$, suppose that $x_n \in \mathcal{H}$ is given. Then select $\lambda_n \in [0, 2]$, and set $x_{n+1} = x_n + \lambda_n(Tx_n + e_n - x_n)$, where $e_n \in \mathcal{H}$.

Theorem 5.2 *Let $(x_n)_{n\geq 0}$ be an arbitrary orbit of Algorithm 5.1. Then $(x_n)_{n\geq 0}$ converges weakly to a fixed point x of T if*

(i) $(\|e_n\|)_{n\geq 0} \in \ell^1$, $T - \operatorname{Id}$ *is demiclosed at* 0, *and* $(\lambda_n)_{n\geq 0}$ *lies in* $[\delta, 2-\delta]$ *for some* $\delta \in \,]0,1[$.

The convergence is strong if any of the following assumptions is added:

(ii) $\underline{\lim}\, d_{\operatorname{Fix} T}(x_n) = 0$.

(iii) $\operatorname{int} \operatorname{Fix} T \neq \varnothing$.

(iv) T *is demicompact at* 0.

(v) $(\exists \chi \in \,]0,1])(\forall n \in \mathbb{N}) \ \ \|Tx_n - x_n\| \geq \chi d_{\operatorname{Fix} T}(x_n)$. *In this case, the convergence estimate* (61) *holds with* $S = \operatorname{Fix} T$.

Proof. It is clear that Algorithm 5.1 is a special case of Algorithm 4.1 and that Theorem 5.2 will follow from Theorem 4.3 if "$T - \operatorname{Id}$ is demiclosed at 0 and $(\lambda_n)_{n\geq 0}$ lies in $[\delta, 2-\delta]$" $\Rightarrow \mathfrak{W}(x_n)_{n\geq 0} \subset \operatorname{Fix} T$. To show this, take $x \in \mathfrak{W}(x_n)_{n\geq 0}$, say $x_{k_n} \rightharpoonup x$. Since $(\forall n \in \mathbb{N})$ $\delta \leq \lambda_{k_n} \leq 2 - \delta \Rightarrow \lambda_{k_n}(2 - \lambda_{k_n}) \geq \delta^2$, it follows from Proposition 4.2(ii) that $(T - \operatorname{Id})x_{k_n} \to 0$. The demiclosedness of $T - \operatorname{Id}$ at 0 therefore implies $x \in \operatorname{Fix} T$. □

Remark 5.3 For T firmly nonexpansive and $\lambda_n \equiv 1$, Theorem 5.2(i) is stated in [49, Rem. 5.6.4].

5.2. Successive approximations with a nonexpansive operator

Let $R\colon \mathcal{H} \to \mathcal{H}$ be a nonexpansive operator with $\operatorname{dom} R = \mathcal{H}$ and $\operatorname{Fix} R \neq \varnothing$. Then a fixed point of R can be obtained via Theorem 5.2 with $T = (\mathrm{Id} + R)/2$, which is firmly nonexpansive [39, Thm. 12.1] (hence in $\mathfrak{T}$ by Proposition 2.2 and, furthermore, nonexpansive so that $T - \mathrm{Id}$ is demiclosed on $\mathcal{H}$ by Remark 2.6) with $\operatorname{Fix} R = \operatorname{Fix} T$. This substitution amounts to halving the relaxations in Algorithm 5.1 and leads to

Algorithm 5.4 At iteration $n \in \mathbb{N}$, suppose that $x_n \in \mathcal{H}$ is given. Then select $\lambda_n \in [0,1]$, and set $x_{n+1} = x_n + \lambda_n(Rx_n + e_n - x_n)$, where $e_n \in \mathcal{H}$.

A direct application of Theorem 5.2 would require the relaxation parameters to be bounded away from 0 and 1. We show below that the nonexpansivity of R allows for a somewhat finer relaxation strategy.

Theorem 5.5 *Let $(x_n)_{n\geq 0}$ be an arbitrary orbit of Algorithm 5.4. Then $(x_n)_{n\geq 0}$ converges weakly to a fixed point of R if*

(i) $(\lambda_n\|e_n\|)_{n\geq 0} \in \ell^1$ *and* $\big(\lambda_n(1-\lambda_n)\big)_{n\geq 0} \notin \ell^1$.

The convergence is strong if any of the following assumptions is added:

(ii) $\underline{\lim}\, d_{\operatorname{Fix} R}(x_n) = 0$.

(iii) $\operatorname{int} \operatorname{Fix} R \neq \varnothing$.

(iv) *R is demicompact at* 0.

Proof. $R - \mathrm{Id}$ is demiclosed on $\mathcal{H}$ by Remark 2.6 and an inspection of the proof of Theorem 5.2 shows that it is sufficient to demonstrate that

$$Rx_n - x_n \to 0. \tag{67}$$

By Proposition 4.2(ii) $\sum_{n\geq 0}\lambda_n(1-\lambda_n)\|Rx_n - x_n\|^2 < +\infty$. Hence, $\sum_{n\geq 0}\lambda_n(1-\lambda_n) = +\infty \Rightarrow \underline{\lim}\,\|Rx_n - x_n\| = 0$. However,

$$(\forall n \in \mathbb{N})\ \ Rx_{n+1} - x_{n+1} = Rx_{n+1} - Rx_n + (1-\lambda_n)(Rx_n - x_n) - \lambda_n e_n \tag{68}$$

and therefore

$$\begin{aligned}(\forall n \in \mathbb{N})\ \ \|Rx_{n+1} - x_{n+1}\| &\leq \|Rx_{n+1} - Rx_n\| + (1-\lambda_n)\|Rx_n - x_n\| + \lambda_n\|e_n\| \\ &\leq \|x_{n+1} - x_n\| + (1-\lambda_n)\|Rx_n - x_n\| + \lambda_n\|e_n\| \\ &\leq \|Rx_n - x_n\| + 2\lambda_n\|e_n\|. \end{aligned} \tag{69}$$

Consequently, as $\sum_{n\geq 0}\lambda_n\|e_n\| < +\infty$, Lemma 3.1(ii) secures the convergence of the sequence $(\|Rx_n - x_n\|)_{n\geq 0}$ and (67) is established. □

Remark 5.6 When $e_n \equiv 0$, Theorem 5.5 is related to several known results:

- The weak convergence statement corresponds to the Hilbert space version of [40, Cor. 3] (see also [65]).
- For constant relaxation parameters, strong convergence under condition (ii) covers the Hilbert space version of [58, Cor. 2.2] and strong convergence under condition (iv) corresponds to the Hilbert space version of [58, Cor. 2.1].
- Strong convergence under condition (iii) was obtained in [53] and [64] by replacing "$(\lambda_n(1-\lambda_n))_{n\geq 0} \notin \ell^1$" by "$\lambda_n \equiv 1$" in (i).

Remark 5.7 Theorem 5.5 authorizes nonsummable error sequences. For instance, for n large, suppose that $\|e_n\| \leq (1+\sqrt{1-1/n})/n^{\kappa}$, where $\kappa \in \,]0,1]$, and that the relaxation rule is $\lambda_n = (1-\sqrt{1-1/n})/2$. Then $\sum_{n\geq 0}\|e_n\|$ may diverge but $\sum_{n\geq 0}\lambda_n\|e_n\| < +\infty$ and $\sum_{n\geq 0}\lambda_n(1-\lambda_n) = +\infty$.

5.3. Gradient method

In the error-free case ($e_n \equiv 0$), it was shown in [24] that convergence results could be derived from Theorem 5.5 for a number of algorithms, including the Forward-Backward and Douglas-Rachford methods for finding a zero of the sum of two monotone operators, the prox method for solving variational inequalities, and, in particular, the projected gradient method. Theorem 5.5 therefore provides convergence results for perturbed versions of these algorithms. As an illustration, this section is devoted to the case of the perturbed gradient method. A different analysis of the perturbed gradient method can be found in [51].

Consider the unconstrained minimization problem

$$\text{Find } x \in \mathcal{H} \text{ such that } f(x) = \bar{f}, \text{ where } \bar{f} = \inf f(\mathcal{H}). \tag{70}$$

The standing assumption is that $f\colon \mathcal{H} \to \mathbb{R}$ is a continuous convex function and that the set S of solutions of (70) is nonempty, as is the case when f is coercive; it is also assumed that f is differentiable and that, for some $\alpha \in \,]0,+\infty[$, $\alpha\nabla f$ is firmly nonexpansive (it follows from [6, Cor. 10] that this is equivalent to saying that ∇f is $(1/\alpha)$-Lipschitz, i.e., that $\alpha\nabla f$ is nonexpansive).

Algorithm 5.8 Fix $\gamma \in \,]0,2\alpha]$ and, at iteration $n \in \mathbb{N}$, suppose that $x_n \in \mathcal{H}$ is given. Then select $\lambda_n \in [0,1]$ and set $x_{n+1} = x_n - \lambda_n\gamma\big(\nabla f(x_n) + e_n\big)$, where $e_n \in \mathcal{H}$.

Theorem 5.9 *Let $(x_n)_{n\geq 0}$ be an arbitrary orbit of Algorithm 5.8. Then $(x_n)_{n\geq 0}$ converges weakly to point in S if $(\lambda_n\|e_n\|)_{n\geq 0} \in \ell^1$ and $\big(\lambda_n(1-\lambda_n)\big)_{n\geq 0} \notin \ell^1$.*

Proof. Put $R = \mathrm{Id} - \gamma\nabla f$. Then

$$\begin{aligned}(\forall (x,y) \in \mathcal{H}^2)\ \|Rx - Ry\|^2 &= \|x-y\|^2 - 2\gamma\langle x-y \mid \nabla f(x) - \nabla f(y)\rangle \\ &\quad + \gamma^2\|\nabla f(x) - \nabla f(y)\|^2 \\ &\leq \|x-y\|^2 - \gamma(2\alpha-\gamma)\|\nabla f(x) - \nabla f(y)\|^2. \end{aligned}\tag{71}$$

Hence R is nonexpansive and Algorithm 5.8 is a special case of Algorithm 5.4. As $\operatorname{Fix} R = (\nabla f)^{-1}(\{0\}) = S$, the claim follows from Theorem 5.5(i). □

Remark 5.10 Strong convergence conditions can be derived from Theorem 5.5(ii)-(iv). Thus, it follows from item (ii) that weak convergence in Theorem 5.9 can be improved to strong convergence if we add the correctness condition [23], [48]:

$$\lim f(x_n) = \bar{f} \quad \Rightarrow \quad \underline{\lim}\, d_S(x_n) = 0. \tag{72}$$

Indeed, by convexity

$$(\forall x \in S)(\forall n \in \mathbb{N}) \;\; 0 \leq f(x_n) - \bar{f} \leq \langle x_n - x \mid \nabla f(x_n) \rangle \leq \sup_{l \geq 0} \|x_l - x\| \cdot \|\nabla f(x_n)\|. \tag{73}$$

Consequently, with the same notation as in the above proof, it follows from (72) that (67) $\Leftrightarrow \nabla f(x_n) \to 0 \Rightarrow f(x_n) \to \bar{f} \Rightarrow \underline{\lim}\, d_S(x_n) = 0$.

5.4. Inconsistent convex feasibility problems

Let $(S_i)_{i \in I}$ be a finite family of nonempty closed and convex sets in $\mathcal{H}$. A standard convex programming problem is to find a point in the intersection of these sets. In instances when the intersection turns out to be empty, an alternative is to look for a point which is closest to all the sets in a least squared distance sense, i.e., to minimize the *proximity function*

$$f = \frac{1}{2} \sum_{i \in I} \omega_i d_{S_i}^2, \quad \text{where} \;\; (\forall i \in I)\; \omega_i > 0 \;\; \text{and} \;\; \sum_{i \in I} \omega_i = 1. \tag{74}$$

The resulting problem is a particular case of (70). We shall denote by S the set of minimizers of f over $\mathcal{H}$ and assume that it is nonempty, as is the case when one of the sets in $(S_i)_{i \in I}$ is bounded since f is then coercive. Naturally, if $\bigcap_{i \in I} S_i \neq \varnothing$, then $S = \bigcap_{i \in I} S_i$.

To solve the (possibly inconsistent) convex feasibility problem (70)/(74), we shall use the following parallel projection algorithm.

Algorithm 5.11 At iteration $n \in \mathbb{N}$, suppose that $x_n \in \mathcal{H}$ is given. Then select $\lambda_n \in [0, 2]$ and set $x_{n+1} = x_n + \lambda_n \big(\sum_{i \in I} \omega_i (P_{S_i} x_n + e_{i,n}) - x_n \big)$, where $(e_{i,n})_{i \in I}$ lies in $\mathcal{H}$.

Theorem 5.12 *Let $(x_n)_{n \geq 0}$ be an arbitrary orbit of Algorithm 5.11. Then $(x_n)_{n \geq 0}$ converges weakly to point in S if $(\lambda_n \| \sum_{i \in I} \omega_i e_{i,n} \|)_{n \geq 0} \in \ell^1$ and $\big(\lambda_n (2 - \lambda_n)\big)_{n \geq 0} \notin \ell^1$.*

Proof. We have $\nabla f = \sum_{i \in I} \omega_i (\mathrm{Id} - P_{S_i})$. Since the operators $(P_{S_i})_{i \in I}$ are firmly nonexpansive by Proposition 2.2, so are the operators $(\mathrm{Id} - P_{S_i})_{i \in I}$ and, in turn, their convex combination ∇f. Hence, Algorithm 5.11 is a special case of Algorithm 5.8 with $\alpha = 1$, $\gamma = 2$, and $(\forall n \in \mathbb{N})\; e_n = \sum_{i \in I} \omega_i e_{i,n}$. The claim therefore follows from Theorem 5.9. ☐

Remark 5.13 Let us make a couple of comments.

- Theorem 5.12 extends [18, Thm. 4], where $e_{i,n} \equiv 0$ and the relaxations parameters are bounded away from 0 and 2 (see also [27] where constant relaxation parameters are assumed).

- Algorithm 5.8 allows for an error in the evaluation of each projection. As noted in Remark 5.7, the average projection error sequence $(\sum_{i\in I}\omega_i e_{i,n})_{n\geq 0}$ does not have to be absolutely summable.

Remark 5.14 Suppose that the problem is consistent, i.e., $\bigcap_{i\in I} S_i \neq \varnothing$.

- If $e_{i,n} \equiv 0$, $\lambda_n \equiv 1$, and $\omega_i = 1/\operatorname{card} I$, Theorem 5.12 was obtained in [4] (see also [66, Cor. 2.6] for a different perspective).
- If I is infinite (and possibly uncountable), a more general operator averaging process for firmly nonexpansive operators with errors is studied in [35] (see also [16] for an error-free version with projectors).
- If the projections can be computed exactly, a more efficient weakly convergent parallel projection algorithm to find a point in $\bigcap_{i\in I} S_i$ is that proposed by Pierra in [59], [60]. It consists in taking T in Algorithm 5.1 as the operator defined in (14) with $(\forall i \in I)\ T_i = P_{S_i}$ and relaxations parameters in $]0,1]$. The large values achieved by the parameters $(L(x_n))_{n\geq 0}$ result in large step sizes that significantly accelerate the algorithm, as evidenced in various numerical experiments (see Remark 6.2 for specific references). This type of extrapolated scheme was first employed in the parallel projection method of Merzlyakov [52] to solve systems of affine inequalities in $\mathbb{R}^N$; the resulting algorithm was shown to be faster than the sequential projection algorithms of [1] and [54]. An alternative interpretation of Pierra's algorithm is the following: it can be obtained by taking T in Algorithm 5.1 as the subgradient projector defined in (9), where f is the proximity function defined in (74). A generalization of Pierra's algorithm will be proposed in Section 6.1.

5.5. Proximal point algorithm

Many optimization problems – in particular (70) – reduce to the problem of finding a zero of a monotone operator $A\colon \mathcal{H} \to 2^{\mathcal{H}}$, i.e., to the problem

$$\text{Find } x \in \mathcal{H} \text{ such that } 0 \in Ax. \tag{75}$$

It will be assumed henceforth that $0 \in \operatorname{ran} A$ and that A is maximal monotone.

The following algorithm, which goes back to [50], is known as the (relaxed) inexact proximal point algorithm.

Algorithm 5.15 At iteration $n \in \mathbb{N}$, suppose that $x_n \in \mathcal{H}$ is given. Then select $\lambda_n \in [0,2]$, $\gamma_n \in]0,+\infty[$, and set $x_{n+1} = x_n + \lambda_n\big((\mathrm{Id} + \gamma_n A)^{-1}x_n + e_n - x_n\big)$, where $e_n \in \mathcal{H}$.

Theorem 5.16 *Let $(x_n)_{n\geq 0}$ be an arbitrary orbit of Algorithm 5.15. Then $(x_n)_{n\geq 0}$ converges weakly to point in $A^{-1}0$ if $(\|e_n\|)_{n\geq 0} \in \ell^1$, $\inf_{n\geq 0}\gamma_n > 0$, and $(\lambda_n)_{n\geq 0}$ lies in $[\delta, 2-\delta]$, for some $\delta \in]0,1[$.*

Proof. It follows from Proposition 2.2 that Algorithm 5.15 is a special case of Algorithm 4.1. Moreover, $(\forall n \in \mathbb{N})\ \operatorname{Fix}(\mathrm{Id} + \gamma_n A)^{-1} = A^{-1}0$. Hence, in view of Theorem 4.3(i), we need to show $\mathfrak{W}(x_n)_{n\geq 0} \subset A^{-1}0$. For every $n \in \mathbb{N}$, define $y_n = (\mathrm{Id} +$

$\gamma_n A)^{-1}x_n$ and $v_n = (x_n - y_n)/\gamma_n$, and observe that $(y_n, v_n) \in \operatorname{gr}A$. In addition, since $\inf_{n\geq 0} \lambda_n(2-\lambda_n) \geq \delta^2$, it follows from Proposition 4.2(ii) that $x_n - y_n \to 0$. Therefore, since $\inf_{n\geq 0} \gamma_n > 0$, we get $v_n \to 0$. Now take $x \in \mathfrak{W}(x_n)_{n\geq 0}$, say $x_{k_n} \rightharpoonup x$. Then $y_{k_n} \rightharpoonup x$ and $v_{k_n} \to 0$. However, as A is maximal monotone, $\operatorname{gr}A$ is weakly-strongly closed, which forces $0 \in Ax$. □

Remark 5.17 Theorem 5.16 can be found in [28, Thm. 3] and several related results can be found in the literature.

- The unrelaxed version (i.e., $\lambda_n \equiv 1$) is due to Rockafellar [68, Thm. 1]. There, it was also proved that $x_{n+1} - x_n \to 0$. This fact follows immediately from Proposition 4.2(iii).
- Perturbed proximal point algorithms are also investigated in [3], [12], [14], [44], [46], and [55].

Remark 5.18 As shown in [42], an orbit of the proximal point algorithm may converge weakly but not strongly to a solution point. In this regard, two comments should be made.

- Strong convergence conditions can be derived from Theorem 4.3(ii)-(v). Thus, the convergence is strong in Theorem 5.16 in each of the following cases:
  - $\sum_{n\geq 0} \|(\mathrm{Id} + \gamma_n A)^{-1}x_n - x_n\|^2 < +\infty \Rightarrow \underline{\lim}\, d_{A^{-1}0}(x_n) = 0$. This condition follows immediately from item (ii). For accretive operators in nonhilbertian Banach spaces and $\lambda_n \equiv 1$, a similar condition was obtained in [55, Sec. 4].
  - $\operatorname{int} A^{-1}0 \neq \varnothing$. This condition follows immediately from item (iii) and can be found in [55, Sec. 6].
  - $\operatorname{dom} A$ is boundedly compact. This condition follows from item (iv) if $(\gamma_n)_{n\geq 0}$ contains a constant subsequence and, more generally, from the argument given in the proof of Theorem 6.9(iv).

  Additional conditions will be found in [12] and [68].
- A relatively minor modification of the proximal point algorithm makes it strongly convergent without adding any specific condition on A. See [71] and Remark 4.5 for details.

6. APPLICATIONS TO BLOCK-ITERATIVE PARALLEL ALGORITHMS

6.1. The algorithm

A common feature of the algorithms described in Section 5 is that $(\forall n \in \mathbb{N})\ \operatorname{Fix} T_n = S$. These algorithms therefore implicitly concern applications in which the target set S is relatively simple. In many applications, however, the target set is not known explicitly but merely described as a countable (finite or countably infinite) intersection of closed convex sets $(S_i)_{i\in I}$ in $\mathcal{H}$. The underlying problem can then be recast in the form of the *countable convex feasibility problem*

$$\text{Find } x \in S = \bigcap_{i\in I} S_i. \tag{76}$$

Here, the tacit assumption is that for every index $i \in I$ it is possible to construct relatively easily at iteration n an operator $T_{i,n} \in \mathfrak{T}$ such that $\operatorname{Fix} T_{i,n} = S_i$. Thus, S is not dealt with directly but only through its supersets $(S_i)_{i\in I}$. In infinite dimensional spaces, a classical method fitting in this framework is Bregman's periodic projection algorithm [11] which solves (76) iteratively in the case $I = \{1, \ldots, m\}$ via the sequential algorithm

$$(\forall n \in \mathbb{N})\;\; x_{n+1} = P_{S_{n \,(\text{mod}\, m)+1}} x_n. \tag{77}$$

As discussed in Remark 5.14, an alternative method to solve this problem is Auslender's parallel projection scheme [4]

$$(\forall n \in \mathbb{N})\;\; x_{n+1} = \frac{1}{m}\sum_{i=1}^{m} P_{S_i} x_n. \tag{78}$$

Bregman's method utilizes only one set at each iteration while Auslender's utilizes all of them simultaneously and is therefore inherently parallel. In this respect, these two algorithms stand at opposite ends in the more general class of parallel block-iterative algorithms, where at iteration n the update is formed by averaging projections of the current iterate onto a block of sets $(S_i)_{i\in I_n \subset I}$. The practical advantage of such a scheme is to provide a flexible means to match the computational load of each iteration to the distributed computer resources at hand.

The first block parallel projection algorithm in a Hilbert space setting was proposed by Ottavy [56] with further developments in [15] and [22]. Variants and extensions of (77) and (78) involving more general operators such as subgradient projectors, nonexpansive and firmly nonexpansive operators have also been investigated [5], [13], [17], [36], [63], [72] and unified in the form of block-iterative algorithms at various levels of generality in [8], [19], [21], and [43]. For recent extensions of (77) in other directions, see [67] and the references therein.

Building upon the framework developed in [8], a general block-iterative scheme was proposed in [45, Algo. 2.1] to bring together the algorithms mentioned above. An essentially equivalent algorithm was later devised in [23, Algo. 7.1] within a different framework. The following algorithm employs yet another framework, namely the $\mathfrak{T}$ operator class, and, in addition, allows for errors in the computation of each operator.

Algorithm 6.1 Fix $(\delta_1, \delta_2) \in \,]0,1[^2$ and $x_0 \in \mathcal{H}$. At every iteration $n \in \mathbb{N}$,

$$x_{n+1} = x_n + \lambda_n L_n \left(\sum_{i\in I_n} \omega_{i,n}(T_{i,n}x_n + e_{i,n}) - x_n\right) \tag{79}$$

where:

① $\emptyset \neq I_n \subset I$, I_n finite.

② $(\forall i \in I_n)$ $T_{i,n} \in \mathfrak{T}$ and $\operatorname{Fix} T_{i,n} = S_i$.

③ $(\forall i \in I_n)$ $e_{i,n} \in \mathcal{H}$ and $e_{i,n} = 0$ if $x_n \in S_i$.

④ $(\forall i \in I_n)$ $\omega_{i,n} \in [0,1]$, $\sum_{i \in I_n} \omega_{i,n} = 1$, and

$$(\exists\, j \in I_n) \begin{cases} \|T_{j,n}x_n - x_n\| = \max\limits_{i \in I_n} \|T_{i,n}x_n - x_n\| \\ \omega_{j,n} \geq \delta_1. \end{cases}$$

⑤ $\lambda_n \in [\delta_2/L_n, 2 - \delta_2]$, where

$$L_n = \begin{cases} \dfrac{\sum_{i \in I_n} \omega_{i,n} \|T_{i,n}x_n - x_n\|^2}{\|\sum_{i \in I_n} \omega_{i,n} T_{i,n}x_n - x_n\|^2} & \text{if } x_n \notin \bigcap_{i \in I_n} S_i \text{ and } \sum_{i \in I_n} \omega_{i,n}\|e_{i,n}\| = 0, \\ 1 & \text{otherwise.} \end{cases}$$

Remark 6.2 The incorporation of errors in the above recursion calls for some comments.

- The vector $e_{i,n}$ stands for the error made in computing $T_{i,n}x_n$. With regard to the convergence analysis, the global error term at iteration n is $\lambda_n \sum_{i \in I_n} \omega_{i,n} e_{i,n}$. Thus, the individual errors $(e_{i,n})_{i \in I_n}$ are naturally averaged and can be further controlled by the relaxation parameter λ_n.

- If $e_{i,n} \equiv 0$, Algorithm 6.1 essentially relapses to [23, Algo. 7.1] and [45, Algo. 2.1]. If we further assume that at every iteration n the index set I_n is a singleton, then it reduces to the exact $\mathfrak{T}$-class sequential method of [9, Algo. 2.8].

- If some index $j \in I_n$ it is possible to verify that $\|T_{j,n}x_n - x_n\| \neq \max_{i \in I_n} \|T_{i,n}x_n - x_n\|$, the associated error $e_{j,n}$ can be neutralized by setting $\omega_{j,n} = 0$.

- Suppose that $\sum_{i \in I_n} \omega_{i,n}\|e_{i,n}\| = 0$, meaning that for each selected index i, either $T_{i,n}x_n$ is computed exactly or the associated error $e_{i,n}$ is neutralized (see previous item). Then extrapolated relaxations up to $(2 - \delta_2)L_n$ can be used, where L_n can attain very large values. In numerical experiments, this type of extrapolated overrelaxations has been shown to induce very fast convergence [20], [21], [25], [37], [52], [60], [61].

6.2. Convergence

Let us first recall a couple of useful concepts.

Definition 6.3 [19] The control sequence $(I_n)_{n\geq 0}$ in Algorithm 6.1 is *admissible* if there exist strictly positive integers $(M_i)_{i\in I}$ such that

$$(\forall (i,n) \in I \times \mathbb{N}) \;\; i \in \bigcup_{k=n}^{n+M_i-1} I_k. \tag{80}$$

Definition 6.4 [8, Def. 3.7] Algorithm 6.1 is *focusing* if for every index $i \in I$ and every generated suborbit $(x_{k_n})_{n\geq 0}$,

$$\begin{cases} i \in \bigcap_{n\geq 0} I_{k_n} \\ x_{k_n} \rightharpoonup x \\ T_{i,k_n}x_{k_n} - x_{k_n} \to 0 \end{cases} \quad \Rightarrow \quad x \in S_i. \tag{81}$$

The notion of a focusing algorithm can be interpreted as an extension of the notion of demiclosedness at 0. Along the same lines, it is convenient to introduce the following extension of the notion of demicompactness at 0.

Definition 6.5 Algorithm 6.1 is *demicompactly regular* if there exists an index $i \in I$ such that, for every generated suborbit $(x_{k_n})_{n\geq 0}$,

$$\begin{cases} i \in \bigcap_{n\geq 0} I_{k_n} \\ \sup_{n\geq 0} \|x_{k_n}\| < +\infty \\ T_{i,k_n}x_{k_n} - x_{k_n} \to 0 \end{cases} \quad \Rightarrow \quad \mathfrak{S}(x_{k_n})_{n\geq 0} \neq \varnothing. \tag{82}$$

Such an index is an *index of demicompact regularity.*

The most relevant convergence properties of Algorithm 6.1 are summarized below. This theorem appears to be the first general result on the convergence of inexact block-iterative methods for convex feasibility problems.

Theorem 6.6 *Suppose that $S \neq \varnothing$ in* (76) *and let $(x_n)_{n\geq 0}$ be an arbitrary orbit of Algorithm 6.1. Then $(x_n)_{n\geq 0}$ converges weakly to a point in S if*

(i) $\big(\lambda_n \| \sum_{i\in I_n} \omega_{i,n} e_{i,n} \|\big)_{n\geq 0} \in \ell^1$, *Algorithm 6.1 is focusing, and the control sequence $(I_n)_{n\geq 0}$ is admissible.*

The convergence is strong if any of the following assumptions is added:

(ii) *Algorithm 6.1 is demicompactly regular.*

(iii) $\operatorname{int} S \neq \varnothing$.

(iv) *There exists a suborbit $(x_{k_n})_{n\geq 0}$ and a sequence $(\chi_n)_{n\geq 0} \in \ell_+ \setminus \ell^2$ such that*

$$(\forall n \in \mathbb{N}) \;\; \max_{i\in I_{k_n}} \|T_{i,k_n}x_{k_n} - x_{k_n}\|^2 \geq \chi_n d_S(x_{k_n}). \tag{83}$$

Proof. We proceed in several steps. Throughout the proof, y is a fixed point in S and $\beta(y) = \sup_{n\geq 0} \|x_n - y\|$. If $\beta(y) = 0$, all the statements are trivially true; we therefore assume otherwise.

Step 1: Algorithm 6.1 is a special instance of Algorithm 4.1.
Indeed, for every $n \in \mathbb{N}$, we can write (79) as

$$x_{n+1} = x_n + \lambda_n(T_n x_n + e_n - x_n), \tag{84}$$

where

$$e_n = \sum_{i\in I_n} \omega_{i,n} e_{i,n} \tag{85}$$

and

$$T_n x_n = x_n + L_n \left(\sum_{i\in I_n} \omega_{i,n} T_{i,n} x_n - x_n \right). \tag{86}$$

It follows from the definition of L_n in ⑤ that the operator T_n takes one of two forms, namely

$$\begin{cases} T_n \colon x \mapsto \sum_{i\in I_n} \omega_{i,n} T_{i,n} x & \text{if } \sum_{i\in I_n} \omega_{i,n} \|e_{i,n}\| \neq 0 \\ T_n \colon x \mapsto x + L\big(x, (T_{i,n})_{i\in I_n}, (\omega_{i,n})_{i\in I_n}\big) \left(\sum_{i\in I_n} \omega_{i,n} T_{i,n} x - x \right) & \text{otherwise,} \end{cases} \tag{87}$$

where the function L is defined in (15). In view of ②, ④, Proposition 2.4, and Remark 2.5, we conclude that in both cases $T_n \in \mathfrak{T}$.

Step 2: $S \subset \bigcap_{n\geq 0} \operatorname{Fix} T_n$.
It follows from (76), (80), ②, and Proposition 2.4 that

$$S = \bigcap_{i\in I} S_i = \bigcap_{n\geq 0} \bigcap_{i\in I_n} S_i \subset \bigcap_{n\geq 0} \bigcap_{\substack{i\in I_n \\ \omega_{i,n}>0}} \operatorname{Fix} T_{i,n} = \bigcap_{n\geq 0} \operatorname{Fix} T_n. \tag{88}$$

Step 3: $(\|x_{n+1} - x_n\|)_{n\geq 0} \in \ell^2$.
The claim follows from Step 1, (85), and Proposition 4.2(iii) since ⑤ $\Rightarrow \overline{\lim}\, \lambda_n < 2$.

Step 4: $\lim \max_{i\in I_n} \|T_{i,n} x_n - x_n\| = 0$.
To see this, we use successively ⑤, (86), and the inequality $L_n \geq 1$ to derive

$$\begin{aligned} (\forall n \in \mathbb{N}) \;\; \lambda_n(2 - \lambda_n)\|T_n x_n - x_n\|^2 &\geq \frac{\delta_2^2}{L_n} \|T_n x_n - x_n\|^2 \\ &= \delta_2^2 L_n \left\| \sum_{i\in I_n} \omega_{i,n} T_{i,n} x_n - x_n \right\|^2 \\ &\geq \delta_2^2 \left\| \sum_{i\in I_n} \omega_{i,n} T_{i,n} x_n - x_n \right\|^2. \end{aligned} \tag{89}$$

By virtue of Step 1 and Proposition 4.2(i), $(x_n)_{n\geq 0}$ is a quasi-Fejér sequence of Type I relative to S and therefore $\beta(y) < +\infty$. Moreover, ② implies that, for every $n \in \mathbb{N}$, $(T_{i,n})_{i\in I_n}$ lies in $\mathfrak{T}$ and $y \in \bigcap_{i\in I_n} \operatorname{Fix} T_{i,n}$. Hence, we can argue as in (17) to get

$$(\forall n \in \mathbb{N}) \quad \left\| \sum_{i\in I_n} \omega_{i,n} T_{i,n} x_n - x_n \right\| \geq \frac{1}{\beta(y)} \sum_{i\in I_n} \omega_{i,n} \|T_{i,n}x_n - x_n\|^2 \geq \frac{\delta_1}{\beta(y)} \max_{i\in I_n} \|T_{i,n}x_n - x_n\|^2, \tag{90}$$

where the second inequality is deduced from ④. Now, since Proposition 4.2(ii) implies that $\lim \lambda_n(2-\lambda_n)\|T_n x_n - x_n\|^2 = 0$, it follows from (89) that $\lim \| \sum_{i\in I_n} \omega_{i,n} T_{i,n} x_n - x_n\| = 0$ and then from (90) that $\lim \max_{i\in I_n} \|T_{i,n}x_n - x_n\| = 0$.

Step 5: $\mathfrak{W}(x_n)_{n\geq 0} \subset S$.

Fix $i \in I$ and $x \in \mathfrak{W}(x_n)_{n\geq 0}$, say $x_{k_n} \rightharpoonup x$. Then it is enough to show $x \in S_i$. By (80), there exist an integer $M_i > 0$ and a strictly increasing sequence $(p_n)_{n\geq 0}$ in $\mathbb{N}$ such that

$$(\forall n \in \mathbb{N}) \quad k_n \leq p_n \leq k_n + M_i - 1 \quad \text{and} \quad i \in I_{p_n}. \tag{91}$$

Hence, by Cauchy-Schwarz,

$$(\forall n \in \mathbb{N}) \quad \|x_{p_n} - x_{k_n}\| \leq \sum_{l=k_n}^{k_n+M_i-2} \|x_{l+1} - x_l\| \leq \sqrt{M_i - 1}\sqrt{\sum_{l\geq k_n} \|x_{l+1} - x_l\|^2} \tag{92}$$

and Step 3 yields $x_{p_n} - x_{k_n} \to 0$. Thus, $x_{p_n} \rightharpoonup x$, while $i \in \bigcap_{n\geq 0} I_{p_n}$ and, by Step 4, $T_{i,p_n}x_{p_n} - x_{p_n} \to 0$. Therefore, (81) forces $x \in S_i$.

Step 6: Weak convergence.

Combine Steps 1, 2, and 5, Theorem 4.3(i), and (85).

Step 7: Strong convergence.

(ii): Let i be an index of demicompact regularity. According to (80), there exists a strictly increasing sequence $(k_n)_{n\geq 0}$ in $\mathbb{N}$ such that $i \in \bigcap_{n\geq 0} I_{k_n}$, where i is an index of demicompact regularity, and Step 4 implies $T_{i,k_n}x_{k_n} - x_{k_n} \to 0$. Since $(x_{k_n})_{n\geq 0}$ is bounded, (82) yields $\mathfrak{S}(x_{k_n})_{n\geq 0} \neq \varnothing$. Therefore, strong convergence results from Step 5 and Theorem 3.11. (iii) follows from Step 1 and Theorem 4.3(iii). (iv) follows from Theorem 4.3(ii). Indeed, using (89), (90), and (83), we get

$$\left(\frac{\beta(y)}{\delta_1\delta_2}\right)^2 \sum_{n\geq 0} \lambda_{k_n}(2-\lambda_{k_n})\|T_{k_n}x_{k_n} - x_{k_n}\|^2 \geq \sum_{n\geq 0} \chi_n^2 d_S(x_{k_n})^2. \tag{93}$$

Hence, since $\sum_{n\geq 0} \lambda_n(2-\lambda_n)\|T_n x_n - x_n\|^2 < +\infty$ by Proposition 4.2(ii) and $(\chi_n)_{n\geq 0} \notin \ell^2$ by assumption, we conclude that $\underline{\lim}\, d_S(x_{k_n}) = 0$. □

6.3. Application to a mixed convex feasibility problem

Let $(f_i)_{i\in I^{(1)}}$ be a family of continuous convex functions from $\mathcal{H}$ into $\mathbb{R}$, $(R_i)_{i\in I^{(2)}}$ a family of firmly nonexpansive operators with domain $\mathcal{H}$ and into $\mathcal{H}$, and $(A_i)_{i\in I^{(3)}}$ a family of maximal monotone operators from $\mathcal{H}$ into $2^{\mathcal{H}}$. Here, $I^{(1)}$, $I^{(2)}$, and $I^{(3)}$ are possibly empty, countable index sets.

In an attempt to unify a wide class of problems, we consider the *mixed convex feasibility problem*

$$\text{Find } x \in \mathcal{H} \text{ such that } \begin{cases} (\forall i \in I^{(1)}) & f_i(x) \leq 0 \\ (\forall i \in I^{(2)}) & R_i x = x \\ (\forall i \in I^{(3)}) & 0 \in A_i x, \end{cases} \tag{94}$$

under the standing assumption that it is consistent. Problem (94) can be expressed as

$$\text{Find } x \in S = \bigcap_{i \in I} S_i, \quad \text{where } I = I^{(1)} \cup I^{(2)} \cup I^{(3)}$$

$$\text{and } (\forall i \in I) \ S_i = \begin{cases} \operatorname{lev}_{\leq 0} f_i & \text{if } i \in I^{(1)} \\ \operatorname{Fix} R_i & \text{if } i \in I^{(2)} \\ A_i^{-1} 0 & \text{if } i \in I^{(3)}. \end{cases} \tag{95}$$

In this format, it is readily seen to be a special case of problem (76) (the closedness and convexity of S_i in each case follows from well known facts).

To solve (94), we shall draw the operators $T_{i,n}$ in Algorithm 6.1 from a pool of subgradient projectors, firmly nonexpansive operators, and resolvents. Since such operators conform to ② in Algorithm 6.1 by Proposition 2.2, this choice is legitimate.

Algorithm 6.7 In Algorithm 6.1 set for every $i \in I$ the operators $(T_{i,n})_{n \geq 0}$ as follows.

- If $i \in I^{(1)}$, $(\forall n \in \mathbb{N})$ $T_{i,n} = G_{f_i}^{g_i}$, where g_i is a selection of ∂f_i (see (9)).
- If $i \in I^{(2)}$, $(\forall n \in \mathbb{N})$ $T_{i,n} = R_i$.
- If $i \in I^{(3)}$, $(\forall n \in \mathbb{N})$ $T_{i,n} = (\mathrm{Id} + \gamma_{i,n} A_i)^{-1}$, where $\gamma_{i,n} \in \,]0, +\infty[$.

The next assumption will ensure that Algorithm 6.7 is well behaved asymptotically.

Assumption 6.8 The subdifferentials $(\partial f_i)_{i \in I^{(1)}}$ map bounded sets into bounded sets and, for every $i \in I^{(3)}$ and every strictly increasing sequence $(k_n)_{n \geq 0}$ in $\mathbb{N}$ such that $i \in \bigcap_{n \geq 0} I_{k_n}$, $\inf_{n \geq 0} \gamma_{i,k_n} > 0$.

Theorem 6.9 *Suppose that $S \neq \emptyset$ in (95) and let $(x_n)_{n \geq 0}$ be an arbitrary orbit of Algorithm 6.7. Then $(x_n)_{n \geq 0}$ converges weakly to a point in S if*

(i) $\left(\lambda_n \|\sum_{i \in I_n} \omega_{i,n} e_{i,n}\|\right)_{n \geq 0} \in \ell^1$, *Assumption 6.8 is satisfied, and the control sequence $(I_n)_{n \geq 0}$ is admissible.*

The convergence is strong if any of the following assumptions is added:

(ii) *For some $i \in I^{(1)}$ and some $\eta \in \,]0, +\infty[$, $\operatorname{lev}_{\leq \eta} f_i$ is boundedly compact.*

(iii) *For some* $i \in I^{(2)}$, R_i *is demicompact at* 0.

(iv) *For some* $i \in I^{(3)}$, $\operatorname{dom} A_i$ *is boundedly compact.*

Proof. Since Algorithm 6.7 is a special case of Algorithm 6.1, we shall derive this theorem from Theorem 6.6. (i): It suffices to show that Assumption 6.8 implies that Algorithm 6.7 is focusing. To this end, fix $i \in I$ and a suborbit $(x_{k_n})_{n\geq 0}$ such that $i \in \bigcap_{n\geq 0} I_{k_n}$, $x_{k_n} \rightharpoonup x$, and $T_{i,k_n}x_{k_n} - x_{k_n} \to 0$. According to (81), we must show $x \in S_i$. Let us consider three cases.

(a) $i \in I^{(1)}$: we must show $f_i(x) \leq 0$. Put $\alpha = \overline{\lim} f(x_{k_n})$. Then, since f is weak lower semicontinuous, $f_i(x) \leq \alpha$ and it is enough to show that $\alpha \leq 0$. Let us extract from $(x_{k_n})_{n\geq 0}$ a subsequence $(x_{l_{k_n}})_{n\geq 0}$ such that $\lim f_i(x_{l_{k_n}}) = \alpha$ and $(\forall n \in \mathbb{N})$ $f_i(x_{l_{k_n}}) > 0$ (if no subsequence exists then clearly $\alpha \leq 0$). Then, by Assumption 6.8, $T_{i,k_n}x_{k_n} - x_{k_n} \to 0 \Rightarrow G_{f_i}^{g_i}x_{l_{k_n}} - x_{l_{k_n}} \to 0 \Rightarrow \lim f_i(x_{l_{k_n}})/\|g_i(x_{l_{k_n}})\| = 0 \Rightarrow \lim f_i(x_{l_{k_n}}) = 0$, where the last implication follows from the boundedness of $(x_{l_{k_n}})_{n\geq 0}$ and Assumption 6.8. Therefore $\alpha \leq 0$.

(b) $i \in I^{(2)}$: we must show $x \in \operatorname{Fix} R_i$. Certainly $T_{i,k_n}x_{k_n} - x_{k_n} \to 0 \Rightarrow (R_i - \operatorname{Id})x_{k_n} \to 0$. Since R_i is firmly nonexpansive, it is nonexpansive. $R_i - \operatorname{Id}$ is therefore demiclosed by Remark 2.6 and the claim ensues.

(c) $i \in I^{(3)}$: we must show $(x, 0) \in \operatorname{gr} A_i$. For every $n \in \mathbb{N}$, define $y_n = (\operatorname{Id} + \gamma_{i,k_n}A_i)^{-1}x_{k_n}$ and $v_n = (x_{k_n} - y_n)/\gamma_{i,k_n}$. Then $\big((y_n, v_n)\big)_{n\geq 0}$ lies in $\operatorname{gr} A_i$ and $T_{i,k_n}x_{k_n} - x_{k_n} \to 0 \Rightarrow y_n - x_{k_n} \to 0 \Rightarrow y_n \rightharpoonup x$. On the other hand, Assumption 6.8 ensures that $y_n - x_{k_n} \to 0 \Rightarrow v_n \to 0$. Since $\operatorname{gr} A_i$ is weakly-strongly closed, we conclude that $(x, 0) \in \operatorname{gr} A_i$.

Let us now show that the three advertised instances of strong convergence yield demicompact regularity and are therefore consequences of Theorem 6.6(ii). Let us fix $i \in I$, a closed ball B, and a suborbit $(x_{k_n})_{n\geq 0}$ such that $i \in \bigcap_{n\geq 0} I_{k_n}$, B contains $(x_{k_n})_{n\geq 0}$, and $T_{i,k_n}x_{k_n} - x_{k_n} \to 0$. We must show $\mathfrak{S}(x_{k_n})_{n\geq 0} \neq \varnothing$. (ii): As shown in (a), $\overline{\lim} f(x_{k_n}) \leq 0$ and therefore the tail of $(x_{k_n})_{n\geq 0}$ lies in the compact set $B \cap \operatorname{lev}_{\leq \eta} f_i$. (iii) is clear. (iv): Define $(y_n)_{n\geq 0}$ as in (c) and recall that $y_n - x_{k_n} \to 0$. Hence $(y_n)_{n\geq 0}$ lies in some closed ball B' and $\mathfrak{S}(x_{k_n})_{n\geq 0} = \mathfrak{S}(y_n)_{n\geq 0}$. Moreover,

$$(\forall n \in \mathbb{N}) \;\; y_n \in \operatorname{ran} (\operatorname{Id} + \gamma_{i,k_n}A_i)^{-1} = \operatorname{dom} (\operatorname{Id} + \gamma_{i,k_n}A_i) = \operatorname{dom} A_i. \tag{96}$$

Hence, $(y_n)_{n\geq 0}$ lies in the compact set $B' \cap \operatorname{dom} A_i$ and the desired conclusion ensues. □

Remark 6.10 To place the above result in its proper context, a few observations should be made.

- Theorem 6.9 combines and, through the incorporation of errors, generalizes various results on the convergence of block-iterative subgradient projection (for $I^{(2)} = I^{(3)} = \varnothing$) and firmly nonexpansive iteration (for $I^{(1)} = I^{(3)} = \varnothing$) methods [8], [19], [21], [22], [45].

- For $I^{(1)} = I^{(2)} = \emptyset$, the resulting inexact block-iterative proximal point algorithm appears to be new. If, in addition, $I^{(3)}$ is a singleton Theorem 6.9(i) reduces to Theorem 5.16; if we further assume that $\lambda_n \equiv 1$, Theorem 6.9 captures some convergence properties of Rockafellar's inexact proximal point algorithm [68].

- Concerning strong convergence, although we have restricted ourselves to special cases of Theorem 6.6(ii), it is clear that conditions (iii) and (iv) in Theorem 6.6 also apply here. At any rate, these conditions are certainly not exhaustive.

- To recover results on projection algorithms, one can set $(f_i)_{i\in I^{(1)}} = (d_{S_i})_{i\in I^{(1)}}$, $(R_i)_{i\in I^{(2)}} = (P_{S_i})_{i\in I^{(2)}}$, and $(A_i)_{i\in I^{(3)}} = (N_{C_i})_{i\in I^{(3)}}$, where N_{C_i} is the normal cone to S_i.

7. PROJECTED SUBGRADIENT METHOD

The algorithms described in Section 4–6 are quasi-Fejér of Type I. In this section, we shall investigate a class of nonsmooth constrained minimization methods which are quasi-Fejér of Type II. As we shall find, the analysis developed in Section 3 will also be quite useful here to obtain convergence results in a straightforward fashion.

Throughout, $f\colon \mathcal{H} \to \mathbb{R}$ is a continuous convex function, C is a closed convex subset of $\mathcal{H}$, and $\bar{f} = \inf f(C)$. Under consideration is the problem

$$\text{Find } x \in C \text{ such that } f(x) = \bar{f} \tag{97}$$

under the standing assumption that its set S of solutions is nonempty, as is the case when $\operatorname{lev}_{\leq\eta} f \cap C$ is nonempty and bounded for some $\eta \in \mathbb{R}$.

In nonsmooth minimization, the use of projected subgradient methods goes back to the 1960's [70]. Our objective here is to establish convergence results for a class of relaxed, projected approximate subgradient methods in Hilbert spaces. Recall that, given $\epsilon \in [0, +\infty[$, the ϵ-subdifferential of f at $x \in \mathcal{H}$ is obtained by relaxing (5) as follows

$$\partial_\epsilon f(x) = \big\{u \in \mathcal{H} \mid (\forall y \in \mathcal{H})\ \ \langle y - x \mid u\rangle + f(x) \leq f(y) + \epsilon\big\}. \tag{98}$$

Algorithm 7.1 At iteration $n \in \mathbb{N}$, suppose that $x_n \in \mathcal{H}$ is given. Then select $\alpha_n \in [0, +\infty[$, $\epsilon_n \in [0, +\infty[$, $\lambda_n \in [0, 2]$, $u_n \in \partial_{\epsilon_n} f(x_n)$, and set

$$x_{n+1} = x_n - \alpha_n u_n + \lambda_n\big(P_C(x_n - \alpha_n u_n) - x_n + \alpha_n u_n\big). \tag{99}$$

Let us open the discussion with some basic facts about this algorithm.

Proposition 7.2 *Let $(x_n)_{n\geq 0}$ be an arbitrary orbit of Algorithm 7.1 such that*

$$\left(\alpha_n^2\|u_n\|^2 + 2\alpha_n\big(\bar{f} - f(x_n)\big)^+ + 2\alpha_n\epsilon_n\right)_{n\geq 0} \in \ell^1. \tag{100}$$

Then

(i) $(x_n)_{n\geq 0}$ *is quasi-Fejér of Type II relative to* S.

(ii) $\big(\lambda_n(2-\lambda_n)d_C(x_n-\alpha_n u_n)^2\big)_{n\geq 0} \in \ell^1$.

(iii) *If* $(\alpha_n)_{n\geq 0} \notin \ell^1$, $\lim \epsilon_n = 0$, *and* $\lim \alpha_n \|u_n\|^2 = 0$, *then* $\underline{\lim}\, f(x_n) \leq \bar{f}$.

Proof. (i): Fix $x \in S$ and set $(\forall n \in \mathbb{N})\ y_n = x_n - \alpha_n u_n$. Then (98) yields

$$(\forall n \in \mathbb{N})\ \|y_n - x\|^2 = \|x_n - x\|^2 + \alpha_n^2\|u_n\|^2 - 2\alpha_n\langle x_n - x \mid u_n\rangle \leq \|x_n - x\|^2 + \alpha_n^2\|u_n\|^2 + 2\alpha_n\big(\bar{f} - f(x_n) + \epsilon_n\big). \tag{101}$$

On the other hand, since $x \in S \subset C = \operatorname{Fix} P_C$ and $P_C \in \mathfrak{T}$ by Proposition 2.2, Proposition 2.3(ii) yields

$$(\forall n \in \mathbb{N})\ \|x_{n+1} - x\|^2 \leq \|y_n - x\|^2 - \lambda_n(2-\lambda_n)d_C(y_n)^2. \tag{102}$$

Altogether

$$(\forall n \in \mathbb{N})\ \|x_{n+1} - x\|^2 \leq \|x_n - x\|^2 - \lambda_n(2-\lambda_n)d_C(y_n)^2 + \alpha_n^2\|u_n\|^2 + 2\alpha_n\big(\bar{f} - f(x_n)\big) + 2\alpha_n\epsilon_n. \tag{103}$$

In view of (100), $(x_n)_{n\geq 0}$ satisfies (3). (ii) follows from (103), (100), and Lemma 3.1(iii). (iii): We derive from (103) that

$$(\forall n \in \mathbb{N})\ \alpha_n\big(2(f(x_n) - \bar{f}) - 2\epsilon_n - \alpha_n\|u_n\|^2\big) \leq \|x_n - x\|^2 - \|x_{n+1} - x\|^2. \tag{104}$$

Hence,

$$\sum_{n\geq 0} \alpha_n\big(2(f(x_n) - \bar{f}) - 2\epsilon_n - \alpha_n\|u_n\|^2\big) \leq \|x_0 - x\|^2 \tag{105}$$

and, since $\sum_{n\geq 0}\alpha_n = +\infty$ by assumption, we get $\underline{\lim}\, 2(f(x_n) - \bar{f}) - 2\epsilon_n - \alpha_n\|u_n\|^2 \leq 0$. The remaining assumptions impose $\underline{\lim}\, f(x_n) \leq \bar{f}$. □

The following theorem describes a situation in which weak and strong convergence can be achieved in Algorithm 7.1.

Theorem 7.3 *Let* $(x_n)_{n\geq 0}$ *be an arbitrary orbit of Algorithm 7.1. Then* $(x_n)_{n\geq 0}$ *converges weakly to a point in* S *and* $\lim f(x_n) = \bar{f}$ *if*

(i) $(\alpha_n)_{n\geq 0} \in \ell^\infty \setminus \ell^1$, $(\epsilon_n)_{n\geq 0} \in \ell^1$, *and, furthermore, there exist* $\delta \in\,]0,1[$, $\kappa \in\,]0,+\infty[$, $(\beta_n)_{n\geq 0} \in \ell_+ \cap \ell^1$, *and* $(\gamma_n)_{n\geq 0} \in \ell_+ \cap \ell^1$ *such that*

$$(\forall n \in \mathbb{N})\ \bar{f} - \gamma_{n+1} \leq f(x_{n+1}) \leq f(x_n) - \kappa\alpha_n\|u_n\|^2 + \beta_n \tag{106}$$

and $(\forall n \in \mathbb{N})\ \delta \leq \lambda_n \leq 2 - \delta$.

The convergence is strong if any of the following assumptions is added:

(ii) $\operatorname{int} S \neq \varnothing$.

(iii) *C is boundedly compact.*

(iv) *For some* $z \in C$, $\operatorname{lev}_{\leq f(z)} f$ *is boundedly compact.*

Proof. Set $\zeta = \bar{f} - \sup_{n\geq 0} \gamma_n$. Then (106) yields

$$(\forall n \in \mathbb{N}) \;\; 0 \leq f(x_{n+1}) - \zeta \leq \big(f(x_n) - \zeta\big) - \kappa\alpha_n\|u_n\|^2 + \beta_n. \tag{107}$$

Hence, it follows from Lemma 3.1(ii) that $(f(x_n))_{n\geq 0}$ converges and from Lemma 3.1(iii) that $(\alpha_n\|u_n\|^2)_{n\geq 0} \in \ell^1$. Hence, since for every $n \in \mathbb{N} \smallsetminus \{0\}$ $\bar{f} - f(x_n) \leq \gamma_n$,

$$\sum_{n\geq 1} \alpha_n\|u_n\|^2 + 2\big(\bar{f} - f(x_n)\big)^+ + 2\epsilon_n \leq \sum_{n\geq 1} \alpha_n\|u_n\|^2 + 2\gamma_n + 2\epsilon_n < +\infty \tag{108}$$

and we deduce from the boundedness of $(\alpha_n)_{n\geq 0}$ that (100) holds. In view of Proposition 7.2(i), Proposition 3.2(i), and Theorem 3.8, to establish weak convergence to a solution, we must show $\mathfrak{W}(x_n)_{n\geq 0} \subset S$. Fix $x \in \mathfrak{W}(x_n)_{n\geq 0}$, say $x_{k_n} \rightharpoonup x$. Note that the assumptions on $(\lambda_n)_{n\geq 0}$ and Proposition 7.2(ii) imply that $\lim d_C(x_n - \alpha_n u_n) = 0$. On the other hand, since $(\alpha_n)_{n\geq 0} \in \ell^\infty$ and $\lim \alpha_n\|u_n\|^2 = 0$, we have $\lim \alpha_n\|u_n\| = 0$ and, therefore, $x_{k_n} - \alpha_{k_n} u_{k_n} \rightharpoonup x$. However, since C is convex, d_C is weak lower semicontinuous and it follows that $d_C(x) \leq \underline{\lim}\, d_C(x_{k_n} - \alpha_{k_n} u_{k_n}) = 0$. As C is closed, we obtain $x \in C$ and, in turn, $\bar{f} \leq f(x)$. The weak lower semicontinuity of f then yields

$$\bar{f} \leq f(x) \leq \underline{\lim}\, f(x_{k_n}) = \lim f(x_n) \leq \bar{f}, \tag{109}$$

where the last inequality is furnished by Proposition 7.2(iii). Consequently $f(x) = \bar{f} = \lim f(x_n)$. Since $x \in C$, we have thus shown $x \in S$, which completes the proof of (i).

Let us now prove the strong convergence statements. By virtue of Theorem 3.11, since $\mathfrak{W}(x_n)_{n\geq 0} \subset S$, it is enough to show $\mathfrak{S}(x_n)_{n\geq 0} \neq \varnothing$. (ii) Apply Proposition 3.10. (iii): As seen above, $P_C x_n - x_n \to 0$. On the other hand, P_C is demicompact (see Remark 2.6). Hence $\mathfrak{S}(x_n)_{n\geq 0} \neq \varnothing$. (iii): Let B be a closed ball containing $(x_n)_{n\geq 0}$. Since $\lim f(x_n) = \bar{f} \leq f(z)$, the tail of $(x_n)_{n\geq 0}$ lies in the compact set $B \cap \operatorname{lev}_{\leq f(z)} f$ and therefore $\mathfrak{S}(x_n)_{n\geq 0} \neq \varnothing$. □

Remark 7.4 A few comments are in order.

- Theorem 7.3 generalizes [26, Prop. 2.2], in which $\dim \mathcal{H} < +\infty$, $C = \mathcal{H}$ (hence $\gamma_n \equiv 0$), and $\beta_n \equiv 0$.
- The second inequality in (106) with $\beta_n \equiv 0$ was interpreted in [26] as an approximate Armijo rule.
- The first inequality in (106) is trivially satisfied with $\gamma_n \equiv 0$ if $(x_n)_{n\geq 0}$ lies in C. This is true in particular when $\lambda_n \equiv 1$.

Remark 7.5 Suppose that $\inf f(C) > \inf f(\mathcal{H})$ and fix $(\gamma_n)_{n\geq 0} \in \ell_+ \cap \ell^2 \smallsetminus \ell^1$. A classical subgradient projection method is described by the recursion [70]

$$x_0 \in C \text{ and } (\forall n \in \mathbb{N}) \ x_{n+1} = P_C\left(x_n - \frac{\gamma_n}{\|u_n\|}u_n\right), \text{ where } u_n \in \partial f(x_n). \tag{110}$$

This algorithm is readily seen to be a particular implementation of Algorithm 7.1 with $\epsilon_n \equiv 0$, $\lambda_n \equiv 1$, and $\alpha_n = \gamma_n/\|u_n\|$. Moreover, $(\alpha_n\|u_n\|)_{n\geq 0} \in \ell^2$ and, since $(f(x_n))_{n\geq 0}$ lies in $[\bar{f}, +\infty[$, (100) holds. Consequently, Proposition 7.2(i) asserts that any sequence $(x_n)_{n\geq 0}$ conforming to (110) is quasi-Fejér of Type II relative to S. Moreover, if ∂f maps the bounded subsets of C to bounded sets, then $(\alpha_n)_{n\geq 0} \notin \ell^1$ and $\lim \alpha_n\|u_n\|^2 = 0$. As a result, Proposition 7.2(iii) implies that $\underline{\lim}\, f(x_n) = \bar{f}$. Thus, if C is boundedly compact, we can extract a subsequence $(x_{k_n})_{n\geq 0}$ such that $\lim f(x_{k_n}) = \bar{f}$ and which converges strongly to some point $x \in C$. Hence, $x \in \mathfrak{S}(x_n)_{n\geq 0} \cap S$ and it follows from Theorem 3.11 that $x_n \to x$. For $\dim \mathcal{H} < +\infty$ and $C = \mathcal{H}$, this result was established in [26, Prop. 5.1]. One will also find in [2] a convergence analysis of the ϵ-subgradient version of (110).

Remark 7.6 A third method for solving (97) is Polyak's subgradient method [62], which assumes that $\bar{f}$ is known. It proceeds by alternating a relaxed subgradient projection onto $\operatorname{lev}_{\leq \bar{f}} f$ with a projection onto C and can therefore be regarded as a special case of Algorithm 6.1 with two sets.

REFERENCES

1. S. Agmon, The relaxation method for linear inequalities, *Canadian Journal of Mathematics* **6** (1954) 382–392.
2. Ya. I. Alber, A. N. Iusem, and M. V. Solodov, On the projected subgradient method for non-smooth convex optimization in a Hilbert space, *Mathematical Programming* **81** (1998) 23–35.
3. P. Alexandre, V. H. Nguyen, and P. Tossings, The perturbed generalized proximal point algorithm, *RAIRO Modélisation Mathématique et Analyse Numérique* **32** (1998) 223–253.
4. A. Auslender, *Méthodes Numériques pour la Résolution des Problèmes d'Optimisation avec Contraintes* Doctoral thesis (Université de Grenoble, France, 1969).
5. J.-B. Baillon, *Comportement asymptotique des contractions et semi-groupes de contractions* Doctoral thesis (Université Paris 6, France, 1978).
6. J.-B. Baillon and G. Haddad, Quelques propriétés des opérateurs angle-bornés et n-cycliquement monotones, *Israel Journal of Mathematics* **26** (1977) 137–150.
7. H. H. Bauschke, Projection algorithms: Results and open problems, *these proceedings*.
8. H. H. Bauschke and J. M. Borwein, On projection algorithms for solving convex feasibility problems, *SIAM Review* **38** (1996) 367–426.
9. H. H. Bauschke and P. L. Combettes, A weak-to-strong convergence principle for Fejér-monotone methods in Hilbert spaces, *Mathematics of Operations Research* to appear.
10. H. H. Bauschke, F. Deutsch, H. Hundal, and S. H. Park, Fejér monotonicity and weak convergence of an accelerated method of projections, in: M. Théra, ed., *Constructive, Experimental, and Nonlinear Analysis* (Canadian Mathematical Society, 2000) 1–6.

11. L. M. Brègman, The method of successive projection for finding a common point of convex sets, *Soviet Mathematics - Doklady* **6** (1965) 688–692.
12. H. Brézis and P. L. Lions, Produits infinis de résolvantes, *Israel Journal of Mathematics* **29** (1978) 329–345.
13. F. E. Browder, Convergence theorems for sequences of nonlinear operators in Banach spaces, *Mathematische Zeitschrift* **100** (1967) 201–225.
14. R. E. Bruck and S. Reich, Nonexpansive projections and resolvents of accretive operators in Banach spaces, *Houston Journal of Mathematics* **3** (1977) 459–470.
15. D. Butnariu and Y. Censor, Strong convergence of almost simultaneous block-iterative projection methods in Hilbert spaces, *Journal of Computational and Applied Mathematics* **53** (1994) 33–42.
16. D. Butnariu and S. D. Flåm, Strong convergence of expected-projection methods in Hilbert spaces, *Numerical Functional Analysis and Optimization* **16** (1995) 601–636.
17. Y. Censor and A. Lent, Cyclic subgradient projections, *Mathematical Programming* **24** (1982) 233–235.
18. P. L. Combettes, Inconsistent signal feasibility problems: Least-squares solutions in a product space, *IEEE Transactions on Signal Processing* **42** (1994) 2955–2966.
19. P. L. Combettes, Construction d'un point fixe commun à une famille de contractions fermes, *Comptes Rendus de l'Académie des Sciences de Paris* **I320** (1995) 1385–1390.
20. P. L. Combettes, The convex feasibility problem in image recovery, in: P. Hawkes, ed., *Advances in Imaging and Electron Physics* **95** (1996) 155–270.
21. P. L. Combettes, Convex set theoretic image recovery by extrapolated iterations of parallel subgradient projections, *IEEE Transactions on Image Processing* **6** (1997) 493–506.
22. P. L. Combettes, Hilbertian convex feasibility problem: Convergence of projection methods, *Applied Mathematics and Optimization* **35** (1997) 311–330.
23. P. L. Combettes, Strong convergence of block-iterative outer approximation methods for convex optimization, *SIAM Journal on Control and Optimization* **38** (2000) 538–565.
24. P. L. Combettes, Fejér-monotonicity in convex optimization, in: C. A. Floudas and P. M. Pardalos, eds., *Encyclopedia of Optimization* (Kluwer, Boston, MA, 2001).
25. P. L. Combettes and H. Puh, Extrapolated projection method for the euclidean convex feasibility problem, research report (City University of New York, 1993).
26. R. Correa and C. Lemaréchal, Convergence of some algorithms for convex minimization, *Mathematical Programming* **62** (1993) 261–275.
27. A. R. De Pierro and A. N. Iusem, A parallel projection method for finding a common point of a family of convex sets, *Pesquisa Operacional* **5** (1985) 1–20.
28. J. Eckstein and D. P. Bertsekas, On the Douglas-Rachford splitting method and the proximal point algorithm for maximal monotone operators, *Mathematical Programming* **55** (1992) 293–318.
29. I. I. Eremin, The relaxation method of solving systems of inequalities with convex functions on the left sides, *Soviet Mathematics - Doklady* **6** (1965) 219–222.
30. I. I. Eremin, Methods of Fejér approximations in convex programming, *Mathematical Notes of the Academy of Sciences of the USSR* **3** (1968) 139–149.
31. I. I. Eremin and V. D. Mazurov, *Nonstationary Processes of Mathematical Program-*

ming (Nauka, Moscow, 1979).
32. Yu. M. Ermol'ev, On the method of generalized stochastic gradients and quasi-Fejér sequences, *Cybernetics* **5** (1969) 208–220.
33. Yu. M. Ermol'ev, Stochastic quasigradient methods, in: Yu. Ermol'ev and R. J.-B. Wets, eds., *Numerical Techniques for Stochastic Optimization* 141–185 (Springer-Verlag, New York, 1988).
34. Yu. M. Ermol'ev and A. D. Tuniev, Random Fejér and quasi-Fejér sequences, *Theory of Optimal Solutions – Akademiya Nauk Ukrainskoĭ SSR Kiev* **2** (1968) 76–83; translated in: *American Mathematical Society Selected Translations in Mathematical Statistics and Probability* **13** (1973) 143–148.
35. S. D. Flåm, Successive averages of firmly nonexpansive mappings, *Mathematics of Operations Research* **20** (1995) 497–512.
36. U. García-Palomares, Parallel-projected aggregation methods for solving the convex feasibility problem, *SIAM Journal on Optimization* **3** (1993) 882–900.
37. U. M. García-Palomares and F. J. González-Castaño, Incomplete projection algorithms for solving the convex feasibility problem, *Numerical Algorithms* **18** (1998) 177–193.
38. A. Genel and J. Lindenstrauss, An example concerning fixed points, *Israel Journal of Mathematics* **22** (1975) 81–86.
39. K. Goebel and W. A. Kirk, *Topics in Metric Fixed Point Theory* (Cambridge University Press, Cambridge, 1990).
40. C. W. Groetsch, A note on segmenting Mann iterates, *Journal of Mathematical Analysis and Applications* **40** (1972) 369–372.
41. L. G. Gubin, B. T. Polyak, and E. V. Raik, The method of projections for finding the common point of convex sets, *USSR Computational Mathematics and Mathematical Physics* **7** (1967) 1–24.
42. O. Güler, On the convergence of the proximal point algorithm for convex minimization, *SIAM Journal on Control and Optimization* **29** (1991) 403–419.
43. K. C. Kiwiel, Block-iterative surrogate projection methods for convex feasibility problems, *Linear Algebra and Its Applications* **215** (1995) 225–259.
44. K. C. Kiwiel, A projection-proximal bundle method for convex nondifferentiable minimization, in: M. Théra and R. Tichatschke, eds., *Ill-Posed Variational Problems and Regularization Techniques* Lecture Notes in Economics and Mathematical Systems **477**, 137–150 (Springer-Verlag, New York, 1999).
45. K. C. Kiwiel and B. Łopuch, Surrogate projection methods for finding fixed points of firmly nonexpansive mappings, *SIAM Journal on Optimization* **7** (1997) 1084–1102.
46. B. Lemaire, Coupling optimization methods and variational convergence, in: K. H. Hoffmann, J. B. Hiriart-Urruty, C. Lemaréchal, and J. Zowe, eds., *Trends in Mathematical Optimization* Schriftenreihe zur Numerischen Mathematik **84**, 163–179 (Birkhäuser, Boston, MA, 1988).
47. B. Lemaire, Bounded diagonally stationary sequences in convex optimization, *Journal of Convex Analysis* **1** (1994) 75–86.
48. E. S. Levitin and B. T. Polyak, Constrained minimization methods, *USSR Computational Mathematics and Mathematical Physics* **6** (1966) 1–50.
49. B. Martinet, *Algorithmes pour la Résolution de Problèmes d'Optimisation et de Min-*

imax Doctoral thesis (Université de Grenoble, France, 1972).
50. B. Martinet, Détermination approchée d'un point fixe d'une application pseudo-contractante. Cas de l'application prox, *Comptes Rendus de l'Académie des Sciences de Paris* **A274** (1972) 163–165.
51. B. Martinet, Perturbation des méthodes d'optimisation – Applications, *RAIRO Analyse Numérique* **12** (1978) 153–171.
52. Y. I. Merzlyakov, On a relaxation method of solving systems of linear inequalities, *USSR Computational Mathematics and Mathematical Physics* **2** (1963) 504–510.
53. J. J. Moreau, Un cas de convergence des itérées d'une contraction d'un espace hilbertien, *Comptes Rendus de l'Académie des Sciences de Paris* **A286** (1978) 143–144.
54. T. S. Motzkin and I. J. Schoenberg, The relaxation method for linear inequalities, *Canadian Journal of Mathematics* **6** (1954) 393–404.
55. O. Nevanlinna and S. Reich, Strong convergence of contraction semigroups and of iterative methods for accretive operators in Banach spaces, *Israel Journal of Mathematics* **32** (1979) 44–58.
56. N. Ottavy, Strong convergence of projection-like methods in Hilbert spaces, *Journal of Optimization Theory and Applications* **56** (1988) 433-461.
57. W. V. Petryshyn, Construction of fixed points of demicompact mappings in Hilbert space, *Journal of Mathematical Analysis and Applications* **14** (1966) 276–284.
58. W. V. Petryshyn and T. E. Williamson, Strong and weak convergence of the sequence of successive approximations for quasi-nonexpansive mappings, *Journal of Mathematical Analysis and Applications* **43** (1973) 459–497.
59. G. Pierra, Méthodes de projections parallèles extrapolées relatives à une intersection de convexes, research report (INPG, Grenoble, France, 1975).
60. G. Pierra, Decomposition through formalization in a product space, *Mathematical Programming* **28** (1984) 96–115.
61. G. Pierra and N. Ottavy, New parallel projection methods for linear varieties and applications, research report, presented at the XIIth International Symposium on Mathematical Programming, Tokyo, 1988.
62. B. T. Polyak, Minimization of unsmooth functionals, *USSR Computational Mathematics and Mathematical Physics* **9** (1969) 14–29.
63. E. Raik, Fejér type methods in Hilbert space, *Eesti NSV Teaduste Akadeemia Toimetised – Füüsika-Matemaatika* **16** (1967) 286–293.
64. E. Raik, A class of iterative methods with Fejér-monotone sequences, *Eesti NSV Teaduste Akadeemia Toimetised – Füüsika-Matemaatika* **18** (1969) 22–26.
65. S. Reich, Weak convergence theorems for nonexpansive mappings in Banach spaces, *Journal of Mathematical Analysis and Applications* **67** (1979) 274–276.
66. S. Reich, A limit theorem for projections, *Linear and Multilinear Algebra* **13** (1983) 281–290.
67. S. Reich and A. J. Zaslavski, Convergence of generic infinite products of nonexpansive and uniformly continuous operators, *Nonlinear Analysis* **36** (1999) 1049–1065.
68. R. T. Rockafellar, Monotone operators and the proximal point algorithm, *SIAM Journal on Control and Optimization* **14** (1976) 877–898.
69. D. Schott, Weak convergence of iterative methods generated by strongly Fejér monotone mappings, *Rostocker Mathematisches Kolloquium* **51** (1997) 83–96.

70. N. Z. Shor, *Minimization Methods for Non-Differentiable Functions* (Springer-Verlag, New York, 1985).
71. M. V. Solodov and B. F. Svaiter, Forcing strong convergence of proximal point iterations in a Hilbert space, *Mathematical Programming* **87** (2000) 189–202.
72. P. Tseng, On the convergence of products of firmly nonexpansive mappings, *SIAM Journal on Optimization* **2** (1992) 425–434.

Inherently Parallel Algorithms in Feasibility and Optimization and their Applications
D. Butnariu, Y. Censor and S. Reich (Editors)

ON THEORY AND PRACTICE OF ROW RELAXATION METHODS

Achiya Dax[a]

[a]Hydrological Service, P.O.B. 6381, Jerusalem 91063, Israel

This paper presents, analyzes, and tests a row relaxation method for solving the regularized linear programming problem

$$\begin{aligned} &\text{minimize } \ 1/2\|\mathbf{x}\|_2^2 + \alpha \mathbf{c}^T \mathbf{x} \\ &\quad \text{subject to } \ A\mathbf{x} \geq \mathbf{b}, \end{aligned}$$

where α is a given positive constant. The proposed method is based on the observation that the dual of this problem has the form

$$\begin{aligned} &\text{maximize } \ \mathbf{b}^T\mathbf{y} - 1/2\|A^T\mathbf{y} - \alpha\mathbf{c}\|_2^2 \\ &\qquad\quad \text{subject to } \ \mathbf{y} \geq \mathbf{0}, \end{aligned}$$

and if $\mathbf{y}_\alpha$ solves the dual then the point $\mathbf{x}_\alpha = A^T\mathbf{y}_\alpha - \alpha\mathbf{c}$ provides the unique solution of the primal. Maximizing the dual objective function by changing one variable at a time results in an effective row-relaxation method which is suitable for solving large sparse problems. One aim of this paper is to clarify the convergence properties of the proposed scheme. Let $\mathbf{y}_k$ denote the estimated dual solution at the end of the k-the iteration, and let $\mathbf{x}_k = A^T\mathbf{y}_k - \alpha\mathbf{c}$ denote the corresponding primal estimate. It is proved that the sequence $\{\mathbf{x}_k\}$ converges to $\mathbf{x}_\alpha$, while the sequence $\{\mathbf{y}_k\}$ converges to a point $\mathbf{y}_\alpha$ that solves the dual. The only assumption which is needed in order to establish these claims is that the feasible region is not empty.

Yet perhaps the more striking features of the algorithm are related to temporary situations in which it attempts to solve an inconsistent system. In such cases the sequence $\{\mathbf{y}_k\}$ obeys the rule

$$\mathbf{y}_k = \mathbf{u}_k + k\mathbf{v},$$

where $\{\mathbf{u}_k\}$ is a fast converging sequence and $\mathbf{v}$ is a fixed vector that satisfies $A^T\mathbf{v} = \mathbf{0}$ and $\mathbf{b}^T\mathbf{v} > 0$. So the sequence $\{\mathbf{x}_k\}$ is almost unchanged for many iterations. The paper ends with numerical experiments that illustrate the effects of this phenomenon and the close links with Kaczmarz's method.

1. INTRODUCTION

In this paper we present, analyze, and test a row relaxation method for solving the regularized linear programming problem

$$\begin{aligned}&\text{minimize } P(\mathbf{x}) = 1/2\|\mathbf{x}\|_2^2 + \alpha\mathbf{c}^T\mathbf{x}\\&\text{subject to } A\mathbf{x} \geq \mathbf{b},\end{aligned} \tag{1.1}$$

where α is a preassigned positive real number, $\|\cdot\|_2$ denotes the Euclidean norm, A is a real $m \times n$ matrix, $\mathbf{b} = (b_1, \ldots, b_m)^T \in \mathbb{R}^m$, $\mathbf{c} = (c_1, \ldots, c_n)^T \in \mathbb{R}^n$, and $\mathbf{x} = (x_1, \ldots, x_n)^T \in \mathbb{R}^n$ is the vector of unknowns. The rows of A are denoted by $\mathbf{a}_i^T$, $i = 1, \ldots, m$. This way the inequality $A\mathbf{x} \geq \mathbf{b}$ can be rewritten as

$$\mathbf{a}_i^T\mathbf{x} \geq \mathbf{b}_i \qquad i = 1, \ldots, m.$$

The discussion of the inconsistent case, when the system $A\mathbf{x} \geq \mathbf{b}$ has no solution, is deferred until Section 4. In all the other sections it is assumed without saying that the feasible region is not empty. This "feasibility assumption" ensures that (1.1) has a unique solution, $\mathbf{x}_\alpha$. The search for algorithms that solve (1.1) is motivated by three reasons. First, it is this type of problem which is solved at each iteration of the proximal point algorithm for solving the linear programming problem

$$\begin{aligned}&\text{minimize } \mathbf{c}^T\mathbf{x}\\&\text{subject to } A\mathbf{x} \geq \mathbf{b}.\end{aligned} \tag{1.2}$$

See [24, 27, 31, 35, 36, 68, 69, 70, 72] for detailed discussions of the proximal point algorithm. In this connection it is worthwhile mentioning that when solving linear programming problems the proximal point algorithm coincides with the Augmented Lagrangian method [27, 36, 67, 69, 72].

The second motivation for replacing (1.2) with (1.1) lies in the following important observation, which is due to Mangasarian and Meyer [59]. Assume that the closed convex set

$$\mathbb{S} = \{\ \mathbf{x} \mid \mathbf{x} \text{ solves } (1.2)\ \}$$

is not empty. Then there exists a **threshold value**, $\alpha^* > 0$, such that $\mathbf{x}_\alpha = \mathbf{x}_{\alpha^*}$ $\forall\, \alpha \geq \alpha^*$, and $\mathbf{x}_{\alpha^*}$ is the unique solution of the problem

$$\begin{aligned}&\text{minimize } 1/2\|\mathbf{x}\|_2^2\\&\text{subject to } \mathbf{x} \in \mathbb{S}.\end{aligned} \tag{1.3}$$

In other words, if α exceeds α^* then $\mathbf{x}_\alpha$ coincides with the minimum norm solution of (1.2). Unfortunately there is no effective way to obtain an a priori estimate of α^* (see below). Hence in practice (1.1) is repeatedly solved with increasing values of α. See, for example, [54–58, 71].

A further insight into the nature of (1.1) is gained by writing this problem in the form

$$\begin{aligned}&\text{minimize } 1/2\|\mathbf{x} + \alpha\mathbf{c}\|_2^2\\&\text{subject to } A\mathbf{x} \geq \mathbf{b}.\end{aligned} \tag{1.4}$$

This presentation indicates that $\mathbf{x}_\alpha$ is essentially the Euclidean projection of $-\alpha\mathbf{c}$ on the feasible region. Consequently, as α moves from 0 to ∞ the projection point, $\mathbf{x}_\alpha$, moves continuously from $\mathbf{x}_0$ to $\mathbf{x}_{\alpha^*}$. The point $\mathbf{x}_0$ denotes the unique solution of the problem

$$\begin{array}{ll}\text{minimize} & 1/2\|\mathbf{x}\|_2^2 \\ \text{subject to} & A\mathbf{x} \geq \mathbf{b}.\end{array} \tag{1.5}$$

Moreover, as shown in [24], there exists a finite number of **break points**,

$$\beta_0 = 0 < \beta_1 < \beta_2 < \ldots < \beta_t,$$

such that for $j = 0, 1, \ldots, t-1$,

$$\mathbf{x}_\alpha = \mathbf{x}_{\beta_j} + ((\alpha - \beta_j)/(\beta_{j+1} - \beta_j))(\mathbf{x}_{\beta_{j+1}} - \mathbf{x}_{\beta_j}) \qquad \forall\, \alpha \in [\beta_j, \beta_{j+1}],$$

and $\mathbf{x}_\alpha = \mathbf{x}_{\beta_t}$ when $\alpha \geq \beta_t$. In other words, the **proximal path** $\{\mathbf{x}_\alpha \mid \alpha \geq 0\}$ is composed from a finite number of consecutive line segments that connect the points $\mathbf{x}_{\beta_j}$, $j = 0, 1, \ldots, t$. Each line segment lies on a different face of the feasible region, and directed at the projection of $-\mathbf{c}$ on that face. (It is possible to "cross" the feasible region but this may happen only once.) Of course the last break point, β_t, equals the Mangasarian-Meyer threshold value, α^*. If the projection of $-\mathbf{c}$ on the j-th face is $\mathbf{0}$ then the j-th line segment $\{\mathbf{x}_\alpha \mid \beta_j \leq \alpha \leq \beta_{j+1}\}$ turns to be a singleton. In this case $[\beta_j, \beta_{j+1}]$ is called a stationary interval. This geometric interpretation provides a simple explanation for the existence of α^* (see [24] for the details). Yet at the same time it clarifies why it is not possible to derive a realistic a-priori estimate of α^*: It is difficult to anticipate the number of stationary intervals and their size. Also a slight change in the shape of the feasible region (or $\mathbf{c}$) may cause an enormous change in the value of α^*.

A third motivation for replacing (1.2) with (1.1) comes from the following fact. The dual of (1.1) has the form

$$\begin{array}{ll}\text{maximize} & D(\mathbf{y}) = \mathbf{b}^T\mathbf{y} - 1/2\|A^T\mathbf{y} - \alpha\mathbf{c}\|_2^2 \\ \text{subject to} & \mathbf{y} \geq \mathbf{0}\end{array} \tag{1.6}$$

and both problems are solvable. Moreover, let $\mathbf{x}_\alpha \in \mathbb{R}^n$ denote the unique solution of (1.1) and let $\mathbf{y}_\alpha \in \mathbb{R}^m$ be any solution of (1.6). Then these vectors are related by the equalities

$$\mathbf{x}_\alpha = A^T\mathbf{y}_\alpha - \alpha\mathbf{c}, \tag{1.7}$$

$$\mathbf{y}_\alpha^T(A\mathbf{x}_\alpha - \mathbf{b}) = 0, \tag{1.8}$$

and

$$D(\mathbf{y}_\alpha) = P(\mathbf{x}_\alpha). \tag{1.9}$$

Consequently the primal-dual inequality

$$D(\mathbf{y}) \leq P(\mathbf{x}) \tag{1.10}$$

holds whenever $\mathbf{y} \geq 0$ and $A\mathbf{x} \geq \mathbf{b}$. The proof of these facts is easily verified by writing down the Karush-Kuhn-Tucker optimality conditions of the two problems (see [24] for the details). Note that while (1.1) has a unique solution, $\mathbf{x}_\alpha$, the dual problem (1.6) might posses infinitely many solutions. Yet the rule (1.7) enables us to retrieve $\mathbf{x}_\alpha$ from any dual solution.

The simplicity of the dual problem opens the way for a wide range of methods. One class of methods is "active set" algorithms which reach the solution of (1.6) in a finite number of iterations, e.g. [2, 3, 11, 12, 17, 32, 45, 60, 63, 64]. A second class of methods consists of relaxation methods which descend from the classical SOR method or other splitting methods. See, for example, [13, 14, 27, 46, 47, 53, 54, 55, 56]. In this paper we consider a row-relaxation scheme that belongs to the second class.

In practice it is convenient to replace (1.6) with an equivalent minimization problem

$$\begin{array}{ll} \text{minimize} & F(\mathbf{y}) = 1/2\|A^T\mathbf{y} - \alpha\mathbf{c}\|_2^2 - \mathbf{b}^T\mathbf{y} \\ \text{subject to} & \mathbf{y} \geq \mathbf{0}. \end{array} \tag{1.11}$$

Hence the proposed row relaxation scheme is actually aimed at solving this problem. The main idea is very simple: The objective function is minimized by changing one variable at a time. The basic iteration is composed of m steps where the i-th step, $i = 1, \ldots, m$, considers the i-th row of A. Let $\mathbf{y} = (y_1, \ldots, y_m)^T \geq \mathbf{0}$ denote the current estimate of the solution at the beginning of the i-th step, and let $\mathbf{x} = A^T\mathbf{y} - \alpha\mathbf{c}$ denote the corresponding primal solution. Then at the i-th step y_i alone is changed in an attempt to reduce the objective function value and all the other variables are kept fixed. The details of the i-th step are as follows.

a) Calculate $\hat{\theta} = (b_i - \mathbf{a}_i^T\mathbf{x})/\mathbf{a}_i^T\mathbf{a}_i$.

b) Calculate $\delta = \max\{\ -y_i\ ,\ \omega\hat{\theta}\ \}$.

c) Set $y_i := y + \delta$ and $\mathbf{x} := \mathbf{x} + \delta\mathbf{a}_i$.

The value of ω is fixed before the iterative process starts. It is a preassigned relaxation parameter that satisfies $0 < \omega < 2$. The symbol := denotes arithmetic assignment. That is, $y_i := y_i + \delta$ means "set the new value of y_i to be $y_i + \delta$ ". It is assumed for simplicity that $\mathbf{a}_i \neq \mathbf{0}$ for $i = 1, \ldots, m$, so the algorithm is well defined. The algorithm may start with any pair of points that satisfy

$$\mathbf{y} \geq \mathbf{0} \quad \text{and} \quad \mathbf{x} = A^T\mathbf{y} - \alpha\mathbf{c},$$

and these relations are kept throughout the iterative process.

Observe that $\hat{\theta}$ is the unique minimizer of the one parameter quadratic function

$$f(\theta) = F(\mathbf{y} + \theta\mathbf{e}_i) = 1/2\|\theta\mathbf{a}_i + \mathbf{x}\|_2^2 - \theta b_i - \mathbf{b}^T\mathbf{y}, \tag{1.12}$$

where $\mathbf{e}_i$ denotes the i-th column of the $m \times m$ identity matrix. The change in y_i during the i-th step is δ, and this change results in the inequality

$$f(0) - f(\delta) \geq 1/2\delta^2\mathbf{a}_i^T\mathbf{a}_i(2 - \omega)/\omega. \tag{1.13}$$

In other words, a change in y_i always leads to a strict reduction in the objective function value. To verify (1.13) we note that

$$\delta = \nu\hat{\theta}$$

for some nonnegative parameter ν. The value of ν may vary from step to step but it always satisfying

$$0 \leq \nu \leq \omega.$$

The resulting change in the objective function value is, therefore,

$$f(0) - f(\delta) = 1/2\nu(2-\nu)\hat{\theta}^2\mathbf{a}_i^T\mathbf{a}_i = 1/2\delta^2\mathbf{a}_i^T\mathbf{a}_i(2-\nu)/\nu. \tag{1.14}$$

So the inequality (1.13) is a direct consequence of the fact that the function $h(\nu) = (2-\nu)/\nu$ is monotonously decreasing in the interval $(0, \omega]$.

The reason that we use a relaxation parameter, $0 < \omega < 2$, lies in the observation that $\omega\hat{\theta}$ is the change of y_i during the i-th step of the SOR iteration for solving the linear system

$$AA^T\mathbf{y} = \alpha A\mathbf{c} + \mathbf{b}. \tag{1.15}$$

For this reason the proposed algorithm can be viewed as a modified SOR method in which the variables are restricted to stay nonnegative.

The appeal of the scheme a) – c) lies in its simplicity and the small amount of working space that it requires: It needs only two vectors, $\mathbf{x}$ and $\mathbf{y}$, in addition to the data A and $\mathbf{b}$. In fact, the only information that we need during the i-th step are $\mathbf{x}$, y_i, $\mathbf{a}_i$ and b_i. Of course, if A is a sparse matrix then only the nonzero entries of $\mathbf{a}_i$ have to be stored, while the computations of the scalar products $\mathbf{a}_i^T\mathbf{x}$ and $\mathbf{a}_i^T\mathbf{a}_i$ are easily adapted to take advantage of this fact. Furthermore, in some applications there is no need to store the entries of A and $\mathbf{b}$. Instead at the i-th step $\mathbf{a}_i^T\mathbf{x}$ and b_i are computed from their definitions. These features make the algorithm suitable for solving large sparse problems. Algorithms that have such properties are often called "row generation" methods or "row action" methods. We prefer the term "row relaxation" methods because of their links with the SOR method. The reader is referred to [5] and [9] for detailed surveys of row action methods and the special environment that characterize their use.

Let $\mathbf{y}_k$, $k = 1, 2, 3, \ldots$, denotes the current estimate of the dual solution at the end of the k-th iteration, and let $\mathbf{x}_k = A^T\mathbf{y}_k - \alpha\mathbf{c}$ denote the corresponding primal vector. A major aim of this paper is to clarify the convergence properties of the sequences $\{\mathbf{x}_k\}$ and $\{\mathbf{y}_k\}$. The interest that we have in this issue is amplified by the close links between the new scheme and several existing methods (see Sections 6 and 7).

The paper is organized as follows. It starts by showing that the sequence $\{\mathbf{x}_k\}$ converges toward $\mathbf{x}_\alpha$, the unique solution of (1.1). Then, in Section 3, it is proved that the sequence $\{\mathbf{y}_k\}$ also converges. The only assumption which is needed in order to establish these claims is that the feasible region is not empty. The question raised in Section 4 is what happens when this assumption fails to hold. It is conjectured that the sequence $\{\mathbf{x}_k\}$ still converges! The key for understanding the behaviour of the proposed algorithm lies in its close links with Kaczmarz's method. There exists an iteration index, k^*, that has the following property: At the "final stage", when $k \geq k^*$, the behaviour of our algorithm is similar to that of a Kaczmarz's method which is aimed at solving the active equalities at $\mathbf{x}_\alpha$. This resemblance is clearly illustrated in the numerical experiments of Section 5. Yet the more striking features which are exposed in our experiments are related to the following phenomena: In some cases the algorithm spends thousands of iterations before reaching the "final stage". A close inspection of this phenomenon reveals highly surprising features of the algorithm. The explanation of these features lies in the behaviour of Kaczmarz's method when applied to solve an inconsistent system (see below).

We shall finish our introduction with a brief overview of Kaczmarz's method and its main features. This overview is essential for better understanding of our method. Assume for a moment that the system $A\mathbf{x} = \mathbf{b}$ is solvable. Then Kaczmarz's method is aimed at solving the minimum norm problem

$$\begin{aligned} &\text{minimize } \ 1/2\|\mathbf{x}\|_2^2 \\ &\text{subject to } \ A\mathbf{x} = \mathbf{b}. \end{aligned} \tag{1.16}$$

The last problem is a special case of (1.1). Hence its dual takes the form

$$\text{maximize } \ \mathbf{b}^T\mathbf{y} - 1/2\|A^T\mathbf{y}\|_2^2, \tag{1.17}$$

and both problems are solvable. Moreover, let $\hat{\mathbf{y}} \in \mathbb{R}^m$ solve the dual then the point $\hat{\mathbf{x}} = A^T\hat{\mathbf{y}}$ solves the primal. Maximizing the dual objective function by changing one variable at a time results in a row relaxation scheme that resembles our method. The difference is that here the dual variables are not restricted to stay nonnegative. Hence there is no need to store and update the dual variables. Consequently the i-th step of the basic iteration, $i = 1, \dots, m$, is simplified as follows:

a*) Set $\hat{\theta} = (b_i - \mathbf{a}_i^T\mathbf{x})/\mathbf{a}_i^T\mathbf{a}_i$.

b*) Set $\delta = \omega\hat{\theta}$.

c*) Set $\mathbf{x} := \mathbf{x} + \delta\mathbf{a}_i$.

Note that for $\omega = 1$ the point $\mathbf{x} + \delta\mathbf{a}_i$ is the Euclidean projection of $\mathbf{x}$ on the hyperplane $\{\mathbf{u} \mid \mathbf{a}_i^T\mathbf{u} = b_i\}$. This geometric interpretation has motivated the original method of Kaczmarz [44], which concentrates on the case when $m = n$ and $\omega = 1$. Later the method has been found as a useful tool for solving large sparse systems that arise in the field of medical image reconstruction from projections [34, 41]. In this field the method is often known under the name ART (Algebraic Reconstruction Technique). Let $\hat{\mathbf{x}}_k$ denote the current estimate of the solution at the end of the k-th iteration of Kaczmarz's method, $k = 1, 2, \dots$. Tanabe [74] has proved that the sequence $\{\hat{\mathbf{x}}_k\}$ always converges, where "always" means that rank(A) can be smaller than n, and that $A\mathbf{x} = \mathbf{b}$ can be inconsistent. The relation between Kaczmarz's method and the SOR method has been observed by Björck and Elfving [4]: Let the sequence $\{\hat{\mathbf{y}}_k\}$ be generated by the SOR method for solving the linear system

$$AA^T\mathbf{y} = \mathbf{b}. \tag{1.18}$$

If the starting points satisfy $\hat{\mathbf{x}}_0 = A^T\hat{\mathbf{y}}_0$ then the equality

$$\hat{\mathbf{x}}_k = A^T\hat{\mathbf{y}}_k \tag{1.19}$$

holds for all k. If the system $A\mathbf{x} = \mathbf{b}$ is solvable then the same is true for (1.18). In this case the sequence $\{\hat{\mathbf{y}}_k\}$ converges to a point $\hat{\mathbf{y}}$ that solves both (1.18) and (1.17). So the primal sequence, $\hat{\mathbf{x}}_k = A^T\hat{\mathbf{y}}_k$, $k = 1, 2, \ldots$, converges to the point $\hat{\mathbf{x}} = A^T\hat{\mathbf{y}}$, which is the unique solution of (1.16). Moreover, the sequence $\{\|\hat{\mathbf{x}}_k - \hat{\mathbf{x}}\|_2\}$ decreases monotonously.

Perhaps the more surprising properties of Kaczmarz's method are related to the case when the systems $A\mathbf{x} = \mathbf{b}$ and $AA^T\mathbf{y} = \mathbf{b}$ are inconsistent. The reader is referred to Dax [16] for a detailed discussion of this situation. It is shown there that in this case the sequence $\{\hat{\mathbf{y}}_k\}$ satisfies the rule

$$\hat{\mathbf{y}}_k = \hat{\mathbf{u}}_k + k\hat{\mathbf{v}} \tag{1.20}$$

where $\{\hat{\mathbf{u}}_k\}$ is a converging sequence and $\hat{\mathbf{v}} \in \text{Null}(A^T)$ is a fixed vector. Now the equality $\hat{\mathbf{x}}_k = A^T\hat{\mathbf{y}}_k = A^T\hat{\mathbf{u}}_k$ explains why the sequence $\{\hat{\mathbf{x}}_k\}$ converges. Moreover, let $\hat{\mathbf{u}}$ denote the limit point of the sequence $\{\hat{\mathbf{u}}_k\}$ and define $\hat{\mathbf{b}} = AA^T\hat{\mathbf{u}}$. Then the sequence $\{\hat{\mathbf{u}}_k\}$ is obtained by applying the SOR method for solving the system

$$AA^T\mathbf{u} = \hat{\mathbf{b}}. \tag{1.21}$$

This means that the primal sequence $\{\hat{\mathbf{x}}_k\}$ is actually gained by applying Kaczmarz's method for solving the linear system

$$A\mathbf{x} = \hat{\mathbf{b}}. \tag{1.22}$$

So the sequence $\{\hat{\mathbf{x}}_k\}$ converges toward $\hat{\mathbf{x}}$, which is the unique solution of the problem

$$\begin{aligned} &\text{minimize } \ 1/2\|\mathbf{x}\|_2^2 \\ &\text{subject to } \ A\mathbf{x} = \hat{\mathbf{b}}, \end{aligned} \tag{1.23}$$

while the sequence $\{\|\hat{\mathbf{x}}_k - \hat{\mathbf{x}}\|_2\}$ decreases monotonously toward zero. It is important to note that although $\hat{\mathbf{y}}$ depends on the choice of the starting point, $\hat{\mathbf{y}}_0$, the vectors $\hat{\mathbf{v}}$, $\hat{\mathbf{b}}$, and $\hat{\mathbf{x}}$ are not effected by this choice. In fact, these three vectors are completely determined by A, $\mathbf{b}$, and the iteration matrix of the SOR method.

Let $\hat{\mathbf{x}}^{(k,i)}$ denote the current primal solution at the end of the i-th step of the k-th iteration of Kaczmarz's method, $i = 1, \ldots, m$, $k = 1, 2 \ldots$. Starting the iteration from the first row is just a matter of convention. Indeed it is possible to assume that the basic iteration starts from the i-th row. This argument shows that for each row index i the subsequences $\hat{\mathbf{x}}^{(k,i)}$, $k = 1, 2, \ldots$, converges to a point $\hat{\mathbf{x}}^{(i)} \in \mathbb{R}^n$. Of course if the system $A\mathbf{x} = \mathbf{b}$ is solvable then all the m subsequence share the same limit point. Yet when solving an inconsistent system these limit points are not necessarily equal!

2. PRIMAL CONVERGENCE

In this section we prove that the sequence $\{\mathbf{x}_k\}$ converges to $\mathbf{x}_\alpha$, the unique solution of (1.1). We shall start by introducing some useful notations. Let $\mathbf{y}^{(k,i)}$ denote the current estimate of the dual solution at the beginning of the i-th step of the k-th iteration, $i = 1, \ldots, m$, $k = 1, 2, 3, \ldots$. This way

$$\mathbf{y}_k = \mathbf{y}^{(k,m+1)} = \mathbf{y}^{(k+1,1)}$$

and $\mathbf{y}_0 = \mathbf{y}^{(1,1)}$ is an arbitrary starting point that satisfies $\mathbf{y}_0 \geq \mathbf{0}$. The corresponding estimates of the primal solution are denoted as $\mathbf{x}^{(k,i)}$. These vectors satisfy the relations

$$\mathbf{x}^{(k,i)} = A^T\mathbf{y}^{(k,i)} - \alpha\mathbf{c},$$

and

$$\mathbf{x}_k = \mathbf{x}^{(k,m+1)} = \mathbf{x}^{(k+1,1)}.$$

The corresponding values of $F(\mathbf{y})$ are defined as $F_k = F(\mathbf{y}_k)$ and $F^{(k,i)} = F(\mathbf{y}^{(k,i)})$. The optimal value of $F(\mathbf{y})$ is denoted by $F_\alpha = F(\mathbf{y}_\alpha)$ where $\mathbf{y}_\alpha \in \mathbb{R}^m$ is any solution of (1.11). With these notations at hand the inequality (1.13) can be rewritten in the form

$$F^{(k,i)} - F^{(k,i+1)} = 1/2\|\mathbf{y}^{(k,i)} - \mathbf{y}^{(k,i+1)}\|_2^2 \mathbf{a}_i^T\mathbf{a}_i(2-\omega)/\omega. \tag{2.1}$$

Summarizing this inequality over an iteration gives

$$F_{k-1} - F_k \geq \|\mathbf{y}_{k-1} - \mathbf{y}_k\|_2^2\rho^2, \tag{2.2}$$

where ρ is the positive constant

$$\rho = (1/2(2-\omega)\omega^{-1} \min_i \mathbf{a}_i^T\mathbf{a}_i)^{1/2}. \tag{2.3}$$

The sequences $\{F^{(k,i)}\}$ and $\{F_k\}$ are, therefore, monotonously decreasing and bounded from below by F_α. Consequently these sequences converge,

$$\lim_{k\to\infty} \|\mathbf{y}_k - \mathbf{y}_{k+1}\|_2 = 0, \tag{2.4}$$

and the limit

$$\lim_{k\to\infty} \|\mathbf{y}^{(k,i)} - \mathbf{y}^{(k,i+1)}\|_2 = 0 \tag{2.5}$$

holds for any row index i, $i = 1, \ldots, m$.

Theorem 1. *The sequence $\{\mathbf{x}_k\}$ converges to $\mathbf{x}_\alpha$, the unique solution of (1.1).*

Proof. Let $\tilde{\mathbf{x}}$ be any feasible point. Then $F(\mathbf{y})$ can be expressed in the form

$$F(\mathbf{y}) = 1/2\|A^T\mathbf{y} - \alpha\mathbf{c} - \tilde{\mathbf{x}}\|_2^2 + (A\tilde{\mathbf{x}} - \mathbf{b})^T\mathbf{y} - P(\tilde{\mathbf{x}}),$$

while the inequalities $A\tilde{\mathbf{x}} \geq \mathbf{b}$, $\mathbf{y}^{(k,i)} \geq \mathbf{0}$ and $(A\tilde{\mathbf{x}} - \mathbf{b})\mathbf{y}^{(k,i)} \geq 0$ imply that $F(\mathbf{y}^{(k,i)}) \geq 1/2\|A^T\mathbf{y}^{(k,i)} - \alpha\mathbf{c} - \tilde{\mathbf{x}}\|_2^2 - P(\tilde{\mathbf{x}})$.

Hence the decreasing property of the sequence $\{F^{(k,i)}\}$ indicates that the sequences $\{A^T\mathbf{y}^{(k,i)} - \alpha\mathbf{c} - \tilde{\mathbf{x}}\}$ and $\{A^T\mathbf{y}^{(k,i)} - \alpha\mathbf{c}\}$ are bounded. This proves that the sequence $\{\mathbf{x}_k\}$ is bounded and that it has at least one cluster point, $\mathbf{x}^*$.

Our next step is to show that $\mathbf{x}^*$ is a feasible point. Let $\{\mathbf{y}_{k_j}\}$ be a subsequence of $\{\mathbf{y}_k\}$ such that

$$\lim_{j\to\infty} (A^T\mathbf{y}_{k_j} - \alpha\mathbf{c}) = \mathbf{x}^*. \tag{2.6}$$

Assume by contradiction that $\mathbf{a}_i^T\mathbf{x}^* < b_i$ for some row index i. Then the limits

$$\lim_{j\to\infty} \mathbf{e}_i^T(\mathbf{y}^{(k_j,i+1)} - \mathbf{y}^{(k_j,i)}) = \lim_{j\to\infty} \omega(b_i - \mathbf{a}_i^T\mathbf{x}^{(k_j,i)})/\mathbf{a}_i^T\mathbf{a}_i = \omega(b_i - \mathbf{a}_i^T\mathbf{x}^*)/\mathbf{a}_i^T\mathbf{a}_i,$$

and

$$\lim_{j\to\infty} (F^{(k_j,i)} - F^{(k_j,i+1)}) = 1/2\omega(2-\omega)(\mathbf{a}_i^T\mathbf{x}^* - b_i)^2/\mathbf{a}_i^T\mathbf{a}_i,$$

contradict the fact that the sequence $\{F^{(k,i)}\}$ is bounded from below.

A similar argument ensures the existence of an iteration index, k^*, that has the following property: Let i be a row index such that $\mathbf{a}_i^T\mathbf{x}^* > b_i$. Then $k_j \geq k^*$ implies that the i-th component of $\mathbf{y}^{(k_j,i+1)}$ is zero. This property leads to the limit

$$\lim_{j\to\infty} (A\mathbf{x}^* - \mathbf{b})^T\mathbf{y}_{k_j} = 0. \tag{2.7}$$

The proof is concluded by showing that $\mathbf{x}^*$ solves (1.1). Using the Karush-Kuhn-Tucker optimality condition of this problem, it is sufficient to prove the existence of a vector $\mathbf{y}^* \in \mathbb{R}^m$ that satisfies

$$\mathbf{y}^* \geq \mathbf{0}, \quad A^T\mathbf{y}^* = \alpha\mathbf{c} + \mathbf{x}^*, \quad \text{and} \quad (A\mathbf{x}^* - \mathbf{b})^T\mathbf{y}^* = 0. \tag{2.8}$$

For this purpose we consider the bounded variables least squares problem

$$\begin{aligned} &\text{minimize} \ \ 1/2\|E^T\mathbf{y} - \mathbf{h}\|_2^2 \\ &\text{subject to} \ \ \mathbf{y} \geq \mathbf{0}, \end{aligned} \tag{2.9}$$

where

$$E = [A, \mathbf{r}] \in \mathbb{R}^{m\times(n+1)}, \quad \mathbf{r} = A\mathbf{x}^* - \mathbf{b}, \quad \text{and} \quad \mathbf{h} = \begin{pmatrix} \alpha\mathbf{c} + \mathbf{x}^* \\ 0 \end{pmatrix} \in \mathbb{R}^{n+1}.$$

The last problem is always solvable (e.g. [22]) while the limits (2.6) and (2.7) indicate that

$$\lim_{j\to\infty} \|E^T\mathbf{y}_{k_j} - \mathbf{h}\|_2 = 0.$$

Hence any vector $\mathbf{y}^* \in \mathbb{R}^m$ that solves (2.9) must satisfy $E^T\mathbf{y}^* = \mathbf{h}$. That is $\mathbf{y}^*$ solves (2.8). □

Let i be a row index such that $\mathbf{a}_i^T\mathbf{x}_\alpha > b_i$. Then the Euclidean distance from $\mathbf{x}_\alpha$ to the hyperplane $\{\mathbf{x} \mid \mathbf{a}_i^T\mathbf{x} = b_i\}$ is $d_i = (\mathbf{a}_i^T\mathbf{x}_\alpha - b_i)/\|\mathbf{a}_i\|_2$. Since the sequence $\{\mathbf{x}_k\}$ converges to $\mathbf{x}_\alpha$, there exists an iteration index, k_i, such that

$$\|\mathbf{x}^{(k,i)} - \mathbf{x}_\alpha\|_2 \le 1/2 d_i \quad \forall\, k \ge k_i.$$

In other words, once k exceeds k_i then the inequality

$$\mathbf{a}_i^T\mathbf{x}^{(k,i)} - b_i \ge 1/2(\mathbf{a}_i^T\mathbf{x}_\alpha - b_i)$$

always holds. The last inequality means that the attempted change in y_i during the i-th step is negative, while the size of the attempted change exceeds $1/2\omega d_i/\|\mathbf{a}_i\|_2$. Consequently y_i must "hit" its boundary in a finite number of steps. Moreover, once y_i "hits" its boundary it will stay there forever. This observation leads to the following important conclusion.

Corollary 2. *There exists an iteration index k^* that has the following property: If $k \ge k^*$ and i is a row index such that $\mathbf{a}_i^T\mathbf{x}_\alpha > b_i$, then $\mathbf{e}_i^T\mathbf{y}_k = 0$.*

A further insight into the nature of our iteration is gained by applying the identity

$$\begin{aligned} F(\mathbf{y}_k) &= 1/2\|A^T\mathbf{y}_k - \alpha\mathbf{c} - \mathbf{x}_\alpha\|_2^2 + (A\mathbf{x}_\alpha - \mathbf{b})^T\mathbf{y}_k - 1/2\|\mathbf{x}_\alpha\|_2^2 - \alpha\mathbf{c}^T\mathbf{x}_\alpha \\ &= 1/2\|\mathbf{x}_k - \mathbf{x}_\alpha\|_2^2 + (A\mathbf{x}_\alpha - \mathbf{b})^T\mathbf{y}_k - P(\mathbf{x}_\alpha). \end{aligned} \tag{2.10}$$

Now the last corollary can be recasted to show that in the "final stage", when $k \ge k^*$, the distance $\|\mathbf{x}_k - \mathbf{x}_\alpha\|_2$ decreases monotonously to zero.

Corollary 3. *Once k exceeds k^* then the following relations hold.*

$$(A\mathbf{x}_\alpha - \mathbf{b})^T\mathbf{y}_k = 0, \tag{2.11}$$
$$F(\mathbf{y}_k) = 1/2\|\mathbf{x}_k - \mathbf{x}_\alpha\|_2^2 - P(\mathbf{x}_\alpha), \tag{2.12}$$
$$F(\mathbf{y}_k) - F(\mathbf{y}_\alpha) = 1/2\|\mathbf{x}_k - \mathbf{x}_\alpha\|_2^2, \tag{2.13}$$

and

$$\|\mathbf{x}_k - \mathbf{x}_\alpha\|_2 \ge \|\mathbf{x}_{k+1} - \mathbf{x}_\alpha\|_2. \tag{2.14}$$

We shall finish this section with a preliminary discussion of the question whether the sequence $\{\mathbf{y}_k\}$ converges. Observe that (2.13) implies the limit

$$\lim_{k\to\infty} F(\mathbf{y}_k) = F_\alpha. \tag{2.15}$$

Therefore any cluster point of the sequence $\{\mathbf{y}_k\}$, if it exists, solves (1.1). The existence of a cluster point is ensured whenever the sequence $\{\mathbf{y}_k\}$ is bounded. One condition that ensures this property is the **Slater constraint qualification**. This condition requires the existence of a point $\hat{\mathbf{x}} \in \mathbb{R}^n$ such that $A\hat{\mathbf{x}} > \mathbf{b}$. Recall that the strict inequality $A\hat{\mathbf{x}} > \mathbf{b}$ implies the existence of a positive constant $\nu > 0$ such that $A\hat{\mathbf{x}} - \mathbf{b} \ge \nu\mathbf{e}$ where $\mathbf{e} = (1, 1, \ldots, 1)^T \in \mathbb{R}^m$. In this case the identity

$$F(\mathbf{y}) = 1/2\|A^T\mathbf{y} - \alpha\mathbf{c} - \hat{\mathbf{x}}\|_2^2 + (A\hat{\mathbf{x}} - \mathbf{b})^T\mathbf{y} - P(\hat{\mathbf{x}})$$

results in the inequalities

$$P(\hat{\mathbf{x}}) + F(\mathbf{y}_k) \geq (A\hat{\mathbf{x}} - \mathbf{b})^T \mathbf{y}_k \geq \nu \|\mathbf{y}_k\|_1,$$

so the decreasing property of the sequence $\{F_k\}$ indicates that the sequence $\{\mathbf{y}_k\}$ is bounded.

Another simplifying assumption is that the rows of A are linearly independent. In this case $F(\mathbf{y})$ is a strictly convex function, so (1.11) has a unique minimizer. The strict convexity of $F(\mathbf{y})$ means that the level set $\{\mathbf{y} \mid F(\mathbf{y}) \leq F(\mathbf{y}_0)\}$ is bounded. Therefore $\{\mathbf{y}_k\}$ is a bounded sequence that has at least one cluster point. On the other hand any cluster point of this sequence solves (1.11). Hence the fact that (1.11) has a unique solution implies that all the sequence converges to this point. Moreover, using Corollary 2 we see that actually only the rows $\{\mathbf{a}_i \mid \mathbf{a}_i^T \mathbf{x}_\alpha = b_i\}$ need to be linearly independent. However, when solving practical problems neither of the above conditions is ensured to exist.

3. DUAL CONVERGENCE

In this section we prove that the sequence $\{\mathbf{y}_k\}$ converges. Let $\mathbb{I}$ be a subset of $\mathcal{M} = \{1, 2, \ldots, m\}$. Let $\tilde{\mathbb{I}}$ denote the complement set of $\mathbb{I}$ with respect to $\mathcal{M}$. That is $\tilde{\mathbb{I}} = \{i \mid i \in \mathcal{M} \text{ and } i \notin \mathbb{I}\}$. Then $\mathbb{I}$ is said to have the **infinity property** if the sequence $\{\mathbf{y}_k\}$ includes an infinite number of points $\mathbf{y}_k = (y_1, \ldots, y_m)^T$ whose components obey the rule:

$$y_i > 0 \text{ when } i \in \mathbb{I} \text{ and } y_i = 0 \text{ when } i \in \tilde{\mathbb{I}}.$$

Since $\mathcal{M}$ has a finite number of subsets, the number of subsets that have the infinity property is also finite. Let $\mathbb{I}_1, \mathbb{I}_2, \ldots, \mathbb{I}_{\ell^*}$ denote the subsets of $\mathcal{M}$ that have the infinity property. Let $\tilde{\mathbb{I}}_\ell$, $\ell = 1, 2, \ldots, \ell^*$, denote the complement of $\mathbb{I}_\ell$ with respect to $\mathcal{M}$. Then for every index ℓ, $\ell = 1, 2, \ldots, \ell^*$, we define $\mathbb{Y}_\ell$ to be the set of all the points $\mathbf{y}_k = (y_1, \ldots, y_m)^T$ that satisfy the rule:

$$y_i > 0 \text{ when } i \in \mathbb{I}_\ell \quad \text{and} \quad y_i = 0 \text{ when } i \in \tilde{\mathbb{I}}_\ell.$$

This definition implies that $\mathbb{Y}_\ell$ is composed of an infinite number of points from the sequence $\{\mathbf{y}_k\}$. Note also that only a finite number of points from the sequence $\{\mathbf{y}_k\}$ do not belong to $\bigcup_{\ell=1}^{\ell^*} \mathbb{Y}_\ell$. Now Corollary 3 can be strengthened as follows.

Corollary 4. *There exists an iteration index k^* that has the following properties:*

a) *If $k \geq k^*$ then $\mathbf{y}_k \in \mathbb{Y}_\ell$ for some index ℓ, $\ell \in \{1, 2, \ldots, \ell^*\}$.*

b) *If $k \geq k^*$ then $\mathbf{y}_k$ satisfies (2.11)–(2.14).*

Corollary 5. *For each index ℓ, $\ell = 1, 2, \ldots, \ell^*$, there exists a point $\mathbf{z}_\ell = (z_1, \ldots, z_m)^T \in \mathbb{R}^m$ that solves (1.11) and satisfies the rule:*

$$z_i \geq 0 \text{ when } i \in \mathbb{I}_\ell \quad \text{and} \quad z_i = 0 \text{ when } i \in \tilde{\mathbb{I}}_\ell.$$

The last corollary is easily verified by following the proof that (2.8) is solvable, using $\mathbf{z}_\ell$ instead of $\mathbf{y}^*$ and assuming that the subsequence $\{\mathbf{y}_{k_j}\}$ belongs to $\mathbb{Y}_\ell$. Also from Corollary 2 we conclude that

$$\mathbf{a}_i^T \mathbf{x}_\alpha = b_i \quad \forall \quad i \in \mathbb{I}_\ell.$$

These equalities can be rewritten as

$$A_\ell \mathbf{x}_\alpha = \mathbf{b}_\ell,$$

where the matrix A_ℓ is obtained from A by deleting the rows whose indices belong to $\tilde{\mathbb{I}}_\ell$. Similarly the vector $\mathbf{b}_\ell$ is obtained from $\mathbf{b}$ by deleting all the components of $\mathbf{b}$ which correspond to indices $i \in \tilde{\mathbb{I}}_\ell$. The next lemma uses the equality $A_\ell \mathbf{x}_\alpha = \mathbf{b}_\ell$ to estimate the distance between $\mathbf{x}_k$ and $\mathbf{x}_\alpha$. For this purpose we need bounds on the singular values of A_ℓ. Let η_ℓ denote the largest singular value of A_ℓ and define $\eta = \max_{i=1,\ldots,\ell^*} \eta_\ell$. Similarly we let σ_ℓ denote the smallest singular value of A_ℓ which differs from zero, and define $\sigma = \min_{\ell=1,\ldots,\ell^*} \sigma_\ell$.

Lemma 6. *Let k be an iteration index such that $k \geq k^*$. Then $\mathbf{y}_k \in \mathbb{Y}_\ell$ for some index ℓ, $\ell \in \{1, 2, \ldots, \ell^*\}$, and*

$$||A_\ell \mathbf{x}_k - \mathbf{b}_\ell||_2/\eta \leq ||\mathbf{x}_k - \mathbf{x}_\alpha||_2 \leq ||A_\ell \mathbf{x}_k - \mathbf{b}_\ell||_2/\sigma. \tag{3.1}$$

Proof. Since $\mathbf{z}_\ell$ solves (1.6) the primal solution satisfies $\mathbf{x}_\alpha = A^T \mathbf{z}_\ell - \alpha \mathbf{c}$ and

$$\mathbf{x}_k - \mathbf{x}_\alpha = (A^T \mathbf{y}_k - \alpha \mathbf{c}) - (A^T \mathbf{z}_\ell - \alpha \mathbf{c}) = A^T \mathbf{y}_k - A^T \mathbf{z}_\ell \in \text{Range}(A_\ell^T).$$

The last observation implies the bounds

$$\sigma_\ell ||\mathbf{x}_k - \mathbf{x}_\alpha||_2 \leq ||A_\ell(\mathbf{x}_k - \mathbf{x}_\alpha)||_2 \leq \eta_\ell ||\mathbf{x}_k - \mathbf{x}_\alpha||_2,$$

while the equality $A_\ell \mathbf{x}_\alpha = \mathbf{b}_\ell$ leads to

$$||A_\ell \mathbf{x}_k - \mathbf{b}_\ell||_2/\eta_\ell \leq ||\mathbf{x}_k - \mathbf{x}_\alpha||_2 \leq ||A_\ell \mathbf{x}_k - \mathbf{b}_\ell||_2/\sigma_\ell.$$

□

In the next proof we apply the positive constants

$$\mu = ||A||_F = \left(\sum_{i=1}^{m} ||\mathbf{a}_i^T||_2^2 \right)^{1/2}$$

and $\beta = 1 + |1 - \omega|/\omega$, where $0 < \omega < 2$ is our relaxation parameter.

Theorem 7. *There exists a positive constant $0 < \lambda < 1$ such that*

$$F_{k+1} - F_\alpha \leq \lambda(F_k - F_\alpha) \qquad \forall\, k \geq k^*. \tag{3.2}$$

Proof. Let k be an iteration index such that $k \geq k^*$. Then $\mathbf{y}_k \in \mathbb{Y}_\ell$ for some index ℓ, $\ell \in \{1, \ldots, \ell^*\}$. Let $\gamma > 0$ be a positive constant that satisfies both $\gamma < 1/(\rho\sigma)$ and $\gamma < (1/2\sigma)/(\eta\beta\mu\|A^T\|_2)$, where ρ is defined by (2.3). Now there are two cases to consider. The first part of the proof concentrates on the case when

$$\|\mathbf{y}_{k+1} - \mathbf{y}_k\|_2 \geq \gamma\|A_\ell\mathbf{x}_k - \mathbf{b}_\ell\|_2. \tag{3.3}$$

In this case Lemma 6 followed by (3.3) and (2.2) gives

$$\begin{aligned} F_k - F_\alpha &= 1/2\|\mathbf{x}_k - \mathbf{x}_\alpha\|_2^2 \leq 1/2\|A_\ell\mathbf{x}_k - \mathbf{b}_\ell\|_2^2/\sigma^2 \leq \\ &\leq 1/2\|\mathbf{y}_{k+1} - \mathbf{y}_k\|_2^2/(\sigma\gamma)^2 \leq (F_k - F_{k+1})/(\rho\sigma\gamma)^2. \end{aligned}$$

To simplify our notations we introduce the constant

$$\tau = 1/(\rho\sigma\gamma)^2.$$

This results in the inequalities

$$\begin{aligned} &F_k - F_\alpha \leq \tau(F_k - F_{k+1}), \\ &F_k - F_\alpha \leq \tau(F_k - F_\alpha) - \tau(F_{k+1} - F_\alpha), \end{aligned}$$

and

$$F_{k+1} - F_\alpha \leq [(\tau - 1)/\tau](F_k - F_\alpha). \tag{3.4}$$

On the other hand the definitions of τ and γ ensures that $\tau > 1$, so $(\tau - 1)/\tau$ is a positive constant that satisfies $0 < (\tau - 1)/\tau < 1$.

The second part of the proof considers the case when

$$\|\mathbf{y}_{k+1} - \mathbf{y}_k\|_2 < \gamma\|A_\ell\mathbf{x}_k - \mathbf{b}_\ell\|_2. \tag{3.5}$$

Since $k \geq k^*$ there exists an index p, $p \in \{1, \ldots, \ell^*\}$, such that $\mathbf{y}_{k+1} \in \mathbb{Y}_p$. In this case the proof is carried out by showing that (3.5) leads to the inequality

$$\|A_p\mathbf{x}_{k+1} - \mathbf{b}_p\|_2 \leq \mu\beta\gamma\|A^T\|_2\|A_\ell\mathbf{x}_k - \mathbf{b}_\ell\|_2. \tag{3.6}$$

Then, when (3.6) is verified, on one hand we have

$$\begin{aligned} F_{k+1} - F_\alpha &= 1/2\|\mathbf{x}_{k+1} - \mathbf{x}_\alpha\|_2^2 \leq 1/2\|A_p\mathbf{x}_{k+1} - \mathbf{b}_p\|_2^2/\sigma^2 \leq \\ &\leq 1/2(\mu\beta\gamma\|A^T\|_2/\sigma)^2\|A_\ell\mathbf{x}_k - \mathbf{b}_\ell\|_2^2; \end{aligned}$$

while on the other hand

$$F_k - F_\alpha = 1/2\|\mathbf{x}_k - \mathbf{x}_\alpha\|_2^2 \geq 1/2\|A_\ell\mathbf{x}_k - \mathbf{b}_\ell\|_2^2/\eta^2.$$

Hence combining these inequalities gives

$$F_{k+1} - F_\alpha \leq (\mu\beta\gamma\eta\|A^T\|_2/\sigma)^2(F_k - F_\alpha) < 1/4(F_k - F_\alpha), \tag{3.7}$$

where the last inequality comes form the definition of γ.

It is left to prove that (3.6) indeed holds. To clarify the main idea behind this claim we first concentrates on the simpler case when $\omega = 1$. Recall that all the components of the vector $A_p\mathbf{x}_{k+1} - \mathbf{b}_p$ correspond to indices $i \in \mathbb{I}_p$. Yet for each index $i \in \mathbb{I}_p$ the change in $\mathbf{x}$ during the i-th step of the k-th iteration results in the equality $\mathbf{a}_i^T\mathbf{x}^{(k,i+1)} = b_i$; while in the rest of the iteration $\mathbf{x}$ makes a restricted movement. More precisely,

$$\begin{aligned}|\mathbf{a}_i^T\mathbf{x}_{k+1} - b_i| &= |\mathbf{a}_i^T(\mathbf{x}_{k+1} - \mathbf{x}^{(k,i+1)}) + \mathbf{a}_i^T\mathbf{x}^{(k,i+1)} - b_i| \\ &= |\mathbf{a}_i^T(\mathbf{x}_{k+1} - \mathbf{x}^{(k,i+1)})| \leq \|\mathbf{a}_i\|_2\|\mathbf{x}_{k+1} - \mathbf{x}^{(k,i+1)}\|_2 \\ &= \|\mathbf{a}_i\|_2\|A^T\mathbf{y}_{k+1} - A^T\mathbf{y}^{(k,i+1)}\|_2 \\ &\leq \|\mathbf{a}_i\|_2\|A^T\|_2\|\mathbf{y}_{k+1} - \mathbf{y}^{(k,i+1)}\|_2 \\ &\leq \|\mathbf{a}_i\|_2\|A^T\|_2\|\mathbf{y}_{k+1} - \mathbf{y}_k\|_2 \\ &< \|\mathbf{a}_i\|_2\|A^T\|_2\gamma\|A_\ell\mathbf{x}_k - \mathbf{b}_\ell\|_2,\end{aligned}$$

were the last inequality is due to (3.5). Now (3.6) is a direct consequence of the resulting bound on $|\mathbf{a}_i^T\mathbf{x}_{k+1} - b_i|$.

The treatment of the more general case, when $\omega \neq 1$, is done in a similar manner. Here for each index $i \in \mathbb{I}_p$ the point

$$\tilde{\mathbf{x}}^{(k,i+1)} = \mathbf{x}^{(k,i+1)} + ((1-\omega)/\omega)\mathbf{e}_i^T(\mathbf{y}^{(k,i+1)} - \mathbf{y}^{(k,i)})\mathbf{a}_i$$

satisfies the equality

$$\mathbf{a}_i^T\tilde{\mathbf{x}}^{(k,i+1)} = b_i.$$

This results in the relations

$$\begin{aligned}|\mathbf{a}_i^T\mathbf{x}_{k+1} - b_i| &= |\mathbf{a}_i^T(\mathbf{x}_{k+1} - \mathbf{x}^{(k,i+1)}) + (\mathbf{a}_i^T\tilde{\mathbf{x}}^{(k,i+1)} - b_i) - \\ &\quad - (\mathbf{a}_i^T\mathbf{a}_i(1-\omega)/\omega)\mathbf{e}_i^T(\mathbf{y}^{(k,i+1)} - \mathbf{y}^{(k,i)})| \\ &\leq |\mathbf{a}_i^T(\mathbf{x}_{k+1} - \mathbf{x}^{(k,i+1)})| + |\mathbf{a}_i^T\mathbf{a}_i(1-\omega)/\omega| \cdot |\mathbf{e}_i^T(\mathbf{y}^{(k,i+1)} - \mathbf{y}^{(k,i)})| \\ &\leq \|\mathbf{a}_i\|_2\|A^T\|_2\|\mathbf{y}^{(k+1)} - \mathbf{y}^{(k)}\|_2 + \|\mathbf{a}_i\|_2^2|(1-\omega)/\omega| \cdot \|\mathbf{y}^{(k+1)} - \mathbf{y}^{(k)}\|_2 \\ &\leq \|\mathbf{a}_i\|_2\|A^T\|_2(1 + |1-\omega|/\omega)\|\mathbf{y}^{(k+1)} - \mathbf{y}^{(k)}\|_2 \\ &= \|\mathbf{a}_i\|_2\|A^T\|_2\beta\|\mathbf{y}^{(k+1)} - \mathbf{y}^{(k)}\|_2 \\ &< \|\mathbf{a}_i\|_2\|A^T\|_2\beta\gamma\|A_\ell\mathbf{x}_k - \mathbf{b}_\ell\|_2,\end{aligned}$$

which confirm the validity of (3.6).

Summarizing our proof we see that either (3.4) or (3.7) holds. Hence in both cases we have $F_{k+1} - F_\alpha \leq \lambda(F_k - F_\alpha)$, where $\lambda = \max\{\ 1/4,\ \ (\tau-1)/\tau\ \}$. □

Theorem 8. *The sequence $\{\mathbf{y}_k\}$ is a Cauchy sequence. Therefore it converges to a point $\mathbf{y}_\alpha$ that solves (1.6). Moreover, define*

$$\xi = \lambda^{1/2} \qquad \textit{and} \qquad \pi = (F_{k^*} - F_\alpha)^{1/2}/\rho.$$

Then $0 < \xi < 1$ and the bound

$$\|\mathbf{y}_{k^*+p} - \mathbf{y}_\alpha\|_2 \leq \pi\xi^p/(1-\xi) \tag{3.8}$$

holds for any positive integer p.

Proof. An alternative way to write (3.2) is

$$F_{k^*+p} - F_\alpha \le \lambda(F_{k^*+p-1} - F_\alpha)$$

where p is an arbitrary positive integer. Using induction on p we obtain that

$$F_{k^*+p} - F_\alpha \le \lambda^p(F_{k^*} - F_\alpha),$$
$$F_{k^*+p} - F_{k^*+p+1} \le F_{k^*+p} - F_\alpha \le \lambda^p(F_{k^*} - F_\alpha),$$

and

$$(F_{k^*+p} - F_{k^*+p+1})^{1/2} \le (\lambda^{1/2})^p(F_{k^*} - F_\alpha)^{1/2}.$$

On the other hand (2.2) can be rewritten as

$$\|\mathbf{y}_{k^*+p} - \mathbf{y}_{k^*+p+1}\|_2 \le (F_{k^*+p} - F_{k^*+p+1})^{1/2}/\rho.$$

Hence by combining the above relations we deduce that

$$\|\mathbf{y}_{k^*+p} - \mathbf{y}_{k^*+p+1}\|_2 \le \xi^p\pi. \tag{3.9}$$

Now for any positive integer q we have the relations

$$\|\mathbf{y}_{k^*+p} - \mathbf{y}_{k^*+p+q}\|_2 = \left\|\sum_{j=1}^{q}(\mathbf{y}_{k^*+p+j-1} - \mathbf{y}_{k^*+p+j})\right\|_2$$
$$\le \sum_{j=1}^{q} \|\mathbf{y}_{k^*+p+j-1} - \mathbf{y}_{k^*+p+j}\|_2$$
$$\le \sum_{j=1}^{q} \xi^{p+j-1}\pi = \pi\xi^p \sum_{j=1}^{q} \xi^{j-1} \le \pi\xi^p \sum_{j=1}^{\infty} \xi^{j-1} = \pi\xi^p/(1-\xi),$$

or

$$\|\mathbf{y}_{k^*+p} - \mathbf{y}_{k^*+p+q}\|_2 \le \pi\xi^p/(1-\xi). \tag{3.10}$$

The last inequality means that $\{\mathbf{y}_k\}$ is a Cauchy sequence. Consequently this sequence converges and the limit point, $\mathbf{y}_\alpha$, solves (1.6). A further use of (3.10) gives

$$\begin{aligned}\|\mathbf{y}_{k^*+p} - \mathbf{y}_\alpha\|_2 &= \|\mathbf{y}_{k^*+p} - \mathbf{y}_{k^*+p+q} + \mathbf{y}_{k^*+p+q} - \mathbf{y}_\alpha\|_2 \\ &\le \|\mathbf{y}_{k^*+p} - \mathbf{y}_{k^*+p+q}\|_2 + \|\mathbf{y}_{k^*+p+q} - \mathbf{y}_\alpha\|_2 \\ &\le \pi\xi^p/(1-\xi) + \|\mathbf{y}_{k^*+p+q} - \mathbf{y}_\alpha\|_2.\end{aligned}$$

Thus by letting q to approach infinity we conclude that (3.8) holds. □

The inequality (3.8) indicates that at the final stage, when $k \ge k^*$, the sequence $\{\mathbf{y}_k\}$ converges at a linear (geometric) rate. Yet the actual rate of convergence at the final stage depends on the properties of the active constraint matrix at $\mathbf{x}_\alpha$. Assume for simplicity that the limit point $\mathbf{x}_\alpha$ satisfies

$$\mathbf{a}_i^T\mathbf{x}_\alpha = b_i \quad \text{for} \quad i = 1, \ldots, \ell, \quad \text{and} \quad \mathbf{a}_i^T\mathbf{x}_\alpha > b_i \quad \text{for} \quad i = \ell + 1, \ldots, m, \tag{3.11}$$

for some row index ℓ. Let $\tilde{A} \in \mathbb{R}^{\ell \times n}$ denote the corresponding active constraints matrix which is composed of the first ℓ rows of A. Similarly we use $\tilde{\mathbf{b}}$ and $\tilde{\mathbf{y}}_\alpha$ to denote the ℓ-vectors which are composed of the first ℓ components of $\mathbf{b}$ and $\mathbf{y}_\alpha$. Then as k exceeds k^* the last $m - \ell$ components of $\mathbf{y}_k$ are always zeros. Hence the behaviour of our method is similar to that of the SOR method for solving the linear system

$$\tilde{A}\tilde{A}^T\mathbf{s} = \tilde{\mathbf{b}} + \alpha\tilde{A}\mathbf{c}, \tag{3.12}$$

where here $\mathbf{s} \in \mathbb{R}^\ell$ denote the vector of unknowns. Of course the two methods are not necessarily identical (otherwise there was no need in this section). Nevertheless, if the truncated limit point $\tilde{\mathbf{y}}_\alpha$ has a small number of zero components then it is reasonable to expect that the two methods will exhibit a similar rate of convergence. Moreover, as we have seen, there are close links between the SOR method for solving (3.12) and Kaczmarz's method for solving the system

$$\tilde{A}\mathbf{x} = \tilde{\mathbf{b}} + \alpha\tilde{A}\mathbf{c}. \tag{3.13}$$

Hence when $k \geq k^*$ the primal sequence is expected to converge at about the same speed as Kaczmarz's method for solving (3.13). Indeed the experiments of Tables 1 and 2 clearly illustrate this point.

It should be noted however, that the "final" rate of convergence is not necessarily the major factor that determines the overall number of iterations. In other words, there is no guarantee for k^* to be small. Thus, as Tables 4–6 show, in some cases most of the iterations are spent before the "final" active set is reached.

4. THE INCONSISTENT CASE

Here, and only in this section, it is assumed that the system $A\mathbf{x} \geq \mathbf{b}$ is inconsistent. In other words, there is no $\mathbf{x} \in \mathbb{R}^n$ such that $A\mathbf{x} \geq \mathbf{b}$. We shall start with a brief overview of the basic facts that characterize such a situation. Let

$$\mathbb{U} = \{\, \mathbf{u} \mid \mathbf{u} = A\mathbf{x} - \mathbf{z}, \quad \mathbf{x} \in \mathbb{R}^n, \quad \mathbf{z} \in \mathbb{R}^m, \quad \text{and} \quad \mathbf{z} \geq \mathbf{0} \,\}$$

denote the set of all points $\mathbf{u} \in \mathbb{R}^m$ for which the system $A\mathbf{x} \geq \mathbf{u}$ is solvable. An alternative way to write $\mathbb{U}$ is

$$\mathbb{U} = \{\mathbf{u} \mid \mathbf{u} = H\mathbf{h}, \quad \mathbf{h} \geq \mathbf{0}\},$$

where

$$H = [A, -A, -I] \in \mathbb{R}^{m\times(2n+m)} \quad \text{and} \quad \mathbf{h} \in \mathbb{R}^{2n+m}.$$

This presentation shows that $\mathbb{U}$ is a closed convex cone [10, 22, 64]. Let $\mathbb{V}$ denote the polar cone of $\mathbb{U}$. That is

$$\mathbb{V} = \{\, \mathbf{v} \mid \mathbf{v}^T\mathbf{u} \leq 0 \quad \forall \mathbf{u} \in \mathbb{U} \,\}.$$

Then, clearly, $\mathbb{V}$ is a closed convex cone. Moreover, as observed in [25], $\mathbb{V}$ can be rewritten in the form

$$\mathbb{V} = \{\mathbf{v} \mid A^T\mathbf{v} = \mathbf{0} \text{ and } \mathbf{v} \geq \mathbf{0}\}.$$

Since $\mathbb{U}$ and $\mathbb{V}$ are **mutually polar cones**, any vector $\mathbf{b} \in \mathbb{R}^m$ has a unique **polar decomposition** of the form

$$\mathbf{b} = \hat{\mathbf{u}} + \hat{\mathbf{v}}, \quad \hat{\mathbf{u}} \in \mathbb{U}, \quad \hat{\mathbf{v}} \in \mathbb{V}, \quad \text{and } \hat{\mathbf{u}}^T\hat{\mathbf{v}} = 0. \tag{4.1}$$

The existence and the uniqueness of the polar decomposition are due to Moreau [61], who established this decomposition for any pair of mutually polar cones in a general Hilbert space. For further discussion of polar cones and their properties see, for example, [1, 51, 61, 64, 66, 73].

The polar decomposition (4.1) indicates that the system $A\mathbf{x} \geq \mathbf{b}$ is solvable if and only if $\mathbf{b} \in \mathbb{U}$. Otherwise, when this system is inconsistent, $\hat{\mathbf{v}} \neq \mathbf{0}$ and

$$\mathbf{b}^T\hat{\mathbf{v}} = (\hat{\mathbf{u}} + \hat{\mathbf{v}})^T\hat{\mathbf{v}} = \hat{\mathbf{v}}^T\hat{\mathbf{v}} > 0.$$

In other words, either the system $A\mathbf{x} \geq \mathbf{b}$ has a solution $\mathbf{x} \in \mathbb{R}^n$, or the system

$$A^T\mathbf{v} = \mathbf{0}, \quad \mathbf{v} \geq \mathbf{0}, \quad \text{and} \quad \mathbf{b}^T\mathbf{v} > 0,$$

has a solution $\mathbf{v} \in \mathbb{R}^m$, but never both. The last statement is Gale's theorem of the alternative (e.g. [20, 26, 33, 52]). Note also that the objective function of (1.11) satisfies $F(\theta\hat{\mathbf{v}}) = F(\mathbf{0}) - \theta\hat{\mathbf{v}}^T\hat{\mathbf{v}}$. This equality indicates that the system $A\mathbf{x} \geq \mathbf{b}$ is solvable if and only if $F(\mathbf{y})$ is bounded from below on $\mathbb{R}^m_+ = \{\mathbf{y} \mid \mathbf{y} \in \mathbb{R}^m \text{ and } \mathbf{y} \geq \mathbf{0}\}$.

Let us turn now to examine the behaviour of the algorithm when the system $A\mathbf{x} \geq \mathbf{b}$ happens to be inconsistent. In order to have a close look at the sequence $\{\mathbf{y}_k\}$ we make the simplifying assumption that only one subset of $\mathcal{M}$ has the "infinity" property. Let $\mathbb{I}$ denote that subset of $\mathcal{M}$. Then our assumption implies the existence of an iteration index k^* that has the following property: If $k \geq k^*$ then the components of $\mathbf{y}_k$ satisfy

$$\mathbf{e}_i^T\mathbf{y}_k > 0 \text{ when } i \in \mathbb{I} \text{ and } \mathbf{e}_i^T\mathbf{y}_k = 0 \text{ when } i \notin \mathbb{I}, \tag{4.2}$$

where, as before, $\mathbf{e}_i$ denotes the i-th column of the $m \times m$ identity matrix. Recall that the i-th dual variable can be changed only during the i-th step of the basic iteration. Therefore for each row index i, $i = 1, \ldots, m$, the subsequence $\mathbf{y}^{(k,i)}$, $k = k^* + 1, k^* + 2, \ldots$, is also satisfying the above condition. In other words, once k exceeds k^* then for $i \in \mathbb{I}$ the i-th dual variable never "hits" its bound, while for $i \notin \mathbb{I}$ the i-th dual variable is always zero. Also there is no loss of generality in assuming that $\mathbb{I} = \{1, 2 \ldots, \tilde{\ell}\}$ for some row index $\tilde{\ell}$. (This assumption is not essential, but it helps to simplify our notations.) Now (4.2) is rephrased as follows: If $k \geq k^*$ then the components of $\mathbf{y}_k = (y_1, \ldots, y_m)^T$ satisfy

$$y_i > 0 \text{ for } i = 1, \ldots, \tilde{\ell}, \quad \text{and} \quad y_i = 0 \text{ for } i = \tilde{\ell} + 1, \ldots, m. \tag{4.3}$$

Let $\tilde{A}$ denote the $\tilde{\ell} \times n$ matrix which is composed of the first $\tilde{\ell}$ rows of A. Let $\tilde{\mathbf{b}}$ and $\tilde{\mathbf{y}}_k$ denote the corresponding $\tilde{\ell}$-vectors which are composed of the first $\tilde{\ell}$ components of

$\mathbf{b}$ and $\mathbf{y}_k$, respectively. Then for $k \geq k^*$ the point $\tilde{\mathbf{y}}_{k+1}$ is obtained from $\tilde{\mathbf{y}}_k$ by the SOR iteration for solving the linear system

$$\tilde{A}\tilde{A}^T\mathbf{y} = \tilde{\mathbf{b}} + \alpha\tilde{A}\mathbf{c}. \tag{4.4}$$

Observe that the system (4.4) is inconsistent. Hence the resemblance between (4.4) and (1.18) implies that for $k \geq k^*$ the sequence $\{\tilde{\mathbf{y}}_k\}$ obeys the rule

$$\tilde{\mathbf{y}}_k = \tilde{\mathbf{u}}_k + (k - k^*)\tilde{\mathbf{v}}, \tag{4.5}$$

where $\{\tilde{\mathbf{u}}_k\}$ is a converging sequence and $\tilde{\mathbf{v}} \in \text{Null}(\tilde{A}^T)$ is a fixed vector. Furthermore, since $\tilde{\mathbf{y}}_k \geq \mathbf{0}$ for $k \geq k^*$, the components of $\tilde{\mathbf{v}}$ must be nonnegative. In addition, the fact that $\{F_k\}$ is a strictly decreasing sequence shows that $\tilde{\mathbf{b}}^T\tilde{\mathbf{v}} > 0$. So $\tilde{\mathbf{v}}$ satisfies

$$\tilde{\mathbf{v}} \geq \mathbf{0} \qquad \tilde{A}^T\tilde{\mathbf{v}} = \mathbf{0}, \qquad \text{and} \qquad \tilde{\mathbf{b}}^T\tilde{\mathbf{v}} > 0. \tag{4.6}$$

Note that the corresponding sequence of primal points satisfies

$$\mathbf{x}_k = A^T\mathbf{y}_k - \alpha\mathbf{c} = \tilde{A}^T\tilde{\mathbf{y}}_k - \alpha\mathbf{c} = \tilde{A}^T\tilde{\mathbf{u}}_k - \alpha\mathbf{c}.$$

Therefore, since $\{\tilde{\mathbf{u}}_k\}$ is a converging sequence, the sequence $\{\mathbf{x}_k\}$ also converges. Moreover, the relationship between the SOR method and Kaczmarz's method indicates that for $k \geq k^*$ the sequence $\{\mathbf{x}_k\}$ is actually generated by applying Kaczmarz's method for solving the inconsistent system

$$\tilde{A}\mathbf{x} = \tilde{\mathbf{b}} + \alpha\tilde{A}\mathbf{c}. \tag{4.7}$$

The next theorem summarizes our main findings.

Theorem 9. *Assume that there exists an iteration index k^* and a subset $\mathbb{I}$ for which (4.2) holds. In this case the sequence $\{\mathbf{x}_k\}$ converges while the dual points obey the rule*

$$\mathbf{y}_k = \mathbf{u}_k + (k - k^*)\mathbf{v} \qquad \forall\, k \geq k^* \tag{4.8}$$

where $\{\mathbf{u}_k\}$ is a converging sequence and $\mathbf{v} \in \mathbb{R}^m$ is a fixed vector that satisfies

$$\mathbf{v} \geq 0, \qquad A^T\mathbf{v} = \mathbf{0}, \qquad \textit{and} \qquad \mathbf{b}^T\mathbf{v} > 0. \tag{4.9}$$

Moreover, for each row index i the subsequence $\mathbf{x}^{(k,i)}$, $k = 1, 2, 3, \ldots,$ converges, but the limit point of these subsequences are not necessarily equal.

Preliminary experiments that we have done suggest that the sequence $\{\mathbf{x}_k\}$ converges in spite of the fact that the feasible region is empty. This property forces the sequence $\{\mathbf{y}_k\}$ to satisfy (4.8) and (4.9). However, proving or disproving these conjectures is not a simple task. So this issue is left for future research.

5. NUMERICAL EXPERIMENTS

In this section we provide the results of some experiments with the proposed row relaxation scheme for solving (1.1). All the computations were carried out on a VAX 9000-210 computer at the Hebrew University computation center. The algorithm was programmed in FORTRAN using double precision arithmetic. The test problems that we have used are similar to those of Mangasarian [54]. The matrix A is fully dense $m \times n$ matrix whose entries are random numbers from the interval $[-1, 1]$. (The random numbers generator is of uniform distribution.) The vectors $\mathbf{b}$ and $\mathbf{c}$ are defined in a way that makes the point

$$\mathbf{e} = (1, 1, \ldots, 1)^T \in \mathbb{R}^n$$

a solution of (1.2). The components of $\mathbf{b}$ are defined by the rule

$$\mathbf{b}_i = \mathbf{a}_i^T\mathbf{e} \quad \text{when} \quad \mathbf{a}_i^T\mathbf{e} > 0, \quad \text{and} \quad b_i = 2\mathbf{a}_i^T\mathbf{e} - 1 \quad \text{when} \quad \mathbf{a}_i^T\mathbf{e} \leq 0.$$

The vector $\mathbf{c}$ is obtained from the equality

$$\mathbf{c} = A^T\mathbf{y}^*,$$

where the components of $\mathbf{y}^* = (y_1^*, \ldots, y_m^*)^T$ satisfy the rule

$$y_i^* = 1 \quad \text{when} \quad \mathbf{a}_i^T\mathbf{e} > 0, \quad \text{and} \quad y_i^* = 0 \quad \text{when} \quad \mathbf{a}_i^T\mathbf{e} \leq 0.$$

The number of active constraints at the point $\mathbf{e}$ is denoted by the integer ℓ. Note that ℓ is also the number of positive components in $\mathbf{y}^*$.

The **starting points** in the experiments of Tables 1, 4, and 5 are always

$$\mathbf{y}_0 = (0, 0, \ldots, 0)^T \in \mathbb{R}^m \quad \text{and} \quad \mathbf{x}_0 = A^T\mathbf{y}_0 - \alpha\mathbf{c} = -\alpha\mathbf{c}.$$

As before we use $\mathbf{y}_k$ and $\mathbf{x}_k = A^T\mathbf{y}_k - \alpha\mathbf{c}$ to denote the current estimate of the solution at the end of the k-th iteration, $k = 1, 2, \ldots$. The progress of these points was inspected by watching the parameters ρ_k and η_k. The definition of these parameters relies on the vectors $(A\mathbf{x}_k - \mathbf{b})_+$ and $(A\mathbf{x}_k - \mathbf{b})_-$ whose components are $\max\{0, \mathbf{a}_i^T\mathbf{x}_k - b_i\}$ and $\min\{0, \mathbf{a}_i^T\mathbf{x}_k - b_i\}$, respectively. The first parameter,

$$\rho_k = \|(A\mathbf{x}_k - \mathbf{b})_-\|_\infty,$$

measures the constraints violation at $\mathbf{x}_k$. The second one,

$$\eta_k = \mathbf{y}_k^T(A\mathbf{x}_k - \mathbf{b})_+ / \max\{1, \|\mathbf{y}_k\|_1\},$$

measures the optimality violation at $\mathbf{y}_k$. One motivation for using these criteria comes from the optimality conditions (2.8), which indicate that $\mathbf{x}_k$ solves (1.1) if and only if $\rho_k = 0$ and $\mathbf{y}_k^T(A\mathbf{x}_k - \mathbf{b})_+ = 0$. A second motivation lies in the identities

$$P(\mathbf{x}_k) - D(\mathbf{y}_k) = \mathbf{y}_k^T(A\mathbf{x}_k - \mathbf{b}) = \mathbf{y}_k^T(A\mathbf{x}_k - \mathbf{b})_+ + \mathbf{y}_k^T(A\mathbf{x}_k - \mathbf{b})_-,$$

which show that $\mathbf{y}_k^T(A\mathbf{x}_k - \mathbf{b})_+$ is an upper bound on the primal-dual gap. The smaller is ρ_k the better is the bound. If α attains a large value then $|P(\mathbf{x}_\alpha)|$, $|D(\mathbf{y}_\alpha)|$, and $\|\mathbf{y}_\alpha\|$, are also expected to have large values. The division of $\mathbf{y}_k^T(A\mathbf{x}_k - \mathbf{b})_+$ by $\max\{1, \|\mathbf{y}_k\|_1\}$ is needed, therefore, to neutralize the effect of a large α.

The figures in Tables 1, 4, 5, and 6 provide the number of iterations which are needed to satisfy the **stopping condition**

$$\max\{\rho_k, \eta_k\} < \delta. \tag{5.1}$$

Except for Table 4, all the experiments were carried out with $\delta = 10^{-10}$. The experiments of Table 4 where carried out with $\delta = 10^{-4}$. The **starred figures** in our tables denote problems in which $\mathbf{x}_\alpha$, the solution of (1.1), fails to solve the original linear programming problem (1.2). This observation was concluded by inspecting the parameter

$$\tau_k = (\mathbf{c}^T\mathbf{x}_k - \mathbf{c}^T\mathbf{e})/\mathbf{c}^T\mathbf{e}.$$

The experiments in Tables 1 and 2 were carried out with various values of the relaxation parameter, ω. The other experiments, which are described in Tables 4–6, were conducted by using **a fixed relaxation parameter** $\omega = 1.6$.

The reading of **Table 1** is quite straightforward. Here each row refers to a different test problem. Let us take for example the case when (1.1) is defined with $m = 50$ $n = 50$ and $\alpha = 1$. In this case counting the number of active constraints at the point $\mathbf{e}$ has shown that $\ell = 30$. The problem has been solved five times, using five different values of ω. Thus, in particular, for $\omega = 1.0$ our algorithm requires 105 iterations in order to satisfy (5.1). Yet for $\omega = 1.2$ only 57 iterations were needed to solve the same problem. Note that when $m = 80$, $n = 50$, and $\alpha = 1$, the solution of (1.1) fails to solve (1.2).

The results presented in Table 1 reveal an interesting feature of our test problems: A major factor that effects the number of iterations is the ratio between ℓ and n. If ℓ is considerably smaller (or larger) than n then the algorithm enjoys a fast rate of convergence. However, when ℓ is about n the algorithm suffers from slow rate of convergence. The explanation of this phenomenon lies in the close links between our algorithm and the Kaczmarz-SOR method. As we have seen, the "final" active set is reached within a finite number of iterations, k^*. Then, as k exceeds k^*, the behaviour of our method is similar to that of the Kaczmarz-SOR method for solving the system (3.13)–(3.12). In order to demonstrate this relationship we have tested the method of Kaczmarz on similar problems.

The experiments with Kaczmarz's method are described in Table 2. These runs are aimed at solving a consistent linear system of the form

$$\tilde{A}\mathbf{x} = \tilde{\mathbf{b}}, \tag{5.2}$$

where $\tilde{A}$ is a random $\ell \times n$ matrix (which is generated in the same way as A) and $\tilde{\mathbf{b}} = \tilde{A}\mathbf{e}$. Let $\tilde{\mathbf{x}}_k$ denote the current estimate of the solution at the end of the k-th iteration of Kaczmarz's method. The figures in **Table 2** provide the number of iterations which are required to satisfy the **stopping condition**

$$\|\tilde{A}\tilde{\mathbf{x}}_k - \tilde{\mathbf{b}}\|_\infty \le 10^{-10}. \tag{5.3}$$

Here the starting point is always $\mathbf{x}_0 = \mathbf{0}$. The reading of Table 2 is similar to that of Table 1. For example, when $\ell = 30$, $n = 50$, and $\omega = 1.4$, the method of Kaczmarz's requires 87 iterations to satisfy (5.3).

A look at Table 2 shows that Kaczmarz's method possesses the same anomaly as our method: If ℓ is considerably smaller (or larger) than n then Kaczmarz's method enjoys a fast rate of convergence. However, when ℓ is about n the method of Kaczmarz suffers from a slow rate of convergence. The reason for this anomaly is revealed in **Table 3**, which examines the eigenvalues of $\tilde{A}\tilde{A}^T$. Recall that Kaczmarz's method for solving (5.2) is essentially the SOR method for solving the system

$$\tilde{A}\tilde{A}^T\mathbf{y} = \tilde{\mathbf{b}}. \tag{5.4}$$

Indeed a comparison of Table 2 with Table 3 suggests that slow rate of convergence is characterized by the existence of small nonzero eigenvalues. This apparently causes the iteration matrix of the SOR method to have eigenvalues with modulus close to 1. Eigenvalues and condition numbers of random matrices (with elements from standard normal distribution) have been studied by several authors. The reader is referred to Edelman [28–30] for detailed discussions of this issue and further references.

A second observation that stems from Tables 1–3 is about the relaxation parameter, ω. We see that if the SOR method has a rapid rate of convergence, i.e. ℓ is considerably smaller (or larger) than n, then the value of ω has a limited effect on the number of iterations. On the other hand, if the SOR method has a slow rate of convergence, i.e. ℓ is about n, then the use of "optimal" ω results in a considerable reduction in the number of iterations.

Another interesting feature is revealed in **Tables 4 and 5**. These experiments are aimed to investigate the effect of α on the number of iterations. So here we have used a fixed relaxation parameter $\omega = 1.6$, and a fixed starting point $\mathbf{y}_0 = \mathbf{0} \in \mathbb{R}^m$. The reading of Tables 4 and 5 is quite simple. For example, from Table 5 we see that for $m = 20$, $n = 20$, and $\alpha = 0.1$ the algorithm requires 64 iterations to terminate, and the solution point solves (1.1) but not (1.2). Similarly, at the same row, when $m = 20$, $n = 20$, and $\alpha = 1$, the algorithm requires 77 iterations to terminate and the limit point solves both (1.1) and (1.2). Hence in this example, when $m = 20$ and $n = 20$, the threshold value, α^*, lies somewhere between $\alpha = 0.1$ and $\alpha = 1$.

In practice α^* is not known in advance. This suggests the use of a large α in order to ensure that the solution of (1.1) also solves (1.2). Indeed, the ability to obtain a solution of (1.2) is the main motivation behind the early SOR methods for solving (1.1), e.g. [54]–[58]. In these methods the difficulty in determining an appropriate value of α is resolved by repeated solutions of (1.1) with increasing values of α. Later methods overcome this difficulty by applying the proximal point algorithm with increasing values of α, e.g. [27] and [75]. So both ways require the solution of (1.1) with a large value of α. However, a look at Tables 4 and 5 shows that the use of a large α can cause a drastic increase in the number of iterations. The larger is the ratio m/n the larger is the increase. The difference between Table 4 and Table 5 lies in the value of the termination criterion δ. A comparison of these tables shows that the "final" rate of convergence is unchanged in spite of the increase in the overall number of iterations! For example, when $m = 200$, $n = 20$, and $\alpha = 1000$ our algorithm requires 24160 iterations to satisfy (5.1) with $\delta = 10^{-4}$. Yet only 12 further iterations are needed to satisfy (5.1) with $\delta = 10^{-10}$.

Let ℓ_k denote the number for positive components in the vector $\mathbf{y}_k$. Let $\mathbf{d}_k = \mathbf{y}_{k+1} - \mathbf{y}_k$ denote the difference vector between two consecutive iterations. Let $\tilde{\mathbf{d}}_k = \mathbf{d}_k/\|\mathbf{d}_k\|_2$ denote the corresponding unit vector in the direction of $\mathbf{d}_k$. In order to find the reasons behind the increase in the number of iterations we have watched the values of the parameters

$$\ell_k, \qquad ||\mathbf{d}_k||_2, \qquad \tilde{\mathbf{d}}_k^T\tilde{\mathbf{d}}_{k-1}, \qquad \text{and} \qquad ||A^T\tilde{\mathbf{d}}_k||_2.$$

This inspection has exposed a highly interesting phenomenon: If α has a large value (i.e. $\alpha \geq 100$) then in the first iterations ℓ_k is rapidly increased toward m. After that, in the rest of the iterative process, ℓ_k is gradually decreased toward its final value, which is about $m/2$. In the way down, when ℓ_k is gradually decreased, the algorithm is often performing several iterations having the same value of ℓ_k and the same dual active set. In other words, there exists an index set $\mathbb{I} \subseteq \mathcal{M}$ for which (4.2) holds for many consecutive iterations. In such a case, when both $\mathbf{y}_k$ and $\mathbf{y}_{k+1}$ satisfy (4.2), the point $\mathbf{y}_{k+1}$ is obtained from $\mathbf{y}_k$ by the SOR iteration for solving (4.4). Moreover, as our records show, situations in which $\mathbf{y}_k$ is "trapped" at the same index set, $\mathbb{I}$, for many consecutive iterations occurs at the following two cases.

Case 1: Here ℓ_k is considerably large than n while the linear system (4.4) is inconsistent. In this case the SOR iterations obey the rule (4.5). Since ℓ_k is larger than n, the sequence $\{\tilde{\mathbf{u}}_k\}$ converges rapidly, so after a few iterations $\mathbf{d}_k$ is actually a fixed vector, $\tilde{\mathbf{v}}$, that belongs to Null($\tilde{A}^T$). Moreover, since $\mathbf{e}_i^T\mathbf{d}_k = 0$ whenever $i \notin \mathbb{I}$, $\tilde{\mathbf{v}} \in$ Null(A^T). Consequently the primal sequence $\{\mathbf{x}_k\}$ is actually "stuck" at the same point for several consecutive iterations. However, by Gale's theorem of the alternative, here $\tilde{\mathbf{v}}$ must has at least one negative component (which can be quite small). So eventually, after several iterations, one component of $\mathbf{y}_k$ hits its bound, ℓ_k is reduced, and so forth.

Case 2: Here ℓ_k is about n, so the SOR method which is associated with the matrix $\tilde{A}\tilde{A}^T$ has a slow rate of convergence. The sequence $\{\mathbf{y}_k\}$ changes, therefore, very slowly. So, again, several iterations are needed in order to move from one active set to the next.

The two situations which are described above (especially the first one) are the main reasons behind the drastic increase in the number of iterations that occurs when using a large value of α. However, eventually the algorithm reaches the final active set, for which the system (4.4) is solvable. Recall also that the "final" active set is actually determined by $\mathbf{x}_\alpha$, while the last point remains unchanged when α exceeds the threshold value α^*. This explains why the "final" rate of convergence is almost unchanged as α increases.

The effect of a large α on the number of iterations is similar to that of starting the algorithm from the point $\mathbf{y}_0 = (\beta, \beta, \ldots, \beta)^T \in \mathbb{R}^m$, using a large value for β. This fact is illustrated in Table 6, from which we see that a large β may cause a dramatic increase in the number of iterations. The reasons for this phenomenon are, again, the two situations described above.

The proposed row relaxation scheme was also used to solve **large sparse problems**. These tests were conducted exactly as in the dense case. The only difference is that now A is a large sparse matrix which is defined in the following way: Each row of A has only three nonzero elements that have random locations. That is, the column indices of the nonzero elements are random integers between 1 and n. The values of the nonzero elements of A are, again, random numbers from the interval $[-1, 1]$. The experiments that we have done indicate that the algorithm is capable to solve efficiently large problems of this type. For example, when $m = n = 3000$, $\alpha = 100$, and $\omega = 1.6$ the algorithm requires 445 iterations to satisfy (5.9). Similarly, when $m = n = 30000$, $\alpha = 100$, and $\omega = 1.6$ the algorithm reached the solution point within 1188 iterations. It is also interesting to note that the parameters m/n, α, and ω effect the number of iterations in the same way as in dense problems. Thus, for example, when m is about $2n$ the algorithm suffers from slow rate of convergence.

6. EXTENSIONS

The description of the basic iteration is aimed to keep it as simple as possible. This helps to see the links with the SOR method and clarifies our analysis. Yet many of the observations made in this paper remain valid when using more complicated schemes. Below we outline a few ways of possible extensions. Some of these options have already discussed in the literature so we will be rather brief here.

Variable relaxation parameters: Some descriptions of Kaczmarz's method allow the use of a different relaxation parameter for each row, e.g. [5, 9, 37]. A further extension is achieved by allowing the use of a different relaxation parameter for each iteration. The only restriction is that all the parameters belong to a preassigned interval $[\mu, \nu]$ such that $0 < \mu < \nu < 2$.

Types of control: The term "control" refers to the order in which the algorithm chooses the rows (dual variables) during the iterative process. The current algorithm uses "cyclic control". However, it is easily modified to allow other types of control; such as "almost cyclic control" or the "Gauss-Southwell control", see [5, 9, 49].

Block iterative methods: The relation with the SOR method suggests the use of block iterative methods. Here the basic iteration consists of p steps where each step deals with a "block" of rows (a group of dual variables). This can be done by following the methods of [7, 13, 15].

Linear equalities and interval constraints: Assume for example that the i-th inequality is replaced with the equality $\mathbf{a}_i^T\mathbf{x} = b_i$. Then by writing this equality as two inequalities, $\mathbf{a}_i^T\mathbf{x} \geq b_i$ and $-\mathbf{a}_i^T\mathbf{x} \geq -b_i$, we obtain that the corresponding dual variable is not restricted to stay nonnegative. A similar simplification is achieved in the case of interval constraints of the form $b_i - d_i \leq \mathbf{a}_i^T\mathbf{x} \leq b_i + d_i$. The need for solving large sparse systems of interval constraints arises in the field of medical image reconstruction, e.g. [6, 8, 38, 39].

Simple bounds on the dual variables: The algorithm is almost unchanged when the inequality $\mathbf{y} \geq \mathbf{0}$ is replaced with $\boldsymbol{\ell} \leq \mathbf{y} \leq \mathbf{u}$, e.g. [18].

7. RELATIONS WITH OTHER METHODS, PROBLEMS, AND CONVERGENCE RESULTS

We shall start with a few words on Hildreth's quadratic programming procedure [42, 51], which is aimed at solving a problem of the form

$$\begin{aligned}&\text{minimize} \ \ {}^{1}\!/\!_{2}\mathbf{z}^T Q\mathbf{z} + \mathbf{q}^T\mathbf{z}\\ &\text{subject to} \ \ B\mathbf{z} \geq \mathbf{b},\end{aligned} \tag{7.1}$$

where Q is an $n \times n$ symmetric positive definite matrix, B is an $m \times n$ matrix, $\mathbf{q} \in \mathbb{R}^n$, $\mathbf{b} \in \mathbb{R}^m$, and $\mathbf{z} \in \mathbb{R}^n$ denotes the vector of unknowns. Since Q is assumed to be positive definite, it has a symmetric decomposition of the form $Q = R^T R$ where R is an $n \times n$ invertible matrix. Using R it is possible to define a new vector of unknowns $\mathbf{x} = R\mathbf{z}$. This transformation turns (7.1) into the regularized linear programming problem

$$\begin{aligned}&\text{minimize} \ \ {}^{1}\!/\!_{2}\mathbf{x}^T\mathbf{x} + \mathbf{c}^T\mathbf{x}\\ &\text{subject to} \ \ A\mathbf{x} \geq \mathbf{b},\end{aligned} \tag{7.2}$$

where $A = BR^{-1}$ and $\mathbf{c} = R^{-T}\mathbf{q}$. The last observation shows that the dual of (7.1) has the form

$$\begin{aligned}&\text{maximize} \ \ \mathbf{b}^T\mathbf{y} - {}^{1}\!/\!_{2}\|A^T\mathbf{y} - \mathbf{c}\|_2^2\\ &\text{subject to} \ \ \mathbf{y} \geq \mathbf{0},\end{aligned} \tag{7.3}$$

and if $\mathbf{y}^*$ solves the dual then the point $\mathbf{z}^* = Q^{-1}(B^T\mathbf{y}^* - \mathbf{q})$ solves (7.1). Hildreth [42] has proposed to solve (7.3) by changing one variable at a time. Hence from this point of view our method is a special case of Hildreth's procedure. However the original method of Hildreth [42] lacks two ingredients that characterize a typical row-action scheme. First it does not form the matrix A and use its rows one at a time. Instead it computes and uses the matrix $H = AA^T = BQ^{-1}B^T$ and the vector $\mathbf{h} = BQ^{-1}\mathbf{q} - \mathbf{b}$. So actually Hildreth's method solves the problem

$$\begin{aligned}&\text{minimize} \ \ {}^{1}\!/\!_{2}\mathbf{y}^T H\mathbf{y} - \mathbf{h}^T\mathbf{y}\\ &\text{subject to} \ \ \mathbf{y} \geq \mathbf{0},\end{aligned} \tag{7.4}$$

by changing one variable at a time while keeping the dual variables nonnegative. Second, unlike our method Hildreth's procedure does not update or use the primal vectors, $\mathbf{x}_k$, $k = 1, 2 \ldots$. Instead the computation of the primal solution, $\mathbf{z}^*$, is deferred to the end of the iterative process, after a dual solution has already found.

The idea of casting Hildreth's procedure as a low-storage row-relaxation scheme is due to Herman and Lent [39] and Lent and Censor [46] who used this method for solving large problems that arise in diagnostic radiology. Lent and Censor [46] have considered the feasibility problem

$$\begin{aligned}&\text{minimize} \ \ {}^{1}\!/\!_{2}\mathbf{x}^T\mathbf{x}\\ &\text{subject to} \ \ A\mathbf{x} \geq \mathbf{b},\end{aligned} \tag{7.5}$$

which is a special case of (1.1). Indeed if $\mathbf{c} = \mathbf{0}$ or $\alpha = 0$ then our method coincides with the Lent-Censor scheme. Lent and Censor [46] use Hoffman's Lemma to show that the sequence $\{\mathbf{x}_k\}$ converges toward $\mathbf{x}^*$, the unique solution of (7.5). Our method of proof replaces this tool with the observation that the least squares problem (2.9) is always solvable. A further discussion of Hildreth's method is given by Iusem and De Pierro [43] who prove the existence of an iteration index, k^*, and a positive constant, $0 < \lambda < 1$, such that

$$\|\mathbf{x}_{k+1} - \mathbf{x}^*\|_2^2 \leq \lambda \|\mathbf{x}_k - \mathbf{x}^*\|_2^2 \qquad \forall k \geq k^*.$$

This observation together with (2.13), outlines an alternative way for deriving Theorem 7.

Let M be an $m \times m$ symmetric positive semidefinite matrix and let $\mathbf{q} \in \mathbb{R}^m$ be a given vector. The need for solving large quadratic programming problems of the form

$$\begin{aligned} &\text{minimize } \ {}^1\!/\!_2\mathbf{y}^T M\mathbf{y} - \mathbf{q}^T\mathbf{y} \\ &\text{subject to } \ \mathbf{y} \geq \mathbf{0} \end{aligned} \tag{7.6}$$

arises in several applications. In particular, a point $\mathbf{y} \in \mathbb{R}^m$ solves the last problem if and only if it solves the corresponding symmetric **linear complementarity problem**; which is to find $\mathbf{y} \in \mathbb{R}^m$ that satisfies

$$\mathbf{y} \geq \mathbf{0}, \qquad M\mathbf{y} \geq \mathbf{q}, \qquad \text{and} \qquad \mathbf{y}^T(M\mathbf{y} - \mathbf{q}) = 0. \tag{7.7}$$

Observe that (7.7) constitutes the Karush-Kuhn-Tucker optimality conditions for (7.6). The use of SOR methods for solving (7.6) has been initiated by the works of Cryer [14], Mangasarian [53], and Cottle, Golub and Sacher [13]. Many other SOR methods are discussed in the survey paper of Lin and Pang [47]. Note also that (7.6) is a special case of (1.11) in which $\mathbf{c} = \mathbf{0}$. Conversely, (1.11) is a special case of (7.6) in which $M = AA^T$ and $\mathbf{q} = \alpha A\mathbf{c} - \mathbf{b}$. Hence the observations that we have made about the behaviour of the sequence $\{\mathbf{y}_k\}$ remain valid when applying the SOR method to solve (7.6).

The existence of effective relaxation methods for solving large systems of linear equations and quadratic programs has raised the question of whether similar methods can be used to handle linear programming problems, e.g. Oettli [62]. The answer was given by Mangasarian [54] whose method is aimed at solving the regularized linear programming problem

$$\begin{aligned} &\text{minimize } \ {}^1\!/\!_2\varepsilon\mathbf{x}^T\mathbf{x} + \mathbf{c}^T\mathbf{x} \\ &\text{subject to } \ A\mathbf{x} \geq \mathbf{b}, \end{aligned} \tag{7.8}$$

where $\varepsilon > 0$ is a positive constant. The dual of this problem has the form

$$\begin{aligned} &\text{maximize } \ \mathbf{b}^T\mathbf{y} - {}^1\!/\!_2\|A^T\mathbf{y} - \mathbf{c}\|_2^2/\varepsilon \\ &\text{subject to } \ \mathbf{y} \geq \mathbf{0}, \end{aligned} \tag{7.9}$$

and if $\mathbf{y}^*$ solves (7.9) then $\mathbf{x}^* = (A^T\mathbf{y}^* - \mathbf{c})/\varepsilon$ solves (7.8). The scheme proposed by Mangasarian [54] is essentially an SOR iteration that attempts to solve the linear system

$$AA^T\mathbf{y} = A\mathbf{c} + \varepsilon\mathbf{b}, \tag{7.10}$$

while keeping the variables nonnegative. The initial version of Mangasarian [54] computes and uses the matrix AA^T. If A is a large sparse matrix then AA^T is often much denser than A. This drawback has led Mangasarian [55] to suggest a modified version that does not require the computation of the product AA^T. For detailed discussions of Mangasarian's SOR method see [27, 54–58, 75]. Let $\{\mathbf{y}_k\}$ and $\{\tilde{\mathbf{y}}_k\}$ denote the sequences of dual points which are generated by our method and Mangasarian's method, respectively. Then in exact arithmetic, when $\alpha = 1/\varepsilon$ and $\mathbf{y}_0 = \tilde{\mathbf{y}}_0/\varepsilon$,

$$\mathbf{y}_k = \tilde{\mathbf{y}}_k/\varepsilon \qquad \text{for} \quad k = 1, 2, \ldots. \tag{7.11}$$

However, the sparsity-preserving version [55] is not a row-relaxation scheme that uses one row at a time. Moreover, unlike our method it does not store or update the primal vector $\mathbf{x}_k$. This constitutes a major difference between the two schemes. The reason that we prefer to deal with (1.1) instead of (7.8) lies in the simple geometric interpretation of (1.1). Former reports on the performance of Mangasarian's SOR method are very encouraging and enthusiastic. See, for example, [54]–[57] and [27]. None of these papers mentions any complication or difficulty. Our experiments reveal a different picture. It is shown that the rate of convergence at the final stage can be arbitrarily slow. Furthermore, the replacing of (1.2) with (1.1) requires the use of a large α, but this may lead to erratic behaviour in which thousands of iterations are spent before reaching the final stage.

Mangasarian [53, 54] has proved that if the sequence $\{\tilde{\mathbf{y}}_k\}$ has an accumulation point, $\mathbf{y}^*$, then this point solves (7.9), or (7.6) if the algorithm is aimed at solving this problem. Thus, if the Slater constraints qualification condition holds, then the sequence $\{\tilde{\mathbf{y}}_k\}$ is bounded and has an accumulation point that solves (7.9). His algorithm does not use the primal sequence, $\{\mathbf{x}_k\}$, so there is no claim about the convergence of this sequence. Pang [65] and Lin and Pang [47] have proved that if (7.6) is solvable then the sequence $\{M\tilde{\mathbf{y}}_k\}$ converges to some vector $M\mathbf{y}^*$ where $\mathbf{y}^*$ solves (7.6). This result can be used to show that the sequence $\{\mathbf{x}_k\}$ converges, but the proof given in Section 2 is simpler and shorter.

The question of whether the sequence $\{\mathbf{y}_k\}$ always converges (under the feasibility assumption) has been left open for a long time. It is a fundamental issue that attracted the attention of several authors [13, 14, 18, 42, 46–50, 53, 54, 65]. The answer has been found only recently by Luo and Tseng [48, 49, 50]. However, as these authors admit, the proof is quite intricate and lengthy. The basic difficulty here is to show that the sequence $\{F_k\}$ converges at a geometric rate [49, 50]. The proof given in Section 3 introduces new arguments which establish this fact in a simple transparent way.

The algorithm proposed in this paper is closely related to several other row-relaxation schemes which are aimed at solving various types of least norm problems. Let us take for example the regularized ℓ_1 problem

$$\text{minimize} \quad {}^1\!/\!_{2}\varepsilon\|\mathbf{x}\|_2^2 + \|A\mathbf{x} - \mathbf{b}\|_1. \tag{7.12}$$

The motivation for replacing the original ℓ_1 problem

$$\text{minimize} \quad \|A\mathbf{x} - \mathbf{b}\|_1 \tag{7.13}$$

with (7.12) is similar to that of replacing (1.2) with (1.1): The dual of (7.12) has the form

$$\begin{aligned} &\text{maximize} \quad \varepsilon\mathbf{b}^T\mathbf{y} - {}^1\!/\!_2\|A^T\mathbf{y}\|_2^2 \\ &\text{subject to} \quad -\mathbf{e} \le \mathbf{y} \le \mathbf{e}, \end{aligned} \tag{7.14}$$

and if $\mathbf{y}_\varepsilon$ solves (7.14) then the vector $\mathbf{x}_\varepsilon = A^T\mathbf{y}_\varepsilon/\varepsilon$ solves (7.12). Moreover, there exists a threshold value $\varepsilon^* > 0$ such that $\varepsilon \in (0, \varepsilon^*]$ implies that $\mathbf{x}_\varepsilon$ is the minimum-norm solution of (7.13). Maximizing the dual objective function by changing one variable at a time results in a simple row-relaxation scheme that resembles Kaczmarz's method. See [18] for the details. Similar row-relaxation schemes have been developed for least squares problems [21, 40], interval constraints problems [6, 8, 38, 39], and ℓ_∞ problems [19, 23]. The resemblance between these methods and the current algorithm suggests that all these methods share similar properties. Indeed many arguments and results from the current analysis are easily adapted to these methods. The advantage of using (1.1) as a "test case" lies in the simplicity of this problem, the existence of a simple geometric interpretation, the existence of a simple dual problem, and a simple row relaxation scheme. This simplicity enables us to see several important features which are not always transparent in other problems or schemes.

8. CONCLUDING REMARKS

Row-relaxation methods have been proved as a useful tool for solving various types of large sparse problems. This suggests that there is a place for such a method when handling large linear programs. The algorithm proposed in this paper can be viewed as a row-relaxation version of Mangasarian's SOR method. Former reports on the performance of this method have been very encouraging. Our experiments reveal a different picture. For example, while the theory supports the use of a large α, in practice this may cause a drastic increase in the number of iterations. Hence an attempt has been made to clarify the major factors that affect the behaviour of the method.

The convergence of the sequence $\{\mathbf{x}_k\}$ implies that the "final" active set is reached within a finite number of iterations, k^*. The "final" rate of convergence is similar to that of the Kaczmarz-SOR method for solving the "final" system (3.12)–(3.13). This feature is illustrated in the experiments of Tables 1–3. However, as Tables 4–6 show, in some cases most of the computational effort is spent before the final stage. The explanation of this phenomenon lies in the behaviour of the Kaczmarz-SOR method when applied to solve an inconsistent system of linear equations. We have seen that in certain situations our algorithm follows the rule (4.5) for several iterations. This mode of behaviour is the reason behind some surprising features of the algorithm.

Nevertheless, in spite of the above drawbacks, the proposed row-relaxation scheme is of clear practical value. It is a simple algorithm that avoids complicate operations such as matrix factorizations or line-searches. It uses a minimal amount of storage, and it is easily adapted to take advantage of sparsity. In fact it requires less computer storage than any other method for solving linear programming problems. It is hoped that the observations made in this paper will help in conducting successful applications of such methods.

REFERENCES

1. A. Ben-Israel, Linear equations and inequalities on finite dimensional, real or complex, vector spaces: A unified Theory *Journal of Mathematical Analysis and Applications* **27** (1969), pp. 367–389.

2. A. Bjorck, A direct method for sparse least squares problems with lower and upper bound, *Numerische Mathematik* **54** (1988), pp. 19–32.
3. A. Bjorck, *Numerical Methods for Least Squares Problems*, SIAM, Philadelphia, 1996.
4. A. Bjorck and T. Elfving, Accelerated projection methods for computing pseudoinverse solutions of systems of linear equations, *BIT* **19** (1979), pp. 145–163.
5. Y. Censor, Row-action methods for huge and sparse systems and their applications, *SIAM Review* **23** (1981), pp. 444–466.
6. Y. Censor, An Automatic relaxation method for solving interval linear inequalities, *J. Math. Anal. and Applic.* **106** (1985), pp. 19–25.
7. Y. Censor, Parallel application of block-iterative methods in medical imaging and radiation therapy, *Mathematical Programming* **42** (1988), pp. 307–325.
8. Y. Censor and A. Lent, An iterative row-action method for interval convex programming, *J. Optim. Theory and Applic.* **34** (1981), pp. 321–353.
9. Y. Censor and S.A. Zenios, *Parallel Optimization, Theory, Algorithms, and Applications*, Oxford University Press, 1997.
10. P.G. Ciarlet, *Introduction to numerical linear algebra and optimisation*, Cambridge University Press, Cambridge, 1989.
11. D.I. Clark and M.R. Osbrorne, On linear restricted and interval least-squares problems, *IMA Journal of Numerical Analysis* **8** (1988), pp. 23–36.
12. T.F. Coleman and L.A. Hulbert, A direct active set algorithm for large sparse quadratic programs with simple bounds, *Mathematical Programming* **45** (1989), pp. 373–406.
13. R.W. Cottle, G.H. Golub and R.S. Sacher, On the solution of large, structured linear complementarity Problems: The block partitioned case, *Applied Mathematics and Optimization* **4** (1978), pp, 348-363.
14. C.W. Cryer, The solution of a quadratic programming problem using systematic overrelaxation, *SIAM Journal of Control* **9** (1971), pp. 385–392.
15. M. D'apuzzo and M.A. De Rosa, A parallel block row-action method for solving large sparse linear systems on distributed memory multiprocessors, *Concurrency: Practice and Experience* **6** (1994), pp. 69–84.
16. A. Dax, The convergence of linear stationary iterative processes for solving singular unstructured systems of linear equations, *SIAM Review* **32** (1990), pp. 611–635.
17. A. Dax, On computational aspects of bounded linear least squares problem, *ACM Transactions on Mathematical Software* **17** (1991), pp. 64–73.
18. A. Dax, A row relaxation method for large ℓ_1 problems, *Linear Algebra and Its Applications* **156** (1991), pp. 793–818.
19. A. Dax, A row relaxation method for large minimax problems, *BIT* **33** (1993), pp. 262–276.
20. A. Dax, The relationship Between Theorems of the Alternative, Least Norm Problems, Steepest Descent Directions, and Degeneracy: A Review, *Annals of Operations Research* **46** (1993), pp. 11–60.
21. A. Dax, On row relaxation methods for large constrained least squares problems, *SIAM J. Sci. Comput.* **14** (1993), pp. 570–584.
22. A. Dax, An elementary proof of Farkas' lemma, *SIAM Review* **39** (1997), pp. 503–507.
23. A. Dax, A proximal point algorithm for minimax problems, *BIT* **37** (1997), pp. 600–

622.
24. A. Dax, Geometric interpretation of the proximal point algorithm, Tech. Rep., Hydrological Service of Israel, 1998.
25. A. Dax, The smallest correction of an inconsistent system of linear inequalities, Tech. Rep., Hydrological Service of Israel, 1998.
26. A. Dax and V.P. Sreedharan, On theorems of the alternative and duality, *Journal of Optimization Theory and Applications* **94** (1997).
27. R. De Leone and O.L. Mangasarian, Serial and parallel solution of large scale linear programs by augmented Lagrangian successive overrelaxation, in: A. Kurzhanski, K. Neuwmann and D. Pallaschke, eds, in *Optimization, Prallel Processing and Applications*, Lecture Notes in Economics and Mathematical Sytems No. 304, (Springer-Verlage, Berline 1988), pp. 103–124.
28. A. Edelman, Eigenvalues and condition numbers of random matrices, *SIAM Journal on Matrix Analysis and Applications* **9** (1988), pp. 543–560.
29. A. Edelman, The distribution and moments of the smallest eigenvalue of a random matrix of Wishart type, *Linear Algebra and Its Applications* **159** (1991), pp. 55–80.
30. A. Edelman, On the distribution of a scaled condition number, *Mathematics of Computation* **58** (1992), pp. 185–190.
31. M.C. Ferris, Finite termination of the proximal point algorithm, *Mathematical Programming* **50** (1991), pp. 359–366.
32. R. Fletcher and M.P. Jackson, Minimization of a quadratic function of many variables subject only to lower and upper bounds, *J. Inst. Maths. Applics.* **14** (1974), pp. 159–174.
33. D. Gale, *The Theory of Linear Economic Models*, McGraw-Hill, New York, 1960.
34. R. Gordon, R. Bender and G.T. Herman, Algebraic reconstruction techniques (ART) for three-dimensional electron microscopy and X-ray photography, *J. Theoret. Biol.* **29** (1970), pp. 471–481.
35. O. Güler, On the convergence of the proximal point algorithm for convex minimization, *SIAM Journal on Control and Optimization* **29** (1991), pp. 403–419.
36. O. Güler, Augmented Lagrangian algorithm for linear programming, *Journal of Optimization Theory and Applications* **75** (1992), pp. 445–470.
37. M. Hanke and W. Niethammer, On the acceleration of Kaczmarz's method for inconsistent linear systems, Linear Algebra and Its Applications **130** (1990), pp. 83–98.
38. G.T. Herman, A relaxation method for reconstructing object from noisy x-rays, *Math. Prog.* **8** (1975), pp. 1–19.
39. G.T. Herman and A. Lent, A family of iterative quadratic optimization algorithms for pairs of inequalities, with applications in diagnostic radiology, *Math. Prog. Study* **9** (1978), pp. 15–29.
40. G. T. Herman, A. Lent and H. Hurwitz, A storage-efficient algorithm for finding the regularized solution of a large, inconsistent system of equations, *J. Inst. Maths. Applics.* **25** (1980), pp. 361–366.
41. G.T. Herman, A. Lent and S.W. Rowland, Art: mathematics and applications, *J. Theoret. Biol.* **42** (1973), pp. 1–32.
42. C. Hildreth, A quadratic programming procedure, *Naval Res. Logist. Quart.* **4** (1957), pp. 79–85; Erratum, Ibid., p. 361.

43. A.N. Iusem and A. De Pierro, On convergence properties of Hildreth's quadratic programming algorithm, *Mathematical Programming* **47** (1990), pp. 37–51.
44. S. Kaczmarz, Angenäherte Auflösung von Systemen linearer Gleichungen, *Bull. Acad. Polon. Sci. Lett. A* **35** (1937), pp. 355–357.
45. C.L. Lawson and R.J. Hanson, *Solving Least Squares Problems*, Second edition, SIAM, Philadelphia, 1995.
46. A. Lent and Y. Censor, Extensions of Hildreth's row-action method for quadratic programming, *SIAM Journal on Control and Optimization* **18** (1980), pp. 444–454.
47. Y.Y. Lin and J.S. Pang, Iterative methods for large convex quadratic programs: A survey, *SIAM Journal on Control and Optimization* **25** (1987), pp. 383–411.
48. Z.Q. Luo and P. Tseng, On the convergence of a matrix splitting algorithm for the symmetric monotone linear complementarity problem, *SIAM Journal on Control and Optimization* **29** (1991), pp. 1037–1060.
49. Z.Q. Luo and P. Tseng, On the convergence of the coordinate descent method for convex differentiable minimization, *Journal of Optimization Theory and Applications* **72** (1992), pp. 7–35.
50. Z.Q. Luo and P. Tseng, Error bounds and convergence analysis of feasible descent methods: a general approach, *Annals of Operations Research* **46** (1993), pp. 157–178.
51. D.G. Luenberger, *Optimization by Vector Space Methods*, John Wiley, New York, 1969.
52. O.L. Mangasarian, *Nonlinear Programming*, McGraw-Hill, New York, 1969.
53. O.L. Mangasarian, Solution of symmetric linear complementarity problems by iterative methods, *Journal of Optimization Theory and Applications* **22** (1977), pp. 465–485.
54. O.L. Mangasarian, Iterative solution of linear programs, *SIAM Journal on Numerical Analysis* **18** (1981), pp. 606–614.
55. O.L. Mangasarian, Sparsity-prserving SOR algorithms for separable quadratic and linear program, *Computers and Operations Research* **11** (1984), pp. 105–112.
56. O.L. Mangasarian and R. De Leone, Parallel successive overrelaxation methods for symmetric linear complementarity problems and linear programs, *Journal of Optimization Theory and Applications* **54** (1987), pp. 437–446.
57. O.L. Mangasarian and R. De Leone, Error bounds for strongly convex programs and (super) linearly convergent iterative schemes for the least 2-norm solution of linear programs, *Applied Mathematics and Optimization* **17** (1988), pp. 1–14.
58. O.L. Mangasarian and R. De Leone, Parallel gradient projection successive overrelaxation for symmetric linear complementarity problems and linear programs, *Annals of Operations Research* **14** (1988), pp. 41–59.
59. O.L. Mangasarian and R.R. Meyer, Nonlinear perturbation of linear programs, *SIAM Journal on Control and Optimization* **17** (1979), pp. 745–752.
60. J.J. More and G. Toraldo, Algorithms for bound constrained quadratic programming problems, *Numerische Mathematik* **55** (1989), pp. 377–400.
61. J.J. Moreau, Decomposition orthogonale d'un espace hilbertien selon deux cones mutuellement polaires, *C.R. Acad. Sci. Paris* **255** (1962), pp. 238–240.
62. W. Oettli, An iterative method, having linear rate of convergence, for solving a pair of dual linear programs, *Mathematical Programming* **3** (1972), pp. 302–311.

63. D.P. O'Leary, A generalized conjugate gradient algorithm for solving a class of quadratic programming problems, *Linear Algebra and Its Applications* **34** (1980), pp. 371–399.
64. M.R. Osborne, *Finite Algorithms in Optimization and Data Analysis*, John Wiley and Sons, Chichester, 1985.
65. J.S. Pang, More results on the convergence of iterative methods for the symmetric linear complementarity problem, *Journal of Optimization Theory and Applications* **49** (1986), pp. 107–134.
66. R. T. Rockafellar, *Convex Analysis*, Princeton University Press, Princeton, NJ, 1970.
67. R. T. Rockafellar, The multiplier method of Hestenes and Powell applied to convex programming, *Journal of Optimization Theory and Applications* **12** (1973), pp. 555–562.
68. R. T. Rockafellar, Monotone operators and the proximal point algorithm, *SIAM Journal on Control and Optimization* **14** (1976), pp. 877–898.
69. R. T. Rockafellar, Augmented Lagrangian and applications of the proximal point algorithms in convex programming, *Mathematics of Operations Research* **1** (1976), pp. 97–116.
70. R. Setiono, Interior proximal point algorithm for linear programs, *Journal of Optimization Theory and Applications* **74** (1992), pp. 425–444.
71. R. Setiono, Interior dual least 2-norm algorithm for linear programs, *SIAM Journal on Control and Optimization* **31** (1993), pp. 875–899.
72. R. Setiono, Interior dual proximal point algorithm for linear programs, *European Journal of Operational Research* **77** (1994), pp. 96–110.
73. J. Stoer and C. Witzgall, *Convexity and Optimization in Finite Dimensions, Part 1*, Springer-Verlag, Berlin, Germany, 1970.
74. K. Tanabe, Projection method for solving a singular system of linear equations and its applications, *Numerische Mathematik* **17** (1971), pp. 203–214.
75. S.J. Wright, Implementing proximal point methods for linear programming, *Journal of Optimization Theory and Applications* **65** (1990), pp. 531–554.

Table 1: Solving random test problems

Problem Description				Number of iterations				
m	n	ℓ	α	$\omega = 1.0$	$\omega = 1.2$	$\omega = 1.4$	$\omega = 1.6$	$\omega = 1.8$
10	20	5	1	17	18	29	53	113
20	20	10	1	31	29	44	77	173
40	20	19	1	533	355	218	97	206
46	20	22	1	31079	3380	2229	3861	1174
80	20	42	1	80	67	61	91	92
200	20	109	1	20	24	27	41	109
20	50	9	1	27	20	34	58	127
50	50	30	1	105	57	61	98	210
80	50	50	1	6260*	4128*	2540*	1346*	549*
80	50	50	10	44232	29744	19254	11579	5220
100	50	61	1	681	529	305	236	244
200	50	112	1	73	86	81	67	144
400	50	212	1	47	47	50	73	151

Table 2: Solving random $\ell \times n$ linear systems with Kaczmarz's method.

Problem Description		Number of iterations				
ℓ	n	$\omega = 1.0$	$\omega = 1.2$	$\omega = 1.4$	$\omega = 1.6$	$\omega = 1.8$
5	20	35	18	28	48	109
10	20	53	31	43	66	147
20	20	8284	5555	3574	2031	654
40	20	67	55	54	64	98
100	20	11	11	12	14	28
10	50	21	23	35	61	138
30	50	172	103	87	119	245
50	50	31921	21510	13997	8260	3584
100	50	72	59	53	59	80
200	50	16	16	19	25	40
400	50	7	7	8	10	18

Table 3: The eigenvalues of AA^T when A is a random $\ell \times n$ matrix

ℓ	n	Number of nonzero eigenvalues which are smaller than γ					
		$\gamma = \ell + n$	$\gamma = 10$	$\gamma = 1$	$\gamma = 0.1$	$\gamma = 0.01$	$\gamma = 0.001$
5	20	5	4	0			
10	20	10	7	0			
20	20	20	15	4	1	1	0
40	20	20	9	0			
100	20	20	0				
10	50	10	3	0			
30	50	30	12	1	0		
50	50	50	23	7	2	1	0
100	50	50	10	0			
200	50	50	0				
400	50	50	0				

Table 4: The effect of α (using $\delta = 10^{-4}$).

Problem Description		Number of iterations					
m	n	$\alpha = 0.01$	$\alpha = 0.1$	$\alpha = 1$	$\alpha = 10$	$\alpha = 100$	$\alpha = 1000$
10	20	19	21	24	28	32	36
20	20	28*	27*	32	40	57	81
40	20	38*	39*	44	98	561	5072
46	20	28*	25*	860	336	1262	8005
80	20	42	31	43	141	1173	11529
200	20	11	13	29	233	2394	24160
20	50	23	24	27	30	31	36
50	50	32*	35*	43	51	65	78
80	50	35*	41*	234*	4123	5590	19587
100	50	42*	80*	109	258	1885	18208
200	50	31	23	45	299	2868	28995
400	50	15	18	59	512	5368	54119

Table 5: The effect of α (using $\delta = 10^{-10}$).

Problem Description		Number of iterations					
m	n	$\alpha = 0.01$	$\alpha = 0.1$	$\alpha = 1$	$\alpha = 10$	$\alpha = 100$	$\alpha = 1000$
10	20	47	47	53	57	61	65
20	20	67*	64*	77	84	101	124
40	20	92*	93*	97	151	614	5125
46	20	70*	70*	3861	855	2202	8943
80	20	122	79	91	189	1221	11577
200	20	23	27	41	244	2404	24172
20	50	54	55	58	60	62	67
50	50	79*	83*	98	106	117	131
80	50	84*	95*	1346*	11579	13040	26995
100	50	104*	198*	236	397	2021	18343
200	50	60	46	67	321	2896	29018
400	50	32	31	73	525	5382	54134

Table 6: The effect of the starting point $\mathbf{y}_0 = (\beta, \beta, \ldots, \beta)^T$.

Problem Description			Number of iterations					
m	n	α	$\beta = 0.01$	$\beta = 0.1$	$\beta = 1$	$\beta = 10$	$\beta = 100$	$\beta = 1000$
10	20	1	53	50	51	52	55	58
20	20	1	77	77	80	83	92	100
40	20	1	97	97	100	117	275	2023
46	20	1	3862	3856	1341	9408	9797	12294
80	20	1	91	100	65	119	962	7124
200	20	1	40	37	30	233	2402	22544
20	50	1	58	58	57	59	59	63
50	50	1	98	95	95	99	97	105
80	50	1	1392*	1517*	1645*	1696*	1715*	1528*
80	50	10	11578	11572	11516	11806	11944	11681
100	50	1	245	242	306	384	241	276
200	50	1	68	68	63	327	1917	17982
400	50	1	72	66	62	522	4937	48960

Inherently Parallel Algorithms in Feasibility and Optimization and their Applications
D. Butnariu, Y. Censor and S. Reich (Editors)

FROM PARALLEL TO SEQUENTIAL PROJECTION METHODS AND VICE VERSA IN CONVEX FEASIBILITY: RESULTS AND CONJECTURES

Alvaro R. De Pierro[a*]

[a]State University of Campinas,
Department of Applied Mathematics,
CP 6065, CEP 13081-970, Campinas, SP, Brazil

We present in this article a brief account on the relationship between parallel and sequential projection methods for solving the convex feasibility problem. It is known that fully parallel methods 'solve' the problem in the infeasible case, computing a least squares solution, whenever it exists. We are mainly concerned with the fact that this property can be extended to sequential, or block parallel methods, using a suitable underrelaxation. We show that classical results on nonexpansive mappings could be used to prove some of these properties and we present several conjectures and new results.

1. INTRODUCTION

We are concerned with the general mathematical problem known as convex feasibility (CF). Given a family of nonempty closed convex sets $\{C_i\}_{i\in I}$, in a Hilbert space H, where I is a set of indices, the convex feasibility problem is described by

$$Find\ x \in C = \cap_{i\in I} C_i. \tag{1}$$

We will consider only the case where the family is finite, that is, $I = [1, m]$, and H is a Hilbert space of finite dimension n (although most of the results presented here are valid in infinite dimension after small modifications and probably substituting strong by weak convergence). $< \cdot >$ will denote the inner product in H and $||\cdot||$, the corresponding norm.

The CF problem appears as the mathematical model in many areas of application like image reconstruction [23], signal processing [11], electron microscopy [12], speckle interferometry [18], topography [32] and others (see [14] for an extended list of these applications in the more general setting of set theoretical estimation). The main common feature of these applications is that the physical problem consists of finding a function that satisfies some known properties. Those properties are mathematically expressed by sets, usually convex, but not always (see, for example, [19] on the phase retrieval problem)

*Work partially supported by CNPq grants nº 301699/81 and 201487/92-6. Current address: University of California, Los Angeles, Department of Mathematics, 405 Hilgard Avenue, 90095-1555, Los Angeles, California.

and many of them are inverse or ill-posed problems [41]. A natural way to solve (1) is to generate a sequence by computing the othogonal projections onto the sets, because this approach decomposes the original problem into simpler ones. The resulting algorithm is known in the literature as POCS (from Projections Onto Convex Sets) and its relaxed version is defined as follows.

If $x^0 \in H$, and in general for $k = 0, 1, 2, ...$, if $x_0^k = x^k$, define for $i = 1, \ldots, m$

$$x_i^k = x_{i-1}^k + \lambda(P_i x_{i-1}^k - x_{i-1}^k) = P_i^\lambda(x_{i-1}^k), \tag{2}$$

and $x^{k+1} = x_m^k$, where P_i denotes the orthogonal projection onto the convex set C_i and λ is a real number in $(0, 1]$. P_i^λ is the relaxed version of P_i. It is well known that, if C is nonempty, the sequence above converges to a point in the intersection [22] (weakly in general)(in the nonempty case, this is true for $\lambda \in (0, 2)$). A recent (and nice) survey on this kind of methods and their main properties can be found in [5]. When the convex sets C_i are hyperplanes, (2) reduces to the well-known ART (Algebraic Reconstruction Techniques) algorithm [23]. Also, it is worth mentioning, that, in many real situations, it makes sense to look, not for any solution in C, but for one with particular properties: minimum norm, maximum entropy, ..., etc (see [7], [17]). Although we are not dealing in this article with algorithms for those specific tasks, many of the properties of (2) (as well as of (3)) are shared by many members of this larger family of algorithms.

Around the end of the 70's and beginning of the 80's, parallel computers appeared in the market, giving an impulse to the search for parallel methods for solving practical problems. Projection methods as (2) are particularly suitable for parallelization. Taking averages of the directions defined by the projections, gives rise to the following algorithm. If $x^0 \in H$, for $k = 0, 1, 2, ...$,

$$x^{k+1} = x^k + \frac{\gamma}{m}\sum_{i=1}^{m}(P_i x^k - x^k); \tag{3}$$

where γ is a positive real number in $(0, 2)$.

Convergence properties of (3) can be found in [15] and, for the linear case, in [16]. For $\gamma = 1$, algorithm (3) is also analized in [36]. In [34], a useful interpretation of (3) was given in such a way that the algorithm can be reduced to the sequential case with only two sets defined in a special product space.

If the problems we are dealing with have noisy data, and that is the standard situation, (1) could be empty, and it is worth analyzing what happens with algorithms (2) and (3) in this case. It can be proven (see [15] and [10] for a proof in a more general setting) that, the fully parallel algorithm (3), converges (weakly in infinite dimension) to a solution of the optimization problem

$$minimize_x \; F(x) = \sum_{i=1}^{m} ||P_i x - x||^2, \tag{4}$$

provided that the solution exists.

Regarding the behavior of (2) when C is empty, a deep analysis of this behavior when $\lambda = 1$, can be found in [6]. Also, in [4], (3) is analyzed in detail when $\gamma = 1$ (see Theorem 6.3).

Remark 1: All the results (and conjectures) in this article also apply when, instead of the parallel algorithm (3), we consider, any fixed convex combination of the projections.

Back to our own experience in tomography and inverse problems, the numerical computation of maximum likelihoods in emission tomography ([40], [37]), triggered in the 80's an intense search for accelerated versions of the EM (expectation maximization) algorithm [38], considered the best option for maximizing Poisson Likelihoods arising in the field. The EM is a fully parallel scaled-gradient type method like (3), that takes a large number of iterations to achieve high likelihood values. The solution for the problem came through taking the opposite direction of (2)→(3); that is, deparallelizing the simultaneous EM, and achieving an almost two orders of magnitude speed-up with the so called OS-EM [25] (from Ordered Subsets-Expectation Maximization). OS-EM achieves in few iterations the needed likelihood values, but does not approximate the maximum itself. At the same time, inspired in work by Herman and Meyer [24], that applied ART to the emission tomography problem, we [8] introduced the appropriate relaxation parameters to OS-EM, in order to obtain convergence to the maximum. This renewed our interest in the original orthogonal projections problem. So, regarding our CF problem, the main question we would like to have answered is:

(Q) What is the relationship between algorithm (2) and the solutions of problem (4)?

Remark 2: It is worth noting here that the behavior of partially parallel (or block) algorithms is the same (from the point of view of the nonfeasibility) as the sequential one (2). That is, when substituting in (2) $(P_i x_{i-1}^k - x_{i-1}^k)$ by a convex combination of projections $\sum_{i \in S_l} \omega_i (P_i x^k - x^k)$ (with $\sum_{i \in S_l} \omega_i = 1$, $\omega_i > 0$ and S_l a subset of the integer interval $[1, m]$), the assertions of this article remain essentially the same (relaxed convex combinations of projections are still strongly nonexpansive, etc).

There are two possible approaches when searching for an answer to (Q) (or two kinds of ways to answer it)(x_i^k's defined as for (2)).

Approach A

Analyze the convergence behavior for a fixed λ, and a given starting point x^0, that is, consider the m limits (if they exist), for $i = 1, \ldots, m$

$$x_i(\lambda) = lim_{k \to \infty} x_i^k, \tag{5}$$

and then consider

$$x_i^* = lim_{\lambda \to 0} x_i(\lambda), \tag{6}$$

if this limit exists. In Section 3 we present an example for which (5) exists for every fixed $\lambda \in (0,1]$, but the limit in (6) does not.

Approach B

Analyze the convergence behavior of

$$x_i^k = x_{i-1}^k + \lambda_k (P_i x_{i-1}^k - x_{i-1}^k), \tag{7}$$

where

$$\lambda_k \underset{k\to\infty}{\longrightarrow} 0, \tag{8}$$

and

$$\sum_{k=0}^{\infty} \lambda_k = +\infty. \tag{9}$$

The Approach A is more related with fixed point theory of nonexpansive operators, and there is a deep study of the case $\lambda = 1$ in [6]. In [13], convergence to the solution of (4) is proven for the case of linear equations (ART), and blocks of equations. The Approach B is closely related with decomposition techniques in optimization. As a matter of fact, the parameters slowly tending to zero appear as a clear necessity in nondifferentiable optimization [39]. There are several articles dealing with the more general problem

$$minimize_x \; F(x) = \sum_{i=1}^{m} f_i(x), \tag{10}$$

most of them in the context of neural networks training and the backpropagation algorithm (see [30], [31] and references therein). In those articles, convergence results are obtained for the sequence $\{F(x^k)\}$ assuming stronger conditions on the relaxation parameters, like

$$\sum_{k=0}^{\infty} {\lambda_k}^2 < +\infty, \tag{11}$$

and/or some stepsize rule. In [29], a convergence result applicable to (7) in the case of affine spaces, is given using condition (11).

In Section 3 we present our main conjectures regarding the behavior of the sequence (2), after analyzing a family of examples supporting our points of view. In Section 4, we present some general results that are used to deduce a simple convergence proof for the affine spaces case when used the Approach B. Also in Section 4, convergence results are given for the polyhedral case using the Approach A. In the next Section, we introduce some boundedness and fixed point results, needed for the remainder of the article. Section 5 summarizes some conclusions, pointing out some research directions. .

2. PRELIMINARY RESULTS AND SOME USEFUL CONSEQUENCES

As pointed out in the previous Section, the Approach A is related with fixed points of nonexpansive mappings. For more information on this subject, we refer to the well-known monograph by Goebel and Reich [20]. So, in the first part of this section, we present some known results. In the second part we describe some boundedness results.

2.1. Nonexpansive Mappings

Let T be a selfmapping of H.

Definition 2.1.1 T is called nonexpansive if

$$||Tx - Ty|| \leq ||x - y||, \quad for\ all\ x, y \in H. \tag{12}$$

Definition 2.1.2 [9] T is called strongly nonexpansive if for every pair of sequences (x_n), (y_n),

$$\boxed{\begin{array}{l}(x_n - y_n) \text{ is bounded,} \\ ||x_n - y_n|| - ||Tx_n - Ty_n|| \to 0 \Rightarrow (x_n - y_n) - (Tx_n - Ty_n) \to 0\end{array}}$$

Proposition 2.1.1 Products of strongly nonexpansive mappings are strongly nonexpansive.
Proof. [9] Lemma 2.1. □

Proposition 2.1.2 Relaxed projections (relaxation in (0,1]) and products of relaxed projections are strongly nonexpansive.
Proof. Let P^λ be a relaxed projection associated with the projection P. If $x_n - y_n$ is bounded, using the fact that P is nonexpansive,

$$\begin{aligned} ||P^\lambda x_n - P^\lambda y_n|| &\leq (1-\lambda)||x_n - y_n|| + \lambda||x_n - y_n|| = ||x_n - y_n|| \\ &\leq (1-\lambda)||x_n - y_n|| + \lambda||x_n - y_n|| = ||x_n - y_n||. \end{aligned} \tag{13}$$

So, if $||x_n - y_n|| - ||P^\lambda x_n - P^\lambda y_n|| \to 0$, then

$$||x_n - y_n||^2 - ||P^\lambda x_n - P^\lambda y_n||^2 \to 0. \tag{14}$$

Now,

$$\begin{aligned} &||x_n - y_n||^2 - ||P^\lambda x_n - P^\lambda y_n||^2 = ||(x_n - y_n) - (P^\lambda x_n - P^\lambda y_n)||^2 + \\ &\quad 2\langle x_n - P^\lambda x_n, P^\lambda x_n - P^\lambda y_n\rangle + 2\langle y_n - P^\lambda y_n, P^\lambda y_n - P^\lambda x_n\rangle. \end{aligned} \tag{15}$$

Expanding the second and third terms:

$$\begin{aligned} &\langle x_n - P^\lambda x_n, P^\lambda x_n - P^\lambda y_n\rangle + \langle y_n - P^\lambda y_n, P^\lambda y_n - P^\lambda x_n\rangle \\ &= \lambda^2[\langle x_n - Px_n, Px_n - Py_n\rangle + \langle y_n - Py_n, Py_n - Px_n\rangle] + \\ &\quad \lambda(1-\lambda)\langle x_n - Px_n - y_n + Py_n, x_n - y_n\rangle; \end{aligned} \tag{16}$$

but $\langle x_n - Px_n - y_n + Py_n, x_n - y_n\rangle = ||x_n - y_n||^2 - \langle Px_n - Py_n, x_n - y_n\rangle \geq 0$, and the first part of the result follows. The second is a consequence of Proposition 2.1.1. □
Proposition 2.1.2 follows also easily from Proposition 2.1 in [9].

In the following, $Fix(T)$ denotes the set of fixed points of the mapping T.

Proposition 2.1.3 If T is a strongly nonexpansive selfmapping of H, then:
(i) Fix $(T) \neq \emptyset$ if and only if $(T^n x)$ converges weakly to some fixed point of T, for all $x \in H$.
(ii) Fix $(T) = \emptyset$ if and only if $lim_n ||T^n x|| = +\infty$, for all $x \in H$.
Proof. [9], Corollaries 1.3 and 1.4. □

2.2. Boundedness

We present in this Section boundedness results from [2], [35] and [1] (yet unpublished) and their consequences. These results provide theoretical foundation for the convergence of the projection sequences in the polyhedral case. The results are in finite dimension, and for projections onto affine sets (intersections of finite number of hyperplanes).

Theorem 2.2.1 For every finite family of affine sets there exists a number r such that, for every sequence (x^k) of projections on the family,

$$||x^k|| \leq ||x^0|| + r \tag{17}$$

Proof. [2], [35]. □

A strengthened version of the theorem above for the relaxed case, is the following [1]:

Theorem 2.2.2 Let $\mathcal{A}$ be a finite family of affine sets in H. Then for every point x there exists a bounded convex set O containing x which is preserved under all projections on members of the family.
(Theorem 2.2.2 clearly implies Theorem 2.2.1, upon taking $x^0 = x$, since every sequence of projections is contained in O .)
Proof. Let D be the union of all sequences of projections (that is, the set of all the elements of the sequences) starting at x, and let $O = conv(D)$ be its convex hull.Let $z \in O$, $A \in \mathcal{A}$, P be the projection operator onto A, and let $y = Pz$. We have to show that $y \in O$. Since $z \in O$, we have $z = \Sigma_i \alpha_i d_i$, where $d_i \in D$, $\alpha_i \geq 0$ for all i, and$\Sigma_i \alpha_i = 1$. But then $y = \Sigma_i \alpha_i P d_i$, and Pd_i all belong to D. Hence $y \in O$. □

Corollary 2.2.3 Any sequence of underrelaxed projections onto a finite family of affine sets is bounded. Moreover, the bound is independent of the value of the relaxation parameter.
Proof. Let x^0 be the first point of the sequence, and O the convex set whose existence is guaranteed by Theorem 2.2.2, which contains x^0. Since O is convex and invariant under projections, it is closed also under underrelaxed projections onto members of $\mathcal{H}$. Relaxation is a particular convex combination; so, any relaxed sequence will be contained in O. □

Corollary 2.2.4 Any sequence of underrelaxed projections onto a finite family of polyhedra is bounded, independently of the value of the relaxation parameter.

Proof. A trivial consequence of the fact that projections onto polyhedra are projections onto vertices, edges or faces (that are determined by affine spaces), and each polyhedron has only a finite number of them. □

3. EXAMPLES, COUNTEREXAMPLES AND CONJECTURES

In this Section we analyze a simple example in the plane, or family of examples, that, in our opinion, contains all the possible failures of convergence. Divergence of orthogonal projections methods only occurs in 'asymptotic situations'.

Example. Let

$$C_1 = \{(x_1, x_2)^t \in R^2 \;/\; x_2 = 1\}, \tag{18}$$

$$C_2^\alpha = \{(x_1, x_2)^t \in R^2 \;/\; x_2 = \alpha\} \tag{19}$$

and,

$$C_3 = \{(x_1, x_2)^t \in R^2 \;/\; x_1 x_2 \geq 1,\; x_1, x_2 > 0\}. \tag{20}$$

For $\alpha \in (-1, 0]$, the LS solution for the problem with the three sets C_1, C_2^α and C_3, exists, and the set of solutions is the half line

$$HL^\alpha = \{(x_1, x_2)^t \in R^2 \;/\; x_1 \geq \frac{2}{1+\alpha},\; x_2 = \frac{1+\alpha}{2}\}. \tag{21}$$

The assertion above is easy to verify by computing for every point in HL^α the gradient of (4) that equals zero. Also trivial to verify is the fact that, for $\alpha \leq -1$, there is not any least squares solution (the starting value of the half line tends to $+\infty$ as α approaches -1).

Now, for $\alpha = 0$ and no relaxation ($\lambda = 1$ in (2)), we get the counterexample showing that convergence of successive projections depends on the order [26]. For example, the sequence $P_3P_1P_2$ converges (for every starting point), but $P_3P_2P_1$ diverges to $+\infty$. This, in spite of the existence of LS solutions.

Now, let us analyze the behavior of the sequences of projections, but with relaxation, that is, $P_3^\lambda P_2^\lambda P_1^\lambda$ and $P_3^\lambda P_1^\lambda P_2^\lambda$. Projection onto C_3 involves computing the zeroes of a $4th$ degree polynomial equation, so, it is better to perform the analysis in terms of fixed points, and use Proposition 2.1.3 (i). It is clear that, if $P_2^\lambda P_1^\lambda$ has a fixed point lying in C_3, it will be a fixed point of $P_3^\lambda P_2^\lambda P_1^\lambda$. On the other hand, it is easy to prove that those are the only fixed points (if a fixed point $x = (x_1, x_2)^t \notin C_3$, then, either $y = P_2^\lambda P_1^\lambda x \in C_3$ or $(P_3^\lambda y)_2 > x_2$, implying that $x \neq P_3^\lambda y$). The same being true for $P_3^\lambda P_1^\lambda P_2^\lambda$. Fixed points of $P_2^\lambda P_1^\lambda$ can be analyzed in one dimension, for a fixed value of x_1. In what follows, fixed points of $P_2^\lambda P_1^\lambda$ will be denoted by x^{F_2} ($x^{F_2} = (x_1^{F_2}, x_2^{F_2})^t$) and fixed points of $P_1^\lambda P_2^\lambda$ by x^{F_1} ($x^{F_1} = (x_1^{F_1}, x_2^{F_1})^t$).

Then, fixed points (x_2 coordinate) of $P_2^\lambda P_1^\lambda$ and $P_1^\lambda P_2^\lambda$ (x^{F_1}) satisfy the linear system

$$\begin{cases} (1-\lambda)x_2^{F_1} - x_2^{F_2} &= -\alpha\lambda \\ x_2^{F_1} + (\lambda-1)x_2^{F_2} &= \lambda \end{cases} \tag{22}$$

and the unique solution is given by

$$x_2^{F_1} = \frac{1+\alpha(1-\lambda)}{2-\lambda}, \tag{23}$$

and

$$x_2^{F_2} = \frac{\alpha+1-\lambda}{2-\lambda}. \tag{24}$$

The condition for the fixed points of $P_1^\lambda P_2^\lambda$ $(x_1^{F_1}, x_2^{F_1})^t$ to lie in C_3 (and $P_3^\lambda P_1^\lambda P_2^\lambda x^k$ to be convergent) is

$$x_1^{F_1} \geq \frac{2-\lambda}{1+\alpha(1-\lambda)} > 0. \tag{25}$$

that is

$$1+\alpha(1-\lambda) > 0. \tag{26}$$

And for $P_2^\lambda P_1^\lambda$

$$x_1^{F_2} \geq \frac{2-\lambda}{1+\alpha-\lambda} > 0. \tag{27}$$

So,

$$\lambda < 1+\alpha. \tag{28}$$

Condition (26) is always true for $\lambda \in (0,1]$ and $\alpha \in [-1,0]$, so, in this case, there are fixed points of $P_3^\lambda P_1^\lambda P_2^\lambda$ for every λ in the interval, but, if $\alpha = -1$, the limit in (6) does not exist (a counterexample for the existence of (6) when the limits in (5) exist for each positive λ). Also, for $\alpha = -2$, we get convergence of (2) for $\lambda \in (\frac{1}{2}, 1)$, but divergence for smaller λ's. From (28), we deduce that, if $\alpha \neq -1$, it is always possible to choose λ small enough such that $P_3^\lambda P_2^\lambda P_1^\lambda x^k$ is convergent. That is, λ must be taken smaller as the gap (distance) between the sets C_3 and C_2^α increases. If $\alpha = -1$, there is not any positive λ for which convergence is achieved. One conclusion is that underrelaxing can be seen as shifting the convex sets towards the LS solution points, whenever they exist, avoiding asymptotic '*dangers*', and ensuring convergence in (6) and (7). Of course, for different orderings, possibly different intervals of λ will produce convergence.

The previous analysis of the examples, that, in our opinion, contain the worst possible cases of behavior (*asymptotic situations*), led us to the following related conjecture.

Conjecture I. Existence of a LS solution for the CF problem is the necessary and sufficient condition for Approaches A and B to find it, in other words; the LS solution exists if and only if the limits in (6) and (7) both exist and solve problem (4).

Finally, it could be asked why A and B, if Conjecture I states that, from the point of view of the LS solution, they are essentially the same. The answer to this question comes from the fact that, if the solution is not unique, choices of one of them by A and B are different. Observation plus several MATLAB experiments led us to

Conjecture II. The limit in (6) is the orthogonal projection of the starting point onto the set of LS solutions, whenever it exists.

Applying (7) to our example ($\alpha = 0$ and the sequence $P_3^\lambda P_2^\lambda P_1^\lambda x^k$), with $\lambda_k = \frac{1}{k}$, starting at the point $(2,0)^t$, the algorithm generates a sequence increasing in the first coordinate, and the projectio of the starting point onto the solution set is $(2, \frac{1}{2})$. So, although B seems to be sometimes more practical, A would mean that small values of λ may compensate giving more stable solutions, especially when dealing with inverse problems.

4. CONVERGENCE RESULTS

4.1. Polyhedra. A

When the convex sets in (1) are polyhedra, because of Corollary 2.2.4 , the sequence generated by (2) is bounded (independently of λ). So, Proposition 2.1.3 means that, for a given positive λ, $x^k(\lambda)$ converges to some $x^*(\lambda)$, the fixed point corresponding to the relaxed projection onto the $m-th$ convex set, that is, $x^*(\lambda) = x_m(\lambda)$. So, if $x_{i-1}^*(\lambda)$ denotes the fixed point corresponding to the $i-th$ projection, then

$$x^*(\lambda) = x^*(\lambda) + \lambda \sum_{i=1}^{m} (P_i x_{i-1}^*(\lambda) - x_{i-1}^*(\lambda)), \tag{29}$$

or,

$$\sum_{i=1}^{m} (P_i x_{i-1}^*(\lambda) - x_{i-1}^*(\lambda)) = 0. \tag{30}$$

Now, from (2), we get that

$$x_i^*(\lambda) = x_{i-1}^*(\lambda) + \lambda(P_i x_{i-1}^*(\lambda) - x_{i-1}^*(\lambda)). \tag{31}$$

Taking limits for λ tending to zero, and using the fact that $P_i x_{i-1}^*(\lambda) - x_{i-1}^*(\lambda)$ is bounded, we deduce that limit points in (6) are all the same and, because of (30), they satisfy

$$\sum_{i=1}^{m} (P_i x^* - x^*) = 0. \tag{32}$$

Therefore we have proven

Theorem 4.1.1 If the convex sets in (1) are polyhedra, the limits points of (6) are solutions of (4).

4.2. Some General Results for B.

This Subsection presents convergence results related with Approach B, equation (7), and always assuming that the sequence is bounded, and, of course, the function bounded below. The results are valid for the general problem (10), so, we will use f_i instead of $||P_i x - x||^2$ and $\nabla f_i(x)$ instead of $2(x - P_i x))$; that is, for continuously differentiable convex functions. The following Lemma is trivial.

Lemma 4.2.1 The difference between iterates $x_i^k - x_{i-1}^k$ $(x^{k+1} - x^k)$ tends to zero.

Proof. It is a consequence of (7) and the fact that $\nabla f_i(x_{i-1}^k)$ is bounded. □

By the way, we prefer to think (intuition plus experiments) that Lemma 4.2.1 is always true for (7), even if there is no LS solution. In other words, following closely [6] (Conjecture 5.2.7) (we are not alone!) we have

Conjecture III. $x_i^k - P_i x_i^k$ is always bounded.

And the sequence of iterates would tend to zero always. Moreover, if Conjecture III is true, then, asymptotically, the objective function will probably be decreasing, independently of the existence of LS solutions. That makes a difference between a general function F like (10), and (4). So, proving Conjecture III, could be the first big step towards proving the previous ones. The definition of relaxed POCS is geometrical, producing the fact that, whenever the relaxation parameter is less or equal than one, the function f_i in the decomposition is decreasing and its directional derivative negative.

Next, another result that shows the necessity of slow decrease of the relaxation sequence.

Theorem 4.2.2 If the sequence generated by (7) is convergent, the limit solves the LS problem (4).

Proof. Expanding from x^0, we have that

$$x^{k+1} = x^0 + \sum_{l=0}^{k} \lambda_l a^l, \tag{33}$$

where

$$a^l = -\sum_{i=1}^{m} \nabla f_i(x) \tag{34}$$

Taking limits for $k \to +\infty$ the series $\sum_{l=0}^{\infty} \lambda_l a^l$ is convergent and the x_i^l's tend to the same limit, say x^*, as $l \to \infty$, because of Lemma 4.1.1. In the same way $a^l \to a^* = \sum_{i=1}^{m} \nabla f_i(x^*)$. If, for some j, $a_j^* > 0$, there exists $a > 0$ and l_0, such that for $l > l_0$, $a_j^l > a > 0$. Therefore $\sum_{l>l_0} \lambda_l a_j^l > a \sum_{l>l_0} \lambda_l$, a contradiction with (11). The same argument applies for $a_j^l < 0$. So, a^* should be zero and x^* a minimum. □

Now, let us go a little bit further after a couple of auxiliary results.

Lemma 4.2.3 If the sequence is bounded and there exists a real number γ such that, $F(x^k) > \gamma > min_x F(x)$, for every k, then, $F(x^{k+1}) < F(x^k)$, for all k large enough.

Proof. Because of Lemma 4.2.1 and the continuity,

$$\nabla F(x^k) - (-d^k)_{k \to \infty} \to 0, \tag{35}$$

where $d^k = -\sum_{i=1}^m \nabla f_i(x_{i-1}^k)$. From (35), it is easy to deduce, using convexity, the hypothesis and the continuity of the derivatives, that, for all k large enough, there exists $\eta > 0$ such that

$$-\nabla F(\theta^k)^t d^k \geq \eta > 0, \tag{36}$$

for θ^k lying in the segment between x^k and x^{k+1}. Now, using Taylor's expansion, we get that

$$F(x^k) - F(x^{k+1}) = -\lambda_k \nabla F(\theta^k)^t d^k \geq \lambda_k \eta, \tag{37}$$

and the result holds. □

Proposition 4.2.4 If the sequence is bounded, there is a limit point that is a minimum.
Proof. If it is not true, then there exists γ positive, such that Lemma 4.2.3 holds. So, using (37), we get that

$$F(x^0) - F(x^l) = \sum_{k=0}^{l} [F(x^k) - F(x^{k+1})] \geq \eta \sum_{k=0}^{l} \lambda_k. \tag{38}$$

The left hand side of (38) is nonnegative and bounded; but, when l tends to ∞, the right hand side is unbounded because of (9), a contradiction. This means that there exists a subsequence $\{x^{k_l}\}$ converging to a point x^* that is a minimum, because of convexity. □

Theorem 4.2.4 If the sequence defined by (7) is bounded, every accumulation point is a solution of (4).
Proof. Define

$$AC = \{set\ of\ limit\ points\}. \tag{39}$$

AC is a connected set because of Ostrowski's Theorem (limit points of a sequence such that the difference between iterates tends to zero) [33], closed (obvious) and bounded because the sequence is bounded. Define now, for each γ the (nonempty) set

$$L_\gamma = \{x : F(x) \ \geq \ \gamma \ > \ minF(x)\}. \tag{40}$$

Suppose that $AC \ \cap \ L_{\bar{\gamma}}$ is nonempty for some $\bar{\gamma}$, that is, there exists a limit point $\bar{x} \in ACL(\bar{\gamma}) = AC \ \cap \ L_{\bar{\gamma}}$ and $F(\bar{x}) > F(x^*)$ (x^* is a minimum, that exists because of Proposition 4.2.4). Clearly $ACL(\gamma) \neq \emptyset$ for every $\gamma < \bar{\gamma}$. Now consider a neighborhood B of AC, such that, $x^k \in B$ for all k large enough and $F(x^{k+1}) < F(x^k)$ if $x^k \ \in \ B \cap L_{\bar{\gamma}}$ (such a neighborhood exists because of the same arguments of Lemma 4.2.3). Let $d_{\bar{\gamma}}$ be the distance between x^* and $L_{\bar{\gamma}}$ (positive because the function is continuous) and let $dl^k = -\lambda_k \sum_{i=1}^m \nabla f_i(x_{i-1}^k)$. For a given p and for all k large enough, $||dl^k|| \ < \ \frac{d_{\bar{\gamma}}}{p}$, p a positive integer. Because of the fact that x^* is a limit point, there are points of the sequence as close as necessary to x^*, say $\frac{d_{\bar{\gamma}}}{p}$. So, considering that $\bar{x} \ \in \ ACL(\bar{\gamma})$, there exists x^{k_1} that belongs to $B \ \cap \ L_{\bar{\gamma}}$, and the distance $dist(x^k, CL_{\bar{\gamma}}) \ < \ \frac{d_{\bar{\gamma}}}{p}$, where CL

stands for the complement set of L (that is, the convex set $CL_{\bar{\gamma}} = \{x : F(x) < \bar{\gamma}\}$), for every $k \geq k_1$, because inside $B \cap L_{\bar{\gamma}}$ the sequence is always decreasing. But this can be proven for every given p, so, if there is an accumulation point in $ACL(\bar{\gamma})$, $\bar{x}$, it should be $F(\bar{x}) = \bar{\gamma}$. Also, $\bar{\gamma}$ is an arbitrary positive number and the preceding rationale is valid for every $\gamma < \bar{\gamma}$. That means that every accumulation point should be a minimum. □

4.3. Affine Spaces. B

If the convex sets are affine spaces, we prove in the following, that convergence of (7) is a direct consequence of the results in the previous sections. Let us consider the problem

$$Ax = b, \tag{41}$$

where A is a matrix with blocks of rows A_i, x and b vectors of appropriate dimensions, and b_i the corresponding blocks of b. Then

$$P_i^\lambda x = x + \lambda A_i^+(b_i - A_i x), \tag{42}$$

where A^+ denotes the pseudoinverse of the matrix A [21].

Theorem 4.3.1 If the convex sets in (1) are affine subspaces, the whole sequence defined by (7) converges to a solution of (4). Moreover, the limit is the projection of the starting point onto the set of LS solutions.

Proof. From Corollary 2.2.3, we know that the sequence is bounded. Using Theorem 4.2.4 every accumulation point is a solution of (4). On the other hand, from (42) and some elementary linear algebra,

$$P_i^\lambda x_{i-1}^k - x_{i-1}^k \in R(A_i^t) \subseteq R(A^t), \tag{43}$$

where R stands for the range of the matrix. So, every limit point x^* is such that $x^* - x^0$ belongs to $R(A^t)$, and x^* is a solution because of Theorem 4.2.4. These are the Kuhn-Tucker conditions for x^* to be the projection of x^0 onto the solution set. But this projection is unique, meaning that the whole sequence converges to it. □

Acknowledgements. Fruitful discussions with my friend (in mathematics, paddling and mountaineering) Nir Cohen have been important inspiration for my work. Dan Butnariu, Yair Censor and Simeon Reich were essential for a pleasant return to orthogonal projection methods (parallel). Margarida always indispensable in my battles against Latex. The referee's criticisms were crucial to improve the first version of the paper.

5. CONCLUDING REMARKS

As we stated at the beginning, we made a short trip through orthogonal projection methods for solving the convex feasibility problem when there is no solution. We filled some gaps, leaving to the interested reader some others, necessary to prove the main conjectures. The extension of Theorem 4.1.1 to the existence of (6) (and not just the limit points) is another immediate task. Essentially, to prove that the necessary and sufficient condition for the existence of a LS solution is to prove boundedness of (2) for every (uniformly) relaxation parameter λ. In [15] it is proven that for the fully parallel algorithm (3), the assertion is true, and some sufficient conditions for the existence of a solution are given (one bounded set, realization of the distance for each two sets, etc). Short steps can be given in the direction of our conjectures by proving boundedness for these particular cases. Finally, we remind, as in the Introduction, that, in our opinion, similar results and conjectures are (essentially) valid when dealing with algorithms based on measures different from the usual 2-norm, once again sequential (block parallel) versus fully parallel ones.

REFERENCES

1. R. Aharoni and A.R. De Pierro, Reaching the least square approximation for a family of convex sets, preprint.
2. R. Aharoni, P. Duchet and B. Wajnryb, Successive projections on hyperplanes, *Journal of Mathematical Analysis and Applications* **103** (1984) 134–138 1984.
3. M. Avriel, *Nonlinear Programing: Analysis and Methods* (Englewood Cliffs, NJ, Prentice Hall, 1976).
4. H.H. Bauschke and J.M. Borwein, Dykstra's alternating projection algorithm for two sets, *Journal of Approximation Theory* **79** (1994) 418–443.
5. H.H. Bauschke and J.M. Borwein, On projection algorithms for solving convex feasibility problems, *SIAM Review* **38** (1996) 367–426.
6. H.H. Bauschke and J.M. Borwein and A.S. Lewis, On the method of cyclic projections for convex sets in Hilbert space, in: Y. Censor and S. Reich, eds, *Recent Developments in Optimization Theory and Nonlinear Analysis, Contemporary Mathematics 204*, (1997) 1–38.
7. L.M. Bregman, The method of successive projection for finding a common point of convex sets, *Soviet Mathematics Doklady* **6** (1965) 688–692.
8. J.A. Browne and A.R De Pierro, A row-action alternative to the EM algorithm for maximizing likelihoods in emission tomography, *IEEE Transactions on Medical Imaging* **15** (1996) 687–699.
9. R.E. Bruck and S. Reich, Nonexpansive projections and resolvents of accretive operators in Banach spaces, *Houston Journal of Mathematics* **3** (1977) 459–470.
10. D. Butnariu, A.N. Iusem and R.S. Burachik, Iterative methods of solving stochastic convex feasibility problems and applications, *Computational Optimization and Applications* **15** (2000) 269–307.
11. J.A. Cadzow, Signal enhancement-A composite property mapping algorithm, *IEEE Transactions on Acoustics, Speech and Signal Processing* **36** (1988) 49–62.
12. J.M. Carazo and J.L. Carrascosa, Information recovery in missing angular data cases: an approach by the convex projection method in three dimensions, *Journal of Microscopy* **145** (1987) 23–43.
13. Y. Censor, P.P.B. Eggermont and D. Gordon, Strong underrelaxation in Kaczmarz's method for inconsistent systems, *Numerische Mathematik* **41** (1983) 83–92.
14. P.L. Combettes, The foundations of set theoretic estimation, *Proceedings of the IEEE*

81 (1993) 182–208.
15. A.R. De Pierro and A.N. Iusem, A parallel projection method of finding a common point of a family of convex sets, *Pesquisa Operacional* **5** (1985) 1–20.
16. A.R. De Pierro and A.N. Iusem, A simultaneous projection method for linear inequalities, *Linear Algebra and its Applications* **64** (1985) 243–253.
17. A.R. De Pierro and A.N. Iusem, A relaxed version of Bregman's method for convex programming, *Journal of Optimization Theory and Applications* **51** (1986) 421–440.
18. S. Ebstein, Stellar speckle interferometry energy spectrum recovery by convex projections, *Applied Optics* **26** (1987) 1530–1536.
19. J.R. Fienup, Phase retrieval algorithms: a comparison, *Applied Optics* **21** (1982) 2758–2769.
20. K. Goebel and S. Reich, *Uniform Convexity, Hyperbolic Geometry and Nonexpansive Mappings*, Monographs and Textbooks in Pure and Applied Mathematics **83** (Marcel Dekker, New York, 1984).
21. G.H. Golub and C.F. Van Loan, *Matrix Computations* (Johns Hopkins University Press, Baltimore, 3rd edition, 1996).
22. L.G. Gubin, B.T. Polyak and E.V. Raik, The method of projections for finding a common point of convex sets, *U.S.S.R. Computational Mathematics and Mathematical Physics* **7** (1967) 1–24.
23. G.T. Herman, *Image Reconstruction from Projections: the Fundamentals of Computerized Tomography* (New York, Academic, 1980).
24. G.T. Herman and L.B. Meyer, Algebraic reconstruction techniques can be made computationally efficient, *IEEE Transactions on Medical Imaging* **12** (1993) 600–609.
25. H.M. Hudson and R.S. Larkin, Accelerated image reconstruction using ordered subsets of projection data, *IEEE Transactions on Medical Imaging* **13** (1994) 601–609.
26. A.N. Iusem, Private Communication.
27. A.N. Iusem and A.R. De Pierro, On the set of weighted least squares solutions of systems of convex inequalities, *Commentationes Mathematicae Universitatis Carolina* **25** (1984) 667–678.
28. A.N. Iusem and A.R. De Pierro, On the convergence of Han's method for convex programming with quadratic objective, *Mathematical Programming* **52** (1991) 265–284.
29. Z.Q. Luo, On the convergence of the LMS algorithm with adaptive learning rate for linear feedforward networks, *Neural Computing* **3** (1991) 226–245.
30. Z.Q. Luo and P. Tseng, Analysis of an approximate gradient projection method with applications to the backpropagation algorithm, *Optimization Methods and Software* **4** (1994) 85–101.
31. O.L. Mangasarian and M.V. Solodov, Serial and parallel backpropagation convergence via monotone perturbed optimization, *Optimization Methods and Software* **4** (1994) 103–116.
32. W. Menke, Applications of the POCS inversion method to interpolating topography and other geophysical fields, *Geophysical Research Letters* **18** (1991) 435–438.
33. A.M. Ostrowski, *Solution of Equations in Euclidean and Banach Spaces* (New York, Academic Press, 1973).
34. G. Pierra, Decomposition through formalization in a product space, *Mathematical*

Programming **28** (1984) 96–115.
35. R. Meshulam, On products of projections, *Discrete Mathematics* **154** (1996) 307–310.
36. S. Reich, A limit theorem for projections, *Linear and Multilinear Algebra* **13** (1983) 281–290.
37. A.J. Rockmore and A. Macovski, A maximum likelihood approach to emission image reconstruction from projections, *IEEE Transactions on Nuclear Science* **23** (1976) 1428–1432.
38. L.A. Shepp and Y. Vardi, Maximum likelihood reconstruction for emission tomography, *IEEE Transactions on Medical Imaging* **1** (1982) 113–121.
39. N.Z. Shor, *Minimization Methods for Non-Differentiable Functions* (Springer-Verlag, Berlin, Heidelberg, Germany, 1985).
40. M.M. Ter-Pogossian, M. Raiche and B.E. Sobel, Positron Emission Tomography, *Scientific American* **243** (1980) 170–181.
41. A.N. Tikhonov and A.V. Goncharsky (eds), *Ill-posed Problems in the Natural Sciences* (MIR Publishers 1987).

Inherently Parallel Algorithms in Feasibility and Optimization and their Applications
D. Butnariu, Y. Censor and S. Reich (Editors)

ACCELERATING THE CONVERGENCE OF THE METHOD OF ALTERNATING PROJECTIONS VIA A LINE SEARCH: A BRIEF SURVEY

F. Deutsch[a]

[a]Department of Mathematics, The Pennsyvania State University, University Park, PA 16802, USA

We give a brief survey of some ways of accelerating the convergence of the iterative method of alternating projections via a line search.

1. INTRODUCTION

In the next section, we shall describe a general iterative method for determining (asymptotically) the best approximation to any given element in a Hilbert space from the set of fixed points of a prescribed bounded linear operator. This idea turns out to be quite fundamental and contains the *method of alternating projections* (or MAP for brevity) as a special case. But it is well-known that the MAP itself has found application in at least ten different areas of mathematics. These include: (1) solving *linear equations* (Kaczmarz [50], Tanabe [69], Herman, Lent, and Lutz [46], Eggermont, Herman, and Lent [34]); (2) the *Dirichlet problem* (Schwarz [64]) which has in turn inspired the "domain decomposition" industry; (3) *probability and statistics* (Wiener and Masani [72], Salehi [63], Burkholder and Chow [13], Burkholder [12], Rota [62], Dykstra [33], Breiman and Friedman [9]); (4) computing *Bergman kernels* (Skwarczynski [65,66], Ramadanov and Skwarczynski [57,58]); (5) *approximating multivariate functions by sums of univariate ones* (Golomb [41], von Golitschek and Cheney [70], Deutsch [28]); (6) *least change secant updates* (Powell [56], Dennis [25], Dennis and Schnabel [26], Dennis and Walker [27]); (7) *multigrid methods* (Braess [8], Gatski, Grosch, and Rose [37,38], Gilbert and Light [40], and Xu and Zikatanov [73]); (8) *conformal mapping* (Wegmann [71]); (9) *image restoration* (Youla [74], Youla and Webb [75]); (10) *computed tomography* (Smith, Solmon, and Wagner [68], Hamaker and Solmon [44], Censor and Herman [19], and Censor [15].) (See Deutsch [29] for a more detailed description of these ten areas and of what is contained in the above-cited references.)

2. THE METHOD OF ALTERNATING PROJECTIONS

Throughout this paper, unless explicitly stated otherwise, H will always denote a (real) Hilbert space with inner product $\langle \cdot, \cdot \rangle$ and norm $\| \cdot \|$. By a *subspace* of H we will mean a nonempty *linear subspace.* If $T : H \to H$ is a bounded linear operator, the set of *fixed points* of T is the set defined by $\operatorname{Fix} T := \{x \in H \mid Tx = x\}$. Clearly, $\operatorname{Fix} T$ is a closed subspace of H, which is nonempty since $0 \in \operatorname{Fix} T$.

Definition 2.1 *T is called:*

- **nonexpansive** *iff* $\|T\| \leq 1$;
- **nonnegative** *iff* $\langle Tx, x\rangle \geq 0$ *for every* $x \in H$;
- **asymptotically regular** *iff* $T^{n+1}x - T^n x \to 0$ *for every* $x \in H$.

If M is a closed subspace of H, we denote the orthogonal projection onto M by P_M. In other words, for every $x \in H$, $P_M(x)$ is the (unique) closest point in M to x: $\|x - P_M(x)\| = \inf_{y \in M} \|x - y\|$. It is well-known and easy to verify that P_M is self-adjoint and idempotent (i.e., $P_M^2 = P_M$) and hence asymptotically regular, nonexpansive, and nonnegative.

Theorem 2.2 *Let T be a bounded linear operator on H and M a closed subspace of H. Of the following three statements, the first two are equivalent and each implies the third:*

1. $\lim_n \|T^n x - P_M x\| = 0$ *for each* $x \in H$;
2. $M = \operatorname{Fix} T$ *and* $T^n x \to 0$ *for each* $x \in M^\perp$;
3. $M = \operatorname{Fix} T$ *and T is asymptotically regular.*

Moreover, if T is nonexpansive, then all three statements are equivalent. In general, the first and last statements are not equivalent.

A direct elementary proof of Theorem 2.2 was given in [4]. A related result can be deduced from a general theorem of Baillon, Bruck, and Reich [1, Theorem 1.1]. Namely, if T is nonexpansive and asymptotically regular, then $\{T^n x\}$ converges to a fixed point of T for each $x \in H$. By applying the mean ergodic theorem for contractions (see [60] or [61, pp. 407-410]), it follows that $T^n x \to P_M x$ for each $x \in H$, where $M = \operatorname{Fix} T$. This provides another proof of the implication (3) $\Rightarrow$ (1) in the case where T is nonexpansive.
From Theorem 2.2 we immediately obtain the following corollary.

Corollary 2.3 *Let T be nonexpansive and $M = \operatorname{Fix} T$. Then*

$$\lim_{n\to\infty} \|T^n x - P_M x\| = 0 \text{ for every } x \in H$$

if and only if T is asymptotically regular.

The natural question raised by this corollary is "which nonexpansive T are asymptotically regular?" One such class of operators are those that are nonexpansive, nonnegative, and self-adjoint. More generally, any operator that can be expressed as the composition of such operators is asymptotically regular.

Theorem 2.4 *Let $T_1, T_2, \ldots, T_k$ be nonexpansive, nonnegative, and self-adjoint linear operators on H, $T := T_1T_2\cdots T_k$, and $M = \text{Fix}\,T$. Then (T is asymptotically regular and hence)*

$$\lim_{n\to\infty} \|T^n x - P_M x\| = 0 \text{ for every } x \in H.$$

Theorem 2.4 is implicit in Halperin [43] and explicit in Smarzewski [67] (see also [4, Theorem 2.5]). Another proof of Theorem 2.4, kindly suggested by Simeon Reich, runs along the following lines. If T is a nonexpansive, nonnegative, and self-adjoint linear operator, then $\|x - Tx\|^2 \leq \|x\|^2 - \|Tx\|^2$ for each $x \in H$. (This fact was also used in the proofs given in [43], [67], and [4].) Thus the operator T is *strongly nonexpansive* in the sense of Bruck and Reich [11]. Since the composition of strongly nonexpansive operators is strongly nonexpansive [11, Proposition 1.1], and a strongly nonexpansive mapping that has a fixed point is asymptotically regular [11, Corollary 1.1], an application of Corollary 2.3 yields the result.

Theorem 2.4 is more encompassing than one might suspect at first glance. Indeed, as a consequence of a general result of Hundal [47], we have that *every finite rank operator with norm strictly less than one on an infinite-dimensional Hilbert space H is a composition of finitely many orthogonal projections! Hence, every finite rank operator on an infinite-dimensional Hilbert space is, up to a constant factor, a composition of finitely many orthogonal projections.*

Using the observation that the fixed point set of a composition of orthogonal projections is the intersection of the ranges of these projections, we immediately deduce the following two consequences of Theorem 2.4.

Corollary 2.5 (von Neumann-Halperin MAP) *Let $M_1, M_2, \ldots, M_k$ be closed subspaces in H and $M = \cap_1^k M_i$. Then, for every $x \in H$,*

$$\lim_{n\to\infty} \|(P_{M_k}P_{M_{k-1}}\cdots P_{M_1})^n x - P_M x\| = 0.$$

Corollary 2.6 (Symmetric version of the MAP) *Let $M_1, M_2, \ldots, M_k$ be closed subspaces in H and $M = \cap_1^k M_i$. Then, for every $x \in H$,*

$$\lim_{n\to\infty} \|(P_{M_1}P_{M_2}\cdots P_{M_k}P_{M_{k-1}}\cdots P_{M_1})^n x - P_M x\| = 0.$$

Corollary 2.5 was proved by von Neumann [54] for the case $k = 2$ and by Halperin [43] for any $k \geq 2$. We should mention that Bruck and Reich [11, Theorem 2.1] have extended Halperin's theorem to uniformly convex spaces.

In keeping with the spirit of this conference, we should note that the MAP and the symmetric MAP are iteration schemes that are "parallelizable" (by using the product space approach of Pierra [55], see also Bauschke and Borwein [2]). Related to this, Lapidus [52, Proposition 26] has established the following parallelizable version of the MAP (see also Reich [59, Theorem 1.7] for a generalization of Lapidus's result to uniformly convex spaces).

Theorem 2.7 *Let $M_1, M_2, \ldots, M_m$ be closed subspaces of the Hilbert space H and let $M = \cap_1^m M_i$. If $0 < \lambda_i < 1$ and $\sum_1^m \lambda_i = 1$, then for each $x \in H$,*

$$\lim_n \left\| \left(\sum_1^m \lambda_i P_{M_i} \right)^n x - P_M x \right\| = 0.$$

3. ACCELERATING THE MAP

According to Corollary 2.3, an iterative algorithm for determining a fixed point for a nonexpansive asymptotically regular mapping T on H can be described as follows: given any $x \in H$, set $x_0 = x$ and

$$x_n = Tx_{n-1} \ (= T^n x) \quad \text{for every } n \geq 1. \tag{1}$$

Then the sequence $\{x_n\}$ converges to $P_{\mathrm{Fix}\,T}x$. We will refer to this algorithm as the "ordinary" algorithm as opposed to the "accelerated" algorithm that we will define below.

However, even in the special case where T is a composition of a finite number of orthogonal projections (i.e., the MAP), it is well-known that convergence of this sequence can be arbitrarily slow for some problems (see Franchetti and Light [35] and Bauschke, Borwein, and Lewis [3]). For *rate of convergence* results for the MAP, see Smith, Solmon, and Wagner [68], Kayalar and Weinert [51], and Deutsch and Hundal [31]. Thus it is of both theoretical and practical importance to find ways of accelerating the convergence of the algorithm described in equation (1). We will be particularly interested in a line search method that, in the case when T is the composition of orthogonal projections, goes back at least to Gubin, Polyak, and Raik [42], and was also studied by Gearhart and Koshy [39]. The idea is that if at the nth stage of an algorithm, we are at a point x, we obtain the next point by finding the "best" point on the line joining x with Tx. This suggests the following definition.

Definition 3.1 *The* **accelerated mapping** A_T *of T is defined on H by the relation*

$$A_T(x) := t(x)Tx + (1 - t(x))x,$$

where the scalar $t(x)$ is given by

$$t(x) := \begin{cases} \langle x, x - Tx\rangle / \|x - Tx\|^2 & \text{if } x \notin \mathrm{Fix}\,T, \\ 1 & \text{if } x \in \mathrm{Fix}\,T. \end{cases}$$

The motivation for this definition is provided by the following result of [4].

Lemma 3.2 *Let T be a nonexpansive and $M = \mathrm{Fix}\,T$. Then $A_T(x)$ is the (unique) point on the line through x and Tx which is closest to $P_M x$. In other words, $A_T(x)$ is the "best" point on the line through x and Tx if our object is to get as close to the desired limiting point $P_M x$ as possible.*

Since [4] has not been published yet, we will outline the proofs of some of its main results. In particular, the main step in the proof of the preceding lemma is to establish the following identity for each $x \in H$, $y \in M$, and $t \in \mathbb{R}$:

$$\|A_T(x) - y\|^2 = \|tTx + (1-t)x - y\|^2 - (t - t(x))^2 \|Tx - x\|^2.$$

With the accelerated mapping, we can now define an "accelerated algorithm". It is described as follows: given $x \in H$, set $x_0 = x, x_1 = Tx_0$, and

$$x_n = A_T(x_{n-1}) \quad (= A_T^{n-1}(Tx)) \text{ for every } n \geq 2.$$

A natural question that can now be posed is: When does the accelerated sequence $\{A_T^{n-1}(Tx)\}$ converge at least as fast to $P_M x$ as does the original sequence $\{T^n x\}$? As we will see below, the answer is that it often does. However, in general we do *not* know whether the accelerated sequence even converges! Nevertheless, we can state the following positive result in this direction.

Theorem 3.3 ([5]). *If T is nonexpansive and $M = \operatorname{Fix} T$, then for each $x \in H$ the accelerated sequence $\{A_T^{n-1}(Tx)\}$ converges* **weakly** *to $P_M x$. In particular, when $\dim H$ is finite, the accelerated algorithm converges in norm to $P_M x$.*

The best possible scenario for the accelerated algorithm is the following result of Hundal [48].

Proposition 3.4 *If $\dim H \leq 3$, M_1 and M_2 are closed subspaces of H, $M = M_1 \cap M_2$, and $T = P_{M_2} P_{M_1}$, then $A_T(Tx) = P_M x$ for every $x \in H$. In other words, the accelerated MAP converges in (at most)* **two** *steps.*

This result should be contrasted with the obvious fact that if M_1 and M_2 are two distinct nonorthogonal one-dimensional subspaces in the Euclidean plane, then the ordinary MAP starting from any $x \neq 0$ is never a finite algorithm.

In higher dimensions, the preceding proposition does not hold in general. For example, we have another result of Hundal [48]:

Proposition 3.5 *If $\dim H \geq 4$, there exist subspaces M_1, M_2 and $x \in H$ such that if $T = P_{M_2} P_{M_1}$ and $M = M_1 \cap M_2$, then $A_T^{n-1}(Tx) \neq P_M x$ for every n. In other words, the accelerated algorithm does not converge in a finite number of steps.*

The proof is based on first choosing an orthonormal set $\{e_1, e_2, e_3, e_4\}$ in H and setting $M_1 = \operatorname{span}\{u_1, u_2\}$, $M_2 = \operatorname{span}\{e_1, e_3\}$, where $u_1 = \frac{1}{\sqrt{2}}(e_1 + e_2)$, and $u_2 = \frac{1}{2}(\sqrt{3}e_3 + e_4)$. If $M = M_1 \cap M_2$ and $T = P_{M_2} P_{M_1}$, we see that $M = \{0\}$ and $T^n(\alpha e_1 + \beta e_3) = (\frac{1}{2})^n \alpha e_1 + (\frac{3}{4})^n \beta e_3$ for every n. A simple induction shows that if $x = \alpha e_1 + \beta e_3$ with $\alpha \neq 0$ and $\beta \neq 0$, then $A_T^n(Tx) \neq 0$ for each n. Since $P_M x = 0$, this proves the proposition.

The main positive result that we know is the following.

Theorem 3.6 ([4, Theorem 3.20]) *Let T be nonexpansive, nonnegative, and self-adjoint, and let $M = \operatorname{Fix} T$. Then for each $x \in H$ and $n \in \mathbb{N}$,*

$$\|A_T^{n-1}(Tx) - P_M x\| \leq \|T^n x - P_M x\|.$$

In other words, the accelerated algorithm is at least as fast as the original.

The proof of this result is somewhat lengthy and involved. We give only an outline. First one can prove by induction that it suffices to verify that

$$\|T^{m-1}(A_T(y))\| \leq \|T^m y\| \tag{2}$$

for each $y \in M^{\perp}$ with $\|y\| = 1$ and every $m \in \mathbb{N}$ with $m \geq 2$. For such y and m, set $N = \operatorname{span}\{y, Ty, T^2y, \ldots, T^m y\}$ and $S = P_N T P_N$, and note that S is a compact, self-adjoint, and nonexpansive linear operator with $N \subset M^{\perp}$. Then apply the spectral theorem for compact self-adjoint operators (see [20, p. 47]) to obtain an orthonormal set of n eigenvectors $\{v_1, v_2, \ldots, v_n\}$ of S such that

$$Sx = \sum_1^n \lambda_i \langle x, v_i\rangle v_i, \quad x \in H,$$

where λ_i is the (nonzero) eigenvalue corresponding to v_i. Using the fact that T is nonnegative and S is nonexpansive, it can be seen that $0 < \lambda_i \leq 1$. Since the possibility λ_i leads to $Tv_i = v_i$, or $v_i \in M \cap M^{\perp} = \{0\}$, which is absurd, we must have

$0 < \lambda_i < 1$ for every i.

Next an induction shows that

$$T^j y = S^j y = \sum_{i=1}^n \langle y, v_i\rangle \lambda_i^j v_i \tag{3}$$

for $j = 1, 2, \ldots, m$, and

$$T^{m-1}(A_T(y)) = \sum_{i=1}^n \langle y, v_i\rangle \lambda_i^{m-1}[1 - (1 - \lambda_i)t(y)]v_i. \tag{4}$$

Using (3) and (4), we see that (2) holds iff

$$\sum_1^n \langle y, v_i\rangle^2 \lambda_i^{2m-2}[1 - (1 - \lambda_i)t(y)]^2 \leq \sum_1^n \langle y, v_i\rangle^2 \lambda_i^{2m},$$

which, after some algebra, may be rewritten as

$$q(t(y)) \leq 0, \tag{5}$$

where

$$q(t) = \alpha t^2 - 2\beta t + \gamma, \qquad \alpha = \sum_1^n \langle y, v_i\rangle^2 \lambda_i^{2m-2}(1 - \lambda_i)^2,$$

$$\beta = \textstyle\sum_1^n \langle y, v_i\rangle^2 \lambda_i^{2m-2}(1 - \lambda_i), \quad \gamma = \displaystyle\sum_1^n \langle y, v_i\rangle^2 \lambda_i^{2m-2}(1 - \lambda_i^2).$$

It is easy to see that $0 < \alpha < \beta < \gamma$, $\beta = \frac{1}{2}(\alpha + \gamma)$, and the zeros of q are given by $t_{min} = \alpha^{-1}(\beta - \sqrt{\beta^2 - \alpha\gamma})$ and $t_{max} = \alpha^{-1}(\beta + \sqrt{\beta^2 - \alpha\gamma})$. It follows that $t_{min} = 1$ and

$t_{max} = \frac{\gamma}{\alpha} > 1$. Since q has a positive leading coefficient, $q(t) \leq 0$ iff $t_{min} \leq t \leq t_{max}$, i.e., $1 \leq t \leq \frac{\gamma}{\alpha}$. Thus to prove that $q(t(y)) \leq 0$, it suffices to prove that

$$1 \leq t(y) \leq \frac{\gamma}{\alpha}. \tag{6}$$

Using (3), one deduces $t(y) \geq 1$. Finally, one can verify the right side of (6) by using the fact that the function

$$h(t) = \frac{\sum_{i=1}^{n} \langle y, v_i \rangle^2 \lambda_i^t (1 - \lambda_i)}{\sum_{j=1}^{n} \langle y, v_j \rangle^2 \lambda_j^t (1 - \lambda_j)^2}$$

is increasing for $t \geq 0$, and hence $h(0) \leq h(2m-2)$.

In particular, taking $T = Q^*Q$ in the preceding theorem, where $Q = P_{M_k} P_{M_{k-1}} \cdots P_{M_1}$, we immediately obtain the next corollary.

Corollary 3.7 *If* $T = P_{M_1} P_{M_2} \cdots P_{M_k} P_{M_{k-1}} \cdots P_{M_1}$ *and* $M = \cap_1^k M_i$, *then for each* $x \in H$,

$$\|A_T^{n-1}(Tx) - P_M x\| \leq \|T^n x - P_M x\| \text{ for each } n \in \mathbb{N}.$$

In other words, the "accelerated symmetric MAP" is at least as fast as the "symmetric MAP".

For the (nonsymmetric) MAP, the analogous result is valid in the case of two subspaces, but not in general. This is a consequence of the next two results from [4].

Theorem 3.8 *If* $T = P_{M_2} P_{M_1}$ *and* $M = M_1 \cap M_2$, *then for each* $x \in H$ *and* $n \in \mathbb{N}$,

$$\|A_T^{n-1}(Tx) - P_M x\| \leq \|T^n x - P_M x\|.$$

That is, the accelerated MAP is at least as fast as the MAP in the case of **two** *subspaces.*

Example 3.9 *If* $\dim H \geq 5$, *there exist five 2-dimensional subspaces* $M_1, M_2, \ldots, M_5$ *in* H *with* $M := \cap_1^5 M_i = \{0\}$ *and a point* $x_0 \in H$ *such that letting* $T = P_{M_5} P_{M_4} P_{M_3} P_{M_2} P_{M_1}$, *we have*

$$\|A_T^{n-1}(Tx_0) - P_M x_0\| > \|T^n x_0 - P_M x_0\| = 0 \text{ for each } n \geq 3.$$

In particular, the accelerated MAP is slower *than the MAP for this example.*

In fact, let $\{e_1, e_2, \ldots, e_5\}$ be an orthonormal set in H, $M_1 = \text{span}\{e_2, e_3\}$, $M_2 = \text{span}\{e_2 + e_4, e_3 + e_5\}$, $M_3 = \text{span}\{e_4, e_5\}$, $M_4 = \{e_1 + e_4, e_2 + e_5\}$, and $M_5 = \text{span}\{e_1, e_2\}$. Letting $T = P_{M_5} P_{M_4} P_{M_3} P_{M_2} P_{M_1}$, it is easy to verify that $M := \text{Fix}\, T = \{0\}$, $\|T\| = \frac{1}{4}$, and

$$Tx = \frac{1}{4}\langle x, e_2 \rangle e_1 + \frac{1}{4}\langle x, e_3 \rangle e_2$$

for each $x \in H$. Setting $x_0 = 4e_3$, it follows that $Tx_0 = e_2$, $T^2 x_0 = \frac{1}{4} e_1$, and $T^n x_0 = 0$ for each $n \geq 3$. Since $P_M x_0 = 0$, it suffices to show that $z_n := A_T^n(Tx_0) \neq 0$ for each $n \geq 0$. Since the range of T is M_5, it follows that

$$z_n = \alpha_n e_1 + \beta_n e_2 \quad (n = 0, 1, 2, \ldots) \tag{7}$$

for some scalars α_n, β_n. If $z_n \neq 0$ for each n, then it follows that for all $n \geq 3$,

$$\|A_T^{n-1}(Tx_0) - P_M x_0\| = \|A_T^{n-1}(Tx_0)\| = \|z_{n-1}\| > 0 = \|T^n x_0\| = \|T^n x_0 - P_M x_0\|,$$

which shows that the accelerated MAP is *slower* that the MAP beginning with the third iterate. Thus it sufffices to verify that $z_n \neq 0$ for each n.

One can show that

$$Tz_n = \frac{1}{4}\beta_n e_1 \quad (n = 0, 1, \ldots,), \tag{8}$$

$$z_{n+1} = 0 \Leftrightarrow \beta_n = 0, \tag{9}$$

and

$$\beta_{n+1}\beta_n = \alpha_{n+1}\left(\frac{1}{4}\beta_n - \alpha_n\right) \quad (n = 0, 1, 2, \ldots,). \tag{10}$$

In particular, if $\beta_n \neq 0$, then

$$\beta_{n+1} = \alpha_{n+1}\left(\frac{1}{4} - \mu_n\right), \text{ where } \mu_n = \frac{\alpha_n}{\beta_n}. \tag{11}$$

If the result were false, let n_0 denote the smallest integer such that $z_{n_0+1} = 0$. By (9), $\beta_{n_0} = 0$. A direct computation reveals that $\alpha_0 = 0, \beta_0 = 1$, and $\alpha_1 = \frac{4}{17}, \beta_1 = \frac{1}{17}$. Hence $n_0 \geq 2$. Using (11), one can deduce that $\beta_n \neq 0$ for all $0 \leq n \leq n_0 - 2$. By considering the set

$$Q^* = \left\{\frac{p}{q} \mid p, q \text{ integers with } p \text{ even and } q \text{ odd}\right\},$$

one can prove that $\mu_n \in Q^*$ for $0 \leq n \leq n_0 - 1$. In particular, $\mu_{n_0-1} \neq \frac{1}{4}$. Then (11) implies that

$$0 = \beta_{n_0} = \alpha_{n_0}\left(\frac{1}{4} - \mu_{n_0-1}\right),$$

which implies $\alpha_{n_0} = 0$. But this shows that $z_{n_0} = 0$, which is a contradiction.

4. RELATED WORK

We mention here some work related to that of the preceding section. In his doctoral thesis, Dyer [32] considers two methods of accelerating the Kaczmarz algorithm for solving the linear system of equations $Ax = b$ in $\mathbb{R}^n$. Both are line searches. If T denotes the composition of projections in the Kaczmarz case, then one of the methods chooses the $n + 1$st point x_{n+1} on the line joining the points $T^i x_n$ and $T^{i+1}x_n$ in such a way as to minimize the residual $\|Ax - b\|$. Brezinski and Redivo-Zagalia [10] posed a hybrid method by combining two iterative schemes for solving a linear system of equations. Given two approximations x' and x'' of a solution to the linear system $Ax = b$, they used the point $\alpha x' + (1 - \alpha)x''$ on the line joining x' and x'' to minimize the new residual $r = b - A(\alpha x' + (1 - \alpha)x'')$. Censor, Eggermont, and Gordon [16] made a case for choosing small relaxation parameters in the Kaczmarz method for solving linear systems of equations. Hanke and Niethammer [45] analyzed this method with small parameters, and proposed accelerating the method using Chebyshev semi-iteration. Björk and Elfving [6] had earlier proposed a symmetric variant of this method which, however, required more computation per cycle. Mandel [53] showed that *strong* underrelaxation for solving a system of linear inequalities in a cyclical fashion provides better estimates for the rate of convergence than does relaxation. Dax [23] described and analyzed a line search acceleration technique for solving linear systems of equations. If y_k is the element at hand at the beginning of the $k+1$st cycle, his idea was to perform a certain number of iterations of the basic iteration which determines a point z_k. Then a line search is performed along the line through the points y_k and z_k to determine the next point y_{k+1}.

In the case of finding the best approximation in the intersection of a finite collection of closed convex sets, it is known that the MAP may not converge or even if it does, the limit may not be the correct element of the intersection. Moreover, for in this general convex sets case, the convergence generally depends on the initial point and there are "tunnelling" effects that slow the convergence. In an attempt to alleviate these difficulties, Crombez [21,22] proposed and studied a variant of the MAP. Iusem and De Pierro [49] gave an accelerated version of Cimmino's algorithm for solving the convex feasibility problem in finite dimensions. This was similar to an algorithm given by Censor and Elfving [18,17] for solving linear inequalities.

De Pierro and Lópes [24] proposed a line search method for accelerating the finite-dimensional symmetric linear complementarity problem. Garcia-Palomares [36] exhibited a superlinearly convergent projection algorithm for solving the convex inequality problem under certain assumptions. The method involves choosing appropriate supersets of the sets which comprise the intersection, and projecting onto these supersets rather than the original sets. However, the projection onto each superset requires solving a quadratic programming problem.

5. RATE OF CONVERGENCE

In this section, we give some rate of convergence theorems for both the original and the accelerated algorithms. Recall that by Corollary 2.3, we have that if T is nonexpansive and asymptotically regular and $M = \operatorname{Fix} T$, then

$$T^n x \to P_M x \quad \text{for each } x \in H.$$

Now we would like to compare how fast this convergence is relative to the accelerated algorithm.

We first note that

$$\|T^n - P_M\| = \|T^n(I - P_M)\| = \|T^n P_{M^\perp}\| = \|(TP_{M^\perp})^n\|,$$

where we used the fact that T commutes with $P_{M^\perp}$ (see [4]), and hence

$$\|T^n x - P_M x\| \le \|(TP_{M^\perp})^n\| \, \|x\| \le \|TP_{M^\perp}\|^n \|x\| = c(T)^n \|x\| \text{ for each } x \in H,$$

where

$$c(T) := \|TP_{M^\perp}\|.$$

Note that in the case when $T = P_M P_N$, then $c(T) = c(M,N)$ is the cosine of the angle between the subspaces M and N and $c(T) < 1$ if and only if $M^\perp + N^\perp$ is closed (see Deutsch [30], Theorem 13). More generally, if $T = P_{M_k} P_{M_{k-1}} \cdots P_{M_1}$, then Bauschke, Borwein, and Lewis [3] showed that $c(T) < 1$ if and only if $M_1^\perp + M_2^\perp + \cdots + M_k^\perp$ is closed.

Theorem 5.1 (Comparison of Rates; [4], Theorem 3.16) *For any $x \in H$ and $n \in \mathbb{N}$,*

$$\|T^n x - P_M x\| \leq \|(TP_{M^\perp})^n\| \, \|x - P_M x\| \leq c(T)^n \|x - P_M x\|$$

and

$$\|A_T^{n-1}(Tx) - P_M x\| \leq \rho_n(x) c(T)^n \|x - P_M x\|,$$

where $\{\rho_n(x)\}$ is a decreasing sequence of nonnegative real numbers with $\rho_1(x) \leq 1$.

Note that while this theorem shows that the upper bound for the accelerated algorithm is no larger than that of the ordinary algorithm, it does *not* state that the accelerated algorithm is at least as fast as the ordinary one. In fact, Example 3.9 above shows that the *opposite* is sometimes true!

With additional conditions on T, a faster rate of convergence is possible.

Theorem 5.2 (Accelerated Convergence Rate; [4], Theorem 3.29) *Let T be nonexpansive, nonnegative, and self-adjoint, and $M = \operatorname{Fix} T$. Then for each $x \in H$ and $n \in \mathbb{N}$,*

$$\|A_T^{n-1}(Tx) - P_M x\| \leq \frac{c(T)^n}{[2 - c(T)]^{n-1}} \|x - P_M x\|.$$

6. OPEN PROBLEMS

Recall (Corollary 2.3) that if T is nonexpansive and asymptotically regular on H and $M = \operatorname{Fix} T$, then

$$\lim_{n\to\infty} \|T^n x - P_M x\| = 0 \text{ for each } x \in H.$$

Problem 6.1 *Must the accelerated algorithm for T also converge? That is, must*

$$\lim_{n\to\infty} \|A_T^{n-1}(Tx) - P_M x\| = 0 \textit{ for each } x \in H?$$

From [5] we know that the sequence $\{A_T^{n-1}(Tx)\}$ always converges **weakly** to $P_M x$, but whether the convergence must also be strong is still open.

The answer is affirmative, i.e.,

$$\lim_{n\to\infty} \|A_T^{n-1}(Tx) - P_M x\| = 0,$$

when any one of the following conditions hold:

1. H is finite-dimensional.
2. T is nonexpansive, nonnegative, and self-adjoint (Theorem 3.6), e.g., if $Q = P_{M_k} P_{M_{k-1}} \cdots P_{M_1}$ and $T = Q^* Q$.

3. $T = P_{M_2}P_{M_1}$ is the composition of two orthogonal projections (Theorem 3.8).

4. $c(T) < 1$ (Theorem 4.1). In particular, if $T = P_{M_k}P_{M_{k-1}} \cdots P_{M_1}$ and $M_1^{\perp} + M_2^{\perp} + \cdots + M_1^{\perp}$ is closed, then $c(T) < 1$ (Bauschke, Borwein, and Lewis [3]).

There is an iterative algorithm, called *Dykstra's algorithm* (Boyle and Dykstra [7]), that asymptotically obtains the best approximation $P_C(x)$ to x from the set $C = \cap_1^k C_i$, where each C_i is a *closed convex* subset of H, using only information about the best approximations onto the individual sets C_i. In the particular case when the convex sets are actually *linear subspaces*, then Dykstra's algorithm yields the same iterates that the MAP does, but not in general. Since it is known that there are numerous important problems that can be treated by Dykstra's algorithm that are not of the form that can be treated by the MAP, it is natural to pose the following problem.

Problem 6.2 *Is there an acceleration scheme for Dykstra's algorithm?*

REFERENCES

1. J.-B. Baillon, R. E. Bruck, and S. Reich, On the asymptotic behavior of nonexpansive mappings and semigroups in Banach spaces, *Houston J. Math.* **4** (1978) 1–9.
2. H. H. Bauschke and J. M. Borwein, On the convergence of von Neumann's alternating projection algorithm for two sets, *Set-Valued Analysis* **1** (1993) 185–212.
3. H. H. Bauschke, J. M. Borwein, and A. S. Lewis, The method of cyclic projections for closed convex sets in Hilbert space, in *Recent Developments in Optimization Theory and Nonlinear Analysis*, Contemporary Mathematics **204**, Amer. Math. Soc., Providence, R.I., 1997, 1–38.
4. H. Bauschke, F. Deutsch, H. Hundal, S.-H. Park, Accelerating the convergence of the method of alternating projections, preprint (July, 1999), 33 pages.
5. H. Bauschke, F. Deutsch, H. Hundal, S.-H. Park, Fejér monotonicity and weak convergence of an accelerated method of projections, in *Constructive, Experimental, and Nonlinear Analysis* (M. Théra, ed.), Amer. Math. Soc. CMS Proceedings series, to appear.
6. A. Björck and T. Elfving, Accelerated projection methods for computing pseudoinverse solutions of systems of linear equations, BIT **19** (1979) 145–163.
7. J. P. Boyle and R. L. Dykstra, A method for finding projections onto the intersection of convex sets in Hilbert spaces, in *Advances in Order Restricted Statistical Inference*, Lecture Notes in Statistics, Springer-Verlag, New York, 1985, 28–47.
8. D. Braess, The contraction number of a multigrid method for solving the Poisson equation, *Numer. Math.* **37** (1981) 387–404.
9. L. Breiman and J. H. Friedman, Estimating optimal transformations for multiple regression and correlation, *J. Amer. Statist.* **80** (1985) 580–598.
10. C. Brezinski and M. Redivo-Zagalia, Hybrid procedures for solving linear systems, *Numer. Math.* **67** (1994) 1–19.
11. R. E. Bruck and S. Reich, Nonexpansive projections and resolvents of accretive operators in Banach spaces, *Houston Jour. Math.* **3** (1977) 459–470.
12. D. L. Burkholder, Successive conditionial expectations of an integrable function, *Ann. Math. Statist.* **33** (1962) 887–893.

13. D. L. Burkholder and Y. S. Chow, Iterates of conditional expectation operators, *Proc. Amer. Math. Soc.* **12** (1961) 887–893.
14. Y. Censor, Row-action methods for huge and sparse systems and their applications, *SIAM Rev.* **23** (1981) 444–466.
15. Y. Censor, Parallel application of block-iterative methods in medical imaging and radiation therapy, *Math. Programming* **42** (1988) 307–325.
16. Y. Censor, P. P. B. Eggermont, and D. Gordon, Strong underrelaxation in Kaczmarz's method for inconsistent systems, *Numer. Math.* **41** (1983) 83–92.
17. Y. Censor and T. Elfving, A nonlinear Cimmino algorithm, Report 33, Department of Mathematics, University of Haifa, 1981.
18. Y. Censor and T. Elfving, New methods for linear inequalities, *Linear Algebra Applics.* **42** (1982) 199–211.
19. Y. Censor and G. T. Herman, On some optimization techniques in image reconstruction from projections, *Appl. Numer. Math.* **3** (1987) 385–391.
20. J. B. Conway, *A Course in Functional Analysis*, second edition, Springer-Verlag, New York, 1990.
21. G. Crombez, Finding projections onto the intersection of convex sets in Hilbert spaces. II, preprint (September, 1997), 22 pages.
22. G. Crombez, Improving the speed of convergence in the method of projections onto convex sets, preprint (April, 2000), 27 pages.
23. A. Dax, Line search acceleration of iterative methods, *Linear Alg. Applics.* **130** (1990) 43–63.
24. A. R. DePierro and J. M. Lópes, Accelerating iterative algorithms for symmetric linear complementarity problems, *Intern. J. Computer Math.* **50** (1994) 35–44.
25. J. E. Dennis, Jr., On some methods based on Broyden's secant approximation to the Hessian, in *Numerical Methods for Non-linear Optimization*, Academic Press, London, 1972, 19–34.
26. J. E. Dennis, Jr., and R. B. Schnabel, Least change secant updates for quasi-Newton methods, *SIAM Review* **21** (1979) 443–459.
27. J. E. Dennis, Jr., and H. F. Walker, Convergence theorems for least-change secant update methods, *SIAM J. Numer. Anal.* **18** (1981) 949–987.
28. F. Deutsch, Von Neumann's alternating method: The rate of convergence, in *Approximation Theory IV* (C. K. Chui, L. L. Schumaker, and J. D. Ward, eds.), Academic Press, New York, 1983, 427–434.
29. F. Deutsch, The method of alternating orthogonal projections, in *Approximation Theory, Spline Functions and Applications* (S. P. Singh, ed.), Kluwer Academic Publ., The Netherlands, 1992, 105–121.
30. F. Deutsch, The angle between subspaces of a Hilbert space, in *Approximation Theory, Spline Functins and Applications*, (S. P. Singh, ed.), Kluwer Academic Publ., The Netherlands, 1995, 107–130.
31. F. Deutsch and H. Hundal, The rate of convergence for the method of alternating projections, II, *J. Math. Anal. Applic.* **205** (1997) 381–405.
32. J. Dyer, Acceleration of the convergence of the Kaczmarz method and iterated homogeneous transformations, *doctoral dissertation*, University of California, Los Angeles, 1965.

33. R. L. Dykstra, An algorithm for restricted least squares regression, *J. Amer. Statist. Assoc.* **78** (1983) 837–842.
34. P. P. B. Eggermont, G. T. Herman, and A. Lent, Iterative algorithms for large partitioned linear systems, with applications to image reconstruction, *Linear Algebra Applics.* **40** (1981) 37–67.
35. C. Franchetti and W. Light, On the von Neumann alternating algorithm in Hilbert space, *J. Math. Anal. Appl.* **114** (1986) 305–313.
36. U. M. Garcia-Palomares, A superlinearly convergent projection algorithm for solving the convex inequality problem, *Operations Research Letters* **22** (1998) 97–103.
37. T. B. Gatski, C. E. Grosch, and M. E. Rose, A numerical study of the two-dimensional Navier-Stokes equations in vorticity-velocity variables, *J. Comput. Physics* **48** (1982) 1–22.
38. T. B. Gatski, C. E. Grosch, and M. E. Rose, The numerical solution of the Navier-Stokes equations for 3-dimensional, unsteady, incompressible flows by compact schemes, *J. Comput. Physics* **82** (1988) 289–329.
39. W. B. Gearhart and M. Koshy, Acceleration schemes for the method of alternating projections, *J. Comp. Applied Math.* **26** (1989) 235–249.
40. J. Gilbert and W. A. Light, Multigrid methods and the alternating algorithm, in *Algorithms for Approximation*, Oxford University Press, New York, 1987, 447–458.
41. M. Golomb, Approximation by functions of fewer variables, in *On Numerical Approximation* (R. E. Langer, ed.), University of Wisconsin Press, Madison, Wisconsin, 1959, 275–327.
42. L. G. Gubin, B. T. Polyak, and E. V. Raik, The method of projections for finding the common points of convex sets, *USSR Comp. Math. Phys.* **7** (1967) 1–24.
43. I. Halperin, The product of projection operators, *Acta. Sci. Math. (Szeged)* **23** (1962) 96–99.
44. C. Hamaker and D. C. Solmon, The angles between null spaces of X-rays, *J. Math. Anal. Appl.* **62** (1978) 1–23.
45. M. Hanke and W. Niethammer, On the acceleration of Kaczmarz's method for inconsistent linear systems, *Linear Algebra Applics.* **130** (1990) 83–98.
46. G. T. Herman, A. Lent, and P. H. Lutz, Iterative relaxation methods for image reconstruction, *Communications of the ACM* **21** (1978) 152–158.
47. H. Hundal, Compact operators as products of projections, this volume.
48. H. Hundal, private communication (August, 2000).
49. A. N. Iusem and A. R. De Pierro, Convergence results for an accelerated nonlinear Cimmino algorithm, *Numer. Math.* **49** (1986) 367–378.
50. S. Kaczmarz, Angenäherte Auflösung von Systemen linearer Gleichungen, *Bull. Internat. Acad. Pol. Sci. Lett.* **A35** (1937) 355–357.
51. S. Kayalar and H. Weinert, Error bounds for the method of alternating projections, *Math. Control Signals Systems* **1** (1988) 43–59.
52. M. L. Lapidus, Generalization of the Trotter–Lie formula, *Integral Equations Operator Theory* **4** (1981) 366–415.
53. J. Mandel, Convergence of the cyclical relaxation method for linear inequalities, *Math. Programming* **30** (1984) 218–228.
54. J. von Neumann, *Functional Operators-Vol. II. The Geometry of Orthogonal Spaces*,

Annals of Math. Studies # 22, Princeton University Press, 1950. (This is a reprint of mimeographed lecture notes first distributed in 1933.)
55. G. Pierra, Decomposition through formalization in a product space, *Math. Programming* **28** (1984) 96–115.
56. M. J. D. Powell, A new algorithm for unconstrained optimization, in *Nonlinear Programming* (J. B. Rosen, O. L. Mangasarian, and K. Ritter, eds.), Academic Press, New York, 1970.
57. I. P. Ramadanov and M. L. Skwarczyński, An angle in $L^2(C)$ determined by two plane domains, *Bull. Polish Acad. Sci. Math.* **32** (1984) 653–659.
58. I. P. Ramadanov and M. L. Skwarczyński, An angle in $L^2(C)$ determined by two plane domains, in *Constructive Theory of Functions '84*, Sofia, Bulgaria, 726–730.
59. S. Reich, A limit theorem for projections, *Linear and Multilinear Algebra* **13** (1983) 281–290.
60. F. Riesz and B. v. Sz. Nagy, Über Kontraktionen des Hilbertschen Raumes, *Acta. Sci. Math. Szeged* **10** (1943) 202–205.
61. F. Riesz and B. Sz. Nagy, *Functional Analysis*, Frederick Unger Pub. Co, New York, 1955.
62. G.-C. Rota, An "alternierende Verfahren" for general positive operators, *Bull. Amer. Math. Soc.* **68** (1962) 95–102.
63. H. Salehi, On the alternating projections theorem and bivariate stationary stochastic processes, *Trans. Amer. Math. Soc.* **128** (1967) 121–134.
64. H. A. Schwarz, Ueber einen Grenzübergang durch alternirendes Verfahren, *Vierteljahrsschrift der Naturforschenden Gesellschaft in Zürich* **15** (1870) 272–286.
65. M. Skwarczyński, Alternating projections in complex analysis, *Complex Analysis and Applications '83*, 1985, Sofia, Bulgaria.
66. M. Skwarczyński, A general description of the Bergman projection, *Annales Polonici Math.* **46** (1985).
67. R. Smarzewski, Iterative recovering of orthogonal projections, preprint (December, 1996).
68. K. T. Smith, D. C. Solmon, and S. L. Wagner, Practical and mathematical aspects of the problem of reconstructing objects from radiographs, *Bull. Amer. Math. Soc.* **83** (1977) 1227–1270.
69. K. Tanabe, Projection method for solving a singular system of linear equations and its applications, *Numer. Math.* **17** (1971) 203–214.
70. M. von Golitschek and E. W. Cheney, On the algorithm of Diliberto and Straus for approximating bivariate functions by univariate ones, *Num. Funct. Anal. Approx.* **1** (1979) 341–363.
71. R. Wegmann, Conformal mapping by the method of alternating projections, *Numer. Math.* **56** (1989) 291–307.
72. N. Wiener and P. Masani, The prediction theory of multivariate stochastic processes, II: The linear predictor, *Acta Math.* **93** (1957) 95–137.
73. J. Xu and L. Zikatanov, The method of alternating projections and the method of subspace corrections in Hilbert space, preprint (June, 2000), 24 pages.
74. D. C. Youla, Generalized image restoration by the method of alternating orthogonal projections, *IEEE Trans. Circuits Syst.* **CAS-25** (1978) 694–702.

75. D. C. Youla and H. Webb, Image restorations by the method of convex projections: Part 1-Theory, *IEEE Trans. Medical Imaging* **MI-1** (1982) 81–94.

Inherently Parallel Algorithms in Feasibility and Optimization and their Applications
D. Butnariu, Y. Censor and S. Reich (Editors)

PICO: AN OBJECT-ORIENTED FRAMEWORK FOR PARALLEL BRANCH AND BOUND*

Jonathan Eckstein[a], Cynthia A. Phillips[b], and William E. Hart[b]

[a]MSIS Department, Faculty of Management and RUTCOR, Rutgers University, 640 Bartholomew Road, Piscataway, NJ 08854-8003, USA

[b]Optimization and Uncertainty Estimation Department, Computation, Computers, and Math Center, Sandia National Laboratories, Mail Stop 1110, P.O. Box 5800, Albuqurque, NM 87185-1110, USA

This paper describes the design of PICO, a C++ framework for implementing general parallel branch-and-bound algorithms. The PICO framework provides a mechanism for the efficient implementation of a wide range of branch-and-bound methods on an equally wide range of parallel computing platforms. We first discuss the basic architecture of PICO, including the application class hierarchy and the package's serial and parallel layers. We next describe the design of the serial layer, and its central notion of manipulating subproblem states. Then, we discuss the design of the parallel layer, which includes flexible processor clustering levels and communication rates, various load balancing mechanisms, and a non-preemptive task scheduler running on each processor. We close by describing the application of the package to a simple branch-and-bound method for mixed integer programming, along with computational results on the ASCI Red massively parallel computer.

1. INTRODUCTION

This paper describes PICO (*P*arallel *I*nteger and *C*ombinatorial *O*ptimizer), an object-oriented framework for parallel implementation of branch-and-bound algorithms. Parts of PICO are based on CMMIP [7–10], a parallel branch-and-bound code for solving mixed integer programming problems on the now-obsolete CM-5 parallel computer. Although CMMIP exhibited excellent scalability to large numbers of processors, its design had a number of limitations. First, it implemented only one specific branch-and-bound algorithm for a single (if fairly general) class of problems; adapting it to more specialized classes of problems or to use more advanced algorithmic techniques, such as branch and cut, proved awkward. Second, CMMIP was designed to showcase certain properties of the

*This work was performed in part at Sandia National Laboratories. Sandia is a multiprogram laboratory operated by Sandia corporation, a Lockheed Martin Company, for the United States Department of Energy under Contract DE-AC04-94AL85000. This work was also supported in part by National Science Foundation Grant CCR-9902092. We would like to thank ILOG, Inc. and DASH Optimization, Inc. for making their linear programming solvers available to us.

CM-5, whose communication network was fast relative to is processors, with specialized hardware and operating system support for particular kinds of interprocessor communication. To run efficiently on systems with less specialized communication capabilities, CMMIP had to be significantly restructured, as in [9].

By contrast, PICO is meant to be a very general parallel branch-and-bound environment. Using object-oriented techniques, the parallel search "engine" is cleanly separated from the details of the application and computing platform. This approach allows the same underlying parallel search code to be used on a wide variety of branch-and-bound applications, ranging from those not requiring linear programming bounds to branch-and-cut methods. The basic search engine also has a large number of run-time parameters that allow the user to control the quantity and pattern of interprocessor communication. On systems with relatively slow, unsophisticated communication abilities, such as networks of workstations, these parameters can be "tuned" so that the code attempts a relatively low level of interprocessor communication. For massively parallel processor (MPP) supercomputers with efficient hardware and software communication support, the parameters can be adjusted to make full use of the available communication bandwidth. A key design goal is that, in such MPP environments, PICO retain and extend the level of scalability exhibited by CMMIP.

Flexible software environments, sometimes called "shells," for constructing branch-and-bound algorithms are not a new idea. Broadly, prior research in this area divides into two main categories. On the one hand, there are a number of packages aimed at serial implementation of sophisticated linear-programming-based branch-and-bound methods, like branch and cut or branch and price. Perhaps the most popular of these environments is MINTO [26], and another noteworthy contribution is the ABACUS object-oriented branch-and-cut environment [17]. PICO bases some of its basic class hierarchy structure on ABACUS.

On the other hand, there have also been a number of tools for parallel implementation of general branch-and-bound algorithms, such as PUBB [30,31], BoB [4], and PPBB-Lib [34]. These efforts stem primarily from the computer science community, and emphasize parallel implementation, but appear to be designed primarily for applications with simple bounding procedures not based on linear programming.

More recently, there have been efforts at parallel implementation of advanced linear-programming-based branching methods. Some recent contributions and works in progress include PARINO [21], SYMPHONY [27], and BCP [1]. SYMPHONY and BCP, which are similar to one another, are broadly extensible libraries, but their current design does not emphasize scalability to large numbers of processors.

The primary goal of PICO is to eventually combine capabilities similar to all of these tools with the scalability and flexibility of the work-distribution scheme of CMMIP, along with additional adjustments to accomodate a large variety of hardware platforms. PICO allows a wide range of branch-and-bound methods, linear-programming-based and otherwise, to use the same basic parallel search engine. This sharing occurs at the link level, without requiring recompilation. While branch-and-cut capabilities are not yet present, PICO's design should allow them to be added cleanly, without major changes to the components already developed.

The literature of parallel branch and bound is vast, and it is not possible to give a comprehensive review here. Two fairly comprehensive but not particularly recent surveys may be found in [14] and [20, Chapter 8]; [5] is a more recent but less comprehensive survey.

The remainder of this paper describes the design of the current components of PICO. Section 2 describes PICO's overall design, including its class hierarchy and the separation of the package into serial and parallel layers. Section 3 discusses the design of the serial layer, which contains a number of novel features not present in earlier branch-and-bound "shells," including the ability to use variable search "protocols," and the key notion of subproblem state. Section 4 describes the parallel layer, and how to migrate an application from the serial to the parallel layer. The parallel layer implements a compound work distribution scheme that generalizes CMMIP's, but can run on general hardware platforms. We also discuss the parallel layer's use of multiple threads of control, which are arbitrated by a non-preemptive "stride" scheduler, and the issue of terminating the computation. Section 5 describes a sample application of PICO to mixed integer programming, without cutting planes, and gives preliminary computational results on the "Janus" massively parallel computer, which consists of thousands of Pentium-II processors. Section 6 presents conclusions and outlines future development plans.

2. THE GENERAL DESIGN OF PICO

PICO is a C++ class library, providing a hierarchically-organized set of capabilities that users may combine and extend to form their own applications. As with ABACUS, extending the core capabilities of PICO requires the development of derived classes that express the additional required functionality. This design is in some sense more demanding than interfaces like MINTO, which simply require that the user provide auxilary functions that are linked into the executable. However, we believe that the class library approach is more powerful and flexible, allowing the use of multiple inheritance, which is critical to PICO's design.

Figure 1 shows a simplified conceptual design, or *inheritance tree* of the library; elements with solid boundaries have been completed or are in an advanced state of development, while those with dashed boundaries are in the planning or early development stages. At the root of the inheritance tree is the *PICO core*, which provides basic capabilities for describing and parallelizing branch-and-bound algorithms. Branch-and-bound methods that have specialized bounding procedures not requiring direct use of linear programming can be defined as direct descendents of the PICO core. Currently there is one such algorithm, for solving binary knapsack problems.

The PICO MIP package extends the PICO core by providing generic capabilities for solving mixed integer programs, using standard LP solvers to solve their linear programming relaxations, as will be described in Section 5. For specialized MIP applications, the PICO MIP can itself be extended and refined by, for example, employing application-specific branching rules, fathoming rules, and heuristic methods for generating incumbent solutions. For example, it is straightforward to extend the PICO MIP to include LP-based approximation algorithms for scheduling problems, using the LP relaxation available at each node. We have exploited this flexibility for applications like the PANTEX production planning problem, which addresses a difficult scheduling problem within the U.S.

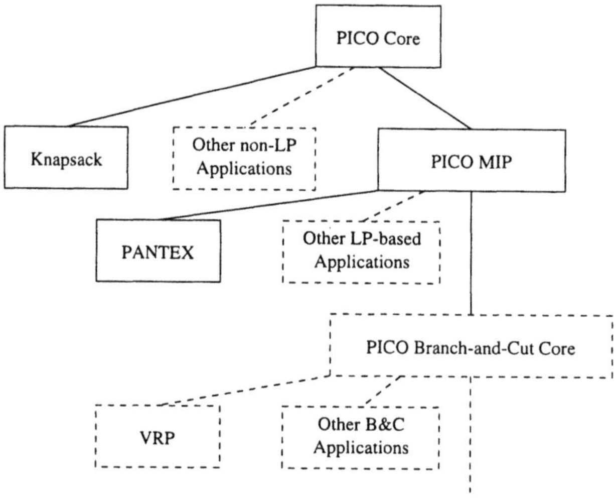

Figure 1. The current conceptual inheritance tree for PICO, in simplified form. Dashed lines indicate components in the planning or early development stages.

Department of Energy (DOE). This application will be discussed in a separate paper; a brief description of the modeling issues may be found in [12], which is a slightly expanded (but older) version of this paper.

We plan to extend the PICO hierarchy by creating a generic branch-and-cut capability that extends PICO MIP. This generic branch-and-cut could then be extended and refined as needed to handle specific applications such as the traveling salesman problem (TSP) or vehicle routing problem (VRP).

PICO consists of two "layers," the *serial layer* and the *parallel layer*. The serial layer provides an object-oriented means of describing branch-and-bound algorithms, with essentially no reference to parallel implementation. The serial layer's design, described in Section 3, has some novel features, and we expect it to be useful in its own right. For users uninterested in parallelism, or simply in the early stages of algorithm development, the serial layer allows branch-and-bound methods to be described, debugged, and run in a familiar, serial development environment.

The parallel layer contains the core code necessary to create parallel versions of serial applications. To parallelize a branch-and-bound application developed with the serial layer, the user simply defines new classes derived from both the serial application and the parallel layer. A fully-operational parallel application only requires the definition of a few additional methods for these derived classes, principally to tell PICO how to pack application-specific problem and subproblem data into MPI message buffers, and later unpack them.

Any parallel PICO application constructed in this way inherits the full capabilities

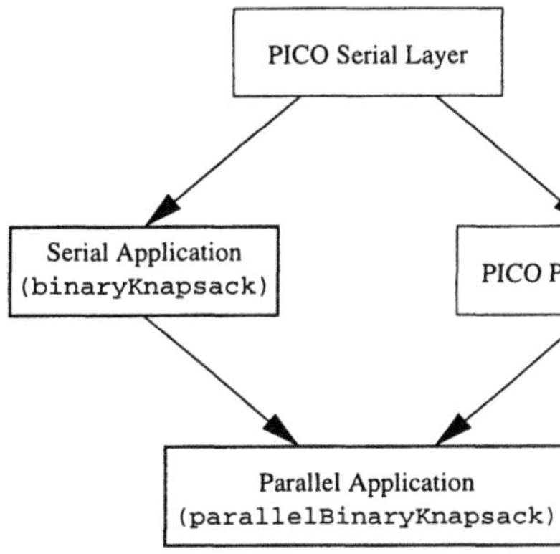

Figure 2. The conceptual relationships of PICO's serial layer, the parallel layer, a serial application (in this case, `binaryKnapsack`), and the corresponding parallel application (in this case, `parallelBinaryKnapsack`).

of the parallel layer, including a wide range of different parallel work distribution and load balancing strategies, and user-configurable levels of interprocessor communication. Application-specific refinements to the parallelization can then be added by the user, but are not required. Section 4 describes the parallel layer, and Figure 2 shows the conceptual relationship between the two layers, a serial application, and its parallelization.

PICO's parallel layer was designed using a distributed-memory computation model, which requires message passing to communicate information between processors. We expect that this design will be effective on a wider range of systems than a design based on a shared-memory model. Although it is always possible to emulate distributed memory and message passing on hardware with memory-sharing capabilities, it is much more difficult to do the reverse. Emulating shared memory without hardware support may involve significant loss of efficiency and low-level control. Furthermore, shared memory, either hardware-supported or emulated, becomes rarer, more expensive, or both as the number of processors increases. Aside from portability considerations, we are particularly interested in the using PICO on DOE's massively parallel systems, for which distributed-memory approaches have proven particularly effective.

The parallel layer uses the MPI [32] standard application interface to pass messages between processors. There are currently two portable interface standards for constructing message-passing programs, MPI and PVM [13]. We selected MPI because it is designed to be customized for maximum performance on MPP systems like Janus, the ASCI Red supercomputer. The design of PVM stresses the ability to operate on heterogeneous platforms, at some sacrifice in performance.

3. THE SERIAL LAYER

This section describes the design of PICO's serial layer. Subsection 3.1 introduces PICO's fundamental object classes, and outlines how an application is constructed by deriving application classes from the fundamental ones. Subsection 3.2 discusses the notion of *subproblem states*, which constitute PICO's basic mechanism for interacting with an application. Different variations on the basic branch-and-bound theme may be specified by manipulating subproblem states in different ways or by storing the subproblems in various kinds of *pools*, as described in Subsection 3.3. Finally, Subsection 3.4 catalogs the various parameters and virtual methods that control the serial layer.

3.1. Object-oriented design

To define a serial branch-and-bound algorithm, a PICO user extends two principal classes in the PICO serial layer, `branching` and `branchSub`. The `branching` class stores global information about a problem instance and contains methods that implement various kinds of serial branch-and-bound algorithms, as described below. The `branchSub` class stores data about each subproblem in the branch-and-bound tree, and it contains methods that perform generic operations on subproblems. This basic organization is borrowed from ABACUS [17], but it is more general, since there is no assumption that cutting planes or linear programming are involved.

For example, our binary knapsack solver defines a class `binaryKnapsack`, derived from `branching`, to describe the capacity of the knapsack and the possible items to be placed in it. We also define a class `binKnapSub`, derived from `branchSub`, which describes the status of the knapsack items at nodes of the branching tree (*i.e.* included, excluded, or undecided); this class describes each node of the branch-and-bound tree. Each object in a subproblem class like `binKnapSub` contains a pointer back to the corresponding instance of the global class, in this case `binaryKnapsack`. Through this pointer, each subproblem object can find global information about the branch-and-bound problem. Finally, both `branching` and `branchSub` are derived from a common base class, `picoBase`, which mainly contains common symbol definitions and run-time parameter objects. Figure 3 illustrates the basic class hierarchy for a serial PICO application.

3.2. Subproblem states

A novel feature of PICO, even at the serial level, is that subproblems remember their *state*. Each subproblem progresses through as many as six of these states, `boundable`, `beingBounded`, `bounded`, `beingSeparated`, `separated`, and `dead`, as illustrated in Figure 4.

A subproblem always comes into existence in state `boundable`, meaning that little or no bounding work has been done for it, although it still has an associated bound value; typically, this bound value is simply inherited from the parent subproblem. Once PICO starts work on bounding a subproblem, its state becomes `beingBounded`, and when the bounding work is complete, the state becomes `bounded`.

Once a problem is in the `bounded` state, PICO may decide to branch on it, a process also called "separation" or "splitting." At this point, the subproblem's state becomes `beingSeparated`. Once separation is complete, the state becomes `separated`, at which point the subproblem's children may be created. Once the last child has been created, the

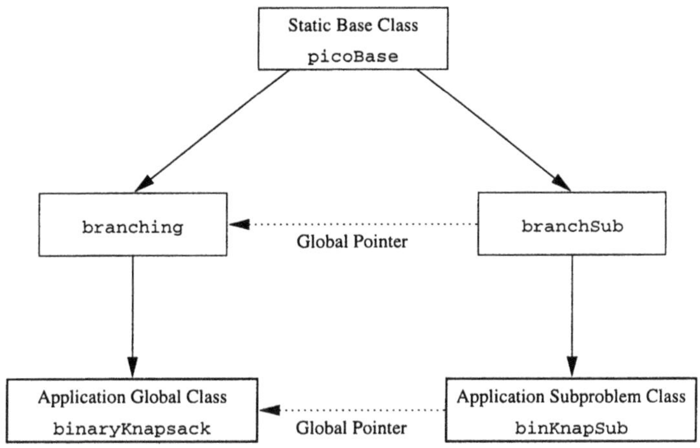

Figure 3. Basic class hierarcy for a serial PICO application (in this case, **binaryKnapsack**, with corresponding subproblem class **binKnapSub**).

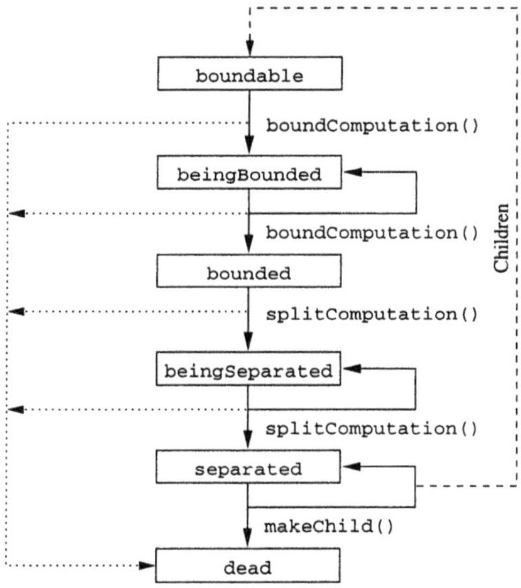

Figure 4. PICO's subproblem state transition diagram. It is possible that a single application of **boundComputation** may take a subproblem from the **boundable** state, through **beingBounded**, to **bounded**. Similarly, a single use of **splitComputation** may move a subproblem from **bounded**, through **beingSeparated**, to **separated**.

subproblem's state becomes `dead`, and it may be deleted from memory. Subproblems may also become `dead` at earlier points in their existence, because they have been fathomed or represent portions of the search space containing no feasible solutions.

Class `branchSub` contains three abstract virtual methods, namely `boundComputation`, `splitComputation`, and `makeChild`, that are responsible for applying these state transitions to subproblems. PICO's search framework interacts with applications primarily through these methods; defining a PICO branch-and-bound application essentially consists of providing definitions for these three operators for the application subproblem class (*e.g.* `binKnapSub`).

The `boundComputation` method's job is to move the subproblem to the `bounded` state, updating the data member `bound` to reflect the computed value. The `boundComputation` method is allowed to pause an indefinite number of times, leaving the subproblem in the `beingBounded` state. The only requirement is that any subproblem will eventually become `bounded` after some finite number of applications of `boundComputation`. This flexibility allows PICO to support branch-and-bound variants where one can suspend bounding one subproblem, set it aside, and turn one's attention to another subproblem or task in the meantime. The subproblem's bound, reflected in the data member `bound`, may change at each step of this process.

The `splitComputation` method's job is similar to `boundComputation`'s, but it manages the separation process. Eventually it must change the problem state to `separated`, set the data member `totalChildren` to the number of child subproblems. Before that, however, it is allowed to return an indefinite number of times with the problem left in the `beingSeparated` state. This feature allows PICO to implement branch-and-bound methods where the work in separating a subproblem is substantial and might need to be paused to attend to some other subproblem or task. The subproblem's bound can be updated by `splitComputation` if the separation process yields additional information about it.

Finally, `makeChild` returns a single child of the subproblem it is applied to, which must be in the `separated` state. After the last child has been made, the subproblem becomes `dead`.

In addition to `boundComputation`, `splitComputation`, and `makeChild`, several additional virtual methods must to be defined to complete the specification of a branch-and-bound application, as described in Subsection 3.4.

3.3. Pools, handlers, and the search framework

PICO's serial layer orchestrates serial branch-and-bound search through a module called the "search framework" (literally, `branching::searchFramework`). The search framework acts as an attachment point for two user-specifiable objects, the *pool* and the *handler*, whose combination determines the exact "flavor" of branch and bound implemented.

The pool object dictates how the currently active subproblems are stored and accessed, which effectively determines the branch-and-bound search order. Currently, there are three kinds of pool: a heap sorted by subproblem bound[2], a stack, and a FIFO queue. If

[2]Objects of type `branchSub` have a member called `integralityMeasure` which may be used by the application to measure how far a subproblem is from being completely feasible (that is, from having `candidateSolution` yield `TRUE`; see Subsection 3.4 and Table 1). If two subproblems have identical

the user specifies the heap pool, then PICO will follow a best-first search order; specifying the stack pool results in a depth-first order, and specifying the queue results in a breadth-first order. For particular applications, however, users may implement additional kinds of pools, thus specifying other search orders.

Critically, at any instant in time, the subproblems in the pool may in principle represent any mix of states: for example, some might be `boundable`, and others `separated`. This feature gives the user flexibility in specifying the *bounding protocol*, which is a separate issue from the search order; the "handler" object implements a particular bounding protocol.

To illustrate what a bounding protocol is, consider the usual branch-and-bound method for mixed integer programming as typically described by operations researchers: one removes a subproblem from the currently active pool, and computes its linear programming relaxation bound. If the bound is strong enough to fathom the subproblem, it is discarded. Otherwise, one selects a branching variable, creates two child subproblems, and inserts them into the pool. This type of procedure is an example of what is often called "lazy" bounding (see for instance [6]), because it views the bounding procedure as something time-consuming (like solving a large linear program) that should be delayed if possible. In the PICO framework, lazy bounding is implemented by keeping all subproblems in the active pool in the `boundable` state.

An alternative approach, common in work originating from the computer science community, is usually called "eager" bounding (again, see [6] for an example of this terminology). Here, all subproblems in the pool have already been bounded. One picks a subproblem out of the pool, immediately separates it, and then forms and bounds each of its children. Children whose bounds do not cause them to be fathomed are returned to the pool.

Lazy and eager bounding each have their own advantages and disadvantages, and the best choice may depend on both the application and the implementation environment. Typically, implementors seek to postpone the most time-consuming operations in the hope that the discovery of a better incumbent solution will make them unnecessary. So, if the bounding operation is much more time-consuming than separation, lazy bounding is most appealing. If the bounding operation is very quick, but separation more difficult, then eager bounding would be more appropriate. Eager bounding may save some memory since leaf nodes of the search tree can be processed without entering the pool, but has a larger task granularity, resulting in somewhat less potential for parallelism.

Because PICO's serial layer stores subproblem states and lets the user specify a handler object, it gives users the freedom to specify lazy bounding, eager bounding, or other protocols. The search framework routine simply extracts subproblems from the pool and passes them to the handler until the pool becomes empty. Currently, there are three possible handlers, `eagerHandler`, `lazyHandler`, and `hybridHandler`, although the user is free to write additional handlers if greater flexibility is required.

The `eagerHandler` and `lazyHandler` methods implement eager and lazy bounding respectively by trying to keep subproblems in the bounded and boundable states, respectively. Problems that become fathomed or `dead` anywhere in the process of applying the

---

bounds, the one with the lower `integralityMeasure` will be placed higher in the heap, since it presumably is more likely to lead to an improved incumbent solution.

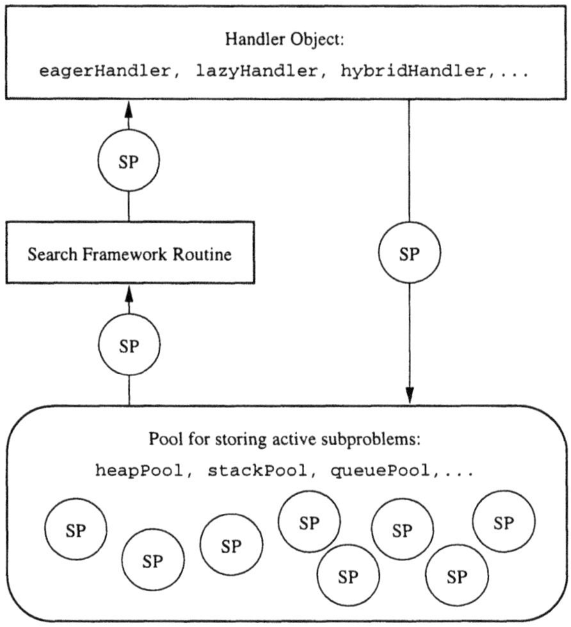

Figure 5. The search framework, pool, and handler. Each "SP" indicates a branch-and-bound subproblem.

boundComputation and splitComputation methods are immediately discarded. Further, to permit users to pause the bounding or separation processes, any subproblem that remains in the beingBounded state after the application of boundComputation, or in the beingSeparated state following the application of splitComputation, is immediately returned to the pool.

The hybridHandler implements a strategy that is somewhere between eager and lazy bounding, and is perhaps the most simple and natural given PICO's concept of subproblem states. Given any subproblem, hybridHandler performs a single application of either boundComputation, splitComputation, or makeChild, to try to advance the subproblem one transition through the state diagram. If the subproblem's state is boundable or beingBounded, it applies boundComputation once. If the subproblem's state is bounded or beingSeparated, it applies splitComputation once. Finally, if the state is separated, the handler performs one call to makeChild, and inserts the resulting subproblem into the pool. It discards any subproblems becoming fathomed or dead at any point in this process.

The combination of multiple handlers, multiple pool implementations, and the user's freedom in implementing boundComputation and splitComputation create considerable flexibility in the kinds of branch-and-bound methods that the serial layer can implement. Figure 5 depicts the relationship of the search framework, pool, and handler.

3.4. Serial layer virtual methods and run-time parameters

In addition to boundComputation, splitComputation, and makeChild, there are a number of additional abstract virtual methods in classes branching and branchSub that the user must define in order to fully describe an application of the PICO serial layer. The two classes also have a large number of other virtual methods that may be overridden at the user's option. Table 1 describes all the required virtual methods and the more commonly-overridden optional ones. The most noteworthy are candidateSolution, updateIncumbent, and incumbentHeuristic, all members of branchSub.

The candidateSolution method tells the PICO search handlers whether a bounded subproblem needs to be separated at all. If this method returns TRUE, PICO assumes that the computed bound is in fact the objective value of the best feasible solution within the portion of the search space corresponding to the subproblem. If this bound is better than the current incumbent, the handler calls updateIncumbent to replace the current incumbent with the solution corresponding to the subproblem. By default, updateIncumbent simply stores the corresponding objective value; in most applications, the user will override this function to also store an application-specific representation of the solution for possible later output.

The incumbentHeuristic method provides a way for the user to specify a heuristic that takes a subproblem for which candidateSolution returns FALSE and attempts to perturb it into a feasible solution. In applications with linear-programming-based bounds, for example, incumbentHeuristic might try to round the fractional variables found in the linear programming relaxation. If it succeeds in finding a solution better than the incumbent, the heuristic should call updateIncumbent. PICO's handlers only call incumbentHeuristic for a subproblem if the method haveIncumbentHeuristic returns TRUE. The default implementation of haveIncumbentHeuristic returns FALSE.

Required virtual method definitions: class `branching`

`readIn`	Read problem instance data from the command line and/or data file.
`blankSub`	Construct an empty subproblem.

Required virtual method definitions: class `branchSub`

`setRootComputation`	Turn a blank subproblem into the root problem.
`boundComputation`	Compute (perhaps only partially) the bound of a subproblem in the `boundable` or `beingBounded` state.
`splitComputation`	Separate (perhaps only partially) a subproblem in the `bounded` or `beingSeparated` state.
`makeChild`	Create a child of a subproblem in the `separated` state.
`candidateSolution`	Return `TRUE` if a `bounded` subproblem does not need further separation.

Optional virtual method definitions: class `branching` (selected)

`preprocess`	Preliminary computation before starting to search
`aPrioriBound`	"Quick and dirty" bound on the best possible solution (*e.g.* for knapsack, the sum of all item values).
`initialGuess`	Initial heuristic feasible solution value (*e.g.* for knapsack, the result of a simple greedy heuristic).
`haveIncumbentHeuristic`	Return `TRUE` if there is a heuristic for forming possible feasible solutions from `bounded` subproblems.
`serialPrintSolution`	Write incumbent solution to an output stream.

Optional virtual method definitions: class `branchSub` (selected)

`incumbentHeuristic`	Attempt to produce a feasible solution from the current (bounded) subproblem.
`updateIncumbent`	Store a new incumbent.

Table 1
Virtual method members of the `branching` and `branchSub` classes. Derived classes should also have their own constructors and destructors.

PICO also provides a general mechanism for specifying run-time parameters. Table 2 shows most of the parameters that control the operation of serial PICO. Technically, the parameters are static objects that are members of the base class `picoBase`. User applications can add an unlimited number of their own run-time parameters, so long as their names are different from those in `picoBase`.

Name	Meaning	Default
`statusPrintFrequency`	Number of subproblems to bound between status printouts	1000
`depthFirst`	Use a stack as the pool, causing depth-first search	`FALSE`
`breadthFirst`	Use a queue for the pool, causing breadth-first search	`FALSE`
`lazyBounding`	Use the lazy bounding handler	`FALSE`
`eagerBounding`	Use the eager bounding handler	`FALSE`
`relTolerance`	A subproblem may be fathomed if its bound is within this factor of the incumbent objective value	10^{-7}
`absTolerance`	A subproblem may be fathomed if its bound is within this absolute distance of the incumbent objective value	0
`validateLog`	Causes quality-control output to be dumped to a file for later analysis	`FALSE`

Table 2
Run-time parameters defined in the static base class `picoBase`, which control the generic operation of the serial layer. The default pool is a heap, which causes best-first search, and the default handler is `hybridHandler`.

4. THE PARALLEL LAYER

We now turn our attention to the parallel layer. Subsection 4.1 describes the basic class hierarchy of the parallel layer and how applications derive from it.

Subsection 4.2 then describes how PICO partitions the set of available processors into *clusters*. Within each cluster, work is managed in a roughly "master-slave" manner by a single processor called the *hub*. However, the degree of "micromanagement" attempted by the hub can be controlled by various run-time parameters. Subsection 4.3 describes the allocation of work within a cluster, along with the notion of *subproblem tokens*. Tokens are small data structures that contain enough information to schedule the bounding of a subproblem, but not enough to actually compute the bound. PICO is much freer about communicating tokens between processors than it is about communicating complete subproblems.

If the number of processors is relatively small, it may be sufficient, depending on how long it takes to process each subproblem, to have a single cluster spanning all available processors. In this case, one hub processor essentially manages the entire search. However, as the number of processors grows, this single hub would become overloaded, and multipler clusters are required. Subsection 4.4 describes how the search workload is balanced between multiple clusters, which is much the same as in [8,10].

Subsection 4.5 describes how and why the PICO parallel layer manages multitasking on each processor. In particular, Subsection 4.5.1 describes the basic scheduling algorithm. The tasks managed by the algorithm are called *threads*, and they are organized into *groups*. Subsection 4.5.2 describes the different kinds of threads and their group divisions. Subsection 4.5.3 outlines how the scheduler is initialized and the search first gets under way, and Subsections 4.5.4 through 4.5.11 then describe each thread provided by the basic parallel layer. Finally, Subsection 4.6 describes the proper detection of the end of the search. When there are multiple clusters, detecting determination is a trickier problem than it might first appear; we adapt some standard techniques from the literature of parallel asynchronous algorithms [24] to detect termination reliably.

4.1. Parallel object-oriented design

The parallel layer's capabilities are encapsulated in the classes `parallelBranching` and `parallelBranchSub`, which have the same function as `branching` and `branchSub`, respectively, except that they perform parallel search of the branch-and-bound tree. Both are derived from a common, static base class `parallelPicoBase`, whose function is similar to `picoBase`, containing mainly common symbol and run-time parameter definitions. Furthermore, each of `parallelBranching` and `parallelBranchSub` is derived from the corresponding class in the serial layer.

To turn a serial application into a parallel application, one must define two new classes. The first is derived from `parallelBranching` and the serial application global class. In the knapsack example, for instance, we defined a new class `parallelBinaryKnapsack` which has both `parallelBranching` and `binaryKnapsack` as `virtual public` base classes. We call this class the *global parallel class.* For each problem instance, the information in the global parallel class is replicated on every processor.

The global parallel class's basic inheritance pattern is repeated for parallel subproblem objects. In the knapsack case, we defined a parallel subproblem class `parBinKnapSub` to have `virtual public` base classes `binKnapSub` and `parallelBranchSub`. As with the serial subproblems, each instance of `parBinKnapSub` has a `parallelBinaryKnapsack` pointer that allows it to locate global problem information. Figure 6 depicts the entire inheritance structure for the parallel knapsack application.

Once this basic inheritance pattern is established, the parallel application automatically combines the description of the application coming from the serial application (in the knapsack case, embodied in `binaryKnapsack` and `binKnapSub`) with the parallel search capabilities of the the parallel layer. For the parallel application to function, however, a few more methods must be defined, as summarized in Table 3.

First, the parallel application global and subproblems classes both require constructors and destructors. However, these methods are essentially trivial to define: the destructors

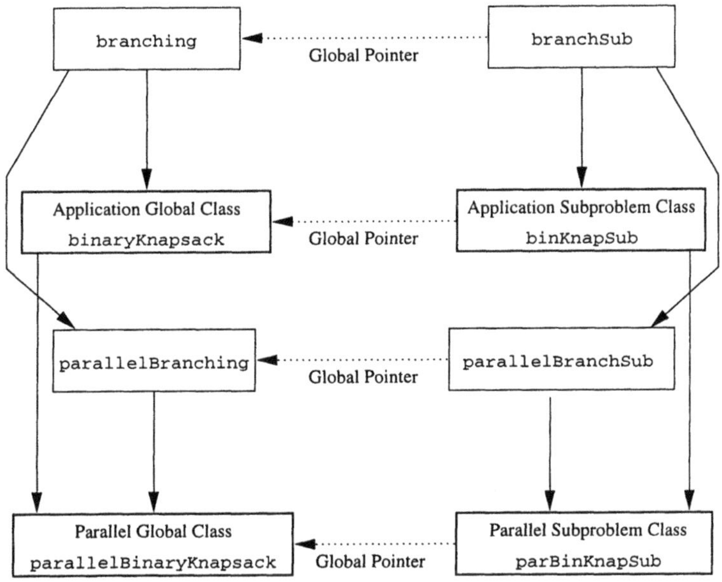

Figure 6. Inheritance structure of the parallel knapsack application. Other parallel applications are similar.

may have empty bodies, and the constructors may simply invoke the constructors for their underlying classes. For technical reasons, the user must define two related methods, **blankParallelSub** in the global class, and **makeParallelChild** in the subproblem class. These methods fulfill the same roles as **blankSub** and **makeChild**, respectively, but in a parallel setting. Typically, these methods do nothing but call the constructor for the parallel subproblem class.

The packing and unpacking virtual methods are of greater interest. The parallel global and subproblem objects each require the definition of two methods, **pack** and **unpack**. The **pack** method is responsible for packing all the application-specific data in an instance of the class into a buffer suitable for sending between processors using the MPI datatype **MPI_PACKED**. The **unpack** method is responsible for unpacking the same data from an MPI receive buffer, and reconstituting the data members of the class instance. The parallel global class **pack** and **unpack** methods are used to distribute the global problem definition before starting the tree search, while the subproblem class **pack** and **unpack** methods are used to send subproblems from one processor to another.

Optionally, the user may also define a **packChild** method, whose functionality is equivalent to **makeChild** followed by a **pack** on the resulting subproblem. This method is discussed further in Subsection 4.5.8.

In addition to the **pack** and **unpack** methods, the parallel application must define one

Required virtual method definitions: class `parallelBranching`

`pack`	Pack application-specific global problem information into a buffer.
`unpack`	Unpack application-specific global problem information from a buffer.
`spPackSize`	Estimate the maximum buffer space needed to pack the application-specific portion of one subproblem.
`blankParallelSub`	Construct an empty subproblem.

Required virtual method definitions: class `parallelBranchSub`

`pack`	Pack application-specific subproblem data into a buffer.
`unpack`	Unpack application-specific subproblem data from a buffer.
`parallelMakeChild`	Construct a single child of the current subproblem, which must be in the `separated` state. Similar to `makeChild`, but returns an object of type `parallelBranchSub`.

Table 3
Abstract virtual methods of the `parallelBranching` and `parallelBranchSub` classes.

additional packing-related method, `spPackSize`. This method, a member of the global parallel class, should return an integer giving the maximum number of bytes required to buffer the application-specific data for a single subproblem. It is allowed to use any information in the global parallel class, and will not be called until the global information has been replicated on all processors. The parallel layer uses this method when allocating buffer space for incoming subproblem information.

We now describe the operation of the parallel layer. The layer is very flexible, but as a result it is also quite complex. For reasons of space, our description is somewhat abbreviated.

4.2. Processor clustering

PICO's parallel layer employs a generalized form of the processor organization used by the later versions of CMMIP [8,10]. Processors are organized into *clusters*, each with one *hub* processor and one or more *worker* processors. The hub processor serves as a "master" in work-allocation decisions, whereas the workers are in some sense "slaves," doing the actual work of bounding and separating subproblems. Unlike CMMIP, however, the degree of control that the hub has over the workers may be varied by a number of run-time parameters, and may not be as tight as a classic "master-slave" system. Further, the hub processor has the option of simultaneously functioning as a worker; CMMIP only permitted this kind of function overlap in clusters consisting of just one processor.

Three run-time parameters, all defined in `parallelPicoBase`, govern the partitioning of processors into clusters: `clusterSize`, `numClusters`, and `hubsDontWorkSize`. First PICO finds the size k of a "typical" cluster via the formula

$$k = \min\left\{\texttt{clusterSize}, \max\left\{\left\lfloor\frac{\overline{p}}{\texttt{numClusters}}\right\rfloor, 1\right\}\right\},$$

where $\overline{p}$ is the total number of processors. Thus, k is the smaller of the cluster sizes that would be dictated by `clusterSize` and `numClusters`. Processors are then gathered into clusters of size k, except that if k does not evenly divide $\overline{p}$, the last cluster will be of size $\overline{p} \bmod k$, which will be between 1 and $k-1$. In clusters whose size is greater than or equal to `hubsDontWorkSize`, the hub processor is "pure," that is, it does not simultaneously function as a worker. In clusters smaller than `hubsDontWorkSize`, the hub processor is also a worker. The rationale for this arrangement is that, in very small clusters, the hub will be lightly loaded, and its spare CPU cycles should be used to help explore the branch-and-bound tree. If a cluster is too big, however, using the hub simultaneously as a worker may unacceptably slow the hub's response to messages from its workers, slowing down the entire cluster. In this case, a "pure" hub is more advantageous.

The value of `hubsDontWorkSize` must be at least 2, so it is impossible to form a cluster with no workers.

4.3. Tokens and work distribution within a cluster

Unlike some "master-slave" implementations of branch and bound, each PICO worker maintains its own pool of active subproblems. This pool may be any of the kinds of pools described in Subsection 3.3, although all workers use the same pool type. Depending on how various run-time parameters are set, however, the pool might be very small, in the extreme case never holding more than one subproblem. Each worker processes its pool in the same general manner as the serial layer: it picks subproblems out of the pool and passes them to a search handler until the pool is empty. There are currently three parallel search handlers, called `eagerHandler`, `lazyHandler`, and `hybridHandler`, which behave in a similar manner to their respective serial counterparts, but with the additional ability to *release* subproblems from the worker to the hub.

For simplicity throughout the remainder of Subsection 4.3, we describe these handlers as if a single cluster spanned all the available processors; in Subsection 4.4, we will amend our description for the case of more than one cluster.

4.3.1. Random release of subproblems

The parallel version of `eagerHandler` decides whether to release a subproblem as soon as it has become `bounded`. The parallel version of `lazyHandler` and `hybridHandler` make the release decision when they create a subproblem. The decision is a random one, with the probability of release controlled by run-time parameters. Released subproblems do not return to the local pool; instead, the worker cedes control over these subproblems to the hub. Eventually, the hub may send control of the subproblem back to the worker, or to another worker.

If the release probability is 100%, then every subproblem is released, and control of subproblems is always returned to the hub at a certain point in their lifetimes (at creation for `lazyHandler` and `hybridHandler`, and upon reaching the `bounded` state for eagerHandler). In this case, the hub and its workers function like a standard "master-slave" system. When the probability is lower, the hub and its workers are less tightly coupled. The release probability is controlled by the run-time parameters minScatterProb, targetScatterProb, and maxScatterProb. The use of three different parameters, instead of a single one, allows the release probability to be sensitive to a worker's load. Basically, if the worker appears to have a fraction $1/w(c)$ of the total work in the cluster, where c denotes the cluster and $w(c)$ the total number of workers in the cluster, then the worker uses the value targetScatterProb. If it appears to have less work, then a smaller value is used, but no smaller than minScatterProb; if it appears to have more work, it uses a larger value, but no larger than maxScatterProb.

In addition to randomly releasing subproblems, each worker automatically releases the first startScatter problems it produces; the startScatter parameter defaults to 5. This feature is intended to speed up the initial dispersion of work.

4.3.2. Subproblem tokens

When a subproblem is released, only a small portion of its data, called a *token* [28,7], is actually sent to the hub. The subproblem itself may move to a secondary pool, called the *server pool*, that resides on the worker. A token consists of only the information needed to identify a subproblem, locate it in the server pool, and schedule it for execution. On a typical 32-bit processor, a token requires 48 bytes of storage, much less than the full data for a subproblem in most applications. Since the hub receives only tokens from its workers, as opposed to entire subproblems, these space savings translate into reduced storage requirements and communication load at the hub.

When making tokens to represent new, boundable subproblems, the parallel version of lazyHandler and hybidHandler take an extra shortcut. Instead of creating a new subproblem with parallelMakeChild and then making a token that points to it, they simply create a token pointing to the parent subproblem, with a special field, whichChild, set to indicate that the token is not for the subproblem itself, but for its children. Optionally, a single token can represent multiple children. If every child of a separated subproblem has been released, the subproblem is moved from the worker pool to the server pool.

4.3.3. Hub operation and hub-worker interaction

Workers that are not simultaneously functioning as hubs periodically send messages to their controlling hub processor. These messages contain blocks of released subproblem tokens, along with data about the workload in the worker's subproblem pool, and other miscellaneous status information.

The hub processor maintains a pool of subproblem tokens that it has received from workers. Again, this pool may be any one of the pools described in Subsection 3.3. Each time it learns of a change in workload status from one of its workers, the hub reevaluates the work distribution in the cluster. The hub tries to make sure that each worker has a sufficient quantity of subproblems, and optionally, that they are of sufficient quality (that is, with bounds sufficiently sufficiently far from the incumbent). Quality balancing is controlled by the run-time parameter qualityBalance, which is TRUE by default. Workload quantity evaluation is via the run-time parameter workerSPThreshHub; if a worker appears to have fewer than this many subproblems in its local pool, the hub judges it "deserving" of more subproblems. If quality balancing is activated, a worker is also judged deserving if the best bound in its pool is worse than the best bound in the hub's pool by a factor exceeding the parameter `qualityBalanceFactor`. Of the workers that deserve work, the hub designates the one with fewest subproblems as being most deserving, unless this number exceeds `workerSPThreshHub`; in that case, the workers are ranked in reverse order of the best subproblem bound in their pools.

As long as there is a deserving worker and the hub's token pool is nonempty, the hub picks a subproblem token from its pool and sends it to the most deserving worker. The message sending the subproblem may not go directly to that worker, however; instead, it goes to the worker that originally released the subproblem. When that worker receives the token, it forwards the necessary subproblem information to the target worker, much as in [7,8,10,28]. This process will be described in more detail in Subsection 4.5.8.

Subproblem dispatch decision are made one at a time; if a token in the hub pool represents several problems, the hub splits it into two, with one token representing one subproblem, and the other any remaining subproblems. It sends the former token and retains the latter for the time being.

However, when a single activation of the hub logic results in multiple dispatch messages to be sent from the hub to the same worker, the hub attempts to pack them, subject to an overall buffer length limit, into a single MPI message, saving system overhead.

If the subproblem release probability is set to 100%, and `workerSPThreshHub` is set to 1, the cluster will function like a classic master-slave system. The hub will control essentially all the active subproblems, and send them to workers whenever those workers become idle. Less extreme parameter settings will reduce the communication load substantially, however, at the cost of possibly greater deviation from the search order that would have been followed by a serial implementation. Also, setting `workerSPThreshHub` larger than 1 helps to reduce worker idleness by giving each worker a "buffer" of subproblems to keep it busy while messages are in transit or the hub is attending to other workers.

The best setting of the parameters controlling the degree of hub-worker communication depends on both the application and the hardware, and may require some tuning, but the scheme has the advantage of being highly flexible without any need for reimplementation or recompilation.

In addition to sending subproblems, the hub periodically broadcasts overall workload information to its workers, so the workers know the approximate relation of their own workloads to other workers'. This information allows each worker to adjust its probability of releasing subproblems appropriately.

4.3.4. Rebalancing

If the probability of workers releasing their subproblems is set too low, or the search process is nearing completion, workers in a cluster may have workloads that are seriously out of balance, yet the hub's token pool may be empty. In this case, the hub has no work to send to underloaded workers. To prevent such difficulties, there is a secondary mechanism, called "rebalancing," by which workers can send subproblem tokens to the hub. If a worker detects that it has a number of subproblems exceeding a user-specifiable factor `workerMaxLoadFactor` times the average load in the cluster, it selects a block of subproblems in its local pool and releases them to the hub. The hub can then redistribute these subproblems to other workers.

4.4. Work distribution between clusters

With any system-application combination, there will be a limit to the size of a cluster that can operate efficiently, even if its hub does not have any worker responsibilities. Depending on the application and the hardware, the hub may simply not be able to keep up with all the messages from its workers, or it may develop excessively long queues of incoming messages.

At this point, one option is to adjust the PICO's run-time parameters to reduce the amount of intra-cluster communication, but if communication is reduced too much, the hub may have difficulty maintaining a proper load balance in the cluster. To be able to use all the available processors, it may then be necessary to partition the system into multiple clusters. Another reason for such partitioning is that particular classes of applications may simply perform better with the more randomized search pattern that results from multiple clusters [8,10].

PICO's method for distributing work between clusters resembles CMMIP's [8,10], with some additional generality: there are two mechanisms for transferring work between clusters, *scattering* and *load balancing*. Scattering comes into play when subproblems are released by the handlers. If there are multiple clusters, and a worker has decided to release a subproblem, the handler makes a second random decision as to whether the subproblems should be released to the worker's own hub or to a cluster chosen at random. This random decision is controlled by the apparent workload of the cluster relative to the entire system, and the parameters `minNonLocalScatterProb`, `targetNonLocalScatterProb`, and `maxNonLocalScatterProb`. When choosing the cluster to scatter to, the probability of picking any particular cluster is proportional to the number of workers it contains, and the worker's own cluster is not excluded. When subproblems are forced to be released through the `startScatter` mechanism, they are always scattered to a random cluster.

To supplement scattering, PICO also uses a form of "rendezvous" load balancing that resembles CMMIP's [8,10]; [23] and [18] also contain earlier, synchronous applications of the same basic idea. This procedure also has the important side effect of gathering and distributing global information on the amount of work in the system, which in turn facilitates control of the scattering process, and is also critical to termination detection in the multi-hub case.

Critical to the operation of the load balancing mechanism is the concept of the *workload* at a cluster c at time t, which we define as

$$L(c,t) \quad = \quad \sum_{P \in C(c,t)} |\overline{z}(c,t) - z(P,c,t)|^{\rho}.$$

Here, $C(c,t)$ denotes the set of subproblems that c's hub knows are controlled by the cluster at time t, $\overline{z}(c,t)$ represents the incumbent value known to cluster c's hub at time t, and $z(P,c,t)$ is the best bound on the objective value of subproblem P known to cluster c's hub at time t. The exponent ρ is either 0, 1, or 2, at the discretion of the user, much as in CMMIP. If $\rho = 0$, only the number of subproblems in the cluster matters. Values of $\rho = 1$ or $\rho = 2$ give progressively higher "weight" to subproblems farther from the incumbent. The default value of ρ is 1.

The rendezvous load balancing mechanism organizes all the cluster hub processors into a balanced tree whose radix (branching factor) is determined by the parameter `loadBalTreeRadix`, with a default value of 2. Periodically, messages "sweep" through this entire tree, from the leaves to the root, and then back down to the leaves. These sweeps are organized into repeating "rounds," each consisting of three sweeps, *synchronization*, *survey*, and *balance*, as follows:

Synchronization Sweep: Each hub does not start participating in the synchronization sweep until one of two conditions is met: either its cluster appears to be idle, or the total CPU time consumed by the load balancing process is less than a parameter-specified fraction `loadBalCPUFrac` of the total elapsed CPU time at the hub. These conditions prevent the load balancing process from "hogging" resources at the hub, but encourage timely load balancing or termination detection if work seems to be running out. Once the conditions are met, each hub makes sure that it has received a message from all of its child hubs (if any). Once all such messages have been received, the hub then sends a message to its parent, unless it is at the root of the tree. Once the root receives messages from all its children, it initiates a broadcast down the tree that the rest of the load balancing process may proceed. This technique is designed so that the survey sweep, which follows immediately, can start in a roughly synchronized way.

Survey Sweep: Each processor waits to receive workload information from its children, if any. It adds these workloads to its own current workload, and forwards the result up the tree. The root is then able to compute an approximate total workload for the system, which it broadcasts down the tree. The messages in this sweep also contain information on the incumbent values $\bar{z}(c, t)$ used to compute the cluster workloads, and other miscellaneous information that is aggregated as the messages pass up the tree. If there is any mismatch between the incumbent values used at the various clusters, or a similar mismatch between the hub and any worker within a cluster, the survey is repeated immediately. Such a mismatch means that a new incumbent value is currently being broadcast (as will be described in Subsection 4.5.7), and some processors have not had an opportunity to prune their subproblem pools to reflect this new incumbent. The aggregate workload information is therefore inconsistent, and must be gathered again. Once consistent information has been gathered, the balance sweep may begin.

Balance Sweep: First, each processor determines whether its cluster should be a *donor* of work, a *receiver* of work, or (typically) neither. Donors are clusters whose workload exceeds the average by a factor of at least `loadBalDonorFac`, while receivers must be below the average by at least `loadBalReceiverFac`. A single message sweep of the tree then counts the total number of donors d and receivers r, and also assigns a unique donor and receiver number. The first $y = \min\{d, r\}$ donors and receivers then "pair up" via a rendezvous procedure involving $2y$ point-to-point messages; see [15, Section 6.3] or [8,10] for a more detailed description of this process. Within each pair, the donor sends a single message containing subproblem tokens to the receiver. Thus, the sweep messages are followed by a possible additional $3y$ point-to-point messages. After these messages, if any, the entire load balance round process repeats, starting with another synchronization sweep. The set of subproblem tokens transferred within each donor-reciever pair is chosen to place the loads of the two corresponding clusters in approximately the same ratio as their numbers of workers, although the message is subject to a maximum buffer length restriction.

Under certain conditions, including at least once at the end of every run, a *termination check* sweep is substituted for the balance sweep. This mechanism will be described in Subsection 4.6.

Peer-to-peer load balancing mechanisms are frequently classified as either "work stealing," that is, initiated by the receiver, or "work sharing," that is, initiated by the donor. The rendezvous method is neither; instead, donors and receivers efficiently locate one another on an equal basis, possibly across a large collection of processors.

4.5. Thread and scheduler architecture

From the preceding discussion, it is clear that the parallel layer requires each processor to perform a certain degree of multitasking. CMMIP handled multitasking by combining user-level interrupts for the highest-priority tasks with an *ad hoc* round-robin scheme for the remaining ones. The former mechanism was not portable, and the latter lacked flexibility (for example, a hub could not simultaneously serve as a worker and still control other worker processors). Instead, PICO defines a *thread* of control for each required task on a processor, and manages these threads through a *scheduler* module. The threads share a common global memory space through a pointer to the instance of `parallelBranching` being solved.

PICO is not truly multithreaded code, however. We do not use POSIX or other standard thread packages, and, on each processor, PICO appears to the operating system as only a single thread of control. The PICO scheduler is non-preemptive, much like the schedulers in the Macintosh and Windows $3.x$ operating systems: the scheduler explicitly calls each thread as a subroutine, and the thread restores control to the scheduler, only when it is ready, through a standard subroutine return. There are several reasons why we took this approach:

- Threads and "thread-safe" system libraries are not available in all environments, especially on MPPs, which often have customized operating systems. In particular, many versions of MPI are not thread-safe, or make their own use of threads. If PICO used "true" threads, it would become less portable.

- The approach simplified debugging and development.

- Since PICO's tasks can be trusted to cooperate, a non-preemptive discipline should be adequate to control them.

- Since threads only return control at times of their choosing, they can leave global data structures in a known state, and there is no need to worry about memory access conflicts and locks.

- The approach allowed us to use our own customized scheduling algorithm.

PICO contains a general-purpose scheduler, which is designed to be useful for other applications as well. We now describe the scheduling algorithm. Further detail for an earlier but similar version of the scheduler may be found in [11].

4.5.1. The scheduling algorithm

At any given time, each thread is in one of three states, *ready*, *waiting*, or *blocked.* Only threads in the ready state are allowed to run. Threads in the waiting state are waiting for the arrival of a particular kind of message, as identified by an MPI tag. The scheduler periodically tests for message arrivals, and changes thread states from waiting back to ready as necessary. Threads in the blocked state are waiting for some event other than a message arrival. The scheduler periodically polls these threads by calling their `ready` virtual methods; once a blocked thread's `ready` method returns `TRUE`, the scheduler changes the thread back to the ready state.

Threads are organized into *groups*, each group having its own priority. Group priorities are absolute, in the sense that the scheduler only runs threads from the highest priority group that contains ready threads. Only if all higher-priority groups have no ready threads will the scheduler permit thread in lower groups to run.

Each group may use one of two scheduling disciplines. The first is a simple round-robin scheme, in which ready threads are selected in a repeating cyclic order. The second possibility is a variant of *stride scheduling* [35,19]. This scheme allows the user to specify the approximate fraction of CPU resources that should be allocated to each thread.

In the stride scheduling discipline, each thread i has a *bias* b_i that specifies its importance. Let R denote the *ready list*, the set of ready threads in the highest nonempty group, then the fraction of the CPU devoted to thread $i \in R$ will be approximately $b_i / \sum_{j \in R} b_j$. Each thread also has a value v_i which specifies its current position in the run queue. The scheduler runs the ready thread with the lowest v_i, and when the thread returns, updates $v_i \leftarrow v_i + u/b_i$, where u is the amount of time just used by the thread.

All stride-scheduled threads start with $v_i = 0$. When a waiting or blocked thread is about to enter the ready list again, its v_i is reset to $\max\{v_i, v_* + k_i\}$, where $v_* = \min_{j \in R}\{v_j\}$, and $k_i \in \Re$ may be user-specified. To prevent numerical precision problems, a constant is periodically subtracted from all the v_i, $i \in R$, so that $v_* = 0$.

4.5.2. Thread group organization and thread types

The threads used by PICO belong to two broad categories: *message-triggered* threads and *compute* threads. There are two thread groups: the message-triggered threads occupy the higher-priority group, which uses round-robin scheduling, and the compute threads make up the lower-priority "base" group, which uses stride scheduling.

A message-triggered thread typically spends most of its time waiting for messages. When a message with the right tag arrives, the scheduler changes the thread's state from waiting to ready. Since message-triggered threads are in the high-priority group, they tend to run soon after they become ready. Once it runs, the thread processes the message, along with all other received messages with the same tag value. The thread then issues a nonblocking receive for another message, changes its state back to waiting, and returns.

Compute threads are usually in the ready state, but may be in the blocked state if they have exhausted all their available work. These threads are scheduled in the proportional-share manner described above, so long as no message-triggered threads need to run. By default, all of PICO's compute threads contain logic to actively manage their *granularity*, that is, the amount of time u they consume before returning to the scheduler. There is a run-time parameter `timeSlice` which specifies an ideal time quantum for compute threads. Compute threads try to manage the amount of work they do at each invocation so that the average value of u is approximately equal to `timeSlice`. The best value of `timeslice` depends on the hardware, the MPI implementation, and the application. A very small value means that message-triggered threads will run soon after their messages arrive, giving fast communication response, whereas a large value will minimize the overhead expended on the scheduler and entering and exiting compute threads. Ideally, one attempts to balance these two goals; in preliminary testing, we have had good results with a value approximately 20 times the time the scheduler needs to check for arriving messages.

4.5.3. Beginning the parallel search

To read in a problem instance, the parallel layer uses the `readAndBroadcast` method. `ReadAndBroadcast` first uses the `readIn` method, inherited from the serial application, to construct the global class information on a single processor. It then uses the the global class `pack` method to copy this information to a buffer, which it then broadcasts. All other processors receive the broadcast, and then use the global class `unpack` method to construct their replicas of the global class object.

Once a problem has been created and replicated on all processors, the application calls the `parallelSearch` method to search for a solution. Before starting the scheduler, this routine first calls the `preprocess` virtual method on all processors. By redefining `preprocess` for the parallel global class, the application may parallize its preprocessing procedure; Section 5 gives an example of this technique.

`ParallelSearch` next initializes the scheduler, creating the thread groups and calling the virtual method `placeTasks` to create the threads and place them the groups. The default version of `placeTasks` should suffice for many applications, and we describe below all the threads that it creates. If the application requires additional threads, it can redefine `placeTasks` to call the default `placeTasks`, and then create and place any additional threads. Again, we will present an example in Subsection 5.

Once the scheduler has been initialized, the first worker in the first cluster creates a blank subproblem, calls `makeRoot` to turn it into the root problem, and then inserts the root problem into its local pool. On all processors, `parallelSearch` then calls the scheduler to begin running all the threads. On each processor, the scheduler then runs until some thread sets the scheduler's global termination flag, after which `parallelSearch` exits.

We now describe all the threads created by the default version of `placeTasks`, as also depicted in Figure 7.

4.5.4. The worker thread

The *worker thread* is a compute thread that is present on every worker processor. This thread simply extracts subproblems from the worker's local pool and passes them to the search handler. If the local pool is empty, the worker thread enters the blocked state.

The worker thread attempts to regulate its granularity by adjusting the number of subproblems it passes to the handler before returning to the scheduler. For applications

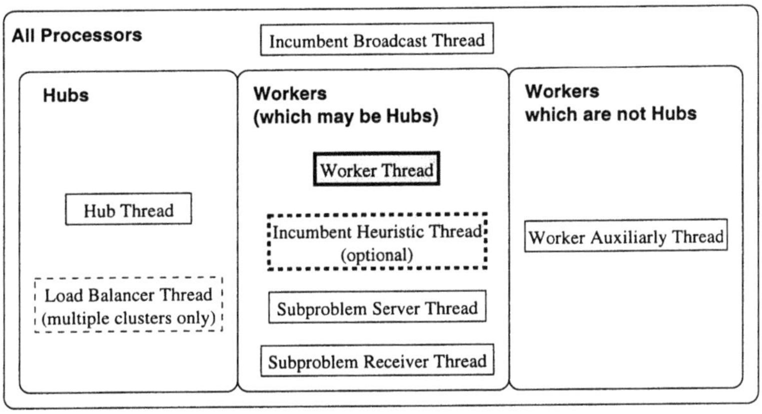

Figure 7. Threads used by the PICO core. Compute threads are shaded; all other threads are message-triggered.

in which the bounding or separation procedure is very time consuming, and may need to be interrupted to allow message-triggered threads to run, the application may modify this granularity-adjustment scheme. Essentially, the virtual methods `boundComputation` and `splitComputation` can set the argument `controlParam` to some value proportional to the amount of work they have done. On subsequent calls, the granularity-control algorithm will pass, via this same argument, a suggested amount of work to perform.

The worker thread is also responsible for pruning the local subproblem pool and server pool on its processor. If running on a processor that is also a hub, the worker thread also calls the method `parallelBranching::activateHub` before returning control to the scheduler. This call allows the hub logic to respond to any changes in the cluster's load resulting from the work just performed, and is described in more detail in Subsection 4.5.6.

4.5.5. The incumbent heuristic thread

The *incumbent heuristic thread* is a second, optional compute thread. It is only created on worker processors, and only if the both the application's `haveIncumbentHeuristic` virtual method returns `TRUE` and the run-time parameter `useIncumbentThread` is set to `TRUE`. This thread's task is to search for better incumbent solutions. The algorithm and granularity-adjustment procedure used are entirely application-specific.

When any of the parallel search handlers move a subproblem to the `bounded` state, and the incumbent heuristic thread exists, they call the `feedToIncumbentThread` virtual method, a member of the `parBranchSub` class. This method can decide whether the subproblem partial solution is of interest to the incumbent heuristic, and, if so, can copy any necessary data to the incumbent heuristic's application-specific internal data structures. There is also a `quickIncumbentHeuristic` virtual method which may be called for any `bounded` subproblem. This method is meant for a "quick and dirty" procedure not requiring a separate thread (such as filling out a knapsack by the greedy method, or rounding up a fractional solution to a set covering problem).

On worker processors, the scheduler uses stride scheduling to arbitrate between the worker and incumbent heuristic threads. The biases of these threads are controlled by the run-time parameters `boundingPriorityBias` and `incSearchPriorityBias`, respectively. In the near future, we plan to add a feature whereby these biases may be dynamically adjusted throughout the course of a run. This technique will allow applications to make heavy use of the incumbent heuristic early in a run, when it is important to locate good incumbents, and then phase it out as the run progresses and it is better to concentrate on proving optimality of the current incumbent.

4.5.6. The hub thread and `activateHub`

The *hub thread* is a message-triggered thread that runs on hub processors, and listens for messages with the tag `hubTag`. These messages may originate from any worker in the system. These messages contain workload status information, tokens of subproblems that are being released or scattered to the hub, and/or acknowledgements of receipt of subproblems dispatched from the hub.

When it awakens, the hub thread processes the contents of all `hubTag` messages received since its last invocation, making the requisite changes in hub logic data structures. It then calls the method `activateHub`.

Calling `activateHub` triggers all the functions of the hub, including pruning the hub's pool of active tokens, distributing subproblems to any deserving workers, and possibly sending messages to workers informing them of the workload distribution in the cluster. When an event occurs that might alter the workload situation in the cluster, `activateHub` may be called by any thread running on a hub processor, and not just the hub thread.

4.5.7. The incumbent broadcast thread

The *incumbent broadcast thread* is a message-triggered thread that runs on all processors, and listens for *incumbent broadcast* messages. Each processor stores both the best objective value it currently knows for the incumbent, `incumbentValue`, and the rank of the processor that generated that value, `incumbentSource`. PICO's incumbent broadcasting scheme is similar to CMMIP's: when a new incumbent is found, PICO uses the `parallelBranching` class `signalIncumbent` method to begin the broadcast. The incumbent broacasting procedure organizes all processors into a balanced tree rooted at the initiating processor. The tree's radix may be specified by a run-time parameter; the default is 32. The broadcast messages contain the objective value of the newly-found incumbent, and the processor number of the tree root.

When the incumbent thread receives an incumbent broadcast message, it compares the message's objective value to the incumbent value now known at the current processor. If the received value is worse, the incumbent thread does not attempt to continue the broadcast. If it is better, the broadcast continues. If the two values are equal, it then compares the processor rank of the processor initiating the broadcast to `incumbentSource`. Only if the broadcast root is smaller than `incumbentSource` will the thread attempt to continue the broadcast. This procedure guarantees that if several processors simultaneously try to broadcast incumbents, that one of the broadcasts with the best value will reach all processors, while the others will be aborted.

If the broadcast should continue, the incumbent broadcast thread updates the local values of `incumbentValue` and `incumbentSource` to match those in the message, and then forwards this information along the broadcast tree. It sets flags forcing the worker thread to become ready, and then prune the server and local worker pools. The incumbent broadcast thread also sets a similar flag to force the hub thread, if present, to perform pruning of the hub pool.

4.5.8. The subproblem server thread

The *subproblem server* is a message-triggered thread that runs on all workers, and listens for work dispatch messages from the hubs. These messages contain subproblem tokens, along with the respective processor ranks of workers to which the subproblems should be delivered. The subproblem server thread's task is to deliver the full information about the subproblems to the specified workers.

For every subproblem token in each dispatch message received since the last time the thread was invoked, the subproblem server thread also decodes the rank w of the corresponding target worker. The thread then processes each token-worker pair as follows: first, it checks the bound on the token to make sure that the problem cannot be fathomed because of some recently broadcast incumbent value. If the subproblem can be fathomed, it sends an acknowledgement message to the originating hub indicating that the subproblem was properly received, but does not bother actually trying to send the subproblem data to the worker. If, as usual, the subproblem cannot be fathomed, the thread then calls the method `parallelBranching::deliverSP` to deliver the subproblem data. Note that if a hub is also a worker, and it wishes to send a subproblem it created itself to some other worker, the hub simply calls the `deliverSP` method directly, rather than sending a dispatch message to itself.

The `deliverSP` method first matches the token with the corresponding subproblem P on the worker; this step is very efficient, because the token contains the memory address of P. `DeliverSP` then separately handles four possible cases, depending on whether the token is a "child" or "self" token, and whether the target worker w is the same as the current processor p, or some other worker. "Self" tokens refer directly to some subproblem in the server pool, while "child" tokens refer to a child of such a subproblem. The cases are handled as follows;

1. If the token is a self token, and $w = p$, P is transferred from the server pool to the local worker pool, and is marked as "delivered."

2. If the token is a self token, and $w \neq p$, the server thread uses the application's subproblem `pack` method to pack P into a buffer, and sends this buffer to w. P is then deleted from the server pool.

3. If the token is a child token, and $w = p$, the server thread uses `parallelMakeChild` to extract a child P' of P, and places P' in the the local worker pool, marking it as delivered. If P has no children left, it is deleted.

4. If the token is a child token, and $w \neq p$, the server thread uses the application's `packChild` method to pack a child P' of P into a buffer. The `packChild` method has the default implementation of creating a child with `makeParallelChild`, placing it

in a buffer with `pack`, and then deleting it. However, the application is free to substitute a more efficient, application-dependent implementation. The buffer is sent to w, and P is deleted if it has no children left.

The messages sent in cases 2 and 4 have MPI tag `deliverSPTag`. If more than one `deliverSPTag` message is destined for a single recipient worker, PICO attempts to pack them, subject to a maximum buffer size restriction, into a single MPI message.

After a subproblem is marked as delivered on a worker that is not a hub, an acknowledgement for that subproblem is included in the worker's next communication to its hub. If the worker is itself a hub, the acknowledgement is entered directly into the hub data structures.

4.5.9. The subproblem receiver thread

The *subproblem receiver* thread is a message-triggered thread present on all workers, and listens for messages with the tag `deliverSPTag`. For each subproblem in the message, the thread calls `blankParallelSub` and then the application's subproblem `unpack` method to recreate the data structures for the subproblem, which it then marks as delivered. If the subproblem cannot be fathomed, the thread inserts it into the local worker pool.

4.5.10. The worker auxiliary thread

The *worker auxiliary thread* is a message-triggered thread that exists only on workers that are not also hubs. It listens for messages with the MPI tag `workerTag`, which are sent by hubs to their workers. Each of these messages can contain one of three possible "signals" from the hub to the worker:

Load Information Signal: In this case, the message contains information about cluster and system-wide workloads. The worker auxiliary thread copies this information to the worker's local data structures.

Termination Check Signal: This signal indicates PICO is double-checking whether the system is really ready to terminate. The worker auxiliary thread immediately replies with a message containing a count of the total messages the worker has sent. The reason for this procedure will be described in Subsection 4.6 below.

Terminate Signal: The hub has determined that the branch-and-bound search has terminated. The worker auxiliary thread sets the scheduler's global termination flag. When the thread exits, the scheduler terminates, and the call to `parallelSearch` returns.

4.5.11. The load balancer thread

The *load balancer thread* orchestrates the load balancing scheme described in Subsection 4.4. It runs on every hub processor. It contains finite-state-machine logic to move all hub processors approximately synchronously through the various message sweeps and other operations described in Subsection 4.4. It is basically a message-triggered thread that listens for various kinds of messages, depending on what phase of the load balancing procedure it is currently in. One exception is that it may enter the blocked state when at the beginning of the synchronize sweep, waiting for conditions to be right for another round of load balancing. The scheduler unblocks it when the sweep is allowed to start.

The load balancer thread is responsible for terminating the search computation, as described below. If there is only one cluster, then no load balancing is necessary, and termination is the thread's only function. In this case, it immediately puts itself in the termination check sweep mode, first listening for termination check reply messages from the workers.

4.6. Termination detection

Proper detection of termination can be a tricky issue in asynchronous distributed-memory computations. CMMIP's termination procedure relied on specific properties of the CM-5's communication hardware and operating system, and could not be generalized to PICO.

In parallel branch and bound, it is important to terminate as soon as, but not before, there are no active subproblems left to be bounded or separated anywhere in the system. In some implementations of MPI, it is also important that when a processor calls `MPI_Finalize` to terminate its computation, that it have received all messages sent to it by other processors (except any that were successfully cancelled via `MPI_Cancel`). If not, `MPI_Finalize` may "hang" or generate system errors.

So, for PICO to be able to terminate, all worker subproblem pools and hub token pools must be empty, and all sent messages must be received. We call this situation *quiescence.*

4.6.1. The case of a single cluster

If there is only one cluster, quiesence is relatively straightforward to detect. The hub knows which subproblems it has assigned to which workers, and through the delivery marking and acknowledgement mechanism, which of these problems have been received. It also has recent workload information from each of its workers. Furthermore, the workload information reported by workers contains counts of total messages sent and received, so hubs can also detect messages in transit.

Once a worker becomes *idle* — that is, it has no more subproblems in its active pool — it reports its workload to its hub immediately. If it is idle and receives a message of any kind, it resends its idle report to the hub, with updated message send and receive counts included.

Suppose that the following conditions hold:

1. The hub has an empty token pool.

2. All the workers have reported themselves idle.

3. All subproblems dispatched from the hub have been acknowledged as delivered.

4. All processors agree about the objective value of the incumbent, and which processor stores the incumbent.

5. The total counts of message sends and receives appear to match when summed over all processors.

In this situation, no more work can possibly arise in the normal operation of the system, and no more messages can be sent by any of the PICO core threads. However, there is still a possiblity of premature termination if any application-specific threads have messages in transit; even though the total count of messages sent and received may appear equal, messages may still be in transit due to the phenomenon of "aliasing," as described in the next subsection. To check for this possibility, the hub sends a termination check signal to all workers. The workers' replies to these messages wake up the hub's load balancing thread, which double-checks the message counts, and terminates the computation if appropriate. This process is described in more detail immediately below.

4.6.2. Multiple clusters

When there are multiple clusters, properly detecting quiescence is more difficult, even if no application-specific threads are present. Basically, termination is detected at the end of the load balancing survey sweep, when the total workload information summed over all clusters is distributed to all hub processors. If all clusters have no active subproblems and the global counts of sent and received messages match, then it is likely that the search can terminate. Note that since the messages used by the load balancing process itself follow a fixed pattern of accumulating and broadcasting along a tree, their arrival patterns are predicatable, and they do not need to be taken into account in the message counting mechanism.

However, even if the survey sweep detects that all clusters appear to be idle and the total counts of sent and received messages match, it is still possible that the system is not quiescent. We call such a state *pseudoquiescence.* The reason is that it is not possible to sample the message send and receive counts from all processors at exactly the same time. Thus, a message can contribute to the total reception count collected by the sweep, without yet contributing to the total send count. The reception of such a message can then masquerade as the reception of another message whose send operation is included in the count but has not been received even by the end of the sweep. Such "aliasing" can cause premature detection of termination. If such a message contained a scattered subproblem, then PICO might terminate with an incomplete proof of optimality, or possibly an incorrect solution. This phenomenon can also occur if there is only a single cluster, but there are application-specific threads that send interprocessor messages.

To prevent such premature termination, we use a variant of the "four counter" method due to Mattern [24], which appears to be the most efficient technique available (the name is misleading, since it is shown in the original reference [24] that the method can be implemented with only three counters). In PICO's case, the procedure works as follows: at the end of the survey sweep, the load balancer threads on all hubs detect pseudoquiescence. Instead of proceeding to the balance sweep, they substitute a termination check sweep instead.

At the start of the termination check sweep, each hub sends a termination check signal message to all its workers (except itself, if it is also a worker). The worker auxiliary threads on these workers respond with their total message sent counts. Once all its workers have responded, each hub adds together the send counts for its entire cluster, including itself. The sweep messages now proceed, adding this information recursively up the cluster hub tree. The overall message count sum forms at the the cluster tree root, and is broadcast back down the tree. If the total message sent count collected by the termination check sweep is the same as that collected by the immediately preceeding survey sweep, then aliasing is impossible, and the system was actually quiescent at the end of the last survey sweep; see [24] for a proof. In this case, the load balancer thread on each hub sends a termination signal message to all its cluster's workers, and then sets the scheduler termination flag. If the counts do not match, then true termination has not occurred, and the load balancer thread simply commences another round of load balancing.

Note that if there is only one cluster, the "tree" used in the termination check "sweep" just consists of a single node and no edges. If the termination check fails in the one-cluster case, the load balancer simply starts another termination check sweep, listening for termination check reply messages from the workers. The signal to send these messages will come from the hub thread the next time it detects possible termination.

5. A SIMPLE APPLICATION TO MIXED INTEGER PROGRAMMING

To demonstrate how the PICO core can be used, we now describe the application of the PICO class library to the solution of general mixed integer programming problems, without the use of cutting planes. This application is the "PICO MIP" referred to in Section 2 and Figure 1.

We stress that this application is not yet meant to be a completely state-of-the-art MIP solver, as it is lacking a number of features present in the best commercial codes. At this point, we present it to illustrate how the PICO core can be easily extended to include additional, advanced features for applications, and to illustrate the degree of parallelism that PICO can acheive.

Technically, the PICO MIP can solve any problem that can be expressed in the industry-standard MPS format. For convenience, we will assume in our discussion that the problem being solved is to find $x \in \Re^n$ satisfying

$$\begin{array}{ll} \min & c^{\mathsf{T}}x \\ \text{S.T.} & Ax = b \\ & \ell \le x \le u \\ & x_j \text{ integer } \forall\, j \in Z, \end{array} \tag{1}$$

where $c \in \Re^n$, $b \in \Re^m$, A is an $m \times n$ matrix, $\ell \in [-\infty, +\infty)^n$, $u \in (-\infty, +\infty]^n$, $\ell \le u$, and $Z \subseteq \{1, \ldots, n\}$ is a nonempty set of indices of variables that are required to take whole-number values. Note that inequality constraints can easily be accomodated in this formulation by introducing slack variables, as is the case with most linear programming solver software.

In the standard branch-and-bound algorithm for this problem, the root problem is simply (1) with the integrality constraints removed. The remaining subproblems are similar, but with some of the lower bounds ℓ_j increased or upper bounds u_j decreased, for $j \in Z$. Let $\ell(P)$ and $u(P)$ denote the lower and upper bound vectors for any given subproblem P. The bound $z(P)$ for subproblem P is obtained by solving the corresponding linear program, yielding some linear programming solution $x(P)$. The value $z(P) = c^{\mathsf{T}}x(P)$ is a lower bound on the objective value of any solution x of (1) that has $\ell(P) \le x \le u(P)$. If

$x_j(P)$ is integer (to within some specified numerical tolerance) for all $j \in Z$, then $x(P)$ represents a feasible solution to (1) that dominates all other solutions with $\ell(P) \le x \le u(P)$.

If there exists any $j \in Z$ for which $x_j(P)$ is not integer, then the subproblem must be separated. We select some such j, call it $j(P)$, and create a *down child* subproblem P^- with $u_{j(P)}(P^-) = \lfloor x_j(P) \rfloor$ and an *up child* P^+ with $\ell_{j(P)}(P^+) = \lceil x_j(P) \rceil$.

5.1. Pseudocosts

One of the keys to making this "textbook" branch-and-bound method work efficiently in practice is to make a good choice of the branching variable $j(P)$ from among the set $J(P)$ of indices $j \in Z$ for which $x_j(P)$ is not integer.

The modeler may specify branching "priorities" to aid in this decision: for example, variables specifying whether or not a particular production plant is to be built would have higher priority than variables specifying which products would be made in each plant. However, priorities are not available for all problems, and often there are many eligible variables with the same priority.

To choose among branching variables, we use a time-tested technique employing *pseudocosts* [2]. At any time t, let $\mathcal{K}(t)$ denote the collection of all subproblems P for which $z(P)$ has been computed. Then define

$$\mathcal{S}_j^+(t) \quad = \quad \left\{ P \in \mathcal{K}(t) \;\middle|\; P^+ \in \mathcal{K}(t),\ j(P) = j \right\}.$$

The "up" pseudo-cost of variable x_j, $j \in Z$, at any time t such that $\mathcal{S}_j^+(t) \neq \emptyset$, is

$$\phi_j^+(t) = \left(\frac{1}{\left|\mathcal{S}_j^+(t)\right|} \right) \sum_{P \in \mathcal{S}_j^+(t)} \left(\frac{z(P^+) - z(P)}{\lceil x_j(P) \rceil - x_j(P)} \right).$$

This quantity attempts to measure how rapidly the subproblem optimal objective value increases, on average, as x_j is forced upward. We define the "down" pseudocost in a similar way, but this time tracking how the objective value changes as variables are forced downward:

$$\begin{aligned} \mathcal{S}_j^-(t) &= \left\{ P \in \mathcal{K}(t) \;\middle|\; P^- \in \mathcal{K}(t),\ j(P) = j \right\}. \\ \phi_j^-(t) &= \left(\frac{1}{\left|\mathcal{S}_j^-(t)\right|} \right) \sum_{P \in \mathcal{S}_j^-(t)} \left(\frac{z(P^-) - z(P)}{x_j(P) - \lfloor x_j(P) \rfloor} \right). \end{aligned}$$

The method for choosing a branch variable is similar to CMMIP's: for each $j \in J(P)$, we calculate a "score" and branch on the variable maximizing the score. To compute the score, we use the pseudocosts to estimate the respective degradations D_j^+ and D_j^- in the objective value for the up and down children, via:

$$\begin{aligned} D_j^+ &= \phi_j^+(t)\left(\lceil x_j(P) \rceil - x_j(P)\right) \\ D_j^- &= \phi_j^-(t)\left(x_j(P) - \lfloor x_j(P) \rfloor\right). \end{aligned}$$

The score is then computed by

$$\sigma_j \quad = \quad \alpha_0 Q_j + \alpha_1 \min\{D_j^+, D_j^-\} + \alpha_2 \max\{D_j^+, D_j^-\},$$

where Q_j is the priority of variable j and α_0, α_1, and α_2 are specified via run-time parameters. Typically, α_0 is chosen very large, so that priority is the overriding consideration. Also, one typically sets $\alpha_2 = 0$, or at any rate $\alpha_2 \leq \alpha_1/10$. Thus, after priority, the next most important consideration is trying to simultaneously "push up" the bounds of both child subproblems.

A critical issue is what to do when $\mathcal{S}_j^+(t) = \emptyset$ or $\mathcal{S}_j^-(t) = \emptyset$. Here, we take a different approach than CMMIP, shown to be superior by Linderoth [21]. Every time the algorithm encounters an index $j \in J(P) \subseteq Z$ that has not been fractional in any prior subproblem solution, it "probes" — that is, computes the objective values for — the subproblems $\hat{P}_j^+$ and $\hat{P}_j^-$ that would result if j were the branching variable. These subproblems do not necessarily appear in the search tree unless j is later chosen to be the branching variable, but are immediatly incorporated into the set $\mathcal{K}(t)$, and thus into the pseudocost calculations above, so there will be at least one element present in each of the sets $\mathcal{S}_j^+(t)$ and $\mathcal{S}_j^-(t)$. If either of the subproblems $\hat{P}_j^+$ or $\hat{P}_j^-$ is infeasible, we narrow the bounds of the variable accordingly, and set the corresponding pseudocost to infinity. If both directions are infeasible for any variable, the current subproblem has no integer solutions and is marked `dead`. To prevent infinite pseudocosts from persisting indefinitely, the following rule is used to combine finite and infinite pseudocost observations: once a variable has a finite pseudocost observation, all previous infinite pseudocosts are treated as a parameter `infeasFactor` times that finite pseudocost and any subsequent infeasible branches found during branching are treated as `infeasfactor` times the current pseudocost.

During the initialization of pseudocosts for a variable, we adjust the bound of the subproblem to (in the case of minimization) the minimum objective value of the two branches, if that is higher than the current bound. This refinement means that each subproblem must also store its parent's LP bound, which may not equal its parent's subproblem bound, in order to calculate of future pseudocosts.

5.2. Other serial aspects of the algorithm

Our algorithm incorporates several other features that are standard in "industrial-strength" MIP solvers. Before starting the search, we run a preprocessor, based on that in MINTO and PARINO [29,26,21]. This preprocessor removes some redundant constraints and fixes the values of some variables, if it can deduce the values they must take in the optimal solution. Variable x_j's value is fixed by setting $\ell_j = u_j$.

The algorithm also applies a standard "locking" procedure after solving the linear program associated with each subproblem. Let $\overline{z}(t)$ denote the objective value of the incumbent at time t. If the absolute value of the reduced cost of a nonbasic variable x_j, $j \in Z$, exceeds $\overline{z}(t) - z(P)$, then x_j may be fixed at its present value in all of P's descendents. This procedure is valid because any descendant solution with a different but still integral value of x_j would necessarily be fathomed. Again, the locking is accomplished by setting $\ell_j = u_j$.

There are a number of other features that are now becoming common in commercial MIP solvers, but are not yet present in our algorithm: cutting plane methods to improve the linear programming bounds, various kinds of rounding heuristics to obtain feasible solutions from subproblem solutions $x(P)$ that do not meet the integrality constraints, and repeated application of the preprocessor at branch-and-bound nodes. We plan to add these features in later implementations or derived applications. Of these features, only an incumbent heuristic was present in CMMIP; we plan to implement a more sophisticated heuristic.

5.3. Serial implementation

Mapping the MIP branch-and-bound algorithm to the PICO serial layer classes and virtual functions was fairly straightforward. Class `MILP` is derived from `branching`, and contains representations of the data b, ℓ, u, c, A, and Z describing and instance of (1). It also contains tables required to calculate and update the pseudocosts $\phi_j^+(t)$ and $\phi_j^-(t)$.

Subproblems are represented by the class `MILPNode`, which is derived from `branchSub`. Essentially, a subproblem P is completely described by the two n-vectors $\ell(P)$ and $u(P)$. In addition, however, each subproblem object stores a compacted representation of a corresponding linear programming basis. For problems in the `boundable` state, this basis corresponds to the optimal solution of the parent problem. For the `bounded` and `separated` states, it describes the optimal solution to the problem itself.

The `preprocess` method for class `MILP` executes the preprocessing procedure, and `MILP`'s `readIn` method reads an MPS or LP format data file into the `MILP` data structures. `MILPNode`'s version of the `boundComputation` method uses a standard linear programming package to calculate $z(P)$ and $x(P)$ for a subproblem. The linear programming solver is encapsulated in a special interface class, allowing different LP packages to be specified at compile time. At present, we are using ILOG, Inc.'s CPLEX 6.x packages, but we have also built encapsulations for DASH Optimization's XPRESS-MP package and the public-domain solver SOPLEX [36].

Except for the root problem, `boundComputation` always begins from the optimal basis of the parent problem, which greatly speeds the calculations. If the last subproblem solved was not the parent of the current subproblem, the parent's basis is reloaded into the LP solver. However, if the parent was the last problem run, then the reload is skipped, further speeding up the bounding of the subproblem. This favorable situation is called a *warm start*. For both root and non-root problems, the user can specify whether the optimization is via primal simplex, dual simplex, or a barrier method. The default for non-root problems is dual simplex.

Once a subproblem's linear program has been solved, `boundComputation` executes the reduced-cost-based variable locking procedure and identifies the set $J(P)$ of variables violating the integrality constraints. The `candidateSolution` method for `MILPNode` simply returns `TRUE` if $J(P) = \emptyset$, and otherwise `FALSE`.

The `splitComputation` method for `MILPNode` scans $J(P)$ for any indices j that have not appeared in $J(P')$ for any prior subproblem P'. For each such index, it computes the objective value of the probe subproblems $\hat{P}_j^+$ and $\hat{P}_j^-$. This procedure may be significantly more time consuming than the original bound computation itself, but should become increasingly rare as the computation progresses. After this probing process is complete, $\mathcal{S}_j^+(t), \mathcal{S}_j^-(t) \neq \emptyset$ for all $j \in J(P)$. `SplitComputation` then calculates the scores σ_j for all $j \in J(P)$ and chooses the branching variable index $j(P)$ to maximize σ_j.

Finally, MILPNode's `makeChild` method creates child subproblems. It creates a fresh subproblem and copies the bound information $\ell(P)$ and $u(P)$ to the child, modifying the bound $\ell_{j(P)}(P^+)$ for the up child, and $u_{j(P)}(P^-)$ for the down child. `MakeChild` also copies

P's optimal basis information to the child.

5.4. Parallel implementation

To make a parallel version of the MIP algorithm, we used the same procedure described in Figure 6 and of Subsection 4.1. We defined a parallel global class `parMILP` with `parallelBranching` and `MILP` as `virtual public` base classes. Further, we also defined a parallel subproblem class `parMILPNode` with `parallelBranchSub` and `MILPNode` as `virtual public` base classes. We also provided straightforward implementations of the constructors and destructors for these classes, along with the virtual methods described in Table 3. These definitions are sufficient to provide a working parallel implementation. Since there is at present no incumbent heuristic in the serial application, there is currently no incumbent heuristic thread in the parallel version.

However, we chose to extend the basic parallelization in a two ways, both relating to pseudocosts. We expect this situation to be an example of a standard pattern: pseudocosts consitute a type of information that is not part of the incumbent or active subproblem pool, but is nevertheless global, in the sense that it is not localized within a particular subproblem. Such global information typically needs some kind of special treatment in a parallel implementation.

Consider how the default parallelization provided by PICO would operate in the case of the MIP algorithm we have just described. The pseudocost tables, needed to calculate $\phi_j^+(t)$ and $\phi_j^-(t)$, data members of `MILP`, will by default be maintained completely independently on each processor. Initially, the first worker in the first cluster solves the root problem P_0, while the other workers remain idle because there is no incumbent heuristic. Typically, the set $J(P_0)$ of the root's integrality-violating variable indices will be large. To initialize the pseudocost information needed to split the root problem, the first worker must solve an additional $2\,|J(P_0)|$ linear programs (albeit from a good starting basis). During this time, all other processors will remain essentially idle, although the work could easily be partitioned into $2\,|J(P_0)|$ independent tasks.

Once the search tree starts to grow, and other workers become busy, a second source of inefficiency would arise. Because the pseudocost tables are maintained separately and independently on each processor, the pseudocost probing operation will be performed whenever a variable x_j, $j \in Z$, is detected to be fractional for the first time *on a given processor*. Thus, probing for any particular variable might occur as many as $\overline{w}$ times, where $\overline{w}$ is the total the number of worker processors, as opposed to once in the serial layer implementation.

To obtain more parallelism at the outset of the algorithm, we designed the `preprocess` routine in `parMILP` so that it functions differently from `MILP`'s. Recall that the parallel search calls `preprocess` before running the scheduler; furthermore, this call is executed on every processor. The parallel version of the preprocessor, `parMILP::preprocess`, starts by first calling the serial version `MILP::preprocess`, to eliminate redundant constraints and fix variables. This calculation is done redundantly on all processors.

Instead of returning at this point, however, the parallel MIP preprocessor now instructs all processors to solve the root problem's linear program. Again, this operation is done in parallel and redundantly on all processors. The preprocessor then identifies the set of integrality-violating variable indices $J(P_0)$. Initializing the pseudocosts for these variables requires solving $2|J(P_0)|$ linear programs. The preprocessor partitions these linear programs into $\overline{p}$ roughly equal-sized groups, each of size approximately $2|J(P_0)|/\overline{p}$. In parallel, without redundancy, each of the $\overline{p}$ processors solves the problems in one of these groups. The preprocessor then makes the combined results of these calculations collectively available to all processors via an `MPI_Allgather` communication operation. The preprocessor then returns. In this manner, the work required to separate the root problem is significantly parallelized.

`ParMILPNode`'s version of `makeRoot` sets the state of the root problem to `bounded` instead of the usual value of `boundable`, since the work of bounding the root problem has already been performed. When the first worker first processes the root problem, it immediately performs separation and chooses a branch variable, a rapid operation since all the necessary pseudocost information is available.

To limit possible redundancy in initializing the pseudocost information for indices $j \notin J(P_0)$, we employ a second strategy. Whenever a worker probes to initialize the pseudocost data for a variable x_j, it places the resulting information in a special buffer, as well as in its regular pseudocost tables. As soon as all newly-fractional variables have been probed for a given subproblem P, the worker broadcasts the buffer to all other workers, as recommended in [21]. The buffer is then reset to empty. Upon reception, all other workers incorporate this information into their own pseudocost tables, making it unnecessary for them to probe any of the variables in $J(P)$ in the future. Otherwise, however, pseudocost information is maintained completely separately by each processor.

Although substantial interprocessor communication is involved, each of the broadcast operations may prevent as many as $2(\overline{w}-1)$ redundant linear program solutions. Each pseudocost pair is broadcast along a balanced tree consisting of all workers, with the originating worker at the root; the radix of this tree is controlled by the run time parameter `pCostTreeRadix`, which defaults to 2.

To receive and forward the messages required for pseudocost broadcasts, we introduce one additional thread, the *pseudocost broadcast thread*. This message-triggered thread listens for incoming pseudocost data and incorporates this information into the local pseudocost tables. If the current processor is not a leaf of the tree for the broadcast in question, it forwards the broadcast to its children. To include this thread in the scheduler, `parMILP` overrides `parallelBranching`'s default implementation of the virtual method `placeTasks`. The substitute implementation first calls the original implementation, in order to create all the standard threads. It then creates an additional thread object (of type `pCostCastObj`) and inserts it into the message-triggered, higher-priority thread group.

It is possible, under this scheme, that some variables may still be probed redundantly: several processors could encounter the same newly-fractional variable at about the same time, with some beginning to probe before the broadcast from the first one reaches them.

The code also includes an option whereby pseudocost information for an index $j \in Z$, may be broadcast by processor p not only the first time processor p encounters a fractional value of $x_j(P)$, but the first k times, where k is set by a run-time parameter. This generalization allows the code to better deal with problems where pseudocosts behave in an "unstable" way, but such problems appear to be rare in practice. Even with this generalization, the approach is considerably simpler than CMMIP's [8,10].

Finally, we note that the current version of the parallel MIP application does not pause either the bounding or separation, that is, `boundComputation` always completes bounding a subproblem, leaving it in the `bounded` state, and `splitComputation` always completes the separation process, leaving a subproblem in the `separated` state. In the future, if we observe situations where invocations of message-triggered threads are unacceptably delayed by very long bounding or separation operations by the worker thread, we could alter our approach. For example, `boundComputation` could return, leaving a subproblem in the `beingBounded` state, after a fixed number of dual simplex pivots. Applying `boundComputation` again would resume the computation. A similar procedure could be applied when evaluating probe subproblems in `splitComputation`.

5.5. Preliminary computational results

To illustrate the parallelism PICO can attain, we now describe the performance of the PICO MIP on the "Janus" ASCI Red supercomputer at Sandia National Laboratories [25]. This system consists of 4,536 nodes, each with two 333-megahertz Pentium II processors and 256 megabytes of RAM. By default, one processor on each node functions as a compute processor, and the second as a communications processor for interacting with the internal network. Optionally, in what is called "virtual node mode," the two processors on each node can be used as compute processors, each with 128 megabytes of RAM. The nodes are linked by an extremely fast $76 \times 32 \times 2$ communications grid, which also includes some "service" nodes that do not directly run user programs. We have measured the system's end-to-end delivery time for a 256-byte message at about 18 microseconds. The system implements "space sharing", rather than time sharing; typically, each active job has full control of some subset of the processing nodes.

In this section we describe the solution of some moderate-difficulty MIP problems from the MIPLIB [3], using between 1 and 128 Janus processors. For our initial experiments, we selected six problems, all solvable on a single processor using the basic branch-and-bound algorithm described above, but still requiring a substantial amount of tree search. Table 4 describes these problems.

Tables 5 through 7 show the solution of these problems using the PICO MIP. The single-processor runs use only the serial layer, and all the other runs use the parallel layer. The parallel runs were configured with a `clusterSize` of 32, so only runs above 32 processors had multiple hubs. `HubsDontWorkSize` was at its default value of 10, so the hubs in the 2 through 8 processor runs doubled as workers, while those in all larger configurations were "pure". The `hybridHandler` bounding protocol was used, in combination with the `heapPool` subproblem pool, which implements best-first search. The scattering parameters were set so that an average of 50% of newly-created subproblems were released to the hubs; 67% of the time, subproblem releases were forced to go to the local hub, and the remaining 33% of the time they were sent to a randomly-chosen hub. Note that the Janus system can sustain higher communications levels than these parameter settings imply.

The $\overline{p}$ columns in the tables represent the number of Janus compute nodes. The remaining data in each table are averaged over $\min\{\overline{p}, 5\}$ runs. The "nodes" column gives the total number of subproblems bounded, and the run times are in seconds. The "speedup" column displays the speedup relative to the corresponding one-processor run. The "idle" column is the percent of the total processor time spent in an idle state, and the "scheduler"

Problem	Binary Variables	General Integer Variables	Continuous Variables	Rows	Application
bell3a	39	32	52	133	Fiber optic network design
lseu	89			28	(Unknown)
misc07	259		1	212	(Unknown)
mod008	319			6	Machine loading
qiu	48		792	1192	Fiber optic network design
stein45	45			331	(Unknown)

Table 4
Description of the test problems.

column is the percent of total processor time devoted to scheduler overhead. Figures 8 through 10 display the same information graphically on a log-log scale. Each "+" data point represents a run of the code, and dashed lines connect the average run times for each processor configuration. The straight dashed line represents an "ideal" situation in which the speedup factor on $\overline{p}$ processors would be exactly $\overline{p}$.

In many of the problems, the size of the search tree inflates fairly dramatically as one moves from the serial to the parallel version of the algorithm. This inflation occurs because a parallel algorithm cannot follow a strict best-first ordering, and the current implementation lacks an incumbent heuristic for general MIP problems. If a high quality incumbent is unavailable for a significant fraction of the run, a parallel algorithm can spend significant amounts of time investigating noncritical subproblems. Experience from [7,10] suggests that even a crude incumbent heuristic can greatly dampen such tree inflation; in the future, we plan to add a (more sophisticated) heuristic to the PICO MIP.

After the initial tree inflation phenomenon, speedups generally remain fairly linear, with some "noise" and gentle degradation, until about 48-64 processors, after which they "tail off." In the near future, we plan to experiment with more difficult problems, larger total numbers of processors, and larger cluster sizes.

One problem, qiu, exhibits large oscillations in node counts, and thus in speedup. We have tentatively traced this phenomenon to an interaction between numerical instabilities in the problem and the pseudocost initialization procedure. Essentially, during the initial pseudocost probing, CPLEX finds different approximately optimal solutions depending on the starting basis. In the case of qiu, these differences lead to significant relative variations in the initial "down" pseudocosts, which are all very similar and on the order of 10^{-5}. These variations can in turn translate into significant changes in the size of the tree search.

Problem	$\overline{p}$	Nodes	Time	Speedup	Idle	Scheduler
bell3a	1	53,975	171.8	1.0	0.0%	0.0%
bell3a	2	53,978	91.7	1.9	0.1%	1.0%
bell3a	3	53,957	62.6	2.7	0.1%	1.0%
bell3a	4	54,521	48.4	3.5	0.6%	1.0%
bell3a	6	53,969	31.1	5.5	0.0%	1.0%
bell3a	8	53,941	23.8	7.2	0.2%	1.3%
bell3a	12	54,518	17.2	10.0	5.1%	1.7%
bell3a	16	54,623	12.6	13.6	3.2%	1.6%
bell3a	24	54,059	8.6	20.0	3.3%	1.2%
bell3a	32	54,090	6.1	28.2	1.6%	1.6%
bell3a	48	53,809	4.6	37.3	9.5%	2.2%
bell3a	64	54,044	3.7	46.4	10.2%	2.7%
bell3a	96	54,082	3.6	47.7	13.3%	0.6%
bell3a	128	54,515	2.1	81.8	17.5%	0.0%
lseu	1	12,143	26.2	1.0	0.0%	0.0%
lseu	2	25,337	27.0	1.0	0.0%	1.1%
lseu	3	16,453	12.0	2.2	0.6%	0.8%
lseu	4	27,462	14.5	1.8	0.0%	1.2%
lseu	6	30,591	10.9	2.4	0.4%	0.9%
lseu	8	16,704	4.5	5.8	0.4%	1.3%
lseu	12	45,312	8.8	3.0	6.1%	1.4%
lseu	16	30,356	4.8	5.5	8.8%	2.1%
lseu	24	34,129	3.8	6.9	9.6%	0.5%
lseu	32	35,228	3.4	7.7	20.6%	1.2%
lseu	48	40,990	2.4	10.9	9.3%	0.0%
lseu	64	35,051	2.0	13.1	10.2%	0.0%
lseu	96	41,758	2.0	13.1	13.3%	0.0%
lseu	128	41,970	1.3	20.2	25.4%	0.0%

Table 5
Computational results for bell3a and lseu.

Problem	$\bar{p}$	Nodes	Time	Speedup	Idle	Scheduler
misc07	1	46,251	722.9	1.0	0.0%	0.0%
misc07	2	52,589	430.1	1.7	0.1%	0.3%
misc07	3	68,648	369.4	2.0	0.1%	0.3%
misc07	4	63,144	261.8	2.8	0.2%	0.3%
misc07	6	97,717	282.5	2.6	0.2%	0.3%
misc07	8	104,550	234.2	3.1	0.3%	0.3%
misc07	12	106,427	191.5	3.8	7.7%	1.1%
misc07	16	47,415	66.9	10.8	6.0%	0.9%
misc07	24	75,285	73.2	9.9	7.7%	0.9%
misc07	32	71,232	54.1	13.4	4.5%	0.7%
misc07	48	31,376	27.1	26.7	24.2%	1.5%
misc07	64	26,185	24.2	29.9	32.9%	2.0%
misc07	96	30,313	23.8	30.4	39.9%	2.3%
misc07	128	22,613	28.6	25.3	60.7%	3.2%
mod008	1	23,695	134.1	1.0	0.0%	0.0%
mod008	2	25,814	71.1	1.9	0.0%	0.6%
mod008	3	24,274	43.9	3.1	0.0%	0.7%
mod008	4	25,952	35.4	3.8	0.0%	0.6%
mod008	6	26,050	24.2	5.5	0.7%	0.8%
mod008	8	28,088	19.3	6.9	0.0%	0.5%
mod008	12	23,563	12.5	10.7	6.1%	1.6%
mod008	16	22,494	9.0	14.9	4.5%	1.1%
mod008	24	24,841	6.9	19.4	3.8%	1.5%
mod008	32	25,787	5.5	24.4	4.0%	0.0%
mod008	48	25,004	4.2	31.9	9.6%	0.5%
mod008	64	24,879	3.4	39.4	10.5%	0.0%
mod008	96	24,804	2.8	47.9	14.1%	0.0%
mod008	128	24,265	2.9	46.2	16.8%	0.0%

Table 6
Computational results for misc07 and mod008.

Problem	$\overline{p}$	Nodes	Time	Speedup	Idle	Scheduler
qiu	1	14,899	1,938.3	1.0	0.0%	0.0%
qiu	2	51,574	3,034.9	0.6	0.4%	0.1%
qiu	3	15,689	670.2	2.9	0.7%	0.1%
qiu	4	109,911	2,909.3	0.7	0.3%	0.1%
qiu	6	17,784	410.0	4.7	1.7%	0.1%
qiu	8	76,666	1,146.7	1.7	0.5%	0.1%
qiu	12	69,430	852.3	2.3	8.0%	0.9%
qiu	16	65,549	630.6	3.1	7.8%	0.8%
qiu	24	55,425	342.9	5.7	10.7%	0.8%
qiu	32	29,387	195.2	9.9	18.0%	1.1%
qiu	48	20,488	120.7	16.1	34.3%	2.0%
qiu	64	21,091	110.2	17.6	40.4%	2.2%
qiu	96	23,084	107.9	18.0	48.9%	2.6%
qiu	128	21,719	102.1	19.0	56.8%	3.0%
stein45	1	64,329	389.4	1.0	0.0%	0.0%
stein45	2	62,773	196.1	2.0	0.7%	0.6%
stein45	3	68,613	140.9	2.8	1.4%	0.6%
stein45	4	61,514	95.4	4.1	0.8%	0.6%
stein45	6	63,274	66.1	5.9	1.3%	0.6%
stein45	8	63,157	49.4	7.9	1.8%	0.6%
stein45	12	75,170	41.7	9.3	8.5%	1.4%
stein45	16	70,854	30.0	13.0	8.4%	1.3%
stein45	24	70,012	21.2	18.4	11.5%	1.4%
stein45	32	72,510	17.3	22.5	15.9%	1.3%
stein45	48	73,773	12.9	30.2	22.2%	1.6%
stein45	64	77,105	11.8	33.0	26.0%	1.7%
stein45	96	77,872	9.7	40.1	35.3%	2.1%
stein45	128	77,172	8.6	45.3	44.3%	2.6%

Table 7
Computational results for qiu and stein45.

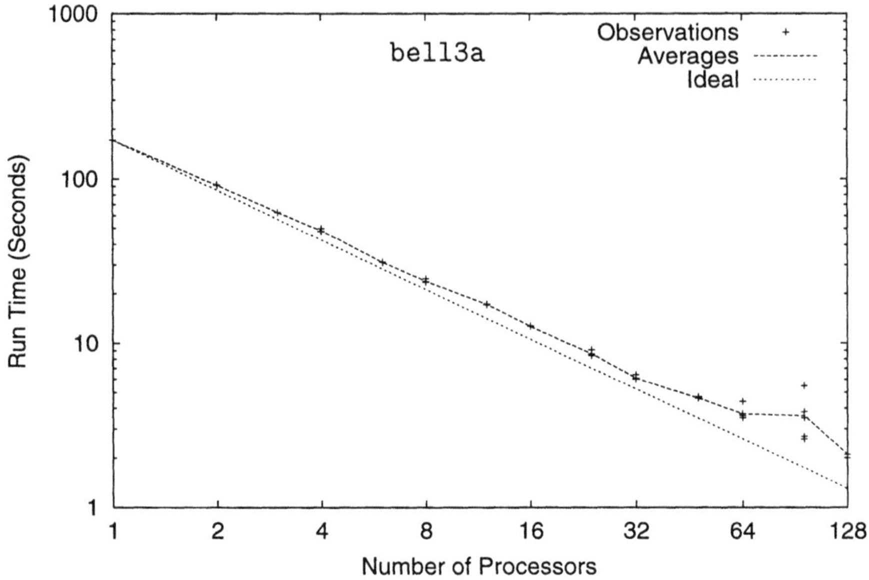

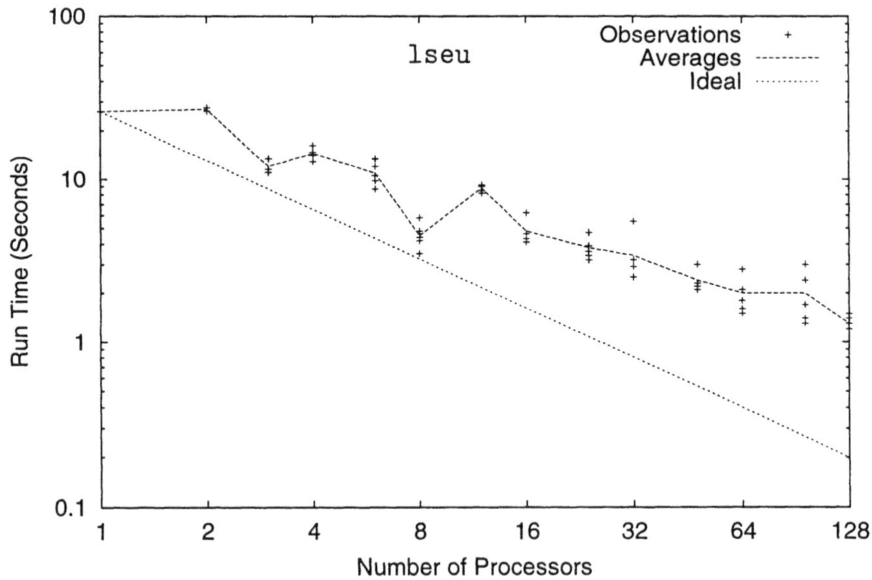

Figure 8. Computational results for `bell3a` and `lseu`.

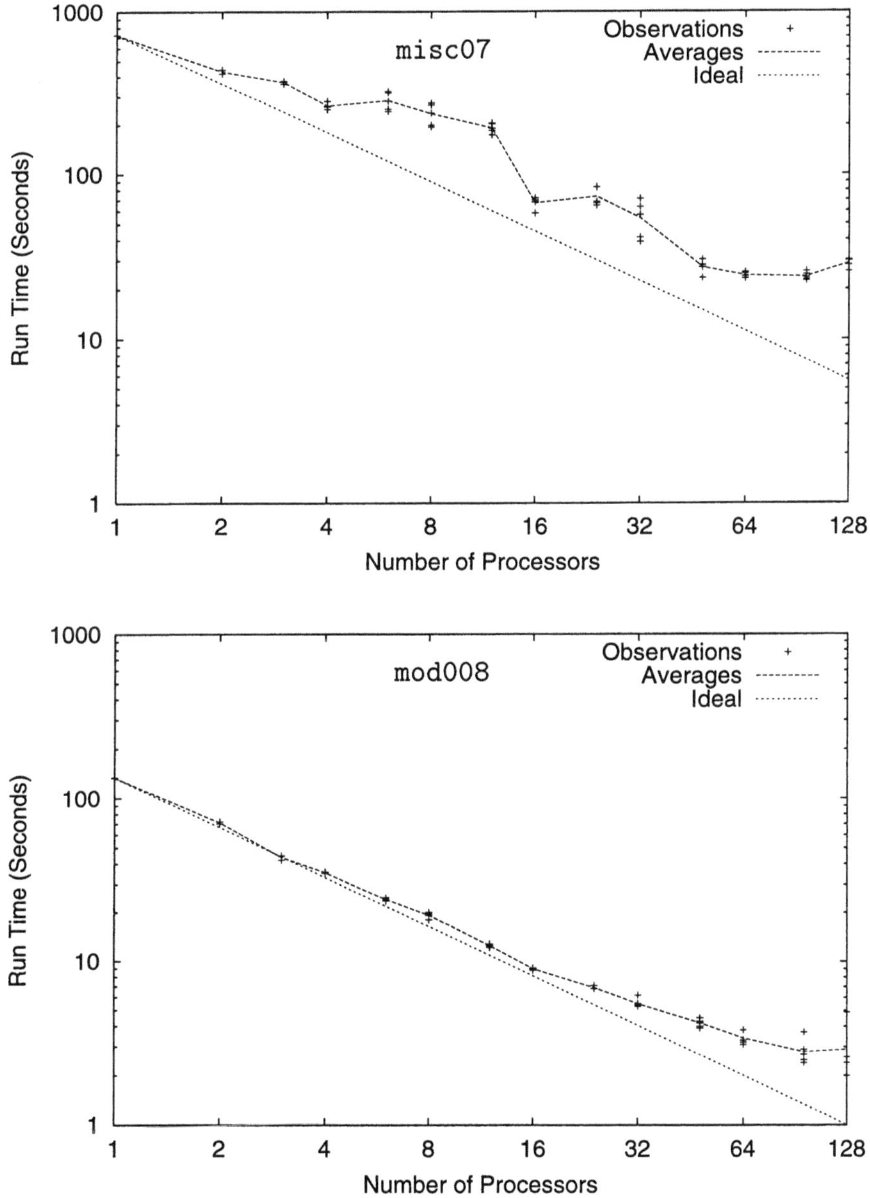

Figure 9. Computational results for misc07 and mod008.

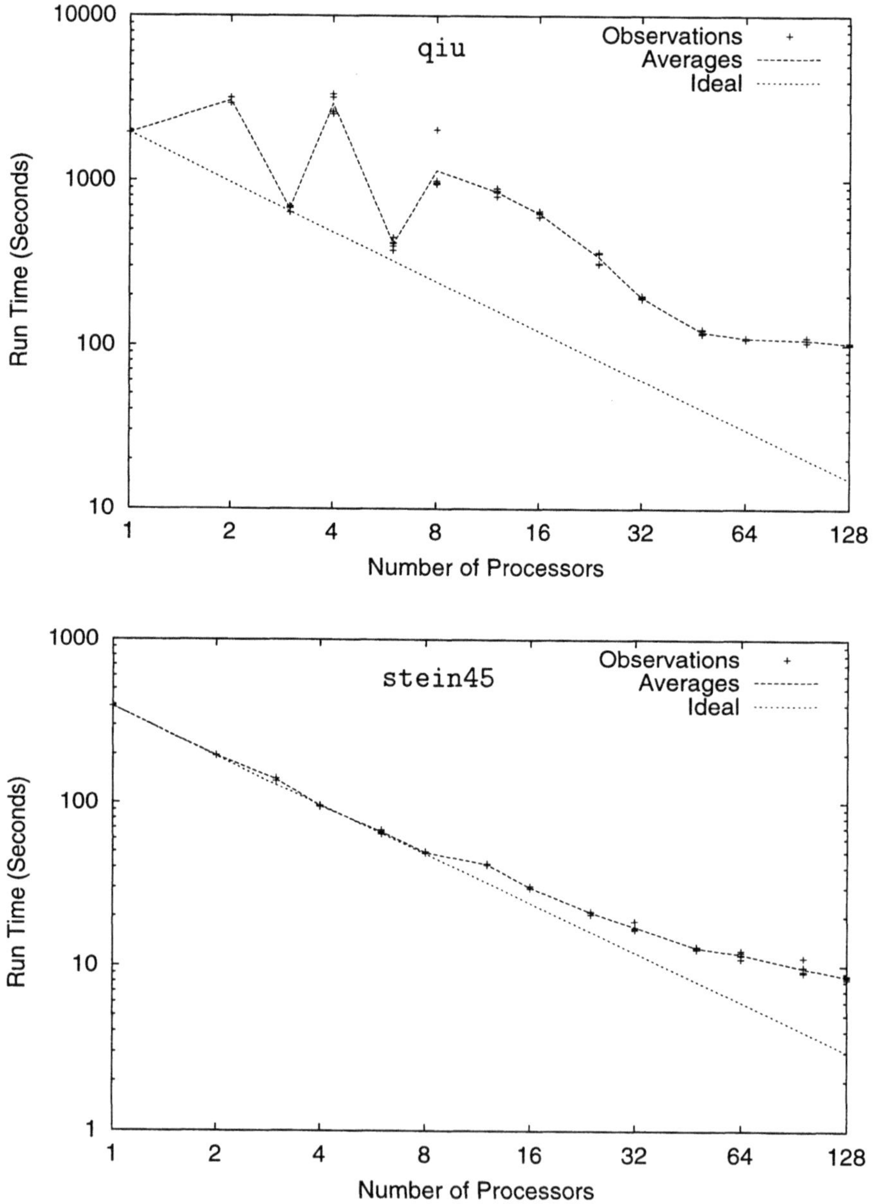

Figure 10. Computational results for `qiu` and `stein45`.

6. CONCLUSION AND FUTURE DEVELOPMENT PLANS

We have described a flexible, object-oriented approach to implementing parallel branch-and-bound algorithms, including a simple application to general mixed integer programming. Limited, preliminary computational testing on a small set of moderately difficult MIP's reveals some initial inflation of the search tree, most likely due to the absence of an incumbent heuristic, followed by fairly linear speedups through 32-48 processors.

The innovations of this work include:

- A novel object-oriented approach to describing branch-and-bound algorithms, using transition operators acting on subproblems moving through a state graph.
- The ability to describe both the search order and bounding protocol in a modular way.
- The division of the class library into serial and parallel layers.
- A continuously adjustable degree of communication between the hub and worker processors within a master-slave cluster.
- Use of stride scheduling to manage concurrent tasks within each processor executing the parallel branch-and-bound method.

In future, we plan to carefully investigate the performance of PICO on various processor configurations, refining its work distribution algorithm, so that the PICO core can be configured to operate efficiently on harder problems and larger processor configurations than described here. We also plan to refine the MIP application by including a parallel incumbent heuristic, as well as adding some other modern features including node-level preprocessing and cutting planes. The flexible underpinnings provided by the PICO core should make these enhancements relatively easy. Forthcoming papers will also describe some more specific applications of PICO, including to the PANTEX scheduling problem.

REFERENCES

1. R. Anbil, F. Barahona, L. Ladányi, R. Rushmeier, and J. Snowdon, *IBM Makes Advances in Airline Optimization*, Research Report RC21465(96887), IBM T. J. Watson Research Center, Yorktown Heights, NY, 1999.
2. E. M. L. Beale, Branch and bound methods for mathematical programming systems, in P. L. Hammer, E. L. Johnson, and B. H. Korte, eds., Discrete Optimization II, *Annals of Discrete Mathematics* **5** (North-Holland, Amsterdam, 1979) 201–219.
3. R. E. Bixby, S. Ceria, C. M. McZeal, and M. W. P. Savelsbergh, *An Updated Mixed Integer Programming Library: MIPLIB 3.0*, Technical Report 98-3, Department of Computational and Applied Mathematics, Rice University, 1998.
4. M. Benachouche, V.-D. Cung, S. Dowaji, B. Le Cun, T. Mautor, and C. Roucairol, Building a parallel branch and bound library, in: *Solving Combinatorial Optimization Problems in Parallel, Lecture Notes in Computer Science* **1054** (Springer, Berlin, 1996) 201–231.

5. J. Clausen, Parallel search-based methods in optimization, in: J. Wasniewski, J. Dongarra, K. Madsen, and D. Olesen, eds., *Applied Parallel Computing: Industrial-Strength Computation and Optimization: Third International Workshop, PARA '96 Proceedings* (Springer, Berlin, 1996) 176–185.
6. J. Clausen and M. Perregaard, On the best search strategy in parallel branch-and-bound: best-first search versus lazy depth-first search, *Annals of Operations Research* **90** (1999) 1–17.
7. J. Eckstein, Parallel branch-and-bound algorithms for general mixed integer programming on the CM-5, *SIAM Journal on Optimization* **4** (1994) 794–814.
8. J. Eckstein, Control strategies for parallel mixed integer branch and bound, *Proceedings of Supercomputing '94* (IEEE Computer Society Press, Los Alamitos, CA, 1994) 41–48.
9. J. Eckstein, How much communication does parallel branch and bound need?, *INFORMS Journal on Computing* **9** (1997) 15–29.
10. J. Eckstein, Distributed versus centralized storage and control for parallel branch and bound: mixed integer programming on the CM-5, *Computational Optimization and Applications* **7** (1997) 199–220.
11. J. Eckstein, W. E. Hart, and C. A. Phillips, Resource management in a parallel mixed integer programming package, *Proceedings of the Intel Supercomputer Users Group Conference*, Albuquerque, NM, (1997), `http://www.cs.sandia.gov/ISUG97/papers/Eckstein.ps`.
12. J. Eckstein, W. E. Hart, and C. A. Phillips, *PICO: an Object-Oriented Framework for Parallel Branch and Bound*, report SAND2000-3000, Sandia National Laboratories (Albuquerque, NM, 2000).
13. A. Geist, A. Beguelin, J. Dongarra, W. Jiang, R. Manchek, and V. Sunderam, *PVM: Parallel Virtual Machine: A Users' Guide and Tutorial for Networked Parallel Computing* (MIT Press, Cambridge, MA, 1994).
14. B. Gendron and T. G. Crainic, Parallel branch-and-bound algorithms: survey and synthesis, *Operations Research* **42** (1994) 1042–1066.
15. W. D. Hillis, *The Connection Machine*, (MIT Press, Cambridge, MA, 1985).
16. C. S. Horstmann, *Mastering C++: an Introduction to C++ and Object-Oriented Programming for C and Pascal Programmers* (John Wiley, New York, 1996).
17. M. Jünger and S. Thienel, Introduction to ABACUS — a branch-and-cut system, *Operations Research Letters* **22** (1998) 83–95.
18. G. Karypis and V. Kumar, Unstructured tree search on SIMD parallel computers: a survey of results, *Proceedings of Supercomputing '92*, (IEEE Computer Society Press, Los Alamitos, CA, 1992) 452–462.
19. M. Kim, H. Lee, and J. Lee, A proportional-share scheduler for multimedia applications, *Proceedings of the IEEE International Conference on Multimedia Computing and Systems '97* (IEEE Computer Society Press, Los Alamitos, CA, 1997) 484–491.
20. V. Kumar, A. Grama, A. Gupta, and G. Karypis, *Introduction to Parallel Computing: Design and Analysis of Algorithms*, (Benjamin/Cummings, Redwood City, CA, 1994).
21. J. T. Linderoth, *Topics in Parallel Integer Optimization*, Ph.D. thesis, Department of Industrial and Systems Engineering, Georgia Institute of Technology, 1998.
22. R. Lüling and B. Monien, Load balancing for distributed branch and bound algo-

rithms, *Proceedings of the International Parallel Processing Symposium* (IEEE Computer Society Press, Los Alamitos, CA, 1992) 543–549.

23. A. Mahanti and C. J. Daniel, A SIMD approach to parallel heuristic search, *Artificial Intelligence* **60** (1993) 243–282.
24. F. Mattern, Algorithms for distributed termination detection, *Distributed Computing* **2** (1987) 161–175.
25. T. G. Mattson and G. Henry, An overview of the Intel TFLOPS supercomputer, *Intel Technical Journal* Q1 1998, `http://developer.intel.com/technology/itj/q11998/articles/art_1.htm`, 1998.
26. G. L. Nemhauser, M. W. P. Savelsbergh, and G. C. Sigismondi, MINTO, a mixed integer optimizer, *Operations Research Letters* **15** (1994) 47–58.
27. T. K. Ralphs and L. Ladányi, *SYMPHONY User's Manual: Preliminary Draft*, `http://branchandcut.org/SYMPHONY/man/man.html`, 2000.
28. V. J. Rayward-Smith, S. A. Rush, and G. P. McKeown, Efficiency considerations in the implementation of parallel branch and bound, *Annals of Operations Research* **43** (1993) 123–145.
29. M. W. P. Savelsbergh, Preprocessing and probing techniques for mixed integer programming problems. *ORSA Journal on Computing* **6** (1994) 445–454.
30. Y. Shinano, M. Higaki, and R. Hirabayashi, Generalized utility for parallel branch and bound algorithms, *Proceedings of the 1995 Seventh IEEE Symposium on Parallel and Distributed Processing* (IEEE Computer Society Press, Los Alamitos, CA, 1995) 392–401.
31. Y. Shinano, K. Harada, and R. Hirabayashi, Control schemes in a generalized utility for parallel branch and bound, *Proceedings of the 1997 Eleventh International Parallel Processing Symposium* (IEEE Computer Society Press, Los Alamitos, CA, 1997) 621–627.
32. M. Snir, S. W. Otto, S. Huss-Lederman, D. W. Walker, and J. Dongarra, *MPI: The complete reference* (MIT Press, Cambridge, MA, 1996).
33. B. Stroustrup, *The C++ Programming Language* (Addison-Wesley, Reading, MA, 1991).
34. S. Tschöke and T. Polzer, *Portable Parallel Branch-and-Bound Library PPBB-Lib User Manual, Library Version 2.0*, Department of Computer Science, University of Paderborn, `http://www.uni-paderborn.de/~ppbb-lib`, 1996.
35. C. A. Waldspurger and W. E. Weihl, *Stride scheduling: Deterministic proportional-share resource management*, Technical Memorandum MIT/LCS/TM-528, MIT Laboratory for Computer Science (Cambridge, MA, 1995).
36. R. Wunderling, *Paralleler und Objektorientierter Simplex*, Ph.D. thesis and technical report TR 96-06, Konrad-Zuse-Zentrum für Informationstechnik, Berlin, 1996.

Inherently Parallel Algorithms in Feasibility and Optimization and their Applications
D. Butnariu, Y. Censor and S. Reich (Editors)

APPROACHING EQUILIBRIUM IN PARALLEL

Sjur Didrik Flåm[a*]

[a]Economics Department, Bergen University and Norwegian School of Economics and Business Administration

Motivated by non-cooperative games we study repeated interaction among non-communicating agents, each dealing with his block of variables, each moving merely on the basis of his marginal payoff and its most recent change. Adjustment or play thus unfolds in parallel. Constraints are accommodated. The main issue is convergence to (Nash) equilibrium.

Key words: gradient systems, Newton's method, parallel computation, noncooperative games, Nash equilibrium, repeated play, differential inclusions, viability, monotonicity, asymptotic stability.

1. INTRODUCTION

Arguably the objects most often sought for in applied science are roots of equations (or equilibria of dynamical systems). Focus is here on constrained instances $0 = m(x), x \in X$, where m maps a finite product $\mathbb{E} = \Pi_{i \in I}\mathbb{E}_i$ of Euclidean spaces into itself, and the feasible set $X = \Pi_{i \in I} X_i$ equals a corresponding product, each component X_i being a nonempty closed convex part of $\mathbb{E}_i$. The said equation then assumes system form

$$0 = m_i(x), \ x_i \in X_i \text{ for all } i \in I, \tag{1}$$

$m_i : \mathbb{E} \to \mathbb{E}_i$ denoting the component map of m that corresponds to x_i.

Let I be a fixed list of different processors or individuals i, each taking on the task to control and monitor merely his block variable x_i. Suppose that i only records the value $m_i(x)$ and its development to date; he neither talks to other agents $j \neq i$ about x_i, nor does he get to observe their realized values $m_j(x)$ or choices x_j. Then, taking existence of solutions for granted, *how can such non-communicating, non-informed agents ultimately come to solve* (1)?

Motivation for this question is found in Section 2, which argues that agents typically need time to learn and adapt. Sections 3, 4 and 5 therefore look at dynamics that may lead to solutions of (1) as steady states. Section 6 and 7 turn such dynamics into iterative procedures. By incorporating constraints this paper adds to [13] but is self-contained.

2. NONCOOPERATIVE GAMES

For motivation suppose in this section that each $i \in I$ seeks to maximize own payoff $\pi_i(x_i, x_{-i})$ with respect to his decision $x_i \in X_i$ given the choice $x_{-i} := (x_j)_{j \in I \setminus i}$ of the

*I thank Ruhrgas for generous support and Bjørn Sandvik for valuable assistance.

rivals. A strategy profile $x \in X$ is then declared a *Nash equilibrium* iff stable against unilateral deviations, that is, precisely when

$$x_i \in \arg\max_{X_i} \pi_i(\cdot, x_{-i}) \text{ for all } i.$$

Presumably, each function $x \mapsto \pi_i(x) = \pi_i(x_i, x_{-i})$ is finite-valued continuous on an open set containing X, and concave in x_i with *marginal payoff* $M_i(x) := \partial\pi_i(x)/\partial x_i$ locally bounded and well defined as the partial superdifferential of convex analysis [19]. Let $N_i(x_i) := \{x_i^* \in \mathbb{E}_i : \langle x_i^*, X_i - x_i\rangle \leq 0\}$ denote the *normal cone* of X_i at the latter's member x_i. The players will then reside at equilibrium iff $0 \in M_i(x) - N_i(x_i)$ for all i. So, defining $m_i(x)$ as the *smallest-norm element* in $M_i(x) - N_i(x_i)$, Nash equilibrium obtains when (1) holds.

Diverse disciplines - including mathematics, system theory, ecology, computer science, and economics - show strong interest in multi-agent interaction [3], [5], [16], [18]. Noncooperative game theory has thus become common ground and Nash equilibrium a key concept [17], [25]. That concept is intricate though, because all parties seemingly hold full rationality and perfect foresight. Real agents possibly have neither - and could also lack experience, knowledge, or memory [6], [20]. If so, they need time to learn and adapt [14], [21], [26].

Manifold mechanisms then come into operation. *Imitation*, being one mechanism, presumes much similarity, visibility, or communication among the concerned parties [22], [24]. *Introspection*, being another, becomes most effective when agents have extended memories or global knowledge about the game [9]. These facts favor *prospection* which, while local in nature, makes virtually no demands on similarity, visibility, communication, memory, or knowledge [12]. Our challenge is therefore to explain whether and why prospective agents eventually may appear smarter and better informed than they really are. In other words: *If the concerned parties know and communicate little, can they solve* (1) *over time?* To come to grips with this question we shall design processes intended to depict modes of repeated play which, while proceeding in parallel, are decentralized, non-coordinated, and non-communicative.

3. GRADIENT PROCESSES

Motivated by noncooperative games, described above, we begin by considering the continuous-time, steepest-ascent, projected-gradient method

$$\frac{dx_i}{dt} =: \dot{x}_i \in M_i(x) - N_i(x_i) \text{ almost everywhere (a.e.) } \forall i. \tag{2}$$

Suppose henceforth that each correspondence $x \mapsto M_i(x)$ is upper semicontinuous with nonempty compact convex values on X, and such that $0 \in M_i(x) - N_i(x_i)$ implies $0 = m_i(x)$. Let $M(x) := \Pi_{i \in I} M_i(x)$.

Proposition 1. (Viability and convergence of the projected gradient method)

(a) *Suppose X is bounded. Then there exists at least one equilibrium, and from any initial point $x(0) \in X$ emanates an infinite, absolutely continuous trajectory $0 \leq t \mapsto x(t) \in X$ satisfying* (2).

(b) *Suppose a ball $\mathcal{B} \subseteq X$ centered at $\bar{x}$ is such that*

$$\sup_{g \in M(x)} \langle x - \bar{x}, g \rangle < 0 \text{ for all } x \in \mathcal{B} \diagdown \bar{x}. \tag{3}$$

Then each infinite solution trajectory $x(\cdot)$ of (2), starting at any $x(0) \in \mathcal{B}$, will converge to $\bar{x}$. Moreover, $\bar{x}$ must be an equilibrium.

Proof. (a) Let $\mathbb{B}$ be a closed ball centered at origin with radius so large that $M(X) \subseteq \mathbb{B}$. Write $N(x) := \Pi_{i \in I} N_i(x_i)$ and consider the truncated dynamics

$$\dot{x} \in \mathcal{M}(x) := [M(x) - N(x)] \cap \mathbb{B} \text{ a.e.}$$

Let $T(x) = cl\{\mathbb{R}_+(X - x)\}$ denote the *tangent cone* of X at $x \in X$. Since $\mathcal{M}$ is upper semicontinuous on X, with $\mathcal{M}(x)$ compact convex and $\mathcal{M}(x) \cap T(x)$ nonempty at any $x \in X$, the first statement now derives from Nagumo's viability theorem ; see [2], 3.3.4. Existence of an equilibrium - that is, the availability of a root of $\mathcal{M}(x)$ - is guaranteed by Kakutani's fixed point theorem [2].

(b) (3) makes the function $0 \le t \mapsto L(t) := \frac{1}{2}\|x(t) - \bar{x}\|^2$ Lyapunov near $\bar{x}$. Indeed, when $x(0) \in \mathcal{B}$ we get (omitting explicit mention of time):

$$\dot{L} = \langle x - \bar{x}, \dot{x} \rangle \in \langle x - \bar{x}, M(x) - N(x) \rangle \le \langle x - \bar{x}, M(x) \rangle \le 0 \text{ a.e.},$$

the last inequality being strict when $x \neq \bar{x}$. Thus $\|x(t) - \bar{x}\|$ tends monotonically downwards to a real number $r \ge 0$. This tells that $x(t) \in \mathcal{B}$ and $r \le \|x(t) - \bar{x}\| \le \|x(0) - \bar{x}\|$ for all $t \ge 0$. Let

$$\mu := \sup\{\langle x - \bar{x}, g \rangle : x \in X, \ g \in M(x), \quad r \le \|x - \bar{x}\| \le \|x(0) - \bar{x}\|\}.$$

If r were positive, the upper semicontinuity of $x \mapsto \sup_{g \in M(x)} \langle x - \bar{x}, g \rangle$ on X would imply $\dot{L} \le \mu < 0$, whence the absurdity $L(t) \searrow -\infty$. Thus $r = 0$. The limit $\bar{x}$ of a trajectory must be invariant; that is, $0 \in M(\bar{x}) - N(\bar{x})$, and the last assertion follows immediately.■

In the game setting described above introduce $\pi(\hat{x}, x) := \sum_{i \in I} \pi_i(\hat{x}_i, x_{-i})$ on $X \times X$ to get a partial superdifferential $\frac{\partial}{\partial \hat{x}}\pi(\hat{x}, x)\,|_{\hat{x}=x} =: M(x)$ with closed graph on X. In fact, $g^n \in M(x^n)$ implies, by definition, that $\pi(\hat{x}, x^n) \le \pi(x^n, x^n) + \langle \hat{x} - x^n, g^n \rangle$ for all $\hat{x} \in X$. So, if $x^n \to x$ in X, and $g^n \to g$, the resulting limit inequality, namely $\pi(\hat{x}, x) \le \pi(x, x) + \langle \hat{x} - x, g \rangle, \forall \hat{x}$ near X, tells that $g \in M(x)$. Since M thus has closed graph and is uniformly bounded on X, it must be upper semicontinuous there.

Often there are serious drawbacks with (2). First, the monotonicity assumption (3) may fail. Second, convergence may come slowly, if at all. These features lead us to consider methods that exploit not merely M but its derivative as well. So, suppose henceforth that $x \mapsto M(x)$ *be single-valued, continuously differentiable with M' non-singular near and on X.*

Before leaving (2) aside we remark parenthetically that discrete-time simulation of that process could proceed iteratively at stages $k = 0, 1, \ldots$ by

$$x_i^{k+1} = P_i\left[x_i^k + s_{ik} g_i^k\right] + e_i^k \text{ for all } i. \tag{4}$$

Here P_i denotes the orthogonal projection in $\mathbb{E}_i$ onto X_i; the number $s_{ik} \geq 0$ is an appropriate step-size; $g_i^k \in M_i(x^k)$ (or, in the context of noncooperative games, g_i^k is an approximate supergradient of $\pi_i(\cdot, x_{-i}^k)$ at the current point x_i^k); and finally, e_i^k accounts for errors stemming from inexact projection. Process (4), being an inaccurate *explicit* Euler integration of (2), demands little knowledge, rationality, or expertise from any agent i. The convergence analysis - found in [7], [10], [11], [12] - invites use of stochastic approximation theory [4]. Format (2) also calls for *implicit* Euler integration in the form of so-called *prox methods* or associated splitting procedures; see [9].

4. HEAVY-BALL PROCESSES

Two objectives structure the subsequent development. First, we want to preserve the appealing, decentralized nature of (2) - and of its discretized version (4). Second, it is desirable to dispense with the often ill-suited monotonicity assumption (3). To these ends, when $x \in X$, the alternative dynamics

$$\dot{x}_i \in M_i(x) - N_i(x_i) + \lambda_i \dot{M}_i(x) \text{ a.e. for all } i, \tag{5}$$

will guide us. Presumably, the time derivatives $\dot{M}_i(x) := \frac{d}{dt} M_i(x(t)), i \in I$, are well defined a.e. Clearly, system (5) embodies some inertia, some heavy-ball momentum, or second-order extrapolation via the terms $\dot{M}_i(x) = M_i'(x)\dot{x}$. Given a game, (5) mirrors that each player i moves in the projected direction $M_i(x) - N_i(x_i)$ of steepest payoff ascent modified somewhat by how swiftly that direction changes.

The numbers λ_i are positive and rather large so as to mitigate the slow-down of (2) near stationary points. System (5) has been studied in [13] as a model of *unconstrained* noncooperative play (with all $N_i \equiv 0$). When interaction is indeed constrained, let $m_i(x)$ denote the smallest-norm element of $M_i(x) - N_i(x_i)$.

Proposition 2. (Convergence of an extrapolative system) *Suppose the absolutely continuous (and viable) trajectory $0 \leq t \mapsto x(t) \in X$ solves*

$$\dot{x}_i = m_i(x) + \lambda_i \dot{M}_i(x) \text{ a.e. for all } i. \tag{6}$$

If all λ_i are sufficiently large, and $0 \leq t \mapsto m_i(x(t))$ is absolutely continuous with $\dot{m}_i(x(t)) = M_i'(x(t))\dot{x}(t)$ a.e. for all i, then any accumulation point $\bar{x}$ of the trajectory must be an equilibrium. In particular, when $\bar{x}$ is isolated, it follows that $x(t) \to \bar{x}$.

Proof. Denote by λ the diagonal matrix having λ_i everywhere along the diagonal in block i. Then (6) can be rewritten succinctly as $\dot{x} = m(x) + \lambda M'(x)\dot{x}$ a.e., that is,

$$\dot{x} = [I - \lambda M'(x)]^{-1} m(x) \text{ a.e.} \tag{7}$$

Let $L(t) := \frac{1}{2} \|m(x(t))\|^2$ and, for brevity, write $m = m(x)$, $M' = M'(x)$ to have

$$\dot{L} = \langle m, M'\dot{x} \rangle = \left\langle m, M' [I - \lambda M']^{-1} m \right\rangle = \left\langle m, \lambda^{-1} \lambda M' [I - \lambda M']^{-1} m \right\rangle \text{ a.e.}$$

Now use the matrix identity $[I - \lambda M']^{-1} - \lambda M' [I - \lambda M']^{-1} = I$ to get

$$\dot{L} = \left\langle m, \lambda^{-1} \left\{ -I + [I - \lambda M']^{-1} \right\} m \right\rangle \text{ a.e.}$$

Thus, for λ sufficiently large it obtains a.e. that $m(x) \neq 0 \Rightarrow \dot{L} < 0$. This completes the proof. ∎

The preceding proof used the chain rule $\dot{M}(x) = M'(x)\dot{x}$. But by assumption no individual i has enough data to calculate $M_i'(x)\dot{x}$, employing that rule.

When $\lambda = 0$, we obtain the selected version $\dot{x} = m(x)$ of (2). Also, if $X_i = \mathbb{E}_i$ so that $N_i \equiv 0$ for all i, then employing large diagonal elements, (7) resembles a scaled version of $\dot{x} = -M'(x)^{-1}M(x)$. The latter is Newton's method - well known to enjoy superb stability. To wit, the function $\mathcal{L}(t) := \frac{1}{2}\|M(x(t))\|^2$ then evolves by $\dot{\mathcal{L}} = -2\mathcal{L}$ so that $\mathcal{L}(t) = \mathcal{L}(0)e^{-2t} \searrow 0$. This observation reiterates that Newton's method yields second-order local convergence. That method does, however, completely destroy the indispensable, decentralized character of (2). Besides, there is no social institution or procedure that takes Newton steps. Such steps cannot depict reasonable or plausible behavior of agents who are loosely coordinated, weakly communicating, or scantly informed.

5. FEASIBILITY

Many situations allow choices or states that transiently are infeasible. To deal with such instances we inquire here how the feasible set X can be made attractive and absorbing. Dynamics (5) and (6) tell nothing about that serious issue. Let the projection $P_S(y)$ comprise those points in $S \subseteq \mathbb{E}$ which are closest to $y \in \mathbb{E}$.

Definition 1. (Aiming towards a set) *Let $\mathcal{V}$ and S be nonempty closed subsets of the same Euclidean space. We say that $\mathcal{V}$ aims towards S at $y \notin S$ if*

$$\sup_{v\in\mathcal{V}} \inf_{s\in P_S(y)} \langle v, y-s\rangle \leq 0.$$

Lemma 1. (A simplified test on aiming) *A convex set $\mathcal{V}$ aims towards S at $y \notin S$ iff there exists $\bar{s} \in convP_S(y)$ such that*

$$\sup_{v\in\mathcal{V}} \langle v, y-\bar{s}\rangle \leq 0. \tag{8}$$

Proof. The set $P_S(y)$ is nonempty compact whence so is its convex hull $convP_S(y)$. Thus, for any given $v \in \mathcal{V}$ we have

$$\inf_{s\in P_S(y)} \langle v, y-s\rangle = \min_{s\in P_S(y)} \langle v, y-s\rangle = \min_{s\in convP_S(y)} \langle v, y-s\rangle .$$

Let $\bar{s}$ be any point in $convP_S(y)$ which minimizes the lower semicontinuous function $s \mapsto \sup_{v\in\mathcal{V}} \langle v, y-s\rangle$ on that set. Since $\mathcal{V}$ is convex, we get by the Lop-sided minimax theorem [1] that $\sup_{v\in\mathcal{V}} \langle v, y-\bar{s}\rangle =$

$$\min_{s\in convP_S(y)} \sup_{v\in\mathcal{V}} \langle v, y-s\rangle = \sup_{v\in\mathcal{V}} \min_{s\in convP_S(y)} \langle v, y-s\rangle = \sup_{v\in\mathcal{V}} \inf_{s\in P_S(y)} \langle v, y-s\rangle \leq 0. \ ∎$$

Lemma 2. (Absorbing sets) *Suppose a convex-valued correspondence $(t,y) \mapsto V(t,y) \subset \mathbb{E}$ aims towards a closed convex set S at any $y \notin S$. Then an absolutely continuous function $0 \leq t \mapsto y(t) \in \mathbb{E}$, satisfying $\dot{y}(t) \in V(t, y(t))$ a.e. when $y(t) \notin S$, cannot leave S after having hit that set.*

Proof. Let $d_S(y) := \min_{s\in S} \|y - s\|$ denote the distance to S. When $y \notin S$, it holds that $\nabla d_S(y) = \frac{y-P_S(y)}{\|y-P_S(y)\|}$. Since $y \mapsto d_S(y)$ is Lipschitz continuous, $t \mapsto d_S(y(t))$ must be absolutely continuous. So, omitting mention of time, we have via (8) that

$$\dot{d}_S(y) = \langle \nabla d_S(y), \dot{y}\rangle \leq \sup_{v\in V(t,y)} \left\langle \frac{y - P_S(y)}{\|y - P_S(y)\|}, v \right\rangle \leq 0 \text{ a.e. when } y \notin S.$$

Pick any instant $t_1 > 0$ such that $y(t_1) \notin S$. Suppose the preceding time

$$t_0 := \max\{t \in [0, t_1] : y(t) \in S\}$$

is well defined. This produces the contradiction $0 < d_S(y(t_1)) = d_S(y(t_1)) - d_S(y(t_0)) = \int_{t_0}^{t_1} \dot{d}_S(y(t)) \leq 0$. ■

Lemma 3. (Finite hitting time) *Suppose* $0 \leq t \mapsto y(t)$ *is absolutely continuous and*

$$\dot{y}(t) \in -\partial f(y(t)) \text{ for almost every tsuch that } f(y(t)) > 0. \tag{9}$$

Here f is finite-valued convex, with $f(y(0)) > 0$, and is - as standing assumption - supposed differentiable almost everywhere along $\{y(t) : f(y(t)) > 0\}$. Then the minimal hitting time $T := \min\{t > 0 : f(y(t)) \leq 0\}$ must be finite under each of the following two conditions:

(I):	$\{f < 0\}$	*is nonempty bounded,*
(II):	$\inf\{\|g\| : g \in \partial f(y) : 0 < f(y) \leq f(y(0))\}$	*is positive.*

Proof. Since $t \mapsto y(t)$ is absolutely continuous, so is the function $t \mapsto f(y(t))$. It holds a.e. on $[0, T)$ that

$$\frac{d}{dt} f(y(t)) = -\|g(t)\|^2 \text{ for some } g(t) \in \partial f(y(t)).$$

If $\{f < 0\}$ is nonempty bounded, there exists a positive number ε such that $C := \{-\varepsilon \leq f \leq f(y(0))\}$ is nonempty compact and $0 \notin \partial f(C)$. Since $\partial f(C)$ is compact as well, all subgradients $G \in \partial f(C)$ must satisfy $\|G\| \geq \gamma$ for some positive constant γ. Consequently, $\frac{d}{dt} f(y(t)) \leq -\gamma^2$ a.e. while $f > 0$. This takes care of case (I). Case (II) is now argued straightforwardly. ■

Lemma 4. (Absorption in finite time) *Let* $0 \leq t \mapsto y(t)$ *be absolutely continuous and satisfy* (9). *Under each of the following two conditions $y(t)$ will hit $\{f \leq 0\}$ in finite time and remain there.*

(I) There is a finite family of convex real-valued functions $f_j, j \in J$, all defined on the same Euclidean space, such that $\cap_{j\in J}\{f_j < 0\}$ is nonempty bounded, and $f := \max_j f_j$, or $f := \max_j f_j^+$, or $f := \sum_j f_j^+$.

(II) S is a closed convex set, and $f := \varphi \circ d_S$ with $\varphi : \mathbb{R} \to \mathbb{R}$ convex increasing, satisfying $\varphi(r) = 0 \Leftrightarrow r = 0$, and $0 \notin \partial\varphi(0)$.

Proof. Let $S := \{f \leq 0\}$. Aiming obtains in all cases because

$$g \in \partial f(y) \text{ and } \bar{s} = P_S(y) \Rightarrow f(y) + \langle g, \bar{s} - y\rangle \leq f(\bar{s}) \leq 0$$

so that $f(y) > 0$ yields $\langle -g, y - \bar{s}\rangle \leq -f(y) < 0$. Lemma 2 says that absorption is guaranteed. It remains to argue that $y(\cdot)$ hits S in finite time. For any instance mentioned under (I) Lemma 3,(II) applies. Finally, for instance (II), when $y \notin S$, we have $\nabla d_S(y) = \frac{y-P_S(y)}{\|y-P_S(y)\|}$ so that $\|\nabla d_S(y)\| = 1$. Consequently, $\frac{d}{dt}\varphi \circ d_S(y(t)) = -\|g(t)\|^2$ for some $g(t) \in \partial\varphi(d(y(t))) \geq \inf \partial\varphi(0) > 0$. Hence Lemma 3, (II) applies once again.■

Collecting these auxiliary results we get:

Proposition 3. (Feasibility after finite time) *For each $i \in I$ suppose at least one of the following two conditions holds:*
(I) There is a finite family of convex functions $f_{ij} : \mathbb{E}_i \to \mathbb{R}$, $j \in J(i)$, with $X_i := cl \cap_{j\in J(i)} \{f_{ij} < 0\}$ nonempty bounded, and $f_i := \sum_{j\in J(i)} f_{ij}^+$, or $f_i := \max_{j\in J(i)} f_{ij}^+$.
(II) X_i is a closed convex set, $f_i := \varphi_i \circ d_{X_i}$ with $\varphi_i : \mathbb{R} \to \mathbb{R}$ convex increasing, $\varphi_i(r) = 0 \Leftrightarrow r = 0$, and $0 \notin \partial\varphi_i(0)$.

Then, if for each i there exists an absolutely continuous function $0 \leq t \mapsto x_i(t)$ such that a.e.

$$x_i(t) \notin X_i \Rightarrow \dot{x}_i(t) \in -\partial f_i(x_i(t))$$

this ensures that $x(t) = [x_i(t)]_{i\in I} \in X$ after finite time. In any case, if $x_i(t) \in bdX_i$ then $\partial f_i(x_i(t)) \subseteq N_i(x_i(t))$. ■

Proposition 3 suggests use of the dynamics $\dot{x}_i \in M_i(x) + \lambda_i \dot{M}_i(x)$ on $intX_i$ and $\dot{x}_i \in -\partial f_i(x_i)$ outside X_i. Making such a choice we must specify what drives x_i at bdX_i. Unless moving transversally, x_i evolves there in a so-called sliding mode [15], [23]. To capture that phenomenon let, in the setting of Proposition 3, $\mathcal{N}_i(x_i) := conv\left[\limsup_{y_i\to x_i} \partial f_i(y_i)\right]$. The correspondence $x_i \mapsto \mathcal{N}_i(x_i)$ so constructed is upper semicontinuous with nonempty compact convex values. Moreover, $\mathcal{N}_i(x_i) = 0$ for $x_i \in intX_i$, $\mathcal{N}_i(x_i) = \partial f_i(x_i)$ for $x_i \notin X_i$, and $\mathcal{N}_i(x_i) \subset N_i(x_i)$ for $x_i \in bdX_i$. To subtract normal components that are directed outwards, if any, let

$$\dot{x}_i \in \mathbf{1}_i(x_i)\left[M_i(x) + \lambda_i \dot{M}_i(x)\right] - \mathcal{N}_i(x_i) \text{ a.e.} \tag{10}$$

Here the usual indicator $x_i \mapsto \mathbf{1}_i(x_i)$ equal 1 on X_i and 0 elsewhere. It suffices for the motion of (10) that each agent i continuously observes and appropriately reacts to his current data x_i, $M_i(x)$, $\dot{M}_i(x)$, $\mathcal{N}_i(x_i)$. One may argue that (10), for any initial $x(0)$, admits an infinitely extendable solution. Taking such a solution for granted we have:

Proposition 4. (Absorption and viability) *Under the hypotheses of Proposition 3 any infinitely extendable trajectory of (10) hits X in finite time and lives there forever after.* ■

Naturally we want (10) to be stationary at every equilibrium. For that purpose suppose that

$$m_i(x) \in M_i(x) - \mathcal{N}_i(x_i)\ \forall i \in I, \forall x \in X. \tag{11}$$

For verification of (11) let $r_i := \sup_{x\in X} \|M_i(x)\|$, this number being finite by assumption. Condition (11) then holds, in the setting of Proposition 3 (I) and (II), if respectively

(I) There is $B(\hat{y}_i, \rho_i) \subseteq \cap_{j\in J(i)} \{f_{ij} < 0\}$ such that $-\max_{j\in J(i)} f_{ij}(\hat{y}_i)/\rho_i \geq r_i$.
(II) $\varphi_i'(0+) \geq r_i$.

6. DISCRETE-TIME APPROXIMATION

The continuous-time format of (10) suits analysis (and analog computers as well). But interaction often happens at discrete times. Such situations fit best to iterative processes (and to digital computers). The state $x^k := (x_i^k)$ that prevails at the beginning of stage k will then be updated sequentially by some specified rule

$$x_i^{k+1} = x_i^k + s_k A_i(k, x^k, x^{k-1}) \text{ for all } i, \tag{12}$$

the initial x^0, x^{-1} being determined by factors not discussed here. The numbers $s_k > 0$, $k = 0, 1, \ldots$, called stepsizes, are predetermined subject to $s_k \searrow 0$ and $\sum s_k = +\infty$. They define an internal clock that starts at time $0 =: \tau_0$ and shows accumulated lapse $\tau_k := s_0 + \cdots + s_{k-1}$ at the onset of stage k. On that clock time ticks dwindle because $\tau_{k+1} - \tau_k = s_k \to 0$, and the horizon lies infinitely distant because $\tau_k \to +\infty$. Interpolating between consecutive points x^k and x^{k+1} we get a piecewise linear approximation

$$\chi(t) := \begin{cases} x^0 & \text{when } t \leq 0, \\ x^k + \frac{t-\tau_k}{\tau_{k+1}-\tau_k}(x^{k+1} - x^k) & \text{when } t \in [\tau_k, \tau_{k+1}], k = 0, 1, \ldots \end{cases}$$

Note that $\chi(\tau_k) = x^k$. Since

$$\dot{\chi}(t) = \frac{x^{k+1} - x^k}{\tau_{k+1} - \tau_k} \text{ for all } t \in (\tau_k, \tau_{k+1}), \tag{13}$$

(12) takes the equivalent form

$$\dot{\chi}(t) = A(k, \chi(\tau_k), \chi(\tau_{k-1})) \text{ when } t \in (\tau_k, \tau_{k+1}), \tag{14}$$

with $\chi(\cdot)$ continuous and $A := (A_i)$. This reformulation leads us to analyze convergence of (12) by means of the corresponding differential equation (10) or (14). For that purpose let $\omega\{x^k\}$ denote the set of all cluster points of $\{x^k\}$, and similarly let

$$\omega\{\chi\} := \{\lim \chi(t_l) : \text{for some sequence } t_l \to +\infty\}. \tag{15}$$

Lemma 5. (Coincidence of ω-limits) *Suppose* $s_k A(k, x^k, x^{k-1}) \to 0$. *Then* $\omega\{x^k\} = \omega\{\chi\}$.

Proof. Evidently, $\omega\{x^k\} \subseteq \omega\{\chi\}$. Conversely, given a cluster point $x = \lim \chi(t_l)$, emerging as some sequence $t_l \to +\infty$, consider stages $k(l) := \max\{k : \tau_k \leq t_l\}$. The assumption tells that $x^{k(l)} \to x$. ■

Lemma 6. (Convergence of approximate differential inclusions) *Let* B_i *be the closed unit ball centered at the origin in* $\mathbb{E}_i$. *Consider a sequence of differential inclusions defined as follows: For each* i *let*

$$\dot{\chi}_i^k \in \mathbf{1}_i(\chi_i^k + y_i^k)\left[M_i(\chi^k + y^k) + \lambda_i \dot{M}_i(\chi^k + z^k)\right] - \mathcal{N}_i(\chi^k + y^k) + r_k B_i \tag{16}$$

with sequences
$\{\chi^k(0)\}$, $\{z^k(0)\}$ *bounded in* $\mathbb{E}$;
$\{\dot\chi^k\}$, $\{\dot z^k\}$ *bounded in* L_∞;
$\{y^k\}$,$\{z^k\}$, $\{r_k\}$ *bounded in* L_∞ *and converging to* 0 *a.e.*
Then there is a subsequence of $\{\chi^k\}$ *which converges uniformly on bounded intervals to an absolutely continuous function* $\chi^\infty : \mathbb{R} \to \mathbb{E}$ *satisfying* (10). *Moreover, we can take the corresponding subsequence of* $\{\dot\chi^k\}$ *to converge weakly to* $\dot\chi^\infty$.

Proof. The presumed essential boundedness of $\{(\dot\chi^k, \dot z^k)\}$ makes the sequence$\{(\chi^k, z^k)\}$ equicontinuous on any bounded interval. Since $\{(\chi^k(0), z^k(0))\}$ is bounded, there exists, by the Arzela-Ascoli theorem, a subsequence of $\{(\chi^k, z^k)\}$ that converges uniformly on bounded intervals to a continuous function pair (χ^∞, z^∞). Moreover, because $\{(\dot\chi^k, \dot z^k)\}$ is bounded in L_∞, whence in L_2, we can take $(\dot\chi^k, \dot z^k)$ to converge weakly along the said subsequence to a limit denoted $\frac{d}{dt}(\chi^\infty, z^\infty)$ - with apologies for temporary abuse of notation. Passing to the limit in

$$(\chi^k(t), z^k(t)) = (\chi^k(0), z^k(0)) + \int_0^t (\dot\chi^k, \dot z^k)$$

we obtain by weak convergence that $(\dot\chi^\infty, \dot z^\infty) = \frac{d}{dt}(\chi^\infty, z^\infty)$ a.e. It remains only to check whether the limiting function χ^∞ solves (10). For this purpose rewrite (16) on the corresponding integral form: For each $i \in I$ and $t > 0$

$$\chi_i^k(t) \in \chi_i^k(0) + \int_0^t \left\{\mathbf{1}_i(\chi_i^k + y_i^k)\left[M_i(\chi^k + y^k) + \lambda_i \dot M_i(\chi^k + z^k)\right] - \mathcal{N}_i(\chi^k + y^k) + r_k B_i\right\}.$$

Recall that M is continuously differentiable. Pass therefore to the limit along the distinguished subsequence to arrive at (10). ■

7. REACHING EQUILIBRIUM

Motivated by format (10) as well as Propositions 2 and 4 we suggest that decentralized adaptations be modelled as follows. Iteratively, at stage $k = 0, 1, ...$ and for all $i \in I$, let

$$x_i^{k+1} = x_i^k + s_k \left\{ \begin{array}{l} m_i(x^k) + \lambda_i \left[M_i(x^k) - M_i(x^{k-1})\right]/s_{k-1} \ \text{ if } x_i^k \in X_i, \\ -\mathcal{N}_i(x_i) \ \text{ otherwise.} \end{array} \right\} \tag{17}$$

For the convergence analysis we *invoke henceforth the hypotheses of Proposition 2 and qualification* (11).

Theorem 1. (Convergence of the discrete-time, heavy-ball method) *Suppose* M' *is non-singular Lipschitz continuous on* X, *that set by assumption being bounded and containing only isolated solutions to* (1). *If the sequence* $\{x^k\}$ *generated by* (17) *becomes feasible after finite time and* $\{(x^{k+1} - x^k)/s_k\}$ *remains bounded, then, provided all* λ_i *are sufficiently large,* x^k *converges to a solution of* (1).

Proof. *We claim that* $\omega\{\chi\}$, *as defined in* (15), *is invariant under* (10). Suppose it really is. Then, for some solution $\bar x$ to (1) we get $\omega\{\chi\} = \{\bar x\}$ because singletons are the only invariant strict subsets of X (by Propositions 2 and 4). Thus, while assuming $\omega\{\chi\}$ invariant, we have shown via Lemma 5 that $\omega\{x^k\} = \omega\{\chi\} = \{\bar{x}\}$ whence $x^k \to \bar{x}$ as desired.

To complete the proof we must verify that asserted invariance. That argument requires a detour. The Lipschitz continuity of M' (with modulus L) yields

$$\begin{aligned} M(x) - M(x^-) &= \int_0^1 \frac{d}{dh} M(x^- + h(x - x^-))dh \\ &= \int_0^1 M'(x^- + h(x - x^-))(x - x^-)dh \\ &= M'(x)(x - x^-) + \int_0^1 \left[M'(x^- + h(x - x^-)) - M'(x)\right](x - x^-)dh \\ &\in M'(x)(x - x^-) + \frac{L}{2}\left\|x - x^-\right\|^2 B. \end{aligned}$$

Therefore, letting $x = x^k$, $x^- = x^{k-1}$, and $\tau_{k-\frac{1}{2}} := \tau_{k-1} + \frac{s_{k-1}}{2}$ we get via (13) that

$$\begin{aligned} \frac{M(x^k) - M(x^{k-1})}{s_{k-1}} &= M'(x^k)\frac{x^k - x^{k-1}}{\tau_k - \tau_{k-1}} + O(s_k) = M'(x^k)\dot{\chi}(\tau_{k-\frac{1}{2}}) + O(s_k) \\ &= M'(\chi(\tau_{k-\frac{1}{2}}))\dot{\chi}(\tau_{k-\frac{1}{2}}) + O(s_k) = \dot{M}(\chi(\tau_{k-\frac{1}{2}})) + O(s_k). \end{aligned}$$

As a result, again invoking (13), we see that when $x^k \in X$, system (17) can eventually be rewritten

$$\dot{\chi}(t) = m(\chi(\tau_k)) + \lambda\dot{M}(\chi(\tau_{k-\frac{1}{2}})) + O(s_k) \text{ for } t \in (\tau_k, \tau_{k+1}). \tag{18}$$

We now shift the initial time 0 of the process χ backwards to gain from $s_k \longrightarrow 0^+$. More precisely, introduce a sequence of functions $\chi^k : \mathbb{R} \to \mathbb{E}$, $k = 0, 1, \ldots$ defined by

$$\chi^k(t) := \chi(\tau_k + t). \tag{19}$$

(In particular, $\chi^0 = \chi$.) The sequence so constructed satisfies (16) for appropriate y^k, z^k, r_k. To see this let the integer (counting) function

$$0 \le r \longmapsto \mathcal{K}(r) := \max\{k : \tau_k \le r\}. \tag{20}$$

keep track of the stage. From (18) we get, writing $\hat{k} = \mathcal{K}(\tau_k + t)$ for short,

$$\dot{\chi}^k(t) = \frac{d}{dt}\chi(\tau_k + t) = m(\chi(\tau_{\hat{k}})) + \lambda\dot{M}(\chi(\tau_{\hat{k}-\frac{1}{2}}) + O(s_{\hat{k}})$$

so (16) obtains with

$$\begin{aligned} y^k(t) &= \chi(\tau_{\hat{k}}) - \chi^k(t) = x^{\mathcal{K}(\tau_k + t)} - \chi^k(t), \\ z^k(t) &= \chi(\tau_{\hat{k}-\frac{1}{2}}) - \chi^k(t) = \chi(\tau_{\mathcal{K}(\tau_k+t)-\frac{1}{2}}) - \chi^k(t), \text{ and} \\ r_n(t) &= O(s_{\mathcal{K}(\tau_k+t)}) \end{aligned}$$

Since $\{x^k\}$ is bounded, so are $\{\chi^k(0) = x^k\}$ and $\{z^k(0)\}$. Clearly, $r_k \to 0$, and $y^k \to 0$ because $\|y^k(t)\| = \|x^{\mathcal{K}(\tau_k+t)} - \chi^k(t)\| \leq \|x^{\mathcal{K}(\tau_k+t)} - x^{\mathcal{K}(\tau_k+t)+1}\| \to 0$. Also, because $\chi(\tau_{\mathcal{K}(\tau_k+t)-\frac{1}{2}})$ and $\chi^k(t)$ both belong to the triangle $conv\left\{x^{\hat{k}-1}, x^{\hat{k}}, x^{\hat{k}+1}\right\}$, we get

$$\|z^k(t)\| = \left\|\chi(\tau_{\mathcal{K}(\tau_k+t)-\frac{1}{2}}) - \chi^k(t)\right\| \leq diam\ conv\left\{x^{\hat{k}-1}, x^{\hat{k}}, x^{\hat{k}+1}\right\}$$
$$\leq \left\|x^{\hat{k}-1} - x^{\hat{k}}\right\| + \left\|x^{\hat{k}} - x^{\hat{k}+1}\right\| = O(s_{\hat{k}-1}) + O(s_{\hat{k}}) = O(s_{\hat{k}-1}) \to 0.$$

When well defined $\dot{z}^k(t) = -\dot{\chi}^k(t) = -\frac{d}{dt}\chi(\tau_k + t)$. So, we also have

$$\|\dot{z}^k(t)\| = \|x^{\mathcal{K}(\tau_k+t)+1} - x^{\mathcal{K}(\tau_k+t)}\| / s_{\mathcal{K}(\tau_k+t)}$$

uniformly bounded by assumption.

After this detour we return to the claim that $\omega\{\chi\}$ is invariant. Denote by $\chi^k(t,x)$ and $\chi^\infty(t,x)$ the unique state at time t of system (19) and (10) respectively provided it passes through x at time 0. Pick any point $x \in \omega\{\chi\}$ and any time t. We must show that $\chi^\infty(t,x) \in \omega\{\chi\}$. Via Lemma 5 some subsequence $\{x^k = \chi(\tau_k)\}_{k\in K}$ converges to x. If necessary, pass to a subsequence to have that $\chi^k \to \chi^\infty$ in the manner described in Lemma 6. Because $\chi^\infty(0,x^k) = \chi^k(0,x^k) = x^k \to x$ as $k \in K$ tends to infinity, this yields

$$\lim_{k\in K} \|\chi^\infty(t,x^k) - \chi^k(t,x^k)\| = 0. \tag{21}$$

Therefore, since $\chi^\infty(t,x^k) \to \chi^\infty(t,x)$ by continuous dependence on initial conditions, (21) implies

$$\chi^\infty(t,x) = \lim_{k\in K} \chi^\infty(t,x^k) = \lim_{k\in K} \chi^k(t,x^k) = \lim_{k\in K} \chi(\tau_k + t) \in \omega\{\chi\}.$$

The asserted invariance has now been verified and this concludes the proof. ■

(17) is a purely primal procedure. Alternatively, using multipliers for penalization, one might develop primal-dual methods like those in [8].

REFERENCES

1. J.-P. Aubin and I. Ekeland, *Applied Nonlinear Analysis* (John Wiley, N.Y., 1984).
2. J.-P. Aubin, *Viability Theory* (Birkhäuser, Basel, 1991).
3. M. Bardi, T. E. S. Raghavan and T. Parthasarathy, *Stochastic and Differential Games* (Birkhäuser, Basel, 1999).
4. M. Benaim, A dynamical system approach to stochastic approximations, *SIAM J. Control and Optimization* **34**, 437–472 (1996).
5. P. Cardaliaguet, M. Quincampoix, and P. Saint-Pierre, Set-valued numerical analysis for optimal control and differential games, Chapter 4 in [3] (1999).
6. J. Conlisk, Why Bounded Rationality? *Journal of Economic Literature* **34**, 669–700 (1996).
7. Yu. M. Ermoliev and S. P. Uryasiev, Nash equilibrium in n-person games, (in Russian) *Kibernetika* **3**, 85–88 (1982).

8. S. D. Flåm and J. Zowe, A primal-dual method for convex programming, in A. Ioffe, M. Marcus, and S. Reich (eds.) *Optimization and Nonlinear Analysis*, Longman Scientific, Essex 119–129 (1992).
9. S. D. Flåm and A. Antipin, Equilibrium programming using proximal-like methods, *Mathematical Programming* **78**, 29–41(1997).
10. S. D. Flåm and C. Horvath, Network games: Adaptations to Cournot-Nash equilibrium, *Annals of Operations Research* **64**, 179–195 (1996).
11. S. D. Flåm, Approaches to economic equilibrium, *J. Economic Dynamics & Control* **20**, 1505–1522 (1996).
12. S. D. Flåm, Learning equilibrium play: a myopic approach, *Computational Optimization and Applications* **14**, 87–102 (1999).
13. S. D. Flåm, Repeated play and Newton's method, *Int. Game Theory Review* **2**, 141–154 (2001).
14. D. Fudenberg and D. K. Levine, *The Theory of Learning in Games* (The MIT Press, Cambridge, 1998).
15. M. P. Glazos, S. Hui, and S. H. Zak, Sliding modes in solving convex programming problems, *SIAM J. Control & Opt.* **36**, 680–697 (1998).
16. J. Hofbauer and K. Sigmund, *Evolutionary Games and Population Dynamics* (Cambridge University Press, 1998).
17. M. J. Osborne and A. Rubinstein, *A Course in Game Theory* (The MIT Press, Cambridge, 1994).
18. A. Robson, Efficiency in evolutionary games: Darwin, Nash, and the secret handshake, *J. Theoretical Biology* **144**, 379–396 (1990).
19. R.T. Rockafellar, *Convex Analysis* (Princeton University Press, Princeton, 1970).
20. A. Rubinstein, *Modeling Bounded Rationality* (The MIT Press, Cambridge, 1998).
21. L. Samuelson, *Evolutionary Games and Equilibrium Selection* (The MIT Press, Cambridge, 1997).
22. K. H. Schlag, Why imitate, and if so, how? *J. Economic Theory* **78**, 130–156 (1998).
23. V. I. Utkin, *Sliding Modes in Control and Optimization*, (Springer-Verlag, Berlin, Germany, 1992).
24. F. Vega-Redondo, The evolution of Walrasian behavior, *Econometrica* **61**, 57–84 (1993).
25. J. W. Weibull, *Evolutionary Game Theory* (The MIT Press, Cambridge, 1995).
26. H. P. Young, *Individual Strategy and Social Structure: An Evolutionary Theory of Institutions* (Princeton University Press, Princeton, N.J., 1998).

Inherently Parallel Algorithms in Feasibility and Optimization and their Applications
D. Butnariu, Y. Censor and S. Reich (Editors)

GENERIC CONVERGENCE OF ALGORITHMS FOR SOLVING STOCHASTIC FEASIBILITY PROBLEMS

Manal Gabour[a], Simeon Reich[b*] and Alexander J. Zaslavski[c]

[a]Department of Mathematics, The Technion-Israel Institute of Technology, 32000 Haifa, Israel

[b]Department of Mathematics, The Technion-Israel Institute of Technology, 32000 Haifa, Israel

[c]Department of Mathematics, The Technion-Israel Institute of Technology, 32000 Haifa, Israel

Given a bounded closed convex subset K of a Banach space X, a complete probability measure space $(\Omega, \mathcal{A}, \mu)$ and a measurable family $\{T_\omega : \omega \in \Omega\}$ of nonexpansive self-mappings of K, we study the asymptotic behavior of the operator $\widetilde{T} : K \to K$ defined by

$$\widetilde{T}x = \int_\Omega T_\omega x d\mu(\omega), \; x \in K.$$

We define an appropriate complete metric space of such families of nonexpansive mappings and establish a generic strong convergence result for the powers of the corresponding operators $\widetilde{T}$. We also study the asymptotic behavior of infinite (random) products of such operators.

1. INTRODUCTION AND PRELIMINARIES

A (generalized) stochastic feasibility problem is the problem of finding almost common fixed points of measurable families of mappings. Such problems arise naturally in many areas of mathematics (see, for example, [1,3,5,11]). Various algorithms for solving stochastic feasibility problems were studied in [1–6,8,11]. In the present paper we propose to solve such problems by an iterative parallel algorithm. We use the generic approach which is common, for example, in the theory of discrete and continuous dynamical systems [7,9,10,13]. Thus, instead of considering a certain convergence property for a single algorithm applied to a single measurable family of mappings, we investigate it for a space of all such families equipped with some natural complete metric and show that this property holds for most of the corresponding algorithms in the sense of Baire's categories. This

*The second author was partially supported by the Israel Science Foundation founded by the Israel Academy of Sciences and Humanities (Grants 293/97 and 592/00), by the Fund for the Promotion of Research at the Technion, and by the Technion VPR Fund. All three authors thank the referee for many helpful comments.

allows us to establish strong convergence without restrictive assumptions on the space and on the operators themselves.

More precisely, let $(\Omega, \mathcal{A}, \mu)$ be a complete probability measure space, $(X, ||\cdot||)$ a Banach space and K a nonempty bounded closed convex subset of X. A mapping $A : K \to K$ is called nonexpansive if $||Ax - Ay|| \leq ||x - y||$ for all $x, y \in K$.

Given a measurable family $\{T_\omega : \omega \in \Omega\}$ of nonexpansive self-mappings of K, we study the asymptotic behavior of the operator $\tilde{T} : K \to K$ defined by

$$\tilde{T}x = \int_\Omega T_\omega x d\mu(\omega), \; x \in K.$$

We construct an everywhere dense G_δ subset $\mathcal{F}$ of a complete metric space of such families and show that for each family $\{T_\omega : \omega \in \Omega\}$ in $\mathcal{F}$ the powers of the corresponding operator $\tilde{T} : K \to K$ converge strongly to the unique fixed point x_T of $\tilde{T} : K \to K$, uniformly on K. If the family has an almost fixed point, then it is unique and coincides with x_T. We also extend our result to the case where a given set is a subset of the fixed points set of all the family involved. This situation encompasses the consistent feasibility problems. Note, however, that our strong convergence results are also applicable to the inconsistent case where there are no common fixed points. We also study the asymptotic behavior of infinite (random) products of the averaging operators $\tilde{T} : K \to K$.

Our paper is organized as follows. The main results are stated in Section 2. Sections 3-5 are devoted to auxiliary results which are needed in the proofs of our main results. These proofs are presented in Sections 6-9.

Denote by $\mathcal{N}$ the set of all nonexpansive mappings $A : K \to K$. For the set $\mathcal{N}$ we consider the metric $\rho_\mathcal{N} : \mathcal{N} \times \mathcal{N} \to [0, \infty)$ defined by

$$\rho_\mathcal{N}(A, B) = \sup\{||Ax - Bx|| : \; x \in K\}, \; A, B \in \mathcal{N}. \tag{1}$$

It is easy to see that the metric space $(\mathcal{N}, \rho_\mathcal{N})$ is complete. Denote by $\mathcal{M}$ the set of all sequences $\{A_t\}_{t=1}^\infty \subset \mathcal{N}$. For the set $\mathcal{M}$ we consider the metric $\rho^s_\mathcal{M} : \mathcal{M} \times \mathcal{M} \to [0, \infty)$ defined by

$$\rho^s_\mathcal{M}(\{A_t\}_{t=1}^\infty, \{B_t\}_{t=1}^\infty) = \sup\{||A_t x - B_t x|| : \; x \in K, \; t = 1, 2, \ldots\},$$
$$\{A_t\}_{t=1}^\infty, \; \{B_t\}_{t=1}^\infty \in \mathcal{M}. \tag{2}$$

It is not difficult to see that the metric space $(\mathcal{M}, \rho^s_\mathcal{M})$ is also complete. The topology generated by this metric on $\mathcal{M}$ will be called the strong topology. In addition to this topology we will also consider the uniformity determined by the base [12]

$$E^w_\mathcal{M}(N, \epsilon) = \{(\{A_t\}_{t=1}, \{B_t\}_{t=1}^\infty) \in \mathcal{M} \times \mathcal{M} :$$
$$||A_t x - B_t x|| \leq \epsilon, \; t = 1, \ldots, N, \; x \in K\}, \tag{3}$$

where N is a natural number and $\epsilon > 0$. It is easy to see that the space $\mathcal{M}$ with this uniformity is metrizable (by a metric $\rho^w_\mathcal{M} : \mathcal{M} \times \mathcal{M} \to [0, \infty)$) and complete. The topology generated by $\rho^w_\mathcal{M}$ on $\mathcal{M}$ will be called the weak topology.

Assume now that $F \subset K$ is a nonempty closed convex subset of K, $P : K \to F$ is nonexpansive and that $Px = x$ for each $x \in F$. Denote by $\mathcal{N}^{(F)}$ the set of all $A \in \mathcal{N}$ such that $Ax = x$ for all $x \in F$. Clearly $\mathcal{N}^{(F)}$ is a closed subset of $(\mathcal{N}, \rho_\mathcal{N})$ and the metric space

$(\mathcal{N}^{(F)}, \rho_{\mathcal{N}})$ is complete. Denote by $\mathcal{M}^{(F)}$ the set of all sequences $\{A_t\}_{t=1}^{\infty} \subset \mathcal{N}^{(F)}$. Clearly $\mathcal{M}^{(F)}$ is a closed subset of $\mathcal{M}$ with the weak topology. In the sequel we will consider the topological subspace $\mathcal{M}^{(F)} \subset \mathcal{M}$ with both the relative weak and strong topologies.

Denote by $\mathcal{N}_\Omega$ the set of all functions

$$\omega \to T_\omega \in \mathcal{N},\ \omega \in \Omega, \tag{4}$$

such that for each $x \in K$ the function $\omega \to T_\omega x$, $\omega \in \Omega$, is measurable. Such functions will be denoted by $(T_\omega)_{\omega\in\Omega}$. For each pair of functions $(T_\omega)_{\omega\in\Omega}$ and $(S_\omega)_{\omega\in\Omega}$ define

$$\rho_{\mathcal{N}_\Omega}((T_\omega)_{\omega\in\Omega}, (S_\omega)_{\omega\in\Omega}) = \sup\{||T_\omega x - S_\omega x|| :\ x \in K,\ \omega \in \Omega\}. \tag{5}$$

It is easy to see that $\rho_{\mathcal{N}_\Omega}$ is a metric on $\mathcal{N}_\Omega$ and that the metric space $(\mathcal{N}_\Omega, \rho_{\mathcal{N}_\Omega})$ is complete. The topology induced on the space $\mathcal{N}_\Omega$ by the metric $\rho_{\mathcal{N}_\Omega}$ will be called the strong topology. On the other hand, the topology induced on $\mathcal{N}_\Omega$ by the pseudometric $\rho^w_{\mathcal{N}_\Omega}$ defined by

$$\rho^w_{\mathcal{N}_\Omega}((T_\omega)_{\omega\in\Omega}, (S_\omega)_{\omega\in\Omega}) = \sup\{||\int_\Omega T_\omega x d\mu(\omega) - \int_\Omega S_\omega x d\mu(\omega)|| :\ x \in K\} \tag{6}$$

will be called the weak topology.

We denote by $\mathcal{M}_\Omega$ the sequence space consisting of all sequences of the form

$$\{(A_{t\omega})_{\omega\in\Omega}\}_{t=1}^{\infty} := \{((A_t)_\omega)_{\omega\in\Omega}\}_{t=1}^{\infty},$$

where

$$((A_t)_\omega)_{\omega\in\Omega} \in \mathcal{N}_\Omega,\ t = 1, 2, \ldots.$$

We equip this sequence space with a metric $\rho_{\mathcal{M}_\Omega}$ defined by

$$\begin{gathered}\rho_{\mathcal{M}_\Omega}(\{(A_{t\omega})_{\omega\in\Omega}\}_{t=1}^{\infty}, \{(B_{t\omega})_{\omega\in\Omega}\}_{t=1}^{\infty}) = \\ \sup\{||A_{t\omega}x - B_{t\omega}x|| :\ t = 1, 2, \ldots,\ \omega \in \Omega,\ x \in K\}, \\ \{(A_{t\omega})_{\omega\in\Omega}\}_{t=1}^{\infty},\ \{(B_{t\omega})_{\omega\in\Omega}\}_{t=1}^{\infty} \in \mathcal{M}_\Omega.\end{gathered} \tag{7}$$

It is easy to see that the metric space $(\mathcal{M}_\Omega, \rho_{\mathcal{M}_\Omega})$ is complete. The topology induced by this metric on $\mathcal{M}_\Omega$ will be called the strong topology. In the sequel we will also consider the closed subset $\mathcal{M}_\Omega^{(F)}$ of $(\mathcal{M}_\Omega, \rho_{\mathcal{M}_\Omega})$ consisting of all sequences $\{(A_{t\omega})_{\omega\in\Omega}\}_{t=1}^{\infty} \in \mathcal{M}_\Omega$ such that

$$A_{t\omega}(x) = x \text{ for each } x \in F,\ t = 1, 2, \ldots \text{ and each } \omega \in \Omega. \tag{8}$$

Finally, we introduce two uniformities for the set $\mathcal{M}_\Omega$. The first uniformity is determined by the following base:

$$\begin{aligned}E^{w^1}_{\mathcal{M}_\Omega}(\epsilon) = \{(\{(S_{t\omega})_{\omega\in\Omega}\}_{t=1}^{\infty}, \{(T_{t\omega})_{\omega\in\Omega}\}_{t=1}^{\infty}) \in \mathcal{M}_\Omega \times \mathcal{M}_\Omega :\\ \rho^w_{\mathcal{N}_\Omega}((S_{t\omega})_{\omega\in\Omega}, (T_{t\omega})_{\omega\in\Omega}) \le \epsilon,\ t = 1, 2, \ldots\},\end{aligned} \tag{9}$$

where $\epsilon > 0$. This uniformity induces on $\mathcal{M}_\Omega$ a topology which we denote by τ_1. Clearly τ_1 is weaker than the strong topology. The second uniformity is determined by the base

$$\begin{aligned}E^{w^2}_{\mathcal{M}_\Omega}(\epsilon, N) = \{(\{(S_{t\omega})_{\omega\in\Omega}\}_{t=1}^{\infty}, \{(T_{t\omega})_{\omega\in\Omega}\}_{t=1}^{\infty} \in \mathcal{M}_\Omega \times \mathcal{M}_\Omega :\\ \rho^w_{\mathcal{N}_\Omega}((S_{t\omega})_{\omega\in\Omega}, (T_{t\omega})_{\omega\in\Omega}) \le \epsilon,\ t = 1, 2, \ldots, N\},\end{aligned} \tag{10}$$

where $\epsilon > 0$ and N is a natural number. This uniformity induces on $\mathcal{M}_\Omega$ a topology which we denote by τ_2. Clearly τ_2 is weaker than τ_1.

For each $(T_\omega)_{\omega\in\Omega} \in \mathcal{N}_\Omega$ define the operator $\widetilde{T} : K \to K$ by

$$\widetilde{T}x = \int_\Omega T_\omega x d\mu(\omega), \; x \in K. \tag{11}$$

Clearly for each $((T_\omega)_{\omega\in\Omega}) \in \mathcal{N}_\Omega$, the operator $\widetilde{T}$ belongs to $\mathcal{N}$. The following two propositions follow easily from the definitions.

Proposition 1.1 *The mapping $((T_\omega)_{\omega\in\Omega}) \to \widetilde{T}$ from $\mathcal{N}_\Omega$ with the weak topology onto $(\mathcal{N}, \rho_\mathcal{N})$ is continuous.*

For each $\{(A_{t\omega})_{\omega\in\Omega}\}_{t=1}^\infty \in \mathcal{M}_\Omega$ define $\{\widetilde{A}_t\}_{t=1}^\infty \in \mathcal{M}$ by

$$\widetilde{A}_t x = \int_\Omega A_{t\omega} x d\mu(\omega), \; x \in K, \; t = 1, 2, \ldots.$$

Proposition 1.2 *The mapping $\{(A_{t\omega})_{\omega\in\Omega}\}_{t=1}^\infty \to \{\widetilde{A}_t\}_{t=1}^\infty$ from $\mathcal{M}_\Omega$ with the topology τ_1 (respectively, τ_2) onto $\mathcal{M}$ with the strong (respectively, weak) topology is continuous.*

2. MAIN RESULTS

In this section we state our main results. Their proofs will be given in subsequent sections.

Theorem 2.1 *There exists a set $\mathcal{F} \subset \mathcal{N}_\Omega$ which is a countable intersection of open (in the weak topology) everywhere dense (in the strong topology) subsets of $\mathcal{N}_\Omega$ so that for each $(A_\omega)_{\omega\in\Omega} \in \mathcal{F}$ there exists $x_A \in K$ such that the following assertions hold:*

1. $\widetilde{A}^n x \to x_A$ as $n \to \infty$, uniformly on K.

2. For each $\epsilon > 0$ there exist a neighborhood $\mathcal{U}$ of $(A_\omega)_{\omega\in\Omega}$ in $\mathcal{N}_\Omega$ with the weak topology and a natural number N such that for each $(B_\omega)_{\omega\in\Omega} \in \mathcal{U}$, each integer $n \geq N$ and each $x \in K$, $||\widetilde{B}^n x - x_A|| \leq \epsilon$.

Theorem 2.2 *There exists a set $\mathcal{F} \subset \mathcal{M}_\Omega$ which is a countable intersection of open (in the topology τ_2) everywhere dense (in the strong topology) subsets of $\mathcal{M}_\Omega$ such that for each $\{(A_{t\omega})_{\omega\in\Omega}\}_{t=1}^\infty \in \mathcal{F}$ the following assertion holds:*

For each $\epsilon > 0$, there exist a neighborhood $\mathcal{U}$ of $\{(A_{t\omega})_{\omega\in\Omega}\}_{t=1}^\infty$ in $\mathcal{M}_\Omega$ with the topology τ_2 and a natural number N such that for each $\{(B_{t\omega})_{\omega\in\Omega}\}_{t=1}^\infty \in \mathcal{U}$, each integer $n \geq N$ and each $x, y \in K$,

$$||\widetilde{B}_n \cdot \ldots \cdot \widetilde{B}_1 x - \widetilde{B}_n \cdot \ldots \cdot \widetilde{B}_1 y|| \leq \epsilon.$$

Theorem 2.3 *There exists a set $\mathcal{F} \subset \mathcal{M}_\Omega$ which is a countable intersection of open (in the topology τ_1) everywhere dense (in the strong topology) subsets of $\mathcal{M}_\Omega$ such that for each $\{(A_{t\omega})_{\omega\in\Omega}\}_{t=1}^\infty \in \mathcal{F}$ the following assertion holds:*

For each $\epsilon > 0$, there exist a neighborhood $\mathcal{U}$ of $\{(A_{t\omega})_{\omega\in\Omega}\}_{t=1}^\infty$ in $\mathcal{M}_\Omega$ with the topology τ_1 and a natural number N such that for each $\{(B_{t\omega})_{\omega\in\Omega}\}_{t=1}^\infty \in \mathcal{U}$, each integer $n \geq N$, each $x, y \in K$ and each $r : \{1, \ldots, n\} \to \{1, 2, \ldots\}$,

$$||\widetilde{B}_{r(n)} \cdot \ldots \cdot \widetilde{B}_{r(1)} x - \widetilde{B}_{r(n)} \cdot \ldots \cdot \widetilde{B}_{r(1)} y|| \leq \epsilon.$$

We denote by $\mathcal{M}_\Omega^{reg}$ the set of all $\{(A_{t\omega})_{\omega\in\Omega}\}_{t=1}^\infty \in \mathcal{M}_\Omega$ for which there exists $x_A \in K$ such that $\tilde{A}_t x_A = x_A$, $t = 1, 2, \ldots$, and by $\bar{\mathcal{M}}_\Omega^{reg}$ the closure of $\mathcal{M}_\Omega^{reg}$ in $\mathcal{M}_\Omega$ with the strong topology. We will consider $\bar{\mathcal{M}}_\Omega^{reg}$ with two relative topologies inherited from $\mathcal{M}_\Omega$: the strong topology and the τ_1 topology. We will refer to these topologies as the strong and weak topologies, respectively.

Theorem 2.4 *There exists a set $\mathcal{F} \subset \bar{\mathcal{M}}_\Omega^{reg}$ which is a countable intersection of open (in the weak topology) everywhere dense (in the strong topology) subsets of $\bar{\mathcal{M}}_\Omega^{reg}$ such that for each $\{(A_{t\omega})_{\omega\in\Omega}\}_{t=1}^\infty \in \mathcal{F}$ the following assertions hold:*

1. There exists $\hat{x} \in K$ such that $\tilde{A}_t \hat{x} = \hat{x}$, $t = 1, 2, \ldots$

2. For each $\epsilon > 0$, there exist a neighborhood $\mathcal{U}$ of $\{(A_{t\omega})_{\omega\in\Omega}\}_{t=1}^\infty$ in $\mathcal{M}_\Omega$ with the topology τ_1 and a natural number N such that for each $\{(B_{t\omega})_{\omega\in\Omega}\}_{t=1}^\infty \in \mathcal{U}$, each integer $n \geq N$, each $x \in K$ and each mapping $r : \{1, \ldots, n\} \to \{1, 2, \ldots\}$,

$$||\tilde{B}_{r(n)} \cdot \ldots \cdot \tilde{B}_{r(1)} x - \hat{x}|| \leq \epsilon.$$

We will consider the space $\mathcal{M}_\Omega^{(F)}$ introduced in Section 1 with three relative topologies inherited from $\mathcal{M}_\Omega$, namely the strong topology, the τ_1 topology and the τ_2 topology. We will refer to these topologies as the strong, relative τ_1 and relative τ_2 topologies, respectively.

Theorem 2.5 *There exists a set $\mathcal{F} \subset \mathcal{M}_\Omega^{(F)}$ which is a countable intersection of open (in the relative τ_2 topology) everywhere dense (in the strong topology) subsets of $\mathcal{M}_\Omega^{(F)}$ such that for each $\{(A_{t\omega})_{\omega\in\Omega}\}_{t=1}^\infty \in \mathcal{F}$ the following assertions hold:*

1. For each $x \in K$ there exists $\lim_{n\to\infty} \tilde{A}_n \cdot \ldots \cdot \tilde{A}_1 x = Px \in F$.

2. For each $\epsilon > 0$ there exist a neighborhood $\mathcal{U}$ of $\{(A_{t\omega})_{\omega\in\Omega}\}_{t=1}^\infty$ in $\mathcal{M}_\Omega^{(F)}$ with the relative τ_2 topology and a natural number N such that for each $\{(B_{t\omega})_{\omega\in\Omega}\}_{t=1}^\infty \in \mathcal{U}$, each integer $n \geq N$ and each $x \in K$,

$$||\tilde{B}_n \cdot \ldots \cdot \tilde{B}_1 x - Px|| \leq \epsilon.$$

Theorem 2.6 *There exists a set $\mathcal{F} \subset \mathcal{M}_\Omega^{(F)}$ which is a countable intersection of open (in the relative τ_1 topology) everywhere dense (in the strong topology) subsets of $\mathcal{M}_\Omega^{(F)}$ such that for each $\{(A_{t\omega})_{\omega\in\Omega}\}_{t=1}^\infty \in \mathcal{F}$ the following assertions hold:*

1. For each $r : \{1, 2, \ldots\} \to \{1, 2, \ldots\}$ and each $x \in K$ there exists

$$\lim_{n\to\infty} \tilde{A}_{r(n)} \cdot \ldots \cdot \tilde{A}_{r(1)} x = P_r x \in F.$$

2. For each $\epsilon > 0$, there exist a neighborhood $\mathcal{U}$ of $\{(A_{t\omega})_{\omega\in\Omega}\}_{t=1}^\infty$ in $\mathcal{M}_\Omega^{(F)}$ with the relative τ_1 topology and a natural number N such that for each $\{(B_{t\omega})_{\omega\in\Omega}\}_{t=1}^\infty \in \mathcal{U}$, each $r : \{1, 2, \ldots\} \to \{1, 2, \ldots\}$, each integer $n \geq N$ and each $x \in K$,

$$||\tilde{B}_{r(n)} \cdot \ldots \cdot \tilde{B}_{r(1)} x - P_r x|| \leq \epsilon.$$

3. AUXILIARY RESULTS FOR THEOREMS 2.2-2.4

In this section we present several lemmas which will be used in the proofs of our main results.

We begin by quoting Lemma 4.2 from [13].

Recall that a set E of operators $A : K \to K$ is uniformly equicontinuous if for any $\epsilon > 0$ there exists $\delta > 0$ such that $||Ax - Ay|| \leq \epsilon$ for all $A \in E$ and all $x, y \in K$ satisfying $||x - y|| \leq \delta$.

Lemma 3.1 *[13]. Assume that E is a nonempty uniformly equicontinuous set of operators $A : K \to K$, N is a natural number and ϵ is a positive number. Then there exists a number $\delta > 0$ such that for each sequence $\{A_t\}_{t=1}^N \subset E$, each sequence $\{B_t\}_{t=1}^N$, where the operators $B_t : K \to K$, $t = 1, \ldots N$, (not necessarily continuous) satisfy*

$$||B_t z - A_t z|| \leq \delta, \quad t = 1, \ldots N, \ z \in K,$$

and each $x \in K$, the following inequality holds:

$$||B_N \cdot \ldots \cdot B_1 x - A_N \cdot \ldots \cdot A_1 x|| \leq \epsilon.$$

Fix $x_* \in K$. Given $\{A_t\}_{t=1}^\infty \in \mathcal{M}$ and $\gamma \in (0, 1)$, we define the mappings $A_{t\gamma} : K \to K$, $t = 1, 2, \ldots$, by

$$A_{t\gamma} x = (1 - \gamma) A_t x + \gamma x_*, \quad x \in K. \tag{12}$$

It is easy to see that $\{A_{t\gamma}\}_{t=1}^\infty \in \mathcal{M}$,

$$||A_{t\gamma} x - A_{t\gamma} y|| \leq (1 - \gamma)||x - y||, \ x, y \in K, \ t = 1, 2, \ldots, \tag{13}$$

and that the set

$$\{\{A_{t\gamma}\}_{t=1}^\infty : \ \{A_t\}_{t=1}^\infty \in \mathcal{M}, \ \gamma \in (0, 1)\}$$

is an everywhere dense subset of $\mathcal{M}$ in the strong topology.

For each $E \subset X$ set

$$\text{rad}(E) = \sup\{||y|| : \ y \in E\}. \tag{14}$$

Lemma 3.2 *Let $\{A_t\}_{t=1}^\infty \in \mathcal{M}$, $\gamma \in (0, 1)$ and let $\epsilon > 0$. Then there exist a natural number N and a neighborhood U of $\{A_{t\gamma}\}_{t=1}^\infty$ in the space $\mathcal{M}$ with the weak topology such that for each $\{B_t\}_{t=1}^\infty \in U$, each $x, y \in K$ and each integer $\tau \geq N$,*

$$||B_\tau \cdot \ldots \cdot B_1 x - B_\tau \cdot \ldots \cdot B_1 y|| \leq \epsilon.$$

Proof. Choose a natural number N such that

$$(1 - \gamma)^N \text{rad}(K) \leq 8^{-1} \epsilon. \tag{15}$$

The inequalities (13) and (15) imply that

$$||A_{N\gamma} \cdot \ldots \cdot A_{1\gamma} x - A_{N\gamma} \cdot \ldots \cdot A_{1\gamma} y|| \leq 4^{-1} \epsilon \text{ for all } x, y \in K. \tag{16}$$

By Lemma 3.1 there exists a neighborhood U of $\{A_{t\gamma}\}_{t=1}^{\infty}$ in the space $\mathcal{M}$ with the weak topology such that for each $\{B_t\}_{t=1}^{\infty} \in U$ and each $x \in K$,

$$||A_{N\gamma} \cdot \ldots \cdot A_{1\gamma}x - B_N \cdot \ldots B_1 x|| \le 4^{-1}\epsilon. \tag{17}$$

Since each one of the operators B_t is nonexpansive, the inequalities (17) and (16) imply that for each $\{B_t\}_{t=1}^{\infty} \in U$, each $x, y \in K$ and each integer $\tau \ge N$,

$$\begin{aligned} ||B_\tau \cdot \ldots \cdot B_1 x - B_\tau \cdot \ldots \cdot B_1 y|| &\le ||B_N \cdot \ldots \cdot B_1 x - B_N \cdot \ldots \cdot B_1 y|| \le \\ ||B_N \cdot \ldots \cdot B_1 x - A_{N\gamma} \cdot \ldots \cdot A_{1\gamma} x|| &+ ||A_{N\gamma} \cdot \ldots \cdot A_{1\gamma} x - A_{N\gamma} \cdot \ldots \cdot A_{1\gamma} y|| + \\ &||A_{N\gamma} \cdot \ldots \cdot A_{1\gamma} y - B_N \cdot \ldots \cdot B_1 y|| \le \epsilon. \end{aligned}$$

Lemma 3.2 is proved.

Lemma 3.3 *Let $\{A_t\}_{t=1}^{\infty} \in \mathcal{M}$, $\gamma \in (0,1)$ and let $\epsilon > 0$. Then there exist a natural number N and a neighborhood U of $\{A_{t\gamma}\}_{t=1}^{\infty}$ in the space $\mathcal{M}$ with the strong topology such that for each $\{B_t\}_{t=1}^{\infty} \in U$, each $x, y \in K$, each integer $\tau \ge N$ and each mapping $r : \{1, \ldots, \tau\} \to \{1, 2, \ldots\}$,*

$$||B_{r(\tau)} \cdot \ldots \cdot B_{r(1)}x - B_{r(\tau)} \cdot \ldots \cdot B_{r(1)}y|| \le \epsilon.$$

Proof. Choose a natural number N such that (15) is valid. Equations (13) and (15) imply that for each mapping $r : \{1, \ldots, N\} \to \{1, 2, \ldots\}$ and each $x, y \in K$,

$$||A_{r(N)\gamma} \cdot \ldots \cdot A_{r(1)\gamma}x - A_{r(N)\gamma} \cdot \ldots \cdot A_{r(1)\gamma}y|| \le 4^{-1}\epsilon. \tag{18}$$

By Lemma 3.1 there exists a neighborhood U of $\{A_{t\gamma}\}_{t=1}^{\infty}$ in the space $\mathcal{M}$ with the strong topology such that for each $\{B_t\}_{t=1}^{\infty} \in U$, each $x \in K$ and each mapping $r : \{1, \ldots, N\} \to \{1, 2, \ldots\}$,

$$||A_{r(N)\gamma} \cdot \ldots \cdot A_{r(1)\gamma}x - B_{r(N)} \cdot \ldots \cdot B_{r(1)}x|| \le 4^{-1}\epsilon. \tag{19}$$

The inequalities (18) and (19) imply that for each $\{B_t\}_{t=1}^{\infty} \in U$, each $x, y \in K$, each integer $\tau \ge N$ and each $r : \{1, \ldots, \tau\} \to \{1, 2, \ldots\}$,

$$\begin{aligned} &||B_{r(\tau)} \cdot \ldots \cdot B_{r(1)}x - B_{r(\tau)} \cdot \ldots \cdot B_{r(1)}y|| \le \\ &||B_{r(N)} \cdot \ldots \cdot B_{r(1)}x - B_{r(N)} \cdot \ldots \cdot B_{r(1)}y|| \le \\ &||B_{r(N)} \cdot \ldots \cdot B_{r(1)}x - A_{r(N)\gamma} \cdot \ldots \cdot A_{r(1)\gamma}x|| + \\ &||A_{r(N)\gamma} \cdot \ldots \cdot A_{r(1)\gamma}x - A_{r(N)\gamma} \cdot \ldots \cdot A_{r(1)\gamma}y|| + \\ &||A_{r(N)\gamma} \cdot \ldots \cdot A_{r(1)\gamma}y - B_{r(N)} \cdot \ldots \cdot B_{r(1)}y|| \le \epsilon. \end{aligned}$$

Lemma 3.3 is proved.

Lemma 3.4 *Let $\{A_t\}_{t=1}^{\infty} \in \mathcal{M}$, $\gamma \in (0,1)$, $\epsilon > 0$, and let $\bar{x} \in K$ satisfy*

$$A_{t\gamma}\bar{x} = \bar{x}, \ t = 1, 2, \ldots \tag{20}$$

Then there exist a natural number N and a neighborhood U of $\{A_{t\gamma}\}_{t=1}^{\infty}$ in the space $\mathcal{M}$ with the strong topology such that for each $\{B_t\}_{t=1}^{\infty} \in U$, each $x \in K$, and each mapping $r : \{1, \ldots, N\} \to \{1, 2, \ldots\}$,

$$||B_{r(N)} \cdot \ldots \cdot B_{r(1)}x - \bar{x}|| \le \epsilon.$$

Proof. Choose a natural number N such that (15) is valid. Equations (13), (15) and (20) imply that for each mapping $r : \{1, \dots, N\} \to \{1, 2, \dots\}$ and each $x \in K$,

$$||A_{r(N)\gamma} \cdot \ldots \cdot A_{r(1)\gamma}x - \bar{x}|| \leq (1-\gamma)^N ||x - \bar{x}|| \leq 2(1-\gamma)^N \mathrm{rad}(K) \leq 4^{-1}\epsilon. \tag{21}$$

By Lemma 3.1, there exists a neighborhood U of $\{A_{t\gamma}\}_{t=1}^{\infty}$ in the space $\mathcal{M}$ with the strong topology such that for each $\{B_t\}_{t=1}^{\infty} \in U$, each $x \in K$ and each mapping $r : \{1, \dots, N\} \to \{1, 2, \dots\}$,

$$||A_{r(N)\gamma} \cdot \ldots \cdot A_{r(1)\gamma}x - B_{r(N)} \cdot \ldots B_{r(1)}x|| \leq 4^{-1}\epsilon. \tag{22}$$

The inequalities (21) and (22) imply that for each $\{B_t\}_{t=1}^{\infty} \in U$, each $x \in K$ and each $r : \{1, \dots, N\} \to \{1, 2, \dots\}$,

$$||B_{r(N)} \cdot \ldots \cdot B_{r(1)}x - \bar{x}|| \leq ||B_{r(N)} \cdot \ldots \cdot B_{r(1)}x - A_{r(N)\gamma} \cdot \ldots \cdot A_{r(1)\gamma}x|| + ||A_{r(N)\gamma} \cdot \ldots \cdot A_{r(1)\gamma}x - \bar{x}|| < \epsilon.$$

This completes the proof of Lemma 3.4.

4. AUXILIARY RESULTS FOR THEOREMS 2.5 AND 2.6

Let $F \subset K$ be a nonempty closed convex subset of K and $P : K \to F$ a nonexpansive mapping such that

$$Px = x, \ x \in F. \tag{23}$$

For each $x \in K$ set

$$d(x, F) = \inf\{||x - y|| : \ y \in F\}.$$

In this section we use the spaces $\mathcal{N}^{(F)}$ and $\mathcal{M}^{(F)}$ defined in the Introduction.

Let $\{A_t\}_{t=1}^{\infty} \in \mathcal{M}^{(F)}$ and $\gamma \in (0, 1)$. For $t = 1, 2, \dots$ define the mappings $A_{t\gamma} : K \to K$ by

$$A_{t\gamma}x = (1-\gamma)A_t x + \gamma Px, \ x \in K. \tag{24}$$

Clearly $\{A_{t\gamma}\}_{t=1}^{\infty} \in \mathcal{M}^{(F)}$. It is easy to see that the set

$$\{\{A_{t\gamma}\}_{t=1}^{\infty} : \ \{A_t\}_{t=1}^{\infty} \in \mathcal{M}^{(F)}, \ \gamma \in (0, 1)\}$$

is everywhere dense in the space $\mathcal{M}^{(F)}$ with the strong topology.

Lemma 4.1 *Let $\{A_t\}_{t=1}^{\infty} \in \mathcal{M}^{(F)}$ and $\gamma \in (0, 1)$. Then for each $x \in K$ and each $t = 1, 2, \dots$, $d(A_{t\gamma}x, F) \leq (1-\gamma)d(x, F)$.*

Proof. Let $x \in K$, $\epsilon > 0$ and let $t \geq 1$ be an integer. There exists $y \in F$ such that

$$||x - y|| < d(x, F) + \epsilon. \tag{25}$$

Then $(1-\gamma)y + \gamma Px \in F$ and by (24),

$$d(A_{t\gamma}x, F) \leq ||A_{t\gamma}x - (1-\gamma)y - \gamma Px|| = ||(1-\gamma)(A_t x - y)|| \leq (1-\gamma)||x - y|| \leq (1-\gamma)(d(x, F) + \epsilon).$$

Since ϵ is an arbitrary positive number we conclude that $d(A_{t\gamma}x, F) \leq (1-\gamma)d(x, F)$. Lemma 4.1 is proved.

Lemma 4.2 *Let $\{A_t\}_{t=1}^{\infty} \in \mathcal{M}^{(F)}$, $\gamma \in (0,1)$ and let $\epsilon > 0$. Then there exist a natural number N, a neighborhood U of $\{A_{t\gamma}\}_{t=1}^{\infty}$ in the space $\mathcal{M}^{(F)}$ with the weak topology and a mapping $Q : K \to F$ such that the following property holds:*

For each $\{B_t\}_{t=1}^{\infty} \in U$ and each $x \in K$,

$$||B_N \cdot \ldots \cdot B_1 x - Qx|| < \epsilon.$$

Proof. Choose a natural number N such that

$$(1-\gamma)^N \text{rad}(K) < 8^{-1}\epsilon. \tag{26}$$

It follows from (26) and Lemma 4.1 that for each $x \in K$,

$$d(A_{N\gamma} \cdot \ldots \cdot A_{1\gamma}x, F) \leq (1-\gamma)^N d(x, F) \leq 2(1-\gamma)^N \text{rad}(K) < 4^{-1}\epsilon.$$

Thus, for each $x \in K$ there exists $Qx \in F$ such that

$$||A_{N\gamma} \cdot \ldots \cdot A_{1\gamma}x - Qx|| < 4^{-1}\epsilon. \tag{27}$$

By Lemma 3.1, there exists a neighborhood U of $\{A_{t\gamma}\}_{t=1}^{\infty}$ in the space $\mathcal{M}^{(F)}$ with the weak topology such that for each $\{B_t\}_{t=1}^{\infty} \in U$ and each $x \in K$,

$$||A_{N\gamma} \cdot \ldots \cdot A_{1\gamma}x - B_N \cdot \ldots \cdot B_1 x|| \leq 4^{-1}\epsilon. \tag{28}$$

The inequalities (27) and (28) imply that for each $\{B_t\}_{t=1}^{\infty} \in U$ and each $x \in K$,

$$||B_N \cdot \ldots \cdot B_1 x - Qx|| \leq ||B_N \cdot \ldots \cdot B_1 x - A_{N\gamma} \cdot \ldots \cdot A_{1\gamma}x|| + ||A_{N\gamma} \cdot \ldots \cdot A_{1\gamma}x - Qx|| \leq 2^{-1}\epsilon.$$

This completes the proof of Lemma 4.2.

Lemma 4.3 *Let $\{A_t\}_{t=1}^{\infty} \in \mathcal{M}^{(F)}$, $\gamma \in (0,1)$ and let $\epsilon > 0$. Then there exist a natural number N and a neighborhood U of $\{A_{t\gamma}\})_{t=1}^{\infty}$ in the space $\mathcal{M}^{(F)}$ with the strong topology such that the following property holds:*

For each $x \in K$ and each mapping $r : \{1, \ldots, N\} \to \{1, 2, \ldots\}$ there exists $Q_r x \in F$ such that for each $\{B_t\}_{t=1}^{\infty} \in U$,

$$||B_{r(N)} \cdot \ldots \cdot B_{r(1)}x - Q_r x|| \leq \epsilon.$$

Proof. Choose a natural number N such that (26) holds. It follows from (26) and Lemma 4.1 that for each $x \in K$ and each $r : \{1, \ldots, N\} \to \{1, 2 \ldots\}$,

$$d(A_{r(N)\gamma} \cdot \ldots \cdot A_{r(1)\gamma}x, F) \leq (1-\gamma)^N d(x, F) < 4^{-1}\epsilon.$$

Thus for each $x \in K$ and each $r : \{1, \ldots, N\} \to \{1, 2, \ldots\}$ we can choose $Q_r x \in F$ such that

$$||A_{r(N)\gamma} \cdot \ldots \cdot A_{r(1)\gamma}x - Q_r x|| < 4^{-1}\epsilon. \tag{29}$$

By Lemma 3.1, there exists a neighborhood U of $\{A_{t\gamma}\}_{t=1}^{\infty}$ in the space $\mathcal{M}^{(F)}$ with the strong topology such that for each $\{B_t\}_{t=1}^{\infty} \in U$, each $x \in K$ and each $r : \{1, \ldots, N\} \to \{1, 2, \ldots\}$,

$$||A_{r(N)\gamma} \cdot \ldots \cdot A_{r(1)\gamma}x - B_{r(N)} \cdot \ldots \cdot B_{r(1)}x|| \leq 4^{-1}\epsilon. \tag{30}$$

The inequalities (29) and (30) imply that for each $\{B_t\}_{t=1}^{\infty} \in U$, each $x \in K$ and each mapping $r : \{1, \ldots, N\} \to \{1, 2, \ldots\}$,

$$||B_{r(N)} \cdot \ldots \cdot B_{r(1)}x - Q_r x|| \leq ||B_{r(N)} \cdot \ldots \cdot B_{r(1)}x - A_{r(N)\gamma} \cdot \ldots \cdot A_{r(1)\gamma}x|| + \\ ||A_{r(N)\gamma} \cdot \ldots \cdot A_{r(1)\gamma}x - Q_r x|| < \epsilon.$$

Lemma 4.3 is proved.

5. AN AUXILIARY RESULT FOR THEOREM 2.1

Let $x_* \in K$. For each $A \in \mathcal{N}$ and $\gamma \in (0, 1)$ define the mapping $A_\gamma : K \to K$ by

$$A_\gamma x = (1 - \gamma)Ax + \gamma x_*, \; x \in K. \tag{31}$$

Let $A \in \mathcal{N}$ and $\gamma \in (0, 1)$. Then

$$||A_\gamma x - A_\gamma y|| \leq (1 - \gamma)||x - y||, \; x, y \in K. \tag{32}$$

By Banach's fixed point theorem, there exists a unique $x_{A\gamma} \in K$ such that

$$A_\gamma x_{A\gamma} = x_{A\gamma} \tag{33}$$

and

$$(A_\gamma)^n x \to x_{A\gamma} \text{ as } n \to \infty, \text{ uniformly on } K. \tag{34}$$

Combining (34) with Lemma 3.2 we obtain the following result.

Lemma 5.1 *Let $A \in \mathcal{N}$, $\gamma \in (0, 1)$ and $\epsilon > 0$. Then there exist a natural number N and a neighborhood U of A_γ in the space $\mathcal{N}$ such that for each $B \in U$, each $x \in K$ and each $t \geq N$, we have $||B^t x - x_{A\gamma}|| \leq \epsilon$.*

6. PROOF OF THEOREM 2.1

Fix $x_* \in K$. For each $(T_\omega)_{\omega \in \Omega} \in \mathcal{N}_\Omega$ and each $\gamma \in (0, 1)$ define $(T_\omega^\gamma)_{\omega \in \Omega} \in \mathcal{N}_\Omega$ by

$$T_\omega^\gamma x = (1 - \gamma)T_\omega x + \gamma x_*, \; x \in K. \tag{35}$$

Clearly, for each $(T_\omega)_{\omega \in \Omega} \in \mathcal{N}_\Omega$, $(T_\omega^\gamma)_{\omega \in \Omega} \to (T_\omega)_{\omega \in \Omega}$, as $\gamma \to 0^+$, in $\mathcal{N}_\Omega$ with the strong topology. Therefore the set

$$\{(T_\omega^\gamma)_{\omega \in \Omega} : \; (T_\omega)_{\omega \in \Omega} \in \mathcal{N}_\Omega, \; \gamma \in (0, 1)\} \tag{36}$$

is an everywhere dense subset of $\mathcal{N}_\Omega$ with the strong topology. It is easy to see that for each $(T_\omega)_{\omega \in \Omega} \in \mathcal{N}_\Omega$, $\gamma \in (0, 1)$ and $x \in K$,

$$\tilde{T}^\gamma x = \int_\Omega T_\omega^\gamma x d\mu(\omega) = \int_\Omega ((1 - \gamma)T_\omega x + \gamma x_*) d\mu(\omega) = \\ \gamma x_* + (1 - \gamma) \int_\Omega T_\omega x d\mu(\omega) = \gamma x_* + (1 - \gamma)\tilde{T}x \tag{37}$$

(see (11)). Lemma 5.1 and (37) now yield the following lemma.

Lemma 6.1 *Let* $(T_\omega)_{\omega\in\Omega} \in \mathcal{N}_\Omega$, $\gamma \in (0,1)$ *and* $\epsilon > 0$. *Then there exist a natural number* N, $\bar{x} \in K$ *and a neighborhood* $\mathcal{U}$ *of* $(T^\gamma_\omega)_{\omega\in\Omega}$ *in* $\mathcal{N}_\Omega$ *with the weak topology such that for each* $(S_\omega)_{\omega\in\Omega} \in \mathcal{U}$, *each integer* $n \geq N$ *and each* $x \in K$, *we have* $||\tilde{S}^n x - \bar{x}|| \leq \epsilon$.

Completion of the proof of Theorem 2.1. Let $(T_\omega)_{\omega\in\Omega} \in \mathcal{N}_\Omega$, $\gamma \in (0,1)$ and let $i \geq 1$ be an integer. By Lemma 6.1, there exist a natural number $N(T,\gamma,i)$, a point $x(T,\gamma,i) \in K$ and an open neighborhood $\mathcal{U}(T,\gamma,i)$ of $(T^\gamma_\omega)_{\omega\in\Omega}$ in $\mathcal{N}_\Omega$ with the weak topology such that for each $(S_\omega)_{\omega\in\Omega} \in \mathcal{U}(T,\gamma,i)$, each integer $n \geq N(T,\gamma,i)$ and each $x \in K$,

$$||\tilde{S}^n x - x(T,\gamma,i)|| \leq 1/i.$$

Define

$$\mathcal{F} = \cap_{q=1}^\infty \cup \{\mathcal{U}(T,\gamma,q) : \ (T_\omega)_{\omega\in\Omega} \in \mathcal{N}_\Omega, \ \gamma \in (0,1)\}.$$

Clearly, $\mathcal{F}$ is a countable intersection of open (in the weak topology) everywhere dense (in the strong topology) subsets of $\mathcal{N}_\Omega$.

Let $(A_\omega)_{\omega\in\Omega} \in \mathcal{F}$ and $\epsilon > 0$. Choose a natural number q such that $4/q < \epsilon$. There exist $\gamma \in (0,1)$ and $(T_\omega)_{\omega\in\Omega} \in \mathcal{N}_\Omega$ such that $(A_\omega)_{\omega\in\Omega} \in \mathcal{U}(T,\gamma,q)$. Then the following property holds:

(P1) For each $(B_\omega)_{\omega\in\Omega} \in \mathcal{U}(T,\gamma,q)$, each integer $n \geq N(T,\gamma,q)$ and each $x \in K$,

$$||\tilde{B}^n x - x(T,\gamma,q)|| \leq 1/q < \epsilon.$$

Since $(A_\omega)_{\omega\in\Omega} \in \mathcal{U}(T,\gamma,q)$, this implies that for each $n \geq N(T,\gamma,q)$ and each $x \in K$,

$$||\tilde{A}^n x - x(T,\gamma,q)|| \leq 1/q.$$

Since ϵ is an arbitrary positive number we conclude that for each $x \in K$ the sequence $\{\tilde{A}^n x\}_{n=1}^\infty$ is a Cauchy sequence which converges in K and

$$|| \lim_{n\to\infty} \tilde{A}^n x - x(T,\gamma,q)|| \leq 1/q < \epsilon.$$

This implies the existence of a point $x_A \in K$ such that for each $x \in K$, $\tilde{A}^n x \to x_A$ as $n \to \infty$. Clearly $||x_A - x(T,\gamma,q)|| \leq 1/q$. When combined with (P1), the latter inequality implies that for each $(B_\omega)_{\omega\in\Omega} \in \mathcal{U}(T,\gamma,q)$, each integer $n \geq N(T,\gamma,q)$ and each $x \in K$,

$$||\tilde{B}^n x - x_A|| \leq 2/q < \epsilon.$$

This completes the proof of Theorem 2.1.

7. PROOFS OF THEOREMS 2.2 AND 2.3

Let $x_* \in K$. For each $\{((A_t)_\omega)_{\omega\in\Omega}\}_{t=1}^\infty = \{(A_{t\omega})_{\omega\in\Omega}\}_{t=1}^\infty \in \mathcal{M}_\Omega$ and each $\gamma \in (0,1)$, define $\{(A^\gamma_{t\omega})_{\omega\in\Omega}\}_{t=1}^\infty \in \mathcal{M}_\Omega$ by

$$A^\gamma_{t\omega} x = (1-\gamma)A_{t\omega}x + \gamma x_*, \ x \in K, \ t = 1,2,\ldots. \tag{38}$$

Clearly, for each $\{(A_{t\omega})_{\omega\in\Omega}\}_{t=1}^\infty \in \mathcal{M}_\Omega$, $\{(A^\gamma_{t\omega})_{\omega\in\Omega}\}_{t=1}^\infty \to \{(A_{t\omega})_{\omega\in\Omega}\}_{t=1}^\infty$, as $\gamma \to 0^+$, in $\mathcal{M}_\Omega$ with the strong topology. Therefore the set

$$\{\{(A^\gamma_{t\omega})_{\omega\in\Omega}\}_{t=1}^\infty : \ \{(A_{t\omega})_{\omega\in\Omega}\}_{t=1}^\infty \in \mathcal{M}_\Omega, \ \gamma \in (0,1)\}$$

is an everywhere dense subset of $\mathcal{M}_\Omega$ with the strong topology. It is easy to see that for each $\{(A_{t\omega})_{\omega\in\Omega}\}_{t=1}^\infty \in \mathcal{M}_\Omega$, $\gamma \in (0,1)$, $x \in K$ and $t \in \{1,2,\dots\}$,

$$\widetilde{A}_t^\gamma x = \int_\Omega A_{t\omega}^\gamma x d\mu(\omega) = \int_\Omega ((1-\gamma)A_{t\omega}x + \gamma x_*)d\mu(\omega) =$$
$$\gamma x_* + (1-\gamma)\int_\Omega A_{t\omega}xd\mu(\omega) = \gamma x_* + (1-\gamma)\widetilde{A}_t x. \quad (39)$$

Lemma 3.2, Proposition 1.2 and (39) now yield the following lemma.

Lemma 7.1 *Let $\{(A_{t\omega})_{\omega\in\Omega}\}_{t=1}^\infty \in \mathcal{M}_\Omega$, $\gamma \in (0,1)$ and $\epsilon > 0$. Then there exist a natural number N and a neighborhood $\mathcal{U}$ of $\{(A_{t\omega}^\gamma)_{\omega\in\Omega}\}_{t=1}^\infty \in \mathcal{M}_\Omega$ with the topology τ_2 such that for each $\{(B_{t\omega})_{\omega\in\Omega}\}_{t=1}^\infty \in \mathcal{U}$, each $x, y \in K$ and each integer $n \geq N$,*

$$||\widetilde{B}_n \cdot \ldots \cdot \widetilde{B}_1 x - \widetilde{B}_n \cdot \ldots \cdot \widetilde{B}_1 y|| \leq \epsilon.$$

Completion of the proof of Theorem 2.2. Let $\{(C_{t\omega})_{\omega\in\Omega}\}_{t=1}^\infty \in \mathcal{M}_\Omega$, $\gamma \in (0,1)$ and let $i \geq 1$ be an integer. By Lemma 7.1, there exist a natural number $N(C,\gamma,i)$ and an open neighborhood $\mathcal{U}(C,\gamma,i)$ of $\{(C_{t\omega}^\gamma)_{\omega\in\Omega}\}_{t=1}^\infty$ in $\mathcal{M}_\Omega$ with the topology τ_2 such that the following property holds:

For each $\{(B_{t\omega})_{\omega\in\Omega}\}_{t=1}^\infty \in \mathcal{U}(C,\gamma,i)$, each integer $n \geq N(C,\gamma,i)$ and each $x, y \in K$,

$$||\widetilde{B}_n \cdot \ldots \cdot \widetilde{B}_1 x - \widetilde{B}_n \cdot \ldots \cdot \widetilde{B}_1 y|| \leq 1/i.$$

Define

$$\mathcal{F} = \cap_{q=1}^\infty \cup \{\mathcal{U}(C,\gamma,q) : \ \{(C_{t\omega})_{\omega\in\Omega}\}_{t=1}^\infty \in \mathcal{M}_\Omega, \ \gamma \in (0,1)\}.$$

Clearly $\mathcal{F}$ is a countable intersection of open (in the topology τ_2) everywhere dense (in the strong topology) subsets of $\mathcal{M}_\Omega$.

Let $\{(A_{t\omega})_{\omega\in\Omega}\}_{t=1}^\infty \in \mathcal{F}$ and $\epsilon > 0$. Choose a natural number q such that $4/q < \epsilon$. There exist $\{(C_{t\omega})_{\omega\in\Omega}\}_{t=1}^\infty \in \mathcal{M}_\Omega$ and $\gamma \in (0,1)$ such that $\{(A_{t\omega})_{\omega\in\Omega}\}_{t=1}^\infty \in \mathcal{U}(C,\gamma,q)$.

Let $\{(B_{t\omega})_{\omega\in\Omega}\}_{t=1}^\infty \in \mathcal{U}(C,\gamma,q)$, $n \geq N(C,\gamma,q)$ and $x, y \in K$. Then

$$||\widetilde{B}_n \cdot \ldots \cdot \widetilde{B}_1 x - \widetilde{B}_n \cdot \ldots \cdot \widetilde{B}_1 y|| \leq q^{-1} < \epsilon,$$

as claimed. Theorem 2.2 is proved.

The next lemma follows from (39), Lemma 3.3 and Proposition 1.2.

Lemma 7.2 *Let $\{(A_{t\omega})_{\omega\in\Omega}\}_{t=1}^\infty \in \mathcal{M}_\Omega$, $\gamma \in (0,1)$ and $\epsilon > 0$. Then there exist a natural number N and a neighborhood $\mathcal{U}$ of $\{(A_{t\omega}^\gamma)_{\omega\in\Omega}\}_{t=1}^\infty$ in $\mathcal{M}_\Omega$ with the topology τ_1 such that for each $\{(B_{t\omega})_{\omega\in\Omega}\}_{t=1}^\infty \in \mathcal{U}$, each $x, y \in K$, each integer $n \geq N$ and each mapping $r : \{1,\dots,n\} \to \{1,2,\dots\}$,*

$$||\widetilde{B}_{r(n)} \cdot \ldots \cdot \widetilde{B}_{r(1)} x - \widetilde{B}_{r(n)} \cdot \ldots \cdot \widetilde{B}_{r(1)} y|| \leq \epsilon.$$

Proof of Theorem 2.3. Let $\{(C_{t\omega})_{\omega\in\Omega}\}_{t=1}^\infty \in \mathcal{M}_\Omega$, $\gamma \in (0,1)$ and let $i \geq 1$ be an integer. By Lemma 7.2 there exist a natural number $N(C,\gamma,i)$ and an open neighborhood $\mathcal{U}(C,\gamma,i)$ of $\{(C_{t\omega}^\gamma)_{\omega\in\Omega}\}_{t=1}^\infty$ in $\mathcal{M}_\Omega$ with the topology τ_1 such that the following property holds:

For each $\{(B_{t\omega})_{\omega\in\Omega}\}_{t=1}^{\infty} \in \mathcal{U}(C,\gamma,i)$, each integer $n \geq N(C,\gamma,i)$, each $x,y \in K$ and each mapping $r : \{1,\ldots,n\} \to \{1,2,\ldots\}$,

$$||\widetilde{B}_{r(n)} \cdot \ldots \cdot \widetilde{B}_{r(1)}x - \widetilde{B}_{r(n)} \cdot \ldots \cdot \widetilde{B}_{r(1)}y|| \leq 1/i.$$

Define

$$\mathcal{F} = \cap_{q=1}^{\infty} \cup \{\mathcal{U}(C,\gamma,q) : \ \{(C_{t\omega})_{\omega\in\Omega}\}_{t=1}^{\infty} \in \mathcal{M}_{\Omega},\ \gamma \in (0,1)\}.$$

Clearly $\mathcal{F}$ is a countable intersection of open (in the topology τ_1) everywhere dense (in the strong topology) subsets of $\mathcal{M}_{\Omega}$.

Let $\{(A_{t\omega})_{\omega\in\Omega}\}_{t=1}^{\infty} \in \mathcal{F}$ and $\epsilon > 0$. Choose a natural number q such that $4/q < \epsilon$. There exist $\{(C_{t\omega})_{\omega\in\Omega}\}_{t=1}^{\infty} \in \mathcal{M}_{\Omega}$ and $\gamma \in (0,1)$ such that $\{(A_{t\omega})_{\omega\in\Omega}\}_{t=1}^{\infty} \in \mathcal{U}(C,\gamma,q)$.

Let $\{(B_{t\omega})_{\omega\in\Omega}\}_{t=1}^{\infty} \in \mathcal{U}(C,\gamma,q)$, $n \geq N(C,\gamma,q)$, $x,y \in K$ and $r : \ \{1,\ldots,n\} \to \{1,2,\ldots\}$. Then

$$||\widetilde{B}_{r(n)} \cdot \ldots \cdot \widetilde{B}_{r(1)}x - \widetilde{B}_{r(n)} \cdot \ldots \cdot \widetilde{B}_{r(1)}y|| \leq q^{-1} < \epsilon.$$

Theorem 2.3 is proved.

8. PROOF OF THEOREM 2.4

Recall that $\mathcal{M}_{\Omega}^{reg}$ is the set of all $\{(A_{t\omega})_{\omega\in\Omega}\}_{t=1}^{\infty} \in \mathcal{M}_{\Omega}$ for which there exists $x_A \in K$ such that

$$\widetilde{A}_t x_A = x_A,\ t = 1,2,\ldots \tag{40}$$

Let $\{(A_{t\omega})_{\omega\in\Omega}\}_{t=1}^{\infty} \in \mathcal{M}_{\Omega}^{reg}$ with $x_A \in K$ satisfying (40) and let $\gamma \in (0,1)$. Define $\{(A_{t\omega}^{\gamma})_{\omega\in\Omega}\}_{t=1}^{\infty} \in \mathcal{M}_{\Omega}$ by

$$A_{t\omega}^{\gamma}(x) = (1-\gamma)A_{t\omega}(x) + \gamma x_A,\ x \in K,\ t = 1,2,\ldots,\ \omega \in \Omega. \tag{41}$$

Clearly for any $x \in K$ and any integer $t \geq 1$ we have

$$\begin{aligned} \widetilde{A}_t^{\gamma} x = \int_{\Omega} A_{t\omega}^{\gamma} x d\mu(\omega) = \int_{\Omega} ((1-\gamma)A_{t\omega}x + \gamma x_A) d\mu(\omega) = \\ \gamma x_A + (1-\gamma)\int_{\Omega} A_{t\omega} x d\mu(\omega) = \gamma x_A + (1-\gamma)\widetilde{A}_t x. \end{aligned} \tag{42}$$

Thus

$$\{(A_{t\omega}^{\gamma})_{\omega\in\Omega}\}_{t=1}^{\infty} \in \mathcal{M}_{\Omega}^{reg} \text{ and } \widetilde{A}_t^{\gamma} x_A = x_A,\ t = 1,2,\ldots \tag{43}$$

It is also easy to see that

$$\{(A_{t\omega}^{\gamma})_{\omega\in\Omega}\}_{t=1}^{\infty} \to \{(A_{t\omega})_{\omega\in\Omega}\}_{t=1}^{\infty} \text{ as } \gamma \to 0^{+} \tag{44}$$

in the strong topology of $\bar{\mathcal{M}}_{\Omega}^{reg}$.

Lemma 3.4, (41), (42), (43) and Proposition 1.2 lead to the following lemma.

Lemma 8.1 *Assume that* $\{(A_{t\omega})_{\omega\in\Omega}\}_{t=1}^{\infty} \in \mathcal{M}_{\Omega}^{reg}$ *with* $x_A \in K$ *satisfying (40), and let* $\gamma \in (0,1)$ *and* $\epsilon > 0$. *Then there exists a natural number* N *and a neighborhood* $\mathcal{U}$ *of* $\{(A_{t\omega}^{\gamma})_{\omega\in\Omega}\}_{t=1}^{\infty}$ *in* $\mathcal{M}_{\Omega}$ *with the topology* τ_1 *such that for each* $\{(B_{t\omega})_{\omega\in\Omega}\}_{t=1}^{\infty} \in \mathcal{U}$, *each* $x \in K$ *and each mapping* $r : \{1,\ldots,N\} \to \{1,2,\ldots\}$,

$$||\widetilde{B}_{r(N)} \cdot \ldots \cdot \widetilde{B}_{r(1)}x - x_A|| \leq \epsilon.$$

Completion of the proof of Theorem 2.4. Let $\{(C_{t\omega})_{\omega\in\Omega}\}_{t=1}^{\infty} \in \mathcal{M}_{\Omega}^{reg}$, $\gamma \in (0,1)$ and let $i \geq 1$ be an integer. By Lemma 8.1 there exist a natural number $N(C,\gamma,i)$ and an open neighborhood $\mathcal{U}(C,\gamma,i)$ of $\{(C_{t\omega}^{\gamma})_{\omega\in\Omega}\}_{t=1}^{\infty}$ in $\mathcal{M}_{\Omega}$ with the topology τ_1 such that the following property holds:

For each $\{(B_{t\omega})_{\omega\in\Omega}\}_{t=1}^{\infty} \in \mathcal{U}(C,\gamma,i)$, each integer $n \geq N(C,\gamma,i)$, each $x \in K$ and each $r : \{1,\ldots,n\} \to \{1,2,\ldots\}$,

$$||\widetilde{B}_{r(n)} \cdot \ldots \cdot \widetilde{B}_{r(1)}x - x(C,\gamma,i)|| \leq 1/i.$$

Define

$$\mathcal{F} = \bar{\mathcal{M}}_{\Omega}^{reg} \cap [\cap_{q=1}^{\infty} \cup \{\mathcal{U}(C,\gamma,q) : \ \{(C_{t\omega})_{\omega\in\Omega}\}_{t=1}^{\infty} \in \mathcal{M}_{\Omega}^{reg}, \ \gamma \in (0,1)\}].$$

Clearly $\mathcal{F}$ is a countable intersection of open (in the weak topology) everywhere dense (in the strong topology) subsets of $\bar{\mathcal{M}}_{\Omega}^{reg}$.

Let $\{(A_{t\omega})_{\omega\in\Omega}\}_{t=1}^{\infty} \in \mathcal{F}$ and $\epsilon > 0$. Choose a natural number q such that $4/q < \epsilon$. There exist $\{(C_{t\omega})_{\omega\in\Omega}\}_{t=1}^{\infty} \in \mathcal{M}_{\Omega}^{reg}$ and $\gamma \in (0,1)$ such that $\{(A_{t\omega})_{\omega\in\Omega}\}_{t=1}^{\infty} \in \mathcal{U}(C,\gamma,q)$. Then the following property holds:

(P2) For each $\{(B_{t\omega})_{\omega\in\Omega}\}_{t=1}^{\infty} \in \mathcal{U}(C,\gamma,q)$, each $n \geq N(C,\gamma,q)$, each $x \in K$ and each mapping $r : \{1,\ldots,n\} \to \{1,2,\ldots\}$,

$$||\widetilde{B}_{r(n)} \cdot \ldots \cdot \widetilde{B}_{r(1)}x - x(C,\gamma,q)|| \leq 1/q.$$

This implies that for each $n \geq N(C,\gamma,q)$, each $x \in K$ and each integer $t \geq 1$,

$$||(\widetilde{A}_t)^n x - x(C,\gamma,q)|| \leq 1/q < \epsilon.$$

Since ϵ is an arbitrary positive number we conclude that there exists $\hat{x} \in K$ such that $\lim_{n\to\infty}(\widetilde{A}_t)^n x = \hat{x}$ for all $t = 1,2,\ldots$ and all $x \in K$, and $||\hat{x} - x(C,\gamma,q)|| \leq 1/q$. Combining the last inequality with (P2) this implies that for each $\{(B_{t\omega})_{\omega\in\Omega}\}_{t=1}^{\infty} \in \mathcal{U}(C,\gamma,q)$, each integer $n \geq N(C,\gamma,q)$, each $x \in K$ and each $r : \{1,\ldots,n\} \to \{1,2,\ldots\}$,

$$||\widetilde{B}_{r(n)} \cdot \ldots \cdot \widetilde{B}_{r(1)}x - \hat{x}|| \leq 2/q < \epsilon.$$

Theorem 2.4 is proved.

9. PROOFS OF THEOREMS 2.5 AND 2.6

Let $\{(A_{t\omega})_{\omega\in\Omega}\}_{t=1}^{\infty} \in \mathcal{M}_{\Omega}^{(F)}$ and $\gamma \in (0,1)$. Define $\{(A_{t\omega}^{\gamma})_{\omega\in\Omega}\}_{t=1}^{\infty} \in \mathcal{M}_{\Omega}^{(F)}$ by

$$A_{t\omega}^{\gamma}x = (1-\gamma)A_{t\omega}x + \gamma Px, \ x \in K, \ t = 1,2,\ldots, \ \omega \in \Omega. \tag{45}$$

Clearly, for any $x \in K$ and any integer $t \geq 1$,

$$\widetilde{A}_t^\gamma x = \int_\Omega A_{t\omega}^\gamma x d\mu(\omega) = \int_\Omega ((1-\gamma)A_{t\omega}x + \gamma Px)d\mu(\omega) =$$
$$\gamma Px + (1-\gamma)\int_\Omega A_{t\omega}xd\mu(\omega) = \gamma Px + (1-\gamma)\widetilde{A}_t x. \tag{46}$$

It is easy to see that

$$\{(A_{t\omega}^\gamma)_{\omega\in\Omega}\}_{t=1}^\infty \to \{(A_{t\omega})_{\omega\in\Omega}\}_{t=1}^\infty \text{ as } \gamma \to 0^+ \tag{47}$$

in the strong topology of $\mathcal{M}_\Omega^{(F)}$.

Lemma 4.2, (45), (46) and Proposition 1.2 now yield the following lemma.

Lemma 9.1 *Assume that* $\{(A_{t\omega})_{\omega\in\Omega}\}_{t=1}^\infty \in \mathcal{M}_\Omega^{(F)}$, $\gamma \in (0,1)$ *and* $\epsilon > 0$. *Then there exist a natural number* N, *a neighborhood* $\mathcal{U}$ *of* $\{(A_{t\omega}^\gamma)_{\omega\in\Omega}\}_{t=1}^\infty$ *in* $\mathcal{M}_\Omega^{(F)}$ *with the relative* τ_2 *topology and a mapping* $Q : K \to F$ *such that the following property holds:*

For each $\{(B_{t\omega})_{\omega\in\Omega}\}_{t=1}^\infty \in \mathcal{U}$ *and each* $x \in K$,

$$||\widetilde{B}_N \cdot \ldots \cdot \widetilde{B}_1 x - Qx|| \leq \epsilon.$$

Lemma 4.3, (45), (46) and Proposition 1.2 imply the following lemma.

Lemma 9.2 *Assume that* $\{(A_{t\omega})_{\omega\in\Omega}\}_{t=1}^\infty \in \mathcal{M}_\Omega^{(F)}$, $\gamma \in (0,1)$ *and* $\epsilon > 0$. *Then there exist a natural number* N *and a neighborhood* $\mathcal{U}$ *of* $\{(A_{t\omega}^\gamma)_{\omega\in\Omega}\}_{t=1}^\infty$ *in* $\mathcal{M}_\Omega^{(F)}$ *with the relative* τ_1 *topology such that the following property holds:*

For each $x \in K$ *and each* $r : \{1,\ldots,N\} \to \{1,2,\ldots\}$, *there exists* $Q_r x \in F$ *such that for each* $\{(B_{t\omega})_{\omega\in\Omega}\}_{t=1}^\infty \in \mathcal{U}$,

$$||\widetilde{B}_{r(N)} \cdot \ldots \cdot \widetilde{B}_{r(1)} x - Q_r x|| \leq \epsilon.$$

Completion of the proof of Theorem 2.5. Let $\{(A_{t\omega})_{\omega\in\Omega}\}_{t=1}^\infty \in \mathcal{M}_\Omega^{(F)}$, $\gamma \in (0,1)$ and let $i \geq 1$ be an integer. By Lemma 9.1, there exist a natural number $N(A,\gamma,i)$ and an open neighborhood $\mathcal{U}(A,\gamma,i)$ of $\{(A_{t\omega}^\gamma)_{\omega\in\Omega}\}_{t=1}^\infty$ in $\mathcal{M}_\Omega$ with the relative τ_2 topology such that the following property holds:

(P3) For each $x \in K$ there exists $Qx \in F$ such that for each $\{(B_{t\omega})_{\omega\in\Omega}\}_{t=1}^\infty \in \mathcal{U}(A,\gamma,i)$ and each $x \in K$,

$$||\widetilde{B}_{N(A,\gamma,i)} \cdot \ldots \cdot \widetilde{B}_1 x - Qx|| \leq 1/i.$$

Define

$$\mathcal{F} = \cap_{q=1}^\infty \cup \{\mathcal{U}(A,\gamma,q) : \ \{(A_{t\omega})_{\omega\in\Omega}\}_{t=1}^\infty \in \mathcal{M}_\Omega^{(F)}, \ \gamma \in (0,1)\}.$$

Clearly $\mathcal{F}$ is a countable intersection of open (in the relative τ_2 topology) everywhere dense (in the strong topology) subsets of $\mathcal{M}_\Omega^{(F)}$.

Let $\{(A_{t\omega})_{\omega\in\Omega}\}_{t=1}^\infty \in \mathcal{F}$ and $\epsilon > 0$. Choose a natural number q such that $4/q < \epsilon$. There exist $\{(C_{t\omega})_{\omega\in\Omega}\}_{t=1}^\infty \in \mathcal{M}_\Omega^{(F)}$ and $\gamma \in (0,1)$ such that $\{(A_{t\omega})_{\omega\in\Omega}\}_{t=1}^\infty \in \mathcal{U}(C,\gamma,q)$. By (P3) the following property holds:

For each $x \in K$, there exists $Qx \in F$ such that for each $\{(B_{t\omega})_{\omega\in\Omega}\}_{t=1}^{\infty} \in \mathcal{U}(C,\gamma,q)$, each $n \geq N(C,\gamma,q)$ and each $x \in K$,

$$||\widetilde{B}_n \cdot \ldots \cdot \widetilde{B}_1 x - Qx|| \leq ||\widetilde{B}_{N(C,\gamma,q)} \cdot \ldots \cdot \widetilde{B}_1 x - Qx|| \leq 1/q < \epsilon/4.$$

Since ϵ is an arbitrary positive number this implies, in particular, that for each $x \in K$ the sequence $\{\widetilde{A}_n \cdot \ldots \cdot \widetilde{A}_1 x\}_{n=1}^{\infty}$ is a Cauchy sequence, there exists

$$\lim_{n\to\infty} \widetilde{A}_n \cdot \ldots \cdot \widetilde{A}_1 x$$

and

$$|| \lim_{n\to\infty} \widetilde{A}_n \cdot \ldots \cdot \widetilde{A}_1 x - Qx|| \leq \epsilon/4.$$

Thus $\lim_{n\to\infty} \widetilde{A}_n \cdot \ldots \cdot \widetilde{A}_1 x \in F$ for all $x \in K$. Set

$$Px = \lim_{n\to\infty} \widetilde{A}_n \cdot \ldots \cdot \widetilde{A}_1 x, \ x \in K.$$

Clearly $||Px - Qx|| \leq \epsilon/4$, $x \in K$.

Therefore, for each $\{(B_{t\omega})_{\omega\in\Omega}\}_{t=1}^{\infty} \in \mathcal{U}(C,\gamma,q)$, each integer $n \geq N(C,\gamma,q)$ and each $x \in K$,

$$||\widetilde{B}_n \cdot \ldots \cdot \widetilde{B}_1 x - Px|| \leq ||\widetilde{B}_n \cdot \ldots \cdot \widetilde{B}^1 x - Qx|| + ||Px - Qx|| \leq \epsilon/2.$$

Theorem 2.5 is proved.

The proof of Theorem 2.6 is analogous to that of Theorem 2.5; instead of Lemma 9.1 we use Lemma 9.2.

REFERENCES

1. D. Butnariu, The expected-projection method: Its behavior and applications to linear operator equations and convex optimization, *J. Applied Analysis* **1** (1995) 95–108.
2. D. Butnariu, Y. Censor and S. Reich, Iterative averaging of entropic projections for solving stochastic convex feasibility problems, *Computational Optimization and Applications* **8** (1997) 21–39.
3. D. Butnariu and S.D. Flam, Strong convergence of expected projection methods in Hilber spaces, *Numerical Functional Analysis and Optimization* **16** (1995) 601–636.
4. D. Butnariu and A.N. Iusem, Local moduli of convexity and their application to finding almost common fixed points of measurable families of operators, *Recent Developments in Optimization Theory and Nonlinear Analysis, Contemporary Mathematics* **204** (1997) 61–91.
5. D. Butnariu and A.N. Iusem, *Totally Convex Functions for Fixed Points Computation and Infinite Dimensional Optimization* (Kluwer, Dordrecht, 2000).
6. D. Butnariu, A.N. Iusem and R.S. Burachik, Iterative methods of solving stochastic convex feasibility problems and applications, *Computational Optimization and Applications* **15** (2000) 269–307.
7. D. Butnariu, S. Reich and A.J. Zaslavski, Generic power convergence of operators in Banach spaces, *Numerical Functional Analysis and Optimization* **20** (1999) 629–650.

8. D. Butnariu and E. Resmerita, The outer Bregman projection method for stochastic feasibility problems in Banach spaces, *Preprint* (2000).
9. F.S. De Blasi and J. Myjak, Sur la convergence des approximations successives pour les contractions non linéaires dans un espace de Banach, *C. R. Acad. Sci. Paris* **283** (1976) 185–187.
10. F.S. De Blasi and J. Myjak, Generic flows generated by continuous vector fields in Banach spaces, *Advances in Math.* **50** (1983) 266–280.
11. W.J. Kammerer and M.Z. Nashed, Iterative methods for best approximate solutions of linear integral equations of the first and second kind, *J. Math. Analysis and Applications* **40** (1972) 547–573.
12. J.L. Kelley, *General Topology* (D. Van Nostrand, New York, 1955).
13. S. Reich and A.J. Zaslavski, Convergence of generic infinite products of nonexpansive and uniformly continuous operators, *Nonlinear Analysis: Theory, Methods and Applications* **36** (1999) 1049–1065.

Inherently Parallel Algorithms in Feasibility and Optimization and their Applications
D. Butnariu, Y. Censor and S. Reich (Editors)

SUPERLINEAR RATE OF CONVERGENCE AND OPTIMAL ACCELERATION SCHEMES IN THE SOLUTION OF CONVEX INEQUALITY PROBLEMS

Ubaldo M. García-Palomares[a*]

[a]Universidad Simón Bolívar,
Dep. Procesos y Sistemas, Apartado 89000, Sartenejas, Caracas 1080-A, Venezuela

This paper proves the existence of an acceleration scheme that introduces a superlinear rate of convergence in the solution of the Convex Inequality Problem **CIP** by projection techniques. This is accomplished by solving the *dual* of a quadratic problem that finds the projection of a point on a group of violated and almost violated constraints. This paper also compares known acceleration schemes by means of defined *attracting directions.* Finally, some remarks are given for possible implementations including the inconsistency case.

1. INTRODUCTION

The Convex Inequality Problem (**CIP**) defined next arises in many applications.

Convex Inequality Problem: Find $x \in C \doteq \{x \in I\!R^n : g(x) \leq 0\}$, where $g(\cdot) : I\!R^n \to I\!R^m$ is a continuous convex function with components $g^1(\cdot), \ldots, g^m(\cdot)$.

A comprehensive study of the theory and usefulness of projection techniques (**PTs**) for solving the more general convex feasibility problem in Hilbertian spaces can be found in the literature (See for instance [1]). (**PTs**) are iterative processes that generate a sequence $\{x_i\}_1^\infty$ which, under mild conditions, converges to some $x \in C$. Although **PTs** have evolved since their early inception by the seminal Von Neumann's contribution [22], the essence of one iteration (with exact projections) can be described as

PT Iteration:
Input: $x_i \notin C \subset I\!R^n, \eta > 0$
1. Choose a closed convex superset $S_i \supset C$
2. Projection: $P_i(x_i) = \arg\min_{y \in S_i} \frac{1}{2}(y - x_i)^T(y - x_i)$ (1)
3. Next point: $x_{i+1} = x_i + \omega_i(P_i(x_i) - x_i)$ for some $\eta \leq \omega_i \leq 2 - \eta$
End of iteration.

Due to their simplicity and robustness **PTs** are among the most successful methods for solving **CIPs**. The computational work of these techniques resides in the solution

*Universidade de Vigo, Dep. Tecnoloxías das Comunicacións, ETS Enxeñeiros de Telecomunicación, 36200 Vigo, España.

of the projection subproblem (1). The smaller this subproblem the faster the iteration. A superlinear rate of convergence has been proved for the primal superlinear algorithm described as

Primal Superlinear Algorithm (PSA) iteration for solving **CIP**
Input: $x_i \notin C \doteq \{x \in I\!R^n : g(x) \leq 0\}, \theta > 0.$

1. $K_i \supseteq \left\{1 \leq k \leq m : g^k(x_i) \geq ||g(x_i)||_\infty - \theta\right\}$

2. $$S_i = \left\{y \in I\!R^n : g^k(x_i) + \nabla g^k(x_i)^T(y - x_i) \leq 0, k \in K_i\right\} \tag{2}$$

3. $$x_{i+1} = \arg\min_{y \in S_i} \frac{1}{2}(y - x_i)^T(y - x_i) \tag{3}$$

End of iteration.

García Palomares suggested the above algorithm in [8, algorithm 3.2] and proved its superlinear convergence in [10]. He proved that

If $int(C) \neq \emptyset$ and if $\nabla g(\cdot)$ is Lipschitz, then the sequence $\{x_i\}_1^\infty$ generated by **PSA** *converges superlinearly to some $x \in C$, i.e.,*

$$\lim_{i\to\infty} \frac{||x_{i+1} - x_i||}{||x_i - x_{i-1}||} = 0, \text{ or equivalently } \lim_{i\to\infty} \frac{||x_{i+1} - x||}{||x_i - x||} = 0. \tag{4}$$

Equation (4) is satisfied by successful algorithms in optimization theory. It is commonly accepted that quasi-Newton methods (superlinear), which need some kind of approximation to second derivatives, are superior to gradient methods (linear). The few iterations needed to achieve convergence generally offsets the computational work per iteration.

We point out that the superset S_i in **PSA** (defined by 2) might consider all violated plus some $\theta-$satisfied constraints. Therefore, the projection problem (3) may be large, especially if compared with row-action methods that only consider one constraint at a time and the projection becomes a trivial step. Most **PTs** often choose the sequence of supersets $\{S_i\}_1^\infty$ mainly to reduce the computational burden due to the projection step. However, this often leads to a (slow) linear rate of convergence, which has been reported by many authors since Gubin's well known paper [16]. In general **PTs** consider projections on a subset of strictly violated constraints, which seems to prevent the superlinear rate of convergence.

Another approach to improve the quality of convergence of **PTs** is by means of acceleration schemes, which are naturally connected to multi-processing environments. Table 1 describes this approach, which we denote as Accelerated Projection Technique **APT**. The key idea embedded in recent papers [7,12,13,18] is to find a factor $\lambda \geq 0$ and a weighting vector $u \geq 0$ such that x_{i+1} given by (5) is the farthest away from x_i without impairing convergence. Although impressive numerical results have been reported [3,7,12–15,25], only Gearhart and Koshy [13] have proven a theoretical improvement to the linear rate of convergence when **APT** is used for finding the intersection of affine subspaces. We recall that underprojection may accelerate convergence in some particular instances [3,17,20].

The main contribution of this paper is found in the next section. Sufficient conditions

Table 1
Accelerated Projection Technique **APT**

Input: The iterate $x_i \notin C$.

1. **FOR** $k = 1, \ldots, p$ **DO**

   (1.1.) Pick appropriate supersets $S_{ik} \supseteq C$.

   (1.2.) Compute $P_k(x_i)$, the projections of x_i on S_{ik}.

   (1.3.) Define the (attracting) directions $d_{ik} = P_k(x_i) - x_i$.

   (1.4.) Pick appropriate values $u_i^k \geq 0, \lambda_i \geq 0$.

2. Let $x_{i+1} = x_i + \lambda_i \sum_{k=1}^{p} u_i^k d_{ik}$. (5)

End of iteration

are given to find weights $u \geq 0$ and a factor $\lambda \geq 0$ that ensure a superlinear rate of convergence for **APTs**. The result is unexpectedly simple (theorem 1). We need: $S_{ik} = \{x \in I\!R^n : g^k(x_i) + \nabla g^k(x_i)^T(x - x_i) \leq 0, k = 1, \ldots, m\}, \{\lambda_i\}_1^\infty \to 1$, and the weights $u^1, \ldots, u^m$ are the dual variables of the projection subproblem (3). The specific algorithm is given in table 2.

The rest of this paper is organized as follows. Section 3 uses the *attracting function* $\varphi(\lambda, u)$ to compare and analyze different acceleration schemes recently proposed in the literature [7,12,18]. It also presents some remarks concerned with possible implementations including the inconsistency case, i.e., $C = \emptyset$.

We end this introduction with a minor note on notation: Lower case Latin letters, except for the range $i \ldots q$ that denote integers, are vectors in $I\!R^n$. Subscripts denote different entities and superscripts denote components, i.e., x_i^k is the k-th component of the vector x_i. Sequences generated by any algorithm, will be denoted by $\{x_i\}_1^\infty, \{S_{ik}\}_1^\infty, \{d_{ik}\}_1^\infty, k = 1, \ldots, m$, and so forth. The projection on a set S_{ik} will be denoted as $P_k(x_i)$. Scalars will be denoted by lower case Greek letters. Unless otherwise stated, $||.||$ refers to the Euclidean norm.

2. OPTIMAL ACCELERATION FOR CIP

In this section we will commit a minor abuse in notation: $\sum_k$ stands for $\sum_{k=1}^m$. We assume that we have computed, either serially or in parallel, $\nabla g^k(x_i), k = 1, \ldots, m$. Let

$$S_{ik} = \{x \in I\!R^n : g^k(x_i) + \nabla g^k(x_i)^T(x - x_i) \leq 0\}, k = 1, \ldots, m. \tag{6}$$

Before proceeding, let us recall well known properties:

P1. $S_{ik} \supset C$. Indeed, if $z \in C$, then $0 \geq g^k(z) \geq g^k(x_i) + \nabla g^k(x)^T(z - x_i)$.

P2. $P_k(x_i) = x_i - \frac{g^k(x_i)}{||\nabla g^k(x_i)||^2} \nabla g^k(x_i)$.

Property **P2** reveals that the direction $d = \sum_k u^k d_{ik}$ in (5) belongs to the cone generated by $\{-\nabla g^1(x_i), \ldots, -\nabla g^m(x_i)\}$. The best direction in this cone will ensure a superlinear rate of convergence as it will be proved below. We need one more definition to clarify the meaning of *best direction*.

Attracting direction: We say that d attracts $x \notin C$ towards C if:

$$[z \in C] \Rightarrow \exists \bar{\lambda} > 0 : ||x + \lambda d - z|| \leq ||x - z|| \text{ for } 0 \leq \lambda \leq \bar{\lambda}. \tag{7}$$

Attracting directions are often obtained from projections on convex sets. It is well known that (See for instance [23,27])

$$[z \in C] \Rightarrow ||x + \lambda(P_C(x) - x) - z||^2 \leq ||x - z||^2 - \lambda(2 - \lambda)||P_C(x) - x||^2$$

Proposition 1 *Let $x \notin C, z \in C$, and let the functions $g^1(.), \ldots, g^m(.)$ that define* **CIP** *be convex and continuously differentiable. The direction $d = -\sum_k u^k \nabla g^k(x)$ attracts x for any fixed nonnegative $u \in \mathbb{R}^m$ such that $\sum_k u^k g^k(x) > 0$.*

Proof: Let $\lambda \geq 0$. By convexity of $g(\cdot)$ we have $||x - \lambda \sum_k u^k \nabla g^k(x) - z||^2 =$

$$\begin{aligned}&= ||x - z||^2 + 2\lambda \textstyle\sum_k u^k \nabla g^k(x)^T (z - x) + \lambda^2 || \sum_k u^k \nabla g^k(x)||^2 \\ &\leq ||x - z||^2 + 2\lambda \textstyle\sum_k u^k (g^k(z) - g^k(x)) + \lambda^2 || \sum_k u^k \nabla g^k(x)||^2 \\ &\leq ||x - z||^2 - \bar{\varphi}(\lambda, u),\end{aligned}$$

where $\bar{\varphi}(\lambda, u) \doteq 2\lambda \sum_k u^k g^k(x) - \lambda^2 || \sum_k u^k \nabla g^k(x)||^2$. If $|| \sum_k u^k \nabla g^k(x)|| = 0$ there is nothing to prove, (7) holds for any $\bar{\lambda} > 0$. Otherwise, we observe that $\bar{\varphi}(\lambda, u)$ is concave with respect to λ and $\bar{\varphi}'_\lambda(0, u) = 2 \sum_k u^k g^k(x) > 0$. It is immediate to prove that

$$\bar{\varphi}(\lambda, u) \geq 0, \text{for } 0 \leq \lambda \leq \bar{\lambda} \doteq 2 \frac{\sum_k u^k g^k(x)}{|| \sum_k u^k \nabla g^k(x)||^2}. \tag{8}$$

As a byproduct we obtain that $\frac{1}{2}\bar{\lambda} = \arg \max_{\lambda \geq 0} \bar{\varphi}(\lambda, u)$.

Note that we may assume $|| \sum_k u^k \nabla g^k(x)|| > 0$, because if this does not hold we can expect $C = \emptyset$, as shown by the following lemma:

Lemma 1 $[C \neq \emptyset, u \geq 0, \sum_k u^k g^k(x) > 0] \Rightarrow \min_{u \geq 0} || \sum_k u^k \nabla g^k(x)|| > 0.$

Proof: If the lemma is not valid, then $C \neq \emptyset$ and there exists $\bar{u} \geq 0 : \sum_k \bar{u}^k g(x) > 0, || \sum_k \bar{u}^k \nabla g^k(x)|| = 0$. From the previous proposition we obtain that $\bar{\varphi}(\lambda, \bar{u}) \leq ||x - P_C(x)||^2$. But we also obtain that $\bar{\varphi}(\lambda, \bar{u}) \to \infty$ as $\lambda \to \infty$, a contradiction.

We now look back at the **APT**. Since $||x_i - \sum_k u^k \nabla g^k(x_i) - z||^2 \leq ||x_i - z||^2 - \bar{\varphi}(1, u)$ for all $z \in C$, it is fully justified to get the optimal acceleration as

$$u_i = \arg \max_{u \geq 0} \bar{\varphi}(1, u) = \arg \max_{u \geq 0} \; 2 \sum_k u^k g^k(x) - || \sum_k u^k \nabla g^k(x)||^2 \,, \tag{9}$$

but this quadratic programming problem happens to be the dual of the strictly convex projection problem (3). They have the same objective value and the minimizer of (3) is given by (10) below (This result is well known in optimization. See for instance [21, p.124]); therefore the dual algorithm described in table 2 exhibits a superlinear rate of convergence. This is formally stated in theorem 1.

Table 2
Dual Superlinear Algorithm

Input: $x_i \notin C \subset I\!R^n, \theta > 0$

1. $K_i \supseteq \left\{1 \leq k \leq m : g^k(x_i) \geq ||g(x_i)||_\infty - \theta\right\}$

2. $u_i = \arg\max\, 2\sum_k u^k g^k(x) - ||\sum_k u^k \nabla g^k(x)||^2 : u_i^k \begin{cases} \geq 0 & k \in K_i, \\ = 0 & k \notin K_i \end{cases}$

3. $x_{i+1} = x_i - \sum_k u_i^k \nabla g^k(x_i)$ (10)

End of iteration.

Theorem 1 *Let $g^k(.) : I\!R^n \to I\!R, k = 1, \ldots, m$ be convex functions with Lipschitz continuous gradients, and assume that $C \doteq \{x \in I\!R^n : g^k(x) \leq 0, k = 1, \ldots, m\}$ has a nonempty interior. The dual algorithm described in table 2 generates a sequence $\{x_i\}_1^\infty$ that converges superlinearly to some $x \in C$.*

Proof: The quadratic optimization problem (table 2, step 2) is the *dual* of the problem $\min_{y \in S_i} \frac{1}{2}||y - x_i||^2$ of the *primal* superlinear algorithm. Therefore, if the initial points coincide, both algorithms generate the same sequence $\{x_i\}_1^\infty$. The theorem follows from [10, Theorem 2].

We point out that a parallel implementation can exhibit a superlinear rate of convergence, which occurs if all gradients $\nabla g^k(x_i), k \in K_i$ are computed in parallel and then the quadratic subproblem (table 2, step 2) is solved. This approach presents two serious drawbacks: *i)* The communication load can surpass the processors computational load, unless the computation of $\nabla g(.)$ involves a significant amount of effort, and *ii)* the quadratic subproblem, as a serial section of the algorithm, undermines the speed up. It must be solved by an efficient (parallel) algorithm.

3. ACCELERATION SCHEMES

In real problems systems are huge and they are split in p blocks. **APT** (Table 1) finds the projection $P_k(x_i)$ on each block, defines $d_{ik} = P_k(x_i) - x_i$ and combines all these directions to generate the next estimate x_{i+1}. It is not relevant to us if this work is performed in parallel. As mentioned in the introduction a lot of attention has been focused on the choice of non negative weights u and the λ factor to accelerate the convergence of block action methods. In this section we analyze the schemes (11,12,13) recently proposed in the open literature [7,11,12,18]. We prove a nice connection among them through (7), the attracting direction concept introduced in the previous section. Despite the excellent numerical results so far obtained for these acceleration schemes, it is argued that a superlinear rate of convergence may be unattainable. We will focus our exposition on exact projections, but we point out that the results can be generalized to inexact or inaccurate projections. The reader can obtain further details in [15].

$$\sum_k u_i^k = 1, \lambda_i = \frac{\sum_k u_i^k ||P_k(x_i) - x_i||^2}{|| \sum_k u_i^k P_k(x_i) - x_i||^2} \tag{11}$$

$$u_i^k = \arg \max_{\sum u^k = 1, \mu \geq 0} \left[\frac{\sum_k u^k ||d_{ik}||^2}{|| \sum_k u^k d_{ik} ||} \right]^2, \lambda_i = 1 \tag{12}$$

$$u_i^k = \arg \max_{u^k \geq 0} \left[2 \sum_k u^k ||d_{ik}||^2 - || \sum_k u^k d_{ik} ||^2 \right], \lambda_i = 1 \tag{13}$$

In the remainder of this section we drop the iteration index i, make $\sum_k$ stand for $\sum_{k=1}^p$, and denote

z as any feasible point, $z \in C$,

x as a non feasible point, $x \notin C$,

$\lambda \geq 0, S_k \supseteq C, d_k \doteq P_k(x) - x, u^k \geq 0, k = 1, \dots, p$.

We recall that $d_k^T(z - x) = d_k^T(z - P_k(x) + d_k) \geq ||d_k||^2$.

Lemma 2 $0 \leq ||x + \lambda \sum_k u^k d_k - z||^2 \leq ||x - z||^2 - \varphi(\lambda, u)$, *where*

$$\varphi(\lambda, u) \doteq 2\lambda \sum_k u^k ||d_k||^2 - \lambda^2 || \sum_k u^k d_k||^2. \tag{14}$$

Proof: $0 \leq ||x + \lambda \sum_k u^k d_k - z||^2$

$= ||x - z||^2 + 2\lambda \sum_k u^k d_k^T(x - z) + \lambda^2 || \sum_k u^k d_k||^2$

$\leq ||x - z||^2 - \varphi(\lambda, u)$.

Lemma 3 *The direction* $\sum_k u^k d_k$ *attracts* x *for any given* $u \geq 0$ *such that* $\sum_k u^k ||d_k|| > 0$.

Proof: It is a replica of proposition 1.

We obtain $\varphi(\lambda, u) \geq 0$ for $0 \leq \lambda \leq \bar{\lambda} \doteq 2 \frac{\sum_k u^k ||d_k||^2}{|| \sum_k u^k d_k||^2}$. (15)

Corollary 1 $\varphi(\omega \bar{\lambda}, u) = 4\omega(1 - \omega) \left[\frac{\sum_k u^k ||d_{ik}||^2}{|| \sum_k u^k d_{ik}||} \right]^2$, *and the maximum value of* $\varphi(., u)$ *occurs at* $\lambda = \frac{1}{2}\bar{\lambda}$.

Proof: It is immediate from the definitions of $\varphi(\lambda, u)$ in (14) and $\bar{\lambda}$ in (15).

Note that $|| \sum_k u^k d_k|| = 0$ probably implies emptiness of the set C.

Lemma 4 $[C \neq \emptyset, \sum_k u^k ||d_k|| > 0] \Rightarrow \min_{u \geq 0} || \sum_k u^k d_k|| > 0$.

Proof: The argument follows almost verbatim from lemma 1.

We point out that when $d_k^T(z - x) = ||d_{ik}||^2$, which occurs when the solution set C and the supersets $S_k, k = 1, \dots, p$ are translated subspaces, x_{i+1} is the closest point to C along $\sum_k u^k d_k$, a desirable property.

We reached the following conclusion:

- $\lambda_i = \arg\max_{\lambda} \varphi(\lambda, u)$ for scheme (11)

- $u_i = \arg \max_{\sum_k u^k=1, u\geq 0} \varphi(\frac{\bar{\lambda}}{2}, u)$ for scheme (12)

- $u_i = \arg\max_{u\geq 0} \varphi(1, u)$ for scheme (13)

We observed that the acceleration schemes 11,12, and 13 are justified. All of them represent some sort of optimal values for the attracting function $\varphi(\lambda, u)$. Given u, (11) finds the best λ, as the maximum along the direction, whereas (12) and (13) find the best u for a given λ. We point out that a superlinear rate can be very hard to achieve. As indicated before we have to consider non violated constraints. It seems that a good strategy may be to start with any acceleration scheme and switch to a superlinear algorithm only at the latest stages of the whole procedure. However, this switching may not be needed for a noisy system.

Numerical tests with several **PTs** versions confirm the usefulness of (11) [12,14,15,25] for its simplicity. On the other end, scheme (12) implies the solution of a difficult quadratic fractional programming problem. For a system of linear inequalities, Kiwiel proposed an algorithm that gives an approximate solution to (12), by solving a sequence of quadratic programming problems. Although computationally (13) is suited to parallel processing [2, Chapter 3] and it is easier than (12), it may degrade the speed-up if it is not efficiently solved. The implementation issue is of fundamental importance and it is, by itself, a subject of search.

A reminiscent of the acceleration schemes of the types (11, 12, 13) can be found in [24]. It is surprising to observe that these schemes have been suggested by different researchers based on different analysis. Scheme (11) was proposed by García-Palomares [9, Proposicion 15] as a way to generate a point x_{i+1} far enough from x_i. It was overlooked by Kiwiel [18, eq. (3.15)]. Combettes suggested it in [6] for the solution of the intersection of affine subspaces. He later generalized its use [7] using Pierra's decomposition approach [26]. Scheme (12) was proposed by Kiwiel to obtain the *deepest surrogate cut.* Scheme (13) was first proposed by García-Palomares and González-Castaño [12]. They developed an attracting function $\varphi(\lambda, u)$ valid for approximate or inaccurate projections. Indeed, they proved that an attracting function can be obtained as long as $d_k^T(z-x) \geq \eta||d_k||^2$, for some $\eta > 0$.

Theorem 1 assumes non emptiness of the interior of the set C. Inaccuracies and noisy data commonly found in real applications may make a system inconsistent, and we should be guarded against this possibility. Inconsistency of a quadratic subproblem (3) obviously implies inconsistency of the larger system. But this test may become superfluous when the constraints are distributed among several processors, or when limited resources oblige the solution of smaller quadratic subproblems. We can also resort to the following non feasibility test adapted from [18]: If ρ_1 is *assumed* to be an upper bound of $||x_1 - P_C(x)||^2$, then the algorithm detects non feasibility when $\sum_{k=1}^{i} \bar{\varphi}(\lambda_k, u_k) > \rho_1$. If we strongly suspect $C = \emptyset$ we should cope with this undesirable condition and dismiss superlinearity.

We can also detect infeasibility with the following linear programming problem:

$$\begin{aligned} u_i, v_i, w_i = \arg\min_{u,v,w} \quad & \textstyle\sum_k (v^k + w^k). \\ \text{such that:} \quad & \textstyle\sum_k u^k \nabla g^k(x_i) + v - w = 0 \\ & \textstyle\sum_k u^k g^k(x_i) = 1 \\ & u^k, v^k, w^k \geq 0 \end{aligned} \tag{16}$$

Observe that if the minimum of (16) is null, then $||\sum_k u_i^k \nabla g^k(x_i)|| = 0$, and from lemmas (1), (4) we may conclude that $C = \emptyset$. More specifically, $S_i = \emptyset$. Otherwise, if a non null minimum is obtained we set $\lambda_i = \dfrac{\sum_k u_i^k g^k(x_i)}{||\sum_k u_i^k \nabla g^k(x_i)||^2}$, as suggested by (8).

To end the feasibility discussion we point out that proximity functions have recently been introduced to measure non feasibility [5,19], and generalized **PTs** converge to a minimum of this function, although quite slowly. The work of Censor et al [4], coupled with the references cited therein, represents a good source of information. Finally, any **CIP** can be easily converted to a feasible system, namely, $\bar{C} \doteq \{x \in I\!R^n : g^k(x) + \gamma_k^2 = 0\}$.

In summary, there is a long way to pave in order to find out the goodness of a superlinear algorithm and its best implementation.

REFERENCES

1. H.H. Bauschke and J.M. Borwein, On projection algorithms for solving convex feasibility problems, *SIAM Review* **38** (1996) 367–426.
2. D. Bertsekas and J.N. Tsitsiklis, *Parallel and distributed computation: Numerical methods* (Prentice Hall, 1989).
3. R. Bramley and A. Sameh, Row projection methods for large nonsymmetric linear systems, *SIAM Journal on Scientific and Statistic Computation* **13** (1992) 168–193.
4. Y. Censor, D. Gordon and R. Gordon, Component averaging: An efficient iterative parallel algorithm for large and sparse unstructured problems, *Parallel Computing,* to appear.
5. P.L. Combettes, Inconsistent signal feasibility problems: least-squares solutions in a product space, *IEEE Transactions on Signal Processing* **SP-42** (1994) 2955–2966.
6. P.L. Combettes, Construction d'un point fixe commun á une famille de contractions fermes, *Comptes Rendus de l'Académie des Sciences de Paris* **Serie I, 320** (1995) 1385–1390.
7. P.L. Combettes, Convex set theoretic image recovery by extrapolated iterations of parallel subgradient projections, *IEEE Transactions on Image Processing* **6-4** (1997) 1–22.
8. U.M. García-Palomares, A class of methods for solving convex systems, *Operation Research Letters* **9** (1990) 181–187.
9. U.M. García-Palomares, Aplicación de los métodos de proyección en el problema de factibilidad convexa: un repaso, *Investigación Operativa* **4-3** (1994) 229–245.
10. U.M. García-Palomares, A superlinearly convergent projection algorithm for solving the convex inequality problem, *Operations Research Letters* **22** (1998) 97–103.
11. U.M. García-Palomares, Relaxation in projection methods, in: P.M. Pardalos and C.A. Floudas, eds., *Encyclopedia of optimization* (Kluwer Academic Publishers) to appear.

12. U.M. García-Palomares and F.J. González-Castaño, Incomplete projection algorithm for solving the convex feasibility problem, *Numerical Algorithms* **18** (1998) 177–193.
13. W.B. Gearhart and M. Koshy, Acceleration schemes for the method of alternating projections, *Journal on Computing and Applied Mathematics* **26** (1989) 235–249.
14. F.J. González-Castaño, Contribución al encaminamiento óptimo en redes de circuitos virtuales con supervivencia mediante arquitecturas de memoria destribuida, *Tesis Doctoral, Universidade de Vigo, Dep. Tecnoloxías das Comunicacións* (1998)
15. F.J. González-Castaño, U.M. García-Palomares, José L. Alba-Castro and J.M. Pousada-Carballo, Fast image recovery using dynamic load balancing in parallel arquitectures, by means of incomplete projections, to appear in *IEEE Transactions on Image Processing*.
16. L.G. Gubin, B.T. Polyak and E.V. Raik, The method of projections for finding the common point of convex sets, *USSR Computational Mathematics and Mathematical Physics* **7** (1967) 1–24
17. G.T. Herman and L.B. Meyer, Algebraic reconstruction techniques can be made computationally efficient, *IEEE Transactions on Medical Imaging* **MI-12** (1993) 600–609
18. K.C. Kiwiel, Block-iterative surrogate projection methods for convex feasibility problems, *Linear Algebra and its Applications* **215** (1995) 225–260.
19. A.N. Iusem and A.R. De Pierro, Convergence results for an accelerated nonlinear Cimmino algorithm, *Numerische Mathematik* **49** (1986) 367–378.
20. J. Mandel, Convergence of the cyclical relaxation method for linear inequalities, *Mahematical Programming* **30-2** (1984) 218–228.
21. O.L. Mangasarian, *Nonlinear Programming*, (Prentice Hall, 1969).
22. J. von Neumann, *Functional Operators, vol. 2* (Princeton University Press, 1950).
23. Z. Opial, Weak convergence of the sequence of successive approximations for nonexpansive mappings, *Bulletin American Mathematical Society* **73** (1967) 591–597.
24. N. Ottavy, Strong convergence of projection like methods in Hilbert spaces, *Journal of Optimization Theory and Applications* **56** (1988) 433–461.
25. H. Özakas, M. Akgül and M. Pinar, A new algorithm with long steps for the simultaneous block projections approach for the linear feasibility problem, *Technical report IEOR 9703, Bilkent University, Dep. of Industrial Engineering, Ankara, Turkey* (1997).
26. G. Pierra, Decomposition through formalization in a product space, *Mathematical Programming* **28-1** (1984) 96–115.
27. D.C. Youla and H. Webb, Restoration by the method of convex projections: Part 1–Theory, *IEEE Transactions on Medical Imaging* **MI-1, 2** (1982) 81–94.

Inherently Parallel Algorithms in Feasibility and Optimization and their Applications
D. Butnariu, Y. Censor and S. Reich (Editors)

ALGEBRAIC RECONSTRUCTION TECHNIQUES USING SMOOTH BASIS FUNCTIONS FOR HELICAL CONE-BEAM TOMOGRAPHY

G. T. Herman[a]*, S. Matej[b], and B. M. Carvalho[c]

[a]Department of Computer and Information Sciences, Temple University, 3rd Floor Wachman Hall, 1805 North Broad Street, Philadelphia, PA 19122-6094, USA

[b]Medical Image Processing Group, Department of Radiology, University of Pennsylvania, 4th Floor Blockley Hall, 423 Guardian Drive, Philadelphia, PA, 19104-6021, USA

[c]Department of Computer and Information Science, University of Pennsylvania, Moore School Building, 200 South 33rd Street, Philadelphia, PA 19104-6389, USA

Algorithms for image reconstruction from projections form the foundations of modern methods of tomographic imaging in radiology, such as helical cone-beam X-ray computerized tomography (CT). An image modeling tool, but one which has been often intermingled with image reconstruction algorithms, is the representation of images and volumes using blobs, which are radially symmetric bell-shaped functions. A volume is represented as a superposition of scaled and shifted versions of the same blob. Once we have chosen this blob and the grid points at which its shifted versions are centered, a volume is determined by the finite set of scaling coefficients; the task of the reconstruction algorithm is then to estimate this set of coefficients from the projection data. This can be done by any of a number of optimization techniques, which are typically iterative.

Block-iterative algebraic reconstruction techniques (ART) are known to have desirable limiting convergence properties (some related to a generalization of least squares estimation). For many modes of practical applications, such algorithms have been demonstrated to give efficacious results even if the process is stopped after cycling through the data only once. In this paper we illustrate that ART using blobs delivers high-quality reconstructions also from helical cone-beam CT data. We are interested in answering the following: "For what variants of block-iterative ART can we simultaneously obtain desirable limiting convergence behavior and good practical performance by the early iterates?"

We suggest a number of approaches to efficient parallel implementation of such algorithms, including the use of footprints to calculate the projections of a blob.

*This work has been supported by NIH Grants HL28438 (for GTH, SM and BMC) and CA 54356 (for SM), NSF Grant DMS96122077 (for GTH), Whitaker Foundation (for SM) and CAPES - Brasília - Brazil (for BMC). We would like to thank Edgar Garduño for his help in producing the images for this paper and Yair Censor and Robert Lewitt for their important contributions at various stages of this research.

1. INTRODUCTION

It was demonstrated over twenty years ago [1] (see alternatively Chapter 14 of [2]) that one can obtain high-quality reconstructions by applying algebraic reconstruction techniques (ART) to cone-beam X-ray computerized tomography (CT) data. The data collection mode used in that demonstration was that of the Dynamic Spatial Reconstructor (DSR [3]), which in addition to having to deal with cone beam data implied that the whole body was not included within the cone beam even radially (and, of course, also not longitudinally) and that only a small subset of the cone-beam projections were taken simultaneously and so the object to be reconstructed (the beating heart) was changing as the data collection progressed. Nevertheless, with the application of ART, high-quality 4D (time-varying 3D) reconstructions of the heart could be obtained. ART was also used for reconstruction from cone-beam X-ray data collected by the Morphometer for the purpose of computerized angiography [4]. In both these data collection devices the X-ray source moved in a circle around the object to be reconstructed.

Helical cone-beam tomography (in which the X-ray source moves in a helix around the object to be reconstructed) is a recent and potentially very useful development, and so it is very hotly pursued (see, e.g., the September 2000 Special Issue of the IEEE Transactions on Medical Imaging, which is devoted to this topic). To justify our interest in adapting ART to this important mode of data collection, we now give a detailed description of our recent and very positive experience with ART using other modes of data collection. Based on this experience we claim that working towards similarly efficacious implementations of ART for helical cone-beam tomography is indeed a worthwhile endeavour. The initial results, reported for the first time in this article, are indeed encouraging.

2. BLOBS FOR RECONSTRUCTION

An image modeling tool, which was described in a general context in [5,6] and utilized in image reconstruction algorithms in [7,8], is the representation of images and volumes using *blobs*, which are radially symmetric bell-shaped functions whose value at a distance r from the origin is

$$b_{m,a,\alpha}(r) = \frac{1}{I_m(\alpha)} \left[\sqrt{1-(r/a)^2}\right]^m I_m\left[\alpha\sqrt{1-(r/a)^2}\right], \tag{1}$$

for $0 \leq r \leq a$ and is zero for $r > a$. In this equation I_m denotes the modified Bessel function of order m, a is the radius of the support of the blob and α is a parameter controlling the blob shape. A volume is represented as a superposition of N scaled and shifted versions of the same blob; i.e., as

$$\overline{f}(x,y,z) = \sum_{j=1}^{N} c_j b_{m,a,\alpha}\left(\sqrt{(x-x_j)^2+(y-y_j)^2+(z-z_j)^2}\right), \tag{2}$$

where $\{(x_j, y_j, z_j)\}_{j=1}^N$ is the set of grid points in the three-dimensional (3D) Euclidean space to which the blob centers are shifted. Once we have chosen these grid points and the specific values of m, a and α, the volume is determined by the finite set $\{c_j\}_{j=1}^N$ of real coefficients; the task of the reconstruction algorithm in this context is to estimate this set of coefficients from the projection data. This can be done by any of a number of optimization techniques, which are typically iterative [9].

An example of the accuraccy that can be obtained by using blobs for volume representation can be seen in Figure 1, where a slice of an approximation of a uniform cylinder of value 1 is displayed using different gray-scale windows. The so-called body-centered cubic (bcc) grid [7] was used with all specific values based on those recommended in [7]. The blob coefficient c_j was set to 1 if the jth grid point was within the specified cylinder and to 0 otherwise. The values of $\overline{f}$ were calculated using (2) for a cubic grid (with finer spacing than the bcc grid). Figure 1 shows computer displays of one slice of this cubic grid. Values away from the surface of the cylinder (a circular perimeter in the given slice) are 0 outside the cylinder and vary between 0.9992 and 1.0014 inside the cylinder, providing us with an approximation which is within 0.15% of the desired value (namely 1.0000). This indicates that volumes can be very accurately aproximated, provided only that we find a suitable set $\{c_j\}_{j=1}^N$. To visualize the nature of these very small inaccuracies, for the display we mapped 1.0014 into intensity 255 (white), and 0.9814 (a) and 0.9914 (b) into intensity 0 (black).

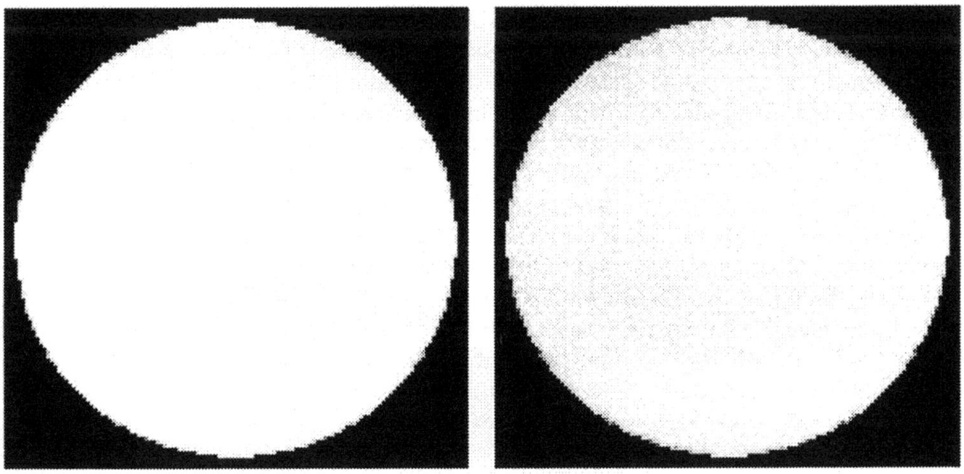

(a) (b)

Figure 1. Interpolated slices of a cylinder on the cubic grid from blobs arranged on a bcc grid using a gray-scale window of [0.9814,1.0014] which is approximately 2% of the total range of values (a) and [0.9914,1.0014] which is approximately 1% of the total range of values (b).

The aim of [7,8] was to study the choices of the grid points and of the parameters m, a and α, combined with implementation of the algorithm to estimate the coefficients, from the point of view of obtaining high quality reconstructions in a reasonable time. In [7] it was demonstrated, in the context of 3D Positron Emission Tomography (PET), that by choosing a not too densely packed bcc grid, a 3D ART (see [10] and also Section 4 below) using blobs can reach comparable or even better quality of reconstructed images than 3D filtered backprojection (FBP) after only one cycle through the projection data (and thereby needing less computer time than the 3D FBP method). They also demonstrated, by simulation experiments, that using blobs in various iterative algorithms leads to substantial improvement in the reconstruction performance in comparison with the traditionally used cubic voxels. These simulation results are consistent with results obtained from physical experiments on a 3D PET scanner [11]. The superior performance of ART with blobs has also been demonstrated in another field of application, namely reconstruction from transmission electron microscopic projections [12–14].

Much of the previous work for the estimation of the $\{c_j\}_{j=1}^N$ was based on generalized projections onto convex sets (POCS) and on generalized distance minimization (GDM), which may be perceived of as the study of various generalizations of De Pierro's modified expectation maximization algorithm for penalized likelihood estimation [15]. A great deal of progress has been made in this direction [16–20], both for simultaneous algorithmic schemes (all measurements are treated at the same time) and row-action algorithmic schemes (the measurements are treated one-by-one, as in ART; see (6) below). In-between block-iterative schemes (to be further discussed in Section 4) have also been investigated, as well as parallel methods to achieve fast reconstructions [21,22] (there will be more about this later as well). Parallel algorithms have been developed for optimizing the parameters (such as the m, a and α) of iterative reconstruction algorithms [23,24]. However, rather than trying to cover all this material, we concentrate on one line of development which took us from innovative basic mathematical development all the way to demonstration of clinical efficacy. (Here we discuss work that took place in our laboratory; there have been many other similar developments in other laboratories.)

As background to this discussion consider [25]. It deals with the following clinically relevant problem. Modern PET scanners provide data from a large acceptance angle. Direct reconstruction of the volume from such data typically requires long reconstruction times. Rebinning techniques, approximating planar sinogram data from the oblique projection data, enable the use of multislice two-dimensional (2D) reconstruction techniques with much lower computational demands; however, rebinning techniques available prior to 1995 often resulted in reconstructions of clinically unacceptable quality.

In 1995 Defrise [26] proposed the very promising method of Fourier rebinning (FORE) which is based on the frequency-distance relation and on the stationary-phase approximation advocated in the context of single photon emission computerized tomography (SPECT) reconstruction [27]. In [25] the performance of FORE for PET with large acceptance angles has been evaluated using a 2D FBP algorithm to reconstruct the individual slices after rebinning. This combination provides reconstructions of quality comparable, for all practical purposes, to those obtained by the 3D version of FBP, but at a more than order of magnitude improvement in computational speed. However, given our positive experience with iterative reconstruction algorithms based on blobs, the question naturally arises: is the 2D FBP algorithm the best choice to apply to the output of FORE? To describe our answer to this question we need to present another independent development in image reconstruction algorithms.

The maximum likelihood (ML) approach to estimate the activity in a cross-section has become popular in PET since it has been shown to provide better images than those produced by FBP. Typically, maximizing likelihood has been achieved using an expectation-maximization (EM) algorithm [28]. However, this simultaneous algorithm has a slow rate of convergence. In [29] a row-action maximum likelihood algorithm (RAMLA) has been presented. It does indeed converge to the image with maximum likelihood and its early iterates increase the likelihood an order of magnitude faster than the standard EM algorithm. Specifically, from the point of view of measuring the uptake in simulated brain phantoms, iterations 1, 2, 3, and 4 of RAMLA perform at least as well as iterations 45, 60, 70, and 80, respectively, of EM. (The computer cost per iteration of the two algorithms is just about the same.) A follow-up simulation study [30] reports that, for the purpose of fully 3D PET brain imaging, an appropriately chosen 3D version of RAMLA using blobs is both faster and superior, as measured by various figures of merit (FOMs) than 3D FBP or 3D ART using blobs.

For this reason, we have decided to compare with other techniques the efficacy of RAMLA as a 2D reconstructor after FORE. In a joint study with Jacobs and Lemahieu [31], we have compared FBP with a variety of iterative methods (in all cases using both pixels and blobs); namely with ART, ML-EM, RAMLA, and ordered subset expectation maximization (OSEM, a block-iterative modification of ML-EM [32]). We have found that, in general, the best performance is obtained with RAMLA using blobs. However, in the meantime, we have found an even better approach, that we have named 2.5D simultaneous multislice reconstruction [33], which takes advantage of the time reduction due to the use of FORE data instead of the original fully 3D data, but at the same time uses 3D iterative reconstruction with 3D blobs. (Thus the reconstruction of the individual slices is coupled and iteration calculations for each particular line are influenced by, and contribute to, several image slices.) The simulation study reported in [33] indicates that 2.5D RAMLA (respectively ART) applied to FORE data is (statistically significantly) superior to 2D RAMLA (respectively ART) applied to FORE data according to each of four PET-related FOMs. For this reason we have started the clinical evaluation of 2.5D RAMLA. Our first published study on this [34] indicates that, as compared to the clinically-used 2D algorithms (OSEM or FBP), the improvements in image quality with 2.5D RAMLA previously seen for simulated data appear to carry over to clinical PET data.

3. BLOBS FOR DISPLAY

There is yet another reason why we believe in the desirableness of using blobs: it not only leads to good reconstructions, but the results of these reconstructions can be displayed in a manner likely to be superior to displays that are obtained based on alternative reconstruction techniques. In this brief section we outline the reasons for this belief.

An often repeated misunderstanding in the radiology literature regarding shaded surface display (SSD) is exemplified by the quote [35]: "The main hallmark of SSD images is the thresholding segmentation that results in a binary classification of the voxels." As can be seen from Figures 1, 3, and 4 of [36], there are ways of segmenting images which succeed where thresholding fails miserably; SSD based on such segmentations will give a more accurate rendition of biological reality than SSD based on thresholding. There is a further assumption in the quote stated above which is also incorrect; namely that thresholding segmentation has to classify whole voxels. This is also not the case provided that we model images using blobs.

One approach to this has been suggested in [5] based on the assumption that there is a fixed threshold t which separates the object of interest from its background. Then the continuous boundary surface of the volume represented by (2) is *exactly*

$$B = \left\{ (x,y,z) \;\middle|\; \sum_{j=1}^{N} c_j b_{m,a,\alpha}\left(\sqrt{(x-x_j)^2 + (y-y_j)^2 + (z-z_j)^2} = t \right) \right\}. \tag{3}$$

Furthermore, if we use (as indeed we always do) blobs which have continuous gradients, then the normal to the surface B at a point (x, y, z) will be parallel to the gradient

$$\nabla \overline{f}(x,y,z) = \sum_{j=1}^{N} c_j \nabla b_{m,a,\alpha}\left(\sqrt{(x-x_j)^2 + (y-y_j)^2 + (z-z_j)^2} \right) \tag{4}$$

of the volume $\overline{f}$ at (x, y, z). Since we have formulas for the $\nabla b_{m,a,\alpha}$ [5], the calculations of the normals to B are also exact and, due to the continuity of the gradient in (4), the normals (and the SSD images) will be smoothly varying (as opposed to surfaces formed from the faces of cubic voxels or any other polygonal surface). An example of such a smooth surface, displayed by the method of "ray-casting" [37], is shown in Figure 2.

4. BLOCK-ITERATIVE RECONSTRUCTION ALGORITHMS

It has been claimed in [38] that applying the simplest form of ART to cone-beam projection data can result in substandard reconstructions, and it has been suggested that a certain alteration of ART leads to improvement. However, besides an illustration of its performance, no properties (such as limiting convergence) of the algorithm have been given. We still need a mathematically rigorous extension of the currently available theory of optimization procedures [9] to include acceptable solutions of problems arising from cone-beam data collection. We discuss this phenomenon in the context of reconstruction using ART with blobs from helical cone-beam data collected according to the geometry of [39].

For this discussion we adopt the notation of [40], because it is natural both for the assumed data collection and for the mathematics that follows. We let M denote the number of times the X-ray source is pulsed as it travels its helical path and L denote the number of lines for which the attenuation line integrals are estimated in the cone-beam for a single pulse. Thus the total number of measurements is LM and we use Y to denote the (column) vector of the individual measurements y_i, for $1 \leq i \leq LM$. We let N denote the number of grid points at which blobs are centered; our desire is to estimate the coefficients $\{c_j\}_{j=1}^{N}$ and thereby define a volume using (2) and even borders within that volume using (3). For $1 \leq i \leq LM$, we let a_{ij} be the integral of the values in the jth blob along the line of the ith measurement (note that these a_{ij} can be calculated analytically for the actual lines along which the data are collected) and we denote by A the matrix whose ijth entry

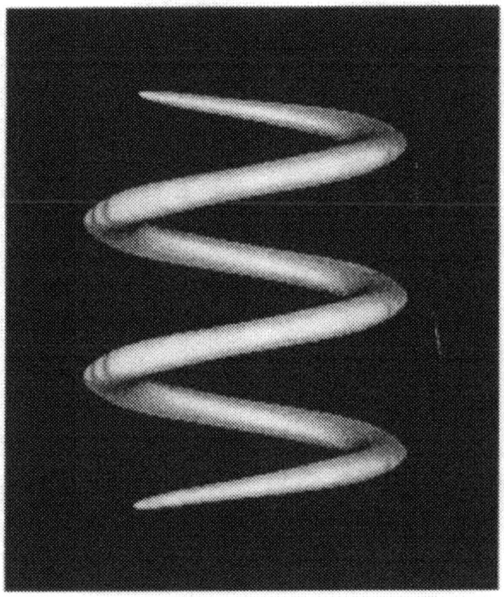

Figure 2. SSD of a surface B, as defined by (3), created using ray-casting.

is a_{ij}. Then, using c to denote the (column) vector whose jth component is c_j, this vector must satisfy the system of approximate equalities:

$$Ac \approx Y. \tag{5}$$

In the notation of [40] the traditional ART procedure for finding a solution of (5) is given by the iterations:

$c^{(0)}$ is arbitrary,

$$c_j^{(n+1)} = c_j^{(n)} + \omega^{(n)} \frac{y_i - \sum_{k=1}^{N} a_{ik} c_k^{(n)}}{\sum_{k=1}^{N} a_{ik}^2} a_{ij}, \qquad \text{for } 1 \leq i \leq N, \qquad n = 0, 1, \ldots, \quad i = n \bmod LM + 1, \tag{6}$$

where $\omega^{(n)}$ is a relaxation parameter. While this procedure has a mathematically well-defined limiting behavior (see, e.g., Theorem 1.1 of [40]), in practice we desire to stop the iterations early for reasons of computational costs. We have found that for the essentially parallel-beam data collection modes of fully 3D PET [41], Fourier rebinned PET [31,33] and transmission electron microscopy [12], one cycle through the data (i.e., $n = LM$) is sufficient to provide us with high quality reconstructions. Although to date we have not been able to achieve similarly high quality reconstructions at the end of one cycle through the data collected using the helical cone-beam geometry, Figure 3 indicates that very high quality reconstructions can be achieved after a small number of cycles through the data.

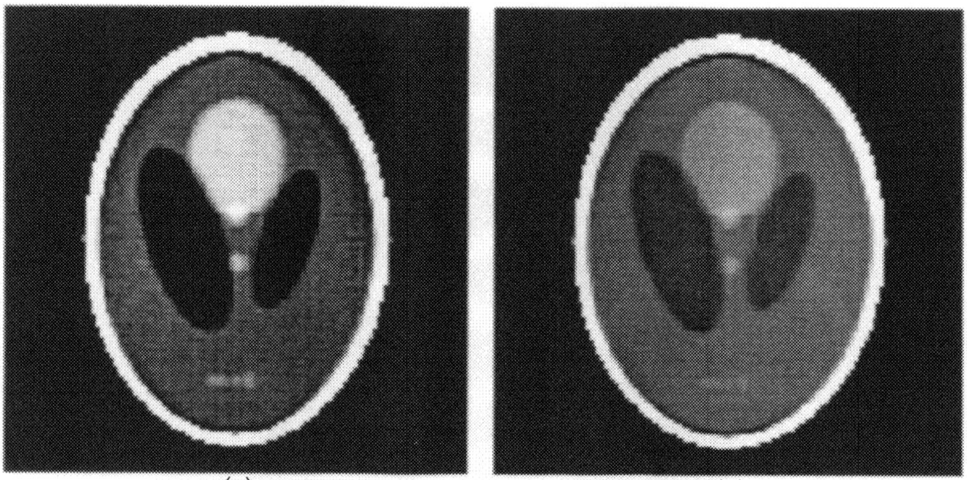

(a) (b)

Figure 3. Slice of the 3D Shepp-Logan phantom [42] reconstructed using ART with blobs (6) shown at gray-scale window settings of [1.00,1.03] which is approximately 1.5% of the total range of values (a) and [0.975,1.055] which is approximately 4% of the total range of values (b). The data were collected using the geometry described in [39] with 64 rows and 128 channels (cone-angle = 9.46°, fan-angle = 21.00°) per projection (i.e., $L = 64 \times 128 =$ 8,192), and 300 projections taken per turn in 2 turns (i.e., $M = 2 \times 300 = 600$). The slice shown is from a $128 \times 128 \times 128$ cubic grid obtained by (2) from a grid of size $N =$ 965,887 (whose points are those points of the bcc grid which is "equivalent" to the cubic grid, as defined in [7], which are inside the cylinder used in Figure 1) with coefficients $\{c_j^{(n)}\}_{j=1}^{N}$ provided by (6) at the end of the seventeenth cycle through the data (i.e., when $n = 17LM =$ 83,558,400). In (6) $c^{(0)}$ was initialized to the zero vector and $\omega^{(n)}$ was chosen to be 0.01 for all n.

This reconstruction compares favorably with those shown in Figure 9 of [43] at matching gray-scale window widths. (One should bear in mind though that the results are not strictly comparable; e.g., we use a helical mode of data collection, while [43] uses a circular path for the X-ray source.) However, we believe that we will be able to further improve on our reconstruction method. We now outline some of the ideas which form the basis of this belief.

Looking at (6) we see that during an iteration the value by which the current estimate of the coefficient of the jth blob is updated is proportional to a_{ij}. Hence, very roughly speaking (and especially if we use very small values of $\omega^{(n)}$), if we initialize the estimate to be the zero vector, then at the end of the first cycle we will get something in which the

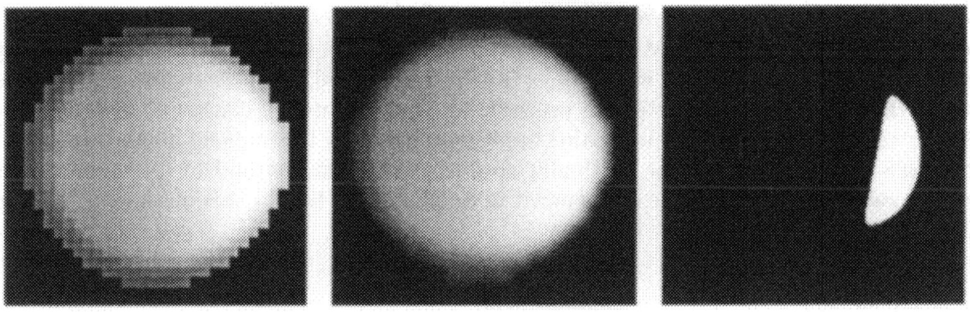

Figure 4. (a) The central one of 113 layers of 27×27 bcc grid points inside a cylinder (see [7]), with the brightness assigned to the jth grid point proportional to the grid coefficient c_j. Other layers above and below are obtained by a sequence of slow rotations around the axis of the cylinder. (b) Calculated values, using (2), of $\overline{f}$ in the same layer as (a) at 200×200 points. (c) Thresholded version of (b) with the threshold t used to define, by (3), the surface B displayed in Figure 2.

jth component is weighted by

$$s_j = \sum_{i=1}^{LM} a_{ij}, \text{ for } 1 \leq j \leq N. \tag{7}$$

For the parallel mode of data collection the values of s_j are nearly the same for all the blobs. However, this is not the case for cone-beam data. Figures 2 and 4 show what happens to the volume $\overline{f}$ of (2) if we set the c_j of (2) to the s_j of (7) for the data collection geometry of [39]: the higher values (near the edge of the cylindrical reconstruction region) reflect the helical path of the X-ray source.

An improvement to ART may result from compensation for this nonuniformity in the iterative formula (the central equation of (6)) by, say, dividing through by s_j. However, there is every reason to approach this problem in a more general context. It is natural to consider instead of the row-action algorithmic scheme (6) its block-iterative version, in which all the measurements taken by one pulse of the X-ray source form a block. A powerful theory is developed for this in [40]. Let Y_i be the L-dimensional vector of those measurements which were taken with the ith pulse of the X-ray source and let A_i be the corresponding submatrix of A. Theorem 1.3 of [40] states that the following block-iterative algorithm has good convergence properties:

$c^{(0)}$ is arbitrary,

$$c^{(n+1)} = c^{(n)} + A_i^T \Sigma^{(n)} \left(Y_i - A_i c^{(n)}\right), \tag{8}$$
$$n = 0, 1, \ldots, \quad i = n \bmod M + 1,$$

where $\Sigma^{(n)}$ is an $L \times L$ relaxation matrix. This theory covers even fully-simultaneous algorithmic schemes (just put all the measurements into a single block). There are also generalizations of the theory which allow the block sizes and the measurement-allocation-to-blocks to change as the iterations proceed [44]. For a thorough discussion of versions of block-iterative ART (and their parallel implementation) we recommend [9] and, specially, Sections 10.4 and 14.4. Our particular special case is the method of Oppenheim [45] whose convergence behaviour is discussed in Section 3 of [40]. In this method $\Sigma^{(n)}$ is an $L \times L$ diagonal matrix whose lth entry is

$$\frac{1}{\sum_{j=1}^{N} a_{[(i-1)L+l]j}} \tag{9}$$

where $i = n \bmod M + 1$ (see (3.4) of [40]).

Potential improvements to the existing theory include the following.

1. *Component-dependent weighting in iterative reconstruction techniques.* The essence of this approach is to introduce in (8) a second ($N \times N$) relaxation matrix $\Delta^{(n)}$ in front of the A_i^T. Then we need to answer the following: For what simple (in the sense of computationally easily implementable) pairs of relaxation matrices $\Sigma^{(n)}$ and $\Delta^{(n)}$ can we simultaneously obtain desirable limiting convergence behavior and good practical performance by the early iterates. Examples of the $\Delta^{(n)}$ to be studied are the diagonal matrix whose jth entry is the reciprocal of the s_j of (7) or, alternatively, proportional to the reciprocal of a similar sum taken over only those measurements which are in the block used in the particular iterative step. In fact the SART method of Andersen and Kak [46], which is used in recent publications such as [43], can be obtained from the method of Oppenheim [40,45] by premultiplication in (8) by exactly such a matrix. To be exact, one obtains SART from the method of Oppenheim by using a $N \times N$ diagonal matrix $\Delta^{(n)}$ whose jth entry is

$$\frac{\lambda}{\sum_{l=1}^{L} a_{[(i-1)L+l]j}} \tag{10}$$

where $i = n \bmod M + 1$ and λ is a relaxation parameter factor (see (2) of [43]). A recently proposed simultaneous reconstruction algorithm which uses j-dependent weighting appears in [47], where it is shown that a certain choice of such weighting leads to substantial acceleration of the algorithm's initial convergence.

2. *A new block-iterative paradigm.* One way of interpreting the algorithm of (8) is as some sort of weighted average of the outcomes of steps of the type indicated in (6), each applied separately to the same $c^{(n)}$ for all the measurements in a particular block [44]. A new paradigm is based on the following idea: since sequential steps of (6) are likely to provide improved reconstructions, why do we not average the output of a *sequence* of steps (over the particular block) of (6)? This approach, which imposes a computational burden similar to the one currently practiced, may be capable of capturing the advantages of both the row-action and the block-iterative (including the fully-simultaneous) algorithms. Preliminary work in this direction [48] includes a proof of convergence to the solution in the case when (5), considered as a system of exact equations, does have a solution. This theory has to be extended

to the (in practice realistic) case of systems of approximate equalities and also to versions of the approach which include component-dependent weighting.

3. *Projections onto fuzzy sets.* The weighted averaging referred to in the previous paragraph raises the question: how are the weights to be determined? This is a particularly relevant question if (5), considered as a system of exact equations, does *not* have a solution, since in such a case algorithms of the type (8) only converge in a cyclic sense (see Theorem 1.3 of [40]). Such algorithms are special cases of the general approach of POCS and so we can reformulate our question: how should the projections onto convex sets be combined if the sets do not have a common point? At an intuitive level the answer is obvious; the more reliable the measurement is, the more weight should be given to its individual influence. Recent work in signal processing using projections onto fuzzy sets [49,50] points a way towards having a mathematically precise formulation and solution of this problem.

In addition to this theoretical work there are some worthwhile heuristic ideas for reconstructions from helical cone-beam data. One promising heuristic approach is the following.

Suppose that we have taken the projection data of an object for which all the blob coefficients c_j are 1. Then, it appears desirable to have a uniform assignment of the blob coefficients after a single step of a modified version of (8), assuming that the initial assignment of the blob coefficients is zero. Assuming that the $\Sigma^{(n)}$ is the identity matrix, we can achieve this aim by choosing $\Delta^{(n)}$ to be a diagonal matrix whose jth entry is inversely proportional to the sum over all lines in the block of the line integral through the jth blob multiplied by the sum of the line integrals through all the blobs. The mathematical expression for this is

$$\sum_{l=1}^{L} \left(a_{[(i-1)L+l]j} \sum_{k=1}^{N} a_{[(i-1)L+l]k} \right) . \tag{11}$$

(Comparing this with (9) and (10), we see that the resulting algorithm can be thought of as a variant of SART [43,46] in which the $\Sigma^{(n)}$ is incorporated into the $\Delta^{(n)}$.) In order for this to work we have to ensure that the value of (11) is not zero. This is likely to demand the forming of blocks which correspond to more than one pulse of the X-ray source.

In addition to investigating the practical efficacy of such heuristic variants of the algorithms on various modes of helical cone-beam data collection (all such investigations can and should be done using the statistically rigorous training-and-testing methodology advocated and used in [12,14,30,31,33,41,51–53]), it is of practical importance to study efficient parallel implementations of them. The following approaches are promising.

1. Distribute the work on multiple processors by exploiting the block structure of the various formulations [9,21].
2. Use ideas from computer graphics for efficient implementation of the forward projection (multiplication by A_i in (8)) and backprojection (multiplication by A_i^T in (8)). One such idea is the use of *footprints*; it is discussed in the next section. For a discussion of the applications of such ideas to cone-beam projections see [43,54] and for an up-to-date survey of hardware available for doing such things in the voxel environment see [55].

3. Assess recent improvements in pipeline image processors and digital signal processors that allow real-time implementation of computation intensive special applications [56–58].

By a combination of these approaches we expect that helical cone-beam reconstruction algorithms can be produced which are superior in image quality to, but are still competitive in speed with, the currently-used approaches.

5. FOOTPRINT ALGORITHMS FOR ITERATIVE RECONSTRUCTIONS USING BLOBS

The central and most computationally demanding operations of any iterative reconstruction algorithm are the forward projection (in (8) this is the estimation of $A_i c^{(n)}$) and the back-projection (in (8) this is the multiplication by A_i^T). To execute these operations efficiently, we employed the so-called footprint algorithm in our previous work on iterative reconstruction from the parallel-beam data [8]. This algorithm, whose basic ideas we outline in this section, can be straightforwardly parallelized and it can be extended to cone-beam data collected using a helical trajectory. In particular we can (and will) use the notation and formalism of (8) for the case of parallel-beam data collection.

We reconstruct a volume as a superimposition of 3D blobs, whose centers are located at specified grid points. Further, the measurements (projection data) are considered to be composed of distinct subsets, where each subset of the projection data is provided by those L lines of integration that are parallel to a vector $(1, \theta_i, \phi_i)$ in spherical coordinates. For each direction (θ_i, ϕ_i), the L lines are perpendicular to the ith 2D "detector plane" and the L-dimensional vector of measurements made by this detector plane is denoted by Y_i. This projection geometry is often called the "parallel-beam X-ray transform" [59]. Although the blob basis functions are localized, they are typically several times larger (in diameter) compared to the classical cubic voxels. Consequently, each blob influences several projection lines from the same 2D detector plane, increasing the computational cost of the algorithm for the evaluation of the line integrals through the 3D volume for each (θ_i, ϕ_i). The following blob-driven footprint algorithm is nevertheless quite effective.

The footprint algorithm is based on constructing the forward projection of the volume onto the 2D detector plane by superimposition of "footprints" (X-ray transforms of individual blobs). The approach is similar to the algorithm called "splatting" [60–62] developed for volume rendering of 3D image data onto a 2D viewing plane. By processing the individual blobs in turn, the ith forward projection of the volume is steadily built up by adding a contribution from each blob for all projection lines from the ith detector plane which intersect this blob. For the spherically-symmetric basis functions, the footprint values are circularly symmetric on the 2D detector plane and are independent of the detector plane orientation. Consequently, only one footprint has to be calculated and this will serve for all orientations. The values of the X-ray transform of the blob are precomputed on a fine grid in the 2D detector plane and are stored in a footprint table. Having the footprint of the generic blob, we compute the contribution of any blob to the given detector plane at any orientation simply by placing the footprint at the appropriate position in the detector plane and adding scaled (by c_j) blob footprint values to the detector plane values. The positions of individual blobs with respect to the detector planes are calculated using a simple incremental approach (for more details on this approach and on sampling requirements of the footprint table, see [8]).

Similarly, in the blob-driven back-projection operation, the same footprint specifies which lines from the detector plane contribute to the given blob coefficient and by what weight. The only difference from the forward projection operation is that, after placing the footprint on the detector plane, the detector plane values from the locations within the footprint are added to the given blob coefficient, where each of the values is weighted by the blob integral value (appropriate elements of the matrix A) stored in the corresponding location of the footprint table. The described forward and back-projection calculations can be straightforwardly parallelized at various levels of the algorithm. For example, this can be done by performing parallel calculations (of both forward and back-projection) for different detector planes (in cases we wish to combine several of them into a single block of the iterative algorithm) or for different sets of projection lines from a given detector plane. The image updates calculated on different processors are then combined at the end of the back-projection operation stage. Alternatively, parallel calculations can be done on different sets of image elements (blobs), in which case the information recombination has to be done in the projection data space after the forward projection stage.

In the outlined blob-driven approach, the computation of the ith forward projection is finished only after sequentially accessing all of the blobs. This approach is well-suited to those iterative reconstruction methods in which the reconstructed image is updated (by the back-projection operation) only after comparisons of all of the forward projected and measured data (simultaneous iterative techniques, e.g., ML-EM [28]) or at least of all data from one or several detector planes (block-iterative algorithms, e.g., (8), block-iterative ART [40], SART [46], OS-EM [32]). In the row-action iterative algorithms (e.g., (6) and other variants of row-action ART [10], RAMLA [29]), which update the reconstructed volume after computations connected with each of the individual projection lines, it is necessary to implement a line-driven approach.

In the line-driven implementation of (6), for each n (which determines i) the forward projection is the calculation of $\sum_{k=1}^{N} a_{ik} c_k^{(n)}$. To do this the footprint algorithm identifies the set of blobs which are intersected by the ith projection line. The centers of these blobs are located in the cylindrical tube with diameter equal to the blob diameter and centered around the ith line. The volume coordinate axis whose direction is closest to the direction of the ith line is determined, and the algorithm sequentially accesses those parallel planes through the volume grid that are perpendicular to this coordinate axis. The cylindrical tube centered around the given line intersects each of the volume planes in an elliptical region. This region defines which blobs from the given plane are intersected by the given line. The unweighted contribution a_{ik} of the kth blob to the ith line is given by the elliptical stretched footprint in the volume plane. (Note that there is something slightly subtle going on here: the stretched footprint is centered where the line crosses the volume plane and not at the individual grid points. That this is all right follows from (2).) The same stretched footprint can be used for all of the parallel volume planes crossed by the ith line and for all of the lines parallel to it. The details of line-driven backprojection are similar.

The line-driven forward and back-projection operations can be calculated in parallel in two different ways. The first is by performing the forward and back-projection calculations in parallel for such lines that are independent, i.e., for which the sets of blobs intersected by them are disjoint. The second is to compute in parallel for the sets of blobs which are located on different volume planes.

In order for it to be useful for helical cone-beam tomography, the methodology described in this section has to be adapted from parallel beams to cone beams. In either case, because of the similarities of the footprint-based projection and back-projection operations (which take up most of the execution time of the reconstruction algorithm) and the operation well-known in computer graphics as "texture mapping," the described projection and back-projection steps can also be implemented by making use of the library functions (say, OpenGL) available on some advanced graphics cards. A very recent paper describing such an approach is [43].

6. SUMMARY

In this article we addressed reconstruction approaches using smooth basis functions, and discussed how these approaches can lead to parallel implementations of algebraic reconstruction techniques for helical cone-beam tomography. Previous works done on row-action methods (such as ART and RAMLA) with blobs have demonstrated that these methods are efficacious for fully 3D image reconstruction in PET and in electron microscopy. We have illustrated in this article that even the most straight-forward implementation of ART gives correspondingly efficacious results when applied to helical cone-beam CT data. However, appropriate generalizations of the block-iterative version of ART have the potential of improving reconstruction quality even more. They also appear to be appropriate for efficient parallel implementations. This is work in progress; our claims await validation by statistically-sound evaluation studies.

REFERENCES

1. M.D. Altschuler, Y. Censor, P.P.B. Eggermont, G.T. Herman, Y.H. Kuo, R.M. Lewitt, M. McKay, H. Tuy, J. Udupa, and M.M. Yau, Demonstration of a software package for the reconstruction of the dynamically changing structure of the human heart from cone-beam X-ray projections, *J. Med. Syst.* **4** (1980) 289–304.
2. G.T. Herman, *Image Reconstruction from Projections: The Fundamentals of Computerized Tomography* (Academic Press, New York, 1980).
3. E.L. Ritman, R.A. Robb, and L.D. Harris, *Imaging Physiological Functions: Experience with the Dynamic Spatial Reconstructor* (Praeger Publishers, New York, 1985).
4. D. Saint-Felix, Y. Trousset, C. Picard, C. Ponchut, R. Romas, A. Rouge, *In vivo* evaluation of a new system for 3D computerized angiography, *Phys. Med. Biol.* **39** (1994) 583–595.
5. R.M. Lewitt, Multidimensional digital image representations using generalized Kaiser-Bessel window functions, *J. Opt. Soc. Amer. A* **7** (1990) 1834–1846.
6. R.M. Lewitt, Alternatives to voxels for image representation in iterative reconstruction algorithms, *Phys. Med. Biol.* **37** (1992) 705–716.
7. S. Matej and R.M. Lewitt, Efficient 3D grids for image reconstruction using spherically-symmetric volume elements, *IEEE Trans. Nucl. Sci.* **42** (1995) 1361–1370.

8. S. Matej and R.M. Lewitt, Practical considerations for 3–D image reconstruction using spherically symmetric volume elements, *IEEE Trans. Med. Imag.* **15** (1996) 68–78.
9. Y. Censor and S.A. Zenios, *Parallel Optimization: Theory, Algorithms, and Applications* (Oxford University Press, New York, 1997).
10. G.T. Herman, Algebraic reconstruction techniques in medical imaging, in: C.T. Leondes, ed., *Medical Imaging Techniques and Applications: Computational Techniques* (Gordon and Breach, Amsterdam, 1998) 1–42.
11. P.E. Kinahan, S. Matej, J.S. Karp, G.T. Herman, and R.M. Lewitt, A comparison of transform and iterative reconstruction techniques for a volume-imaging PET scanner with a large acceptance angle, *IEEE Trans. Nucl. Sci.* **42** (1995) 2281–2287.
12. R. Marabini, G.T. Herman, and J.M. Carazo, 3D reconstruction in electron microscopy using ART with smooth spherically symmetric volume elements (blobs), *Ultramicrosc.* **72** (1998) 53–65.
13. R. Marabini, E. Rietzel, R. Schroeder, G.T. Herman, and J.M. Carazo, Three-dimensional reconstruction from reduced sets of very noisy images acquired following a single-axis tilt schema: Application of a new three-dimensional reconstruction algorithm and objective comparison with weighted backprojection, *J. Struct. Biol.* **120** (1997) 363–371.
14. R. Marabini, G.T. Herman, and J.M. Carazo, Fully Three-Dimensional Reconstruction in Electron Microscopy, in: C. Börgers and F. Natterer, eds., *Computational Radiology and Imaging: Therapy and Diagnostics* (Springer, New York, 1999) 251–281.
15. A.R. De Pierro, A modified expectation maximization algorithm for penalized likelihood estimation in emission tomography, *IEEE Trans. Med. Imag.* **14** (1995) 132–137.
16. D. Butnariu, Y. Censor, and S. Reich, Iterative averaging of entropic projections for solving stochastic convex feasibility problems, *Comput. Optim. Appl.* **8** (1997) 21–39.
17. Y. Censor, A.N. Iusem, and S.A. Zenios, An interior point method with Bregman functions for the variational inequality problem with paramonotone operators, *Math. Programming* **81** (1998) 373–400.
18. Y. Censor and S. Reich, Iterations of paracontractions and firmly nonexpansive operators with applications to feasibility and optimization, *Optimization* **37** (1996) 323–339.
19. Y. Censor and S. Reich, The Dykstra algorithm with Bregman projections, *Comm. Appl. Anal.* **2** (1998) 407–419.
20. L. M. Bregman, Y. Censor, and S. Reich, Dykstra's algorithm as the nonlinear extension of Bregman's optimization method, *Journal of Convex Analysis* **6** (1999) 319–333.
21. Y. Censor, E.D. Chajakis, and S.A. Zenios, Parallelization strategies of a row-action method for multicommodity network flow problems, *Parallel Algorithms Appl.* **6** (1995) 179–205.
22. G.T. Herman, Image reconstruction from projections, *Journal of Real-Time Imaging* **1** (1995) 3–18.
23. I. García, P.M. Ortigosa, L.G. Casado, G.T. Herman, and S. Matej, A parallel implementation of the controlled random search algorithm to optimize an algorithm for reconstruction from projections, in: *Proceedings of Third Workshop on Global Optimization* (The Austrian and the Hungarian Operation Res. Soc., Szeged, Hungary, 1995) 28–32.

24. I. García, P.M. Ortigosa, L.G. Casado, G.T. Herman, and S. Matej, Multidimensional optimization in image reconstruction from projections, in: I.M. Bomze, T. Csendes, T. Horst, and P. Pardalos, eds., *Developments in Global Optimization* (Kluwer Academic Publishers, 1997) 289–299.
25. S. Matej, J.S. Karp, R.M. Lewitt, and A.J. Becher, Performance of the Fourier rebinning algorithm for PET with large acceptance angles, *Phys. Med. Biol.* **43** (1998) 787–795.
26. M. Defrise, A factorization method for the 3D x-ray transform, *Inverse Problems* **11** (1995) 983–994.
27. W. Xia, R.M. Lewitt, and P.R. Edholm, Fourier correction for spatially-variant collimator blurring in SPECT, *IEEE Trans. Med. Imag.* **14** (1995) 100–115.
28. L.A. Shepp and Y. Vardi, Maximum likelihood reconstruction in positron emission tomography, *IEEE Trans. Med. Imag.* **1** (1982) 113–122.
29. J. Browne and A.R. De Pierro, A row-action alternative to the EM algorithm for maximizing likelihoods in emission tomography, *IEEE Trans. Med. Imag.* **15** (1996) 687–699.
30. S. Matej and J.A. Browne, Performance of a fast maximum likelihood algorithm for fully 3D PET reconstruction, In: P. Grangeat and J.-L. Amans, eds., *Three-Dimensional Image Reconstruction in Radiology and Nuclear Medicine* (Kluwer Academic Publishers, 1996) 297–316.
31. F. Jacobs, S. Matej, R.M. Lewitt and I. Lemahieu, A comparative study of 2D reconstruction algorithms using pixels and opmized blobs applied to Fourier rebinned 3D data, In: *Proceedings of the 1999 International Meeting on Fully Three-Dimensional Image Reconstruction in Radiology and Nuclear Medicine* (Egmond aan Zee, The Netherlands, 1999) 43–46.
32. H.M. Hudson and R.S. Larkin, Accelerated image reconstruction using ordered subsets of projection data, *IEEE Trans. Med. Imaging* **13** (1994) 601–609.
33. T. Obi, S. Matej, R.M. Lewitt and G.T. Herman, 2.5D simultaneous multislice reconstruction by series expansion methods from Fourier-rebinned PET data, *IEEE Trans. Med. Imaging* **19** (2000) 474–484.
34. E. Daube-Witherspoon, S. Matej, J.S. Karp, and R.M. Lewitt, Application of the 3D row action maximum likelihood algorithm to clinical PET imaging, In: *1999 Nucl. Sci. Symp. Med. Imag. Conf. CDROM* (IEEE, Seatle, WA, 2000) M12-8.
35. M. Remy-Jardin and J. Remy, Spiral CT angiography of the pulmonary circulation, *Radiology* **212** (1999) 615–636.
36. B.M. Carvalho, C.J. Gau, G.T. Herman, and T.Y. Kong, Algorithms for fuzzy segmentation, *Pattern Anal. Appl.* **2** (1999) 73–81.
37. A. Watt, *3D Computer Graphics, 3rd Edition* (Addison–Wesley, Reading, MA, 2000).
38. K. Mueller, R. Yagel, and J.J. Wheller, Anti-aliased three-dimensional cone-beam reconstruction of low-contrast objects with algebraic methods, *IEEE Trans. Med. Imag.* **18** (1999) 519–537.
39. H. Turbell and P.-E. Danielsson, Helical cone-beam tomography, *Internat. J. Imag. Systems Tech.* **11** (2000) 91–100.
40. P.P.B. Eggermont, G.T. Herman, and A. Lent, Iterative algorithms for large partitioned linear systems, with applications to image reconstruction, *Linear Algebra and*

its Applications **40** (1981) 37–67.
41. S. Matej, G.T. Herman, T.K. Narayan, S.S. Furuie, R.M. Lewitt, and P.E. Kinahan, Evaluation of task-oriented performance of several fully 3D PET reconstruction algorithms, *Phys. Med. Biol.* **39** (1994) 355–367.
42. C. Jacobson, *Fourier methods in 3D-reconstruction from cone-beam data*, PhD Thesis, Department of Electrical Engineering, Linköping University, 1996.
43. K. Mueller and R. Yagel, Rapid 3-D cone-beam reconstruction with the simultaneous algebraic reconstruction technique (SART) using 2-D texture mapping hardware, *IEEE Trans. Med. Imag.* **19** (2000) 1227–1237.
44. R. Aharoni and Y. Censor, Block-iterative projection methods for parallel computation of solutions to convex feasibility problems, *Linear Algebra Appl.* **120** (1989) 165–175.
45. B.E. Oppenheim, Reconstruction tomography from incomplete projections, In: M.M. Ter-Pogossian et al., eds., *Reconstruction Tomography in Diagnostic Radiology and Nuclear Medicine* (University Park Press, Baltimore, 1977) 155–183.
46. A.H. Andersen and A.C. Kak, Simultaneous algebraic reconstruction technique (SART): A superior implementation of the ART algorithm, *Ultrason. Imag.* **6** (1984) 81–94.
47. Y. Censor, D. Gordon, and R. Gordon, Component averaging: An efficient iterative parallel algorithm for large and sparse unstructured problems, *Parallel Computing*, to appear.
48. Y. Censor, T. Elfving, and G.T. Herman, Averaging strings of sequential iterations for convex feasibility problems, *In this volume.*
49. M.R. Civanlar and H.J. Trussel, Digital signal restoration using fuzzy sets, *IEEE Trans. Acoust. Speech Signal Process.* **34** (1986) 919–936.
50. D. Lysczyk and J. Shamir, Signal processing under uncertain conditions by parallel projections onto fuzzy sets, *J. Opt. Soc. Amer. A* **16** (1999) 1602–1611.
51. J.A. Browne and G.T. Herman, Computerized evaluation of image reconstruction algorithms, *Internat. J. Imag. Systems Tech.* **7** (1996) 256–267.
52. M.T. Chan, G.T. Herman, and E. Levitan, A Bayesian approach to PET reconstruction using image-modeling Gibbs priors: Implementation and comparison, *IEEE Trans. Nucl. Sci.* **44** (1997) 1347–1354.
53. M.T. Chan, G.T. Herman, and E. Levitan, Bayesian image reconstruction using image-modeling Gibbs priors, *Internat. J. Imag. Systems Tech.* **9** (1998) 85–98.
54. K. Mueller, R. Yagel, and J.J. Wheller, Fast implementation of algebraic methods for three-dimensional reconstruction from cone-beam data, *IEEE Trans. Med. Imag.* **18** (1999) 538–548.
55. H. Ray, H. Pfister, D. Silver, and T.A. Cook, Ray casting architectures for volume visualization, *IEEE Trans. Visualization Comput. Graph.* **5** (1999) 210–223.
56. C.R. Coggrave and J.M. Huntley, High-speed surface profilometer based on a spatial light modulator and pipeline image processor, *Opt. Engrg.* **38** (1999) 1573–1581.
57. S. Dumontier, F. Luthon and J.-P. Charras, Real-time DSP implementation for MRF-based video motion detection, *IEEE Trans. Image Process.* **8** (1999) 1341–1347.
58. P.N. Morgan, R.J. Iannuzzelli, F.H. Epstein, and R.S. Balaban, Real-time cardiac MRI using DSP's, *IEEE Trans. Med. Imag.* **18** (1999) 649–653.

59. F. Natterer, *The Mathematics of Computerized Tomography* (Chichester: John Wiley & Sons, Inc., 1986).
60. L. Westover, Interactive volume rendering, in: *Proceedings of the Chapel Hill Workshop on Volume Visualization* (Dept. of Computer Science, Univ. of North Carolina, Chapel Hill, N.C., 1989) 9–16.
61. L. Westover, Footprint evaluation for volume rendering, *Computer Graphics (Proc. of ACM SIGGRAPH'90 Conf.)* **24** (1990) 367–376.
62. D. Laur and P. Hanrahan, Hierarchical splatting: A progressive refinement algorithm for volume rendering, *Computer Graphics (Proc. of ACM SIGGRAPH'91 Conf.)* **25** (1991) 285–288.

Inherently Parallel Algorithms in Feasibility and Optimization and their Applications
D. Butnariu, Y. Censor and S. Reich (Editors)
2001 Elsevier Science B.V.

COMPACT OPERATORS AS PRODUCTS OF PROJECTIONS

Hein S. Hundal[a]

[a]Momentum Investment Services, 1981 North Oak Lane, State College, PA 16803 USA

Products of projections appear in the analysis of many algorithms. Indeed, many algorithms are merely products of projections iterated. For example, the best approximation of a point from the intersection of a set of subspaces can be determined from the following parallel algorithm. Define $x_{n+1} = \frac{1}{k}\sum_{i=0}^{k} P_i$ where P_i is the orthogonal projection onto a subspace C_i. Then $\lim_{n\to\infty}(x_n)$ is the best approximation to x_0 in the intersection of the C_i. This algorithm can be represented as a product of projections in a higher dimensional space. When forming hypotheses about a product of projections, the natural question that arises is "Which linear operators can be factored into a product of projections?" This note demonstrates that many bounded linear operators can be represented as scalar multiples of a product of projections. In the case of compact operators with norm strictly less than one, a constructive method for factoring the operator can often be obtained. Furthermore, all compact operators with norm strictly less than one have a simple extension in a higher dimensional Hilbert space and can be represented as a product of projections. Of course, this implies that EVERY compact operator has an extension that is a scalar times a product of projections.

1. INTRODUCTION

Recently, Oikhberg characterized which linear operators could be represented as products of projections in reference [1]. In this paper, we give a constructive proof that all compact operators have an extension into a higher dimensional space that is the product of five projections and a scalar.

2. PRELIMINARY LEMMAS

Lemma 1 *Inequalities:*

1. $\cos(x) \geq 1 - x^2/2$ *for all* $x \in R$.

2. $(1-x)^n \geq 1 - nx$ *for all* $x \leq 2$ *and integers* $n \geq 0$. *(Bernoulli's Inequality)*

Proof:
1) The proof of part a) follows from Taylor's theorem.
2) Bernoulli's inequality is well-known. □

Lemma 2

1. *For all integers* $n \geq 1$, $\cos^n\left(\frac{\pi}{2n}\right) \geq 1 - \frac{\pi^2}{8n}$.

2. $\lim_{n\to\infty} \cos^n(\frac{\pi}{2n}) = 1$.

Proof: By Lemma 1, for $n \geq 1$,

$$1 \geq \cos^n(\frac{\pi}{2n}) \geq \left(1 - \frac{\pi^2}{8n^2}\right)^n \geq 1 - \frac{\pi^2}{8n}.$$

Taking the limits of the right and left hand sides forces $\lim_{n\to\infty} \cos^n(\frac{\pi}{2n})$ to be 1.

□

Remark 1

For the remainder of the paper we will use the following notations:

$$\prod_{i=n}^{1} Q_i = Q_n Q_{n-1} Q_{n-2} \cdots Q_1$$

and for any vector x,

$$[x] := \text{span}(x).$$

Lemma 3 *Let* $\langle u, v\rangle = 0$ *and* $\|u\| > \|v\| > 0$. *Then there exists a positive integer*

$$n < \frac{1}{\left(1 - \frac{\|v\|}{\|u\|}\right)} \frac{\pi^2}{8} + 1 \tag{1}$$

and n *norm one vectors* $x_1, x_2, \ldots, x_n$ *in* $\text{span}(\{u, v\})$ *such that*

$$v = \left(\prod_{i=n}^{1} P_{[x_i]}\right) u.$$

Note: This lemma is very similar to Lemma 2 in reference [1]. In reference [1], the vectors are not necessarily parallel and the bound on n is $c/(1 - \|v\| / \|u\|)$.

Proof: Let $c_i := \cos^i\left(\frac{\pi}{2i}\right)$. Then $c_1 = 0$ and $\lim_{i\to\infty} c_i = 1$ by Lemma 2. So there exists an integer $I \geq 1$ such that

$$c_I < \frac{\|v\|}{\|u\|} \leq c_{I+1}. \tag{2}$$

Let $n = I + 1$. We can bound the value of n with the above inequality. By the definition of c,

$$\cos^I\left(\frac{\pi}{2I}\right) < \frac{\|v\|}{\|u\|}.$$

Applying Lemma 2.1 and some algebra yields

$$\begin{aligned}
1-\frac{\pi^2}{8I} &< \frac{\|v\|}{\|u\|}\\
1-\frac{\|v\|}{\|u\|} &< \frac{\pi^2}{8I}\\
\frac{1}{1-\frac{\|v\|}{\|u\|}} &> \frac{8I}{\pi^2}\\
\frac{1}{1-\frac{\|v\|}{\|u\|}}\frac{\pi^2}{8} &> I.
\end{aligned}$$

But $I=n-1$, so

$$\frac{1}{1-\frac{\|v\|}{\|u\|}}\frac{\pi^2}{8}+1>n.$$

Now we define an angle, $\theta(z)$, and a vector valued function $f(z,i)$ in terms of I. For a key value of z, $f(z,i)$ will be the vector x_i and $\theta(z)$ will be the angle between the x_i vectors. Specifically, for any $z\in[0,1]$ and any integer i, let

$$f(z,i):=\frac{u}{\|u\|}\cos(i\,\theta(z))+\frac{v}{\|v\|}\sin(i\,\theta(z))$$

where

$$\theta(z):=\frac{\pi}{2(I+z)}.$$

The vectors $f(z,i)$ have the following three properties:

1. $f(z,i)\in\text{span}(u,v)$,
2. $\langle f(z,i),f(z,i+1)\rangle=\cos(\theta(z))$, and
3. $\|f(z,i)\|=1$.

Property 1 is obvious. For property 2, use the orthogonality of u and v in the following expansion,

$$\begin{aligned}
\langle f(z,i),f(z,i+1)\rangle &= \Big\langle \frac{u}{\|u\|}\cos(i\,\theta(z))+\frac{v}{\|v\|}\sin(i\,\theta(z))\,,\\
&\qquad \frac{u}{\|u\|}\cos((i+1)\theta(z))+\frac{v}{\|v\|}\sin((i+1)\theta(z))\Big\rangle\\
&= \cos(i\,\theta(z))\cos((i+1)\theta(z))+\\
&\qquad \sin(i\,\theta(z))\sin((i+1)\theta(z))\\
&= \cos(i\,\theta(z)-(i+1)\theta(z))\\
&= \cos(\theta(z)).
\end{aligned}$$

Property 3 also follows from orthogonality of u and v.

$$\begin{aligned}\|f(z,i)\|^2 &= \left\|\frac{u}{\|u\|}\cos(i\,\theta(z))\right\|^2 + \left\|\frac{v}{\|v\|}\sin(i\,\theta(z))\right\|^2 \\ &= \cos(i\,\theta(z))^2 + \sin(i\,\theta(z))^2 \\ &= 1.\end{aligned}$$

In the main part of the proof, we will need the well-known formula for projection onto a one dimensional subspace. For any vector w with norm one,

$$P_{[w]}x = \langle x, w\rangle\, w. \tag{3}$$

Applying formula (3) with $x = f(z,i)$, and $w = f(z,i+1)$ yields

$$\begin{aligned}P_{[f(z,i+1)]}f(z,i) &= \langle f(z,i), f(z,i+1)\rangle\, f(z,i+1) &(4)\\ &= \cos(\theta(z))f(z,i+1). &(5)\end{aligned}$$

Using (5) in an induction argument gives the following formula for the product of these projections.

$$\left(\prod_{i=I}^{1} P_{[f(z,i)]}\right) f(z,0) = \left(\prod_{i=I}^{1} P_{[f(z,i)]}\right) \frac{u}{\|u\|} = \cos^I(\theta(z))f(z,I) \tag{6}$$

Finding a key value for z will complete the proof. Let

$$g(z) = \cos^I(\theta(z))\sin(I\;\theta(z)).$$

Then

$$\begin{aligned}g(0) &= \cos^I\left(\frac{\pi}{2I}\right)\sin\left(I\frac{\pi}{2I}\right) = c_I\\ g(1) &= \cos^I\left(\frac{\pi}{2(I+1)}\right)\sin\left(I\frac{\pi}{2(I+1)}\right)\\ &= \cos^I\left(\frac{\pi}{2(I+1)}\right)\cos\left(\frac{\pi}{2} - I\frac{\pi}{2(I+1)}\right)\\ &= \cos^I\left(\frac{\pi}{2(I+1)}\right)\cos\left(\frac{\pi}{2(I+1)}\right)\\ &= c_{I+1}.\end{aligned}$$

These values of g, continuity of g, and (2) imply there exists a real number z^* such that

$$g(z^*) = \frac{\|v\|}{\|u\|}. \tag{7}$$

Applying (6), (3), the definitions of f and g, and (7) gives

$$\begin{aligned} P_{[v]}\left(\prod_{i=I}^{1} P_{[f(z^*,i)]}\right) u &= P_{[v]}\left(\cos^I(\theta(z^*)) f(z^*, I)\, \|u\|\right) \\ &= \cos^I(\theta(z^*)) \frac{v}{\|v\|} \left\langle \frac{v}{\|v\|}, f(z^*, I) \right\rangle \|u\| \\ &= \cos^I(\theta(z^*)) \sin(I\,\theta(z^*)) \frac{v}{\|v\|}\, \|u\| \\ &= g(z^*) \frac{v}{\|v\|}\, \|u\| \\ &= v. \end{aligned}$$

So, if we let $x_i = f(z^*, i)$ for $i = 1, \ldots, I$, $x_{I+1} = v/\,\|v\|$, and $n = I + 1$, then the lemma holds. □

For the remainder of the paper assume that X is a Hibert space.

Lemma 4 *Consider two sets of vectors* $\{u_i | i = 1, \ldots, m\} \subset X$ *and* $\{v_i | i = 1, \ldots, m\} \subset X$ *with* $m \leq \infty$ *and* $\alpha \in (0, 1)$. *If*

$$\alpha \|u_i\| \geq \|v_i\|, u_i \perp v_i, \text{ and} \tag{8}$$

$$\operatorname{span}\{u_i, v_i\} \perp \operatorname{span}\{u_j, v_j\} \tag{9}$$

whenever $i \neq j$, *then there exists an integer*

$$p < \left(\frac{\pi^2}{8(1-\alpha)} + 1\right) \tag{10}$$

and $x_{ij} \in X$, $i = 1, \ldots, m$, $j = 1, \ldots, p$ *satisfying*

$$x_{ij} \in \operatorname{span}\{u_i, v_i\}, \quad \|x_{ij}\| = 1 \text{ or } 0, \tag{11}$$

$$v_i = \left(\prod_{j=p}^{1} P_{[x_{ij}]}\right) u_i, \quad \text{and} \tag{12}$$

$$v_i = \left(\prod_{j=p}^{1} P_{M_j}\right) u_i \tag{13}$$

where

$$M_j = \operatorname{span}\{x_{ij} | i = 1, \ldots, m\}. \tag{14}$$

Proof: First we define p and the x_{ij} vectors. For every i, one of two following cases holds.

Case 1: $\|v_i\| = 0$.

In this case let $p_i = 1$ and $x_{i1} = 0$. Note that (11) and

$$v_i = \left(\prod_{j=p_i}^{1} P_{[x_{ij}]}\right) u_i \tag{15}$$

hold.

Case 2: $\|v_i\| > 0$.

By (8) and Lemma 3 (with $u = u_i$ and $v = v_i$) there exists a positive integer

$$p_i < \left(\frac{\pi^2}{8(1 - \frac{\|v_i\|}{\|u_i\|})} + 1\right) \le \left(\frac{\pi^2}{8(1-\alpha)} + 1\right),$$

and vectors $x_{i1}, x_{i2}, \ldots, x_{ip_i}$ satisfying (11), and (15).

So in both cases (11) and (15) are satisfied. Furthermore, if we define $p := \max p_i$, then (10) is satisfied. To obtain (12) we define the remaining vectors

$$x_{ij} := x_{ip_i} \quad \text{for} \quad j = p_i + 1, p_i + 2, \ldots, p. \tag{16}$$

Now (12) follows from (15) and (16). It remains to prove (13).

Let M_j be defined by (14). To prove (13) we prove the more general claim:

$$\left(\prod_{j=k}^{1} P_{M_j}\right) u_i = \left(\prod_{j=k}^{1} P_{[x_{ij}]}\right) u_i \tag{17}$$

for $k = 1, 2, \ldots, p$.

We prove the claim by induction on k. For $k = 1$, (17) reduces to

$$P_{M_1} u_i = P_{[x_{i1}]} u_i$$

where $M_j = \text{span}\{x_{lj} | l = 1, 2, \ldots, m\}$. The vectors in $\{x_{lj} | l = 1, 2, \ldots, m\}$ are orthogonal to one another and have norm 0 or 1 by (11) and (9). Therefore,

$$P_{M_j} = \sum_{l=1}^{m} P_{[x_{lj}]}, \quad \text{and} \tag{18}$$

the vector $x_{lj} \in \text{span}\{u_l, v_l\}$ is orthogonal to $\text{span}\{u_i, v_i\}$ whenever $l \ne i$ by (9). So

$$P_{M_1} u_i = \sum_{l=1}^{m} P_{[x_{l1}]} u_i = P_{[x_{i1}]} u_i.$$

Thus the claim holds when $k = 1$.

Similarly, for $p \ge k > 1$,

$$\begin{aligned}
\left(\prod_{j=k+1}^{1} P_{M_j}\right) u_i &= P_{M_{k+1}} \left(\prod_{j=k}^{1} P_{M_j}\right) u_i \\
&= P_{M_{k+1}} \left(\prod_{j=k}^{1} P_{[x_{ij}]}\right) u_i \quad \text{by the inductive hypothesis} \\
&= \sum_{l=m}^{1} P_{[x_{l,k+1}]} \left(\prod_{j=k}^{1} P_{[x_{ij}]}\right) u_i \quad \text{by (18).}
\end{aligned}$$

However, $x_{ik} \in \operatorname{span}\{u_i, v_i\}$ is orthogonal to $x_{l,k+1} \in \operatorname{span}\{u_l, v_l\}$ whenever $l \neq i$. So,

$$\begin{aligned}\left(\prod_{j=k+1}^{1} P_{M_j}\right) u_i &= P_{[x_{i,k+1}]}\left(\prod_{j=k}^{1} P_{[x_{ij}]}\right) u_i \\ &= \left(\prod_{j=k+1}^{1} P_{[x_{ij}]}\right) u_i.\end{aligned}$$

This proves the claim and the lemma. □

Lemma 5 *Consider an operator $T : X \to X$, $m \leq \infty$, and three orthonormal subsets of X: $Y = \{y_1, \dots, y_m\}$, $Z = \{z_1, \dots, z_m\}$, and $W = \{w_1, \dots, w_m\}$. If*

$$\operatorname{span}(W) \subset Y^{\perp} \cap Z^{\perp} \text{ and} \tag{19}$$

$$Tx = \sum_{i=1}^{m} \langle y_i, x \rangle\, \sigma_i z_i \tag{20}$$

where σ_i and β are real numbers satisfying

$$0 \leq \sigma_i \leq \beta < 1, \tag{21}$$

then there exists an integer

$$p < \left(\frac{\pi^2}{4(1-\sqrt{\beta})} + 2\right)$$

and subspaces $M_0, M_1, \dots, M_p$ such that

$$T = \prod_{j=0}^{p} P_{M_j}. \tag{22}$$

Proof: First we apply Lemma 4 with $\alpha = \sqrt{\beta}$, $u_i = y_i$, and $v_i = \alpha w_i$ to obtain subspaces $M_1, M_2, \dots, M_{p_1}$ such that

$$\alpha w_i = v_i = \left(\prod_{j=p_1}^{1} P_{M_j}\right) u_i = \left(\prod_{j=p_1}^{1} P_{M_j}\right) y_i \tag{23}$$

with

$$p_1 < \left(\frac{\pi^2}{8(1-\alpha)} + 1\right). \tag{24}$$

Again we can apply Lemma 4 with $\alpha = \sqrt{\beta}$, $u_i = \alpha w_i$, and $v_i = \sigma_i z_i$ to obtain additional subspaces $M_{p_1+1}, M_{p_1+2}, \dots, M_p$ where

$$\sigma_i z_i = v_i = \left(\prod_{j=p}^{p_1+1} P_{M_j}\right) \alpha w_i \tag{25}$$

with

$$p \leq \left(\frac{\pi^2}{8(1-\alpha)} + 1\right) + p_1. \tag{26}$$

Let $Q = \left(\prod_{j=p}^{1} P_{M_j}\right)$. By (24) and (26),

$$p \leq \left(\frac{\pi^2}{4(1-\sqrt{\beta})} + 2\right).$$

By (23) and (25), $\sigma_i z_i = Q y_i$. Thus $Tx = Qx$ for all $x \in \operatorname{span}(Y)$. The norm of T is bounded by β so T is continuous. Thus $Tx = Qx$ for all $x \in \overline{\operatorname{span}}(Y)$. Now let $M_0 = \overline{\operatorname{span}}(Y)$. Then $TP_{M_0}x = QP_{M_0}x$ for all $x \in X$. But $TP_{M_0^\perp} = 0$. So

$$Tx = TP_{M_0^\perp}x + TP_{M_0}x = TP_{M_0}x = QP_{M_0}x$$

for all $x \in X$. That proves (22) and the lemma. □

3. THEOREMS

Theorem 1 *If $T : X \to X$ is compact, $\|T\| = \beta < 1$, and*

$$\dim(\ker T \cap \ker T^*) \geq \dim(\operatorname{range}(T)), \tag{27}$$

then there exists an integer

$$p \leq \left(\frac{\pi^2}{4(1-\sqrt{\beta})} + 3\right)$$

and subspaces $M_1, M_2, \ldots, M_p$ such that

$$T = \prod_{j=p}^{1} P_{M_j}.$$

Proof: By the singular value decomposition theorem (see [2] pages 204-207), there exist two orthonormal sets $Y = \{y_i\}_{i=1}^m$, $Z = \{z_i\}_{i=1}^m$, real positive numbers $\sigma_1, \sigma_2, \ldots, \sigma_m$, and an integer m, $\infty \geq m \geq 0$, such that

$$Tx = \sum_{i=1}^{m} \langle y_i, x\rangle \, \sigma_i z_i. \tag{28}$$

(If $m = 0$, we use the convention that the empty sum is zero.) Note that $\|T\| \geq \sup_i \sigma_i$, so $0 < \sigma_i \leq \beta$ for all i. Let

$$\begin{aligned} W &= \ker(T) \cap \ker T^* \\ &= \ker(T) \cap \operatorname{range}(T)^\perp. \end{aligned}$$

Obviously, W is a subspace. It is clear that $\dim(W) \geq \dim(\operatorname{range}(T))$. By (28) and the definitions of y_i, σ_i, and z_i, $\dim(\operatorname{range}(T)) = m$. Thus there exists a set of m orthonormal vectors $\{w_1, w_2, \ldots, w_m\}$ in W. Note that $\overline{\text{span}}(Y) = \ker(T)^{\perp}$ and $\overline{\text{span}}(Z) = \text{range}(T)$ so we can apply Lemma 5 to obtain a positive integer

$$p \leq \frac{\pi^2}{4(1-\sqrt{\beta})} + 2$$

and subspaces $M_0, M_1, ..., M_p$, such that

$$T = \prod_{j=p}^{0} P_{M_j}.$$

A shift of indices completes the proof of the theorem. □

Corollary 1 *If $T : X \to X$ is compact and $\|T\| = \beta < 1$, then there exists and integer*

$$p \leq \left(\frac{\pi^2}{4(1-\sqrt{\beta})} + 3\right) \tag{29}$$

and subspaces of X^2: $M_1, M_2, \ldots, M_p$ such that

$$Q = \prod_{j=p}^{1} P_{M_j}.$$

where

$$Q(x, y) = (Tx, 0)$$

Proof: First we show that $Q^*(x, y) = (T^*x, 0)$. For any four vectors x_1, x_2, y_1, and y_2,

$$\begin{aligned}
\langle Q^*(x_1, y_1) - (T^*x_1, 0), (x_2, y_2)\rangle &= \langle Q^*(x_1, y_1), (x_2, y_2)\rangle \\
&\quad - \langle (T^*x_1, 0), (x_2, y_2)\rangle \\
&= \langle (x_1, y_1), Q(x_2, y_2)\rangle - \langle T^*x_1, x_2\rangle \\
&= \langle (x_1, y_1), (Tx_2, 0)\rangle - \langle x_1, Tx_2\rangle \\
&= \langle x_1, Tx_2\rangle - \langle x_1, Tx_2\rangle \\
&= 0.
\end{aligned}$$

Thus $Q^*(x, y) = (T^*x, 0)$. So $\ker(Q^*) = \ker(T^*) \times X$ and $\ker(Q) \cap \ker(Q^*) = (\ker(T) \cap \ker(T^*)) \times X$. Finally,

$$\dim(\ker(Q) \cap \ker(Q^*)) \geq \dim(X) \geq \dim(\text{range}(T)) = \dim(\text{range}(Q)),$$

so we can apply Theorem 1 to Q to prove the corollary. □

Remark 2

A direct consequence of this lemma is that for every compact operator S, the operator $Q : X^2 \to X^2$ defined by

$$Q(x, y) = (Sx/(100\,\|S\|), 0)$$

is expressible as

$$Q = P_{M_5} P_{M_4} P_{M_3} P_{M_2} P_{M_1}.$$

To see this, set $T = S/(100\,\|S\|)$. Then $\|T\| = 1/100 = \beta$ so $p \leq 5$ by (29).

Corollary 2 *If $T : X \to X$ is finite rank, $\|T\| = \beta < 1$, and X is infinite dimensional, then there exists an integer*

$$p \leq \left(\frac{\pi^2}{4(1-\sqrt{\beta})} + 3\right)$$

and subspaces $M_1, M_2, \ldots, M_p$ such that

$$T = \prod_{j=p}^{1} P_{M_j}.$$

Proof: If T has finite rank, T is compact so we can apply the singular value decomposition theorem to obtain

$$Tx = \sum_{i=1}^{m} \langle y_i, x\rangle \, \sigma_i z_i \tag{30}$$

as in (28) with m equal to the rank of T. Computing the kernels and ranges of T gives

$$\begin{aligned} \ker(T) &= \bigcap_{i=1}^{m} (y_i)^{\perp} \\ \text{range}(T) &= \text{span}_{i=1,\ldots,m} z_i \\ \ker(T^*) &= (\text{range}(T))^{\perp} = \bigcap_{i=1}^{m} (z_i)^{\perp}. \end{aligned}$$

So

$$W = (\ker(T)) \cap (\ker(T^*)) = (\bigcap (y_i)^{\perp}) \cap (\bigcap (z_i)^{\perp})$$

has a finite codimension less than $2m$ thus it has infinite dimension. Consequently, $\dim(W) > \dim(\text{range}(T))$ so we can apply Theorem 1 to complete the proof. □

REFERENCES

1. T. Oikhberg, Products of orthogonal projections, *Proc. of Amer. Math. Soc.* **127** (1999) 3659–3669.
2. N. Young, *Introduction to Hilbert Spaces* (Cambridge Univ. Press, New York 1988).

Inherently Parallel Algorithms in Feasibility and Optimization and their Applications
D. Butnariu, Y. Censor and S. Reich (Editors)

PARALLEL SUBGRADIENT METHODS FOR CONVEX OPTIMIZATION

K. C. Kiwiel[a*] and P. O. Lindberg[b]

[a]Systems Research Institute, Newelska 6, 01–447 Warsaw, Poland

[b]Linköping University, S-58183 Linköping, Sweden

We study subgradient methods for minimizing a sum of convex functions over a closed convex set. To generate a search direction, each iteration employs subgradients of a subset of the objectives evaluated at the current iterate, as well as past subgradients of the remaining objectives. A stepsize is selected via ballstep level controls for estimating the optimal value. We establish global convergence of the method. When applied to Lagrangian relaxation of separable problems, the method allows for almost asynchronous parallel solution of Lagrangian subproblems, updating the iterates as soon as new subgradient information becomes available.

1. INTRODUCTION

We consider subgradient methods for the convex constrained minimization problem

$$f_* := \inf \{ f(x) : x \in S \} \quad \text{with} \quad f := \sum_{i=1}^m f^i, \tag{1}$$

where S is a nonempty closed convex set in the Euclidean space $\mathbb{R}^n$ with inner product $\langle \cdot, \cdot \rangle$ and norm $|\cdot|$, each $f^i : \mathbb{R}^n \to \mathbb{R}$ is a convex function and we can find its value and a subgradient $g_{f^i}(x) \in \partial f^i(x)$ at any $x \in S$. The *optimal set* $S_* := \operatorname{Arg\,min}_S f$ may be empty. We assume that for each $x \in \mathbb{R}^n$ we can find $P_S x := \arg\min_S |x - \cdot|$, its orthogonal projection on S.

An important special case arises in Lagrangian relaxation [1], [3, §5.5.3], [8, Chap. XII], where f is the dual function of a primal separable problem of the form

$$\text{maximize} \quad \sum_{i=1}^m \psi_{0i}(z_i) \quad \text{s.t.} \quad z_i \in Z_i,\ i = 1{:}\, m, \quad \sum_{i=1}^m \psi_{ji}(z_i) \geq 0,\ j = 1{:}\, n, \tag{2}$$

with each $Z_i \subset \mathbb{R}^{m_i}$ compact and $\psi_{ji} : Z_i \to \mathbb{R}$ upper semicontinuous. Then, by viewing x as a Lagrange multiplier vector for the coupling constraints of (2), we obtain a dual problem of the form (1) with $S := \mathbb{R}^n_+$ and

$$f^i(x) := \max \{ \psi_{0i}(z_i) + \langle x, \psi_{\cdot i}(z_i) \rangle : z_i \in Z_i \}, \tag{3}$$

*Research supported by the Polish State Committee for Scientific Research under Grant 8T11A00115, and the Swedish Research Council for Engineering Sciences (TFR).

where $\psi_{\cdot i} := (\psi_{1i}, \ldots, \psi_{ni})$; further, any solution $z_i(x)$ of the Lagrangian subproblem in (3) provides a subgradient $g_{f^i}(x) := \psi_{\cdot i}(z_i(x))$ of f^i at x.

We present an extension of the ballstep subgradient method [12] that finds f_* as follows. Given the kth iterate $x^k \in S$, a *target level* f_{lev}^k that estimates f_*, and for each $i \in I := \{1\colon m\}$, the *linearization* of the partial objective f^i collected at iteration $j_i^k \leq k$:

$$f_j^i(\cdot) := f^i(x^j) + \left\langle g_{f^i}(x^j), \cdot - x^j \right\rangle \leq f^i(\cdot) \quad \text{with} \quad g_{f^i}(x^j) \in \partial f^i(x^j), \quad j = j_i^k, \tag{4}$$

our method uses the *aggregate linearization* of the objective f of (1)

$$f_k := \sum_{i=1}^{m} f_{j_i^k}^i \leq f \tag{5}$$

and its halfspace corresponding to the target level f_{lev}^k

$$H_k := \left\{ x : f_k(x) \leq f_{\text{lev}}^k \right\} \tag{6}$$

to approximate the *level set* of f

$$\mathcal{L}_f(f_{\text{lev}}^k) := \left\{ x : f(x) \leq f_{\text{lev}}^k \right\} \subset H_k = \mathcal{L}_{f_k}(f_{\text{lev}}^k).$$

Then, following the original algorithm of [15], the method generates the next iterate

$$x^{k+1} = P_S\left(x^k + t_k\left[P_{H_k}x^k - x^k\right]\right), \tag{7}$$

where t_k is a *relaxation factor* satisfying

$$t_k \in T := [t_{\min}, t_{\max}] \quad \text{for some fixed} \quad 0 < t_{\min} \leq t_{\max} < 2. \tag{8}$$

The targets are chosen via a ballstep strategy that works in groups of iterations. Within each group, the target f_{lev}^k is fixed, and the method attempts to minimize f over a ball around the best point found so far, shifting the ball and lowering the target when sufficient progress occurs, or shrinking the ball and increasing the target upon discovering that it is too low. To assess progress without evaluating the objective f at each iteration, at iteration k the method only uses the value $f(x^{j_k})$ for some $j_k \leq k$ such that $j_k \to \infty$ as $k \to \infty$; in words, f should be evaluated infinitely often at some selected past iterates, but we allow arbitrarily large delays (i.e., $k - j_k$ needn't be bounded).

We show that $\inf_k f(x^{j_k}) = f_*$ if the indices j_i^k in (5) satisfy a mild quasicyclicity condition, under which each f^i is evaluated at least once (i.e., $j_i^k = k$) in a cycle of iterations, and the length of the cycle may grow to infinity (but not too fast; cf. Def. 3.1).

In effect, when applied to Lagrangian relaxation, our method allows for almost asynchronous parallel solution of the Lagrangian subproblems (3), updating the iterates as soon as new subgradient information becomes available. Our interest is not limited to parallel environments, since the method may also be useful in serial settings. Namely, in many applications (see, e.g., [7]) some of the Lagrangian subproblems (3) are easy to solve, and others are much harder. Then it is natural to use subgradient information from the easier subproblems more frequently than from the harder ones.

In the context of Lagrangian relaxation, subgradient schemes similar to (4)–(8) were considered in [4,6,13], and later in [9,17] under the name *interleaved* subgradient methods. The latter need the solution of a single subproblem per iteration. Relative to classical subgradient methods, they save work and may avoid zigzags because the search direction is changed only partially. Convergence was not established in [4,9,13,17]. The convergence result of [6], which requires $f_{\rm lev}^k \equiv f_*$, $S_* \neq \emptyset$ and almost cyclic updates of j_i^k, is easily recovered in our framework. Good results are reported in [9,17] for serial implementations on job-shop scheduling problems. For a parallel implementation with a heuristic choice of $f_{\rm lev}^k$, [6] gives highly encouraging results for the unit commitment problems from power production planning used in [5]. This raises hopes for our method, since ballstep choices seem to be competitive with heuristic ones [11,12].

Our method differs significantly from incremental subgradient methods [2,10,14,16], [3, §6.3.2], which take steps by using individual subgradients of successive f^is.

The paper is organized as follows. In §2 we present our method. Its convergence is analyzed in §3. Several modifications are given in §§4–6.

Our notation is fairly standard. $B(x,r) = \{y : |y-x| \leq r\}$ is the ball with center x and radius r. $d_C(\cdot) := \inf_{y\in C} |\cdot - y|$ is the distance function of a set $C \subset \mathbb{R}^n$.

2. THE BALLSTEP LEVEL ALGORITHM

For theoretical purposes, it is convenient to regard our constrained problem $f_* := \inf_S f$ (cf. (1)) as the unconstrained problem $f_* = \inf f_S$ with the *extended objective* $f_S := f + \imath_S$, where $\imath_S$ is the *indicator function* of S ($\imath_S(x) = 0$ if $x \in S$, ∞ if $x \notin S$).

We first state the simplest extension of the ballstep subgradient method [12]. Its rules will be commented upon below; see [12] for additional motivations.

Algorithm 2.1

Step 0 (*Initiation*). Select an initial $x^1 \in S$, a level gap $\delta_1 > 0$, ballstep parameters $R > 0$, $\beta \in [0,1)$, and stepsize bounds $t_{\min}$, $t_{\max}$ (cf. (8)). Set $f_{\rm rec}^0 = \infty$, $\rho_1 = 0$. Set the counters $k = l = k(1) = 1$ ($k(l)$ denotes the iteration number of the lth change of $f_{\rm lev}^k$).

Step 1 (*Objective evaluation*). Choose a nonempty set $I^k \subset I$, $I^k = I$ if $k = 1$. For each $i \in I^k$, evaluate $f^i(x^k)$, $g_{f^i}(x^k)$ and set $j_i^k = k$. Set $j_i^k = j_i^{k-1}$ for $i \in I \setminus I^k$. Let j_k be the largest $j \leq k$ for which $f(x^j)$ is known. If $f(x^{j_k}) < f_{\rm rec}^{k-1}$, set $f_{\rm rec}^k = f(x^{j_k})$ and $x_{\rm rec}^k = x^{j_k}$, else set $f_{\rm rec}^k = f_{\rm rec}^{k-1}$ and $x_{\rm rec}^k = x_{\rm rec}^{k-1}$.

Step 2 (*Stopping criterion*). If $j_k = k$, $f_k(x^k) = f(x^k)$ and $\nabla f_k = 0$, terminate ($x^k \in S_*$).

Step 3 (*Sufficient descent detection*). If $f_{\rm rec}^k \leq f_{\rm rec}^{k(l)} - \frac{1}{2}\delta_l$, start the next group: set $k(l+1) = k$, $\delta_{l+1} = \delta_l$, $\rho_k := 0$, replace x^k by $x_{\rm rec}^k$ and increase the group counter l by 1.

Step 4 (*Projections*). Set the level $f_{\rm lev}^k = f_{\rm rec}^{k(l)} - \delta_l$. Choose a relaxation factor $t_k \in T$ (cf. (8)). Set $x^{k+1/2} = x^k + t_k(P_{H_k}x^k - x^k)$, $\check\rho_k = t_k(2-t_k)d_{H_k}^2(x^k)$, $\rho_{k+1/2} = \rho_k + \check\rho_k$, $x^{k+1} = P_S x^{k+1/2}$, $\check\rho_{k+1/2} = |x^{k+1} - x^{k+1/2}|^2$, $\rho_{k+1} = \rho_{k+1/2} + \check\rho_{k+1/2}$.

Step 5 (*Target infeasibility detection*). Set the ball radius $R_l = R(\delta_l/\delta_1)^\beta$. If

$$(R_l - |x^{k+1} - x^{k(l)}|)^2 > R_l^2 - \rho_{k+1}, \tag{9}$$

i.e., the target level is too low (see below), then go to Step 6; otherwise, go to Step 7.

Step 6 (*Level increase*). Start the next group: set $k(l+1) = k$, $\delta_{l+1} = \frac{1}{2}\delta_l$, $\rho_k := 0$, replace x^k by $x_{\rm rec}^k$, increase the group counter l by 1 and go to Step 4.

Step 7. Increase k by 1 and go to Step 1.

Remarks 2.2 (i) At Step 0, since $\min_{B(x^1,R)} f_S \geq f(x^1) - R|\nabla f_1|$ from $f \geq f_1$ (cf. (5)), it is reasonable to set $\delta_1 = R|\nabla f_1|$ when R estimates $d_{S_*}(x^1)$. (Conversely, one may use $R = \delta_1/|\nabla f_1|$ when δ_1 estimates $f(x^1) - f_*$.) The values $t_{\min} = t_{\max} = 1$ and $\beta = \frac{1}{2}$ seem to work well in practice [12,11].

(ii) Step 1 shouldn't ignore any f^i forever, i.e., each i should be in I^k infinitely often; not much more is required for convergence (cf. Def. 3.1). For generating suitable target levels, f should be evaluated at infinitely many iterates, i.e., we should have $j_k \to \infty$ as $k \to \infty$. The *record* point x^k_{rec} has the best f-value f^k_{rec} found so far.

(iii) Let us split the iterations into *groups* $K_l := \{k(l){:}\, k(l+1) - 1\}$, $l \geq 1$. Within group l, starting from the record point $x^k = x^{k(l)}_{\text{rec}}$ with $f^{k(l)}_{\text{rec}} = f(x^{k(l)}_{\text{rec}})$, the method aims at the *frozen* target $f^k_{\text{lev}} = f^{k(l)}_{\text{rec}} - \delta_l$. If at least half of the *desired objective reduction* δ_l is achieved at Step 3, group $l+1$ starts with the same $\delta_{l+1} = \delta_l$, but $f^{k(l+1)}_{\text{rec}} \leq f^{k(l)}_{\text{rec}} - \frac{1}{2}\delta_l$ with $x^{k(l+1)} = x^{k(l+1)}_{\text{rec}}$. Alternatively, it starts at Step 6 with $\delta_{l+1} = \frac{1}{2}\delta_l$, $x^{k(l+1)} = x^{k(l+1)}_{\text{rec}}$. Hence (cf. Steps 0 and 1) we have the following basic relations: $\delta_{l+1} \leq \delta_l$, $x^{k(l)} = x^{k(l)}_{\text{rec}} \in S$ and $f^{k(l)}_{\text{rec}} = f(x^{k(l)})$ for all l.

(iv) At Step 4, in view of (4)–(6), we may use the explicit relations

$$x^{k+1/2} = x^k - t_k[f_k(x^k) - f^k_{\text{lev}}]_+ \nabla f_k/|\nabla f_k|^2,$$

where $[\cdot]_+ := \max\{\cdot, 0\}$ and $\nabla f_k = \sum_i g_{f^i}(x^{j^k_i})$, and

$$d_{H_k}(x^k) = [f_k(x^k) - f^k_{\text{lev}}]_+/|\nabla f_k|. \tag{10}$$

The *Fejér* quantity ρ_k is updated for the infeasibility test (9).

(v) At Step 5, the *ball radius* $R_l = R(\delta_l/\delta_1)^\beta \leq R$ is nonincreasing; $R_l \equiv R$ if $\beta = 0$. Ideally, R_l should be of order $d_{S_*}(x^{k(l)})$, and hence shrink as $x^{k(l)}$ approaches S_*.

(vi) Algorithm 2.1 is a *ballstep* method, which in group l attempts to minimize f_S approximately over the ball $B(x^{k(l)}, R_l)$, shifting it when sufficient progress occurs, or increasing the target level otherwise. We show in §3 that (9) implies $f^k_{\text{lev}} = f^{k(l)}_{\text{rec}} - \delta_l < \min_{B(x^{k(l)},R_l)} f_S$, i.e., the target is too low, in which case δ_l is halved at Step 6, f^k_{lev} is increased at Step 4 and x^{k+1} is recomputed. Note that l increases at Step 6, but k does not, so relations like $f^k_{\text{lev}} = f^{k(l)}_{\text{rec}} - \delta_l$ always involve the *current values* of the counters k and l at Step 4.

(vii) At Step 4, $x^{k+1} = x^{k+1/2} = x^k$ if $f_k(x^k) \leq f^k_{\text{lev}}$. A series of such null steps terminates once $f(x^k)$ becomes known. Indeed, if for some $k' \leq k$, $x^k = x^{k-1} = \ldots = x^{k'}$ and $\bigcup_{j=k'}^k I^j = I$, then $f(x^k) = f_k(x^k)$, i.e., $j_k = k$, and hence $f_k(x^k) > f^k_{\text{lev}}$ at Step 4.

3. CONVERGENCE

Throughout this section, we assume that the method does not stop, that $j_k \to \infty$ at Step 1, and that the algorithm is quasicyclic in the following sense.

Definition 3.1 The algorithm is *quasicyclic* if there is an increasing sequence of integers $\{\tau_p\}_{p=1}^\infty$ such that $\tau_1 = 1$, $\sum_{p=1}^\infty (\tau_{p+1} - \tau_p)^{-1} = \infty$ and $I = \bigcup_{k=\tau_p}^{\tau_{p+1}-1} I^k$ for all p.

Our aim is to prove that $\inf_l f(x^{k(l)}) = f_*$. To save space, we only show how to extend the analysis of [12, §3] to the present setting.

Since the proofs of the following results of [12] remain valid, they are omitted.

Lemma 3.2 (cf. [12, Lems. 3.1 and 3.3]) (i) *At Step* 4, *we have*

$$\check{\rho}_k = t_k(2-t_k)d_{H_k}^2(x^k) \le |x^k - \bar{x}|^2 - |x^{k+1/2} - \bar{x}|^2 \qquad \forall \bar{x} \in \mathcal{L}_{f_S}(f_{\text{lev}}^k), \tag{11a}$$

$$\check{\rho}_{k+1/2} = |x^{k+1} - x^{k+1/2}|^2 \le |x^{k+1/2} - \bar{x}|^2 - |x^{k+1} - \bar{x}|^2 \qquad \forall \bar{x} \in S. \tag{11b}$$

(ii) *If* (9) *holds at Step* 5 *then* $f_{\text{rec}}^{k(l)} - \delta_l < \min_{B(x^{k(l)},R_l)} f_S$.

(iii) *If* $f_{\text{rec}}^{k(l)} - \delta_l \le \min_{B(x^{k(l)},R_l)} f_S$ *(e.g.,* (9) *holds), then for* $\alpha := R/\delta_1^\beta$ *we have*

$$f_S(x^{k(l)}) - f_S(\cdot) \le \delta_l \max\{|\cdot - x^{k(l)}|/R_l, 1\} = \max\{|\cdot - x^{k(l)}|\delta_l^{1-\beta}/\alpha, \delta_l\}. \tag{12}$$

As observed in [12, Rems. 3.4], the optimality estimate (12) suggests that Step 2 may employ the stopping criterion $\delta_l \le \epsilon_{\text{opt}}(1 + |f_{\text{rec}}^k|)$ with $\epsilon_{\text{opt}} = 10^{-d}$ to obtain a relative accuracy of about d digits in the optimal value.

To show that $l \to \infty$, we must provide a replacement for [12, Lem. 3.5]. First, it is instructive to see what would happen if l stayed fixed.

Lemma 3.3 *If l stays fixed then* $x^k \to \bar{x} \in \mathcal{L}_{f_S}(f_{\text{lev}}^{k(l)})$ *and* $f(x^k) \to f(\bar{x}) \le f_{\text{lev}}^{k(l)}$.

Proof. Since l is fixed, (9) fails for $k \ge k(l)$, and thus $|x^{k+1} - x^{k(l)}| \le 2R_l$ and $\rho_{k+1} \le R_l^2$. Hence the sequence $\{x^k\}$ is bounded, whereas the facts $\rho_{k(l)} = 0$ (cf. Steps 0, 3, 6) and

$$\rho_{k+1} \ge \rho_k + t_k(2-t_k)d_{H_k}^2(x^k)$$

(cf. Step 4) with $t_k(2-t_k) \ge t_{\min}(2-t_{\max}) > 0$ (cf. (8)) yield

$$t_{\min}(2-t_{\max}) \sum_{k=k(l)}^{\infty} d_{H_k}^2(x^k) \le \sum_{k=k(l)}^{\infty} t_k(2-t_k)d_{H_k}^2(x^k) \le \lim_{k\to\infty} \rho_k \le R_l^2. \tag{13}$$

Since $\{x^k\}$ is bounded and (cf. (4)–(5)) $\nabla f_k = \sum_i g_{f^i}(x^{j_i^k})$, where the subgradient mappings $g_{f^i}(\cdot) \in \partial f^i(\cdot)$ are locally bounded, there is $G < \infty$ such that $|\nabla f_k| \le G$ for all k. Thus $d_{H_k}(x^k) \ge [f_k(x^k) - f_{\text{lev}}^k]_+/G$ (cf. (10)) with $f_{\text{lev}}^k = f_{\text{lev}}^{k(l)}$ and (13) give

$$\overline{\lim_{k\to\infty}} f_k(x^k) \le f_{\text{lev}}^{k(l)}. \tag{14}$$

Next, for each p, the Cauchy-Schwarz inequality implies

$$(\tau_{p+1} - \tau_p)^{-1} \left(\sum_{k=\tau_p}^{\tau_{p+1}-1} d_{H_k}(x^k) \right)^2 \le \sum_{k=\tau_p}^{\tau_{p+1}-1} d_{H_k}^2(x^k). \tag{15}$$

Summing and invoking Definition 3.1 and (13) yield the existence of a subsequence $\{\tau_{p'}\}$ such that $\sum_{k=\tau_{p'}}^{\tau_{p'+1}-1} d_{H_k}(x^k) \to 0$. Hence, since $|x^{k+1} - x^k| \le |x^{k+1/2} - x^k|$ by (11b) with $\bar{x} = x^k$, where $|x^{k+1/2} - x^k| = t_k d_{H_k}(x^k) \le 2d_{H_k}(x^k)$ (cf. Step 4), we have

$$\sum_{k=\tau_{p'}}^{\tau_{p'+1}-1} |x^{k+1} - x^k| \to 0. \tag{16}$$

Extracting a subsequence if necessary, assume $x^{\tau_{p'}} \to \bar{x}$ ($\bar{x} \in S$, since $\{x^k\} \subset S$ and S is closed). Let $K := \{\tau_{p'+1}\}$. By (16) and the triangle inequality, $x^k \xrightarrow{K} \bar{x}$. Similarly, since $I = \bigcup_{k=\tau_{p'}}^{\tau_{p'+1}-1} I^k$ by Def. 3.1, for each $i \in I$ we have (cf. Step 1) $x^{j_i^k} \xrightarrow{K} \bar{x}$, using $j_i^k \geq \tau_{p'}$ for $k \in K$ in (16); thus, since f^i is continuous and g_{f^i} is locally bounded in (4), we obtain

$$f^i_{j_i^k}(x^k) = f^i(x^{j_i^k}) + \left\langle g_{f^i}(x^{j_i^k}), x^k - x^{j_i^k} \right\rangle \xrightarrow{K} f^i(\bar{x}).$$

Hence by (5),

$$f_k(x^k) = \sum_{i \in I} f^i_{j_i^k}(x^k) \xrightarrow{K} \sum_{i \in I} f^i(\bar{x}) = f(\bar{x}).$$

Combining this with $\overline{\lim}_k f_k(x^k) \leq f_{\text{lev}}^{k(l)}$ (cf. (14)) gives $f(\bar{x}) \leq f_{\text{lev}}^{k(l)}$, i.e., $\bar{x} \in \mathcal{L}_{f_S}(f_{\text{lev}}^{k(l)})$. By (11), $|x^{k+1} - \bar{x}| \leq |x^k - \bar{x}|$ for all $k \geq k(l)$, so $x^k \xrightarrow{K} \bar{x}$ yields $x^k \to \bar{x}$ and $f(x^k) \to f(\bar{x})$ by continuity of f. □

We may now show that infinitely many groups are generated.

Lemma 3.4 *We have $l \to \infty$.*

Proof. For contradiction, suppose l stays fixed. By Lemma 3.3, $\lim_k f(x^k) \leq f_{\text{lev}}^{k(l)} = f_{\text{rec}}^{k(l)} - \delta_l$ (cf. Step 4). Hence our assumption that $j_k \to \infty$ and the rules of Step 1 yield $\lim_k f(x^{j_k}) = \lim_k f(x^k)$ and $\lim_k f_{\text{rec}}^k \leq f_{\text{rec}}^{k(l)} - \delta_l$. Thus eventually $f_{\text{rec}}^k \leq f_{\text{rec}}^{k(l)} - \frac{1}{2}\delta_l$ (since $\delta_l > 0$) and Step 3 must increase l, a contradiction. □

We need another simple property of $\{\delta_l\}$ and the quantity $f_\infty := \inf_l f(x^{k(l)})$.

Lemma 3.5 *Either $f_\infty = -\infty$ or $\delta_l \downarrow 0$.*

Proof. Let $\delta_\infty = \lim_{l\to\infty} \delta_l$ ($l \to \infty$ by Lem. 3.4). If $\delta_\infty > 0$ then (cf. Steps 3 and 6) $f(x^{k(l+1)}) \leq f(x^{k(l)}) - \frac{1}{2}\delta_l$ with $\delta_l = \delta_\infty > 0$ for all large l yield $f_\infty = -\infty$. □

We may now prove the main convergence result of this section.

Theorem 3.6 *We have $f_\infty = f_*$, i.e., $f(x^{k(l)}) \downarrow \inf_S f$.*

Proof. Use Lemmas 3.2(iii) and 3.5 in the proof of [12, Thm 3.7]. □

Corollary 3.7 *If $S_* \neq \emptyset$ is bounded (i.e., f_S is coercive) then $\{x^{k(l)}\}$ is bounded and $d_{S_*}(x^{k(l)}) \to 0$. Conversely, if $\{x^{k(l)}\}$ is bounded then $d_{S_*}(x^{k(l)}) \to 0$.*

Proof. Use Theorem 3.6 in the proof of [12, Cor. 3.8]. □

Remarks 3.8 (i) The proof of Theorem 3.6 only requires that $\{R_l\} \subset (0, \infty)$ and $\delta_l/R_l \to 0$ in (12) if $\delta_l \downarrow 0$; cf. [12, Rem. 3.9(i)].

(ii) Our results only require that each f^i be finite convex on S and $g_{f^i}(\cdot) \in \partial f^i(\cdot)$ be locally bounded on S; then f^i is locally Lipschitz continuous on S [12, Rem. 3.9(ii)].

(iii) Our results extend easily to the "true" ballstep version of [12, Lem. 3.10], which additionally projects x^{k+1} on $B(x^{k(l)}, R_l)$ to ensure $\{x^k\}_{k \in K_l} \subset B(x^{k(l)}, R_l)$; this helps in practice [12, Rem. 3.11(iii)].

4. USING A FIXED TARGET LEVEL

We now consider a simplified version of Algorithm 2.1 that employs a fixed target level. The following theorem extends a classical result of B.T. Polyak [15].

Theorem 4.1 *Consider the subgradient iteration described by* (4)–(8) *with a fixed target level* $f_{\rm lev}^k \equiv f_{\rm lev}^1$, $x^1 \in S$ *and* j_i^k *chosen as in Step* 1 *of Algorithm* 2.1 *in a quasicyclic way. Suppose that either* $f_{\rm lev}^1 > f_*$, *or* $f_{\rm lev}^1 = f_*$ *and* $S_* \neq \emptyset$. *Then* $x^k \to \bar{x} \in \mathcal{L}_{f_S}(f_{\rm lev}^1)$ *and* $f(x^k) \to f(\bar{x}) \leq f_{\rm lev}^1$, *where* $\bar{x} \in S_*$ *if* $f_{\rm lev}^1 = f_*$.

Proof. If $f(x^1) \leq f_{\rm lev}^1$ then (5)–(7) yield $x^1 = x^k \in H_k$ for all k, and we may take $\bar{x} = x^1$. So suppose $f(x^1) > f_{\rm lev}^1$. Let $\delta_1 := f(x^1) - f_{\rm lev}^1$ and $R := |x^1 - \tilde{x}|$ for some $\tilde{x} \in S$ such that $f(\tilde{x}) \leq f_{\rm lev}^1$. Then the iteration corresponds to Algorithm 2.1 with Steps 2–3 omitted and $j_k \equiv 1$. Further, l stays fixed at 1. Indeed, if $l = 1$ at Step 5 then $R_l = R$, $f_{\rm rec}^{k(l)} - \delta_l = f_{\rm lev}^1 \geq f(\tilde{x})$ and $f(\tilde{x}) \geq \min_{B(x^{k(l)},R_l)} f_S$, so (9) can't hold due to Lemma 3.2(ii). Therefore, our assertion follows from Lemma 3.3. □

5. ACCELERATIONS

As in [12, §9], we may accelerate Algorithm 2.1 by replacing the linearization f_k with a more accurate model ϕ_k of the essential objective f_S from the family Φ_μ^k defined below.

Definition 5.1 Given $\mu \in (0,1]$, let $\Phi_\mu^k := \{\phi \in \Phi : d_{\mathcal{L}(\phi, f_{\rm lev}^k)}(x^k) \geq \mu d_{H_k}(x^k)\}$, where $\Phi := \{\phi : \mathbb{R}^n \to (-\infty, \infty] : \phi$ is closed proper convex and $\phi \leq f_S\}$, $\mathcal{L}(\phi, \cdot) := \mathcal{L}_\phi(\cdot)$.

A few remarks on this definition are in order.

Remarks 5.2 (i) If $\phi_k \in \Phi$ and $\phi_k \geq f_k$ then $\phi_k \in \Phi_1^k$.

(ii) Let $\hat{f}^k := \max_{j \in J^k} f_j$, where $k \in J^k \subset \{1\colon k\}$. Then $\hat{f}^k \in \Phi_1^k$.

(iii) Note that $\phi_k \in \Phi$ if ϕ_k is the maximum of several accumulated linearizations $\{f_j\}_{j=1}^k$, or their convex combinations, possibly augmented with $\imath_S$ or its affine minorants.

Fixing $\mu \in (0,1]$, suppose at Step 4 of Algorithm 2.1 we choose an objective model $\phi_k \in \Phi_\mu^k$ and replace the halfspace H_k by the level set $\mathcal{L}_k := \mathcal{L}_{\phi_k}(f_{\rm lev}^k)$ for setting

$$x^{k+1/2} = x^k + t_k(P_{\mathcal{L}_k}x^k - x^k) \quad \text{and} \quad \check{\rho}_k = t_k(2 - t_k)d_{\mathcal{L}_k}^2(x^k).$$

We now comment on the properties of this modification.

Remarks 5.3 (i) The results of §§3–4 extend easily to this modification. First, $\mathcal{L}_k$ replaces H_k in Lem. 3.2 [12, Rem. 7.3(i)]. Second, in the proof of Lem. 3.3, we may replace H_k by $\mathcal{L}_k$ in (13) and (15), and use $d_{\mathcal{L}_k}(x^k) \geq \mu d_{H_k}(x^k)$ and $|x^{k+1/2} - x^k| = t_k d_{\mathcal{L}_k}(x^k)$.

(ii) If $\inf \phi_k \geq f_{\rm lev}^k$ (e.g., $\mathcal{L}_k = \emptyset$) then $f_{\rm lev}^k \leq f_*$, so Algorithm 2.1 may go to Step 6 to set $\delta_{l+1} = \min\{\frac{1}{2}\delta_l, f_{\rm rec}^k - f_{\rm lev}^k\}$ (possibly with $\inf \phi_k$ replacing $f_{\rm lev}^k$).

(iii) Simple but useful choices of ϕ_k include the aggregate subgradient strategy [12, Ex. 7.4(v)] and the conjugate subgradient and average direction strategies of [12, Lem. 7.5 and Rem. 7.6] (with $f(x^k)$ replaced by $f_k(x^k)$ and g^k by ∇f_k), possibly combined with subgradient reduction [11, Rem. 8.5].

6. USING DIVERGENT SERIES STEPSIZES

Consider the subgradient iteration with *divergent series stepsizes*

$$x^{k+1} = P_S\left(x^k - \nu_k \nabla f_k\right) \quad \text{with } \nu_k > 0,\ \nu_k \to 0,\ \sum_{k=1}^{\infty} \nu_k = \infty, \tag{17}$$

where the aggregate linearization f_k is generated via (4)–(5) with j_i^k chosen as in Step 1 of Algorithm 2.1. We assume that the algorithm is *almost cyclic*, i.e., there is a positive integer τ such that $I = I^k \cup \cdots \cup I^{k+\tau-1}$ for each k.

Theorem 6.1 *In addition to the above assumptions on the subgradient iteration* (17), *suppose for some $C < \infty$ we have $|g_{f^i}(x^k)| \le C$ for all i and k. Then $\underline{\lim}_k f(x^k) = f_*$.*

Proof. The main idea is to show that the iteration (17) may be viewed as an approximate subgradient method for problem (1), i.e., that for a suitable error tolerance ϵ_k,

$$\nabla f_k \in \partial_{\epsilon_k} f(x^k) := \left\{ g : f(\cdot) \ge f(x^k) - \epsilon_k + \left\langle g, \cdot - x^k \right\rangle \right\}. \tag{18}$$

For each $i \in I$, the subgradient inequality (4) yields

$$f^i(\cdot) \ge f^i_{j_i^k}(\cdot) = f^i(x^k) - \epsilon_i^k + \left\langle g_{f^i}(x^{j_i^k}), \cdot - x^k \right\rangle \tag{19a}$$

with

$$\begin{aligned} \epsilon_i^k &:= f^i(x^k) - f^i_{j_i^k}(x^k) = \left[f^i(x^k) - f^i(x^{j_i^k})\right] - \left\langle g_{f^i}(x^{j_i^k}), x^k - x^{j_i^k} \right\rangle \\ &\le \left\langle g_{f^i}(x^k), x^k - x^{j_i^k} \right\rangle - \left\langle g_{f^i}(x^{j_i^k}), x^k - x^{j_i^k} \right\rangle \\ &\le |g_{f^i}(x^k)||x^k - x^{j_i^k}| + |g_{f^i}(x^{j_i^k})||x^k - x^{j_i^k}| \\ &\le 2C|x^k - x^{j_i^k}|, \end{aligned} \tag{19b}$$

using the subgradient inequality, the Cauchy-Schwarz inequality (twice) and our assumption $|g_{f^i}(x^j)| \le C$. Summing (19) and using (5) gives $\nabla f_k \in \partial_{\epsilon_k} f(x^k)$, i.e., (18), with

$$|\nabla f_k| = \left|\sum\nolimits_{i\in I} g_{f^i}(x^{j_i^k})\right| \le C|I| \tag{20}$$

and

$$\epsilon_k := \sum\nolimits_{i\in I} \epsilon_i^k \le \sum\nolimits_{i\in I} 2C|x^k - x^{j_i^k}| \le 2C|I| \max\nolimits_{i\in I} |x^k - x^{j_i^k}|. \tag{21}$$

By (17), nonexpansiveness of the orthogonal projection P_S and (20), we have

$$|x^{k+1} - x^k| \le \nu_k |\nabla f_k| \to 0. \tag{22}$$

Since $k - j_i^k \le \tau - 1$ for each i by almost cyclicity, (22) and the triangle inequality give $\max_i |x^k - x^{j_i^k}| \to 0$ in (21). Therefore, we have $\nabla f_k \in \partial_{\epsilon_k} f(x^k)$ in (17) with $\sup_k |\nabla f_k| < \infty$ (cf. (20)) and $\epsilon_k \to 0$. Hence the conclusion follows from well-known results on approximate subgradient methods; cf. [3, Ex. 6.3.13]. □

Acknowledgment. We thank an anonymous referee for helpful comments.

REFERENCES

1. J. E. Beasley, Lagrangean relaxation, in: C. R. Reeves (Ed.), *Modern Heuristic Techniques for Combinatorial Problems*, Blackwell Scientific Publications, Oxford, 1993, pp. 243–303.
2. A. Ben-Tal, T. Margalit, A. Nemirovski, The ordered subsets mirror descent optimization method and its use for the positron emission tomography reconstruction problem, *SIAM J. Optim.* , to appear.
3. D. P. Bertsekas, *Nonlinear Programming*, 2nd Edition, Athena Scientific, Belmont, MA, USA, 1999.
4. S. Feltenmark, Using primal structure in dual optimization, in: P. O. Lindberg (Ed.), 4th Stockholm Optimization Days, August 16–17, 1993, Dept. of Mathematics, Royal Institute of Technology, Stockholm, 1993.
5. S. Feltenmark, K. C. Kiwiel, Dual applications of proximal bundle methods, including Lagrangian relaxation of nonconvex problems, *SIAM J. Optim.* **10** (2000) 697–721.
6. S. Feltenmark, P. O. Lindberg, Experiments with partial subgradient updates and application to separable programming, Working paper, Dept. of Mathematics, Royal Institute of Technology, Stockholm (1996).
7. N. Gröwe-Kuska, K. C. Kiwiel, M. P. Nowak, W. Römisch, I. Wegner, Power management in a hydro-thermal system under uncertainty by Lagrangian relaxation, Preprint 99-19, Institut für Mathematik, Humboldt-Univ. Berlin, Berlin, Germany (1999).
8. J.-B. Hiriart-Urruty, C. Lemaréchal, *Convex Analysis and Minimization Algorithms*, Springer-Verlag, Berlin, Germany, 1993.
9. C. A. Kaskavelis, M. C. Caramanis, Efficient Lagrangian relaxation algorithms for industry size job-shop scheduling problems, *IEE Trans. Scheduling Logistics* **30** (1998) 1085–1097.
10. V. M. Kibardin, Decomposition into functions in the minimization problem, *Avtomat. i Telemekh.* **9** (1979) 66–79, English transl. in *Automat. Remote Control* **40** (1980) 1311–1323.
11. K. C. Kiwiel, T. Larsson, P. O. Lindberg, Dual properties of ballstep subgradient methods, with applications to Lagrangian relaxation, Tech. Rep. LiTH-MAT-R-1999-24, Dept. of Mathematics, Linköping Univ., Linköping, Sweden (1999).
12. K. C. Kiwiel, T. Larsson, P. O. Lindberg, The efficiency of ballstep subgradient level methods for convex optimization, *Math. Oper. Res.* **24** (1999) 237–254.
13. P. O. Lindberg, S. Feltenmark, A. Nöu, A. Piria, Using primal structure in large scale dual optimization, in: *Conference on Large Scale Optimization*, Univ. of Florida, Gainsville, FL, USA, 1993.
14. A. Nedić, D. P. Bertsekas, Incremental subgradient methods for nondifferentiable optimization, Tech. rep., Dept. of Electrical Engineering and Computer Science, M.I.T., Cambridge, MA (1999), revised: June 2000; submitted to *SIAM J. Optim.*
15. B. T. Polyak, Minimization of unsmooth functionals, *Zh. Vychisl. Mat. i Mat. Fiz.* **9** (1969) 509–521, English transl. in it USSR Comput. Math. and Math. Phys. **9** (1969) 14–29.
16. M. V. Solodov, S. K. Zavriev, Error-stability properties of generalized gradient-type algorithms, *J. Optim. Theory Appl.* **98** (1998) 663–680.

17. X. Zhao, P. B. Luh, J. Wang, The surrogate gradient algorithm for Lagrangian relaxation method, *J. Optim. Theory Appl.* **100** (1999) 699–712.

Inherently Parallel Algorithms in Feasibility and Optimization and their Applications
D. Butnariu, Y. Censor and S. Reich (Editors)

DIRECTIONAL HALLEY AND QUASI-HALLEY METHODS IN N VARIABLES

Yuri Levin[a]* and Adi Ben-Israel[b]

[a]RUTCOR–Rutgers Center for Operations Research, Rutgers University, 640 Bartholomew Rd, Piscataway, NJ 08854-8003, USA

[b]RUTCOR–Rutgers Center for Operations Research, Rutgers University, 640 Bartholomew Rd, Piscataway, NJ 08854-8003, USA

A directional Halley method for functions f of n variables is shown to converge, at a cubic rate, to a solution. To avoid the second derivative needed in the Halley method we propose a directional quasi-Halley method, with one more function evaluation per iteration than the directional Newton method, but with convergence rates comparable to the Halley method.

1. INTRODUCTION

When describing iterations such as

$$x^{k+1} := \Phi(x^k, d^k) \ , \text{ or } x^{k+1} := \Psi(x^k, x^{k-1}) \ , \quad k = 0, 1, \dots \tag{1}$$

we sometimes denote the current point by x, the next point by x_+, and the previous point by x_-, so that (1) is written simply as

$$x_+ := \Phi(x, d) \ , \text{ or } x_+ := \Psi(x, x_-) \ .$$

Consider a single equation in n unknowns,

$$f(\mathbf{x}) = 0 \ , \quad \text{or} \quad f(x_1, x_2, \dots, x_n) = 0 \ . \tag{2}$$

Under standard assumptions on f and the initial approximation, the *directional Newton method*

$$\mathbf{x}_+ := \mathbf{x} - \frac{f(\mathbf{x})}{\nabla f(\mathbf{x}) \cdot \mathbf{d}} \mathbf{d} \ , \tag{3}$$

converges to a solution at a quadratic rate, for certain directions $\mathbf{d}$ related to the gradient $\nabla f(\mathbf{x})$, see [5, Theorems 2–3]. An important special case of (3) is when $\mathbf{d}$ is along the gradient $\nabla f(\mathbf{x})$, giving

$$\mathbf{x}_+ := \mathbf{x} - \frac{f(\mathbf{x})}{||\nabla f(\mathbf{x})||^2} \nabla f(\mathbf{x}) \ . \tag{4}$$

*The work was supported by DIMACS.

For $n = 1$, method (3) reduces to the scalar Newton method

$$x_+ := x - \frac{f(x)}{f'(x)} \tag{5}$$

with which it shares quadratic convergence. Applying (5) to the function $f(x)/\sqrt{f'(x)}$ we get

$$x_+ := x - \frac{f(x)}{f'(x) - \dfrac{f''(x)f(x)}{2f'(x)}}, \tag{6}$$

the (scalar) *Halley method* with cubic rate of convergence. The *quasi-Halley method* of [1] replaces the second derivative in (6) by a difference $(f'(x) - f'(x_-))/(x - x_-)$,

$$x_+ := x - \frac{f(x)}{f'(x) - \dfrac{(f'(x) - f'(x_-)f(x)}{2(x - x_-)f'(x)}} \tag{7}$$

without losing much in convergence rate, see [1, Theorem 3] where (7) was shown to have order $1 + \sqrt{2}$, and [1, Theorem 5] where the Halley step h and quasi-Halley step q were shown to satisfy $|h - q| = O(|u_-|^3)$ where u_- is the previous Newton step. This shows that sufficiently close to a solution, the Halley and quasi-Halley methods are indistinguishable, as confirmed by numerical experiments.

A Halley method for solving operator equations in Banach space was given by Safiev [7] and Yao [9]. They assume three times differentiable mappings, with bijective first derivatives. Specializing to functions $f : \mathbb{R}^n \to \mathbb{R}^n$, where the second derivative f'' is a tensor, the Halley method of [7] and [9] is of theoretical interest, but difficult to implement.

On the other hand, a Halley method for solving (2), a single equation in n unknowns, is practical. For $f : \mathbb{R}^n \to \mathbb{R}$ we define, in analogy with the directional Newton method (3), the *directional Halley Method* as:

$$\mathbf{x}_+ := \mathbf{x} - \frac{f(\mathbf{x})}{\left(\nabla f(\mathbf{x}) - \dfrac{f(\mathbf{x})}{2\nabla f(\mathbf{x}) \cdot \mathbf{d}} f''(\mathbf{x})\mathbf{d}\right) \cdot \mathbf{d}} \mathbf{d}, \tag{8}$$

where f'' is the Hessian matrix of f. For $n = 1$, (8) reduces to the scalar Halley method (6).

We establish cubic convergence for the method (8) for the directions $\mathbf{d}$ in two cases:

- directions $\mathbf{d}$ nearly constant throughout the iterations, see § 2, Theorem 1, and
- directions $\mathbf{d}$ along the gradient $\nabla f(\mathbf{x})$, in which case (8) becomes

$$\mathbf{x}_+ := \mathbf{x} - \frac{f(\mathbf{x})}{\|\nabla f(\mathbf{x})\|^2 - \dfrac{f(\mathbf{x})}{2\,\|\nabla f(\mathbf{x})\|^2} \nabla f(\mathbf{x}) \cdot f''(\mathbf{x})\, \nabla f(\mathbf{x})} \nabla f(\mathbf{x}), \tag{9}$$

see § 3, Theorem 2.

In § 4 we propose the following *directional quasi-Halley method*

$$\mathbf{x}_+ := \mathbf{x} - \frac{f(\mathbf{x})}{\|\nabla f(\mathbf{x})\|^2 \left(1 - f\left(\mathbf{x} - \frac{f(\mathbf{x})}{\|\nabla f(\mathbf{x})\|^2}\nabla f(\mathbf{x})\right) / f(\mathbf{x})\right)} \nabla f(\mathbf{x}) \,, \tag{10}$$

obtained by approximating the term involving the second derivative in (9)

$$\frac{f(\mathbf{x})}{2\,\|\nabla f(\mathbf{x})\|^2} \nabla f(\mathbf{x}) \cdot f''(\mathbf{x}) \nabla f(\mathbf{x}) \quad \text{by} \quad \frac{\|\nabla f(\mathbf{x})\|^2}{f(\mathbf{x})} f\left(\mathbf{x} - \frac{f(\mathbf{x})}{\|\nabla f(\mathbf{x})\|^2} \nabla f(\mathbf{x})\right) .$$

The advantages of the directional quasi–Halley method (10) include:

- it avoids the second derivative needed in (9), requiring only one more function evaluation per iteration than the directional Newton method (4),
- if both methods (9) and (10) converge, their steps near a solution are approximately equal, see Theorem 3.

In §§ 5–6 we study three directional methods, along the direction of the gradient:

- the directional Newton method (4),
- the directional Halley method (9), and
- the directional quasi-Halley method (10).

We will drop the adjective "directional" when referring to these methods. The corresponding steps, along the gradient $\nabla f(\mathbf{x})$,

$$\text{Newton step, } \mathbf{u} = \frac{-f(\mathbf{x})}{\|\nabla f(\mathbf{x})\|^2} \nabla f(\mathbf{x}), \tag{11a}$$

$$\text{Halley step, } \mathbf{h} = \frac{-f(\mathbf{x})}{\|\nabla f(\mathbf{x})\|^2 - \frac{f(\mathbf{x})}{2\,\|\nabla f(\mathbf{x})\|^2} \nabla f(\mathbf{x}) \cdot f''(\mathbf{x}) \nabla f(\mathbf{x})} \nabla f(\mathbf{x}), \tag{11b}$$

$$\text{quasi-Halley step, } \mathbf{q} = \frac{-f(\mathbf{x})}{\|\nabla f(\mathbf{x})\|^2 \left(1 - f\left(\mathbf{x} - \frac{f(\mathbf{x})}{\|\nabla f(\mathbf{x})\|^2}\nabla f(\mathbf{x})\right) / f(\mathbf{x})\right)} \nabla f(\mathbf{x}), \tag{11c}$$

are compared in § 5. In numerical experiments, reported in § 6, the three methods were applied to randomly generated test problems with polynomials in several variables. In terms of average number of iterations, the Halley and quasi-Halley methods are comparable, and both are superior to the Newton method (4).

2. THE DIRECTIONAL HALLEY METHOD (8) WITH NEARLY CONSTANT DIRECTIONS

In this section we study the convergence of the method (8), for directions $\{\mathbf{d}^i : i = 0, 1, \cdots\}$ that are nearly constant in the sense of condition (16) below. We use the following result, a consequence of the Mean Value Theorem,

Lemma 1. *If* $f : \mathbb{R}^n \to \mathbb{R}$ *is twice differentiable in an open set* S *then for any* $\mathbf{x}, \mathbf{y} \in S$

$$\|\nabla f(y) - \nabla f(\mathbf{x})\| \leq \|\mathbf{y} - \mathbf{x}\| \sup_{0 \leq t \leq 1} \|f''(\mathbf{x} + t(\mathbf{y} - \mathbf{x}))\| .$$

The main tool for proving convergence is the majorizing sequence, due to Kantorovich and Akilov [4], see also [6, Chapter 12.4] and [9], where a majorizing sequence was used to prove cubic convergence for Halley's method in Banach space.

Definition 1. *A sequence* $\{y^k\}$, $y^k \geq 0$, $y^k \in \mathbb{R}$ *for which* $\|\mathbf{x}^{k+1} - \mathbf{x}^k\| \leq y^{k+1} - y^k$, $k = 0, 1, \ldots$ *is called a majorizing sequence for* $\{\mathbf{x}^k\}$.

Note that any majorizing sequence is necessarily monotonically increasing.
We recall the following lemma, proved in [6, Chapter 12.4, Lemma 12.4.1].

Lemma 2. *Let* $\{y^k\}$ *is a majorizing sequence for* $\{\mathbf{x}^k\} \subset \mathbb{R}^n$ *and* $\lim_{k\to\infty} y^k = y^* < \infty$. *Then there exists* $\mathbf{x}^* = \lim_{k\to\infty} \mathbf{x}^k$ *and* $\|\mathbf{x}^* - \mathbf{x}^k\| \leq y^* - y^k$, $k = 0, 1, \ldots$. □

To prove the convergence of (8), we write that iteration as

$$\mathbf{x}^{k+1} := \mathbf{x}^k - \alpha^k f(\mathbf{x}^k)\, \mathbf{v}^k , \text{ where} \tag{12a}$$

$$\mathbf{v}^k := \frac{\mathbf{d}^k}{\nabla f(\mathbf{x}^k) \cdot \mathbf{d}^k} , \text{ and} \tag{12b}$$

$$\alpha^k := \frac{1}{1 - \frac{1}{2} f(\mathbf{x}^k)\, \mathbf{v}^k \cdot f''(\mathbf{x}^k) \mathbf{v}^k} . \tag{12c}$$

Theorem 1. Let $f : \mathbb{R}^n \to \mathbb{R}$ be a three times differentiable function, $\mathbf{x}^0 \in \mathbb{R}^n$, and assume that

$$\sup_{\mathbf{x} \in X_0} \|f''(\mathbf{x})\| = M , \tag{13a}$$

$$\sup_{\mathbf{x} \in X_0} \|f'''(\mathbf{x})\| = N , \tag{13b}$$

where X_0 is defined as

$$X_0 := \{\mathbf{x} := \|\mathbf{x} - \mathbf{x}^0\| \leq (1+q)\, B\} , \tag{14}$$

for q and B given in terms of constants L, T, C that are assumed to satisfy

$$|\nabla f(\mathbf{x}^0) \cdot \mathbf{d}^0| \geq \frac{1}{L} , \tag{15a}$$

$$|f(\mathbf{x}^0)| \leq \frac{B}{L} , \tag{15b}$$

$$T := CLB < \frac{1}{2} , \text{ where } C := \sqrt{M^2 + \frac{2}{3} \frac{N}{L(1 - \frac{1}{2} MLB)}} , \tag{15c}$$

$$q = \frac{1 - \sqrt{1 - 2T}}{1 + \sqrt{1 - 2T}} , \tag{15d}$$

$\mathbf{d}^0$ is the initial direction, and all directions satisfy

$$\mathbf{d}^i \cdot \nabla f(\mathbf{x}^0) \geq \mathbf{d}^0 \cdot \nabla f(\mathbf{x}^0) \ , \ \mathbf{d}^i \in \mathbb{R}^n \ , \ \| \mathbf{d}^i \| = 1 \ , \ i = 0, 1, \ldots \ . \tag{16}$$

Then:
(a) All the points $\mathbf{x}^{k+1} := \mathbf{x}^k - \alpha^k f\left(\mathbf{x}^k\right) \mathbf{v}^k$, $k = 0, 1, \cdots$ lie in X_0.
(b) $\mathbf{x}^* = \lim_{k\to\infty} \mathbf{x}^k$ exists, $\mathbf{x}^* \in X_0$, and $f(\mathbf{x}^*) = 0$.
(c) The order of convengence of the directional Halley method (8) is cubic.

Remark 1. Condition (16) says that all directions $\mathbf{d}^i$ are as close, in angle, to the initial gradient $\nabla f(\mathbf{x}^0)$ as the initial direction $\mathbf{d}^0$,

$$\angle(\mathbf{d}^i, \nabla f(\mathbf{x}^0)) \leq \angle(\mathbf{d}^0, \nabla f(\mathbf{x}^0)) \ , \ \mathbf{d}^i \in \mathbb{R}^n \ , \ \| \mathbf{d}^i \| = 1 \ , \ i = 0, 1, \ldots \ . \tag{17}$$

Equivalently, all directions $\mathbf{d}^i$ are in a circular cone with axis $\nabla f(\mathbf{x}^0)$, generated by $\mathbf{d}^0$. In particular, (16) holds if the directions are fixed

$$\mathbf{d}^i = \mathbf{d}^0 \ , \ i = 1, 2, \cdots \ , \tag{18}$$

in which case, (8) reduces to the scalar Halley method (6) for the function $F(t) := f(\mathbf{x}^0 + t\mathbf{d}^0)$, along the line $\mathbf{L} := \{\mathbf{x}^0 + t\mathbf{d} : t \in \mathbb{R}\}$. The scalar quasi–Halley method (7) can also be used for $F(t)$ along $\mathbf{L}$.

3. A GRADIENT–DIRECTIONAL HALLEY METHOD

To prove the convergence of (9), we write that iteration as

$$\mathbf{x}^{k+1} := \mathbf{x}^k - \alpha^k f\left(\mathbf{x}^k\right) \mathbf{v}^k \ , \ \text{where} \tag{19a}$$

$$\mathbf{v}^k := \frac{\nabla f(\mathbf{x}^k)}{\|\nabla f(\mathbf{x}^k)\|^2} \ , \ \text{and} \tag{19b}$$

$$\alpha^k := \frac{1}{1 - \frac{1}{2} f(\mathbf{x}^k)\, \mathbf{v}^k \cdot f''(\mathbf{x}^k)\mathbf{v}^k} \ . \tag{19c}$$

The following theorem is analogous to Theorem 1. The proof is given in Appendix B.

Theorem 2. Let $f : \mathbb{R}^n \to \mathbb{R}$ be a three times differentiable function, $\mathbf{x}^0 \in \mathbb{R}^n$, and assume that

$$\sup_{\mathbf{x}\in X_0} \|f''(\mathbf{x})\| = M \ , \tag{20a}$$

$$\sup_{\mathbf{x}\in X_0} \|f'''(\mathbf{x})\| = N \ , \tag{20b}$$

where X_0 is defined in (21d) below. Let there exist constants B, L such that

$$\|\nabla f(\mathbf{x}^0)\| \geq \frac{1}{L}, \tag{21a}$$

$$|f(\mathbf{x}^0)| \leq \frac{B}{L}, \tag{21b}$$

$$T := CLB < \frac{1}{2}, \text{ where } C := \sqrt{M^2 + \frac{2}{3}\frac{N}{L(1-\frac{1}{2}MLB)}}, \tag{21c}$$

and finally let $X_0 := \{\mathbf{x} := \|\mathbf{x} - \mathbf{x}^0\| \leq (1+q)\,B\}$, where $q = \dfrac{1-\sqrt{1-2T}}{1+\sqrt{1-2T}}$. (21d)

Then:
(a) All the points $\mathbf{x}^{k+1} := \mathbf{x}^k - \alpha^k f\left(\mathbf{x}^k\right)\,\mathbf{v}^k$, $k = 0, 1, \cdots$ lie in X_0.
(b) $\mathbf{x}^* = \lim_{k\to\infty} \mathbf{x}^k$ exists, $\mathbf{x}^* \in X_0$, and $f(\mathbf{x}^*) = 0$.
(c) The order of convengence of the Halley method (9) is cubic.

4. DIRECTIONAL QUASI-HALLEY ITERATION

Consider the Taylor expansion of $f(x^k - \frac{f(\mathbf{x}^k)}{\|\nabla f(\mathbf{x}^k)\|^2}\nabla f(\mathbf{x}^k))$:

$$\begin{aligned} f\left(x^k - \frac{f(\mathbf{x}^k)}{\|\nabla f(\mathbf{x}^k)\|^2}\nabla f(\mathbf{x}^k)\right) &= f(\mathbf{x}^k) - \nabla f(\mathbf{x}^k)\cdot\frac{f(\mathbf{x}^k)}{\|\nabla f(\mathbf{x}^k)\|^2}\nabla f(\mathbf{x}^k) \\ &+ \frac{(f(\mathbf{x}^k))^2\,\nabla f(\mathbf{x}^k)\cdot f''(\mathbf{x}^k)\,\nabla f(\mathbf{x}^k)}{2\,\|\nabla f(\mathbf{x}^k)\|^4} + O\left(\frac{|f(\mathbf{x}^k)|^2}{\|\nabla f(\mathbf{x}^k)\|^2}\right) \\ &= \frac{f(\mathbf{x}^k)}{\|\nabla f(\mathbf{x}^k)\|^2}\frac{f(\mathbf{x}^k)\,\nabla f(\mathbf{x}^k)\cdot f''(\mathbf{x}^k)\,\nabla f(\mathbf{x}^k)}{2\,\|\nabla f(\mathbf{x}^k)\|^2} + O\left(\|\mathbf{u}^k\|^2\right), \text{ see (11a)}. \end{aligned}$$

Multiplying by $\dfrac{\|\nabla f(\mathbf{x}^k)\|^2}{f(\mathbf{x}^k)}$ we get

$$\frac{\|\nabla f(\mathbf{x}^k)\|^2}{f(\mathbf{x}^k)} f\left(x^k - \frac{f(\mathbf{x}^k)}{\|\nabla f(\mathbf{x}^k)\|^2}\nabla f(\mathbf{x}^k)\right) = \frac{f(\mathbf{x}^k)\,\nabla f(\mathbf{x}^k)\cdot f''(\mathbf{x}^k)\nabla f(\mathbf{x}^k)}{2\,\|\nabla f(\mathbf{x}^k)\|^2} + O(|f(\mathbf{x}^k)|),$$

showing that $\dfrac{f(\mathbf{x}^k)\nabla f(\mathbf{x}^k)\cdot f''(\mathbf{x}^k)\nabla f(\mathbf{x}^k)}{2\,\|\nabla f(\mathbf{x}^k)\|^2}$

can be approximated by $\dfrac{\|\nabla f(\mathbf{x}^k)\|^2}{f(\mathbf{x}^k)} f\left(x^k - \dfrac{f(\mathbf{x}^k)}{\|\nabla f(\mathbf{x}^k)\|^2}\nabla f(\mathbf{x}^k)\right)$

if $|f(\mathbf{x}^k)|$ is sufficiently small. Substituting this approximation to (9), we get the following iteration

$$\mathbf{x}^{k+1} := \mathbf{x}^k - \frac{f(\mathbf{x}^k)}{\|\nabla f(\mathbf{x}^k)\|^2 \left(1 - \frac{f\left(\mathbf{x}^k - \frac{f(\mathbf{x}^k)}{\|\nabla f(\mathbf{x}^k)\|^2}\nabla f(\mathbf{x}^k)\right)}{f(\mathbf{x}^k)}\right)} \nabla f(\mathbf{x}^k)\,,\ k = 0, 1, \ldots \quad (22a)$$

$$= \mathbf{x}^k - \frac{f(\mathbf{x}^k)}{f(\mathbf{x}^k + \mathbf{u}^k) - f(\mathbf{x}^k)}\mathbf{u}^k\,,\ k = 0, 1, \ldots \quad (22b)$$

This **quasi-Halley method** does not reduce to its scalar counterpart (7) for $n = 1$.

5. COMPARISON OF STEPS

The results of this section are extensions of [1, § 6].
We use the Taylor expansion of f, see [2, (8.14.3)],

$$f(\mathbf{x} + \mathbf{u}) = f(\mathbf{x}) + f'(\mathbf{x}) \cdot \mathbf{u} + \frac{1}{2} f''(\mathbf{x}) \cdot \mathbf{u}^{(2)} + \frac{1}{3!} f'''(\boldsymbol{\xi}) \cdot \mathbf{u}^{(3)}\,,$$

for some $\boldsymbol{\xi}$ betwee $\mathbf{x}$ and $\mathbf{x} + \mathbf{u}$, that we write as

$$f(\mathbf{x} + \mathbf{u}) = f(\mathbf{x}) + \nabla f(\mathbf{x}) \cdot \mathbf{u} + \frac{1}{2} \mathbf{u} \cdot f''(\mathbf{x})\mathbf{u} + \frac{1}{3!} f'''(\boldsymbol{\xi}) \cdot \mathbf{u}^{(3)}\,. \quad (23)$$

Two iterative methods are comparable locally if at a given point they produce comparable steps. We will compare the steps in terms of length.

We first compare the steps of the Newton and Halley methods, assuming both steps emanate from the same point $\mathbf{x}^k$, arrived at by Newton's method. This corresponds to a hypothetical situation where at an iterate $\mathbf{x}^k$ of the Newton method we have an option continuing (and making a Newton step) or switching to the Halley method (9), making a *Halley step*

To simplify the writing we denote by f_k the function f evaluated at $\mathbf{x}^k$. Similarly, ∇f_k and f_k'' denote the gradient ∇f and the Hessian f'' evaluated at $\mathbf{x}^k$. The steps to be compared are the Newton step $\mathbf{u}^k$ and the Halley step $\mathbf{h}^k$, see (11a)–(11b), written as

$$\mathbf{u}^k = -\frac{f_k}{\|\nabla f_k\|^2}\nabla f_k \quad \text{and} \quad \mathbf{h}^k = -\frac{f_k}{\|\nabla f_k\|^2 - \frac{f_k \cdot \nabla f_k \cdot f_k'' \cdot \nabla f_k}{2 \cdot \|\nabla f_k\|^2}} \nabla f_k.$$

The next lemma gives a condition for the Newton Halley steps to have the same sign.

Lemma 3. *The steps* $\mathbf{u}^k$ *and* $\mathbf{h}^k$ *have the same sign if and only if*

$$\|\nabla f_k\|^2 > \frac{f_k \nabla f_k \cdot f_k'' \nabla f_k}{2\,\|\nabla f_k\|^2}, \quad (24)$$

in which case

$$\|\mathbf{h}^k\| \geq \|\mathbf{u}^k\| \quad \text{if } f_k \nabla f_k \cdot f_k'' \nabla f_k \geq 0, \quad (25a)$$

$$\|\mathbf{h}^k\| < \|\mathbf{u}^k\| \quad \text{if } f_k \nabla f_k \cdot f_k'' \nabla f_k < 0\,. \quad (25b)$$

Proof. The steps $\mathbf{u}^k$ and $\mathbf{h}^k$ have the same sign if and only if

$$\begin{aligned} \|\nabla f_k\|^2 - \frac{f_k \nabla f_k \cdot f_k'' \nabla f_k}{2\,\|\nabla f_k\|^2} &\geq 0\,, \\ \text{that is } \|\nabla f_k\|^2 &\geq \frac{f_k \nabla f_k \cdot f_k'' \nabla f_k}{2\,\|\nabla f_k\|^2}\,, \\ \text{in which case } \|\nabla f_k\|^2 &\geq \left| \|\nabla f_k\|^2 - \frac{f_k \nabla f_k \cdot f_k'' \nabla f_k}{2\,\|\nabla f_k\|^2} \right| \quad \text{if} \quad f_k \nabla f_k \cdot f_k'' \nabla f_k \geq 0\,, \\ \|\nabla f_k\|^2 &< \left| \|\nabla f_k\|^2 - \frac{f_k \nabla f_k \cdot f_k'' \nabla f_k}{2\,\|\nabla f_k\|^2} \right| \quad \text{if} \quad f_k \nabla f_k \cdot f_k'' \nabla f_k < 0\,. \end{aligned}$$

□

The point $\mathbf{x}^k$ where the steps $\mathbf{u}^k$ and $\mathbf{h}^k$ are compared is arrived at by the Newton method. It is therefore reasonable to assume that the following conditions hold, see [5, eq. 10a-10b]

$$\|f_k''\| \leq M\,, \tag{26a}$$

$$\|\nabla f_k\| \geq 2\,\|\mathbf{u}^k\|\,M,\ k = 1, 2, \dots \tag{26b}$$

Theorem 3. *If conditions (26a) and (26b) hold, then the Newton step* $\mathbf{u}^k$ *and the Halley step* $\mathbf{h}^k$ *have the same sign, and are related by*

$$\frac{2}{3}\,\|\mathbf{u}^k\| \leq \|\mathbf{h}^k\| \leq \frac{4}{3}\,\|\mathbf{u}^k\|\,. \tag{27}$$

Proof.

$$\|\nabla f_k\| \geq 2\,\|\mathbf{u}^k\|\,M = 2\frac{|f_k|}{\|\nabla f_k\|}M, \text{ by } (26b)$$

$$\therefore \|\nabla f_k\|^2 \geq 2\,|f_k|\,M \geq 2\,|f_k|\,\|f_k''\| \geq 2\,|f_k|\,\frac{\nabla f_k \cdot f_k'' \nabla f_k}{\|\nabla f_k\|^2}, \text{ by } (26a).$$

Therefore (24) holds, showing that $\mathbf{u}^k$ and $\mathbf{h}^k$ have the same sign. Then

$$\begin{aligned} \mathbf{u}^k - \mathbf{h}^k &= \frac{f_k}{\|\nabla f_k\|^2 - \frac{f_k \nabla f_k \cdot f_k'' \nabla f_k}{2\|\nabla f_k\|^2}} \nabla f_k - \frac{f_k}{\|\nabla f_k\|^2} \nabla f_k \\ &= \frac{(f_k)^2 \left(\frac{\nabla f_k \cdot f_k'' \nabla f_k}{2\|\nabla f_k\|^2}\right)}{\|\nabla f_k\|^2 \left(\|\nabla f_k\|^2 - \frac{f_k \nabla f_k \cdot f_k'' \nabla f_k}{2\|\nabla f_k\|^2}\right)} \nabla f_k \end{aligned}$$

$$\therefore\ \|\mathbf{u}^k - \mathbf{h}^k\| = \frac{(f_k)^2 \frac{|\nabla f_k \cdot f_k'' \nabla f_k|}{2\|\nabla f_k\|^2}}{\|\nabla f_k\| \left| \|\nabla f_k\|^2 - \frac{f_k \nabla f_k \cdot f_k'' \nabla f_k}{2\|\nabla f_k\|^2} \right|}. \tag{28}$$

From (26a) and (26b) it follows that $|\nabla f_k \cdot f_k'' \nabla f_k| \leq M\,\|\nabla f_k\|^2$, and $\frac{|f_k| M}{2\|\nabla f_k\|^2} \leq \frac{1}{4},\ k = 0, 1, 2, \dots$, which substituted in (28) gives

$$\|\mathbf{u}^k - \mathbf{h}^k\| \leq \frac{(f_k)^2 \frac{M}{2}}{\|\nabla f_k\|^3 \left(1 - \frac{|f_k| M}{2\|\nabla f_k\|^2}\right)} \leq \frac{(f_k)^2 \frac{M}{2}}{\|\nabla f_k\|^3\,|f_k|\,M} \frac{4}{3} = \frac{1}{3} \frac{f_k}{\|\nabla f_k\|} = \frac{\|\mathbf{u}^k\|}{3},$$

proving (27). □

The Halley step $\mathbf{h}^k$ therefore lies in the interval with endpoints $\frac{2}{3}\mathbf{u}^k$ and $\frac{4}{3}\mathbf{u}^k$.
We next compare the Halley step $\mathbf{h}^k$ and the quasi-Halley step (11c)

$$\mathbf{q}^k = -\frac{f(\mathbf{x}^k)}{\|\nabla f(\mathbf{x}^k)\|^2 \left(1 - \frac{f(\mathbf{x}^k+\mathbf{u}^k)}{f(\mathbf{x}^k)}\right)} \nabla f(\mathbf{x}^k) , \; k = 0, 1, \dots \tag{29}$$

evaluated at the same point $\mathbf{x}^k$ where $f(\mathbf{x}^k) \neq 0$. Numerical experience shows that, close to a root to which both methods converge, the Halley step and the quasi-Halley step are very close. This is explained by the the following theorem.

Theorem 4. *Let f have continuous third derivative in the interval X_0, and let*

$$\sup_{\mathbf{x}\in X_0} \|f'''(\mathbf{x})\| = N.$$

If conditions (26a) and (26b) hold, then

$$\|\mathbf{h}^k - \mathbf{q}^k\| \le \frac{4}{27}\frac{N}{M} \|\mathbf{u}^k\|^2 , \tag{30}$$

i.e. the difference between the Halley step $\mathbf{h}^k$ and the quasi-Halley step $\mathbf{q}^k$ is of order $O(\|\mathbf{u}^k\|^2)$, where $\mathbf{u}^k$ is the Newton step at the current point.

Proof. Let $\mathbf{g}_k := \nabla f_k$. Then

$$\begin{aligned}
\mathbf{q}^k - \mathbf{h}^k &= \frac{f_k \mathbf{g}_k}{\|\mathbf{g}_k\|^2}\left(\frac{1}{1 - \frac{f_k \mathbf{g}_k \cdot f_k'' \mathbf{g}_k}{2\|\mathbf{g}_k\|^4}} - \frac{1}{1 - \frac{f(\mathbf{x}^k+\mathbf{u}^k)}{f_k}}\right) \\
&= -\frac{\mathbf{g}_k}{\|\mathbf{g}_k\|^2} \frac{f(\mathbf{x}^k + \mathbf{u}^k) - \frac{(f_k)^2 \mathbf{g}_k \cdot f_k'' \mathbf{g}_k}{2\|\mathbf{g}_k\|^4}}{\left(1 - \frac{f_k \mathbf{g}_k \cdot f_k'' \mathbf{g}_k}{2\|\mathbf{g}_k\|^4}\right)\left(1 - \frac{f(\mathbf{x}^k+\mathbf{u}^k)}{f_k}\right)} \\
&= -\frac{\mathbf{g}_k}{\|\mathbf{g}_k\|^2} \frac{f_k + \mathbf{g}_k \cdot \mathbf{u}^k + \frac{1}{2}\mathbf{u}^k \cdot f_k'' \mathbf{u}^k + \frac{1}{6} f'''(\xi) \cdot (\mathbf{u}^k)^{(3)} - \frac{(f_k)^2 \mathbf{g}_k \cdot f_k'' \mathbf{g}_k}{2\|\mathbf{g}_k\|^4}}{\left(1 - \frac{f_k \mathbf{g}_k \cdot f_k'' \mathbf{g}_k}{2\|\mathbf{g}_k\|^4}\right)\left(1 - \frac{f(\mathbf{x}^k+\mathbf{u}^k)}{f_k}\right)} , \\
&\qquad \text{for some } \xi \text{ between } \mathbf{x}^k + \mathbf{u}^k \text{ and } \mathbf{x}^k , \\
&= -\frac{\mathbf{g}_k}{\|\mathbf{g}_k\|^2} \frac{\frac{1}{6} f'''(\xi) \cdot (\mathbf{u}^k)^{(3)}}{\left(1 - \frac{f_k \mathbf{g}_k \cdot f_k'' \mathbf{g}_k}{2\|\mathbf{g}_k\|^4}\right)\left(1 - \frac{f(\mathbf{x}^k+\mathbf{u}^k)}{f_k}\right)} .
\end{aligned}$$

Note,

$$\begin{aligned}
\frac{|f(\mathbf{x}^k + \mathbf{u}^k)|}{|f_k|} &\le \frac{|\frac{1}{2}\mathbf{u}^k \cdot f_k''(\theta)\,\mathbf{u}^k|}{|f_k|} , \text{ for some } \theta \text{ between } \mathbf{x}^k + \mathbf{u}^k \text{ and } \mathbf{x}^k , \\
&\le \frac{\frac{1}{2} M \|\mathbf{u}^k\|^2}{|f_k|} \le \frac{1}{4} , \text{ since } |f_k| \ge 2M \|\mathbf{u}^k\|^2 .
\end{aligned}$$

$$\therefore\ \left\|\mathbf{h}^k-\mathbf{q}^k\right\| \le \frac{\frac{1}{6}N\left\|\mathbf{u}^k\right\|^3}{\|\mathbf{g}_k\|\left|1-\frac{f_k\,\mathbf{g}_k\cdot f_k''\mathbf{g}_k}{2\|\mathbf{g}_k\|^4}\right|\frac{3}{4}} = \frac{\frac{2}{9}N\left\|\mathbf{u}^k\right\|^3 f_k}{\|\mathbf{g}_k\|\left|f_k-\frac{1}{2}\mathbf{u}^k\cdot f_k''\,\mathbf{u}^k\right|}$$

$$\le \frac{\frac{2}{9}N\left\|\mathbf{u}^k\right\|^4}{|f_k|-\left|\frac{1}{2}\mathbf{u}^k\cdot f_k''\,\mathbf{u}^k\right|}$$

$$\le \frac{\frac{2}{9}N\left\|\mathbf{u}^k\right\|^4}{2M\left\|\mathbf{u}^k\right\|^2-\frac{1}{2}M\left\|\mathbf{u}^k\right\|^2}\,,\ \text{since } |f_k|\ge 2M\left\|\mathbf{u}^k\right\|^2$$

$$= \frac{4}{27}\frac{N}{M}\left\|\mathbf{u}^k\right\|^2\,. \qquad \square$$

The comparison (30) between the Halley step $\mathbf{h}^k$ and the quasi-Halley step $\mathbf{q}^k$ is in terms of the current Newton step $\mathbf{u}^k$. The same comparison in terms of the previous Newton step $\mathbf{u}^{k-1}$ (assuming the current point $\mathbf{x}^k$ is arrived at by the Newton method) gives

$$\left\|\mathbf{h}^k-\mathbf{q}^k\right\| \le \frac{NM}{27\,\|\mathbf{g}_k\|^2}\left\|\mathbf{u}^{k-1}\right\|^4\,.$$

Example 1. As in [5], an arbitrary system of m equations in n unknowns:

$$\mathbf{f}(\mathbf{x}) = \mathbf{0}\,,\ \text{or}\quad f_i(x_1,x_2,\ldots,x_n) = 0\,,\ i=1,\ldots,m \tag{31}$$

can be replaced by an equivalent single equation

$$\sum_{i=1}^{m} f_i^2(\mathbf{x}) = 0\,, \tag{32}$$

which can be solved by both the Halley and the quasi-Halley methods. An example is given in § C.4, Appendix C.

6. NUMERICAL EXPERIENCE

The Halley method (9) and the quasi-Halley method (10) were tested on polynomials in n variables, $n = 2,\cdots,9$, and degree d, $d = 2,\cdots,10$. The *degree* of a polynomial in n variables is the maximal degree of its terms, e.g. $degree(x^2y^3+x^4+y^4) = 5$. For each combination of n and d, 100 random polynomials were generated and solved by both methods, starting with the same initial point $\mathbf{x}^0 = (1,1,\cdots,1)$. For comparison, each such polynomial was also solved by the Newton method (4). The stopping rule in all methods was identical, the norm of the step value is less than 10^{-12}, with an upper bound of 30 iterations. The average numbers of iterations of the directional Newton, Halley and quasi-Halley methods, for 100 random polynomials with degree $d = 2,\cdots,10$ and n variables, $n = 2,\cdots,9$ are tabulated in Table 1.

Figures 1–2 illustrate two typical sections of Table 1.

In Figure 1 the number of variables is fixed at 5. For each degree from 2 to 10, 100 random polynomials were generated and solved by the three methods, recording the average number of iterations as a function of the degree.

Number of variables	2			3			4			5		
Degree	**N**	**H**	**q-H**	**N**	**H**	**q-H**	**N**	**H**	**q-H**	**N**	**H**	**q-H**
2	12.45	11.86	7.87	10.05	10.40	7.93	10.46	11.69	6.82	12.47	11.22	6.92
3	11.60	12.25	8.88	10.75	10.98	7.75	11.05	9.66	6.64	10.45	10.23	6.28
4	13.74	11.16	8.09	13.24	10.70	8.15	10.66	10.73	7.45	11.11	8.18	6.56
5	14.30	11.96	9.65	12.38	11.77	8.79	10.16	7.80	8.02	10.38	8.41	8.85
6	12.94	10.48	9.26	11.92	9.82	9.25	10.58	9.26	8.21	9.06	8.82	7.11
7	13.25	10.22	10.11	11.50	9.16	8.00	10.21	7.36	8.06	10.08	8.46	7.76
8	9.47	10.56	8.90	12.54	9.44	9.75	11.69	8.74	7.98	8.45	7.39	8.77
9	10.86	8.99	9.09	10.33	10.17	9.09	10.00	10.23	9.61	8.34	7.17	7.92
10	11.61	10.04	8.99	10.23	10.76	9.76	9.20	7.83	9.32	8.22	8.54	8.11

Number of variables	6			7			8			9		
Degree	**N**	**H**	**q-H**	**N**	**H**	**q-H**	**N**	**H**	**q-H**	**N**	**H**	**q-H**
2	9.00	7.95	7.23	9.12	8.14	8.37	11.20	10.59	9.13	9.82	10.08	7.77
3	9.64	10.67	6.00	10.88	9.36	7.44	10.03	8.38	6.92	9.24	9.92	7.14
4	8.69	7.18	6.25	9.68	8.85	7.46	10.02	9.15	6.95	8.26	7.48	6.96
5	9.93	9.32	6.57	8.89	7.37	8.43	8.45	6.62	6.55	8.54	7.00	7.40
6	9.35	8.32	7.08	7.65	7.41	6.77	7.98	6.55	6.61	7.70	7.52	5.48
7	7.80	7.47	8.24	7.65	6.03	6.38	7.62	8.37	6.66	8.21	6.89	7.80
8	8.79	6.27	6.91	7.66	7.03	7.64	7.33	7.43	7.51	6.75	4.98	7.34
9	8.21	6.03	7.05	8.47	7.20	7.69	7.74	6.58	6.68	8.25	6.66	7.30
10	7.26	7.53	7.26	8.10	6.70	7.45	6.78	6.35	7.14	6.51	5.66	6.38

Table 1
Comparison of the directional Newton (**N**), Halley (**H**) and quasi-Halley (**q-H**) methods in terms of the average number of iterations for 100 random polynomials with the given degree and number of variables.

In Figure 2 the polynomial degree is fixed at 5. For each number of variables from 2 to 10, 100 random polynomials were generated and solved by the three methods, recording the average number of iterations as a function of the number of variables.

7. CONCLUSIONS

Three methods for solving a single equation in n unknowns are discussed here:

- the Newton method (4),
- the Halley method (9), and
- the quasi-Halley method (10).

The Newton method is of order 2, see [5, Theorem 1], and requires the evaluation of $f(\mathbf{x})$, $\nabla f(\mathbf{x})$ (the *Newton data*) at each iteration. We use this method as our basis for comparison.

The Halley method is of order 3, see Theorem 1 above, but requires the Hessian $f''(\mathbf{x})$ in addition to the Newton data.

The order of the quasi-Halley method is unknown (the related scalar quasi-Halley method (7) has order $1 + \sqrt{2}$, see [1]), however in practice it performs similarly to the Halley method, see the comparison of steps in Theorem 4, and the numerical experience reported in § 6. In terms of work per iteration, the quasi-Halley method requires one

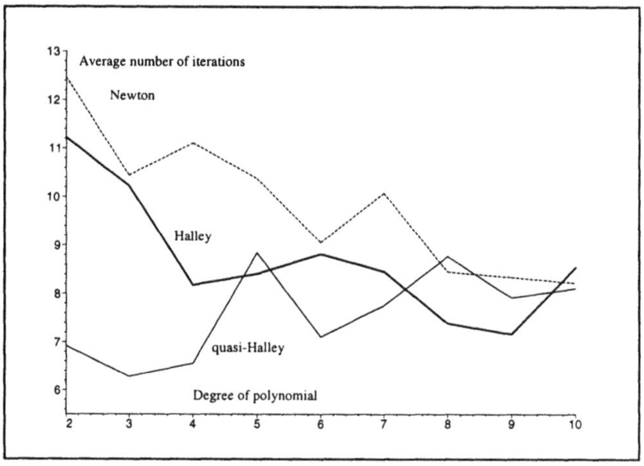

Figure 1. Comparison of the Newton, Halley and quasi-Halley methods for polynomials with 5 variables.

more function evaluation than the Newton method, $f(\mathbf{x}+\mathbf{u})$ where $\mathbf{u}$ is the next Newton step. Another advantage of the quasi-Halley method is that it is amenable for parallel implementation, since the matrix $f''(\mathbf{x})$ is avoided.

REFERENCES

1. A. Ben-Israel, Newton's method with modified functions, *Contemp. Math.* **204** (1997) 39–50.
2. J. Dieudonné, *Foundations of Modern Analysis* (Academic Press, 1960).
3. C.-E. Fröberg, *Numerical Mathematics: Theory and Computer Applications* (Benjamin, 1985).
4. L.V. Kantorovich and G.P. Akilov, *Functional Analysis in Normed Spaces* (Macmillan, New York, 1964).
5. Y. Levin and A. Ben-Israel, Directional Newton methods in n variables, *Math. of Computation*, to appear.
6. J.M. Ortega and W.C. Rheinboldt, *Iterative Solution of Nonlinear Equations in Several Variables* (Academic Press, New York, 1970).
7. R.A. Safiev, The methods of tangent hyperbolas, *Soviet Math. Dokl.* **4** (1963) 482–485.
8. J.F. Traub, *Iterative Methods for the Solution of Equations* (Prentice–Hall, 1964).
9. Q. Yao, On Halley iteration, *Numer. Math.* **81** (1999) 647-677.

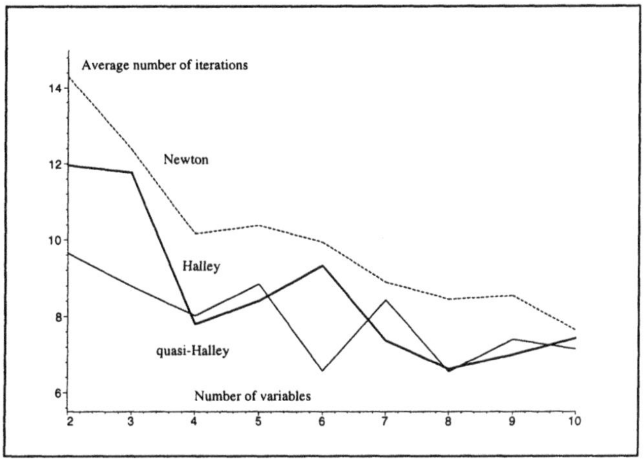

Figure 2. Comparison of the Newton, Halley and quasi-Halley methods for polynomials of degree 5.

APPENDIX A: PROOF OF THEOREM 1

Proof. We construct a majorizing sequence for $\{\mathbf{x}^k\}$ in terms of the auxiliary function

$$\varphi(y) := \frac{C}{2}y^2 - \frac{1}{L}y + \frac{B}{L}\,. \tag{A.1}$$

The quadratic equation $\varphi(y) = 0$ has two roots $r_1 = (1+q)B$, $r_2 = \left(1 + \frac{1}{q}\right)B$, and $0 < r_1 < r_2$. Then $\varphi(y) = \frac{C}{2}(y - r_1)(y - r_2)$, $\varphi'(y) = \frac{C}{2}((y - r_1) + (y - r_2))$, and $\varphi''(y) = C$.

Starting from $y^0 = 0$, apply the scalar Halley iteration (6) to the function $\varphi(y)$ to get

$$\begin{aligned} y^{k+1} &= y^k - \frac{\varphi(y^k)}{\varphi'(y^k) - \frac{1}{2}\frac{\varphi(y^k)\varphi''(y^k)}{\varphi'(y^k)}}\,, \quad k = 0,1,2,\dots \\ &= y^k - \frac{\frac{C}{2}(y^k)^2 - \frac{1}{L}y^k + \frac{B}{L}}{Cy^k - \frac{1}{L} - \frac{C}{2}\frac{\left(\frac{C}{2}(y^k)^2 - \frac{1}{L}y^k + \frac{B}{L}\right)}{Cy^k - \frac{1}{L}}}\,, \quad k = 0,1,2,\dots \end{aligned} \tag{A.2}$$

obtained by substituting $\varphi'(y) = Cy - \dfrac{1}{L}$, $\varphi''(y) = C$.

Next we prove that the sequences $\{\mathbf{x}^k\}$ and $\{y^k\}$, generated by (8) and (A.2) respectively, satisfy for $k = 0, 1, \dots$,

$$|f(\mathbf{x}^k)| \le \varphi(y^k) , \tag{A.3a}$$

$$\|\mathbf{v}^k\| \le -\frac{1}{\varphi'(y^k)} , \tag{A.3b}$$

$$|\alpha^k| \le \frac{1}{1 - \frac{M}{2}\frac{\varphi(y^k)}{(\varphi'(y^k))^2}} , \tag{A.3c}$$

$$\|\mathbf{x}^{k+1} - \mathbf{x}^k\| \le y^{k+1} - y^k . \tag{A.3d}$$

Statement (A.3d) says that $\{y^k\}$ is a majorizing sequence for $\{\mathbf{x}^k\}$. The proof is by induction.
Verification for $k = 0$:

$$\begin{aligned}
|f(\mathbf{x}^0)| &\le \frac{B}{L} = \varphi(y^0) = \varphi(0) ,\\
\|\mathbf{v}^0\| \le L &= -\frac{1}{\varphi'(y^0)} ,\\
|\alpha^0| &= \left|\frac{1}{1 - \frac{1}{2} f(\mathbf{x}^0)\, \mathbf{v}^0 \cdot f''(\mathbf{x}^0)\, \mathbf{v}^0}\right| \le \frac{1}{1 - \frac{1}{2} M \frac{\varphi(y^0)}{(\varphi'(y^0))^2}} ,\\
\|\mathbf{x}^1 - \mathbf{x}^0\| &= \|\alpha^0 \mathbf{v}^0 f(\mathbf{x}^0)\| \le |\alpha^0|\, \|\mathbf{v}^0\|\, |f(\mathbf{x}^0)|\\
&\le \frac{-\frac{\varphi(y^0)}{\varphi'(y^0)}}{1 - \frac{1}{2} M \frac{\varphi(y^0)}{(\varphi'(y^0))^2}} \le \frac{-\frac{\varphi(y^0)}{\varphi'(y^0)}}{1 - \frac{1}{2} C \frac{\varphi(y^0)}{(\varphi'(y^0))^2}} = y^1 - y^0,
\end{aligned}$$

showing that equations (A.3a)-(A.3d) hold for $k = 0$. Suppose (A.3a)-(A.3d) hold for $k \le n$.
Proof of (A.3a) for $n + 1$:

$$\begin{aligned}
\|\mathbf{x}^{n+1} - \mathbf{x}^0\| &= \left\|\sum_{k=0}^{n} \left(\mathbf{x}^{k+1} - \mathbf{x}^k\right)\right\| \le \sum_{k=0}^{n} \left(y^{k+1} - y^k\right) = y^{n+1} - y^0 = y^{n+1} \le r_1 .\\
\therefore\ \mathbf{x}^{n+1} &\in X_0 .\\
\text{Let } a_n(\mathbf{x}) &:= (\mathbf{x} - \mathbf{x}^n)\left(\nabla f(\mathbf{x}^n) \cdot \mathbf{d}^n - \frac{f(\mathbf{x}^n)}{2\nabla f(\mathbf{x}^n) \cdot \mathbf{d}^n}\, \mathbf{d}^n \cdot f''(\mathbf{x}^n)\, \mathbf{d}^n\right) + f(\mathbf{x}^n)\mathbf{d}^n .\\
\therefore\ a_n(\mathbf{x}) &= (\mathbf{x} - \mathbf{x}^{n+1})\left(\nabla f(\mathbf{x}^n) \cdot \mathbf{d}^n - \frac{f(\mathbf{x}^n)}{2\nabla f(\mathbf{x}^n) \cdot \mathbf{d}^n}\, \mathbf{d}^n \cdot f''(\mathbf{x}^n)\, \mathbf{d}^n\right)\\
&\quad + (\mathbf{x}^{n+1} - \mathbf{x}^n)\left(\nabla f(\mathbf{x}^n) \cdot \mathbf{d}^n - \frac{f(\mathbf{x}^n)}{2\nabla f(\mathbf{x}^n) \cdot \mathbf{d}^n}\, \mathbf{d}^n \cdot f''(\mathbf{x}^n)\, \mathbf{d}^n\right) + f(\mathbf{x}^n)\mathbf{d}^n\\
&= (\mathbf{x} - \mathbf{x}^{n+1})\left(\nabla f(\mathbf{x}^n) \cdot \mathbf{d}^n - \frac{f(\mathbf{x}^n)}{2\nabla f(\mathbf{x}^n) \cdot \mathbf{d}^n}\, \mathbf{d}^n \cdot f''(\mathbf{x}^n)\, \mathbf{d}^n\right) .\\
\text{Let } p_n(\mathbf{x}) &:= \frac{\nabla f(\mathbf{x}^n)}{\nabla f(\mathbf{x}^n) \cdot \mathbf{d}^n} \cdot a_n(\mathbf{x})\\
&= (\mathbf{x} - \mathbf{x}^n) \cdot \left(\nabla f(\mathbf{x}^n) - \frac{f(\mathbf{x}^n)}{2\nabla f(\mathbf{x}^n) \cdot \mathbf{d}^n}\, f''(\mathbf{x}^n)\, \mathbf{d}^n\right) + f(\mathbf{x}^n) .
\end{aligned}$$

Note that $p_n(\mathbf{x}^{n+1}) = 0$. On the other hand, $p_n(\mathbf{x}^{n+1})$ can be represented as

$$\begin{aligned}
p_n(\mathbf{x}^{n+1}) &= f(\mathbf{x}^n) + \nabla f(\mathbf{x}^n)\cdot\left(\mathbf{x}^{n+1}-\mathbf{x}^n\right) - \frac{f(\mathbf{x}^n)}{2\nabla f(\mathbf{x}^n)\cdot\mathbf{d}^n}\mathbf{d}^n\cdot f''(\mathbf{x}^n)\left(\mathbf{x}^{n+1}-\mathbf{x}^n\right) \\
&\quad +\frac{1}{2}\left(\mathbf{x}^{n+1}-\mathbf{x}^n\right)\cdot f''(\mathbf{x}^n)\left(\mathbf{x}^{n+1}-\mathbf{x}^n\right) - \frac{1}{2}\left(\mathbf{x}^{n+1}-\mathbf{x}^n\right)\cdot f''(\mathbf{x}^n)\left(\mathbf{x}^{n+1}-\mathbf{x}^n\right) \\
&= f(\mathbf{x}^n) + \nabla f(\mathbf{x}^n)\cdot\left(\mathbf{x}^{n+1}-\mathbf{x}^n\right) + \frac{1}{2}\left(\mathbf{x}^{n+1}-\mathbf{x}^n\right)\cdot f''(\mathbf{x}^n)\left(\mathbf{x}^{n+1}-\mathbf{x}^n\right) \\
&\quad -\frac{1}{2}\left(\frac{f(\mathbf{x}^n)\,\mathbf{d}^n}{\nabla f(\mathbf{x}^n)\cdot\mathbf{d}^n} + \mathbf{x}^{n+1}-\mathbf{x}^n\right)\cdot f''(\mathbf{x}^n)\left(\mathbf{x}^{n+1}-\mathbf{x}^n\right) \\
&= f(\mathbf{x}^n) + \nabla f(\mathbf{x}^n)\cdot\left(\mathbf{x}^{n+1}-\mathbf{x}^n\right) + \frac{1}{2}\left(\mathbf{x}^{n+1}-\mathbf{x}^n\right)\cdot f''(\mathbf{x}^n)\left(\mathbf{x}^{n+1}-\mathbf{x}^n\right) \\
&\quad -\frac{1}{4}\frac{f(\mathbf{x}^n)\,\mathbf{d}^n\cdot f''(\mathbf{x}^n)\,\mathbf{d}^n}{(\nabla f(\mathbf{x}^n)\cdot\mathbf{d}^n)^2}\left(\mathbf{x}^{n+1}-\mathbf{x}^n\right)\cdot f''(\mathbf{x}^n)\left(\mathbf{x}^{n+1}-\mathbf{x}^n\right). \\
\therefore\ 0 &= f(\mathbf{x}^n) + \nabla f(\mathbf{x}^n)\cdot\left(\mathbf{x}^{n+1}-\mathbf{x}^n\right) + \frac{1}{2}\left(\mathbf{x}^{n+1}-\mathbf{x}^n\right)\cdot f''(\mathbf{x}^n)\left(\mathbf{x}^{n+1}-\mathbf{x}^n\right) \\
&\quad -\frac{1}{4}\frac{f(\mathbf{x}^n)\,\mathbf{d}^n\cdot f''(\mathbf{x}^n)\,\mathbf{d}^n}{(\nabla f(\mathbf{x}^n)\cdot\mathbf{d}^n)^2}\left(\mathbf{x}^{n+1}-\mathbf{x}^n\right)\cdot f''(\mathbf{x}^n)\left(\mathbf{x}^{n+1}-\mathbf{x}^n\right).
\end{aligned}$$

Thus,

$$\begin{aligned}
f\left(\mathbf{x}^{n+1}\right) &= f\left(\mathbf{x}^{n+1}\right) - f(\mathbf{x}^n) - \nabla f(\mathbf{x}^n)\cdot\left(\mathbf{x}^{n+1}-\mathbf{x}^n\right) - \frac{1}{2}\left(\mathbf{x}^{n+1}-\mathbf{x}^n\right)\cdot f''(\mathbf{x}^n)\left(\mathbf{x}^{n+1}-\mathbf{x}^n\right) \\
&\quad +\frac{1}{4}\frac{f(\mathbf{x}^n)\,\mathbf{d}^n\cdot f''(\mathbf{x}^n)\,\mathbf{d}^n}{(\nabla f(\mathbf{x}^n)\cdot\mathbf{d}^n)^2}\left(\mathbf{x}^{n+1}-\mathbf{x}^n\right)\cdot f''(\mathbf{x}^n)\left(\mathbf{x}^{n+1}-\mathbf{x}^n\right). \qquad \text{(A.4)}
\end{aligned}$$

So, by induction

$$\begin{aligned}
\left|f\left(\mathbf{x}^{n+1}\right)\right| &\le \frac{N}{6}\left\|\mathbf{x}^{n+1}-\mathbf{x}^n\right\|^3 + \frac{M^2}{4}\frac{|f(\mathbf{x}^n)|}{(\nabla f(\mathbf{x}^n)\cdot\mathbf{d}^n)^2}\left\|\mathbf{x}^{n+1}-\mathbf{x}^n\right\|^2 \\
&\le \left\{\frac{N}{6}\left\|\mathbf{x}^{n+1}-\mathbf{x}^n\right\| + \frac{M^2}{4}\frac{\varphi(y^n)}{(\varphi'(y^n))^2}\right\}\left(y^{n+1}-y^n\right)^2, \text{ by induction} \\
&= \left\{\frac{N}{6}|\alpha^n|\,\|\mathbf{v}^n\|\,|f(\mathbf{x}^n)| + \frac{M^2}{4}\frac{\varphi(y^n)}{(\varphi'(y^n))^2}\right\}\left(y^{n+1}-y^n\right)^2 \\
&\le \left\{\frac{N}{6}\frac{-\frac{\varphi(y^n)}{\varphi'(y^n)}}{1-\frac{M}{2}\frac{\varphi(y^n)}{(\varphi'(y^n))^2}} + \frac{M^2}{4}\frac{\varphi(y^n)}{(\varphi'(y^n))^2}\right\}\left(y^{n+1}-y^n\right)^2 \\
&= \left\{\frac{N}{6}\frac{-\varphi'(y^n)}{1-\frac{M}{2}\frac{\varphi(y^n)}{(\varphi'(y^n))^2}} + \frac{M^2}{4}\right\}\frac{\varphi(y^n)}{(\varphi'(y^n))^2}\left(y^{n+1}-y^n\right)^2.
\end{aligned}$$

The following result is analogous to (A.4), and is proved the same way

$$\varphi\left(y^{n+1}\right) = \frac{C^2}{4}\frac{\varphi(y^n)}{(\varphi'(y^n))^2}\left(y^{n+1}-y^n\right)^2. \qquad \text{(A.5)}$$

For any $y \in [0, r_1]$ the function $\frac{\varphi(y)}{(\varphi'(y))^2}$ is decreasing, since

$$\left(\frac{\varphi(y)}{(\varphi'(y))^2}\right)' = \frac{(r_2 - r_1)^2}{\frac{C}{2}(2y - r_1 - r_2)^3} < 0, \; y \in [0, r_1] ,$$

$$\text{and } \frac{\varphi(y^n)}{(\varphi'(y^n))^2} \le \frac{\varphi(y^0)}{(\varphi'(y^0))^2} .$$

$$\therefore \frac{-\varphi'(y^n)}{1 - \frac{M}{2}\frac{\varphi(y^n)}{(\varphi'(y^n))^2}} \le \frac{-\varphi'(y^0)}{1 - \frac{M}{2}\frac{\varphi(y^0)}{(\varphi'(y^0))^2}} = \frac{1}{L(1 - \frac{M}{2}LB)} . \tag{A.6}$$

Consequently,

$$\begin{aligned} \left|f\left(\mathbf{x}^{n+1}\right)\right| &\le \frac{1}{4}\left(M^2 + \frac{2}{3}\frac{N}{L\left(1 - \frac{MBL}{2}\right)}\right)\frac{\varphi(y^n)}{(\varphi'(y^n))^2}\left(y^{n+1} - y^n\right)^2, \text{ by (A.6)} \\ &= \frac{C^2}{4}\frac{\varphi(y^n)}{(\varphi'(y^n))^2}\left(y^{n+1} - y^n\right)^2 = \varphi\left(y^{n+1}\right), \text{ by (A.5)} . \end{aligned}$$

Proof of (A.3b) for $n+1$:

$$\begin{aligned} \left\|\mathbf{v}^{n+1}\right\| &= \frac{1}{|\nabla f(\mathbf{x}^{n+1}) \cdot \mathbf{d}^{n+1}|} = \frac{1}{\left|\nabla f\left(\mathbf{x}^0\right) \cdot \mathbf{d}^{n+1} - \left(\nabla f\left(\mathbf{x}^0\right) - \nabla f\left(\mathbf{x}^{n+1}\right)\right) \cdot \mathbf{d}^{n+1}\right|} \\ &= \frac{1}{\left|\nabla f\left(\mathbf{x}^0\right) \cdot \mathbf{d}^{n+1}\right| \left|1 - \frac{(\nabla f(\mathbf{x}^0) - \nabla f(\mathbf{x}^{n+1})) \cdot \mathbf{d}^{n+1}}{\nabla f(\mathbf{x}^0) \cdot \mathbf{d}^{n+1}}\right|} \\ &\le \frac{1}{\left|\nabla f\left(\mathbf{x}^0\right) \cdot \mathbf{d}^0\right| \left|1 - \frac{(\nabla f(\mathbf{x}^0) - \nabla f(\mathbf{x}^{n+1})) \cdot \mathbf{d}^{n+1}}{\nabla f(\mathbf{x}^0) \cdot \mathbf{d}^0}\right|}, \text{ by (17)} , \\ &\le \frac{L}{1 - LM\left\|\mathbf{x}^{n+1} - \mathbf{x}^0\right\|}, \text{ by Lemma 2} , \\ &\le \frac{L}{1 - LCy^{n+1}} = \frac{1}{\frac{1}{L} - Cy^{n+1}} = \frac{1}{-\varphi'\left(y^{n+1}\right)} . \end{aligned}$$

Proof of (A.3c) for $n+1$:

$$\begin{aligned} \left|\alpha^{n+1}\right| &= \frac{1}{\left|1 - \frac{1}{2}f\left(\mathbf{x}^{n+1}\right)\mathbf{v}^{n+1} \cdot f''\left(\mathbf{x}^{n+1}\right)\mathbf{v}^{n+1}\right|} \\ &\le \frac{1}{1 - \frac{M}{2}\frac{\varphi(y^{n+1})}{(\varphi'(y^{n+1}))^2}}, \text{ by } (A.3a) \text{ and } (A.3b). \end{aligned}$$

Proof of (A.3d) for $n+1$:

$$\begin{aligned} \left\|\mathbf{x}^{n+2} - \mathbf{x}^{n+1}\right\| &= \left\|\alpha^{n+1}f(\mathbf{x}^{n+1})\mathbf{v}^{n+1}\right\| \le \frac{-\frac{\varphi(y^{n+1})}{\varphi'(y^{n+1})}}{1 - \frac{M}{2}\frac{\varphi(y^{n+1})}{(\varphi'(y^{n+1}))^2}}, \text{ by } (A.3a), (A.3b) \text{ and } (A.3c). \\ &\le \frac{-\frac{\varphi(y^{n+1})}{\varphi'(y^{n+1})}}{1 - \frac{C}{2}\frac{\varphi(y^{n+1})}{(\varphi'(y^{n+1}))^2}} = y^{n+2} - y^{n+1}. \end{aligned}$$

Consequently, (A.3a)–(A.3d) hold for all $n \ge 0$. Since the sequence $\{\mathbf{x}^k\}$ is majorized by the sequence $\{y^k\}$ it follows from Lemma 2 that $\lim_{k\to\infty} \mathbf{x}^k = \mathbf{x}^*$ and $\mathbf{x}^* \in X_0$.

The scalar Halley method has a cubic rate of convergence, [8], and as shown in [9], $|y^{k+1} - y^k| \leq \beta \frac{\theta^{3^k}}{1-\theta^{3^k}}$, where $\theta < 1$. Therefore $\|\mathbf{x}^{k+1} - \mathbf{x}^k\| \leq \beta \frac{\theta^{3^k}}{1-\theta^{3^k}}$, and the sequence $\{\mathbf{x}^k\}$ has at least cubic rate of convergence. □

APPENDIX B: PROOF OF THEOREM 2

Proof. Let the function φ and the sequence $\{y^k\}$ be given by (A.1) and (A.2) respectively, and let $\mathbf{v}^k$ and α^k be defined by (19b)–(19c). We show that $\{y^k\}$ is a majorizing sequence for the sequence $\{\mathbf{x}^k\}$ generated by (9). This is statement (B.1d), proved below for all k. Indeed, we prove for $k = 0, 1, \cdots$,

$$
\begin{aligned}
|f(\mathbf{x}^k)| &\leq \varphi(y^k) , && \text{(B.1a)}\\
\|\mathbf{v}^k\| &\leq -\frac{1}{\varphi'(y^k)} , && \text{(B.1b)}\\
|\alpha^k| &\leq \frac{1}{1 - \frac{M}{2}\frac{\varphi(y^k)}{(\varphi'(y^k))^2}} , && \text{(B.1c)}\\
\|\mathbf{x}^{k+1} - \mathbf{x}^k\| &\leq y^{k+1} - y^k . && \text{(B.1d)}
\end{aligned}
$$

The proof is by induction.
Verification for $k = 0$:

$$
\begin{aligned}
|f(\mathbf{x}^0)| &\leq \frac{B}{L} = \varphi(y^0) = \varphi(0) ,\\
\|\mathbf{v}^0\| \leq L &= -\frac{1}{\varphi'(y^0)} ,\\
|\alpha^0| &= \left| \frac{1}{1 - \frac{1}{2} f(\mathbf{x}^0)\,\mathbf{v}^0 \cdot f''(\mathbf{x}^0)\,\mathbf{v}^0} \right| \leq \frac{1}{1 - \frac{1}{2} M \frac{\varphi(y^0)}{(\varphi'(y^0))^2}},\\
\|\mathbf{x}^1 - \mathbf{x}^0\| &= \|\alpha^0 \mathbf{v}^0 f(\mathbf{x}^0)\| \leq |\alpha^0|\, \|\mathbf{v}^0\|\, |f(\mathbf{x}^0)|\\
&\leq \frac{-\frac{\varphi(y^0)}{\varphi'(y^0)}}{1 - \frac{1}{2} M \frac{\varphi(y^0)}{(\varphi'(y^0))^2}} \leq \frac{-\frac{\varphi(y^0)}{\varphi'(y^0)}}{1 - \frac{1}{2} C \frac{\varphi(y^0)}{(\varphi'(y^0))^2}} = y^1 - y^0,
\end{aligned}
$$

showing that equations (B.1a)-(B.1d) hold for $k = 0$.
Suppose (B.1a)-(B.1d) hold for $k \leq n$.
Proof of (B.1a) for $n + 1$:

$$
\begin{aligned}
\|\mathbf{x}^{n+1} - \mathbf{x}^0\| &= \left\| \sum_{k=0}^{n} \left(\mathbf{x}^{k+1} - \mathbf{x}^k \right) \right\| \leq \sum_{k=0}^{n} \left(y^{k+1} - y^k \right) = y^{n+1} - y^0 = y^{n+1} \leq r_1 .\\
\therefore\ \mathbf{x}^{n+1} &\in X_0 .
\end{aligned}
$$

Let

$$
\begin{aligned}
a_n(\mathbf{x}) &:= (\mathbf{x}-\mathbf{x}^n)\,(\|\nabla f(\mathbf{x}^n)\|^2 \\
&\quad -\frac{f(\mathbf{x}^n)}{2\,\|\nabla f(\mathbf{x}^n)\|^2}\nabla f(\mathbf{x}^n)\cdot f''(\mathbf{x}^n)\,\nabla f(\mathbf{x}^n)) + f(\mathbf{x}^n)\nabla f(\mathbf{x}^n)\,. \\
\therefore\ a_n(\mathbf{x}) &= (\mathbf{x}-\mathbf{x}^{n+1})(\|\nabla f(\mathbf{x}^n)\|^2 \\
&\quad -\frac{f(\mathbf{x}^n)}{2\,\|\nabla f(\mathbf{x}^n)\|^2}\nabla f(\mathbf{x}^n)\cdot f''(\mathbf{x}^n)\,\nabla f(\mathbf{x}^n)) + (\mathbf{x}^{n+1}-\mathbf{x}^n)(\|\nabla f(\mathbf{x}^n)\|^2 \\
&\quad -\frac{f(\mathbf{x}^n)}{2\,\|\nabla f(\mathbf{x}^n)\|^2}\nabla f(\mathbf{x}^n)\cdot f''(\mathbf{x}^n)\,\nabla f(\mathbf{x}^n)) + f(\mathbf{x}^n)\nabla f(\mathbf{x}^n) \\
&= (\mathbf{x}-\mathbf{x}^{n+1})(\|\nabla f(\mathbf{x}^n)\|^2 - \frac{f(\mathbf{x}^n)}{2\,\|\nabla f(\mathbf{x}^n)\|^2}\nabla f(\mathbf{x}^n)\cdot f''(\mathbf{x}^n)\,\nabla f(\mathbf{x}^n))\,.
\end{aligned}
$$

Let

$$
\begin{aligned}
p_n(\mathbf{x}) &:= \frac{\nabla f(\mathbf{x}^n)}{\|\nabla f(\mathbf{x}^n)\|^2}\cdot a_n(\mathbf{x}) \\
&= (\mathbf{x}-\mathbf{x}^n)\cdot\left(\nabla f(\mathbf{x}^n) - \frac{f(\mathbf{x}^n)}{2\,\|\nabla f(\mathbf{x}^n)\|^2} f''(\mathbf{x}^n)\,\nabla f(\mathbf{x}^n)\right) + f(\mathbf{x}^n)\,.
\end{aligned}
$$

Note that $p_n(\mathbf{x}^{n+1}) = 0$. On the other hand, $p_n(\mathbf{x}^{n+1})$ can be represented as

$$
\begin{aligned}
p_n(\mathbf{x}^{n+1}) &= f(\mathbf{x}^n) + \nabla f(\mathbf{x}^n)\cdot(\mathbf{x}^{n+1}-\mathbf{x}^n) - \frac{f(\mathbf{x}^n)}{2\,\|\nabla f(\mathbf{x}^n)\|^2}\nabla f(\mathbf{x}^n)\cdot f''(\mathbf{x}^n)\,(\mathbf{x}^{n+1}-\mathbf{x}^n) \\
&\quad +\frac{1}{2}(\mathbf{x}^{n+1}-\mathbf{x}^n)\cdot f''(\mathbf{x}^n)\,(\mathbf{x}^{n+1}-\mathbf{x}^n) - \frac{1}{2}(\mathbf{x}^{n+1}-\mathbf{x}^n)\cdot f''(\mathbf{x}^n)\,(\mathbf{x}^{n+1}-\mathbf{x}^n) \\
&= f(\mathbf{x}^n) + \nabla f(\mathbf{x}^n)\cdot(\mathbf{x}^{n+1}-\mathbf{x}^n) + \frac{1}{2}(\mathbf{x}^{n+1}-\mathbf{x}^n)\cdot f''(\mathbf{x}^n)\,(\mathbf{x}^{n+1}-\mathbf{x}^n) \\
&\quad -\frac{1}{2}\left(\frac{f(\mathbf{x}^n)\,\nabla f(\mathbf{x}^n)}{\|\nabla f(\mathbf{x}^n)\|^2} + \mathbf{x}^{n+1}-\mathbf{x}^n\right)\cdot f''(\mathbf{x}^n)\,(\mathbf{x}^{n+1}-\mathbf{x}^n) \\
&= f(\mathbf{x}^n) + \nabla f(\mathbf{x}^n)\cdot(\mathbf{x}^{n+1}-\mathbf{x}^n) + \frac{1}{2}(\mathbf{x}^{n+1}-\mathbf{x}^n)\cdot f''(\mathbf{x}^n)\,(\mathbf{x}^{n+1}-\mathbf{x}^n) \\
&\quad -\frac{1}{4}\frac{f(\mathbf{x}^n)\,\nabla f(\mathbf{x}^n)\cdot f''(\mathbf{x}^n)\,\nabla f(\mathbf{x}^n)}{\|\nabla f(\mathbf{x}^n)\|^4}(\mathbf{x}^{n+1}-\mathbf{x}^n)\cdot f''(\mathbf{x}^n)\,(\mathbf{x}^{n+1}-\mathbf{x}^n)\,. \\
\therefore\ 0 &= f(\mathbf{x}^n) + \nabla f(\mathbf{x}^n)\cdot(\mathbf{x}^{n+1}-\mathbf{x}^n) + \frac{1}{2}(\mathbf{x}^{n+1}-\mathbf{x}^n)\cdot f''(\mathbf{x}^n)\,(\mathbf{x}^{n+1}-\mathbf{x}^n) \\
&\quad -\frac{1}{4}\frac{f(\mathbf{x}^n)\,\nabla f(\mathbf{x}^n)\cdot f''(\mathbf{x}^n)\,\nabla f(\mathbf{x}^n)}{\|\nabla f(\mathbf{x}^n)\|^4}(\mathbf{x}^{n+1}-\mathbf{x}^n)\cdot f''(\mathbf{x}^n)\,(\mathbf{x}^{n+1}-\mathbf{x}^n)\,.
\end{aligned}
$$

Thus,

$$
\begin{aligned}
f(\mathbf{x}^{n+1}) &= f(\mathbf{x}^{n+1}) - f(\mathbf{x}^n) - \nabla f(\mathbf{x}^n)\cdot(\mathbf{x}^{n+1}-\mathbf{x}^n) \\
&\quad -\frac{1}{2}(\mathbf{x}^{n+1}-\mathbf{x}^n)\cdot f''(\mathbf{x}^n)\,(\mathbf{x}^{n+1}-\mathbf{x}^n) \\
&\quad +\frac{1}{4}\frac{f(\mathbf{x}^n)\,\nabla f(\mathbf{x}^n)\cdot f''(\mathbf{x}^n)\,\nabla f(\mathbf{x}^n)}{\|\nabla f(\mathbf{x}^n)\|^4}(\mathbf{x}^{n+1}-\mathbf{x}^n)\cdot f''(\mathbf{x}^n)\,(\mathbf{x}^{n+1}-\mathbf{x}^n)\,.
\end{aligned}
$$

So, by induction

$$
|f(\mathbf{x}^{n+1})| \le \frac{N}{6}\left\|\mathbf{x}^{n+1}-\mathbf{x}^n\right\|^3 + \frac{M^2}{4}\frac{|f(\mathbf{x}^n)|}{\|\nabla f(\mathbf{x}^n)\|^2}\left\|\mathbf{x}^{n+1}-\mathbf{x}^n\right\|^2
$$

$$\leq \left\{ \frac{N}{6} \left\| \mathbf{x}^{n+1} - \mathbf{x}^{n} \right\| + \frac{M^2}{4} \frac{\varphi(y^n)}{(\varphi'(y^n))^2} \right\} \left(y^{n+1} - y^n\right)^2, \text{ by induction}$$

$$= \left\{ \frac{N}{6} |\alpha^n| \left\| \mathbf{v}^n \right\| |f(\mathbf{x}^n)| + \frac{M^2}{4} \frac{\varphi(y^n)}{(\varphi'(y^n))^2} \right\} \left(y^{n+1} - y^n\right)^2$$

$$\leq \left\{ \frac{N}{6} \frac{-\frac{\varphi(y^n)}{\varphi'(y^n)}}{1 - \frac{M}{2}\frac{\varphi(y^n)}{(\varphi'(y^n))^2}} + \frac{M^2}{4} \frac{\varphi(y^n)}{(\varphi'(y^n))^2} \right\} \left(y^{n+1} - y^n\right)^2$$

$$= \left\{ \frac{N}{6} \frac{-\varphi'(y^n)}{1 - \frac{M}{2}\frac{\varphi(y^n)}{(\varphi'(y^n))^2}} + \frac{M^2}{4} \right\} \frac{\varphi(y^n)}{(\varphi'(y^n))^2} \left(y^{n+1} - y^n\right)^2.$$

For any $y \in [0, r_1]$ the function $\frac{\varphi(y)}{(\varphi'(y))^2}$ is decreasing, since

$$\left(\frac{\varphi(y)}{(\varphi'(y))^2}\right)' = \frac{(r_2 - r_1)^2}{\frac{C}{2}(2y - r_1 - r_2)^3} < 0, \; y \in [0, r_1].$$

$$\text{Also, } \frac{\varphi(y^n)}{(\varphi'(y^n))^2} \leq \frac{\varphi(y^0)}{(\varphi'(y^0))^2}.$$

$$\therefore \frac{-\varphi'(y^n)}{1 - \frac{M}{2}\frac{\varphi(y^n)}{(\varphi'(y^n))^2}} \leq \frac{-\varphi'(y^0)}{1 - \frac{M}{2}\frac{\varphi(y^0)}{(\varphi'(y^0))^2}} = \frac{1}{L(1 - \frac{M}{2}LB)}.$$

$$\therefore \left|f\left(\mathbf{x}^{n+1}\right)\right| \leq \frac{1}{4}\left(M^2 + \frac{2}{3}\frac{N}{L\left(1 - \frac{MBL}{2}\right)}\right) \frac{\varphi(y^n)}{(\varphi'(y^n))^2} \left(y^{n+1} - y^n\right)^2$$

$$= \frac{C^2}{4} \frac{\varphi(y^n)}{(\varphi'(y^n))^2} \left(y^{n+1} - y^n\right)^2 = \varphi\left(y^{n+1}\right), \text{ by (A.5)}.$$

<u>Proof of (B.1b) for $n+1$:</u>

$$\left\|\mathbf{v}^{n+1}\right\| = \frac{1}{\left\|\nabla f\left(\mathbf{x}^{n+1}\right)\right\|} = \frac{1}{\left\|\nabla f\left(\mathbf{x}^0\right) - \left(\nabla f\left(\mathbf{x}^0\right) - \nabla f\left(\mathbf{x}^{n+1}\right)\right)\right\|}$$

$$= \frac{1}{\left\|\nabla f\left(\mathbf{x}^0\right)\right\| \left\| \frac{\nabla f(\mathbf{x}^0)}{\|\nabla f(\mathbf{x}^0)\|} - \frac{\nabla f(\mathbf{x}^0) - \nabla f(\mathbf{x}^{n+1})}{\|\nabla f(\mathbf{x}^0)\|} \right\|} \leq \frac{L}{1 - LM\left\|\mathbf{x}^{n+1} - \mathbf{x}^0\right\|}, \text{ by lemma 2}$$

$$\leq \frac{L}{1 - LCy^{n+1}} = \frac{1}{\frac{1}{L} - Cy^{n+1}} = \frac{1}{-\varphi'\left(y^{n+1}\right)}.$$

<u>Proof of (B.1c) for $n+1$:</u>

$$\left|\alpha^{n+1}\right| = \frac{1}{\left|1 - \frac{1}{2}f\left(\mathbf{x}^{n+1}\right)\mathbf{v}^{n+1} \cdot f''\left(\mathbf{x}^{n+1}\right)\mathbf{v}^{n+1}\right|}$$

$$\leq \frac{1}{1 - \frac{M}{2}\frac{\varphi(y^{n+1})}{(\varphi'(y^{n+1}))^2}}, \text{ by } (A.3a) \text{ and } (A.3b).$$

Proof of (B.1d) for $n+1$:

$$\left\|\mathbf{x}^{n+2}-\mathbf{x}^{n+1}\right\| = \left\|\alpha^{n+1} f(\mathbf{x}^{n+1})\mathbf{v}^{n+1}\right\| \le \frac{-\frac{\varphi(y^{n+1})}{\varphi'(y^{n+1})}}{1-\frac{M}{2}\frac{\varphi(y^{n+1})}{(\varphi'(y^{n+1}))^2}}, \text{ by } (A.3a), (A.3b) \text{ and } (A.3c).$$

$$\le \frac{-\frac{\varphi(y^{n+1})}{\varphi'(y^{n+1})}}{1-\frac{C}{2}\frac{\varphi(y^{n+1})}{(\varphi'(y^{n+1}))^2}} = y^{n+2}-y^{n+1}.$$

Consequently, (B.1a)–(B.1d) hold for all $n \ge 0$. So by Lemma 2, $\lim_{k\to\infty} \mathbf{x}^k = \mathbf{x}^*$ and $\mathbf{x}^* \in X_0$.

So, the sequence $\{\mathbf{x}^k\}$ is majorized by the sequence $\{y^k\}$ generated by the scalar Halley method. The scalar Halley method has a cubic rate of convergence, [8], and $|y^{k+1}-y^k| \le \beta\frac{\theta^{3^k}}{1-\theta^{3^k}}$, where $\theta < 1$, [9]. So, $\left\|\mathbf{x}^{k+1}-\mathbf{x}^k\right\| \le \beta\frac{\theta^{3^k}}{1-\theta^{3^k}}$, and the sequence $\{\mathbf{x}^k\}$ has at least cubic rate of convergence. □

APPENDIX C: MAPLE PROGRAMS

Note. In the examples below all equations have zero RHS's, so the values of the functions give an indication of the error. Also, the functions use a vector variable x of unspecified dimension, making it necessary to define the dimension, say

```
> x:=array(1..3):
```

before using a function. We use the *linalg* package, so

```
> restart:with(linalg):
```

C.1. The directional Halley Method (8).

The function **HalleyDirNext(f,x,x0,d)** computes the next directional Halley iterate for $f(\mathbf{x})$ at $\mathbf{x}^0$ in the direction $\mathbf{d}$

```
> HalleyDirNext:=proc(f,x,x0,d)
> local val,gr,c,cc,hes;
> val:=eval(subs(x=x0,f)):
> gr:=eval(subs(x=x0,grad(f,x))):
> c:=dotprod(gr,d):
> hes:=eval(subs(x=x0,hessian(f,x))):
> cc:=dotprod(d,evalm(hes &* d)):
> evalm(x0-(val/(c-(cc*val)/(2*c))*d));
> end:
```

C.2. The gradient Halley Method (9).

The function **HalleyGrad(f,x,x0,N)** computes N iterations for $f(\mathbf{x})$ starting at $\mathbf{x}^0$

```
> HalleyGrad:=proc(f,x,x0,N)
> local d,sol,valf; global k;
> k:=0; sol:=array(0..N):sol[0]:=x0:
> valf:=eval(subs(x=x0,f)):
> print(f);
> lprint(Iterate,0):print(sol[0]):
> lprint(function):print(valf):
```

```
> for k from 1 to N do d:=eval(subs(x=sol[k-1],grad(f,x))):
> sol[k]:=HalleyDirNext(f,x,sol[k-1],d):
> valf:=eval(subs(x=sol[k],f)):
> if (sqrt(dotprod(sol[k]-sol[k-1],sol[k]-sol[k-1]))<eps) then
> break fi:
> od:
> lprint(Iterate,k-1):print(sol[k-1]):
> lprint(function):print(valf):
> end:
```

Example C.2. $f(x) = \exp(1 - x_1^2 - x_2) - 1$, $x^0 = (1.0, 1.2)$, 10 *iterations.*

```
> HalleyGrad(exp(1-x[1]-x[2])-1,x,[1.,1.2],3);
```

$e^{(1-x_1-x_2)} - 1$

```
Iterate   0
```

$$[1., \ 1.2]$$

```
Function
```

$$-.6988057881$$

```
Iterate   3
```

$$[.4000000000, \ .6000000000]$$

```
Function
```

$$0$$

C.3. The gradient quasi-Halley Method (10).

The function **QuasiDirNext(f, x, x0, d)** computes the next directional quasi-Halley iterate of $f(\mathbf{x})$ at $\mathbf{x}^0$ in the direction $\nabla f(\mathbf{x}^0)$.

```
> QuasiDirNext:=proc(f,x,x0)
> local val,val1,x1,gr,c; val:=eval(subs(x=x0,f)):
> gr:=eval(subs(x=x0,grad(f,x))):  c:=dotprod(gr,gr):
> x1:=evalm(x0-(val/c)*gr):  val1:=eval(subs(x=x1,f)):
> if (val1<>val) then evalm(x0-val/(c*(1-val1/val))*gr)
> else evalm(x0-val/(c)*gr); fi:
> end:
```

The function **QuasiGrad(f,x,x0,N)** computes N iterations for the function $f(\mathbf{x})$ starting at $\mathbf{x}^0$.

```
> QuasiGrad:=proc(f,x,x0,N)
> local d,sol,valf; global k; k:=0;
> sol:=array(0..N):
> sol[0]:=x0:valf:=eval(subs(x=x0,f)):
> print(f); lprint(Iterate,0):print(sol[0]):lprint(Function):print(valf):
> for k from 1 to N do
> sol[k]:=QuasiDirNext(f,x,sol[k-1]):
> valf:=eval(subs(x=sol[k],f)):
> if (sqrt(dotprod(sol[k]-sol[k-1],sol[k]-sol[k-1]))<eps) then break fi:
> od:
> lprint(Iterate,k-1):print(sol[k-1]):  lprint(Function):print(valf):
> end:
```

Example C.3. $f(x) = x_1^2 - x_2$, $x^0 = (2.1, 1.2)$, 4 *iterations.*

```
> QuasiGrad(x[1]^2-x[2],x,[2.1,1.2],4);
```

$$x_1{}^2 - x_2$$

```
Iterate   0
```

$$[2.1,\ 1.2]$$

```
Function
```

$$3.21$$

```
Iterate   3
```

$$[1.192944003,\ 1.423115393]$$

```
Function
```

$$.1\,10^{-8}$$

C.4. Systems of equations.

The function **SOS(x)** computes the sum of squares of the components of the vector **x**. It is used in some of the functions below, and works better than the MAPLE function `norm(x,2)`^{2} which is not differentiable if any $x_i = 0$.

```
> SOS:=proc(x)
> local n,k;
> n:=vectdim(x):
> sum(x[k]^2,k=1..n);
> end:
```

The function **SystemHalleyGrad(f,x,x0,N)** computes N iterations of the directional Halley method for the sum of squares $\sum_{i=1}^{n} f_i^2(x)$, starting at x^0.

```
> SystemHalleyGrad:=proc(f,x,x0,N)
> local n,F;
> n:=vectdim(x0);F:=SOS(f);print(f);
> HalleyGrad(F,x,x0,N);
> end:
```

The function **SystemQuasiGrad(f,x,x0,N)** computes N iterations of the directional quasi-Halley method for the sum of squares $\sum_{i=1}^{n} f_i^2(x)$, starting at x^0.

```
> SystemQuasiGrad:=proc(f,x,x0,N)
> local n,F;
> n:=vectdim(x0);F:=SOS(f);print(f);
> QuasiGrad(F,x,x0,N);
> end:
```

Example C.4. *Fröberg, p. 186, Example 1*

```
> x:=array(1..3):
> SystemHalleyGrad([x[1]^2-x[1]+x[2]^3+x[3]^5,x [1]^3+x[2]^5-x[2]+x[3]^7,
> x[1]^5+x[2]^7+x[3]^11-x[3]],x,[0.4,0.3,0.2],10):
```

$$[x_1{}^2 - x_1 + x_2{}^3 + x_3{}^5,\ x_1{}^3 + x_2{}^5 - x_2 + x_3{}^7,\ x_1{}^5 + x_2{}^7 + x_3{}^{11} - x_3]$$

$$(x_1{}^2 - x_1 + x_2{}^3 + x_3{}^5)^2 + (x_1{}^3 + x_2{}^5 - x_2 + x_3{}^7)^2 + (x_1{}^5 + x_2{}^7 + x_3{}^{11} - x_3)^2$$

```
Iterate   0
```

$$[.4,\ .3,\ .2]$$

```
Function
```

$$.1357076447$$

```
Iterate   10
```

$$[.002243051296,\ .0002858171153,\ -.0002540074383]$$

```
Function
```

$$.5154938245\,10^{-5}$$

```
> x:=array(1..3):
> SystemQuasiGrad([x[1]^2-x[1]+x[2]^3+x[3]^5,x[ 1]^3+x[2]^5-x[2]+x[3]^7,
> x[1]^5+x[2]^7+x[3]^11-x[3]],x,[0.4,0.3,0.2],10);
```

$$[{x_1}^2 - x_1 + {x_2}^3 + {x_3}^5,\ {x_1}^3 + {x_2}^5 - x_2 + {x_3}^7,\ {x_1}^5 + {x_2}^7 + {x_3}^{11} - x_3]$$

$$({x_1}^2 - x_1 + {x_2}^3 + {x_3}^5)^2 + ({x_1}^3 + {x_2}^5 - x_2 + {x_3}^7)^2 + ({x_1}^5 + {x_2}^7 + {x_3}^{11} - x_3)^2$$

```
Iterate   0
```

$$[.4,\ .3,\ .2]$$

```
Function
```

$$.1357076447$$

```
Iterate   10
```

$$[.0001876563761,\ .4627014469\,10^{-5},\ -.3061094461\,10^{-5}]$$

```
Function
```

$$.3523247963\,10^{-7}$$

Inherently Parallel Algorithms in Feasibility and Optimization and their Applications
D. Butnariu, Y. Censor and S. Reich (Editors)
2001 Elsevier Science B.V.

ERGODIC CONVERGENCE TO A ZERO OF THE EXTENDED SUM OF TWO MAXIMAL MONOTONE OPERATORS

Abdellatif Moudafi[a] and Michel Théra[b]

[a]Université des Antilles et de la Guyane
Département de Mathématiques EA 2431, Pointe-à-Pitre, Guadeloupe and
Département Scientifique Interfacultaires, B.P 7168, 97278 Schoelcher Cedex
Martinique, France

[b]LACO, UMR-CNRS 6090, Université de Limoges
123, avenue Albert Thomas, 87060 Limoges Cedex France

In this note we show that the splitting scheme of Passty [13] as well as the barycentric-proximal method of Lehdili & Lemaire [8] can be used to approximate a zero of the extended sum of maximal monotone operators. When the extended sum is maximal monotone, we generalize a convergence result obtained by Lehdili & Lemaire for convex functions to the case of maximal monotone operators. Moreover, we recover the main convergence results of Passty and Lehdili & Lemaire when the pointwise sum of the involved operators is maximal monotone.

1. INTRODUCTION AND PRELIMINARIES

A wide range of problems in physics, economics and operation research can be formulated as a generalized equation $0 \in T(x)$ for a given set-valued mapping T on a Hilbert space X. Therefore, the problem of finding a zero of T, i.e., a point $x \in X$ such that $0 \in T(x)$ is a fundamental problem in many areas of applied mathematics.

When T is a maximal monotone operator, a classical method for solving the problem $0 \in T(x)$ is the *Proximal Point Algorithm*, proposed by Rockafellar [19] which extends an earlier algorithm established by Martinet [10] for $T = \partial f$, i.e., when T is the subdifferential of a convex lower semicontinuous proper function. In this case, finding a zero of T is equivalent to the problem of finding a minimizer of f.

The case when T is the pointwise sum of two operators A and B is called a *splitting* of T. It is of fundamental interest in large-scale optimization since the objective function splits into the sum of two simpler functions and we can take advantage of this separable structure. For an overview of various splitting methods, we refer to Eckstein [7] and, for earlier contributions on the proximal point algorithm related to this paper, to Brézis & Lions [3], Bruck & Reich [5] and Nevanlinna & Reich [11]. Let us also mention that ergodic convergence results appear in Bruck [4] and in Reich [14], for example.

Using conjugate duality, splitting methods may apply in certain circumstances to the dual objective function. Recall that the general framework of the conjugate duality is the following [18]:

Consider a convex lower semicontinuous function f on the product $H \times U$ of two Hilbert spaces H and U. Define
$L(x,v) := \inf_u\{f(x,u) - \langle u, v\rangle\}$ and $g(p,v) := \inf_x\{L(x,v) - \langle x,p\rangle\}$.
Setting $f_0(x) := f(x,0)$ and $g_0(v) := g(0,v)$, a well-known method to solve $\inf f_0$ is the *method of multipliers* which consists in solving the dual problem $\max g_0$ using the *Proximal Point Algorithm* [20].

It has been observed that in certain situations the study of a problem involving monotone operators leads to an operator that turns out to be larger than the pointwise sum. Consequently, there have been several attempts to generalize the usual pointwise sum of two monotone operators such as for instance the well-known extension based on the Trotter-Lie formula. More recently, in 1994, the notion of *variational sum* of two maximal monotone operators was introduced in [1] by Attouch, Baillon and Théra using the Yosida regularization of operators as well as the graph convergence of operators. Recently, another concept of extended sum was proposed by Revalski and Théra [15] relying on the so-called ε-enlargements of operators.

Our focus in this paper is on finding a zero of the extended sum of two monotone operators when this extended sum is maximal monotone. Since, when the pointwise sum is maximal monotone, the extended sum and the pointwise sum coincides, the proposed algorithm will subsume the classical *Passty scheme* and the *barycentric-proximal method* of Lehdili and Lemaire which are related to weak ergodic type convergence.

Throughout we will assume that X is a real Hilbert space. The inner product and the associated norm will be designated respectively by $\langle\cdot,\cdot\rangle$ and $\|\cdot\|$. Given a (multivalued) operator $A: X \rightrightarrows X$, the *graph* of A is denoted by $\mathrm{Gr}(A) := \{(x,u) \in X \times X \mid u \in Ax\}$, its *domain* by $\mathrm{Dom}\,(A) := \{x \in X \mid\ Ax \neq \emptyset\}$ and its inverse operator is $A^{-1}: X \rightrightarrows X$, $A^{-1}u := \{x \in X \mid u \in Ax\}$, $y \in X$. The operator A is called *monotone* if $\langle y-x, v-u\rangle \geq 0$, whenever $(x,u) \in \mathrm{Gr}\, A$ and $(y,v) \in \mathrm{Gr}\, A$. We denote by $\bar{A}$ the operator $\bar{A}x := \overline{Ax}$, $x \in X$, where the overbar means the norm-closure of a given set. The monotone operator A is said to be *maximal* if its graph is not contained properly in the graph of any other monotone operator from X to X. The graph $\mathrm{Gr}\,(A)$ is a closed subset with respect to the product of the norm topologies in $X \times X$.

Finally, given a maximal monotone operator $A: X \rightrightarrows X$ and a positive λ, recall that the *Yosida regularization* of A of order λ is the operator $A_\lambda := (A^{-1} + \lambda I)^{-1}$, and that the *resolvent* of A of order λ is the operator $J_\lambda^A := (I + \lambda A)^{-1}$, where I is the identity mapping. For any $\lambda > 0$, the Yosida regularization A_λ and the resolvent J_λ^A are everywhere defined single-valued maximal monotone operators.

Let $f : X \to \mathbb{R} \cup \{+\infty\}$ be an extended real-valued lower semicontinuous convex function in X which is *proper* (i.e. the *domain* $\mathrm{dom}\, f := \{x \in X : f(x) < +\infty\}$ of f is non-empty). Given $\varepsilon \geq 0$, the *ε-subdifferential* of f is defined at $x \in \mathrm{dom}\, f$ by:

$$\partial_\varepsilon f(x) := \Big\{u \in X \mid f(y) - f(x) \geq \langle y - x, u\rangle - \varepsilon \quad \text{for every } y \in X \Big\},$$

and $\partial_\varepsilon f(x) := \emptyset$, if $x \notin \mathrm{dom}\, f$. When $\varepsilon = 0$, $\partial_0 f$ is the *subdifferential* ∂f of f, which, as it is well-known, is a maximal monotone operator.

The concept of approximate subdifferential leads to similar enlargements for monotone operators. One which has been investigated intensively in the past few years is the following: for a monotone operator $A : X \rightrightarrows X$ and $\varepsilon \geq 0$, the *ε-enlargement* of A is $A^\varepsilon : X \rightrightarrows X$, defined by $A^\varepsilon x := \{u \in X : \langle y - x, v - u\rangle \geq -\varepsilon \text{ for any } (y, v) \in \mathrm{Gr}\,(A)\}$. A^ε has closed convex images for any $\varepsilon \geq 0$ and due to the monotonicity of A, one has $Ax \subset A^\varepsilon x$ for every $x \in X$ and every $\varepsilon \geq 0$. In the case $A = \partial f$ one has $\partial_\varepsilon f \subset (\partial f)^\varepsilon$ and the inclusion can be strict.

2. GENERALIZED SUMS AND SPLITTING METHODS

We start by recalling different types of sums of monotone operators. We then present two splitting methods for finding a zero of the extended sum.

Let $A, B : X \rightrightarrows X$ be two monotone operators. As usual $A + B : X \rightrightarrows X$ denotes the pointwise sum of A and B: $(A + B)x = Ax + Bx$, $x \in X$. $A + B$ is a monotone operator with $\mathrm{Dom}\,(A + B) = \mathrm{Dom}\,A \cap \mathrm{Dom}\,B$. However, even if A and B are maximal monotone operators, their sum $A + B$ may fail to be maximal monotone.

The above lack of maximality of the pointwise sum inspired the study of possible generalized sums of monotone operators. Recently, the *variational sum* was proposed in [1] using the Yosida approximation. More precisely, let $A, B : X \rightrightarrows X$ be maximal monotone operators and $\mathcal{I} := \left\{(\lambda, \mu) \in \mathbb{R}^2 \,|\, \lambda, \mu \geq 0, \quad \lambda + \mu \neq 0\right\}$.The idea of the variational sum, $A \underset{v}{+} B$, is to take as a sum of A and B the graph-convergence limit (i.e. the Painlevé-Kuratowski limit of the graphs) of $A_\lambda + B_\mu$, $(\lambda, \mu) \in \mathcal{I}$, when $(\lambda, \mu) \to 0$. Namely, $A \underset{v}{+} B$ is equal to

$$\liminf_{\mathcal{F}}(A_\lambda + B_\mu) =$$
$$\left\{(x, y) \,|\, \forall\{(\lambda_n, \mu_n)\} \subset \mathcal{I}, \lambda_n, \mu_n \to 0, \exists (x_n, y_n) \in A_\lambda + B_\mu; (x_n, y_n) \to (x, y)\right\}.$$

By contrast to the pointwise sum, this definition also takes into account the behaviour of the operators at points in the neighborhood of the initial point. It follows from this definition that $\mathrm{Dom}\,(A) \cap \mathrm{Dom}\,(B) \subset \mathrm{Dom}\,(A \underset{v}{+} B)$ and that $A \underset{v}{+} B$ is monotone. It was shown in [1] that if $A + B$ is a maximal monotone operator then, $A + B = A \underset{v}{+} B$. Moreover, the subdifferential of the sum of two proper convex lower semicontinuous functions is equal to the variational sum of their subdifferentials.

Another type of generalized sum was proposed by Revalski and Théra [15] relying on the notion of enlargement: the *extended sum* of two monotone operators $A, B : X \rightrightarrows X$ is defined in [15] for each $x \in X$ by

$$A \underset{e}{+} B(x) = \bigcap_{\varepsilon > 0} \overline{A^\varepsilon x + B^\varepsilon x},$$

where the closure on the right hand side is taken with respect to the weak topology. Evidently, $A + B \subset A \underset{e}{+} B$ and hence, $\mathrm{Dom}\,(A) \cap \mathrm{Dom}\,(B) \subset \mathrm{Dom}\,(A \underset{e}{+} B)$. As shown in [15] (Corollary 3.2), if $A+B$ is a maximal monotone operator then, $A+B = A \underset{e}{+} B$. Furthermore, the subdifferential of the sum of two convex proper lower semicontinuous functions is equal to the extended sum of their subdifferentials ([15], Theorem 3.3.)

Let us now recall two splitting methods for the problem of finding a zero of the sum of two maximal monotone operators with maximal monotone sum. The first one has been proposed by Passty and is based on a regularization of one of the operators. Actually, replacing the problem

$$(\mathcal{P}) \quad \text{find} \quad x \in X \quad \text{such that} \quad 0 \in (A+B)x$$

by

$$(\mathcal{P}_\lambda) \quad \text{find} \quad x \in X \quad \text{such that} \quad 0 \in (A+B_\lambda)x$$

leads to the following equivalent fixed-point formulation

$$\text{find} \quad x \in X \quad \text{such that} \quad x = J_\lambda^A \circ J_\lambda^B x. \tag{1}$$

Indeed, $(\mathcal{P}_\lambda)$ can be transformed in the following way

$$0 \in x - J_\lambda^B x + \lambda A x$$

which, in turn as A is maximal monotone, is equivalent to

$$x = J_\lambda^A \circ J_\lambda^B x.$$

Starting from a given initial point $x_0 \in X$ and iterating the above relation with a variable λ_n tending to zero gives the scheme of Passty:

$$x_n = J_{\lambda_n}^A \circ J_{\lambda_n}^B x_{n-1} \quad \forall n \in \mathbb{R}^*. \tag{2}$$

Another approach called the *barycentric-proximal* method was proposed by Lehdili and Lemaire [8]. It is based on a complete regularization of the two operators under consideration and consists in replacing the problem $(\mathcal{P})$ by the problem $(\mathcal{P}_{\lambda,\mu})$:

$$(\mathcal{P}_{\lambda,\mu}) \quad \text{find} \quad x \in X \quad \textit{such that} \quad 0 \in (A_\lambda + B_\mu)x.$$

By definition of the Yosida approximate, $(\mathcal{P}_{\lambda,\mu})$ can be written as a fixed-point problem, namely

$$\text{find} \quad x \in X \quad \text{such that} \quad x = \frac{\mu}{\lambda+\mu} J_\lambda^A x + \frac{\lambda}{\lambda+\mu} J_\mu^B x. \tag{3}$$

The barycentric-proximal method is nothing but the iteration method for (3) with variable parameters. More precisely, for a given $x_0 \in X$, the iteration is given by:

$$\tilde{x}_n = \frac{\mu_n}{\lambda_n+\mu_n} J_{\lambda_n}^A \tilde{x}_{n-1} + \frac{\lambda_n}{\lambda_n+\mu_n} J_{\mu_n}^B \tilde{x}_{n-1} \quad \forall n \in \mathbb{R}^*. \tag{4}$$

For the sake of simplicity, we suppose that $\lambda_n = \mu_n$ for all $n \in \mathbb{R}^*$.

3. THE MAIN RESULT

In what follows, we show that these methods allow to approximate a solution of the problem

$$(\mathcal{Q}) \quad \text{find} \quad x \in X \quad \text{such that} \quad 0 \in (A \underset{e}{+} B)x,$$

where A and B are two maximal monotone operators.

When $A+B$ is maximal monotone, we recover the results by Lehdili and Lemaire for (4) and Passty for (2). Moreover, in the case of convex minimization our result generalizes a theorem of Lehdili and Lemaire. Indeed, in this setting, the extended and the variational sums coincide.

To prove our main result we need the following variant of Opial's lemma [12]:

Lemma 1 *Let $\{\lambda_n\}_{n\in\mathbb{R}}$ be a sequence of positive reals such that $\sum_{n=1}^{+\infty} \lambda_n = +\infty$ and $\{x_n\}_{n\in\mathbb{R}}$ be a sequence with weighted average z_n given by $z_n := \frac{\sum_{k=1}^{n} \lambda_k x_k}{\sum_{k=1}^{n} \lambda_k}$. Let us assume that there exists a nonempty closed convex subset S of X such that*

- *any weak limit of a subsequence of $\{z_n\}_{n\in\mathbb{R}}$ is in S;*
- *$\lim_{n\to+\infty} \|x_n - u\|$ exists for all $u \in S$.*

Then, $\{z_n\}_{n\in\mathbb{R}}$ weakly converges to an element of S.

Theorem 2 *Let A and B be two maximal monotone operators such that $A \underset{e}{+} B$ is a maximal monotone operator and* $\operatorname{Dom} A \cap \operatorname{Dom} B \neq \emptyset$. *Suppose also that problem $(\mathcal{Q})$ has a solution, i.e, the $S := (A \underset{e}{+} B)^{-1}(0)$ is nonempty. Let $\{x_n\}_{n\in\mathbb{R}}$ (resp. $\{\tilde{x}_n\}_{n\in\mathbb{R}}$) be a sequence generated by (2) (resp. by (4)) and z_n (resp. $\tilde{z}_n$) be the corresponding weighted average. Let us assume that*

$$\sum_{n=1}^{+\infty} \lambda_n^2 < +\infty \quad \text{and} \quad \sum_{n=1}^{+\infty} \lambda_n = +\infty.$$

Then, any weak limit point of a subsequence of $\{z_n\}_{n\in\mathbb{R}}$ (resp. $\{\tilde{z}_n\}_{n\in\mathbb{R}}$) is a zero of the extended sum. Moreover, if A^ϵ is locally bounded [1] *on S, then the whole sequence weakly converges to some zero of the extended sum.*

Proof: Let us show that all the assumptions of Lemma 1 are satisfied for $S = (A \underset{e}{+} B)^{-1}(0)$. First let us remark that thanks to the maximal monotonicity of $A \underset{e}{+} B$, the set S is closed and convex and nonempty (by assumption). Take $(x, y) \in A \underset{e}{+} B$. By definition of the extended sum, it amounts to saying that for each $\epsilon > 0$, $y \in$

[1]Recall that $A : X \rightrightarrows X$ is *locally bounded* if for each point x from the norm-closure of Dom (A) there is a neighborhood U of x such that $A(U)$ is a norm-bounded subset in X

$\overline{A^\epsilon(x) + B^\epsilon(x)}$. Equivalently, for all $\epsilon > 0$, there exists a sequence $\{y_{p,\epsilon}\}$ weakly converging to y, with

$$y_{p,\epsilon} = y_{1,p,\epsilon} + y_{2,p,\epsilon},\ y_{1,p,\epsilon} \in A^\epsilon(x) \text{ and } y_{2,p,\epsilon} \in B^\epsilon(x).$$

Define inductively x_n and v_n by $x_n = J^A_{\lambda_n} v_n$ and $v_n = J^B_{\lambda_n} x_{n-1}$. Equivalently we have,

$$\frac{v_n - x_n}{\lambda_n} \in A(x_n) \quad \text{and} \quad \frac{x_{n-1} - v_n}{\lambda_n} \in B(v_n). \tag{5}$$

Definition of ϵ-enlargments combined to relations (5) yields

$$\left\langle \frac{v_n - x_n}{\lambda_n} - y_{1,p,\epsilon}, x_n - x \right\rangle \geq -\epsilon \quad \text{and} \quad \left\langle \frac{x_{n-1} - v_n}{\lambda_n} - y_{2,p,\epsilon}, v_n - x \right\rangle \geq -\epsilon.$$

In other words

$$\langle v_n - x_n, x_n - x \rangle \geq \lambda_n \langle y_{1,p,\epsilon}, x_n - x \rangle - \lambda_n \epsilon \tag{6}$$

and

$$\langle x_{n-1} - v_n, v_n - x \rangle \geq \lambda_n \langle y_{2,p,\epsilon}, v_n - x \rangle - \lambda_n \epsilon. \tag{7}$$

From (6) and (7), using the general equality

$$2 < a - b, b - c >= ||a - c||^2 - ||a - b||^2 - ||b - c||^2$$

we obtain

$$||x_{n-1} - x||^2 - ||v_n - x||^2 \geq ||x_{n-1} - v_n||^2 + 2\lambda_n \langle y_{2,p,\epsilon}, v_n - x \rangle - 2\lambda_n \epsilon$$

and

$$||v_n - x||^2 - ||x_n - x||^2 \geq ||v_n - x_n||^2 + 2\lambda_n \langle y_{1,p,\epsilon}, x_n - x \rangle - 2\lambda_n \epsilon.$$

By adding the two last inequalities, we infer

$$\begin{aligned} ||x_{n-1} - x||^2 - ||x_n - x||^2 \quad & \geq ||x_{n-1} - v_n||^2 + ||v_n - x_n||^2 + 2\lambda_n \langle y_{2,p,\epsilon}, v_n - x \rangle \\ & +2\lambda_n \langle y_{1,p,\epsilon}, x_n - x \rangle - 4\lambda_n \epsilon \\ & \geq ||v_n - x_n||^2 + 2\lambda_n \langle y_{2,p,\epsilon}, v_n - x \rangle \\ & +2\lambda_n \langle y_{1,p,\epsilon}, x_n - x \rangle - 4\lambda_n \epsilon \\ & = ||v_n - x_n||^2 + 2\lambda_n \langle y_{2,p,\epsilon}, v_n - x \rangle - 2\lambda_n \langle y_{2,p,\epsilon}, x_n - x \rangle \\ & +2\lambda_n \langle y_{p,\epsilon}, x_n - x \rangle - 4\lambda_n \epsilon \\ & = ||v_n - x_n||^2 + 2\lambda_n \langle y_{2,p,\epsilon}, v_n - x_n \rangle \\ & +2\lambda_n \langle y_{p,\epsilon}, x_n - x \rangle - 4\lambda_n \epsilon. \end{aligned}$$

Therefore,

$$||x_{n-1} - x||^2 - ||x_n - x||^2 \ \geq -\lambda_n^2 ||y_{2,p,\epsilon}||^2 + 2\lambda_n \langle y_{p,\epsilon}, x_n - x \rangle - 4\lambda_n \epsilon. \tag{8}$$

After summation and division by $\sum_{k=1}^{n}\lambda_k$, we get

$$2\langle y_{p,\epsilon}, z_n - x\rangle \leq \frac{\|x_0 - x\|^2}{\sum_{k=1}^{n}\lambda_k} + \|y_{2,p,\epsilon}\|^2\frac{\sum_{k=1}^{n}\lambda_k^2}{\sum_{k=1}^{n}\lambda_k} + 4\epsilon. \tag{9}$$

Passing to the limit as $n \to +\infty$ for a subsequence, gives

$$\langle y_{p,\epsilon}, \bar{z} - x\rangle \leq 2\epsilon.$$

Then passing to the limit as $p \to +\infty$ and as ϵ goes to zero yields $\langle y, x - \bar{z}\rangle \geq 0$. Thus, thanks to the maximality of $A \underset{e}{+} B$, we derive that $0 \in (A \underset{e}{+} B)\bar{z}$, that is every weak limit point of a subsequence of $\{z_n\}$ is a solution of (Q). Therefore the first assumption of Lemma 1 is satisfied.
Now taking $x \in (A \underset{e}{+} B)^{-1}(0)$ in (8), we obtain

$$\|x_n - x\|^2 \leq \|x_{n-1} - x\|^2 + \lambda_n^2\|y_{2,p,\epsilon}\|^2 - 2\lambda_n\langle y_{p,\epsilon}, x_n - x\rangle + 4\lambda_n\epsilon \quad \forall\epsilon > 0. \tag{10}$$

Local boundedness of A^ϵ and relation $y_{2,p,\epsilon} \in A^\epsilon(x)$ show that $\{y_{2,p,\epsilon}\}$ is bounded when $p \to +\infty$ and when $\epsilon \to 0$. This combined to the fact that $\{y_{p,\epsilon}\} \to 0$ as $p \to +\infty$ gives

$$\|x_n - x\|^2 \leq \|x_{n-1} - x\|^2 + C\lambda_n^2. \tag{11}$$

So, as $\sum_{n=1}^{+\infty}\lambda_n^2 < +\infty$, it is well-known that $\lim_{n\to+\infty}\|x_n - x\|^2$ exists, that is, the second assumption of Lemma 1 is also verified.

Let us now establish that $\{\tilde{z}_n\}$ weakly converges to some $\tilde{z} \in (A \underset{e}{+} B)^{-1}(0)$. Set $u_n = J_{\lambda_n}^A\tilde{x}_{n-1}, v_n = J_{\lambda_n}^B\tilde{x}_{n-1}$ and take again $(x, y) \in A \underset{e}{+} B$. By definition of the extended sum, for all $\epsilon > 0$, there exists a sequence $\{y_{p,\epsilon}\}$ weakly converging to y and satisfies

$$y_{p,\epsilon} = y_{1,p,\epsilon} + y_{2,p,\epsilon}, \text{with } y_{1,p,\epsilon} \in A^\epsilon(x) \text{ and } y_{2,p,\epsilon} \in B^\epsilon(x).$$

Definition of ϵ-enlargements combined to relation (12)

$$\frac{\tilde{x}_{n-1} - u_n}{\lambda_n} \in A(u_n) \quad \text{and} \quad \frac{\tilde{x}_{n-1} - v_n}{\lambda_n} \in B(v_n) \tag{12}$$

yields

$$\left\langle\frac{\tilde{x}_{n-1} - u_n}{\lambda_n} - y_{1,p,\epsilon}, u_n - x\right\rangle \geq -\epsilon \quad \text{and} \quad \left\langle\frac{\tilde{x}_{n-1} - v_n}{\lambda_n} - y_{2,p,\epsilon}, v_n - x\right\rangle \geq -\epsilon.$$

Equivalently,

$$\langle\tilde{x}_{n-1} - u_n, u_n - x\rangle \geq \lambda_n\langle y_{1,p,\epsilon}, u_n - x\rangle - \lambda_n\epsilon \tag{13}$$

and

$$\langle\tilde{x}_{n-1} - v_n, v_n - x\rangle \geq \lambda_n\langle y_{2,p,\epsilon}, v_n - x\rangle - \lambda_n\epsilon. \tag{14}$$

From (13) and (14), we obtain:

$$\|\tilde{x}_{n-1} - x\|^2 - \|u_n - x\|^2 \geq \|\tilde{x}_{n-1} - u_n\|^2 + 2\lambda_n\langle y_{1,p,\epsilon}, u_n - x\rangle - 2\lambda_n\epsilon$$

and

$$\|\tilde{x}_{n-1} - x\|^2 - \|v_n - x\|^2 \geq \|\tilde{x}_{n-1} - v_n\|^2 + 2\lambda_n \langle y_{2,p,\epsilon}, v_n - x\rangle - 2\lambda_n \epsilon$$

Therefore,

$$\|\tilde{x}_{n-1} - x\|^2 - \|u_n - x\|^2 \geq -\lambda_n^2 \|y_{1,p,\epsilon}\|^2 + 2\lambda_n \langle y_{1,p,\epsilon}, \tilde{x}_{n-1} - x\rangle - 2\lambda_n \epsilon \tag{15}$$

and

$$\|\tilde{x}_{n-1} - x\|^2 - \|v_n - x\|^2 \geq -\lambda_n^2 \|y_{2,p,\epsilon}\|^2 + 2\lambda_n \langle y_{2,p,\epsilon}, \tilde{x}_{n-1} - x\rangle - 2\lambda_n \epsilon \tag{16}$$

As $\tilde{x}_n = \frac{1}{2}(u_n + v_n)$, using convexity of the norm, we obtain

$$-\|\tilde{x}_n - x\|^2 \geq -\frac{1}{2}\|u_n - x\|^2 - \frac{1}{2}\|v_n - x\|^2. \tag{17}$$

Mutiplying (15) and (16) by $\frac{1}{2}$, summing up and adding (17) gives:

$$\|\tilde{x}_{n-1} - x\|^2 - \|\tilde{x}_n - x\|^2 \geq -\frac{1}{2}\lambda_n^2(\|y_{1,p,\epsilon}\|^2 + \|y_{2,p,\epsilon}\|^2) + 2\lambda_n \langle y_p, \tilde{x}_{n-1} - x\rangle - 2\lambda_n \epsilon.$$

Summing the last inequality from $k = 1$ to n and dividing by $\sum_{k=1}^n \lambda_k$, gives

$$2\langle y_{p,\epsilon}, \tilde{z}_n - x\rangle \leq \frac{\|x_0 - x\|^2}{\sum_{k=1}^n \lambda_k} + \frac{1}{2}(\|y_{1,p,\epsilon}\|^2 + \|y_{2,p,\epsilon}\|^2)\frac{\sum_{k=1}^n \lambda_k^2}{\sum_{k=1}^n \lambda_k} + 2\epsilon. \tag{18}$$

We finish the proof by proceeding along the same lines as in the first part.

Corollary 3

- *If $A + B$ is maximal monotone, we recover results by Passty (resp. Lehdili and Lemaire), that is $\{z_n\}$ (resp. $\{\tilde{z}_n\}$) converges weakly to some $\bar{z} \in (A + B)^{-1}(0)$.*

- *If $A \underset{v}{+} B$ and $A \underset{e}{+} B$ are both maximal monotone, we generalize the convergence result of Lehdili and Lemaire to maximal monotone operators, that is, $\{\tilde{z}_n\}$ converges weakly to some $\bar{z} \in (A \underset{v}{+} B)^{-1}(0)$. Indeed, in this case, our proof works without assuming that A^ϵ is locally bounded.*

- *In the case $A = \partial f$, $B = \partial g$, $\{z_n\}$ (resp. $\{\tilde{z}_n\}$) converges weakly to a solution of*

  (OP) find $x \in X$ such that $0 \in \partial(f + g)(x)$

  or equivalently

  find $x \in X$ such that $x \in argmin(f + g)$,

  and we recover the convergence result of Lehdili and Lemaire for (4).

4. EXAMPLES

It is clear that the maximality assumption on $A+B$ is too restrictive. Indeed, since the maximality of $A+B$ is equivalent to the graph convergence of $A_\lambda + B$ towards $A+B$ and we need only the convergence of the zeroes of $A_\lambda + B$ towards the zeroes of $A+B$ and not of the entire graph.

When $A+B$ is not maximal monotone the problem is to know whether the method (2) (resp. (4)) converges to a zero of a maximal continuation of $A+B$. This is illustrated by the following example:

4.1. Example 1 (we refer to [9])

Let us consider the problem (Q) in $\mathbb{R}^3$ of finding the closest point to the origin in the intersection of a two-dimensional subspace L and a cylinder C lying on that subspace. We consider below both cases whether $0 \in L \cap C$ or not. The problem (Q) is then $\min_{x \in L\cap C} \frac{1}{2}\|x\|^2$ which is equivalent to finding a zero of the extended sum of the maximal monotone operators $A = \partial I_C$ and $B = I + \partial I_L$. More precisely to finding a zero of $I + \partial I_{L\cap C}$, where I_C stands for the indicator function of the set C. It is easy to see that $A \underset{e}{+} B$ is maximal (indeed, for any $x \in L \cap C, (A \underset{e}{+} B)x = \{x\} + (L\cap C)^\perp$, a two-dimensional plane orthogonal to the boundary $L \cap C$).

On the other hand, if we want to achieve a decomposition of the constraints, we can consider the pointwise sum of A and B. Then, we get that Dom $(A+B) = L\cap C$ and for any $x \in L \cap C, (A+B)x = \{x\} + L^\perp$, which is the one dimensional line orthogonal to L at x. Two situations may arise: either $0 \in L\cap C$ or not. In both cases $A+B$ is not maximal monotone because $(A+B)x$ is strictly contained in $(A \underset{e}{+} B)x$ and, moreover, in the second case $A+B$ has no zero at all. We show that even in that case, the splitting method (2) (resp. (4)) works and converges to the right point. We analyze directly the second case where $A+B$ has no zero: let $x_{n-1} \in C$ and y_{n-1} be the unique point in L such that: $y_{n-1} = \operatorname{argmin}_{x\in L}\left(\frac{1}{2}\|x\|^2 + \frac{1}{2\lambda_n}\|x - x_{n-1}\|^2\right)$. We can see that y_{n-1} is the projection on L of $\frac{1}{1+\lambda_n}x_{n-1}$ so that:

$$x_n = \mathrm{Proj}_C \circ \mathrm{Proj}_L \left(\frac{1}{1+\lambda_n} x_{n-1}\right). \tag{19}$$

The sequence $\{z_n\}_{n\in\mathbb{R}}$ converges to $y = \mathrm{Proj}_{L\cap C} 0$ which is a zero of the extended sum of A and B.

For completeness, we would like to emphasize that the motivation for the introduction of various generalizations of the concept of sum for operators comes in particular from the theory of Schrödinger equations and problems arising in quantum theory. Let us briefly recall the mathematical setting of these equations:

4.2. Example 2 (see for example [1])

Let $\Omega \subset \mathbb{R}^N$ be an open subset of $\mathbb{R}^N$. Let $V : \Omega \to \mathbb{R}^+$ be a locally integrable function with some singularities (for example, $V(x) = \frac{1}{\|x\|^p}$). The differential operator attached to the Schrödinger equations can be formally written as $Cu = -\Delta u + Vu$. Take

$X = L^2(\Omega), A = \partial\Phi$, and $B = \partial\Psi$ the subdifferential operators of the convex lower semicontinuous proper functions

$$\Phi(u) := \begin{cases} \dfrac{1}{2}\displaystyle\int_\Omega |Du|^2 dx & \text{if} \quad u \in H_0^1(\Omega) \\ +\infty & \text{on} \quad L^2(\Omega)\backslash H_0^1(\Omega) \end{cases}$$

and

$$\Psi(u) := \begin{cases} \dfrac{1}{2}\displaystyle\int_\Omega Vu^2 dx & \text{if} \quad Vu^2 \in L^1(\Omega) \\ +\infty & \text{elsewhere.} \end{cases}$$

$$Au := -\Delta u \quad \text{with} \quad D(A) := H^2(\Omega) \cap H_0^1(\Omega)$$

and

$$Bu := Vu \quad \text{with} \quad D(B) := \Big\{ u \in L^2(\Omega) \,|\, Vu \in L^2(\Omega) \Big\}.$$

The pointwise sum of the maximal monotone operators A and B is a naive approach and is not adapted to the situation. Indeed, if u belongs to $D(A) \cap D(B)$, then $u \in H^2(\Omega)$ and $Vu \in L^2(\Omega)$, these conditions are usually incompatible. On the opposite, the approach developped in this note consists in taking $C = A \underset{e}{+} B$. Since $A = \partial\Phi$ and $B = \partial\Psi$ are subdifferential of convex lower semicontinuous proper function, by ([15], theorem 3.3) $A \underset{e}{+} B = \partial(\Phi + \Psi)$.

At each step of Algorithm (2) (resp. (4)), one has to compute

$$J_\lambda^B u = u - \frac{\lambda V}{1 + \lambda V} u, \tag{20}$$

and then to compute the solution of a linear problem involving the Laplace operator. The latter methods converge to a zero of the extended sum of A and B given by

$$A \underset{e}{+} B = -\Delta u + Vu \quad \text{with} \quad D(A \underset{e}{+} B) = \Big\{ u \in H_0^1 \,|\, Vu^2 \in L^1(\Omega), -\Delta u + Vu \in L^2(\Omega) \Big\}.$$

Acknowledgments. The authors are grateful to the anonymous referee and to Professor S. Reich for their valuable comments and remarks.

REFERENCES

1. H. Attouch, J.-B. Baillon and M. Théra, Variational sum of monotone operators, *J. Convex Anal.* **1** (1994) 1–29.
2. H. Attouch, J.-B. Baillon and M. Théra, Weak solutions of evolution equations and variational sum of maximal monotone operators, *Southeast Asian Bulletin of Mathematics* **19** (1995) 117–126.
3. H. Brézis and P.-L. Lions, Produits infinis de résolvantes, *Israel J. Math.* **29** (1978) 329–345.

4. R. E. Bruck, On the weak convergence of an ergodic iteration for the solution of variational inequalities for monotone operators in Hilbert space, *J. Math. Anal. Appl.* **61** (1977) 159–164.
5. R. E. Bruck and S. Reich, Nonexpansive projections and resolvents of accretive operators in Banach spaces, *Houston J. Math.* **3** (1977) 459–470.
6. R.S. Burachik, A.N. Iusem and B. F. Svaiter, Enlargements of maximal monotone operators with applications to variational inequalities, *Set-Valued Analysis* **5** (1997) 159–180.
7. J. Eckstein, *Splitting Methods for Monotone Operators with Applications to Parallel Optimization*, Ph.D. Thesis, (CICS-TH-140, MIT, 1989).
8. N. Lehdili and B. Lemaire, The barycentric proximal method, *Communications on Applied Nonlinear Analysis* **6** (1999) 29–47.
9. Ph. Mahey and Pham Dinh Tao, Partial regularization of the sum of two maximal monotone operators, *Mathematical Modelling and Numerical Analysis* **27** (1993) 375–392.
10. B. Martinet, *Algorithmes pour la résolution de problèmes d'optimisation et minimax*, Thèse d'Etat (Université de Grenoble, 1972).
11. O. Nevanlinna and S. Reich, Strong convergence of contraction semigroups and of iterative methods for accretive operators in Banach spaces, *Israel J. Math.* **32** (1979) 44–58.
12. Z. Opial, Weak convergence of the sequence of successive approximations for nonexpansive mapping, *Bull. Amer. Math. Soc.* **73** (1967) 591–597.
13. G. B. Passty, Ergodic convergence to a zero of the sum of maximal monotone operators, *J. Math. Anal. Appl.* **72** (1979) 383–390.
14. S. Reich, Constructive techniques for accretive and monotone operators, in: *Applied Nonlinear Analysis*, (Academic Press, New York) (1979) 335–345.
15. J.-P. Revalski and M. Théra, Enlargements and sums of monotone operators, *Nonlinear Analysis, Theory, Methods and Applications*, to appear.
16. J.-P. Revalski and M. Théra, Variational and extended sums of monotone operators, in: M. Théra and R. Tichatschke, eds., *Proceedings of the Conference on Ill-posed Variational Problems and Regularization Techniques* Lecture Notes in Economics and Mathematical Systems, Springer-Verlag, **477** (1999) 229–246.
17. J.-P. Revalski and M. Théra, Generalized Sums of Monotone Operators, *Comptes Rendus de l'Académie des Sciences, Paris* **329**, Série I4 (1999) 979–98.
18. R.T. Rockafellar, *Conjugate duality and optimization* (SIAM, Philadelphia, PA, USA, 1974).
19. R.T. Rockafellar, Monotone operators and the proximal point algorithm, *SIAM J. Control and Optimization* **14** (1975) 877–898.
20. R.T. Rockafellar, Augmented Lagrangians and applications of the proximal point algorithm in convex programming, *Math. Oper. Res.* **1** (1976) 97–116.

Inherently Parallel Algorithms in Feasibility and Optimization and their Applications
D. Butnariu, Y. Censor and S. Reich (Editors)

DISTRIBUTED ASYNCHRONOUS INCREMENTAL SUBGRADIENT METHODS

A. Nedić,[a] D. P. Bertsekas,[b*] and V. S. Borkar[c†]

[a]Massachusetts Institute of Technology, Rm. 35-307,
77 Massachusetts Ave., Cambridge, MA 02139 USA

[b]Massachusetts Institute of Technology, Rm. 35-210,
77 Massachusetts Ave., Cambridge, MA 02139 USA

[c]School of Technology and Computer Science,
Tata Institute of Fundamental Research,
Homi Bhabha Road, Mumbai 400005, India

We propose and analyze a distributed asynchronous subgradient method for minimizing a convex function that consists of the sum of a large number of component functions. This type of minimization arises in a dual context from Lagrangian relaxation of the coupling constraints of large scale separable problems. The idea is to distribute the computation of the component subgradients among a set of processors, which communicate only with a coordinator. The coordinator performs the subgradient iteration incrementally and asynchronously, by taking steps along the subgradients of the component functions that are available at the update time. The incremental approach has performed very well in centralized computation, and the parallel implementation should improve its performance substantially, particularly for typical problems where computation of the component subgradients is relatively costly.

1. INTRODUCTION

We focus on the problem

$$\begin{array}{ll} \text{minimize} & f(x) = f_1(x) + \ldots + f_m(x) \\ \text{subject to} & x \in X, \end{array} \qquad (1)$$

where $f_i : \Re^n \to \Re$ are convex functions and X is a nonempty, closed, and convex subset of $\Re^n$. We are primarily concerned with the case where f is nondifferentiable. A special case of particular interest arises in Lagrangian relaxation where f is the dual function of a primal separable combinatorial problem that is solved by decomposition. Within this

*Research supported by NSF under Grant ACI-9873339.
†Research supported by NSF-KDI under Grant ECS9873451.

context, x is a vector of Lagrange multipliers corresponding to coupling constraints, and the component functions $f_i(x)$ are obtained by solving subproblems corresponding to the separable terms in the cost function and the constraints.

In this paper, we consider the solution of problem (1) using distributed asynchronous versions of the classical subgradient iteration

$$x(t+1) = \mathcal{P}_X \left[x(t) - \alpha(t) \sum_{i=1}^{m} g_i(t) \right]. \tag{2}$$

Here $g_i(t)$ is a subgradient of f_i at $x(t)$, $\alpha(t)$ is a positive stepsize, and $\mathcal{P}_X$ denotes the projection on the set $X \subset \Re^n$.

The most straightforward way to parallelize the above iteration is to use multiple processors to compute in parallel the component subgradients $g_i(t)$. Once all of these components have been computed, they can be collected at a single processor, called the *updating processor*, who will execute the update of the vector $x(t)$ using iteration (2). The updating processor will then distribute/broadcast in some way the new iterate $x(t+1)$ to the subgradient-computing processors who will collectively compute the new subgradients for the subsequent iteration. This parallelization approach is quite efficient as long as the computation of the subgradients (which is parallelized) takes much more time than their addition and the execution of the iteration (2) (which are performed serially). Fortunately, this is typically the case in the principal context of interest to us, i.e., duality and Lagrangian relaxation.

The parallel algorithm just described is mathematically equivalent to the serial iteration (2). It can be termed *synchronous*, in the sense that there is clear division between the computations of successive iterations, i.e., all computation relating to iteration t must be completed before iteration $t+1$ can begin. In this paper we are interested in *asynchronous* versions of the subgradient method (2), where the subgradient components at a given iteration do not necessarily correspond to the same value of x. An example of such a method is

$$x(t+1) = \mathcal{P}_X \left[x(t) - \alpha(t) \sum_{i=1}^{m} g_i(\tau_i(t)) \right], \tag{3}$$

where for all i, we have $\tau_i(t) \leq t$ and the difference $t - \tau_i(t)$ may be viewed as a "delay." Such an iteration may be useful if for some reason (e.g., excessive computation or communication delay) some subgradient components $g_i(t)$ are not available at time t, and to avoid further delay in executing the update of $x(t)$, the most recently computed components $g_i(\tau_i(t))$ are used in (3) in place of the missing components $g_i(t)$.

A more general version of the asynchronous iteration (3) is given by

$$x(t+1) = \mathcal{P}_X \left[x(t) - \alpha(t) \sum_{i \in I(t)} g_i(\tau_i(t)) \right], \tag{4}$$

where $I(t)$ is a nonempty subset of the index set $\{1, \ldots, m\}$, $\tau_i(t)$ satisfies $\tau_i(t) \leq t$ for all t and i, $g_i(\tau_i(t))$ is a subgradient of f_i computed at $x(\tau_i(t))$, and $x(0) \in X$ is an initial point. To visualize the execution of this iteration, it is useful to think of the computing system as consisting of two parts: the *updating system* (US for short), and the *subgradient computing system* (GCS for short) (see Fig. 1).

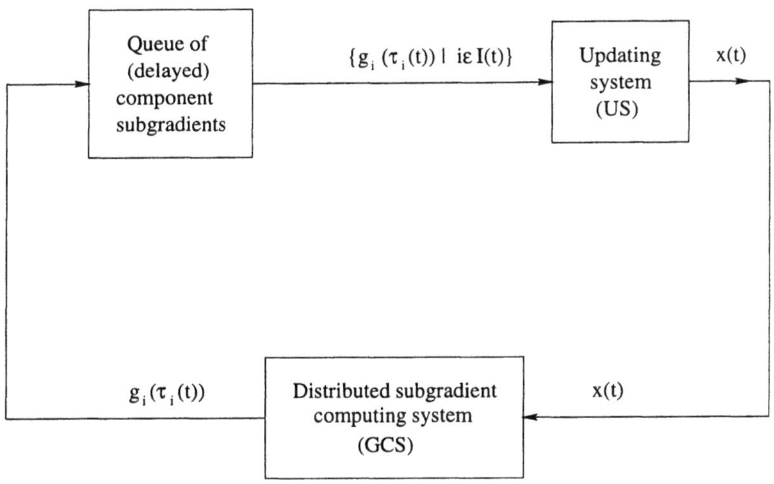

Figure 1. Visualization of the implementation of the asynchronous distributed iteration

$$x(t+1) = \mathcal{P}_X \left[x(t) - \alpha(t) \sum_{i \in I(t)} g_i(\tau_i(t)) \right].$$

This iteration is executed by the updating system (US), using subgradients deposited in a queue by the distributed subgradient computing system (GCS).

The US executes iteration (4) at each time t and delivers the corresponding iterates $x(t)$ to the GCS. The GCS uses the values $x(t)$ obtained from the US, computes subgradient components $g_i(\tau_i(t))$, and deposits them in a queue from where they can be accessed by the US. There is no synchronization between the operations of the CS and the GCS. Furthermore, while the GCS may involve multiple processors that compute subgradient components in parallel, the characteristics of the GCS (e.g., shared memory, message passing, communication architecture, number of processors, synchronization rules, etc.) are not material to the description of our algorithm.

The motivation for considering iteration (4) rather than its special case (3) is twofold. First, it makes sense to keep the US busy with updates while the GCS is computing subgradient components. This is particularly so if the computation of g_i is much more time consuming for some i than for others, thereby creating a synchronization bottleneck.

Second, it appears that updating the value of $x(t)$ as quickly as possible and using it in the calculation of the component subgradients has a beneficial effect in the convergence rate of the subgradient method. This is the main characteristic of *incremental subgradient methods*, which we have studied in [19], and which may be viewed as the special case of iteration (4) where $\tau_i(t) = t$ for all i and t, and the set $I(t)$ consists of a *single* index.

We note that incremental methods for differentiable unconstrained problems have a long tradition, most notably in the training of neural networks, where they are known as *backpropagation methods*. They are related to the Widrow-Hoff algorithm [25] and to stochastic gradient/stochastic approximation methods, and they are supported by several recent convergence analyses Bertsekas [3], Bertsekas and Tsitsiklis [5]–[6], Gaivoronski [8], Grippo [10], Luo [16], Luo and Tseng [17], Mangasarian and Solodov [18], Tseng [23]. It has been experimentally observed that incremental gradient methods often converge much faster than the steepest descent method when far from the eventual limit. However, near convergence, they typically converge slowly because they require a diminishing stepsize [e.g., $\alpha_k = O(1/k)$] for convergence. If α_k is instead taken to be a small enough constant, "convergence" to a limit cycle occurs, as first shown by Luo [16].

The incremental subgradient method was studied first by Kibardin [12], and more recently by Solodov and Zavriev [22], Nedić and Bertsekas [19]–[21], and Ben-Tal, Margalit and Nemirovski [1]; the incremental subgradient method of the form (3) is considered by Zhao, Luh, and Wang [26], Kiwiel and Lindberg [13]; the incremental ϵ-subgradient method is analyzed by Kiwiel [14]; applications of incremental methods can be found in Kaskavelis and Caramanis [11], Ben-Tal, Margalit and Nemirovski [1]. As we have discussed in our papers [19] and [21], incremental subgradient methods exhibit a similar behavior to incremental gradient methods. We believe that the incremental structure that is inherent in our proposed parallel subgradient method (4) results in convergence and rate of convergence characteristics that are similar to those of incremental gradient and subgradient methods. In particular, we expect an enhanced convergence rate over the nonincremental version given by Eq. (3).

In this paper, we will analyze the following version of iteration (4), given by

$$x(t+1) = \mathcal{P}_X \left[x(t) - \alpha(t) g_{i(t)}(\tau(t)) \right]. \tag{5}$$

Here we assume for simplicity that the set $I(t)$ consists of a single element denoted $i(t)$. For $X = \Re^n$, this simplification does not involve an essential loss of generality since an iteration involving multiple component function subgradients may be broken down into several iterations each involving a single component function subgradient. When $X \neq \Re^n$, our analysis can be extended for the more general iteration (4).

The most important assumptions in our analysis are:

(a) The stepsize $\alpha(t)$ is either constant, or is diminishing to 0 and satisfies some common technical conditions such as $\sum_{t=0}^{\infty} \alpha(t) = \infty$ (see a more precise statement later). In the case of a constant stepsize, we only show convergence to optimality within an error which depends on the length of the stepsize.

(b) The "delay" $t - \tau(t)$ is bounded from above by some (unknown) positive integer D, so that our algorithm belongs to the class of *partially asynchronous methods*, as defined by Bertsekas and Tsitsiklis [4].

(c) All the component functions f_i are used with the same "long-term frequency" by the algorithm. Precise methods to enforce this assumption are given later, but basically what we mean is that if $n_i(t)$ is the number of times up to iteration t where a subgradient of the component f_i is used by the algorithm, then the ratios $n_i(t)/t$ should all be asymptotically equal to $1/m$ (as $t \to \infty$).

(d) The subgradients $g_{i(t)}(\tau(t))$ used in the method are bounded.

The restriction (c) can be easily enforced in a number of ways, by regulating the frequency of the indices of subgradient components computed by the subgradient computing system. We will consider one specific approach, whereby we first select a sequence of indexes $\{j(t)\}$ according to one of two rules:

(1) The *cyclic rule* where the sequence $\{j(t)\}$ is obtained by a permutation of each of the periodic blocks $\{1, 2, \ldots, m\}$ in the periodic sequence $\{1, 2, \ldots, m, 1, 2, \ldots, m, \ldots\}$.

(2) The *random rule* where $\{j(t)\}$ is a sequence of independent identically distributed random variables, each taking the values $1, 2, \ldots, m$ with equal probability $1/m$.

Given a sequence $\{j(t)\}$ obtained by the cyclic or the random rule, the sequence $\{i(t)\}$ used in iteration (5) is given by

$$i(t) = j(\pi(t)), \tag{6}$$

where $\pi(\cdot)$ is a permutation mapping that maps the set $\{0, 1, \ldots\}$ into itself, such that for some positive integer T, we have

$$|\pi(t) - t| \leq T, \qquad \forall\, t = 0, 1, \ldots. \tag{7}$$

The permutation mapping $\pi(\cdot)$ captures the asynchronous character of the algorithm, whereby component function subgradients are offered to the updating system in the order of $\{j(\pi(t))\}$, which is different than the order of $\{j(t)\}$ in which their computation was initiated within the subgradient computing system.

A version of the algorithm that does not work is when the component subgradients $g_{i(t)}(\tau(t))$ are normalized by multiplying with $1/||g_{i(t)}(\tau(t))||$, which may be viewed as a weight associated with the component $f_{i(t)}$ at time t. Unless these weights are asymptotically equal, this modification would effectively alter the effective "long-term frequency" by which the components f_i are selected, thereby violating a fundamental premise for the validity of our algorithm.

We note that our proposed parallel algorithms (4) and (5) do not fit the framework of the general algorithmic models of Chapters 6 and 7 of Bertsekas and Tsitsiklis [4], so it is not covered by the line of analysis of that reference. In the latter models, at each time t, only *some of the components of* x are updated using an equation that (within our subgradient method context) would depend on *all components* f_i (perhaps with communication delays). By contrast in the present paper, at each time t, *all components of* x are updated using an equation that involves some of the components f_i. While it is possible to consider alternative asynchronous subgradient methods of the type given in Chapters 6 and 7 of Bertsekas and Tsitsiklis [4], we believe that these methods would not be as well suited to typical subgradient optimization problems, which arise in the context of Lagrangian relaxation and duality.

The proof ideas of the present paper are related to those of parallel asynchronous deterministic and stochastic gradient methods as discussed in Tsitsiklis, Bertsekas, and Athans [24], and Bertsekas and Tsitsiklis [4], as well as the proof ideas of incremental deterministic and randomized subgradient methods as discussed in Nedić and Bertsekas [19]. In particular, the key proof idea is to view the parallel asynchronous method as an iterative method with deterministic or stochastic errors, the effects of which are controlled with an appropriate mechanism, such as stepsize selection. An alternative approach is possible based on differential inclusions that extend the "ODE" approach for the analysis of stochastic approximation algorithms (see Benveniste, Metivier, and Priouret [2], Borkar [7], and Kushner and Yin [15]).

We note also that while a (nonparallel) incremental subgradient method with a diminishing stepsize can be viewed as an ϵ-subgradient method (under some boundedness assumptions) [20], for the algorithm of this paper, this connection is much harder to make and has not been attempted. The reason is that with delays in the calculations of the component subgradients, it is difficult to estimate the ϵ parameter corresponding to the ϵ-subgradient method. Furthermore, for a different stepsize rule (e.g., a constant stepsize), it is not possible to make any fruitful connection with an ϵ-subgradient method.

The paper is structured as follows. In Section 2, we state and discuss our results for the cyclic rule. In Section 3, we do the same for the random selection rule. In Sections 4 and 5, we give the proofs of the convergence results presented in Sections 2 and 3, respectively.

2. CONVERGENCE RESULTS FOR A CYCLIC SELECTION RULE

Throughout the paper, we use f^* and X^* to denote the optimal function value and the optimal solution set for problem (1), respectively. In this section, we present convergence results for the method with the cyclic selection rule, under the following assumption.

Assumption 2.1:

(a) There exists a positive constant C such that for all $i = 1, \ldots, m$

$$||g|| \le C, \qquad \forall\ g \in \partial f_i(x(\tau(t))) \cup \partial f_i(x(t)), \qquad t = 0, 1 \ldots,$$

where $\partial f_i(x)$ denotes the set of subgradients of f_i at a point x.

(b) There exists a positive integer D such that

$$t - \tau(t) \le D, \qquad \forall\ t = 0, 1, \ldots.$$

Note that if the components f_i are polyhedral or if the set X is compact, then Assumption 2.1(a) holds. Assumption 2.1(b) is natural, since our algorithm does not use the value of the bound D.

We now give the convergence result for a constant stepsize.

Proposition 2.1: Let Assumption 2.1 hold. Then, for the sequence $\{x(t)\}$ generated by the method with the cyclic selection rule and the stepsize fixed to some positive scalar α, the following hold:

(a) If $f^* = -\infty$, then

$$\liminf_{t\to\infty} f(x(t)) = -\infty.$$

(b) If f^* is finite, then

$$\liminf_{t\to\infty} f(x(t)) \le f^* + mC^2\left(\frac{1}{2} + m + 2D + T\right)\alpha.$$

Next, we consider a diminishing stepsize that satisfies the following.

Assumption 2.2: The stepsize $\alpha(t)$ is given by

$$\alpha(t) = \frac{r_0}{(l + r_1)^q}, \qquad t = \sigma_l, \sigma_l + 1, \ldots, \sigma_{l+1} - 1, \quad l = 0, 1 \ldots,$$

where r_0, r_1, and q are some positive scalars with $0 < q \le 1$, and the sequence $\{\sigma_l\}$ is increasing and is such that $\sigma_{l+1} - \sigma_l \le S$ for some positive integer S and all l.

For this stepsize we have the following convergence result.

Proposition 2.2: Let Assumptions 2.1 and 2.2 hold. Then, for the sequence $\{x(t)\}$ generated by the method with the cyclic selection rule, we have

$$\liminf_{t\to\infty} f(x(t)) = f^*.$$

The result of Prop. 2.2 can be strengthened in the case where the optimal solution set X^* is nonempty, under a mild additional restriction on the stepsize. This stronger convergence result is given in the next proposition.

Proposition 2.3: Let Assumptions 2.1 and 2.2 hold, where $\frac{1}{2} < q \le 1$ in Assumption 2.2. Also, let X^* be nonempty. Then the sequence $\{x(t)\}$ generated by the method with the cyclic selection rule converges to an optimal solution.

3. CONVERGENCE RESULTS FOR A RANDOM SELECTION RULE

In this section, we present convergence results for the method with the random selection rule. The initial point $x(0) \in X$ and the stepsize $\alpha(t)$ are deterministic. In our analysis we use the following assumption.

Assumption 3.1:

(a) Assumption 2.1 holds.

(b) Assumption 2.2 holds with the scalar q satisfying $\frac{3}{4} < q \leq 1$.

(c) The sequence $\{j(t)\}$ is a sequence of independent random variables each of which is uniformly distributed over the set $\{1, \ldots, m\}$. Furthermore, the sequence $\{j(t)\}$ is independent of the sequence $\{x(t)\}$.

We now give the convergence result for a diminishing stepsize.

Proposition 3.1: Let Assumption 3.1 hold. Then, for the sequence $\{x(t)\}$ generated by the method with the random selection rule, we have with probability 1

$$\liminf_{t \to \infty} f(x(t)) = f^*.$$

When the optimal solution set X^* is nonempty, we can strengthen Prop. 3.1, as shown in the following proposition.

Proposition 3.2: Let Assumption 3.1 hold and let X^* be nonempty. Then the sequence $\{x(t)\}$ generated by the method with the random selection rule converges to some optimal solution with probability 1.

Remark 3.1: If the underlying set X is compact, then it can be shown that Prop. 3.2 holds using a wider range of values for q in the stepsize rule [cf. Assumption 3.1(b)]. In particular, for $\frac{1}{2} < q \leq 1$, the result of Prop. 3.2 is valid, which we discuss in more detail in Section 5 (cf. Remark 5.1).

4. CONVERGENCE PROOFS FOR A CYCLIC SELECTION RULE

In this section and the next one, for notational convenience, we define $\alpha(t) = \alpha(0)$ for $t < 0$, and $t_k = km$ and $x_k = x(t_k)$ for all k. Here we give the proofs for the convergence results of Section 2. The proofs are complicated, so we break them down in several steps. We first provide some estimates of the progress of the method in terms of the distances of the iterates to an arbitrary point in X and in terms of the objective function values. These estimates are given in the subsequent Lemma 4.2. Some preliminary results needed for the proof of Lemma 4.2 and Lemma 5.2 of Section 5 are given in the following lemma.

Lemma 4.1: Let Assumption 2.1 hold. Then, for the iterates $x(t)$ generated by the method, the following hold:

(a) For all $y \in X$ and t, we have

$$\begin{aligned} \|x(t+1) - y\|^2 \leq\ & \|x(t) - y\|^2 - 2\alpha(t)\big(f_{j(t)}(x(t)) - f_{j(t)}(y)\big) \\ & + C^2(1 + 4D)\alpha^2(t - D) \\ & + 2\alpha(t) \sum_{l=1}^{m} \big(\delta^l_{j(t)} - \delta^l_{i(t)}\big)\big(f_l(x(t)) - f_l(y)\big), \end{aligned} \tag{8}$$

where δ^l_i is the Kronecker symbol (i.e., $\delta^l_i = 1$ if $l = i$ and $\delta^l_i = 0$ otherwise).

(b) For all $y \in X$, and N and K with $N \geq K$, we have

$$\sum_{l=1}^{m} \sum_{t=K}^{N} \alpha(t)\big(\delta^l_{j(t)} - \delta^l_{i(t)}\big)\big(f_l(x(t)) - f_l(y)\big) \leq C^2 T \sum_{t=K}^{N} \alpha^2(t-T)$$

$$+ \max\{C, G(y)\} \sum_{t=K}^{N} \big(\alpha(t-T) - \alpha(t+T)\big) \left(C \sum_{r=0}^{t} \alpha(r) + ||x_0 - y||\right)$$

$$+ c(y)\big(\alpha^2(K) + \alpha(K) + \alpha^2(N+1-T) + \beta\alpha(N+1-T)\big)$$

$$+ \left(\alpha(K)||x(K) - y||^2 + \frac{1}{\beta}\alpha(N+1-T)||x(N+1) - y||^2\right), \tag{9}$$

where β is an arbitrary positive scalar, and

$$G(y) = \max\{||g|| \mid g \in \partial f_l(y),\ l = 1, \ldots, m\}, \tag{10}$$

$$c(y) = \max\left\{CT^2\big(C + G(y)\big), \frac{T^2}{2}\big(C^2 + G^2(y)\big)\right\}. \tag{11}$$

Proof: (a) From the definition of $x(t+1)$ [cf. Eq. (5)], the nonexpansion property of the projection, and the subgradient boundedness [cf. Assumption 2.1(a)], we have for all $y \in X$ and t

$$\begin{aligned} ||x(t+1) - y||^2 &\leq ||x(t) - y||^2 - 2\alpha(t) g_{i(t)}(\tau(t))'\big(x(t) - y\big) + C^2\alpha^2(t) \\ &\leq ||x(t) - y||^2 - 2\alpha(t)\big(f_{i(t)}(x(t)) - f_{i(t)}(y)\big) \\ &\quad +4C\alpha(t)||x(t) - x(\tau(t))|| + C^2\alpha^2(t), \end{aligned} \tag{12}$$

where in the last inequality we use

$$g_{i(t)}(\tau(t))'\big(x(t) - y\big) \geq f_{i(t)}(x(t)) - f_{i(t)}(y) - 2C||x(t) - x(\tau(t))||,$$

which can be obtained by using the fact $x(t) = x(\tau(t)) + (x(t) - x(\tau(t)))$, the convexity of $f_{i(t)}$, and the subgradient boundedness. Furthermore, from the relation

$$||x(t) - x(\hat{t})|| \leq C \sum_{s=\hat{t}}^{t-1} \alpha(s), \qquad \forall\, t, \hat{t},\ t \geq \hat{t}, \tag{13}$$

and the facts $t - D \leq \tau(t) \leq t$ for all t, and $\alpha(r) \leq \alpha(t-D)$ for $r = t-D, \ldots, t-1$ and all t, we obtain

$$||x(t) - x(\tau(t))|| \leq C \sum_{r=t-D}^{t-1} \alpha(r) \leq CD\alpha(t-D).$$

By using this estimate and the fact $\alpha(t) \le \alpha(t-D)$ for all t in Eq. (12), we have

$$||x(t+1) - y||^2 \le ||x(t) - y||^2 - 2\alpha(t)\Big(f_{i(t)}(x(t)) - f_{i(t)}(y)\Big) + C^2(1+4D)\alpha^2(t-D),$$

from which, by adding and subtracting $2\alpha(t)\big(f_{j(t)}(x(t)) - f_{j(t)}(y)\big)$, and by using the Kronecker symbol, we obtain Eq. (8).

(b) We introduce the following sets:

$$M_{K,N} = \Big\{t \in \{K,\ldots,N\} \mid j(t) = i(p(t)) \text{ with } p(t) \in \{K,\ldots,N\}\Big\},$$

$$P_{K,N} = \Big\{t \in \{K,\ldots,N\} \mid j(t) = i(p(t)) \text{ with } p(t) < K \text{ or } p(t) > N\Big\}, \tag{14}$$

$$Q_{K,N} = \Big\{t \in \{K,\ldots,N\} \mid i(t) = j(\pi(t)) \text{ with } \pi(t) < K \text{ or } \pi(t) > N\Big\}, \tag{15}$$

where $p(t)$ is the inverse of the permutation mapping $\pi(t)$, i.e., $p(t) = \pi^{-1}(t)$. Note that, since $|\pi(t) - t| \le T$ for all t [cf. Eq. (7)], for the inverse mapping $p(t)$ we have

$$|p(t) - t| \le T, \qquad \forall\, t = 0, 1, \ldots.$$

The set $M_{K,N}$ contains all $t \in \{K,\ldots,N\}$ for which the subgradient $g_{j(t)}$ of $f_{j(t)}$ is used in an update of $x(t)$ at some time between K and N. Similarly, the set $P_{K,N}$ contains all $t \in \{K,\ldots,N\}$ for which the subgradient $g_{j(t)}$ of $f_{j(t)}$ is used in an update $x(t)$ at some time before K or after N [i.e., $j(t) = i(p(t)) \notin \{i(K),\ldots,i(N)\}$]. The set $Q_{K,N}$ contains all $t \in \{K,\ldots,N\}$ for which the subgradient $g_{i(t)}$ of $f_{i(t)}$ is used in an update $x(t)$ at some time between K and N, but the $j(\pi(t))$ corresponding to $i(t)$ does not belong to the set $\{j(K),\ldots,j(N)\}$. By using the above defined sets, we have

$$\sum_{l=1}^{m}\sum_{t=K}^{N} \alpha(t)\Big(\delta^l_{j(t)} - \delta^l_{i(t)}\Big)\Big(f_l(x(t)) - f_l(y)\Big) =$$

$$\sum_{l=1}^{m}\sum_{t\in M_{K,N}} \alpha(t)\Big(\delta^l_{j(t)} - \delta^l_{i(t)}\Big)\Big(f_l(x(t)) - f_l(y)\Big)$$

$$+\sum_{l=1}^{m}\sum_{t\in P_{K,N}} \alpha(t)\delta^l_{j(t)}\Big(f_l(x(t)) - f_l(y)\Big)$$

$$-\sum_{l=1}^{m}\sum_{t\in Q_{K,N}} \alpha(t)\delta^l_{i(t)}\Big(f_l(x(t)) - f_l(y)\Big). \tag{16}$$

Next we estimate each of the terms in the preceding relation. According to the definition of $M_{K,N}$, we have $j(t) = i(p(t))$ for all $t \in M_{K,N}$ [i.e., $g_{j(t)}$ is used at time $p(t)$ with the corresponding step $\alpha(p(t))$], so that

$$\sum_{l=1}^{m}\sum_{t\in M_{K,N}} \alpha(t)\Big(\delta^l_{j(t)} - \delta^l_{i(t)}\Big)\Big(f_l(x(t)) - f_l(y)\Big) =$$

$$\sum_{l=1}^{m}\sum_{t\in M_{K,N}}\delta_{j(t)}^{l}\Big[\alpha(t)\big(f_l(x(t))-f_l(y)\big)-\alpha(p(t))\big(f_l(x(p(t)))-f_l(y)\big)\Big]$$

$$=\sum_{l=1}^{m}\sum_{t\in M_{K,N}}\delta_{j(t)}^{l}\alpha(p(t))\big(f_l(x(t))-f_l(x(p(t)))\big)$$

$$+\sum_{l=1}^{m}\sum_{t\in M_{K,N}}\delta_{j(t)}^{l}\big(\alpha(t)-\alpha(p(t))\big)\big(f_l(x(t))-f_l(y)\big).$$

By using the convexity of each f_l, the subgradient boundedness, the monotonicity of $\alpha(t)$, and the facts $|p(t)-t|\le T$ and $\sum_{l=1}^{m}\delta_{j(t)}^{l}=1$ for all t, from the preceding relation we obtain

$$\sum_{l=1}^{m}\sum_{t\in M_{K,N}}\alpha(t)\big(\delta_{j(t)}^{l}-\delta_{i(t)}^{l}\big)\big(f_l(x(t))-f_l(y)\big)\le C\sum_{t=K}^{N}\alpha(t-T)||x(t)-x(p(t))||$$

$$+\max\{C,G(y)\}\sum_{t=K}^{N}\big(\alpha(t-T)-\alpha(t+T)\big)||x(t)-y||,$$

where $G(y)$ is given by Eq. (10). Furthermore, we have

$$||x(t)-x(p(t))||\le CT\alpha(t-T),$$

$$||x(t)-y||\le C\sum_{r=0}^{t}\alpha(r)+||x_0-y||,$$

where in the first relation we use the monotonicity of $\alpha(t)$ and the fact $|p(t)-t|\le T$, and in the second relation we use Eq. (13). By substituting the last two relations in the preceding inequality, we have

$$\sum_{l=1}^{m}\sum_{t\in M_{K,N}}\alpha(t)\big(\delta_{j(t)}^{l}-\delta_{i(t)}^{l}\big)\big(f_l(x(t))-f_l(y)\big)\le C^2T\sum_{t=K}^{N}\alpha^2(t-T)$$

$$+\max\{C,G(y)\}\sum_{t=K}^{N}\big(\alpha(t-T)-\alpha(t+T)\big)\left(C\sum_{r=0}^{t}\alpha(r)+||x_0-y||\right). \quad (17)$$

Now we consider the second term on the right hand side of Eq. (16). For $t=K,\ldots,N$ we may have $j(t)\notin\{i(K),\ldots,i(N)\}$ possibly at times $t=K,\ldots,K+T-1$ and $t=N+1-T,\ldots,N$. Therefore, from the convexity of each f_l, the subgradient boundedness, and the fact $\sum_{l=1}^{m}\delta_{j(t)}^{l}=1$ for all t, we obtain

$$\sum_{l=1}^{m}\sum_{t\in P_{K,N}}\alpha(t)\delta_{j(t)}^{l}\big(f_l(x(t))-f_l(y)\big)\le C\sum_{t=K}^{K-1+T}\alpha(t)||x(t)-y||+C\sum_{t=N+1-T}^{N}\alpha(t)||x(t)-y||.$$

By using Eq. (13), the triangle inequality, and the monotonicity of $\alpha(t)$, we have

$$\begin{aligned} C\sum_{t=K}^{K-1+T}\alpha(t)\|x(t)-y\| &\le C\sum_{t=K}^{K-1+T}\alpha(t)\big(\|x(t)-x(K)\|+\|x(K)-y\|\big)\\ &\le C^2T^2\alpha^2(K)+CT\alpha(K)\|x(K)-y\|\\ &\le C^2T^2\alpha^2(K)+\frac{\alpha(K)}{2}\big(C^2T^2+\|x(K)-y\|^2\big), \end{aligned}$$

where in the last inequality we exploit the fact $2ab \le a^2 + b^2$ for any scalars a and b. Similarly, it can be seen that

$$\begin{aligned} C\sum_{t=N+1-T}^{N}\alpha(t)\|x(t)-y\| &\le C\sum_{t=N+1-T}^{N}\alpha(t)\big(\|x(t)-x(N+1)\|+\|x(N+1)-y\|\big)\\ &\le C^2T^2\alpha^2(N+1-T)+CT\alpha(N+1-T)\|x(N+1)-y\|\\ &\le C^2T^2\alpha^2(N+1-T)\\ &\quad+\frac{\alpha(N+1-T)}{2}\left(\beta C^2T^2+\frac{1}{\beta}\|x(N+1)-y\|^2\right), \end{aligned}$$

where the last inequality follows from the fact $2ab \le \beta a^2 + \frac{1}{\beta}b^2$ for any scalars a,b, and β with $\beta > 0$. Therefore

$$\begin{aligned} \sum_{l=1}^{m}\sum_{t\in P_{K,N}}\alpha(t)\delta^l_{j(t)}\big(f_l(x(t))-f_l(y)\big) &\le C^2T^2\big(\alpha^2(K)+\alpha^2(N+1-T)\big)\\ &\quad+\frac{C^2T^2}{2}\big(\alpha(K)+\beta\alpha(N+1-T)\big)\\ &\quad+\frac{1}{2}\left(\alpha(K)\|x(K)-y\|^2+\frac{1}{\beta}\alpha(N+1-T)\|x(N+1)-y\|^2\right). \quad (18) \end{aligned}$$

Finally, we estimate the last term in Eq. (16). For $t = K,\ldots,N$ we may have $i(t) \notin \{j(K),\ldots,j(N)\}$ possibly at times $t = K,\ldots,K+T-1$ and $t = N+1-T,\ldots,N$. Therefore, similar to the preceding analysis, it can be seen that

$$\begin{aligned} -\sum_{l=1}^{m}\sum_{t\in Q_{K,N}}\alpha(t)\delta^l_{i(t)}\big(f_l(x(t))-f_l(y)\big) &\le G(y)CT^2\big(\alpha^2(K)+\alpha^2(N+1-T)\big)\\ &\quad+\frac{G^2(y)T^2}{2}\big(\alpha(K)+\beta\alpha(N+1-T)\big)\\ &\quad+\frac{1}{2}\left(\alpha(K)\|x(K)-y\|^2+\frac{1}{\beta}\alpha(N+1-T)\|x(N+1)-y\|^2\right), \quad (19) \end{aligned}$$

where $G(y)$ is given by Eq. (10). By substituting Eqs. (17)–(19) in the relation (16), and by using the definition of $c(y)$ [cf. Eq. (11)], we obtain Eq. (9). **Q.E.D.**

Lemma 4.2: Let Assumption 2.1 hold. Then, for the iterates $x(t)$ generated by the method, the following hold:

(a) For all $y \in X$, and k_0 and $\hat{k}$ with $\hat{k} > k_0$, we have

$$\left(1 - \frac{2}{\beta}\,\alpha(t_{\hat{k}} - W)\right)||x_{\hat{k}} - y||^2 \le \big(1 + 2\alpha(t_{k_0})\big)||x_{k_0} - y||^2$$

$$-2\sum_{k=k_0}^{\hat{k}-1} \alpha(t_k)\big(f(x_k) - f(y)\big) + 2\tilde{C}\sum_{k=k_0}^{\hat{k}-1} \alpha^2(t_k - W)$$

$$+2K(y)\sum_{k=k_0}^{\hat{k}-1}\big(\alpha(t_k - W) - \alpha(t_{k+1} + W)\big)\left(C\sum_{r=0}^{t_{k+1}} \alpha(r) + ||x_0 - y||\right)$$

$$+2c(y)\big(\alpha^2(t_{k_0}) + \alpha(t_{k_0}) + \alpha^2(t_{\hat{k}} - W) + \beta\alpha(t_{\hat{k}} - W)\big), \qquad (20)$$

where $W = \max\{T, D\}$, β is an arbitrary positive scalar,

$$K(y) = mC + m\max\{C, G(y)\},$$

$$\tilde{C} = mC^2\left(\frac{1}{2} + m + 2D + T\right), \qquad (21)$$

and $G(y)$ and $c(y)$ are defined by Eqs. (10) and (11), respectively.

(b) For for all $y \in X$ and $\hat{k} \ge 1$, we have

$$\frac{\sum_{k=0}^{\hat{k}-1} \alpha(t_k) f(x_k)}{\sum_{k=0}^{\hat{k}-1} \alpha(t_k)} \le f(y) + \frac{\big(1 + 2\alpha(0)\big)||x_0 - y||^2}{2\sum_{k=0}^{\hat{k}-1} \alpha(t_k)} + \tilde{C}\frac{\sum_{k=0}^{\hat{k}-1} \alpha^2(t_k - W)}{\sum_{k=0}^{\hat{k}-1} \alpha(t_k)}$$

$$+K(y)\frac{\sum_{k=0}^{\hat{k}-1}\big(\alpha(t_k - W) - \alpha(t_{k+1} + W)\big)\left(C\sum_{r=0}^{t_{k+1}} \alpha(r) + ||x_0 - y||\right)}{\sum_{k=0}^{\hat{k}-1} \alpha(t_k)}$$

$$+c(y)\frac{\big(\alpha^2(0) + \alpha(0) + \alpha^2(t_{\hat{k}} - W) + \beta\alpha(t_{\hat{k}} - W)\big)}{\sum_{k=0}^{\hat{k}-1} \alpha(t_k)}, \qquad (22)$$

where $\beta \ge 2\alpha(0)$.

Proof: (a) By using the convexity of each $f_{j(t)}$, the subgradient boundedness, the monotonicity of $\alpha(t)$, and the following relation [cf. Eq. (13)]

$$||x(t) - x(\hat{t})|| \le C\sum_{s=\hat{t}}^{t-1} \alpha(s), \qquad \forall\, t, \hat{t},\ t \ge \hat{t},$$

we have for any $t \in \{t_k, \dots, t_{k+1}-1\}$

$$f_{j(t)}(x(t)) \geq f_{j(t)}(x_k) + g_{j(t)}(t_k)'\big(x(t) - x_k\big) \geq f_{j(t)}(x_k) - mC^2\alpha(t_k),$$

where $g_{j(t)}(t_k)$ is a subgradient of $f_{j(t)}$ at x_k. By substituting this relation in Eq. (8) [cf. Lemma 4.1] and by summing over $t = t_k, \dots, t_{k+1}-1$, we obtain

$$\begin{aligned} ||x_{k+1} - y||^2 \leq \; & ||x_k - y||^2 - 2\sum_{t=t_k}^{t_{k+1}-1} \alpha(t)\big(f_{j(t)}(x_k) - f_{j(t)}(y)\big) \\ & + mC^2(1 + 2m + 4D)\alpha^2(t_k - D) \\ & + 2\sum_{l=1}^{m}\sum_{t=t_k}^{t_{k+1}-1} \alpha(t)\big(\delta^l_{j(t)} - \delta^l_{i(t)}\big)\big(f_l(x(t)) - f_l(y)\big), \end{aligned} \tag{23}$$

where we also use $\alpha(t_k) \leq \alpha(t_k - D)$ and

$$\sum_{t=t_k}^{t_{k+1}-1} \alpha^2(t-D) \leq m\alpha^2(t_k - D), \qquad \sum_{t=t_k}^{t_{k+1}-1} \alpha(t) \leq m\alpha(t_k),$$

which follow from the monotonicity of $\alpha(t)$ and the fact $t_{k+1} - t_k = m$ for all k.

Now we estimate the second term on the right hand side in the inequality (23). For this we define

$$I_k^+(y) = \Big\{t \in \{t_k, \dots, t_{k+1}-1\} \mid f_{j(t)}(x_k) - f_{j(t)}(y) \geq 0\Big\},$$

$$I_k^-(y) = \{t_k, \dots, t_{k+1}-1\} \setminus I_k^+(y).$$

Since $\alpha(t_k) \geq \alpha(t) \geq \alpha(t_{k+1})$ for all t with $t_k \leq t < t_{k+1}$, we have for $t \in I_k^+(y)$

$$\alpha(t)\big(f_{j(t)}(x_k) - f_{j(t)}(y)\big) \geq \alpha(t_{k+1})\big(f_{j(t)}(x_k) - f_{j(t)}(y)\big),$$

and for $t \in I_k^-(y)$

$$\alpha(t)\big(f_{j(t)}(x_k) - f_{j(t)}(y)\big) \geq \alpha(t_k)\big(f_{j(t)}(x_k) - f_{j(t)}(y)\big).$$

Hence for all t with $t_k \leq t < t_{k+1}$

$$\begin{aligned} \sum_{t=t_k}^{t_{k+1}-1} \alpha(t)\big(f_{j(t)}(x_k) - f_{j(t)}(y)\big) \geq \; & \alpha(t_{k+1}) \sum_{t \in I_k^+(y)} \big(f_{j(t)}(x_k) - f_{j(t)}(y)\big) \\ & + \alpha(t_k) \sum_{t \in I_k^-(y)} \big(f_{j(t)}(x_k) - f_{j(t)}(y)\big) \\ = \; & \alpha(t_k) \sum_{t=t_k}^{t_{k+1}-1} \big(f_{j(t)}(x_k) - f_{j(t)}(y)\big) \\ - \; & \big(\alpha(t_k) - \alpha(t_{k+1})\big) \sum_{t \in I_k^+(y)} \big(f_{j(t)}(x_k) - f_{j(t)}(y)\big). \end{aligned} \tag{24}$$

Furthermore, by using the convexity of each $f_{j(t)}$, the subgradient boundedness, and Eq. (13), we can see that

$$f_{j(t)}(x_k) - f_{j(t)}(y) \leq C\big(||x_k - x_0|| + ||x_0 - y||\big) \leq C\left(C\sum_{r=0}^{t_k}\alpha(r) + ||x_0 - y||\right),$$

and, since the cardinality of $I_k^+(y)$ is at most m, we obtain

$$\sum_{t\in I_k^+(y)}\big(f_{j(t)}(x_k) - f_{j(t)}(y)\big) \leq mC\left(C\sum_{r=0}^{t_k}\alpha(r) + ||x_0 - y||\right).$$

For the cyclic rule we have $\{j(t_k), \ldots, j(t_{k+1} - 1)\} = \{1, \ldots, m\}$, so that

$$\sum_{t=t_k}^{t_{k+1}-1}\big(f_{j(t)}(x_k) - f_{j(t)}(y)\big) = f(x_k) - f(y).$$

By using the last two relations in Eq. (24), we obtain

$$\begin{aligned}\sum_{t=t_k}^{t_{k+1}-1}\alpha(t)\big(f_{j(t)}(x_k) - f_{j(t)}(y)\big) \;\geq\; & \alpha(t_k)\big(f(x_k) - f(y)\big)\\ & -mC\big(\alpha(t_k) - \alpha(t_{k+1})\big)\left(C\sum_{r=0}^{t_k}\alpha(r) + ||x_0 - y||\right),\end{aligned}$$

which when substituted in Eq. (23) yields for all $y \in X$ and k

$$\begin{aligned}||x_{k+1} - y||^2 \;\leq\; & ||x_k - y||^2 - 2\alpha(t_k)\big(f(x_k) - f(y)\big) + mC^2(1 + 2m + 4D)\alpha^2(t_k - D)\\ & +2mC\big(\alpha(t_k) - \alpha(t_{k+1})\big)\left(C\sum_{r=0}^{t_k}\alpha(r) + ||x_0 - y||\right)\\ & +2\sum_{l=1}^{m}\sum_{t=t_k}^{t_{k+1}-1}\alpha(t)\big(\delta_{j(t)}^l - \delta_{i(t)}^l\big)\big(f_l(x(t)) - f_l(y)\big).\end{aligned}$$

By summing these inequalities over $k = k_0, \ldots, \hat{k} - 1$, and by using the facts $t_k < t_{k+1}$, $\alpha(t_k) \leq \alpha(t_k - W)$, $\alpha(t_{k+1}) \geq \alpha(t_{k+1} + W)$, $\alpha^2(t_k - D) \leq \alpha^2(t_k - W)$, we obtain

$$\begin{aligned}||x_{\hat{k}} - y||^2 \;\leq\; & ||x_{k_0} - y||^2 - 2\sum_{k=k_0}^{\hat{k}-1}\alpha(t_k)\big(f(x_k) - f(y)\big)\\ & +mC^2(1 + 2m + 4D)\sum_{k=k_0}^{\hat{k}-1}\alpha^2(t_k - W)\\ & +2mC\sum_{k=k_0}^{\hat{k}-1}\big(\alpha(t_k - W) - \alpha(t_{k+1} + W)\big)\left(C\sum_{r=0}^{t_{k+1}}\alpha(r) + ||x_0 - y||\right)\\ & +2\sum_{k=k_0}^{\hat{k}-1}\sum_{l=1}^{m}\sum_{t=t_k}^{t_{k+1}-1}\alpha(t)\big(\delta_{j(t)}^l - \delta_{i(t)}^l\big)\big(f_l(x(t)) - f_l(y)\big),\end{aligned} \tag{25}$$

with $W = \max\{D, T\}$. Note that the last term in the preceding relation can be written as the sum over $l = 1, \ldots, m$ and over $t = t_{k_0}, \ldots, t_{\hat{k}} - 1$, so that by using Eq. (9) of Lemma 4.1 where $K = t_{k_0}$ and $N = t_{\hat{k}} - 1$, and the monotonicity of $\alpha(t)$, we have

$$\sum_{l=1}^{m} \sum_{t=t_{k_0}}^{t_{\hat{k}}-1} \alpha(t)\left(\delta^l_{j(t)} - \delta^l_{i(t)}\right)\left(f_l(x(t)) - f_l(y)\right) \le C^2 T \sum_{t=t_{k_0}}^{t_{\hat{k}}-1} \alpha^2(t-W)$$

$$+\max\{C, G(y)\} \sum_{t=t_{k_0}}^{t_{\hat{k}}-1} \left(\alpha(t-W) - \alpha(t+W)\right)\left(C\sum_{r=0}^{t}\alpha(r) + ||x_0 - y||\right)$$

$$+c(y)\left(\alpha^2(t_{k_0}) + \alpha(t_{k_0}) + \alpha^2(t_{\hat{k}} - W) + \beta\alpha(t_{\hat{k}} - W)\right)$$

$$+\left(\alpha(t_{k_0})||x(t_{k_0}) - y||^2 + \frac{1}{\beta}\alpha(t_{\hat{k}} - W)||x_{\hat{k}} - y||^2\right). \tag{26}$$

Furthermore, by the monotonicity of $\alpha(t)$, it follows that

$$\sum_{t=t_{k_0}}^{t_{\hat{k}}-1} \alpha^2(t-W) = \sum_{k=k_0}^{\hat{k}-1} \sum_{t=t_k}^{t_{k+1}-1} \alpha^2(t-W) \le m \sum_{k=k_0}^{\hat{k}-1} \alpha^2(t_k - W),$$

$$\sum_{t=t_{k_0}}^{t_{\hat{k}}-1} \left(\alpha(t-W) - \alpha(t+W)\right)\left(C\sum_{r=0}^{t}\alpha(r) + ||x_0 - y||\right) \le$$

$$m \sum_{k=k_0}^{\hat{k}-1} \left(\alpha(t_k - W) - \alpha(t_{k+1} + W)\right)\left(C\sum_{r=0}^{t_{k+1}}\alpha(r) + ||x_0 - y||\right).$$

The desired relation (20) follows by substituting the last two inequalities in Eq. (26) and by using the resulting relation in Eq. (25).

(b) The desired relation (22) follows from Eq. (20), by setting $k_0 = 0$, by dividing with $2\sum_{k=0}^{\hat{k}-1}\alpha(t_k)$, and by using the fact $\beta \ge 2\alpha(0) \ge 2\alpha(t)$ for all t. **Q.E.D.**

Now we prove Prop. 2.1.

Proof of Prop. 2.1: We prove (a) and (b) simultaneously. Since $\alpha(t) = \alpha$ for $t \in (-\infty, \infty)$ [recall that we defined $\alpha(t) = \alpha(0)$ for $t < 0$], from Lemma 4.2(b), we obtain for all $y \in X$ and $\hat{k} \ge 1$

$$\frac{1}{\hat{k}} \sum_{k=0}^{\hat{k}-1} f(x_k) \le f(y) + \frac{(1+2\alpha)||x_0 - y||^2}{2\alpha\hat{k}} + \tilde{C}\alpha + c(y)\frac{1+2\alpha+\beta}{\alpha\hat{k}}.$$

By letting $\hat{k} \to \infty$ and by using the following inequality

$$\liminf_{\hat{k}\to\infty} f(x_{\hat{k}}) \leq \liminf_{\hat{k}\to\infty} \frac{1}{\hat{k}} \sum_{k=0}^{\hat{k}-1} f(x_k),$$

we have for all $y \in X$

$$\liminf_{\hat{k}\to\infty} f(x_{\hat{k}}) \leq f(y) + \tilde{C}\alpha,$$

from which the results (a) and (b) follow by taking the minimum over $y \in X$. **Q.E.D.**

In the proofs of Props. 2.2 and 2.3 we use properties of the stepsize given in the following lemma.

Lemma 4.3: Let the stepsize $\alpha(t)$ satisfy Assumption 2.2. Then we have

$$\lim_{k\to\infty} \frac{\alpha^2(t_k - W)}{\alpha(t_k)} = 0, \qquad \lim_{k\to\infty} \frac{\alpha(t_k - W) - \alpha(t_{k+1} + W)}{\alpha(t_k)} \sum_{t=0}^{t_{k+1}} \alpha(t) = 0,$$

$$\sum_{k=0}^{\infty} \alpha(t_k) = \infty, \qquad \sum_{k=0}^{\infty} \Big(\alpha(t_k - W) - \alpha(t_{k+1} + W)\Big) < \infty,$$

where $t_k = mk$ and W is a nonnegative integer. In addition, for $\frac{1}{2} < q \leq 1$ we have

$$\sum_{k=0}^{\infty} \alpha^2(t_k - W) < \infty, \qquad \sum_{k=0}^{\infty} \Big(\alpha(t_k - W) - \alpha(t_{k+1} + W)\Big) \sum_{t=0}^{t_{k+1}} \alpha(t) < \infty.$$

Proof: Let $0 < q \leq 1$. The stepsize $\alpha(t)$ is smallest when $S = 1$, so that

$$\sum_{k=0}^{\infty} \alpha(t_k) \geq r_0 \sum_{k=0}^{\infty} \frac{1}{(km + r_1)^q} = \infty.$$

Let $\{l_k\}$ be a sequence of nonnegative integers such that for all k

$$\alpha(t_k - W) = \frac{r_0}{(l_k + r_1)^q}. \tag{27}$$

Note that $l_k \to \infty$ as $k \to \infty$. Given the value of $\alpha(t_k - W)$, the values of $\alpha(t_k)$ and $\alpha(t_{k+1} + W)$ are smallest if we decrease the stepsize $\alpha(t)$ at each time t for $t > t_k - W$. Therefore

$$\alpha(t_k) \geq \frac{r_0}{(l_k + W + r_1)^q}, \tag{28}$$

$$\alpha(t_{k+1} + W) \geq \frac{r_0}{(l_k + m + 2W + r_1)^q}, \tag{29}$$

where in the last inequality above we use the fact $t_k = mk$. By combining Eqs. (27) and (28), we see that

$$\lim_{k\to\infty} \frac{\alpha^2(t_k - W)}{\alpha(t_k)} = 0.$$

Next from Eqs. (27) and (29) we obtain

$$\begin{aligned} \alpha(t_k - W) - \alpha(t_{k+1} + W) &= r_0 \frac{(l_k + m + 2W + r_1)^q - (l_k + r_1)^q}{(l_k + r_1)^q (l_k + m + 2W + r_1)^q} \\ &\leq \frac{r_0 q(m + 2W)}{(l_k + r_1)(l_k + W + r_1)^q}, \end{aligned} \tag{30}$$

where in the last inequality above we exploit the facts $l_k + m + 2W + r_1 \geq l_k + W + r_1$ and

$$b^q - a^q = q \int_a^b \frac{dx}{x^{1-q}} \leq \frac{q}{a^{1-q}} \int_a^b dx = \frac{q(b-a)}{a^{1-q}},$$

for all b and a with $b \geq a > 0$, and $0 < q \leq 1$. In particular, the relation (30) implies that

$$\alpha(t_k - W) - \alpha(t_{k+1} + W) \leq \frac{r_0 q(m + 2W)}{(l_k + r_1)^{1+q}}, \tag{31}$$

so that $\sum_{k=0}^{\infty} \big(\alpha(t_k - W) - \alpha(t_{k+1} + W)\big) < \infty$. Furthermore, by combining Eqs. (28) and (30), we obtain for all k

$$\frac{\alpha(t_k - W) - \alpha(t_{k+1} + W)}{\alpha(t_k)} \leq \frac{q(m + 2W)}{l_k + r_1}. \tag{32}$$

Now we estimate $\sum_{t=0}^{t_{k+1}} \alpha(t)$. By using the definition and the monotonicity of $\alpha(t)$, and Eq. (27) we have for all k large enough (so that $t_k - W > 0$)

$$\sum_{t=0}^{t_{k+1}} \alpha(t) \leq \sum_{t=0}^{t_k - W} \alpha(t) + (1 + m + W)\alpha(t_k - W) \leq \sum_{l=0}^{l_k} \frac{S r_0}{(l + r_1)^q} + (1 + m + W)\alpha(t_k - W).$$

Since

$$\sum_{l=0}^{l_k} \frac{1}{(l + r_1)^q} \leq \begin{cases} \frac{1}{r_1} + \ln(l_k + r_1) & \text{if } q = 1, \\ \frac{1}{r_1^q} + \frac{(l_k + r_1)^{1-q}}{1-q} & \text{if } 0 < q < 1, \end{cases} \tag{33}$$

from the preceding relation we obtain

$$\sum_{t=0}^{t_{k+1}} \alpha(t) \leq u_k, \tag{34}$$

where

$$u_k = \begin{cases} O\Big(\ln(l_k + r_1)\Big) & \text{if } q = 1, \\ O\Big((l_k + r_1)^{1-q}\Big) & \text{if } 0 < q < 1. \end{cases} \tag{35}$$

This together with Eq. (32) implies that

$$\lim_{k\to\infty} \frac{\alpha(t_k - W) - \alpha(t_{k+1} + W)}{\alpha(t_k)} \sum_{t=0}^{t_{k+1}} \alpha(t) = 0.$$

Now, let $\frac{1}{2} < q \leq 1$. Then, by using the definition of $\alpha(t)$, we have for K large enough (so that $t_k - W > 0$)

$$\sum_{k=K}^{\infty} \alpha^2(t_k - W) \leq \sum_{t=K}^{\infty} \alpha^2(t - W) \leq \sum_{s=0}^{\infty} \alpha^2(s) \leq \sum_{l=0}^{\infty} \frac{Sr_1^2}{(l + r_1)^{2q}},$$

implying that $\sum_{k=0}^{\infty} \alpha^2(t_k - W)$ is finite. Furthermore, by combining Eqs. (31), (34), and (35), we obtain

$$\alpha(t_k - W) - \alpha(t_{k+1} + W) \sum_{t=0}^{t_{k+1}} \alpha(t) \leq v_k,$$

where

$$v_k = \begin{cases} O\left(\frac{\ln(l_k + r_1)}{(l_k + r_1)^2}\right) & \text{if } q = 1, \\ O\left(\frac{1}{(l_k + r_1)^{2q}}\right) & \text{if } 0 < q < 1. \end{cases}$$

Hence $\sum_{k=0}^{\infty} \alpha(t_k - W) - \alpha(t_{k+1} + W) \sum_{t=0}^{t_{k+1}} \alpha(t)$ is finite. **Q.E.D.**

We are now ready to prove Props. 2.2 and 2.3.

Proof of Prop. 2.2: It suffices to show that $\liminf_{k\to\infty} f(x_k) = f^*$. For this we need the following two relations

$$\liminf_{N\to\infty} b_N \leq \liminf_{N\to\infty} \frac{\sum_{k=0}^{N-1} c_k b_k}{\sum_{k=0}^{N-1} c_k}, \tag{36}$$

$$\limsup_{N\to\infty} \frac{\sum_{k=0}^{N-1} c_k b_k}{\sum_{k=0}^{N-1} c_k} \leq \limsup_{N\to\infty} b_N, \tag{37}$$

which hold for any two scalar sequences $\{b_k\}$ and $\{c_k\}$ with $c_k > 0$ for all k and $\sum_{k=0}^{\infty} c_k = \infty$ (see Lemma 2.1 of Kiwiel [14]). By letting $\hat{k} \to \infty$ in Eq. (22) of Lemma 4.2, by using Lemma 4.3 and the relation (37), where $c_k = \alpha(t_k)$ and

$$b_k = \frac{\tilde{C}\alpha^2(t_k - W)}{\alpha(t_k)} + K(y)\frac{\alpha(t_k - W) - \alpha(t_{k+1} + W)}{\alpha(t_k)} \sum_{t=0}^{t_{k+1}} \alpha(t),$$

from Eq. (22) it can be seen that for all $y \in X$

$$\liminf_{\hat{k}\to\infty} \frac{\sum_{k=0}^{\hat{k}-1} \alpha(t_k) f(x_k)}{\sum_{k=0}^{\hat{k}-1} \alpha(t_k)} \le f(y).$$

By using the relation (36) with $c_k = \alpha(t_k)$ and $b_k = f(x_k)$, from the preceding relation we obtain for all $y \in X$

$$\liminf_{\hat{k}\to\infty} f(x_{\hat{k}}) \le f(y),$$

from which the result follows by taking the minimum over $y \in X$. **Q.E.D.**

Proof of Prop. 2.3: By letting $\beta = 1$ and $y = x^*$ for some $x^* \in X^*$, and by dropping the nonpositive term involving $f(x_k) - f^*$, from Eq. (20) of Lemma 4.2, we obtain for all $x^* \in X^*$ and $\hat{k} > k_0$

$$\Big(1 - 2\alpha(t_{\hat{k}} - W)\Big)||x_{\hat{k}} - x^*||^2 \le \Big(1 + 2\alpha(t_{k_0})\Big)||x_{k_0} - x^*||^2 + 2\tilde{C} \sum_{k=k_0}^{\hat{k}-1} \alpha^2(t_k - W)$$
$$+2K(x^*) \sum_{k=k_0}^{\hat{k}-1} \Big(\alpha(t_k - W) - \alpha(t_{k+1} + W)\Big) \left(C \sum_{r=0}^{t_{k+1}} \alpha(r) + ||x_0 - x^*||\right)$$
$$+2c(x^*)\Big(\alpha^2(t_{k_0}) + \alpha(t_{k_0}) + \alpha^2(t_{\hat{k}} - W) + \alpha(t_{\hat{k}} - W)\Big), \tag{38}$$

As $\hat{k} \to \infty$, by applying Lemma 4.3, from the preceding relation it follows that

$$\limsup_{\hat{k}\to\infty} ||x_{\hat{k}} - x^*|| < \infty,$$

i.e., $\{x_k\}$ is bounded. Furthermore, according to Prop. 2.2, we have

$$\liminf_{k\to\infty} f(x_k) = f^*,$$

which, by the continuity of f and the boundedness of $\{x_k\}$, implies the existence of a subsequence $\{x_{k_j}\} \subset \{x_k\}$ and a point $\hat{x} \in X^*$ such that

$$\lim_{j\to\infty} ||x_{k_j} - \hat{x}|| = 0.$$

Next, in Eq. (38) we let $x^* = \hat{x}$ and $k_0 = k_j$ for some j. Then by first letting $\hat{k} \to \infty$ and then $j \to \infty$, and by using Lemma 4.3 and the fact $x_{k_j} \to \hat{x}$, we obtain

$$\limsup_{\hat{k}\to\infty} ||x_{\hat{k}} - \hat{x}|| = 0.$$

Q.E.D.

5. CONVERGENCE PROOFS FOR A RANDOM SELECTION RULE

In this section we give proofs of Props. 3.1 and 3.2. The proofs rely on the martingale convergence theorem, as stated for example in Gallager [9], p. 256.

Theorem 5.1: (*Martingale Convergence Theorem*) Let $\{Z_k : k = 0, 1, \ldots\}$ be a martingale and assume that there is some positive scalar M such that $E\{Z_k^2\} \leq M$ for all k. Then there is a random variable Z such that, for all sample sequences except a set of probability 0, we have

$$\lim_{k\to\infty} Z_k = Z.$$

In the proofs we also use some properties of $\alpha(t)$ and $x(t)$ that are given, respectively, in Lemmas 5.1 and 5.2 below.

Lemma 5.1: Let Assumption 2.2 hold with $\frac{3}{4} < q \leq 1$. Then we have

$$\sum_{t=0}^{\infty} \alpha^2(t-W) < \infty, \qquad \sum_{t=0}^{\infty} \Delta(t) < \infty, \qquad \sum_{t=0}^{\infty} \alpha(t) = \infty,$$

$$\sum_{t=0}^{\infty} \Delta(t) \sum_{r=0}^{t} \alpha(r) < \infty, \qquad \sum_{t=0}^{\infty} \alpha^2(t) \left(\sum_{r=0}^{t} \alpha(r)\right)^2 < \infty, \tag{39}$$

where $\Delta(t) = \alpha(t-T) - \alpha(t+T)$ and W is a nonnegative integer.

Proof: We show the last relation in Eq. (39). The rest can be shown similar to the proof of Lemma 4.3. Note that $\alpha(t)$ is largest when we change the step every S iterations, i.e., $\sigma_{l+1} - \sigma_l = S$ for all l, so that

$$\alpha(t) \leq \frac{r_0}{(l+r_1)^q}, \qquad t \in \{lS, \ldots, (l+1)S-1\}, \quad l = 0, 1, \ldots,$$

and consequently [cf. Eq. (33)]

$$\sum_{r=0}^{t} \alpha(r) \leq Sr_0 \sum_{k=0}^{l} \frac{1}{(k+r_1)^q} \leq \begin{cases} Sr_0 \left(\frac{1}{r_1} + \ln(l+r_1)\right) & \text{if } q = 1, \\ Sr_0 \left(\frac{1}{r_1^q} + \frac{(l+r_1)^{1-q}}{1-q}\right) & \text{if } q < 1. \end{cases}$$

Therefore $\alpha^2(t) \left(\sum_{r=0}^{t} \alpha(r)\right)^2 \leq w_l$ for $t \in \{lS, \ldots, (l+1)S-1\}$, where w_l is of the order $\frac{\ln^2(l+1+r_1)}{(l+r_1)^2}$ for $q = 1$ and of the order $\frac{1}{(l+r_1)^{4q-2}}$ for $q < 1$. Hence $\sum_{t=0}^{\infty} \alpha^2(t) \left(\sum_{r=0}^{t} \alpha(r)\right)^2$ is finite for $q > \frac{3}{4}$. **Q.E.D.**

Lemma 5.2: Let Assumption 3.1 hold.

(a) For all $y \in X$ and t, we have

$$\begin{aligned}
||x(t+1)-y||^2 \le\ & ||x(t)-y||^2 - \frac{2\alpha(t)}{m}\big(f(x(t))-f(y)\big) + 2\big(z_y(t)-z_y(t-1)\big) \\
& + C^2(1+4D)\alpha^2(t-D) \\
& + 2\alpha(t)\sum_{l=1}^{m}\big(\delta_{j(t)}^l - \delta_{i(t)}^l\big)\big(f_l(x(t))-f_l(y)\big),
\end{aligned} \tag{40}$$

where δ_i^l is the Kronecker symbol, $z_y(-1)=0$, and for $t \ge 0$

$$z_y(t) = \sum_{r=0}^{t}\alpha(r)\left(\frac{1}{m}\big(f(x(r))-f(y)\big) - \big(f_{j(r)}(x(r)) - f_{j(r)}(y)\big)\right). \tag{41}$$

(b) For all $y \in X$, and N and K with $N \ge K$, we have

$$\begin{aligned}
||x(N+1)-y||^2 \le\ & ||x(K)-y||^2 - \frac{2}{m}\sum_{t=K}^{N}\alpha(t)\big(f(x(t))-f(y)\big) \\
& + 2\big(z_y(N)-z_y(K-1)\big) \\
& + C^2\big(1+4D+2T\big)\sum_{t=K}^{N}\alpha^2(t-W) \\
& + 2\max\{G(y),C\}\sum_{t=K}^{N}\Delta(t)\left(C\sum_{r=0}^{t}\alpha(r) + ||x(0)-y||\right) \\
& + 2c(y)\big(\alpha^2(K)+\alpha(K)+\alpha^2(N+1-T)+\alpha(N+1-T)\big) \\
& + 2\big(\alpha(K)||x(K)-y||^2 + \alpha(N+1-T)||x(N+1)-y||^2\big),
\end{aligned}$$

where $W = \max\{D,T\}$, $G(y)$ and $c(y)$ are given by Eqs. (10) and (11), respectively, and $\Delta(t) = \alpha(t-T) - \alpha(t+T)$ for all t.

(c) For all $y \in X$, the sequence $\{z_y(t)\}$ defined by Eq. (41) is a convergent martingale.

Proof: (a) The relation (40) follows from Eq. (8) of Lemma 4.1 by adding and subtracting $\frac{2\alpha(t)}{m}\big(f(x_k)-f(y)\big)$, and by using the definition of $z_y(t)$ [cf. Eq. (41)].

(b) Summing the inequalities (40) over $t = K,\dots,N$ yields for all $y \in X$

$$\begin{aligned}
||x(N+1)-y||^2 \le\ & ||x(K)-y||^2 - \frac{2}{m}\sum_{t=K}^{N}\alpha(t)\big(f(x(t))-f(y)\big) \\
& + 2\big(z_y(N)-z_y(K-1)\big) + C^2(1+4D)\sum_{t=K}^{N}\alpha^2(t-D) \\
& + 2\sum_{t=K}^{N}\sum_{l=1}^{m}\alpha(t)\big(\delta_{j(t)}^l - \delta_{i(t)}^l\big)\big(f_l(x(t))-f_l(y)\big).
\end{aligned}$$

The desired relation follows from the preceding relation, by using Eq. (9) of Lemma 4.1 with $\beta = 1$ and $x_0 = x(0)$.

(c) Let $y \in X$ be fixed. First, we show that the sequence $\{z_y(t)\}$ is a martingale. By using the definition of $z_y(t)$ [cf. Eq. (41)], we have

$$\begin{aligned} E\Big\{z_y(t) \mid z_y(t-1)\Big\} &= z_y(t-1) \\ &\quad + \alpha(t) E\left\{\frac{1}{m}\Big(f(x(t)) - f(y)\Big) - \Big(f_{j(t)}(x(t)) - f_{j(t)}(y)\Big)\right\} \\ &= z_y(t-1), \end{aligned}$$

where in the last equality we use the iterated expectation rule and

$$E\left\{\frac{1}{m}\Big(f(x(t)) - f(y)\Big) - \Big(f_{j(t)}(x(t)) - f_{j(t)}(y)\Big)\Big|\, x(t)\right\} = 0,$$

which follows from the properties of $\{j(t)\}$ [cf. Assumption 3.1(c)]. Hence $z_y(t)$ is indeed a martingale.

To apply the martingale convergence theorem, which guarantees the convergence of $z_y(t)$, we have to prove that $E\{z_y^2(t)\}$ is bounded. From the definition of $z_y(t)$ [cf. Eq. (41)] it follows that

$$E\Big\{z_y^2(t)\Big\} = \sum_{r=0}^{t} \alpha^2(r) E\left\{\left(\frac{1}{m}\Big(f(x(r)) - f(y)\Big) - \Big(f_{j(r)}(x(r)) - f_{j(r)}(y)\Big)\right)^2\right\}. \tag{42}$$

This is because the expected values of the cross terms appearing in $z_y^2(t)$ are equal to 0, which can be seen by using the iterated expectation rule [i.e., by conditioning on the values $x(s)$ and $x(r)$ for $s, r \leq t$] and by exploiting the properties of $j(r)$ [cf. Assumption 3.1(c)]. Furthermore, by using convexity of each f_i, the triangle inequality, and the following relation [cf. Eq. (13)]

$$||x(t) - x(\hat{t})|| \leq C \sum_{s=\hat{t}}^{t-1} \alpha(s), \qquad \forall\, t, \hat{t},\ t \geq \hat{t},$$

for every r we have

$$\begin{aligned} &\left(\frac{1}{m}\Big(f(x(r)) - f(y)\Big) - \Big(f_{j(r)}(x(r)) - f_{j(r)}(y)\Big)\right)^2 \\ &\qquad \leq \frac{2}{m^2}\Big(f(x(r)) - f(y)\Big)^2 + 2\Big(f_{j(r)}(x(r)) - f_{j(r)}(y)\Big)^2 \\ &\qquad \leq 4\Big(\max\{G(y), C\}\Big)^2 ||x(r) - y||^2 \\ &\qquad \leq 4\Big(\max\{G(y), C\}\Big)^2 \left(C\sum_{s=0}^{r} \alpha(s) + ||x(0) - y||\right)^2 \\ &\qquad \leq 8\Big(\max\{G(y), C\}\Big)^2 \left[C^2\left(\sum_{s=0}^{r} \alpha(s)\right)^2 + ||x(0) - y||^2\right]. \end{aligned}$$

By using this inequality and Lemma 5.1 in Eq. (42), we see that $E\{z_y^2(t)\}$ is bounded. Thus, according to the martingale convergence theorem, as $t \to \infty$, the sequence $\{z_y(t)\}$ converges to some random variable with probability 1. **Q.E.D.**

Next we prove Prop. 3.1.

Proof of Prop. 3.1: First we consider the case where f^* is finite. Let $\epsilon > 0$ be arbitrary and let $\hat{y} \in X$ be such that

$$f(\hat{y}) \leq f^* + \epsilon.$$

Fix a sample path, denoted by $\mathcal{P}$, for which the martingale $\{z_{\hat{y}}(t)\}$ is convergent [cf. Lemma 5.2(c)]. From Lemma 5.2(b), where $K = 0$ and $y = \hat{y}$, we have for the path $\mathcal{P}$ and sufficiently large N

$$\begin{aligned}\frac{2}{m}\sum_{t=0}^{N}\alpha(t)\Big(f(x(t)) - f(\hat{y})\Big) \;\leq\; & \Big(1 + 2\alpha(0)\Big)\|x(0) - \hat{y}\|^2 + 2z_{\hat{y}}(N) \\ & + C^2\Big(1 + 4D + 2T\Big)\sum_{t=0}^{N}\alpha^2(t - W) \\ & + 2\max\{C, G(\hat{y})\}\sum_{t=0}^{N}\Delta(t)\left(C\sum_{r=0}^{t}\alpha(r) + \|x(0) - \hat{y}\|\right) \\ & + 2c(\hat{y})\Big(\alpha^2(0) + \alpha(0) + \alpha^2(N + 1 - T) + \alpha(N + 1 - T)\Big),\end{aligned}$$

where we use the fact $z_{\hat{y}}(-1) = 0$ and we take $N \geq k_0$ with k_0 such that $1 - 2m\alpha(k - T) \geq 0$ for all $k \geq k_0$. Since $z_{\hat{y}}(N)$ converges, by dividing the above inequality with $\frac{2}{m}\sum_{t=0}^{N}\alpha(t)$, by letting $N \to \infty$, and by using Lemma 5.1, we obtain

$$\liminf_{N\to\infty}\frac{\sum_{t=0}^{N}\alpha(t)f(x(t))}{\sum_{t=0}^{N}\alpha(t)} \leq f(\hat{y}).$$

By using the facts $f(\hat{y}) \leq f^* + \epsilon$ and

$$\liminf_{N\to\infty} f(x(N+1)) \leq \liminf_{N\to\infty}\frac{\sum_{t=0}^{N}\alpha(t)f(x(t))}{\sum_{t=0}^{N}\alpha(t)}$$

[cf. Eq. (36)], we have for the path $\mathcal{P}$

$$\liminf_{N\to\infty} f(x(N+1)) \leq f^* + \epsilon.$$

Thus $\liminf_{t\to\infty} f(x(t)) \leq f^* + \epsilon$ with probability 1, and since ϵ is arbitrary, it follows that

$$\liminf_{t\to\infty} f(x(t)) = f^*.$$

In the case where $f^* = -\infty$, we choose an arbitrarily large positive integer M and a point $\hat{y} \in X$ such that $f(\hat{y}) \leq -M$, and use the same line of analysis as in the preceding case. **Q.E.D.**

Now we present a proof of Prop. 3.2

Proof of Prop. 3.2: Let d be the dimension of X^*. Then there exist $d+1$ distinct points $y_0, \ldots, y_d \in X^*$ that are in general position, i.e., they are such that the vectors $y_1 - y_0, \ldots, y_d - y_0$ are linearly independent. According to Lemma 5.2(c), for each $s = 0, \ldots, d$ the martingale $z_s(t) = z_{y_s}(t)$ converges, as $t \to \infty$, to some random variable with probability 1.

Now, fix a sample path, denoted by $\mathcal{P}$, for which

$$\liminf_{t \to \infty} f(x(t)) = f^* \tag{43}$$

(cf. Prop. 3.1) and every martingale $z_s(t)$ is convergent. Furthermore, fix an $s \in \{0, \ldots, d\}$. Let K_0 be a positive integer large enough so that

$$1 - 2\alpha(K - T) > 0, \qquad \forall\ K \geq K_0.$$

By using Lemma 5.2(b) with $y = y_s$ and $N > K \geq K_0$, and by dropping the nonnegative term involving $f(x(t)) - f(y_s)$, we obtain

$$\begin{aligned}\Big(1 - 2\alpha(N+1-T)\Big)||x(N+1) - y_s||^2 &\leq \Big(1 + 2\alpha(K)\Big)||x(K) - y_s||^2 \\ &+2\Big(z_s(N) - z_s(K-1)\Big) + C^2\Big(1 + 4D + 2T\Big)\sum_{t=K}^{N} \alpha^2(t - W) \\ &+2\max\{G(y_s), C\}\sum_{t=K}^{N} \Delta(t)\left(C\sum_{r=0}^{t}\alpha(r) + ||x(0) - y_s||\right) \\ &+2c(y_s)\Big(\alpha^2(K) + \alpha(K) + \alpha^2(N+1-T) + \alpha(N+1-T)\Big).\end{aligned}$$

By using Lemma 5.1 and the convergence of $\{z_s(t)\}$, from the preceding relation we obtain

$$\limsup_{N \to \infty} ||x(N+1) - y_s||^2 \leq \liminf_{K \to \infty} ||x(K) - y_s||^2.$$

Hence $\lim_{t \to \infty} ||x(t) - y_s||^2$ exists for the path $\mathcal{P}$ and all $s = 0, \ldots, d$.

Let us define for $s = 0, \ldots, d$

$$\beta_s = \lim_{t \to \infty} ||x(t) - y_s||. \tag{44}$$

Because $y_0, \ldots, y_d$ are distinct points, at most one β_s can be zero, in which case for the path $\mathcal{P}$ we have $\lim_{t \to \infty} ||x(t) - y_s|| = 0$, and we are done. Now, let all the scalars β_s be positive. Without loss of generality, we may assume that all β_s are equal, for otherwise we can use a linear transformation of the space $\Re^n$ such that in the transformed space the values $\hat{\beta}_0, \ldots, \hat{\beta}_d$ corresponding to $\beta_0, \ldots, \beta_d$, respectively, are all equal. Therefore we set $\beta_s = \beta$ for all s, so that every limit point of $\{x(t)\}$ is at the same distance β from each of the points $y_0, \ldots, y_d$. Let $\mathcal{M}_d$ be the d-dimensional linear manifold containing X^*. Because $y_0, \ldots, y_d$ are in general position, the set of all points that are at the same distance from each y_s is an $n-d$-dimensional linear manifold $\mathcal{M}_{n-d}$. Hence all limit points of $\{x(t)\}$ lie in the manifold $\mathcal{M}_{n-d}$.

In view of Eqs. (43) and (44), $\{x(t)\}$ must have a limit point $\tilde{x}$ for which $f(\tilde{x}) = f^*$. Therefore $\tilde{x} \in X^*$ implying that $\tilde{x} \in \mathcal{M}_d \cap \mathcal{M}_{n-d}$, and since $\mathcal{M}_d$ is orthogonal to $\mathcal{M}_{n-d}$, $\tilde{x}$ is the unique limit point of $\{x(t)\}$ for the path $\mathcal{P}$. Hence $\{x(t)\}$ converges to some optimal solution with probability 1. **Q.E.D.**

Remark 5.1: If the set X is compact, then it can be shown that

$$E\left\{z_y^2(t)\right\} \leq 4\left(\max\{G(y), C\}\right)^2 \sup_{x \in X} \|x - y\|^2 \sum_{r=0}^{t} \alpha^2(r).$$

In this case, the martingale $\{z_y(t)\}$ is convergent for $\frac{1}{2} < q \leq 1$ in Assumption 3.1(b), so that the result of Prop. 3.2 holds under Assumption 3.1(b) with $\frac{1}{2} < q \leq 1$ instead of $\frac{3}{4} < q \leq 1$.

REFERENCES

1. A. Ben-Tal, T. Margalit, and A. Nemirovski, The Ordered Subsets Mirror Descent Optimization Method and its Use for the Positron Emission Tomography Reconstruction, *SIAM J. Optim.*, to appear.
2. A. Benveniste, M. Metivier, and P. Priouret, *Adaptive Algorithms and Stochastic Approximations* (Springer-Verlag, New York, 1990).
3. D. P. Bertsekas, A New Class of Incremental Gradient Methods for Least Squares Problems, *SIAM J. on Optimization* **7** (1997) 913–926.
4. D. P. Bertsekas and J. N. Tsitsiklis, *Parallel and Distributed Computation: Numerical Methods* (Prentice-Hall, Inc., 1989).
5. D. P. Bertsekas and J. N. Tsitsiklis, *Neuro-Dynamic Programming* (Athena Scientific, Belmont, Massachusetts, 1996).
6. D. P. Bertsekas and J. N. Tsitsiklis, Gradient Convergence in Gradient Methods, *SIAM J. on Optimization* **3** (2000) 627–642.
7. V. S. Borkar, Asynchronous Stochastic Approximation, *SIAM J. on Optimization* **36** (1998) 840–851.
8. A. A. Gaivoronski, Convergence Analysis of Parallel Backpropagation Algorithm for Neural Networks, *Opt. Meth. and Software* **4** (1994) 117–134.
9. R. G. Gallager, *Discrete Stochastic Processes* (Kluwer Academic Publishers, 1996).
10. L. Grippo, A Class of Unconstrained Minimization Methods for Neural Network Training, *Opt. Meth. and Software* **4** (1994) 135–150.
11. C. A. Kaskavelis and M. C. Caramanis, Efficient Lagrangian Relaxation Algorithms for Industry Size Job-Shop Scheduling Problems, *IIE Trans. on Scheduling and Logistics* **30** (1998) 1085–1097.
12. V. M. Kibardin, Decomposition into Functions in the Minimization Problem, *Automation and Remote Control* **40** (1980) 1311–1323.

13. K. C. Kiwiel and P. O. Lindberg, Parallel Subgradient Methods for Convex Optimization, submitted to the Proceedings of the March 2000 Haifa Workshop "Inherently Parallel Algorithms in Feasibility and Optimization and Their Applications", D. Butnariu, Y. Censor, and S. Reich, Eds., *Studies in Computational Mathematics*, Elsevier, Amsterdam.
14. K. C. Kiwiel, Convergence of Approximate and Incremental Subgradient Methods for Convex Optimization, submitted to *SIAM J. on Optimization.*
15. H. J. Kushner and G. Yin, it Stochastic Approximation Algorithms and Applications (Springer-Verlag, New York, 1997).
16. Z. Q. Luo, On the Convergence of the LMS Algorithm with Adaptive Learning Rate for Linear Feedforward Networks, *Neural Computation* **3** (1991) 226–245.
17. Z. Q. Luo and P. Tseng, Analysis of an Approximate Gradient Projection Method with Applications to the Backpropagation Algorithm, *Opt. Meth. and Software* **4** (1994) 85–101.
18. O. L. Mangasarian and M. V. Solodov, Serial and Parallel Backpropagation Convergence Via Nonmonotone Perturbed Minimization, *Opt. Meth. and Software* **4** (1994) 103–116.
19. A. Nedić and D. P. Bertsekas, Incremental Subgradient Methods for Nondifferentiable Optimization, *Lab. for Info. and Decision Systems Report LIDS-P-2460* (Massachusetts Institute of Technology, Cambridge, MA, 1999).
20. A. Nedić and D. P. Bertsekas, Incremental Subgradient Methods for Nondifferentiable Optimization, submitted to *SIAM J. Optimization.*
21. A. Nedić and D. P. Bertsekas, Convergence Rate of Incremental Subgradient Algorithms, *Lab. for Info. and Decision Systems Report LIDS-P-2475* (Massachusetts Institute of Technology, Cambridge, MA, 2000), to appear in *Stochastic Optimization: Algorithms and Applications*, Eds. S. Uryasev and P. M. Pardalos.
22. M. V. Solodov and S. K. Zavriev, Error Stability Properties of Generalized Gradient-Type Algorithms, *J. of Opt. Theory and Applications* **98** (1998) 663–680.
23. P. Tseng, An Incremental Gradient(-Projection) Method with Momentum Term and Adaptive Stepsize Rule, *SIAM J. on Optimization* **2** (1998) 506–531.
24. J. N. Tsitsiklis, D. P. Bertsekas, and M. Athans, Distributed Asynchronous Deterministic and Stochastic Gradient Optimization Algorithms, *IEEE Trans. on Automatic Control* **AC-31** (1986) 803–812.
25. B. Widrow and M. E. Hoff, Adaptive Switching Circuits, *Institute of Radio Engineers, Western Electronic Show and Convention, Convention Record*, part **4** (1960) 96–104.
26. X. Zhao, P. B. Luh, and J. Wang, Surrogate Gradient Algorithm for Lagrangian Relaxation, *J. of Opt. Theory and Applications* **100** (1999) 699–712.

Inherently Parallel Algorithms in Feasibility and Optimization and their Applications
D. Butnariu, Y. Censor and S. Reich (Editors)

RANDOM ALGORITHMS FOR SOLVING CONVEX INEQUALITIES

B.T. Polyak[a]*

[a]Institute for Control Science, Moscow, Russia

A random subgradient method is proposed for the general convex feasibility problem with very large (or infinite) number of inequalities. Under the strong feasibility assumption the method terminates in a finite number of iterations with probability one. A convergent version of the method for infeasible case is also provided. The algorithm can be applied for solving linear matrix inequalities arising in control. Numerical simulation demonstrates high efficiency of the approach for large dimensional problems.

1. INTRODUCTION

The problem of solving convex inequalities (also known as convex feasibility problem) is one of the basic problems of numerical analysis. It arises in numerous applications, including statistics, parameter estimation, pattern recognition, image restoration, tomography and many others, see e.g. monographs and surveys [1–4] and references therein. Particular cases of the problem relate to solving of linear inequalities and to finding a common point of convex sets. The specific feature of some applications is a huge number of inequalities to be solved, while the dimensionality of the variables is moderate, see e.g. the examples of applied problems in Section 5 below. Under these circumstances many known numerical methods are inappropriate. For instance, finding the most violated inequality may be a hard task; dual methods also can not be applied due to large number of dual variables.

In this paper we focus on simple iterative methods which are applicable for the case of very large (and even infinite) number of inequalities. They are based on projection-like algorithms, originated in the works [5–10]. There are many versions of such algorithms; they can be either parallel or non-parallel (row-action); in the last case the order of projections is usually chosen as cyclical one or the-most-violated one, see [1–4]. All these methods are well suited for the finite (and not too large) number of constraints. The novelty of the method under consideration is its random nature, which allows to treat large dimensional and infinite dimensional cases. Although the idea of exploiting stochastic algorithms for optimization problems with continua of constraints has been known in the literature [14–16], it lead to much more complicated calculations than the proposed method. Another feature of the method is its finite termination property – under the strong feasibility assumption a solution is found after a finite number of steps

*The paper has been supported by NATO Fellowship and RFFI-99-01-00340 grant.

with probability one. The version of a projection method for linear inequalities with this property has been proposed first by V.Yakubovich [11], see also [12]; it was extended to general convex case by V.Fomin [13]. In Section 2 we provide a general result on finite convergence of a random algorithm for a convex feasibility problem. In Section 3 the infeasible case is considered and the convergent version of the algorithm under the lack of feasibility condition is proposed. We discuss various extensions of the algorithms and problems of their implementation in Section 4. Section 5 is devoted to applications, which are different from standard ones. First we deal with numerous problems in systems and control which can be described in terms of linear matrix inequalities (LMIs). The robust quadratic stability is taken as an example; it can be cast as LMIs with very large or infinite number of inequalities. Thus the proposed algorithms happen to be effective tools for their solution. Other applications relate to robust linear inequalities problem and to finding the largest ellipsoid inscribed in a polytope.

2. CONVEX FEASIBILITY ALGORITHM

Consider the general convex feasibility problem: find a point x in the set

$$C = \{x \in X, \qquad f(x,\omega) \leq 0, \qquad \forall \omega \in \Omega\}. \tag{1}$$

Here $X \subset R^n$ is a convex closed set, $f(x,\omega)$ is convex in x for all $\omega \in \Omega$, while Ω is an arbitrary set (finite or infinite). Particular cases of the problem are:

1. *Finite number of inequalities*: $\Omega = \{1, ..., m\}$.
2. *Semiinfinite problem*: $\Omega = [0, T] \subset R^1$.
3. *Finding a common point of convex sets*: $f(x,\omega) = dist(x, C_\omega) = \min_{y \in C_\omega} ||x - y||$, where $C_\omega \subset R^n$ are closed convex sets and $C = \cap_{\omega \in \Omega} C_\omega$. Here and elsewhere $||x||$ denotes the Euclidean norm of a vector.
4. *Linear inequalities:* $f(x,\omega) = a(\omega)^T x - b(\omega)$.

We suppose that a subgradient $\partial_x f(x,\omega)$ is available at any point $x \in X$ for all $\omega \in \Omega$ (we mean an arbitrary subgradient if the set of them is not a singleton). The algorithm has the following structure: at k-th iteration, we generate randomly $\omega_k \in \Omega$; we assume that they are independent identically distributed (i.i.d.) samples from some probabilistic distribution p_ω on Ω. Two key assumptions are done:

Assumption 1 (strong feasibility). The set C is nonempty and contains an interior point:

$$\exists x^* \in C : ||x - x^*|| \leq r \Longrightarrow x \in C.$$

Here $r > 0$ is a constant assumed to be known.

Assumption 2 (distinguishability of feasible and infeasible points). If $x \in X, x \notin C$, then the probability to generate a violated inequality is not vanishing:

$$\Pr\{f(x,\omega) > 0\} > 0.$$

This is the only assumption on the probability distribution p_ω. For instance, if Ω is a finite set and each element of Ω is generated with nonzero probability, then the above assumption holds.

Algorithm 1 (feasible case). *For an initial point $x_0 \in X$ proceed*

$$x_{k+1} = P_X(x_k - \lambda_k \partial_x f(x_k, \omega_k)) \tag{2}$$

$$\lambda_k = \begin{cases} \frac{f(x_k,\omega_k)+r||\partial_x f(x_k,\omega_k)||}{||\partial_x f(x_k,\omega_k)||^2}, & \text{if } f(x_k,\omega_k) > 0 \\ 0, & \text{otherwise.} \end{cases} \tag{3}$$

Here P_X is a projection operator onto X (that is, $||x - P_X(x)|| =$dist (x, X)). Thus a subgradient calculation just for one inequality is performed at each step, and this inequality is randomly chosen among all inequalities Ω. Note that the value of r (radius of a ball in the feasible set) is used in the algorithm; its modification for r unknown will be presented later. To explain the stepsize choice in the Algorithm, let us consider two particular cases.

1. *Linear inequalities.* $f(x,\omega) = a(\omega)^T x - b(\omega), X = R^n$, thus $\partial_x f(x_k, \omega_k) = a_k = a(\omega_k), f(x_k,\omega_k) = a_k^T x_k - b_k, b_k = b(\omega_k)$, and the algorithm takes the form

$$x_{k+1} = x_k - \frac{(a_k^T x_k - b_k)_+ + r||a_k||}{||a_k||^2} a_k$$

for $(a_k^T x_k - b_k)_+ \neq 0$, otherwise $x_{k+1} = x_k$; here $c_+ = \max\{0, c\}$. For $r = 0$ the method coincides with the projection method for solving linear inequalities by Agmon-Motzkin-Shoenberg [6,7].

2. *Common point of convex sets.* $f(x,\omega) = dist(x, C_\omega), C = \cap_{\omega\in\Omega} C_\omega, X = R^n$. Then $\partial_x f(x_k,\omega_k) = (x_k - P_k(x_k))/\rho_k$, where P_k is a projection onto the set $C_k = C_{\omega_k}$ and $\rho_k = ||x_k - P_k(x_k)||$. Thus the algorithm takes the form:

$$x_{k+1} = P_k(x_k) + \frac{r}{\varrho_k}(P_k(x_k) - x_k)$$

provided $x_k \notin C_k$, otherwise $x_{k+1} = x_k$. We conclude, that for $r = 0$ each iteration of the algorithm is the same as for the projection method for finding the intersection of convex sets [9,1].

Having this in mind, the stepsize rule has very natural explanation. Denote y_{k+1} the point which is generated with the same formula as x_{k+1}, but with $r = 0$, suppose also $X = R^n$. Then for linear inequalities y_{k+1} is a projection of x_k onto the half-space $\{x : a_k^T x - b_k \leq 0\}$. Similarly, if we deal with finding a common point of convex sets, y_{k+1} is a projection of x_k onto the set C_k. It is easy to estimate that $||x_{k+1} - y_{k+1}|| = r$. Thus the step in the algorithm is overrelaxed (additively) projection; we make an extra step (at length r) inside the current feasible set.

The idea of additive overrelaxation is due to Yakubovich , who applied such method for linear inequalities [11]. In all above mentioned works the order of sorting out the inequalities was either cyclic or the-most-violated one was taken, in contrast with the random order in the proposed algorithm. Now we formulate the main result on the convergence of the algorithm.

Theorem 1 *Under Assumptions 1, 2, Algorithm 1 finds a feasible point in a finite number of iterations with probability one, i.e. there exists with probability one such N that $x_N \in C$ and $x_k = x_N$ for all $k \geq N$.*

Proof Define

$$\overline{x} = x^* + \frac{r}{||\partial_x f(x_k, \omega_k)||} \partial_x f(x_k, \omega_k)$$

where x^* is an interior point of C, mentioned in Assumption 1. Then, due to Assumption 1, $\overline{x}$ is a feasible point: $\overline{x} \in C$ and, in particular, $f(\overline{x}, \omega_k) \leq 0$. Now, if $f(x_k, \omega_k) > 0$, it follows from the properties of a projection that

$$||x_{k+1} - x^*||^2 \leq ||x_k - x^* - \lambda_k \partial_x f(x_k, \omega_k)||^2 = ||x_k - x^*||^2 +$$

$$+(\lambda_k)^2 ||\partial_x f(x_k, \omega_k)||^2 - 2\lambda_k \partial_x f(x_k, \omega_k)^T (x_k - \overline{x}) - 2\lambda_k \partial_x f(x_k, \omega_k)^T (\overline{x} - x^*).$$

We now consider the last two terms in the inequality above. Due to convexity of $f(x, \omega)$ and to the feasibility of $\overline{x}$, we obtain

$$\partial_x f(x_k, \omega_k)^T (x_k - \overline{x}) \geq f(x_k, \omega_k)$$

while, due to definition of $\overline{x}$

$$\partial_x f(x_k, \omega_k)^T (\overline{x} - x^*) = r||\partial_x f(x_k, \omega_k)||.$$

Thus, we write

$$||x_{k+1} - x^*||^2 \leq ||x_k - x^*||^2 + (\lambda_k)^2 ||\partial_x f(x_k, \omega_k)||^2 - 2\lambda_k (f(x_k, \omega_k) + r||\partial_x f(x_k, \omega_k)||).$$

Now, substituting the value of λ_k (3), we get

$$||x_{k+1} - x^*||^2 \leq ||x_k - x^*||^2 - \frac{(f(x_k, \omega_k) + r||\partial_x f(x_k, \omega_k)||)^2}{||\partial_x f(x_k, \omega_k)||^2} \leq ||x_k - x^*||^2 - r^2.$$

Therefore, if $f(x_k, \omega_k) > 0$, then we obtain

$$||x_{k+1} - x^*||^2 \leq ||x_k - x^*||^2 - r^2. \tag{4}$$

From this formula, we conclude that no more than $M = ||x_0 - x^*||^2 / r^2$ correction steps can be executed. On the other hand, if x_k is infeasible, then due to Assumption 2, there is a non-zero probability to make a correction step. Thus, with probability one, the method can not terminate at an infeasible point. We, therefore, conclude that the algorithm must terminate after a finite number of iterations at a feasible solution.■

It is possible also to estimate the probability of the termination of the iterative process after k iterations; some results of this kind can be found in [23].

Note that the proof of the theorem follows the standard lines of similar proofs for projection-like methods and is based on estimation of the distance to the feasible point x^*. For instance, it differs from the ones in [11] just in minor technical details.

3. ALGORITHM FOR APPROXIMATE FEASIBILITY

In this section, we consider the case when the set C is empty. Then the problem formulation should be changed - we are looking for a point x^* which minimizes some measure of infeasibility on X. The natural choice of such *feasibility indicator function* is

$$F(x) = Ef(x,\omega)_+, \tag{5}$$

where E denotes expectation with respect to p_ω and $f_+ = \max\{f, 0\}$. Then $F(x)$ is convex in x and $F(x) > 0$ on X if inequalities (1) have no feasible solution, otherwise $F(x) = 0$ for any feasible point. Thus the problem is converted into stochastic minimization problem for function (5) on X. There are numerous algorithms for this purpose; we can adopt the general algorithm of Nemirovsky and Yudin [17]. It has the same structure as Algorithm 1 for feasible case - at each iteration we generate randomly one constraint, defined by a sample ω_k, calculate its subgradient in the point x_k and perform the subgradient step with projection onto X. However, the stepsize rule is different and the algorithm also includes averaging.

We assume that $F(x)$ has a minimum point x^* on X and $||\partial_x f(x_k,\omega_k)|| \le \mu$ for all $x \in X, \omega \in \Omega$. Remind that $\partial_x f(x_k,\omega_k)_+ = \partial_x f(x_k,\omega_k)$ if $f(x_k,\omega_k) > 0$ and $\partial_x f(x_k,\omega_k)_+ = 0$ otherwise.

Algorithm 2 (infeasible case). *With an initial point $x_0 \in X$ and $m_{-1} = 0$ proceed*

$$\begin{aligned}
x_{k+1} &= P_X(x_k - \lambda_k \partial_x f(x_k,\omega_k)_+), && (6)\\
\lambda_k &> 0, \quad \lambda_k \longrightarrow 0, \quad \sum \lambda_k = \infty && (7)\\
m_k &= m_{k-1} + \lambda_k && (8)\\
\overline{x}_k &= \frac{m_{k-1}}{m_k}\overline{x}_{k-1} + \frac{\lambda_k}{m_k}x_k. && (9)
\end{aligned}$$

Now we can formulate the result on the convergence of the algorithm.

Theorem 2 *For Algorithm 2 we have $\lim_{k\to\infty} EF(\overline{x}_k) = F(x^*)$. Moreover, for a finite k the following estimate holds*

$$EF(\overline{x}_k) - F(x^*) \le C(k) = \frac{\|x_0 - x^*\|^2 + \mu^2 \sum_{i=0}^{k-1} \lambda_i^2}{2\sum_{i=0}^{k-1} \lambda_i}. \tag{10}$$

Proof Consider the distance from the current point x_{k+1} to the optimal point x^*. By the definition of projection, we have that $\|P_X(x) - x^*\| \le \|x - x^*\|$, for any x, and for any $x^* \in X$, therefore,

$$\|x_{k+1} - x^*\|^2 \le \|x_k - x^*\|^2 - 2\lambda_k (x_k - x^*)^T \partial_x f(x_k,\omega_k)_+ + \lambda_k^2 \|\partial_x f(x_k,\omega_k)_+\|^2. \tag{11}$$

Now, for a convex function $g(x)$, for any x, x^* it holds that $(x - x^*)^T \partial g(x) \ge g(x) - g(x^*)$, hence

$$(x_k - x^*)^T \partial_x f(x_k,\omega_k)_+ \ge f(x_k,\omega_k)_+ - f(x^*,\omega_k)_+.$$

On the other hand, from the boundedness condition on the subgradients we have that $||\partial_x f(x_k, \omega_k)_+||^2 \leq \mu^2$, therefore, (11) writes $||x_{k+1} - x^*||^2 \leq ||x_k - x^*||^2 - 2\lambda_k(f(x_k, \omega_k)_+ - f(x^*, \omega_k)_+) + \lambda_k^2\mu^2$. Denoting $u_k = E||x_k - x^*||^2$, and taking expectation of both sides of the last inequality, we get

$$u_{k+1} \leq u_k - 2\lambda_k \left(EF(x_i) - F(x^*)\right) + \lambda_k^2\mu^2,$$

and summing such inequalities

$$u_k \leq u_0 - 2\sum_{i=0}^{k-1} \lambda_i \left(EF(x_i) - F(x^*)\right) + \mu^2 \sum_{i=0}^{k-1} \lambda_i^2,$$

therefore,

$$\sum_{i=0}^{k-1} \lambda_i \left(EF(x_i) - F(x^*)\right) \leq \frac{1}{2}(u_0 + \mu^2 \sum_{i=0}^{k-1} \lambda_i^2). \tag{12}$$

It is obvious that (8), (9) provide a recurrent version of the Cesaro's mean, therefore, $\overline{x}_k$ is an averaged version of the x_k's:

$$\overline{x}_k = \frac{\sum_{i=0}^k \lambda_i x_i}{\sum_{i=0}^k \lambda_i}.$$

From Jensen inequality for convex functions, we then have

$$F(\overline{x}_k) = F\left(\frac{\sum_{i=0}^k \lambda_i x_i}{\sum_{i=0}^k \lambda_i}\right) \leq \frac{\sum_{i=0}^k \lambda_i F(x_i)}{\sum_{i=0}^k \lambda_i},$$

therefore,

$$EF(\overline{x}_k) - F(x^*) \leq \frac{\sum_{i=0}^k \lambda_i (EF(x_i) - F(x^*)}{\sum_{i=0}^k \lambda_i},$$

and, by (12)

$$EF(\overline{x}_k) - F(x^*) \leq \frac{u_0 + \mu^2 \sum_{i=0}^k \lambda_i^2}{2\sum_{i=0}^k \lambda_i},$$

which proves (10). With the further assumptions on the step sizes $\sum_{i=0}^{\infty} \lambda_i = \infty; \ \ \lambda_i \to 0$, it immediately follows that $EF(\overline{x}_k) \to F(x^*)$.■

The above result in more general form (for arbitrary convex functions in a Banach space) can be found in [17, Section 5.2]; we provided the proof above because it can be simplified and adapted for the problem at hand.

4. EXTENSIONS AND IMPLEMENTATION

In this Section we consider some versions of the above proposed basic algorithms as well as some problems arising in the process of their implementation.

4.1. Nonrandom order of generating inequalities

Note that estimate (4) for Algorithm 1 holds for an arbitrary rule of generating inequalities; the only assumption is that we encounter a violated inequality. Thus we can conclude that the total number of correction steps for an arbitrary strategy of generating inequalities is not more than $M = ||x_0 - x^*||^2/r^2$. If we take the most violated inequality at each iteration, then we automatically deal with a violated inequality, hence for this case the total number of iterations before termination does not exceed M. Similarly, if the number of inequalities is finite and we use the cyclic order, then at each cycle we meet not less than one violated inequality. Hence the number of cycles for termination is not greater than M. Of course such nonrandom rules can be more preferable if the number of inequalities is not large.

4.2. Choice of r

Usually the quantity r in Algorithm 1 is not critical for feasible case, that is we have fast convergence for any r in (3) which is less or equal the true r in Assumption 1. In most cases it was taken to be $10^{-2} \div 10^{-3}$ in numerical simulation. However there are two risky extreme situations. If we take r too small (e.g. $r = 0$) then the method turns into the standard projection algorithm with random control, which has no finite termination property. Indeed, the behavior of Algorithm 1 with $r = 0.01$ and $r = 0$ was really different for many examples. On the other hand one can loose convergence if r is overestimated and chosen to be too large. To avoid such situations we can provide more rigorous algorithm which can be validated for r unknown. Choose a sequence

$$\alpha_k > 0, \qquad \alpha_k \longrightarrow 0, \qquad \sum \alpha_k^2 = \infty$$

($\alpha_k = 1/\sqrt{k}$ is an example). Now let $n(k)$ denote the number of correction steps preceding k-th iteration (i.e. $n(0) = 0, n(k) = n(k-1) + 1$ if $\lambda_k > 0, n(k) = n(k-1)$ otherwise). Then for Algorithm 1 with r replaced by $r_k = \alpha_{n(k)}$ the statement of Theorem 1 remains valid. Indeed, after a finite number of iterations we get $r_k \leq r$ (because $\alpha_k \longrightarrow 0$) and then instead of (4) we obtain

$$||x_{k+1} - x^*||^2 \leq ||x_k - x^*||^2 - r_k^2.$$

The same considerations as at the end of the proof of Theorem 1 guarantee finite termination property with probability one.

4.3. Multiplicative overrelaxation

We have mentioned that Algorithm 1 is a projection method with additive overrelaxation. But we can include multiplicative overrelaxation as well. Instead of formula (3) for stepsize in Algorithm 1 let us take

$$\lambda_k = \begin{cases} \eta \frac{f(x_k,\omega_k) + r||\partial_x f(x_k,\omega_k)||}{||\partial_x f(x_k,\omega_k)||^2}, & \text{if } f(x_k,\omega_k) > 0 \\ 0, & \text{otherwise.} \end{cases} \tag{13}$$

where the multiplier $0 < \eta < 2$ is plugged in. Then, repeating the calculations in the proof of Theorem 1 we get that (4) should be replaced with

$$||x_{k+1} - x^*||^2 \leq ||x_k - x^*||^2 - \eta(2-\eta)r^2$$

and all conclusions of the theorem remain true (with $M = ||x_0 - x^*||^2/\eta(2-\eta)r^2$). Such multiplier $\eta > 1$ accelerates convergence; in practical simulation it was usually taken $\eta = 1.8$.

4.4. Distinguishing feasible and infeasible cases

We have different algorithms for these cases; however in many situations we do not know in advance if the inequalities are feasible or not. If Algorithm 1 is applied for an infeasible problem, it looses convergence - there arise oscillations in some bounded region. Such oscillations usually can be recognized. On the other hand, Algorithm 2 converges for both cases - feasible and infeasible, but it has no finite termination property for feasible inequalities (and its convergence may be slow in contrast with fast convergence of Algorithm 1). These considerations can be exploited in the process of implementation as follows. First, in many applications (see Section 5 below) we deal with uncertain systems having some uncertainty indicator. For the zero value of the indicator the system of inequalities is feasible and it looses feasibility for increasing values of the indicator. Thus we can try to apply Algorithm 1 for some grid of indicator values and observe when it looses convergence. Note that closer is the indicator to its critical value, slower is the convergence of Algorithm 1. Second, if we have just one fixed problem we can try to apply Algorithm 1; the lack of its convergence can be easily recognized. Third, we can start with Algorithm 2 and observe its behavior; if the averaged measure of violation of inequalities tends to zero, we may hope that the problem is feasible and pass to Algorithm 1.

4.5. Choice of probability distribution

There are some types of problems where the probability distribution p_ω is beyond our choice - there is a natural random mechanism which generates the sample ω_k. We encounter such situation in many identification and estimation problems with random inputs. However in the most of the problems the randomization is ensured by our will and the choice of p_ω can vary. The only condition on the distribution is Assumption 2 which is not very restrictive. The natural candidate for p_ω is the uniform distribution on Ω. Such distribution can be easily generated when Ω is a ball in some matrix or vector norm via the technique developed in [24]. If Ω is a finite set, the probabilities of generating each inequality can be adjusted in the process of iterations: we can increase the probabilities of violated constraints and decrease the probabilities of nonviolated ones. Such nonstationary distributions can be covered by an extension of Theorem 1; they can lead to the acceleration of convergence.

4.6. Stopping rule

There is no contradiction between the possible infinite number of inequalities and the finite termination property for Algorithm 1. The serious problem in this case is the stopping rule. How to verify that a candidate solution is a feasible point? If the number of inequalities is finite and moderate (few thousands) we can sequentially check all of them. However, if the number of inequalities is too large or infinite, the only possible approach is a stochastic one. That is, if we do not have a correction step after many random generations of inequalities we can guarantee with the probability close to one (which can be estimated) that the point is feasible.

4.7. Projection on X

The set X in the problem formulation is not specified - it is an arbitrary convex closed set with the only property that the projection onto X can be found explicitly. However the inspection of the proofs of Theorems 1 and 2 confirms that P_X is not necessarily the projection operator, but any operator with the property

$$||x - P_X(y)|| \leq ||x - y||, \qquad \forall x \in X, y \in R^n. \tag{14}$$

Finding of such operator can be easier than constructing of the true projection. Consider a typical example. Let X be the standard simplex:

$$X = \{x : x \geq 0, \ x^T e \leq 1\}, \qquad e = (1, ..., 1).$$

Then the simple operator

$$P_X(x) = \begin{cases} \frac{x_+}{x_+^T e}, & \text{if } x_+^T e > 1, \\ x_+, & \text{otherwise,} \end{cases}$$

has the above property while the true projection is hard to present in an explicit form. Here x_+ is the projection of a vector $x = (x_1, ..., x_n) \in R^n$ onto nonnegative orthant: $x_+ = (x_1^+, ..., x_n^+)$, $x_i^+ = \max\{0, x_i\}$.

4.8. Choice of λ_k

Theorem 2 gives the estimate of accuracy of the approximation after k iterations (10). This provides an opportunity to optimize the choice of stepsizes. Indeed, optimizing $C(k)$ over λ_i, we get

$$\lambda_i = \lambda^* = \frac{||x_0 - x^*||}{\mu} \frac{1}{\sqrt{k}}, \quad i = 0, \ldots, k-1.$$

If the number of iterations is fixed in advance to k, then the best choice are constant step sizes $\lambda^* = \frac{||x_0 - x^*||}{\mu\sqrt{k}}$, which yields

$$EF(\overline{x}_k) - F(x^*) \leq 2\frac{||x_0 - x^*||\mu}{\sqrt{k}}.$$

If we do not fix a-priori the number of iterations, a good choice for the steps is $\lambda_k = \frac{c}{\sqrt{k}}$, which provides asymptotically the same estimate $EF(\overline{x}_k) - F(x^*) = O\left(k^{-1/2}\right)$.

5. APPLICATIONS

5.1. Solving linear matrix inequalities

Linear Matrix Inequalities (LMIs) are one of the most powerful tools for model formulation in various fields of systems and control, see [18]. There exist well developed techniques for solving such inequalities as well as for solving optimization problems subject to such inequalities (Semidefinite Programming) [19,?]. However, in a number of applications (for instance, in robust stabilization and control, see below) the number of LMIs is extremely large (or even infinite) and such problems are beyond the applicability of standard LMI tools. Let us cast these problems in the framework of the above proposed approach.

The space S_m of $m \times m$ symmetric real matrices equipped with the scalar product $\langle A, B\rangle = \text{Tr } AB$ and Frobenius norm becomes a Hilbert space. Then we can define the projection A_+ of a matrix A onto the cone of positive semidefinite matrices. This projection can be found in the explicit form [21]. Indeed, if $A = RDR^T, R^{-1} = R^T$, is eigenvector - eigenvalue decomposition of A and D =diag $(d_1, ..., d_m)$, then

$$A_+ = RD_+R^T, \tag{15}$$

where D_+ =diag $(d_1^+, ..., d_m^+)$ and $d_i^+ = \max\{0, d_i\}$. Linear matrix inequality is the expression of the form $A(x) = A_0 + \sum_{i=1}^n x_i A_i \leq 0$, where $A_i \in S_m, i = 0, 1, ..., n$ are given matrices, $x = (x_1, ..., x_n) \in R^n$ is vector variable and $A \leq 0$ means that A is negative semidefinite. The general system of LMIs can be written as

$$A(x, \omega) = A_0(\omega) + \sum_{i=1}^n x_i A_i(\omega) \leq 0 \qquad \forall \omega \in \Omega. \tag{16}$$

Here Ω is a set of indices which can be finite or infinite. The problem under consideration is to find a $x \in R^n$ which satisfies LMIs (16). Our first goal is to convert these LMIs into a system of convex inequalities. For this purpose, introduce the scalar function

$$f(x, \omega) = ||A_+(x, \omega)|| \tag{17}$$

where $A(x, \omega)$ is given by (16) and A_+ is defined in (15).

Lemma 1 *Matrix inequalities (16) are equivalent to scalar inequalities*

$$f(x, \omega) \leq 0 \qquad \forall \omega \in \Omega.$$

The function $f(x, \omega)$ is convex in x and its subgradient is given by the formula

$$\partial_x f(x, \omega) = \frac{1}{f(x, \omega)} \begin{pmatrix} TrA_1A_+(x, \omega) \\ \vdots \\ TrA_nA_+(x, \omega) \end{pmatrix}$$

if $f(x, \omega) > 0$; otherwise $\partial_x f(x, \omega) = 0$.

These statements can be checked by direct calculations (see also [22]).

Now we are in position to apply Algorithms 1 and 2 for solving (or approximate solving in infeasible case) LMIs (16). At each iteration we deal with one LMI, generated randomly (that is, we generate $\omega_k \in \Omega$ according to some density p_ω on Ω and take LMI $A(x, \omega_k) \leq 0$), and the subgradient step requires one eigenvalue - eigenvector decomposition of the symmetric matrix $A(x_k, \omega_k)$. Thus the iteration is very simple from the computational point of view and the method can be applied to large dimensional problems.

Similar approach to quadratic matrix inequalities is proposed in [23].

Example: *Quadratic stability of interval matrices.* Given an interval system of linear ordinary differential equations $\dot{x} = Ax$ where A is an $n \times n$ interval matrix with entries a_{ij} :

$$|a_{ij} - a_{ij}^0| \leq \rho|s_{ij}|, \qquad i, j = 1, ..., n. \tag{18}$$

Here matrices A^0, S are given and ρ is a scalar parameter, which characterizes the range of uncertainty. The question is if the system is quadratically stable, i.e. if there exists a common quadratic Lyapunov function (Px, x) with $P > 0$ for all systems of the interval family. This is equivalent to solving LMIs

$$P \geq \epsilon I, \quad A^T P + PA \leq 0, \qquad \forall A \in \mathcal{A}, \tag{19}$$

here $\mathcal{A}$ is the set of admissible interval matrices (18), I is the identity matrix, $\epsilon > 0$ while P is the matrix variable. Note that we replaced the condition $P > 0$ with $P \geq \epsilon I$ with arbitrary $\epsilon > 0$ due to homogenous character of the inequalities. At the first glance the problem reads different than LMIs (16) (we deal with matrix variables P, not vector variables x). However such problems can be easily considered in the framework of LMIs, see [18–20]. In our case we introduce the scalar function

$$f(P, A) = ||(A^T P + PA)_+||;$$

it is a convex function in P and its subgradient is

$$\partial_P f(P, A) = \frac{A^T(A^T P + PA)_+ + (A^T P + PA)_+ A}{||(A^T P + PA)_+||} \tag{20}$$

provided that $f(P, A) \neq 0$, otherwise $\partial_P f(P, A) = 0$. Thus the iterative method for solving (19) has the following form. At k-th iteration we have an approximation P_k and generate a random matrix $A_k \in \mathcal{A}$ (say, we can take a uniform distribution on the set of interval matrices, generating i.i.d. sample from such distribution is very simple problem). Then we calculate the subgradient $\partial_P f(P_k, A_k)$ according to (20) and make the subgradient step as in Algorithm 1 or 2 (with projection onto $X = \{P : P \geq \epsilon I\}$).

The results of numerical simulation are reported in [22]. Note that the number of inequalities in (19) is infinite but it is easy to verify that checking of vertex inequalities is sufficient to find a solution. The uniform distribution on vertices of the box (18) was taken to generate random matrices A_k. The total amount of vertex inequalities in $n \times n$ interval family is $N = 2^{n^2}$, thus for $n = 3$ $N = 512$ while for $n = 10$ $N \cong 10^{30}$. In spite of such huge number of inequalities, typically after 50 ($n = 3$) or 400 ($n = 10$) iterations a solution was found in feasible case.

Other applications to control (such as robust design of LQR for uncertain linear systems) are addressed in [23].

5.2. Robust linear inequalities

Suppose we are solving a system of linear inequalities

$$Ax \leq b \tag{21}$$

where the data are uncertain

$$||A - A^0|| \leq \rho_1, \qquad ||b - b^0|| \leq \rho_2 \tag{22}$$

with nominal data A^0, b^0 and levels of uncertainty ρ_1, ρ_2 while $|| \cdot ||$ denote some norms. The problem is to find a point x which satisfies (21) for all admissible values of A_ω, b_ω from (22). Introduce a scalar function

$$f(x, \omega) = ||(Ax - b)_+||$$

here c_+ denotes a projection of a vector $c = (c_1, ..., c_m) \in R^m$ onto nonnegative orthant: $c_+ = (c_1^+, ...c_m^+), c_i^+ = \max(0, c_i)$. Then $f(x, \omega)$ is convex in x and its subgradient can be calculated as

$$\partial_x f(x, \omega) = \frac{A^T(Ax - b)_+}{||(Ax - b)_+||}$$

for $f(x, \omega) \neq 0$ (if $f(x, \omega) = 0$, then $\partial_x f(x, \omega) = 0$). The scheme of the algorithm remains the same. We generate a random sample A_k, b_k, uniformly distributed in the set (22) (this can be done effectively for various matrix and vector norms by means of the technique from [24]). Then for the current approximation x_k we calculate the subgradient $\partial_x f(x_k, \omega_k)$ and make a subgradient step according to Algorithm 1 (if the problem is feasible). Note that the direction of the step in the algorithm is the same as in Block-AMS algorithm in [2, p. 101]; the difference is the random control rule in contrast with cyclic rule in [2] and different stepsize.

Simulation results [22] demonstrate very fast convergence of the method for feasible case.

Another approach to the problem (21), (22) is to deal with rows of (21) separately, then we can apply row-action version of the algorithm. This is simpler; however such approach is possible not for all matrix norms in (22), for instance an operator norm does not permit such simplification.

5.3. Finding inscribed ellipsoid

Consider a polytope in $R^n : Y = \{y : a_i^T y \leq b_i, i = 1, ..., m\}$ and an ellipsoid in the same space $E = \{y : ||P^{-1}(y - c)|| \leq 1\}$ where the vector $c \in R^n$ is its center and the matrix $P > 0$ defines its shape. The following result is known (see, e.g., [18]).

Lemma 2 *The ellipsoid E lies in Y if and only if the following inequalities hold*

$a_i^T c + ||Pa_i|| \leq b_i, i = 1, ..., m.$

Thus we have a system of convex inequalities with respect to ellipsoid characteristics $x = [c, P]$ which are equivalent to the condition $E \subset Y$. Our goal is to find the largest ellipsoid inscribed in Y. The words "the largest" can be specified in various ways; for our purposes the most convenient measure of the size of an ellipsoid is Tr P. We do not discuss here how to solve such optimization problem; instead we can fix some value $\mu > 0$ and require Tr $P \geq \mu$. Hence we arrive to the following feasibility problem with vector - matrix variables: find $x \in X, \quad f_i(x) \leq 0, i = 1, ..., m.$

Here

$$x = [c, P], \qquad X = \{x : P \geq \epsilon I, TrP \geq \mu\}, \qquad f_i(x) = a_i^T c + ||Pa_i|| - b_i,$$

$\epsilon > 0$ is some fixed small number; a solution of the above problem exists, if Y contains an interior point, ϵ is small enough and μ is not too large. The subgradient of $f_i(x)$ can be calculated with no problems:

$$\partial_c f_i(x) = a_i^T, \qquad \partial_P f_i(x) = \frac{Pa_i a_i^T + a_i a_i^T P}{2||Pa_i||},$$

and instead of projection onto X we can perform successive projections onto sets $\{P \geq \epsilon I\}$ and $\{$ Tr $P \geq \mu\}$ (compare example in Section 4.7). With this understanding we can apply the above proposed methods. More details can be found in the paper [25].

6. CONCLUSIONS

Fast and simple algorithm has been proposed for solving a system of convex inequalities which are intractable by standard means due to too large (or infinite) number of inequalities. It guarantees a finite termination of iterations with probability one. Its extension on the infeasible case is also provided. Numerous applications of the algorithm to various problems demonstrate its high efficiency.

Acknowledgment. Some parts of this paper are the results of the joint work with R. Tempo, G. Calafiore, F. Dabbene and P. Gay.

REFERENCES

1. H.H. Bauschke and J.M. Borwein, On projection algorithms for solving convex feasibility problems, *SIAM Review* **38** (1996) 367–426.
2. Y. Censor and S.A. Zenios, *Parallel Optimization: Theory, Algorithms and Applications* (Oxford University Press, New York, 1997).
3. P.L. Combettes, The convex feasibility problem in image recovery, *Advances in Imaging and Electron Physics* **95** (1996) 155–270.
4. U.M. Garcia-Palomares and F.J. Gonzales-Castano, Incomplete projection algorithms for solving the convex feasibility problem, *Numerical Algorithms* **18** (1998) 177–193.
5. S. Kaczmarz, Angenaherte Aufslosung von Systemen linearer Gleichungen, *Bull. Intern. Acad. Polon. Sci., Lett. A* (1937) 355–357; English translation: Approximate solution of systems of linear equations, *Intern. J. Control* **57** (1993) 1269–1271.
6. S. Agmon, The relaxation method for linear inequalities, *Canad. J. Math.* **6** (1954) 382–393.
7. T.S. Motzkin and I. Shoenberg, The relaxation method for linear inequalities, *Canad. J. Math.* **6** (1954) 393–404.
8. I.I. Eremin, The relaxation method of solving systems of inequalities with convex functions on the left sides, *Soviet Math. Dokl.* **6** (1965) 219–222.
9. L.G. Gubin, B.T. Polyak and E.V. Raik, The method of projections for finding the common point of convex sets, *USSR Comp. Math. and Math. Phys.* **7** (1967) 1–24.
10. B.T. Polyak, Minimization of nonsmooth functionals, *USSR Comp. Math. and Math. Phys.* **9** (1969) 14–29.
11. V.A. Yakubovich, Finite terminating algorithms for solving countable systems of inequalities and their application in problems of adaptive systems, *Doklady AN SSSR* **189** (1969) 495–498 (in Russian).
12. V. Bondarko and V.A. Yakubovich, The method of recursive aim inequalities in adaptive control theory, *Intern. J. on Adaptive Contr. and Sign. Proc.* **6** (1992) 141–160.
13. V.N. Fomin, *Mathematical Theory of Learning Processes* (LGU Publ., Leningrad, 1976) (in Russian).
14. N.M. Novikova, Stochastic quasi-gradient method for minimax seeking, *USSR Comp. Math. and Math. Phys.* **17** (1977) 91–99.
15. A.J.Heunis, Use of Monte-Carlo method in an algorithm which solves a set of functional inequalities, *J. Optim. Theory Appl.* **45** (1984) 89–99.

16. S.K. Zavriev, A general stochastic outer approximation method, *SIAM J. Control Optim.* **35** (1997) 1387–1421.
17. A.S. Nemirovsky and D.B. Yudin, *Informational Complexity and Efficient Methods for Solution of Convex Extremal Problems* (Nauka, Moscow, 1978) (in Russian); (John Wiley, NY, 1983) (English translation).
18. S. Boyd, L. El Ghaoui, E. Feron and V. Balakrishnan, *Linear Matrix Inequalities in Systems and Control Theory* (SIAM Publ., Philadelphia, 1994).
19. Yu. Nesterov and A. Nemirovskii, *Interior-Point Polynomial Algorithms in Convex Programming* (SIAM Publ., Philadelphia, 1994).
20. R. Saigal, L. Vandenberghe and H. Wolkowicz (eds), *Handbook of Semidefinite Programming* (Kluwer, Waterloo, 2000).
21. B.T. Polyak, Gradient methods for solving equations and inequalities, *USSR Comp. Math. and Math. Phys.* **4** (1964) 17–32.
22. G. Calafiore and B. Polyak, Fast algorithms for exact and approximate feasibility of robust LMIs, *Proceedings of 39th CDC* (Sydney, 2000) 5035–5040.
23. B.T. Polyak and R. Tempo, Probabilistic robust design with linear quadratic regulators, *Proceedings of 39th CDC* (Sydney, 2000) 1037–1042.
24. G. Calafiore, F. Dabbene and R. Tempo, Randomized algorithms for probabilistic robustness with real and complex structured uncertainty, *IEEE Trans. Autom. Control* (in press).
25. F. Dabbene, P. Gay and B.T. Polyak, Inner ellipsoidal approximation of membership set: a fast recursive algorithm, *Proceedings of 39th CDC* (Sydney, 2000) 209–211.

Inherently Parallel Algorithms in Feasibility and Optimization and their Applications
D. Butnariu, Y. Censor and S. Reich (Editors)

PARALLEL ITERATIVE METHODS FOR SPARSE LINEAR SYSTEMS

Y. Saad[a]*

[a]University of Minnesota, Department of Computer Science and Engineering, 200 Union st., SE, Minneapolis, MN 55455, USA

This paper presents an overview of parallel algorithms and their implementations for solving large sparse linear systems which arise in scientific and engineering applications. Preconditioners constitute the most important ingredient in solving such systems. As will be seen, the most common preconditioners used for sparse linear systems adapt domain decomposition concepts to the more general framework of "distributed sparse linear systems". Variants of Schwarz procedures and Schur complement techniques will be discussed. We will also report on our own experience in the parallel implementation of a fairly complex simulation of solid-liquid flows.

1. INTRODUCTION

One of the main changes in the scientific computing field as we enter the 21st century is a definite penetration of parallel computing technologies in real-life engineering applications. Parallel algorithms and methodologies are no longer just the realm of academia. This trend is due in most part to the maturation of parallel architectures and software. The emergence of standards for message passing languages such as the Message Passing Interface (MPI) [19], is probably the most significant factor leading to this maturation. The move toward message-passing programming and away from large shared memory supercomputers has been mainly motivated by cost. Since it is possible to build small clusters of PC-based networks of workstations, large and expensive supercomputers or massively parallel platforms such as the CRAY T3E become much less attractive. These clusters of workstations as well as some of the medium size machines such as the IBM SP and the SGI Origin 2000, are often programmed in message passing, in most cases employing the MPI communication library. It seems likely that this trend will persist as many engineers and researchers in scientific areas are now familiar with this mode of programming.

This paper gives an overview of methods used for solving "distributed sparse linear systems". We begin by a simple illustration of the main concepts. Assume we have to solve the sparse linear system

$$Ax = b, \tag{1}$$

*Work supported by NSF/CTS 9873236, NSF/ACI-0000443, and by the Minnesota Supercomputer Institute.

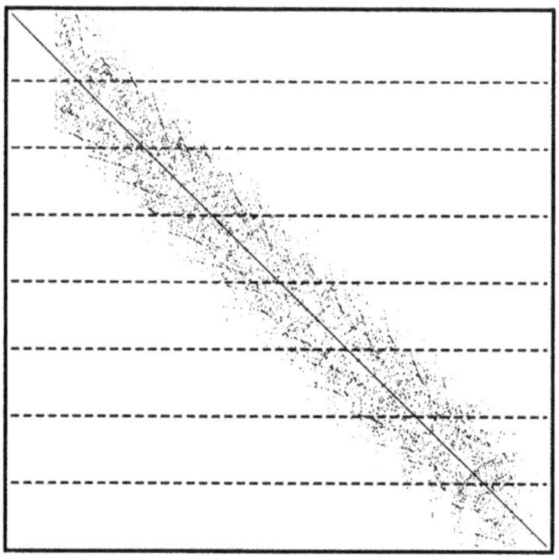

Figure 1. A small sparse matrix partitioned among 8 processors

where A is a large and sparse nonsymmetric real matrix of size n. One such system is depicted in Figure 1 (without the right-hand side). A dot represents a nonzero element in the matrix. The figure shows the simplest possible assignment of the data to processors in which blocks or rows of approximately the same number of contiguous rows are assigned to processors. Though not shown in the figure the right-hand side is to be partitioned in the same way. We refer to this as a distributed linear system - since the equations and right-hand side are (conformally) distributed among processors. The above mapping of the equations to processors is straightforward and there are clearly more efficient ways of partitioning a general sparse linear system using graph partitioners which we will briefly survey in Section 2. Regardless on how the system is partitioned and mapped into processors, the fundamental issue of how to solve such "distributed sparse linear systems" remains identical.

Systems similar to the one shown in the figure can arise, for example, from finite element discretizations of partial differential equations. It is typical in this case to partition the finite element mesh by a graph partitioner and assign a cluster of elements which represent a physical subdomain to each processor. Each processor then assembles only the local equations associated with the elements assigned to it. In other instances, the linear system is only available in assembled form. Here also a graph partitioner is invoked to determine a good way to map pairs of equations-unknowns to processors. In either case, each processor will end up with a set of equations (rows of the linear system) and a vector of the variables associated with these rows. This natural way of distributing a sparse linear system has been adopted by most developers of software for distributed sparse linear systems (see, e.g., [25,3,27,13]) because it is closely tied to a physical viewpoint.

Graph partitioning methods are surveyed in Section 2. We then consider methods for implementing Krylov subspace accelerators (Section 3) and preconditioners (4). In Section 5 an application of these methods to the simulation of liquid-solid flows will be described to illustrate some of the difficulties encountered in real life simulations.

2. GRAPH PARTITIONING

Graph partitioning has been the subject of much research in the last three decades. Initially, these techniques have been mainly used to partition electronic circuits [56,14, 35,11,45,12]. Naturally, graph partitioning for parallel computing relied on some of the methods developed in circuit simulation and other areas. The problem is one of optimization with constraints. Roughly speaking, the ideal partioned graph would be such that each subgraph has about the same number of nodes and the total number of edge cuts is minimum.

If $G = (V, E)$ is the graph to partition, where V is the set of vertices and E is the set of edges, then the problem is to find a set of subsets $S_1, S_2, \ldots, S_p$ of V such that $S_1 \bigcup S_2 \ldots \bigcup S_p = V$ and such that a few requirements related to efficiency of the underlying algorithm are satisfied. Note that subsets can be arbitrary and 'overlapping' is allowed in the sense that for $i \neq j$ $S_i \bigcap S_j$ may be non-empty. Figure 2 shows a 4×3 mesh split among 4 processors, with $S_1 = \{1, 2, 5, 6, \}$, $S_2 = \{3, 4\}$, $S_3 = \{9, 10\}$, $S_4 = \{7, 8, 11, 12\}$.

There are two distinct and independent requirements. The first is a load balancing constraint, since we would like each processor to perform roughly the same amount of work. In practice this is translated by imposing that each S_i contain about the same number of vertices, but we will return to this later. The second requirement is that the number of edge-cuts be as small as possible. An edge-cut is an edge in the graph that links two vertices that are on two different subgraphs. This second requirement is made in order to reduce overhead costs associated with communication. Indeed, edge-cuts represent data that must be exchanged during the (iterative) solution process, and minimizing these edge-cuts will lead to reduced communication costs. In recent years a number of new methods were developed to obtain graph partitioners which attempt to satisfy both requirements. Another requirement, which is implicit, is that the cost of the partitioning algorithm itself must be inexpensive.

Three classes of algorithms have been proposed for partitioning a graph according to the requirements just stated. 'Geometric algorithms' use information about the physical coordinates of the vertices in order to partition the graph [16,41,40,10,58]. These 'mesh-based' techniques have basically been superseded by two other classes of methods that appeared in the mid 1990s. The first is the class of 'spectral partitioners' [42,4,1,20,22,43] which are based on some well-known properties of eigenvectors of so-called Laplaceans of graphs, for details see [42]. Recursive spectral partitioners seem to give the best overall results in terms of the optimization requirements mentioned above. However, the algorithm is rather expensive, and less costly multilevel methods have been developed [21,30] which have become the most common techniques in use today. Among these Metis [30,31] seems to be currently the de-facto standard in graph partitioning. Parallel versions [33] as well as a number of variants [34,28,29] for a number of different applications have been

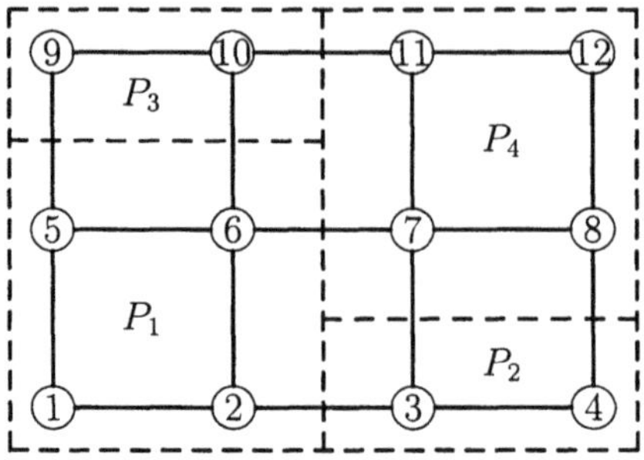

Figure 2. A 5-point matrix partitioned among 4 processors

developed.

Often the two criteria mentioned above for graph partitioning can give poor results in complex situations, since the partitioning will affect convergence properties in a manner that is rather difficult to predict theoretically. An illustration of this is shown in Section 5 where a realistic application of solid-liquid flow simulation is described.

3. DISTRIBUTED KRYLOV SUBSPACE SOLVERS

Figure 3 shows a 'physical domain' viewpoint of a sparse linear system. This representation borrows from the domain decomposition literature – so the term 'subdomain' is often used instead of the more proper term 'subgraph'. Each point (node) belonging to a 'subdomain' is actually a pair representing an equation and an associated unknown. As is often done, we distinguish between three types of unknowns: (1) Interior variables are those that are coupled only with local variables by the equations; (2) Local interface variables are those coupled with non-local (external) variables as well as local variables; and (3) External interface variables are those variables in other processors which are coupled with local variables.

Along with this figure, we can represent the local equations as shown in Figure 4. The local equations do not correspond to contiguous equations in the original system. The matrix represented in the figure can be viewed as a reordered version of the equations associated with a local numbering of the equations/unknowns pairs.

In Figure 4 the rows of the matrix assigned to a certain processor have been split into two parts: a *local* matrix A_i which acts on the local variables and an *interface* matrix U_i which acts on remote variables. A given processor must first receive these remote variables from adjacent processor(s) before performing locally its part of the matrix-vector product. Most of the data structures for the parallel solution of distributed sparse linear systems separate the boundary points from the interior points and list interface nodes last after the interior nodes. This 'local ordering' yields more efficient interprocessor communication, and reduces local indirect addressing during matrix-vector products. The use of block

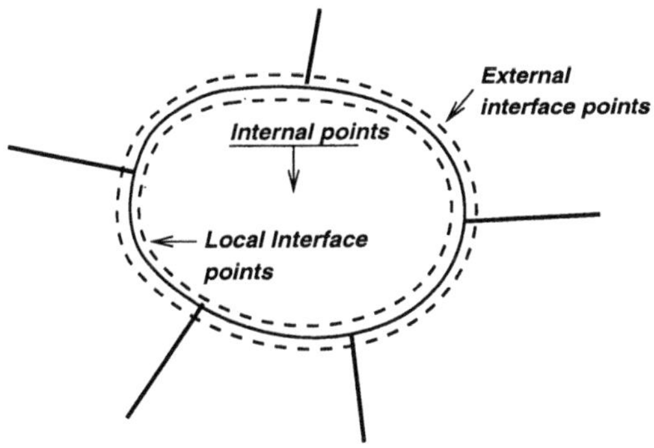

Figure 3. A local view of a distributed sparse matrix.

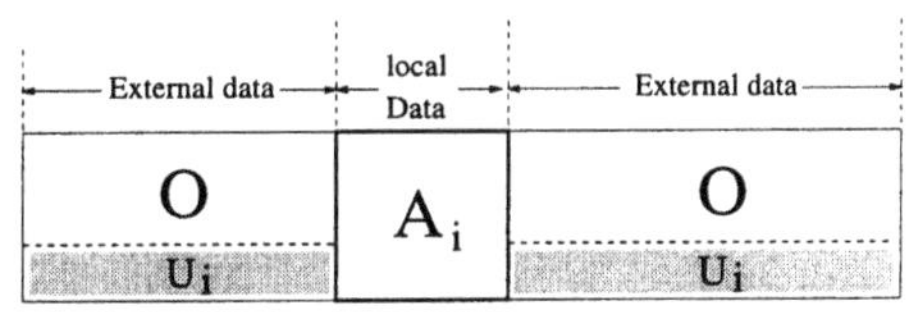

Figure 4. Corresponding local matrix.

sparse row format can also give a sufficient reduction in indirect addressing. The zero blocks in the local matrix of Figure 4 are due to the fact that local internal nodes are not coupled with external nodes.

Thus, each local vector of unknowns x_i is split in two parts: the subvector u_i of internal nodes followed by the subvector y_i of local interface variables. The right-hand side b_i is conformally split in the subvectors f_i and g_i,

$$x_i = \begin{pmatrix} u_i \\ y_i \end{pmatrix} \ ; \quad b_i = \begin{pmatrix} f_i \\ g_i \end{pmatrix} \ .$$

The local matrix A_i residing in processor i as defined above is block-partitioned according to this splitting, leading to

$$A_i = \left(\begin{array}{c|c} B_i & E_i \\ \hline F_i & C_i \end{array} \right) . \tag{2}$$

With this, the local equations can be written as follows.

$$\begin{pmatrix} B_i & E_i \\ F_i & C_i \end{pmatrix} \begin{pmatrix} u_i \\ y_i \end{pmatrix} + \begin{pmatrix} 0 \\ \sum_{j \in N_i} E_{ij} y_j \end{pmatrix} = \begin{pmatrix} f_i \\ g_i \end{pmatrix} \tag{3}$$

The submatrix $E_{ij}y_j$ accounts for the contribution to the local equation from the neighboring subdomain number j and N_i is the set of subdomains that are neighbors to subdomain i. The sum of these contributions, seen on the left side of of (3) is the result of multiplying a certain matrix by the external interface variables. It is clear that the result of this product will affect only the local interface variables as is indicated by the zero in the upper part of the second term in the left-hand side of (3). For practical implementations, the subvectors of external interface variables are grouped into one vector called $y_{i,ext}$ and the notation

$$\sum_{j \in N_i} E_{ij} y_j \equiv U_i y_{i,ext}$$

will denote the contributions from external variables to the local system (3). In effect this represents a local ordering of external variables to write these contributions in a compact matrix form. With this notation, the left-hand side of the (3) becomes

$$w_i = A_i x_i + U_{i,ext} y_{i,ext} \tag{4}$$

Note that w_i is also the local part the matrix-by vector product Ax in which x is a vector which has the local vector components x_i, $i = 1, \ldots, s$. Matrix-vector product operations can be performed in three stages. First the x_i is multiplied by A_i. Then a communication step takes place in which the external data $y_{i,ext}$ is received. In effect this is an exchange operation, since each processor needs to receive the remote interface variables from other processors. In the third and last stage, the external data is multiplied by the local matrix U_i and the result is added to the current result $A_i x_i$.

Since communication is an important part of the matrix-vector product operation, it is useful at the outset to gather the data structure representing the local part of the linear matrix as was just described. In this preprocessing phase it is also important to form any additional data structures required to prepare for the intensive communication that will take place during the solution phase. In particular, each processor needs to know (1) the processors with which it must communicate, (2) the list of interface points and (3) a break-up of this list into pieces of data that must be sent and received to/from the "neighboring processors".

Several packages based on distributed sparse matrices and domain-decomposition type approaches utilize basically the same approaches. A first version of PSPARSLIB [54] was available in 1993-1994 using P4, a precursor to the current MPI, and then CMMD the communication Library on the Connection Machine 5. A number of other packages were developed in the mid-1990s. Among them we mention PETSc [3], Block-Solve [27], Aztec [25], and ParPre [13]. All these packages offer at least a few of the standard "block", i.e., domain-based, preconditioners.

The primary accelerator used in PSPARSLIB is a flexible variant of GMRES called FGMRES [47]. This is a right-preconditioned variant of GMRES which allows variations in the preconditioner at each step. Details on the implementations of parallel Krylov accelerators are given in [50,55,48]. PSPARSLIB uses a reverse communication protocol is used to avoid passing data structures related to the coefficient matrix or the preconditioner. With this implementation, the FGMRES code itself contains no communication calls. All such calls are relegated to either matrix-vector operations which are performed outside the main body of the acceleration code, or dot-product operations which usually rely on special library calls.

4. PRECONDITIONERS

We now turn our attention to preconditioning techniques for distributed sparse matrices. Many of the ideas used in preconditioning distributed sparse matrices are borrowed from the Domain Decomposition literature; see, for example [15,36,57]. The simplest of these use block preconditionings based on the domains, and are termed Schwarz alternating procedures in the Partial Differential Equations literature. Let R_i be a restriction operator which projects a global vector onto its restriction in subdomain i. Algebraically, this is represented by an $n_i \times n$ matrix, consisting of zeros and ones, where n_i is the size of subdomain i. The transpose of R_i represents a prolongation operator from subdomain i to the whole space. Let A_i be the local submatrix associated with subdomain i. In the Block Jacobi (or additive Schwarz) procedure, the global solution is updated as,

$$x := x + \sum_{i=1}^{s} R_i^T A_i^{-1} R_i r$$

where r is the residual vector $b - Ax$. Each of the solves with the matrices A_i is done independently. The method does not specify which technique is used for solving these local systems. It clear that a direct solution method could be used in case each subproblem is small enough.

However, it is common to use iterative solvers for the local systems since these are often fairly large. An iterative solver consists of an accelerator such as the conjugate gradient algorithm, and a preconditioner such as an Incomplete LU (ILU) factorization. For details on such techniques see, for example, [2,5,18,49]. In terms of software, a publically available package called SPARSKIT [46] includes a fair number of standard accelerators and preconditioners. Among the preconditioners available in SPARSKIT, the ILU with threshold (ILUT) reaches a good compromise between simplicity and robustness and is widely used.

In the of domain decomposition methods, however, the accelerator for the global iteration must take into account the fact that the preconditioner may not a constant operator (i.e., that it may change at each step of the outer iteration). Many variations are possible. In general, we found that iterating to solve each of the sub problems accurately is not cost-effective. Instead often a simple forward-backward sweep, with ILU factors obtained from an ILUT preconditioner, often yields the fastest combination. Subdomain partitions may be allowed to overlap. This technique works reasonably well for a small number of overlapping subdomains as was shown in experiments using a purely algebraic form in [44].

This can be extended to block Multiplicative Schwarz procedure, which is a block-Gauss-Seidel iteration in which, likewise, a block is associated with a domain. Here, the algorithm can be described by a simple loop which updates the local solutions for subdomain 1, 2, ..., s in turn, taking into account the latest value reached for the current solution each time. The only difference with the block Jacobi iteration is that the solution and residual are updated immediately after each local correction δ_i. In the block Jacobi case, these local corrections are all computed using the same residual r and then they are added at the same time to the current solution x.

A technique of labeling the domains, known as "multicoloring", has been be exploited to extract parallelism from the Gauss-Seidel procedure, see e.g., [57]. In this procedure, the subdomains are colored such that subdomains that are coupled by an equation are always assigned a different color. One such coloring is illustrated in Figure 5 where the numbers refer to a color. With this coloring, the Gauss-Seidel loop can now proceed differently. Since couplings between vertices of different domains are only possible when these domains have different colors, the Gauss-Seidel sweep can proceed by color. Thus, in Figure 5 all 4 domains of color number 1 can update their variables first at the same time. Then the two domains with color number 2 can simultaneously do their update,

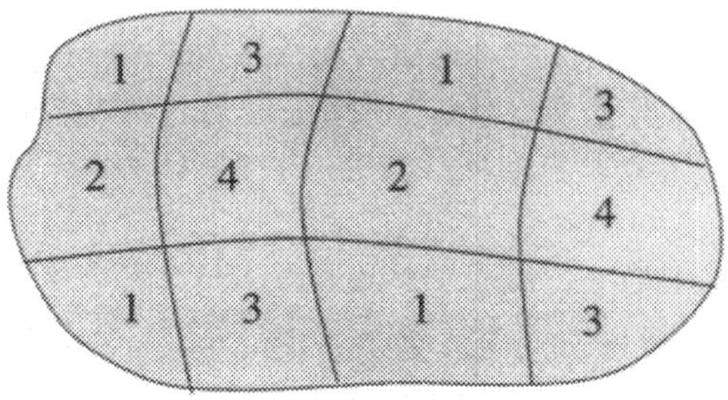

Figure 5. Subdomain coloring for the multicolor Multiplicative Schwarz procedure

and so on. For this example this color-based sweep will require 4 steps instead of $s = 12$ steps for the domain-based sweep. The color-based loop is as follows.

For $color = 1, \ldots, num_{colors}$ Do:
  If `mycolor` $=$ $color$ Do:
    $\delta := A_i^{-1} R_i r$
    $x := x + \delta$
    $r := r - A\delta$
  EndIf
EndDo

Here, `mycolor` represents the color of the node to which the subdomain is assigned. Though this procedure represents an improvement over the standard Gauss-Seidel procedure, it is still sequential with respect to the colors since when one color is active (one of the steps of the color loop), all processors of a different color will be idle. This may lead to substantial loss of efficiency. Several remedies have been proposed, the most common one of which is to further partition each subdomain in each processor and then find a global coloring of the new, finer, partitioning. The goal here is to have all colors of this new partitioning represented in each processor. An example of a second level partitioning of the domain in Figure 5 is shown in Figure 6.

This two-level coloring now presents a different disadvantage, namely that the rate of convergence of the new procedure is usually smaller than that of the original scheme, due to the fact that the subdomains are smaller. Other remedies have been proposed, see e.g., [52] for a discussion. Our own experience is that many of the remedies do not come for free in that they tend to increase the cost of the scheme either due to the increased number of iterations or other overhead.

Distributed versions of the Incomplete LU factorization ILU(0) have been developed and tested on the Connection Machine 5 as early as 1993 [38]. More recently, the idea was generalized to higher order ILU(k) factorizations by Karypis and Kumar [32] and Hysom and Pothen [26]. While the original ILU factorization is sequential in nature, it is fairly easy to devise parallel versions by what amounts to reordering on the global system. The effect of such reorderings can be detrimental to convergence and this has been the subject

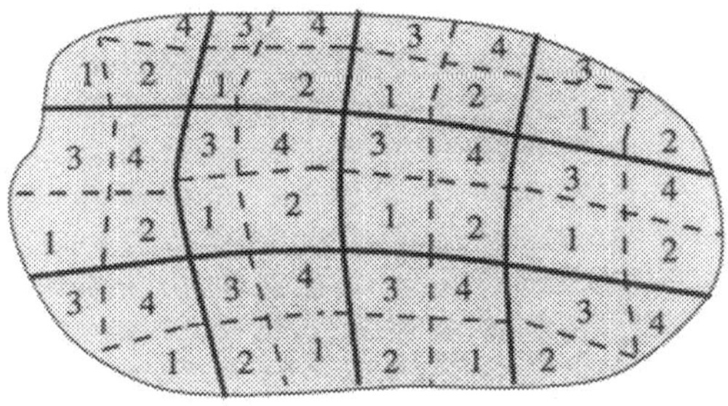

Figure 6. Two-level multicoloring of the subdomains for the multicolor Multiplicative Schwarz procedure

of recent research, see, e.g., [6].

The strategy for unraveling parallelism in ILU is illustrated in Figure 7 in the simplest case of ILU(0). The unknowns associated with the interior nodes do not depend on the other variables. As a result they can be eliminated independently and in parallel in the ILU(0) process. The rows associated with the nodes on the interface of the subdomain will require more attention. Indeed, now the order in which to eliminate these rows is important, since this order will essentially determine the amount of parallelism. At one extreme we can have the rows eliminated one by one in the natural order $1, 2, \ldots, n$ which results in no parallelism. We can also try to reorder the equations in such a way as to maximize parallelism as is done in sparse Gaussian elimination – but this would be too costly for incomplete LU factorizations. A somewhat intermediate solution, is to eliminate the equations using a simple "schedule", i.e., a global order in which to eliminate the rows, set by a priority rule among processors. As an example, we could impose a global order based on the labels of the processors, meaning that between two adjacent processors, the one with a lower number has a higher priority in processing rows than the other. Thus, in our illustration of Figure 7 Processor 10 processes the rows associated with its interior nodes first, then it will wait for those interface rows belonging to processors 2, 4 and 6. Once they are received, by Processor 10, it will eliminate all its rows associated with (all) its interface nodes. Finally, once this is done, the interface rows in Processor 10 that are awaited by Processors 13 and 14 will be sent to them. This sets a global ordering for the elimination process. The local ordering of the rows, i.e., the order in which we process the interface rows in the same processor (e.g. Processor 10) may not be as important. The global order based on PE-number, defines a natural priority graph and parallelism can easily be exploited in a data-driven implementation.

However, it seems rather artificial to use the label of a processor to determine the global order of the elimination. In fact, we can also define a proper order in which to perform the elimination by *replacing the PE-numbers by any labels provided any two neighboring processors have a different label.* The most natural way of doing this is to perform a multicoloring of the subdomains, and use the colors in exactly the same way as above to define an order of the tasks. The colors can be defined to be positive integers as in the

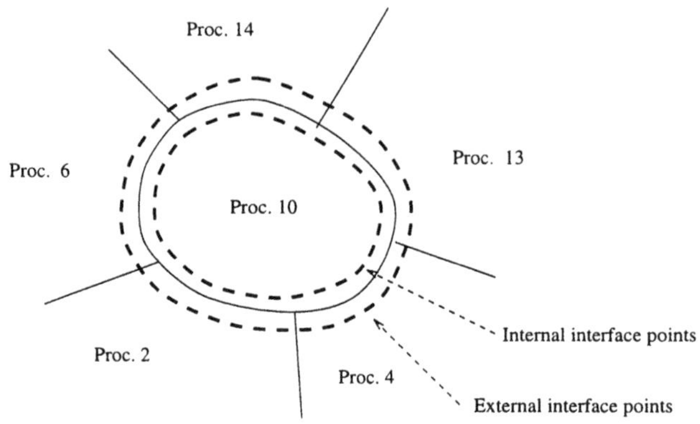

Figure 7. Distributed ILU(0) with a priority rule based on PE numbers

illustration of Figure 5.

It is possible to eliminate the interior variables u_i from each of the local systems (3). Doing so will leave us with a system which involves only interface variables (denoted by y_i (3)), and referred to as a Schur complement system. This system can be solved and then the other variables can be computed. One of the main difficulties with this approach is to precondition the Schur complement system. Indeed, depending on the formulation, this system may be dense or have diagonal blocks that are dense [49]. It is possible to define approximations for this Schur complement from which preconditioners can then be derived [8,7,13]. An alternative is to define preconditioners for the global system based on some approximate solution of the Schur complement system [53,51]. This approach does not require that the Schur complement system be solved exactly since the solution is only used to define a preconditioner. Generally speaking, Schur complement preconditioners are often harder to implement but of better qualily than Schwarz-based preconditioners.

5. AN APPLICATION

The direct numerical simulation of solid-liquid flows is a problem which poses many challenges. Problems of this type arise in various applications such as oil drilling and refining, rheology, and manufacturing processes. The number of particles can range from the tens to the tens of thousands, leading to solutions of nonlinear systems that can be very large and rather difficult to solve. Such simulations gather all of the possible difficulties that can render parallel implementations difficult. First, the linear systems encountered are large as well as ill-conditioned and far from diagonally dominant. A second challenge is that the particles are moving, so the meshes must be recomputed occasionally, and therefore a remapping of the data is also performed. Finally, by the nature of the equations, some of the unknowns are heavily coupled – meaning that some of the equations may involve a relatively large number of unknowns. This may lead to a heavy communication overhead. In the end, it is rather difficult to obtain a performance which scales ideally. On the other hand, note that engineers are often content to simply be able to run these simulations, taking advantage of large memories afforded by distributed memory computers, even if the gain in execution time is moderate.

There are two different formulations currently in use for simulating solid-liquid flows. The Distributed Lagrange-Multiplier (DLM) [17] approach which usually employs structured grids and the Arbitrary Lagrangian-Eulerian (ALE) [23,24] technique. In the ALE formulation, the fluid particles are embedded in the grid and are treated like moving, solid boundaries. One notable difficulty with the ALE formulation arises from a projection step developed by Maury and Glowinski [39] in which the fluid particles are collapsed to a single point. This step affects the variables associated with the particle center (i.e., the velocity vector and angular velocity), as well as the elements coupled with the surface of the particle. When fluid particles are very close together it is likely that one finite element will lie on two different particles or two different particles will have surface elements that share a common edge. As a result, information must be communicated between both particles in order to correctly generate the local system matrices. In situations where many particles are represented on 2 or more subdomains, the communication associated with matrix-vector operations can become a bottleneck in the solution process. This difficulty is not present in the DLM model because the particle grid is independent from the fluid grid.

It is apparent that the mapping of the particles to processors plays an important role in obtaining good speedup and efficiency of a parallel solver. We will discuss a few methods for mapping the fluid particles. The solid-liquid flow simulator used in this discussion is essentially a parallel implementation of a finite element ALE code, originally described in [24] and later modified in [9] to include the particle projection step.

Partitioning of the domain causes no particular difficulty except when dealing with the particles. The domain is first partitioned using a standard partitioner, such as Metis, or a heuristic partitioner based on counting elements assigned to a given processor from left to right (in the domain) and stopping when a certain number of elements are counted. We refer to this second approach as 'bin-counting'.

Once this is done we then need to map information to processors. Four particle partitioning strategies have been tested. Figure 8 shows one of the methods tested, which is a somewhat simplified partitioning of the surface elements of 2 particles that is representative of the problem faced. Numbers at the center of the elements represent the processors to which the elements are mapped. This figure does not show all the information that is affected by the particle partitioning strategy. The 4 methods are described next. For details see [37].

In the methods referred to as **M1** the particle centers and all variables associated with all surface elements duplicated on each processor. This can result in a large amount of overlap. For example, in Figure 8, all information shown would be duplicated on subdomains 1, 2 and 3. In the method labeled **M2**, only the particle centers of subdomains contributing from surface elements would be duplicated. In Figure 8, only the centers would be copied onto subdomains 1, 2 and 3. In order to eliminate communication in the local matrix assembly process, all the information is available locally on each processor. In the third approach (**M3**) no data is overlapped. The particle is placed into a given subdomain based on the weight of the subdomains represented by the surface elements.

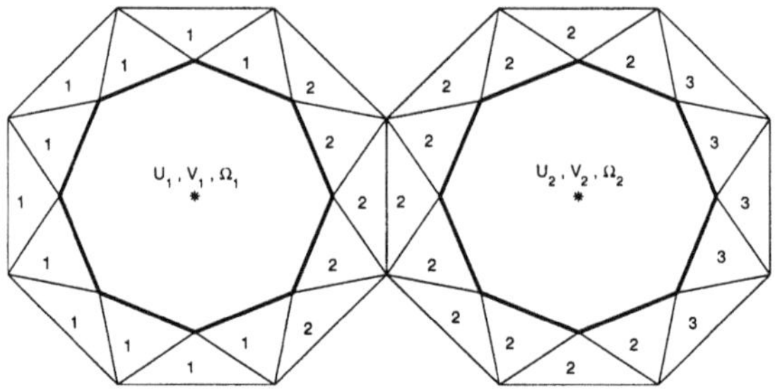

Figure 8. An example of a typical partitioning of the surface elements for two closely spaced particles.

For the situation shown in Figure 8, particle 1 would be placed in subdomain 1 and particle 2 would be placed into subdomain 2. Again, all information necessary to assemble the local matrices is available locally. Finally, the fourth technique (**M4**), is similar to **M3** but the effect of the particle subdomain assignment is further strengthened by reassigning each surface element that has an edge on the particle to have the same domain as the particle center. This is illustrated in Figure 9.

It is also possible to include additional overlap by creating one or more layers of elements at subdomain boundaries. The overlap for one layer is obtained by examining every element that shares a vertex node with element E_i. If the assigned subdomain is different than that assigned to E_i, the element is included in the same domain as E_i. Additional layers of overlap can be included by implementing this idea recursively.

One test case on which we now report consists of 1000 particles in a sedimentation column of 1000 particles, initially at rest in a packed array as shown in Figure 10 for a much smaller number of particles (100). The domain is a rectangular box of width 21.05 cm and length 104.5 cm. The particle and fluid densities are 1.1 and 1.0 $\frac{\text{g}}{\text{cm}^3}$ respectively. The particles are initially configured in a 20×50 array. They are then allowed to fall in a Newtonian fluid under the influence of gravity.

The tests were performed on an IBM-SP2 located at the Minnesota Supercomputing Institute. This machine has several subconfigurations, but the portion used for these experiments consists of 17 nodes, each node having four 222 MHz Power3+ processors sharing 16 GB of memory. The timing of the performance studies was done using a wall-clock timer provided in the MPI implementation (MPI_WTime). The parallel solver for the saddle point problem is an additive Schwarz algorithm using local ILU preconditioning and GMRES for the accelerator. The convergence criterion is a reduction of the initial preconditioned residual be a factor of 10^{-8}. Only results using the Metis partitioner are shown here, a partitioner based on "bin-counting' (mentioned earlier) was also tested elsewhere, see [37].

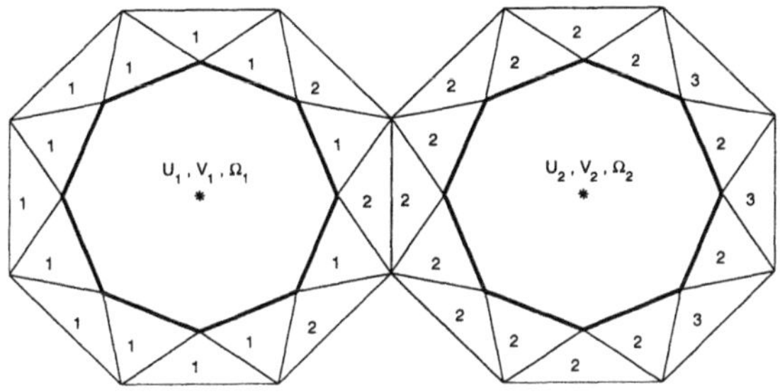

Figure 9. Particle partitioning example from Figure 8 using method M4

We found that, not surprisingly, methods M1 and M2 are generally more robust than methods M3 and M4, since there is more overlap associated with M1 and M2 (including particle centers). An analysis of data communication indicates the amount of data to be communicated becomes significant as the number of processors increases. For Methods **M1** and **M2** and 16 processors, the amount of the total local variables to be communicated with adjacent processors is 50% and 30% on average respectively. The use of an Additive Schwarz algorithm implies that, aside from a few scalar product operations in the GMRES accelerator, the only place that a substantial amount of communication occurs is in the matrix-vector operation. However, the large quantities of data to be averaged result in a bottle neck in the solution process. As one might expect, the amount of data communication is less for Method **M2**, but still substantial. Method **M3** requires only about half as much data communication as Method **M2**, but the performance is more erratic, particularly for 2 processors and Metis partitioning.

6. CONCLUSION

Several strategies exist for implementing preconditioned Krylov subspace methods on parallel computers. For distributed sparse matrices, it seems that the popular concepts in domain decomposition methods have been adapted fairly easily to the situation. In fact, in realistic applications, variants of the additive Schwarz procedure with some overlap seem to be popular choice when the number of processors is not too large. Good efficiency may vary significantly even if the developer pays attention to performance issues related to the implementation on a specific architecture. The two main sources of impediment to performance are poor load balancing and excessive communication overhead. However, another source of difficulty is loss of performance due to the increase in the number of iterations. The behavior of all preconditioners, even as simple as the Schwarz procedure, is not predictable in the realistic situations encountered in solving real-life problems. We have shown just one example example of such an application. As research is still very active in this area, the real challenge is to develop parallel iterative solvers that are not

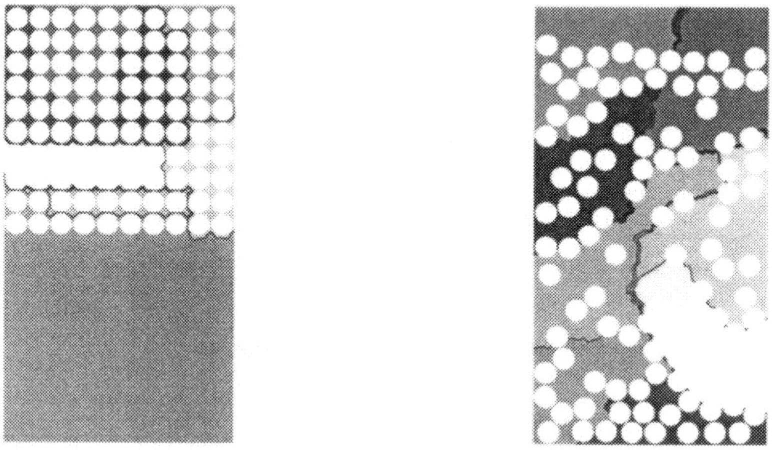

Figure 10. Computational domain for times $t = 0$ and $t = 1.2$ seconds, for a 100 particle problem, partitioned into 8 subdomains by Metis

only 'scalable' to larger numbers of processors but also robust.

Acknowledgment. The application discussed in Section 5 represents joint work with H. G. Choi, Z. Li, and L. Little, and is presented in detail elsewhere [37]. Many thanks to all three, as well as other team members involved in this project, for their very hard work.

REFERENCES

1. C. J. Alpert and S.-Z. Yao, *Spectral partitioning: The more eigenvectors the better*, in Proceedings ACM/IEEE Design Automation Conference, 1995.
2. O. Axelsson, *Iterative Solution Methods* (Cambridge University Press, New York, 1994).
3. S. Balay, W. D. Gropp, L. C. McInnes, and B. F. Smith, *PETSc 2.0 users manual*, Tech. Rep. ANL-95/11 - Revision 2.0.24, Argonne National Laboratory, 1999.
4. S. T. Barnard and H. D. Simon, *A fast multilevel implementation of recursive spectral bisection for partitioning unstructured problems*, in Proceedings of the sixth SIAM conference on Parallel Processing for Scientific Computing, 1993, 711–718.
5. R. Barrett, M. Berry, T. Chan, J. Demmel, J. Donato, J. Dongarra, V. Eijkhout, R. Pozo, C. Romine, and H. van der Vorst, *Templates for the solution of linear systems:*

Table 1
Performance results for 1000 particle sedimentation case using 4 different overlapping strategies.

	M1				M2			
Procs	Its	Time	Speedup	Efficiency	Its	Time	Speedup	Efficiency
2	56	85.6	1.000	1.000	61	91.3	1.000	1.000
4	62	58.9	1.453	0.727	73	65.6	1.392	0.696
8	67	41.9	2.043	0.511	80	44.8	2.038	0.510
16	72	32.2	2.658	0.332	88	33.1	2.758	0.345

	M3				M4			
Procs	Its	Time	Speedup	Efficiency	Its	Time	Speedup	Efficiency
2	74	114.9	1.000	1.000	74	115.1	1.000	1.000
4	147	124.9	0.920	0.460	147	124.6	0.924	0.462
8	89	48.9	2.350	0.587	89	48.8	2.359	0.590
16	112	38.8	2.961	0.370	112	40.3	2.856	0.357

building blocks for iterative methods (SIAM, Philadelphia, 1994).

6. M. Benzi, W. Joubert, and G. Mateescu, *Numerical experiments with parallel orderings for ILU preconditioners*, Electronic Transactions on Numerical Analysis, **8** (1999), 88–114.
7. T. F. Chan and T. P. Mathew, *An application of the probing technique to the vertex space method in domain decomposition*, in Fourth International Symposium on Domain Decomposition Methods for Partial Differential Equations, R. Glowinski, Y. A. Kuznetsov, G. A. Meurant, J. Périaux, and O. Widlund, eds., Philadelphia, PA, 1991, SIAM.
8. ——, *The interface probing technique in domain decomposition*, SIAM J. on Mat.rix Anal. and Appl., **13** (1992), 212–238.
9. H. G. Choi, *Splitting method for the combined formulation of fluid-particle problem.* Submitted to Comp. Meth. Appl. Meth. Engr., 2000.
10. N. Chrisochoides, E. Houstis, and J. Rice, *Mapping algorithms and software environment for data parallel pde iterative solvers*, Journal of Parallel and Distributed Computing, **21** (1994), 75–95.
11. J. Cong and M. L. Smith, *A parallel bottom-up clustering algorithm with applications to circuit partitioning in vlsi design*, in Proc. ACM/IEEE Design Automation Conference, 1993, 755–760.
12. S. Dutt and W. Deng, *VLSI circuit partitioning by cluster-removal using iterative improvement techniques*, in Proc. Physical Design Workshop, 1996.
13. V. Eijkhout and T. Chan, *ParPre a parallel preconditioners package, reference manual for version 2.0.17*, Tech. Rep. CAM Report 97-24, UCLA, 1997.
14. T. B. et al, *Improving the performance of the kernighan-lin and simulated annealing graph bisection algorithm*, in Proc. ACM/IEEE Design Automation Conference, 1989,

775–778.
15. C. Farhat and J. X. Roux, *Implicit parallel processing in structural mechanics*, Computational Mechanics Advances, **2** (1994), 1–124.
16. J. R. Gilbert, G. L. Miller, and S.-H. Teng, *Geometric mesh partitioning: Implementation and experiments*, in Proceedings of International Parallel Processing Symposium, 1995.
17. R. Glowinski, T. W. Pan, T. I. Hesla, and D. D. Joseph, *A distributed lagrange multiplier fictitious domain method for particulate flows*, Int. J. Multiphase Flow, (1999), 755–794.
18. A. Greenbaum, *Iterative Methods for Solving Linear Systems* (SIAM, Philadelpha, PA, 1997).
19. W. Gropp, E. Lusk, and A. Skjellum, *Using MPI: Portable Parallel Programming with the Message Passing Interface* (MIT press, 1994).
20. L. Hagen and A. Kahng, *Fast spectral methods for ratio cut partitioning and clustering*, in Proceedings of IEEE International Conference on Computer Aided Design, 1991, 10–13.
21. B. Hendrickson and R. Leland, *An improved spectral graph partitioning algorithm for mapping parallel computations*, Tech. Rep. SAND92-1460, UC-405, Sandia National Laboratories, Albuquerque, NM, 1992.
22. B. Hendrickson and R. Leland, *Multidimensional spectral load balancing*, in Proceedings of the sixth SIAM Conference on Parallel Processing for Scientific Computing, 1993, 953–961.
23. H. H. Hu, *Direct simulation of flows of solid-liquid mixtures*, Int. J. Multiphase Flow, **22** (1996), 335–352.
24. H. H. Hu, D. D. Joseph, and M. J. Crochet, *Direct simulation of flows of fluid-particle motions*, Theor. Comp. Fluid Dyn., **3** (1992), 285–306.
25. S. A. Hutchinson, J. N. Shadid, and R. S. Tuminaro, *Aztec user's guide. version 1.0*, Tech. Rep. SAND95-1559, Sandia National Laboratories, Albuquerque, NM, 1995.
26. D. Hysom and A. Pothen, *A scalable parallel algorithm for incomplete factor preconditioning*, Tech. Rep. (preprint), Old-Dominion University, Norfolk, VA, 2000.
27. M. T. Jones and P. E. Plassmann, *BlockSolve95 users manual: Scalable library software for the solution of sparse linear systems*, Tech. Rep. ANL-95/48, Argonne National Lab., Argonne, IL., 1995.
28. G. Karypis, R. Aggarwal, V. Kumar, and S. Shekhar, *Multilevel hypergraph partitioning: Application in vlsi domain*, IEEE Transactions on VLSI Systems, **7** (1999).
29. G. Karypis, E.-H. Han, , and V. Kumar, *A hierarchical clustering algorithms using dynamic modeling*, IEEE Computer, Special Issue on Data Analysis and Mining, **32** (1999), 68–75.
30. G. Karypis and V. Kumar, *Multilevel graph partition and sparse matrix ordering*, in Intl. Conf. on Parallel Processing, 1995. Available on WWW at URL http://www.cs.umn.edu/~karypis.
31. G. Karypis and V. Kumar, *Multilevel k-way partitioning scheme for irregular graphs*, Journal of Parallel and Distributed Computing, **48** (1998), 96–129.
32. ——, *Parallel threshold-based ilu factorization*, Tech. Rep. AHPCRC-98-061, Department of Computer Science / Army HPC res. ctr., University of Minnesota, Minneapo-

lis, MN, 1998.

33. ——, *Parallel multilevel k-way partition scheme for irregular graphs*, SIAM Review, **41** (1999), 278–300.
34. ——, *Multilevel k-way hypergraph partitioning*, VLSI Design, (2000). to appear.
35. B. W. Kernighan and S. Lin, *An efficient heuristic procedure for partitioning graphs*, The Bell System Technical Journal, (1970).
36. P. LeTallec, *Domain decomposition methods in computational mechanics*, Computational Mechanics Advances, **1** (1994), 121–220.
37. Z. Li, L. Little, Y. Saad, and H. G. Choi, *Particle partitioning strategies for the parallel computation of solid-liquid flows*. In preparation, 2000.
38. S. Ma and Y. Saad, *Distributed ILU(0) and SOR preconditioners for unstructured sparse linear systems*, Tech. Rep. ahpcrc-94-027, Army High Perf. Comput. Res. Center, University of Minnesota, Minneapolis, MN, 1994.
39. B. Maury and R. Glowinski, *Fluid-particle flow: A symmetric formulation*, C.R. Acad. Sci, Paris, **324** (1997), 1079–1084.
40. G. L. Miller, S.-H. Teng, W. Thurston, and S. A. Vavasis, *Automatic mesh partitioning*, in Sparse Matrix Computations: Graph Theory Issues and Algorithms, A. George, J. R. Gilbert, and J. W.-H. Liu, eds., (An IMA Workshop Volume). Springer-Verlag, New York, NY, 1993.
41. G. L. Miller, S.-H. Teng, and S. A. Vavasis, *A unified geometric approach to graph separators*, in Proceedings of 31st Annual Symposium on Foundations of Computer Science, 1991, 538–547.
42. A. Pothen, H. D. Simon, and K.-P. Liou, *Partitioning sparse matrices with eigenvectors of graphs*, SIAM Journal of Matrix Analysis and Applications, **11** (1990), 430–452.
43. A. Pothen, H. D. Simon, L. Wang, and S. T. Bernard, *Towards a fast implementation of spectral nested dissection*, in Supercomputing '92 Proceedings, 1992, 42–51.
44. G. Radicati di Brozolo and Y. Robert, *Parallel conjugate gradient-like algorithms for solving sparse non-symmetric systems on a vector multiprocessor*, Parallel Computing, 11 (1989), 223–239.
45. B. M. Riess, K. Doll, and F. M. Johannes, *Partitioning very large circuits using analytical placement techniques*, in Proceedings ACM/IEEE Design Automation Conference, 1994, 646–651.
46. Y. Saad, *SPARSKIT: A basic tool kit for sparse matrix computations*, Tech. Rep. 90-20, Research Institute for Advanced Computer Science, NASA Ames Research Center, Moffet Field, CA, 1990.
47. ——, *A flexible inner-outer preconditioned GMRES algorithm*, SIAM J. on Sci. and Stat. Comput., **14** (1993), 461–469.
48. ——, *Krylov subspace methods in distributed computing environments*, in State of the Art in CFD, M. Hafez, ed., 1995, 741–755.
49. ——, *Iterative Methods for Sparse Linear Systems* (PWS publishing, New York, 1996).
50. Y. Saad and A. Malevsky, *PSPARSLIB: A portable library of distributed memory sparse iterative solvers*, in Proceedings of Parallel Computing Technologies (PaCT-95), 3-rd international conference, St. Petersburg, Russia, Sept. 1995, V. E. M. et al., ed., 1995.

51. Y. Saad and M. Sosonkina, *Distributed Schur complement techniques for general sparse linear systems*, Tech. Rep. umsi-97-159, Minnesota Supercomputer Institute, University of Minnesota, Minneapolis, MN, 1997. To appear.
52. Y. Saad and M. Sosonkina, *Non-standard parallel solution strategies for distributed sparse linear systems*, in Parallel Computation: Proc. of ACPC'99, A. U. P. Zinterhof, M. Vajtersic, ed., Lecture Notes in Computer Science, Berlin, 1999, Springer-Verlag, 13–27.
53. Y. Saad, M. Sosonkina, and J. Zhang, *Domain decomposition and multi-level type techniques for general sparse linear systems*, in Domain Decomposition Methods 10, Providence, RI, 1998, American Mathematical Society.
54. Y. Saad and K. Wu, *Parallel sparse matrix library (P_SPARSLIB): The iterative solvers module*, Tech. Rep. 94-008, Army High Performance Computing Research Center, Minneapolis, MN, 1994.
55. ——, *Design of an iterative solution module for a parallel sparse matrix library (P_SPARSLIB)*, in Proceedings of IMACS conference, Georgia, 1994, W. Schonauer, ed., 1995.
56. D. G. Schweikert and B. W. Kernighan, *A proper model for the partitioning of electrical circuits*, in Proc. ACM/IEEE Design Automation Conference, 1972, 57–62.
57. B. Smith, P. Bjørstad, and W. Gropp, *Domain decomposition: Parallel multilevel methods for elliptic partial differential equations* (Cambridge University Press, New-York, NY, 1996).
58. P. Wu and E. N. Houstis, *Parallel adaptive mesh generation and decomposition*, Engineering and Computers, **12** (1996), 155–167.

ON THE RELATION BETWEEN BUNDLE METHODS FOR MAXIMAL MONOTONE INCLUSIONS AND HYBRID PROXIMAL POINT ALGORITHMS

Claudia A. Sagastizábal[a*] and Mikhail V. Solodov[b†]

[a]IMPA, Estrada Dona Castorina 110, Jardim Botânico, Rio de Janeiro, RJ 22460-320, Brazil. On leave from INRIA, BP 105, 78153 Le Chesnay Cedex, France

[b]IMPA–Instituto de Matemática Pura e Aplicada,
Estrada Dona Castorina 110, Jardim Botânico, Rio de Janeiro, RJ 22460-320, Brazil

We demonstrate that bundle methods for computing zeroes of general maximal monotone operators can be cast as a special case in a certain class of hybrid proximal point algorithms. This fact is significant for several reasons. First, it provides an alternative convergence proof, which is technically simple, for serious steps of bundle methods by invoking the corresponding results for hybrid proximal point methods. This includes the linear rate of convergence results, which were not available previously. Second, relating the two methodologies supplies a computationally realistic implementation of hybrid proximal point methods for the most general case, i.e., for operators without any particular structure.

1. INTRODUCTION

Given a maximal monotone operator T on $\Re^n$, we consider the problem:

$$\text{Find } x \in \Re^n \text{ such that } 0 \in T(x)\,. \tag{1}$$

As is well known, this problem is fundamental in mathematics and related sciences. It also subsumes many important practical applications. When T has some specific structure, specialized algorithms can usually be used to solve (1). For example, when $T = \partial f$, the subdifferential of a convex function f, methods of convex optimization come into play, including bundle techniques, see [15, Chapters XIV, XV] and [5, Part II]. When T represents a variational inequality over a closed convex set with a continuous mapping, then certain projection methods can be used, see [14] and [33, Introduction] for references and relevant discussions. However, when T is a general maximal monotone operator, relatively few methods are applicable. Until very recently, only *implicit* procedures based on resolvents (i.e., proximal-type methods) were available for solving (1) in the general

*Research partially supported by CNPq under Grant No. 301800/96-0 and by FAPERJ under Grant No. 150-581/00.
†Research is supported in part by CNPq Grant 300734/95-6, by PRONEX–Optimization, and by FAPERJ.

case. In fact, the bundle method of [8] and the projection method with averaging of [1] appear to be currently the only *explicit* methods for solving general maximal monotone inclusions. To make this remark precise, we note that we consider an algorithm for solving (1) as explicit if:

- It only requires the knowledge of *one* $v \in T(x)$ for each given $x \in \Re^n$ (recall that the ability of an algorithm to work with this minimal information is a well-known practical requirement in nonsmooth optimization, which is an important example of problem (1)).

- All parts of the method are computationally implementable, independently of the structure of T.

- The method converges under no assumptions other than the maximal monotonicity of T and $T^{-1}(0) \neq \emptyset$.

The iterative procedure of [1] is essentially a subgradient-type method with vanishing stepsizes, where convergence is "forced" by Cesàro averaging. Methods of this type have inherently slow (sublinear) convergence even for well-structured problems. In optimization, bundle methods are known to be in practice typically superior to standard subgradient algorithms, e.g., such as in [30]. On the theoretical side, bundle methods also have the potential for the faster linear rate of convergence. It should be mentioned, however, that the method of [1] is developed in a considerably much more general setting than [8] or the present paper (in [1], the Banach space setting is considered, and the problem data can be subject to perturbations).

In this paper we consider bundle methods under the light of inexact proximal point algorithms, namely the hybrid variant of [36], see also [35,37,38]. The insight given by this new interpretation is twofold. First, it provides an alternative convergence proof, which is technically simple, for serious steps of bundle methods by invoking the corresponding results for hybrid proximal point methods. Second, relating the two methodologies supplies a computationally realistic implementation of hybrid proximal point methods for the most general case, i.e., when the operator may not have any special structure.

Our paper is organized as follows. In § 2 we outline the hybrid proximal point algorithm, together with its relevant convergence properties. Some useful theory from [9] and [8] on certain enlargements of maximal monotone operators is reviewed in § 3. Finally, in Section 4 we establish the connection between bundle and hybrid proximal methods and give some new convergence results, including the linear rate of convergence for bundle methods.

2. HYBRID PROXIMAL POINT ALGORITHMS

A classical algorithm applicable to solve (1) is the proximal point method, e.g., [24,29, 6,26,23,18]. This algorithm consists of solving a sequence of subproblems of the following structure:

Given $x^k \in \Re^n$, find $x^{k+1} \in \Re^n$ and $v^{k+1} \in T(x^{k+1})$ such that $0 = v^{k+1} + \mu_k(x^{k+1} - x^k)$,(2) where $\mu_k > 0$ is a regularization parameter. It is quite clear that in this form the proximal point algorithm is purely conceptual (in other words, not implementable and/or not practical). This has to do with the obvious fact that subproblems (2) are structurally as difficult to solve as the original problem (1) itself. Nevertheless, the ideas of the proximal point method are useful for developing more sophisticated (and implementable!) computational algorithms. A well-known example is precisely the family of bundle methods for nonsmooth convex optimization, see [15,20,4,21].

To pass from the domain of conceptual to implementable, one has to start with inexact versions of (2). Perhaps the most natural relaxation of (2) can be stated as follows:

$$\text{Given } x^k \in \Re^n\,,\ \text{find } y^k \in \Re^n \text{ and } v^k \in T(y^k) \text{ such that } 0 = v^k + \mu_k(y^k - x^k) + r^k\,, \quad (3)$$

where $r^k \in \Re^n$ is some error term induced by the approximation method. Traditionally, see [29,39,7,10,13,11], inexact proximal algorithms defined the next iterate as the approximate solution given by (3), i.e., $x^{k+1} = y^k$. To guarantee convergence, error terms were required to satisfy some *"a priori summability"* condition, such as

$$\|r^k\| \le \sigma_k\,, \quad \sum_{k=0}^{\infty} \sigma_k < \infty\,. \quad (4)$$

Recently, the so-called hybrid proximal methods appeared as an alternative proposal aimed to develop more constructive (and hopefully practical) approximation criteria. We refer the reader to [36,35,37,38] for more detailed discussions. Here, we shall only consider the approximation criterion of [36]. Specifically, we regard a pair $(y^k, v^k) \in \Re^n \times \Re^n$ as an acceptable approximation to the solution of the k-th proximal subproblem (2), if it satisfies (3) with

$$\|r^k\| \le \sigma_k \max\{\|v^k\|, \mu_k\|y^k - x^k\|\}\,, \quad \sigma_k \in [0,1)\,. \quad (5)$$

This rule is constructive, in the sense that it is clearly related to the behaviour of the method around the current point for the problem being solved. Furthermore, as opposed to (4), the relaxation parameters σ_k can be kept bounded away from zero (in fact, they can be even fixed), which essentially means that the *relative* error in (3) need not tend to zero. Computationally, the latter is a realistic condition, suitable for applications. Moreover, criterion (5) is weaker than standard conditions, like (4). This can be seen from the fact (see an example in [36]) that a sequence generated according to (3) and (5) with $x^{k+1} = y^k$, need not converge under the assumption $\liminf_{k\to\infty} \sigma_k > 0$. By contrast, a sequence generated using (4) does converge. Nevertheless, approximate solutions satisfying (3) and (5) are "good enough". Specifically, convergence can be forced by introducing a simple (explicit) projection step at no additional computational cost.
The algorithm proposed in [36] is the following.

Algorithm 2.1 Hybrid Projection–Proximal Point Method (HPPM)
Choose any $x^0 \in \Re^n$ *and set* $k := 0$.

Inexact proximal step. *Having* x^k, *choose* $\mu_k \in (0, +\infty)$, $\sigma_k \in [0,1)$, *and solve (approximately) the* k*-th proximal subproblem so that* (3) *and* (5) *are satisfied.*
Stop if $v^k = 0$ *or* $y^k = x^k$. *Otherwise, compute*

Projection step.

$$x^{k+1} = x^k - \langle v^k, x^k - y^k \rangle \|v^k\|^{-2} v^k \,. \tag{6}$$

Set $k := k+1$; and repeat. ∎

Remark 2.1 Since it is not specified how to solve subproblems in order to satisfy (3) and (5), at this stage HPPM is also a conceptual scheme. However, in a number of applications where T has some structure, the constructive nature of (5) allows to solve subproblems by simple and efficient means. Such is the case of *truly* globally convergent Newton-type methods for systems of monotone equations [34] and complementarity problems [32]. The key fact is that under certain natural assumptions, just one step of Newton method is sufficient to solve the k-th proximal subproblem within the error tolerance given by (5). See also [31] for some other possibilities in the context of variational inequalities. Another interesting application is the modified forward-backward splitting method of [40], which can be shown to be a specific realization of the method of [35] (closely related to HPPM) in the case when T is a sum of two maximal monotone operators, one of which is continuous. In § 4, we show how to develop a (constructive) bundle method satisfying (3) and (5) for general maximal monotone T.

Remark 2.2 HPPM is well-defined, because the exact solution of the k-th proximal point subproblem always satisfies (3) and (5) for any $\sigma_k \geq 0$. If $\sigma_k = 0$, only the exact solution is acceptable; otherwise there are many acceptable (approximate) solutions.

Remark 2.3 If $x^k \notin T^{-1}(0)$, it can be seen that (3) and (5) imply that $\langle v^k, x^k - y^k \rangle > 0$. On the other hand, by the monotonicity of T, $\langle v^k, \bar{x} - y^k \rangle \leq 0$ for any $\bar{x} \in T^{-1}(0)$. It follows that the hyperplane $\{x \in \Re^n \mid \langle v^k, x - y^k \rangle = 0\}$ strictly separates x^k from $T^{-1}(0)$. The last step of HPPM, namely (6), projects x^k onto this hyperplane, essentially ensuring global convergence.
The projection step of HPPM plays here the role of "forcing" convergence. It is interesting to note that in the proximal method of [2] a somewhat similar role seems to be played by averaging. We point out, however, that the method of [2] itself is very different from HPPM, which is based on constructive approximation criteria for solving the subproblems.

HPPM preserves all the desirable convergence properties of the classical exact proximal point algorithm. Specifically, assuming that the method does not terminate finitely (which can only occur at the exact solution of (1)), we have the following results.

Theorem 2.1 *[36, Section 2] Suppose that T is a maximal monotone operator, with $T^{-1}(0) \neq \emptyset$, and consider HPPM Algorithm 2.1. Then the following hold:*

(i) The sequence $\{x^k\}$ generated by HPPM is bounded.

(ii) If parameters are chosen so that $\bar{\sigma} := \limsup_{k\to\infty} \sigma_k < 1$ and $\sum_{k=0}^{\infty} \mu_k^{-2} = +\infty$, then $\{x^k\}$ converges to some $\bar{x} \in T^{-1}(0)$.

(iii) If, in addition, there exists $\bar{\mu} > 0$ such that $\mu_k \leq \bar{\mu}$ for all k, and two constants $L > 0$ and $\delta > 0$ such that

$$\|v\| \leq \delta\,, v \in T(x) \quad \Longrightarrow \quad \operatorname{dist}(x, T^{-1}(0)) \leq L\|v\|\,, \tag{7}$$

then $\{x^k\}$ converges to $\bar{x} \in T^{-1}(0)$ R-linearly.

(iv) If, in addition to (7), $\bar{x} \in T^{-1}(0)$ is unique, then the convergence rate is Q-linear.

■

When T is a general maximal monotone operator, up to now HPPM remained as conceptual as the original proximal point method, as so far no techniques have been proposed to solve the proximal point subproblems to satisfy (3) and (5) in this general case. We show in § 4 below how this can be done by the bundle strategy of [8]. Specifically, we demonstrate that a modification of the method of [8] can be viewed as a particular practical realization of HPPM. As a by-product, we obtain an immediate convergence proof for serious steps of the bundle method, including the linear rate of convergence result. We note that the rate of convergence analysis for bundle methods was not previously available.

3. BUNDLE METHODS

We shall assume from now on that the domain of T is the whole space $\Re^n$. From a purely theoretical point of view, this assumption may be deemed restrictive. However, it is important to keep in mind that we are considering implementable algorithms which are developed using just one $v \in T(x)$ for each $x \in \Re^n$. Formally speaking, this precludes T from having a restricted domain. However, in most applications infinite values are not inherent in the problem, but are rather induced by the indicator function of the feasible region. If one is to deal with constraints explicitly (via projections, for example, so that T needs to be evaluated only at feasible points), then bundle methods can be extended to constrained problems, after taking care of the usual technicalities (see also Remark 3.4). For the sake of clarity of presentation, we shall not deal with this issue in the present paper.

3.1. Enlargements of maximal monotone operators

It is known that the ε-subdifferential plays a paramount role in the convergence analysis of bundle methods for nonsmooth optimization. For a general maximal monotone operator, a similar tool is a certain outer approximation of T. More precisely, given some $\varepsilon \geq 0$, the ε-enlargement of T at $x \in \Re^n$ [7] is defined by

$$T^{\varepsilon}(x) := \{u \in \Re^n \mid \langle v - u, y - x\rangle \geq -\varepsilon \text{ for all } y \in \Re^n\,, v \in T(y)\}\,.$$

It is clear that $T^0 = T$, and $T(x) \subset T^{\varepsilon}(x)$ for all $x \in \Re^n\,, \varepsilon \geq 0\,.$

One of the most important properties of T^ε is that, unlike T itself, it is continuous as an operator of both x and ε. Continuity is crucial when constructing practical algorithms, because it ensures the following key property:

$$\text{Given } \{(\varepsilon_k, x^k)\} \to (0, \bar{x} \in T^{-1}(0)), \text{ there exists } \{s^k \in T^{\varepsilon_k}(x^k)\} \text{ such that } s^k \to 0.$$

Information accumulated over the course of iterations is related to the ε-enlargement of T through the following consequence of the weak *transportation formula* [9, Theorem 2.2]:

Lemma 3.1 *[8, Corollary 2.3] Take any $z^i \in \Re^n, w^i \in T(z^i), i = 1, \ldots, j$. Let $x \in \Re^n$ and $\rho > 0$ be such that $\|z^i - x\| \le \rho$ for all $i = 1, \ldots, j$. Let*

$$(\hat{x}, \hat{s}) := (\sum_{i=1}^{j} \lambda_i z^i, \sum_{i=1}^{j} \lambda_i w^i), \quad \sum_{i=1}^{j} \lambda_i = 1, \lambda_i \ge 0, i = 1, \ldots, j.$$

Then it holds that

$$\hat{s} \in T^\varepsilon(\hat{x}) \quad \text{with } \|\hat{x} - x\| \le \rho \text{ and } \varepsilon \le 2\rho M, \quad \text{where } M := \max\{\|w^i\| \mid i = 1, \ldots, j\}.$$

■

For properties and applications of ε-enlargements of maximal monotone operators, as well as other related issues, we refer the reader to [7,9,8,35,37,41,27].

3.2. Towards an implementable algorithm

Before stating our implementable bundle method, which is technically quite involved, we describe a related conceptual scheme, introduced in [9], which is helpful for transmitting some key ideas.

Algorithm 3.1 Conceptual Algorithmic Scheme (CAS)
Choose positive parameters $\tau, R, \varepsilon, \alpha$ and θ, with $\alpha, \theta \in (0,1)$. Choose any $x^0 \in \Re^n$. Set $k := 0$.

Compute search direction. *Stop if $0 \in T(x^k)$.*
Otherwise, find m_k, the smallest *nonnegative integer m, such that*

$$\|s^k\| > \alpha^m \tau, \quad \text{where } s^k := \arg\min\{\|v\| \mid v \in T^{\varepsilon\alpha^m}(x^k)\}.$$

Line-search step. *Find l_k, the* smallest *nonnegative integer l, such that*

$$\langle v^k, s^k \rangle > \theta \|s^k\|^2, \quad \text{where } v^k \in T(y^k), \; y^k := x^k - \alpha^l R s^k.$$

Projection step. *Compute*

$$x^{k+1} = x^k - \langle v^k, x^k - y^k \rangle \|v^k\|^{-2} v^k.$$

Set $k := k + 1$; and repeat. ■

Remark 3.1 In the given form, CAS is a conceptual method because computing the minimal-norm element of $T^{\varepsilon}(x^k)$ (actually, even of $T(x^k)$) is in general prohibitively computationally expensive, and in fact, typically impossible. At the very least, for this one should have available a full functional representation for the set $T^{\varepsilon}(x^k)$ for all ε and x^k, which is certainly too much to wish for. Recalling nonsmooth optimization, one practical requirement in this setting is that algorithms should be developed with the use of only one subgradient at each point. In order to develop stable methods with reliable stopping tests, it is also important to keep track of previous information along iterations, in the form of a *bundle* of information. This is the underlying philosophy in bundle methods for nonsmooth optimization, which constituted a huge progress when compared to subgradient methods. It is precisely the use of bundle methods that allowed efficient numerical solution of complex real-world applications [19,3].

Remark 3.2 CAS can be viewed as an analog of the steepest-descent method for nonsmooth convex optimization. Indeed, if we consider the minimal-norm element v of $\partial f(x)$, then by the optimality condition it holds that $\langle v, u - v\rangle \geq 0$ for all $u \in \partial f(x)$. Hence,

$$\|v\|^2 = \min_{u\in\partial f(x)} \langle v, u\rangle \leq \min_{u\in\partial f(x)} \max_{w\in\partial f(x)} \langle w, u\rangle = \min_{u\in\partial f(x)} f'(x; u)\,,$$

where $f'(x;u)$ stands for the usual directional derivative of f at $x \in \Re^n$ in the direction $u \in \Re^n$. Assuming that x is not a minimizer of f, v is then the direction of maximal descent for f at $x \in \Re^n$, and one can perform a line-search in this direction to decrease the value of the objective function f.

Remark 3.3 It was shown in [9, Proposition 3.2] that CAS is well-defined (m_k and l_k are finite integers for each k), and generates a (finitely or infinitely) convergent sequence, provided $T^{-1}(0) \neq \emptyset$.

Remark 3.4 CAS is related to methods proposed in [16] for solving multi-valued variational inequalities. However, the methods of [16] are not aimed at devising implementable algorithms. On the other hand, [16] can be combined with CAS to produce a constrained version of the conceptual scheme. This constrained version can then serve as the basis for a constrained bundle method to handle the case when T has restricted domain.

4. BUNDLE METHODS AS A HYBRID PROXIMAL POINT ALGORITHM

To make the basic scheme of CAS computationally viable, in [8] a bundle strategy is employed to approximate the sets $T^{\varepsilon}(x^k)$, as follows:

- Polyhedral approximations of $T^{\varepsilon}(x^k)$ are defined using the information that has been accumulated along iterations. The line-search step in CAS is then modified to include an important safeguard. This safeguard essentially verifies whether the current polyhedral approximation is "good enough" or "poor". In the bundle methods terminology, these two situations correspond to serious or null steps, respectively. See Algorithm 4.1 below.

- Having in hand these suitable polyhedral approximations of $T^{\varepsilon}(x^k)$, they can be used to approximate minimal-norm elements of $T^{\varepsilon}(x^k)$. The latter problem can be solved by fast and reliable *quadratic programming* techniques, [22,12]. In nonsmooth optimization there exist very efficient quadratic programming solvers customized specifically for bundle methods, [17]. To keep the size of quadratic programs manageable, the method of [8] uses a selection rule to work with a reduced bundle at every step.

We proceed to formally state a bundle method, which is a modification of that in [8], adapted for our purposes. First, some notation is in order.

- Given $z^i \in \Re^n$ and $w^i \in T(z^i)$, $i = 1, \dots, k$, by Δ_k we denote the bundle of accumulated information, i.e., $\Delta_k := \cup_{i=1}^{k}\{(z^i, w^i)\}$.
- Given a set of points $w^i, i \in I$, by $\mathrm{conv}\{w^i, i \in I\}$ we denote its convex hull, omitting the index set I when it is clear from the context.
- Serious-steps indices (i.e., indices of iterates yielding a significant progress toward a solution) will be collected in the set K_S.

Algorithm 4.1 Bundle Method (BM)

Choose positive parameters $\tau, R, \mathrm{TOL}, \alpha$ *and* θ, *with* $\alpha \in (0,1)$ *and* $\theta \in (1/2, 1)$. *Choose an integer* $\bar{m}$ *such that* $2\alpha^{\bar{m}} R \le \mathrm{TOL}$. *Choose any* $x^0 \in \Re^n$ *and define* $\Delta_{-1} := \emptyset, K_S := \emptyset$. *Set* $k := 0$.

Major iteration. *Compute some* $u^k \in T(x^k)$. *Stop if* $u^k = 0$.
Otherwise, set $\hat{m}_k := 0$; $(z^k, w^k) := (x^k, u^k)$.

Update the bundle. *Set* $\Delta_k := \Delta_{k-1} \cup \{(z^k, w^k)\}$.

Compute direction. *For* $m = \hat{m}_k, \hat{m}_k + 1, \dots,$

compute

$$s := \arg\min\{\|v\| \mid v \in \mathrm{conv}\{w^i\}, (z^i, w^i) \in \Delta_k, i \text{ such that } \|z^i - x^k\| \le \alpha^m R\},$$

until

$$\|s\| > \alpha^m \tau \quad \text{or} \quad m = \bar{m}.$$

Set $s^k := s, \hat{m}_k := m$.

Line-search step. *For* $l = 0, 1, \dots,$

compute

$$y := x^k - \alpha^l R \|s^k\|^{-1} s^k, \quad v \in T(y),$$

until

$$\langle v, s^k \rangle > \theta \min\{\|s^k\|, \|v\|\}^2 \quad \text{or} \quad l = \hat{m}_k.$$

Set $y^k := y, v^k := v, l_k := l$.

Null step.
If

$$\langle v^k, s^k \rangle \le \theta \min\{\|s^k\|, \|v^k\|\}^2,$$

then

set $(z^{k+1}, w^{k+1}) := (y^k, v^k)$, $x^{k+1} := x^k$, $k := k+1$,

go to **"Update the bundle"** *step.*

Otherwise,

Serious step. *Define*

$$x^{k+1} := x^k - \langle v^k, x^k - y^k \rangle \|v^k\|^{-2} v^k, \quad K_S := K_S \cup \{k\}.$$

Set $k := k+1$*; and go to* **"Major iteration"** *step.* ■

Remark 4.1 We note that the criterion employed by our algorithm to accept a serious step is weaker (hence, easier to satisfy) than the criterion in [8], which is $\langle v, s^k \rangle > \theta \|s^k\|^2$. This improvement was made possible precisely due to relating our method to HPPM Algorithm 2.1 (see the proof of Theorem 4.1).

Remark 4.2 For each m, the task of computing direction s is equivalent to solving the quadratic program

$$\min_\lambda \{\| \sum_i \lambda_i w^i \|^2 \mid \lambda_i \ge 0, \sum_i \lambda_i = 1, i \text{ such that } \|z^i - x^k\| \le \alpha^m R, (z^i, w^i) \in \Delta_k\}.$$

Not that the problem constraints have a very special structure (the feasible region is the unit-simplex), which can be exploited by sophisticated quadratic programming solvers.

Remark 4.3 The "Compute direction" step in BM Algorithm 4.1 differs from CAS Algorithm 3.1 in that the sets $T^\varepsilon(x^k)$ in the latter are replaced by their polyhedral approximations in the former. Furthermore, these polyhedral approximations are constructed as a convex hull of certain *selected* elements in the full bundle Δ_k. Thus the method is always working with a *reduced* bundle, which is important for keeping the size of quadratic programs manageable. In addition, to control memory requirements for storing the full bundle Δ_k, one could use bundle compression techniques, similar to those in nonsmooth optimization [15, Chapter XIV].

The following is our main result, proving convergence when there is an infinite number of serious steps.

Theorem 4.1 *Suppose that T is a maximal monotone operator, with $T^{-1}(0) \neq \emptyset$, and consider the sequence of serious steps $\{x^k\}_{k \in K_S}$ generated by BM Algorithm 4.1. Assume further that the set K_S is infinite. Then the following hold:*

(i) The sequence $\{x^k\}_{k\in K_S}$ can be considered as a sequence generated by HPPM Algorithm 2.1. Furthermore, with parameters σ_k and μ_k chosen appropriately, the assumptions of Theorem 2.1(ii) hold, and so $\{x^k\}_{k\in K_S}$ converges to some $\bar{x} \in T^{-1}(0)$.

(ii) If, in addition, assumptions in Theorem 2.1 (iii) [respectively, (iv)] hold, then $\{x^k\}_{k\in K_S}$ converges to $\bar{x} \in T^{-1}(0)$ R-linearly [respectively, Q-linearly].

Proof. Since we assume that BM Algorithm 4.1 performs an infinite number of serious steps, we shall consider only indices k in K_S (note that x^k is not being modified for $k \notin K_S$). BM Algorithm 4.1 accepts a serious step when

$$\langle v^k, s^k\rangle > \theta \min\{\|s^k\|, \|v^k\|\}^2 . \tag{8}$$

Having in mind the framework of HPPM Algorithm 2.1, define

$$\mu_k := (\alpha^{l_k} R)^{-1}\|s^k\| \quad \text{and} \quad r^k := s^k - v^k .$$

With this notation, the line-search step of BM Algorithm 4.1 becomes $y^k = x^k - \mu_k^{-1} s^k$, or, equivalently,

$$\begin{aligned} 0 &= s^k + \mu_k(y^k - x^k) \\ &= v^k + \mu_k(y^k - x^k) + r^k , \quad v^k \in T(y^k) . \end{aligned}$$

Thus, condition (3) in HPPM Algorithm 2.1 is satisfied. To check (5), note that

$$\begin{aligned} \|r^k\|^2 &= \|s^k - v^k\|^2 \\ &= \|s^k\|^2 - 2\langle s^k, v^k\rangle + \|v^k\|^2 \\ &< \|s^k\|^2 - 2\theta \min\{\|s^k\|, \|v^k\|\}^2 + \|v^k\|^2 && \text{[by (8)]} \\ &= \max\{(1-2\theta)\|s^k\|^2 + \|v^k\|^2 , (1-2\theta)\|v^k\|^2 + \|s^k\|^2\} \\ &\le 2(1-\theta)\max\{\|s^k\|, \|v^k\|\}^2 \\ &= 2(1-\theta)\max\{\mu_k\|y^k - x^k\|, \|v^k\|\}^2 \\ &\le \sigma_k^2 \max\{\mu_k\|y^k - x^k\|, \|v^k\|\}^2 . && \text{[for } \sigma_k \ge \sqrt{2(1-\theta)}\text{]} \end{aligned}$$

We conclude that the pair $(y^k, v^k \in T(y^k))$ yielding a serious step in BM Algorithm 4.1 is acceptable in the "Inexact proximal step" of HPPM Algorithm 2.1, where μ_k and σ_k are as chosen above. In particular, Theorem 2.1(i) implies that $\{x^k\}$ is bounded. Obviously, $\{y^k\}$ is then also bounded. Being a maximal monotone operator with domain $\Re^n$, T is bounded on bounded sets [28, Theorem 1]. Hence, there exists some $M > 0$ such that $\|w^i\| \le M$ for all $(z^i, w^i) \in \Delta_k$ and all $k = 1, 2, \ldots$. Since $s^k \in \text{conv}\{w^i\}$ for all k, it follows that $\|s^k\| \le M$. Finally,

$$\mu_k = (\alpha^{l_k} R)^{-1}\|s^k\| \le (\alpha^{\bar{m}} R)^{-1} M =: \bar{\mu} .$$

Therefore, our choice of μ_k satisfies the assumption of Theorem 2.1(iii) (hence, also of Theorem 2.1(ii)). If we take $\sigma_k \in (\sqrt{2(1-\theta)}, 1)$, this choice also satisfies the condition of Theorem 2.1(ii), because $\theta \in (1/2, 1)$. Hence, all the assertions now follow from Theorem 2.1 (ii)-(iv). ■

Note that our implementation of the bundle algorithm has no infinite loops, and so it is always well-defined. We have established that if an infinite number of serious steps is generated, the algorithm falls within the general framework of the hybrid projection-proximal point method, and so all the convergence results, including rates of convergence, readily follow. For completeness, it remains to consider the case when K_S is finite, i.e., there is a last serious step $x^{k_{last}}$, followed by an infinite sequence of null steps.

In our development, we shall need the following technical result.

Lemma 4.1 *[15, Lemma IX.2.1.1] Let $\gamma > 0$ be fixed, and consider two infinite sequences $\{v^j\}$ and $\{s^j\}$ such that for all $j = 1, 2, \ldots$, it holds that*

$$\langle v^k - v^j, s^j \rangle \geq \gamma \|s^j\|^2 \quad \textit{for all } j > k\,. \tag{9}$$

If $\{v^j\}$ is bounded, then $\{s^j\} \to 0$ as $j \to \infty$. ■

Theorem 4.2 *Suppose that T is a maximal monotone operator, with $T^{-1}(0) \neq \emptyset$, and consider BM Algorithm 4.1. Assume further that the set K_S is finite and let k_{last} be the index k yielding the last serious step. Then the following hold:*

(i) The sequence $\{s^k\}$ tends to zero.

(ii) The last serious step $x^{k_{last}}$ is an approximate solution of (1) in the following sense. For each k sufficiently large and each s^k (recall that $s^k = \operatorname{conv}\{w^i\}, (z^i, w^i) \in \Delta_k$), the point $\hat{x}^k = \operatorname{conv}\{z^i\}$ generated by the same convex combination coefficients as s^k, satisfies

$$s^k \in T^{\varepsilon_k}(\hat{x}^k)\,, \quad \|\hat{x}^k - x^{k_{last}}\| \leq \text{TOL}/2 \quad \textit{and} \quad \varepsilon_k \leq M\text{TOL}\,,$$

where $M := \max\{\|w^i\| \mid (z^i, w^i) \in \Delta_k, i \text{ such that } \|z^i - x^{k_{last}}\| \leq \text{TOL}/2\}$.

Proof. [(i)] For all $k \geq k_{last}$, the line-search step in BM always results in a null step. The latter means that

$$\langle v^k, s^k \rangle \leq \theta \min\{\|s^k\|, \|v^k\|\}^2 \quad \text{and} \quad l_k = \hat{m}_k \text{ for all } k \geq k_{last}\,. \tag{10}$$

Observe that $\hat{m}_k$ is set to zero after serious steps, while within a sequence of null steps it can only be increased or stay the same. Hence, $\{\hat{m}_k\}$ is nondecreasing for all $k \geq k_{last}$. Since $\hat{m}_k$ is also bounded above by $\bar{m}$ for all k, it follows that for k sufficiently large, say $k \geq k_1$, $\hat{m}_k$ remains fixed. Taking also into account the second relation in (10), we conclude that

$$l_k = \hat{m}_k = \tilde{m} \leq \bar{m} \text{ for all } k \geq k_1\,. \tag{11}$$

In particular, the line-search step generates y^k satisfying

$$\|y^k - x^{k_{last}}\| = \alpha^{\tilde{m}} R \text{ for all } k \geq k_1\,. \tag{12}$$

Since for all $k \geq k_1$ we have that $\hat{m}_k = \tilde{m}$ in the "Compute direction" step, (12) ensures that the pair (y^k, v^k) appears in the selected bundle used to compute s^j for all $j > k > k_1$. Using the optimality conditions for the minimum-norm problem defining s^j, this means that

$$\langle s^j, v^k - s^j \rangle \geq 0 \text{ for all } j > k > k_1 .$$

Or, equivalently,

$$\langle v^k, s^j \rangle \geq \|s^j\|^2 \text{ for all } j > k > k_1 .$$

Subtracting from the above inequality the first relation in (10) written with $k := j > k_1$, we obtain that

$$\langle v^k - v^j, s^j \rangle \geq (1 - \theta)\|s^j\|^2 \text{ for all } j > k > k_1 .$$

Writing (12) with $k := j > k_1$, since $x^{k_{last}}$ is fixed we have that $\{y^j\}_{j>k_1}$ is bounded, and hence, so is $\{v^j\}_{j>k_1}$. Using Lemma 4.1, we conclude that $\{s^j\} \to 0$.

[(ii)] Note first that if $\tilde{m} < \bar{m}$ in (11), the "Compute direction" step implies that $\|s^k\| > \alpha^{\bar{m}}\tau$, which leads to a contradiction with item (i). Hence, $\tilde{m} = \bar{m}$.
Since $s^k \in \text{conv}\{w^i\}$, we have that

$$s^k = \sum_i \lambda_i^k w^i , \quad \text{with } \lambda_i^k \geq 0 \text{ and } \sum_i \lambda_i^k = 1 .$$

Consider now the corresponding convex combination $\hat{x}^k := \sum_i \lambda_i^k z^i$ and recall that the bundle Δ_k gathers pairs (z^i, w^i) such that $\|z^i - x^{k_{last}}\| \leq \alpha^{\bar{m}} R$. Then Lemma 3.1 gives the final result:

$$s^k \in T^{\varepsilon_k}(\hat{x}^k) \quad \text{with } \|\hat{x}^k - x^{k_{last}}\| \leq \alpha^{\bar{m}} R \leq \text{TOL}/2 \quad \text{and} \quad \varepsilon_k \leq 2\alpha^{\bar{m}} RM \leq \text{TOL}M ,$$

where $M := \max\{\|w^i\| \mid (z^i, w^i) \in \Delta_k , i \text{ such that } \|z^i - x^{k_{last}}\| \leq \alpha^{\bar{m}} R\}$. ∎

As a practical matter, if BM Algorithm 4.1 happens to generate a finite number of serious steps and the approximate solution $x^{k_{last}}$ does not appear satisfactory, then one can decrease the tolerance parameter TOL, increase $\bar{m}$, and proceed. In [8, Lemma 4.2], it is established that if TOL $= 0$ and the number of serious steps is finite, then $x^{k_{last}} \in T^{-1}(0)$, i.e., the exact solution is generated.

5. CONCLUDING REMARKS

We have demonstrated that serious steps of a bundle method for solving maximal monotone inclusions can be viewed as iterations of a hybrid proximal point algorithm. This implies that bundling strategies can be used as a constructive and computationally realistic tool to find acceptable approximate solutions of proximal subproblems. Conversely, our development provides an alternative convergence analysis for bundle methods, including the rate of convergence results, by relating them to the proximal point algorithms. We hope that the exhibited relation adds a further insight into the nature of both methodologies.

Finally, we conjecture that the presented approach can be combined with decomposition principles, such as in [25], in order to develop implementable decomposition algorithms, with a potential for parallel implementation.

REFERENCES

1. Ya.I. Alber. On average convergence of the iterative projection methods, 1997. To be published in *Taiwanese Journal of Mathematics*.
2. Ya.I. Alber. Proximal projection method for variational inequalities and Cesàro averaged approximations, 1998. Preprint of the Technion No. 1051. Haifa, Israel.
3. L. Bacaud, C. Lemaréchal, A. Renaud, and C.A. Sagastizábal. Bundle methods in stochastic optimal power management: a disaggregated approach using preconditioners. *Computational Optimization and Applications*, 2001. Accepted for publication.
4. J.F. Bonnans, J.C. Gilbert, C. Lemaréchal, and C. Sagastizábal. A family of variable metric proximal point methods. *Mathematical Programming* **68** (1995) 15–47.
5. J.F. Bonnans, J.Ch. Gilbert, C. Lemaréchal, and C. Sagastizábal. *Optimisation Numérique: aspects théoriques et pratiques.* (Springer-Verlag, Berlin, 1997).
6. R.E. Bruck and S. Reich. Nonexpansive projections and resolvents of accretive operators in Banach spaces. *Houston Journal of Mathematics* **3** (1977) 459–470.
7. R.S. Burachik, A.N. Iusem, and B.F. Svaiter. Enlargement of monotone operators with applications to variational inequalities. *Set-Valued Analysis* **5** (1997) 159–180.
8. R.S. Burachik, C.A. Sagastizábal, and B. F. Svaiter. Bundle methods for maximal monotone operators. In R. Tichatschke and M. Théra, editors, *Ill-posed variational problems and regularization techniques*, Lecture Notes in Economics and Mathematical Systems, No. **477**, 49–64. (Springer-Verlag, 1999).
9. R.S. Burachik, C.A. Sagastizábal, and B.F. Svaiter. ε-Enlargements of maximal monotone operators: Theory and applications. In M. Fukushima and L. Qi, editors, *Reformulation - Nonsmooth, Piecewise Smooth, Semismooth and Smoothing Methods*, pp. 25–44. (Kluwer Academic Publishers, 1999).
10. J.V. Burke and M. Qian. A variable metric proximal point algorithm for monotone operators. *SIAM Journal on Control and Optimization* **37** (1998) 353–375.
11. R. Cominetti. Coupling the proximal point algorithm with approximation methods. *Journal of Optimization Theory and Applications* **95** (1997) 581–600.
12. R. W. Cottle, J.-S. Pang, and R. E. Stone. *The Linear Complementarity Problem.* (Academic Press, New York, 1992).
13. J. Eckstein. Approximate iterations in Bregman-function-based proximal algorithms. *Mathematical Programming* **83** (1998) 113–123.
14. P.T. Harker and J.-S. Pang. Finite-dimensional variational inequality problems : A survey of theory, algorithms and applications. *Mathematical Programming* **48** (1990) 161–220.
15. J.-B. Hiriart-Urruty and C. Lemaréchal. *Convex Analysis and Minimization Algorithms.* (Springer-Verlag, Berlin, 1993).
16. A.N. Iusem and L.R. Lucambio Pérez. An extragradient-type method for non-smooth variational inequalities. *Optimization*, 2000, to appear.
17. K.C. Kiwiel. A method for solving certain quadratic programming problems arising in nonsmooth optimization. *IMA Journal of Numerical Analysis* **6** (1986) 137–152.
18. B. Lemaire. The proximal algorithm. In J.-P. Penot, editor, *New Methods of Optimization and Their Industrial Use.* International Series of Numerical Mathematics 87, pp. 73–87 (Birkhauser, Basel, 1989).

19. C. Lemaréchal, F. Pellegrino, A. Renaud, and C. Sagastizábal. Bundle methods applied to the unit-commitment problem. In J. Doležal and J. Fidler, editors, *System Modelling and Optimization*, pp. 395–402. (Chapman and Hall, 1995).
20. C. Lemaréchal and C. Sagastizábal. An approach to variable metric bundle methods. In J. Henry and J.-P. Yvon, editors, *Lecture notes in Control and Information Sciences No.* **197**, System Modelling and Optimization, pp. 144–162 (Springer-Verlag, Berlin, 1994).
21. C. Lemaréchal and C. Sagastizábal. Variable metric bundle methods : From conceptual to implementable forms. *Mathematical Programming* **76** (1997) 393–410.
22. Y. Y. Lin and J.-S. Pang. Iterative methods for large convex quadratic programs: A survey. *SIAM Journal on Control and Optimization* **25** (1987) 383–411.
23. F.J. Luque. Asymptotic convergence analysis of the proximal point algorithm. *SIAM Journal on Control and Optimization* **22** (1984) 277–293.
24. B. Martinet. Regularisation d'inequations variationelles par approximations successives. *Revue Française d'Informatique et de Recherche Opérationelle* **4** (1970) 154–159.
25. T. Pennanen. Dualization of generalized equations of maximal monotone type. *SIAM Journal on Optimization* **10** (2000) 809–835.
26. S. Reich. Weak convergence theorems for nonexpansive mappings in Banach spaces. *Journal of Mathematical Analysis and Applications* **67** (1979) 274–276.
27. J.P. Revalski and M. Théra. Variational and extended sums of monotone operators. In R. Tichatschke and M. Théra, editors, *Ill-posed variational problems and regularization techniques*, Lecture Notes in Economics and Mathematical Systems, No. **477**, pp. 229–246. (Springer-Verlag, 1999).
28. R. T. Rockafellar. Local boundedness of nonlinear monotone operators. *Michigan Mathematical Journal* **16** (1969) 397–407.
29. R.T. Rockafellar. Monotone operators and the proximal point algorithm. *SIAM Journal on Control and Optimization* **14** (1976) 877–898.
30. N.Z. Shor. *Minimization methods for nondifferentiable functions.* (Springer-Verlag, Berlin, 1985).
31. M. V. Solodov. A class of globally convergent algorithms for pseudomonotone variational inequalities. In M C. Ferris, O.L. Mangasarian, and J.-S. Pang, editors, *Complementarity: Applications, Algorithms and Extensions.* (Kluwer Academic Publishers, 2001).
32. M. V. Solodov and B. F. Svaiter. A truly globally convergent Newton-type method for the monotone nonlinear complementarity problem. *SIAM Journal on Optimization* **10** (2000) 605–625.
33. M. V. Solodov and P. Tseng. Modified projection-type methods for monotone variational inequalities. *SIAM Journal on Control and Optimization* **34** (1996) 1814–1830.
34. M.V. Solodov and B.F. Svaiter. A globally convergent inexact Newton method for systems of monotone equations. In M. Fukushima and L. Qi, editors, *Reformulation–Nonsmooth, Piecewise Smooth, Semismooth and Smoothing Methods*, pp. 355–369. (Kluwer Academic Publishers, 1999).
35. M.V. Solodov and B.F. Svaiter. A hybrid approximate extragradient–proximal point algorithm using the enlargement of a maximal monotone operator. *Set-Valued Anal-*

ysis **7** (1999) 323–345.
36. M.V. Solodov and B.F. Svaiter. A hybrid projection–proximal point algorithm. *Journal of Convex Analysis* **6** (1999) 59–70.
37. M.V. Solodov and B.F. Svaiter. Error bounds for proximal point subproblems and associated inexact proximal point algorithms. *Mathematical Programming* **88** (2000) 371–389.
38. M.V. Solodov and B.F. Svaiter. An inexact hybrid generalized proximal point algorithm and some new results on the theory of Bregman functions. *Mathematics of Operations Research* **25** (2000) 214–230.
39. P. Tossings. The perturbed proximal point algorithm and some of its applications. *Applied Mathematics and Optimization* **29** (1994) 125–159.
40. P. Tseng. A modified forward-backward splitting method for maximal monotone mappings. *SIAM Journal on Control and Optimization* **38** (2000) 431–446.
41. L. Veselý. Local uniform boundedness principle for families of ε-monotone operators. *Nonlinear Analysis, Theory, Methods and Applications* **24** (1995) 1299–1304.

Inherently Parallel Algorithms in Feasibility and Optimization and their Applications
D. Butnariu, Y. Censor and S. Reich (Editors)
2001 Elsevier Science B.V.

NEW OPTIMIZED AND ACCELERATED PAM METHODS FOR SOLVING LARGE NON-SYMMETRIC LINEAR SYSTEMS: THEORY AND PRACTICE

H. Scolnik[a]* N. Echebest[b] M. T. Guardarucci[b] and M. C. Vacchino[b]

[a]Departamento de Computación, Facultad de Ciencias Exactas y Naturales, Universidad de Buenos Aires, Argentina

[b]Departamento de Matemática, Universidad Nacional de La Plata, Buenos Aires, Argentina

The Projected Aggregation Methods generate the new point x^{k+1} as the projection of x^k onto an "aggregate" hyperplane usually arising from linear combinations of the hyperplanes defined by the blocks. In [13] an acceleration scheme was introduced for algorithms in which an optimized search direction arises from the solution of small quadratic subproblems. In this paper we extend that theory to classical methods like Cimmino's and to the generalized convex combination as defined in [5]. We prove that the resulting new highly parallel algorithms improve the original convergence rate and present numerical results which show their outstanding computational efficiency.
Keywords: projected aggregation methods, row partition strategies, parallel iterative methods.

1. INTRODUCTION

This paper presents an extension of the iterative accelerated block projection method introduced in [13] and further developed in [14] for computing a solution x^* of a non-symmetric linear system $Ax = b$, $A \in \Re^{m\times n}$. Here, we consider a convex combination of block projections instead of optimized ones. We also apply our scheme to the generalized convex combination as defined in [5].

Among the most frequently used iterative methods for solving large, sparse, non-symmetric linear systems $Ax = b$, $A \in \Re^{n\times n}$ are those based on Krylov subspaces techniques ([11]). These methods, when convergent, are extremely fast but may suffer from breakdowns even when using preconditioners (see for instance [3] and [14]). The row projection methods avoid those difficulties ([3]) although in their classical formulations ([6], [10]) convergence is usually very slow.

Block versions ([4],[8],[14]) of these methods consider a partition of $A^t = [A_1^t, A_2^t, .., A_q^t]$ and the corresponding partition of b. For each iterate x^k, an exact or approximate projection onto each block is obtained and a combination of the corresponding directions leads to

*Work supported by the universities of Buenos Aires and La Plata (Project 11/X243), Argentina.

the new iterate. In [8] several possible parallel algorithms are presented and convergence conditions are given. This approach is particularly useful for parallel implementations when the system is appropiately decomposed ([3]). These Projected Aggregation Methods generate the new point x^{k+1} as the projection of x^k onto an "aggregate" hyperplane usually arising from linear combinations of the hyperplanes defined by the blocks. Our aim is to improve the speed of convergence of a particular kind of them by projecting the directions given by the blocks onto the aggregate hyperplane defined in the last iteration.

This paper is organized as follows: In Section 2 we summarize our results concerning acceleration schemes for these sort of algorithms and state some new ones, together with the related convergence theory. In Section 3 we present some numerical results.

2. THE ALGORITHMS: BASIC PROPERTIES AND CONVERGENCE RESULTS

Given a system $Ax = b$, $A \in \Re^{m \times n}$, $m \leq n$, we will assume that it is compatible, and the matrix is not nil.

Let $x^* \in \Re^n$ be a solution of $Ax = b$.

From hereafter $\|x\|$ will denote the Euclidean norm of $x \in \Re^n$. We assume that each row of A has $\|a_i\| = 1$.

Given $M \in \Re^{n \times l}$, $1 \leq l < n$, P_M will denote the orthogonal projector onto $R(M)$, subspace spanned by the columns of M, and $P_{M^\perp}$ the orthogonal projector onto the orthogonal subspace to $R(M)$.

The block version of the projection methods ([1]) consider a partition of A into q row blocks as follows: $A^t = [A_1{}^t, A_2{}^t, \ldots, A_q{}^t]$, $A_i \in \Re^{m_i \times n}$, with $\sum_{i=1}^q m_i = m$, and the corresponding partition of $b^t = (b_1^t, b_2^t, \ldots, b_q^t)$, where $1 \leq q \leq m$.

This yields a partition of the set $\{1, \ldots, m\} = I_1 \cup I_2 \cup, \ldots, \cup I_q$, where I_j contains indices rows of $Ax = b$ belonging to $A_j x = b_j$, for each $j = 1, \ldots, q$.

We denote, for $i = 1, \ldots, q$

$L_i = \{x \in \Re^n : A_i x = b_i, b_i \in \Re^{m_i}\}$

We will assume that each block of the matrix has full rank, that is $rank(A_i) = m_i$.

For every $i = 1, \ldots, q$ the orthogonal projection onto L_i is the mapping $P_i : \Re^n \rightarrow L_i$ given by

$P_i(x) = argmin\{\|x - y\| \ : \ y \in L_i\}$

The basic idea in the parallel projection methods by blocks is given a point x^k, a closer point y_i^k to L_i is computed, and the next iterate is defined by means of a combination of the generated directions $d_i^k = y_i^k - x^k$.

In the special case where y_i^k is the exact projection onto L_i, for all $i = 1, \ldots, q$, we can get the following elementary results ([14]), which are here included because they are useful for understanding and proving the main results.

Lemma 2.1 *Given* x^k, $x^k \neq x^*$. *If* $y_i^k = P_i(x^k)$, $d_i^k = y_i^k - x^k$ *and* $rank(A_i) = m_i$ *for* $i = 1, \ldots, q$, *then*

(i) $d_i^k = A_i^t v_i^k$, *with* $v_i^k \in \Re^{m_i}$, *being*

$v_i^k = (A_i A_i^t)^{-1}(b_i - A_i x^k) = (A_i A_i^t)^{-1} A_i (x^* - x^k)$

(ii) $\|d_i^k\|^2 = (d_i^k)^t (x^* - x^k)$

(iii) If $d^k = \sum_{i=1}^q w_i^k d_i^k$ *then* $(d^k)^t (x^* - x^k) = \sum_{i=1}^q w_i^k \|d_i^k\|^2$.

In the PAM methods the iterative step is defined by means of the projection of the current iterate x^k onto a new hyperplane added to the system arising from a linear combination of the original hyperplanes ([7]).

U. M. García-Palomares in [8] described variations of PAMs for solving structured convex systems, and introduced novel methods (PPAMs) that allow a high degree of parallelism for nonstructured convex systems. The large system is splitted into smaller subsystems, not necessarily disjoint. The parallel algorithms presented in [8], can be described for the particular case of systems of linear equations using our notation and hypotheses, in the following way.

Given an iterate x^k, compute for each block a weighted direction $w_i^k \lambda_i^k d_i^k$ aimed at finding a closer point y_i^k to L_i, considering $d_i^k = A_i^t v_i^k$, under the general scheme for PAMs for solving $A_i x = b_i$, and λ_i^k is defined as

$$\lambda_i^k = argmin_\lambda \|x^k + \lambda d_i^k - x^*\|^2 \tag{2.1}$$

To ensure convergence to a solution of the system define the new iterate x^{k+1}, using a combination of directions d_i^k, $d^k = \sum_{i=1}^q w_i^k \lambda_i^k d_i^k$, and compute $x^{k+1} = x^k + w_k \lambda_k d^k$, where $\eta \leq w_k \leq 2 - \eta$, $\eta > 0$, and

$$\lambda^k = argmin_\lambda \|x^k + \lambda d^k - x^*\|^2 \tag{2.2}$$

The parallel version for linear systems considering a certain choice of $\{w_i^k\}_{i=1}^q$, $\eta \leq w_i^k \leq 2 - \eta$, $\eta > 0$, is described by the iterative step.

Iterative Step (PPAM): Given x^k

Do for $i = 1, \ldots, q$ in parallel

Compute y_i^k a closer point to L_i:

Define $d_i^k = A_i^t v_i^k$.

Define λ_i^k, using (2.1).

Define w_i^k , $\eta \leq w_i^k \leq 2 - \eta$.

End do.

Define $d^k = \sum_{i=1}^q \lambda_i^k w_i^k d_i^k$,

$x^{k+1} = x^k + w_k \lambda_k d^k$, and λ_k from (2.2), and $\eta \leq w_k \leq 2 - \eta$. $\diamond$

Remark 2.1 *In particular if $y_i^k = P_i(x^k)$ then λ_i^k is 1. Furthermore, it is known that $d_i^k = A_i^t v_i^k$, $v_i^k \in \Re^{m_i}$, for all $i = 1, \ldots, q$. It follows that a vector $v^k \in \Re^m$ exists such that $d^k = A^t v^k$. This property is typical of the PAM methods.*

Lemma 2.2 *If $d^k = \sum_{i=1}^q w_i^k d_i^k$, where $d_i^k = P_i(x^k) - x^k$ for all $i = 1, \ldots, q$, and $\lambda_k = argmin_\lambda \|x^k + \lambda d^k - x^*\|^2$, then*

$$\lambda_k = (d^k)^t (x^* - x^k)/\|d^k\|^2 = \sum_{i=1}^q w_i^k \|d_i^k\|^2/\|d^k\|^2 \tag{2.3}$$

If $x^{k+1} = x^k + \lambda_k d^k$, then the sequence $\{x^k\}$ generated by this procedure satisfies

$$\|x^{k+1} - x^*\|^2 = \|x^k - x^*\|^2 - \alpha_k, \text{ where} \tag{2.4}$$

$$\alpha_k = \lambda_k^2 \|d^k\|^2 = (\sum_{i=1}^{q} w_i^k \|d_i^k\|^2)^2 / \|d^k\|^2. \tag{2.5}$$

Proof Considering the solution of (2.2) $\lambda_k = (d^k)^t(x^* - x^k)/\|d^k\|^2$, and using (iii) in Lemma 2.1 its explicit expression follows.

Taking into account that $x^{k+1} = x^k + \lambda_k d^k$, where the combination d^k satisfies (iii) in Lemma 2.1, and considering the expression given in (2.3) for the PAM methods, the result follows. ∘

The convergence of the sequence $\{x^k\}$ generated by this procedure is assured under conditions of Theorem 2.1 and Lemma 2.1 in [8].

It is possible to choose $\{w_i^k\}_{i=1}^q$ in such a way that an optimal combination of the directions $\{d_i^k\}_{i=1}^q$ is obtained in the sense that

$\min_{w \in \Re^q} \|x^k + \sum_{i=1}^q w_i d_i^k - x^*\|$.

This idea has been introduced in [14] ([12]) led to the iterative step, $x^{k+1} = x^k + \sum_{i=1}^q w_i^k d_i^k$, where $w^k \in \Re^q$ is the solution of the convex quadratic problem

$$\min_{u \in \Re^q} \|x^k - x^*\|^2 + 2u^t(D^k)^t(x^k - x^*) + u^t(D^k)^t D^k u, \tag{2.6}$$

where $D^k = [d_1^k, \ldots, d_q^k]$.

Remark 2.2 *Considering our general hypotheses, a direction d_i^k may exist such that is a linear combination of the other directions. Hence, the matrix D^k of problem (2.6) is defined using only linearly independent directions which means that rank$(D^k) = q_k$ with $q_k \leq q$. Therefore, the optimal solution of problem (2.6), by the optimality criteria, satisfies $w^k = ((D^k)^t D^k)^{(-1)}(D^k)^t(x^* - x^k)$. We can get the explicit expression of w^k, considering $(d_i^k)^t(x^* - x^k) = \|d_i^k\|^2$ by (ii) of Lemma 2.1. Thus, the optimal solution of (2.6) is $w^k = ((D^k)^t D^k)^{(-1)}(\|d_1^k\|^2, \|d_2^k\|^2, \ldots, \|d_{q_k}^k\|^2)^t$.*

Within this framework it is useful to take into account the following results given in [14].

Lemma 2.3 *In each iteration k the optimal direction $d^k = D^k w^k$, satisfies*

i) $(D^k)^t D^k w^k = (D^k)^t(x^* - x^k)$, where

$(D^k)^t(x^* - x^k) = (\|d_1^k\|^2, \|d_2^k\|^2, \ldots, \|d_{q_k}^k\|^2)^t$.

ii) $\|d^k\|^2 = (d^k)^t(x^* - x^k) = \sum_{i=1}^{q_k} w_i^k \|d_i^k\|^2$

iii) $(d_i^k)^t d^k = \|d_i^k\|^2$, for all $i = 1, \ldots, q$, and

$\|d^k\| \geq \|d_i^k\|$, *for all* $i = 1, \ldots, q$

iv) The new iterate x^{k+1}, is such that $(d^k)^t(x^{k+1} - x^*) = 0$.

A sequence generated by any algorithm using such a direction satisfies:

Lemma 2.4 *The sequence $\{x^k\}$ generated by the optimal direction defined in (2.6) satisfies*

$\|x^{k+1} - x^*\|^2 = \|x^k - x^*\|^2 - \alpha_k^*$ *where*

$$\alpha_k^* = \sum_{i=1}^{q_k} w_i^k \|d_i^k\|^2 = \|d^k\|^2 \tag{2.7}$$

Furthermore, from the optimality of the linear combination of $\{d_i^k\}_{i=1}^q$ *considered, it is obvious that* $\alpha_k^* \geq \alpha_k$, *where* α_k *is the one defined in (2.5).*

Result(iv) of Lemma 2.3 reflects the geometrical property that from the new iterate $x^k = x^{k-1} + d^{k-1}$ the best direction $x^* - x^k$ lies on the optimal hyperplane used for computing the projection from x^{k-1}.

This led us to consider the possibility of performing from each x^k the next iterative step on such hyperplane, which was optimally chosen at x^{k-1} in the sense described previously.

2.1. An accelerated block projection algorithm

Given an iterate x^k, one possibility is to project the optimal direction d^k, computed by means of (2.6), onto the hyperplane $\{x \in \Re^n \ : \ (d^{k-1})^t(x - x^k) = 0\}$. Such a direction, denoted by $\hat{d}^k$, is $\hat{d}^k = \sum_{i=1}^{q_k} w_i^k P_{v^\perp}(d_i^k)$, where $v = x^k - x^{k-1}$.

The new iterate $\hat{x}^{k+1}$ could be defined as the point $x^k + \hat{\lambda}\hat{d}^k$ corresponding to the solution of the problem

$$\min_{\lambda} \|x^k + \lambda \hat{d}^k - x^*\|^2 \tag{2.8}$$

In this case we obtained the following result ([14]):

Lemma 2.5 *If* $\hat{x}^{k+1} = x^k + \hat{\lambda}\hat{d}^k$, *where* $\hat{d}^k = P_{v^\perp}(d^k)$, $v = x^k - x^{k-1}$, *and* $\hat{\lambda}$ *is defined by (2.8), then* $\|\hat{x}^{k+1} - x^*\|^2 = \|x^k - x^*\|^2 - \hat{\alpha}_k$, *where*

$$\hat{\alpha}_k = (\|d^k\|^2/\|\hat{d}^k\|^2)\alpha_k^*, \tag{2.9}$$

with α_k^* *given in (2.7), is defined by the optimal iterative step.*

In order to explain the accelerated convergence features of a procedure which at the *kth* iterate, $k \geq 1$, uses the previous direction $\hat{d}^k = P_{v^\perp}(d^k)$, it was necessary to obtain several results based on considering that the previous step $v = x^k - x^{k-1}$, satisfies the conditions:

(C1) $v^t(x^* - x^k) = 0$.

(C2) $(d_i^{k-1})^t v = \|d_i^{k-1}\|^2$, for $i = 1, \ldots, q$.

In particular, note that the iterative step $v = d^{k-1}$, defined in x^{k-1} using (2.6), satisfies both conditions due to (iii) and (iv) of Lemma 2.3.

Lemma 2.6 *If at* x^k, $k \geq 1$, $x^k \neq x^*$, *we consider the optimal direction* $d^k = D^k w^k$ *defined by (2.6), and the previous step* $v = x^k - x^{k-1}$ *satisfying the conditions (C1) and (C2), then*

(i) $\|P_{v^\perp}(d^k)\| > 0$.

(ii) $\|P_{v^\perp}(d^k)\| < \|d_k\|$.

Such a result allowed us to prove the following

Theorem 2.1 *If* $\hat{x}^{k+1} = x^k + \hat{\lambda}\hat{d}^k$*, is defined as in (2.8) at kth iterate,* $k \geq 1$*, where* $\hat{d}^k = P_{v^\perp}(d^k)$*, and* $v = x^k - x^{k-1}$ *satisfies (C1) and (C2), then*

$$\|\hat{x}^{k+1} - x^*\|^2 = \|x^k - x^*\|^2 - \hat{\alpha}_k, \quad with \quad \hat{\alpha}_k > \alpha_k^*.$$

With the aim of arriving at the new accelerated procedure, we take into account Theorem 2.1, where the used direction belongs to the subspace defined by

$$[P_{v^\perp}(d_1^k), P_{v^\perp}(d_2^k), \ldots, P_{v^\perp}(d_q^k)]$$

Then, it was natural to choose $\hat{d}^k$ at x^k, as the best combination of $\{P_{v^\perp}(d_i^k)\}_{i=1}^q$ such that the distance between $x^{k+1} = x^k + \hat{d}^k$ and x^* is minimized.

This idea led to define the iterative step, in the following way.

Given x^k, $k \geq 1$, $x^k \neq x^*$, the next iterate $x^{k+1} = x^k + \hat{D}^k w^k$, where $w^k \in \Re^{q_k}$ is the solution of the quadratic problem

$$\min_{u \in \Re^{q_k}} \|x^k + \hat{D}^k u - x^*\|^2, \tag{2.10}$$

where $v = x^k - x^{k-1}$, $\hat{D}^k = [P_{v^\perp}(d_1^k), P_{v^\perp}(d_2^k), \ldots, P_{v^\perp}(d_{q_k}^k)]$, and q_k is the number of linearly independent directions in $\{P_{v^\perp}(d_i^k)\}_{i=1}^q$.

At x^0, define $x^1 = x^0 + d^0$, where d^0 is defined as in (2.6).

Now we will describe the iterative step used for defining the Accelerated Block Algorithm (ACPAM).

We shall use the notation $Q_0 = I_n$, $Q_k = P_{v^\perp}$, $v = x^k - x^{k-1}$ for $k \geq 1$, and $q_k = rank(\hat{D}^k)$ for all $k \geq 0$.

Iterative Step (ACPAM) : Given x^k, Q_k

> **Do for $i = 1, \ldots, q$ in parallel**
> **Compute** $y_i^k = P_i(x^k)$
> **Define** $d_i^k = y_i^k - x^k$.
> **Define** $\hat{d}_i^k = Q_k(d_i^k)$.
> **End do.**
> **Define** $x^{k+1} = x^k + \hat{d}^k$, **where** $\hat{d}^k = \hat{D}^k w^k$
> $w^k = argmin_{u \in \Re^{q_k}} \|x^k + \hat{D}^k u - x^*\|^2$,
> $\hat{D}^k = [\hat{d}_1^k, \hat{d}_2^k, \ldots, \hat{d}_{q_k}^k]$.
> **Set** $v = \hat{d}^k$
> $k = k + 1$. ◇

It was proved the iterative step of ACPAM is well defined and that the sequence generated satisfies the conditions *(C1)* and *(C2)* needed for proving Theorem 2.1.

Hence, it was possible to established the key result

Theorem 2.2 *The sequence* $\{x^k\}$ *generated by ACPAM, satisfies*

$$\|x^{k+1} - x^*\|^2 = \|x^k - x^*\|^2 - \hat{\alpha}_k^*, \quad with \quad \hat{\alpha}_k^* > \alpha_k^*. \tag{2.11}$$

Now we will describe the complete Accelerated Block Algorithm.

Algorithm ACPAM (is ALG2 in [14]).

Step 0. *Split the matrix into blocks by rows using the method described in [14],* $A^t = [A_1^t, A_2^t, \ldots, A_q^t]$, *and the corresponding partition of* $b^t = (b_1^t, \ldots, b_q^t)$, *obtaining for each block* $i = 1, \ldots, q$, *with* m_i *rows, the matrix* $(A_i A_i^t)^{-1} = (L_{m_i}^t D_{mi} L_{m_i})^{-1}$

Main Step. *Given the starting point* $x^0 \in \Re^n$, $\epsilon > 0$.

k=0

While ($\|Ax^k - b\| > \epsilon$) **do**

For each $i = 1, \ldots, q$,

$d_i^k = A_i^t(A_i A_i^t)^{-1}(-r_i^k)$, *where* $r_i^k = A_i x_k - b_i$

Define $\hat{d}_i^k = Q_k(d_i^k)$.

Set $x^{k+1} = x^k + \hat{d}^k$, *where* $\hat{d}^k = \hat{D}^k w^k$

$w^k = ((\hat{D}^k)^t \hat{D}^k)^{(-1)}(\|d_1^k\|^2, \|d_2^k\|^2, \ldots, \|d_{q_k}^k\|^2)^t$

$\hat{D}^k = [\hat{d}_1^k, \hat{d}_2^k, \ldots, \hat{d}_{q_k}^k]$.

Set $v = \hat{d}^k$

$k = k + 1$

End while;

End procedure.$\diamond$

2.2. Acceleration of Block Cimmino algorithm

Our proposal is to show that, under the hypotheses of Lemma 2.2, if we define the iterate x^{k+1} using (2.2), in the direction $\tilde{d}^k$ which is the projection $P_{v^\perp}(\sum_{i=1}^q w_i^k d_i^k)$, with $\sum w_i^k = 1$ and v the previous direction $\tilde{d}^{k-1}$, the convergence rate of (2.4) is also accelerated.

Let $\{w_i > 0\}_{i=1}^q$ be such that $\sum_{i=1}^q w_i = 1$, and $\tilde{d}^k = \sum_{i=1}^q w_i \tilde{d}_i^k$, being $\tilde{d}_i^k = Q_k(d_i^k)$, if Q_k denotes the projector onto the subspace orthogonal to the previous step. In particular $Q_0 = I_n$, where I_n the identity matrix.

The new iterate x^{k+1} is defined as in the general algorithm presented by García-Palomares, but now using the new direction $\tilde{d}^k$.

This scheme, considering $y_i^k = P_i(x^k)$ for each block $i = 1, \ldots, q$, is described by the iterative step.

Iterative Step (ACIMM) : Given x^k, **and** Q_k

Do for $i = 1, \ldots, q$ **in parallel**

Compute $y_i^k = P_i(x^k)$

Define $d_i^k = y_i^k - x^k$**.**

Compute $\tilde{d}_i^k = Q_k(d_i^k)$**.**

End do.

Define $\tilde{d}^k = \sum_{i=1}^q w_i \tilde{d}_i^k$ **,**

$x^{k+1} = x^k + \lambda_k \tilde{d}^k$**,** λ_k defined by (2.2). $\diamond$

This algorithm leads to the iterative step, $x^{k+1} = x^k + \lambda_k \tilde{d}^k$, where λ_k is the solution of the quadratic problem

$$\min_{\lambda \in \Re} \lambda^2 \|\tilde{d}^k\|^2 - 2\lambda(\tilde{d}^k)^t(x^* - x^k) \tag{2.12}$$

whose solution is $\lambda_k = \frac{(\tilde{d}^k)^t(x^*-x^k)}{\|\tilde{d}^k\|^2}$. We will be able to prove $\lambda_k = \frac{\sum_{i=1}^q w_i \|d_i^k\|^2}{\|\tilde{d}^k\|^2}$, considering the next results.

Lemma 2.7 *In each iteration k, $x^k \neq x^*$, the new iterate x^{k+1} of ACIMM is such that $(\tilde{d}^k)^t(x^{k+1} - x^*) = 0$.*

Proof As a consequence of the definition of x^{k+1} and λ_k
$(\tilde{d}^k)^t(x^* - x^{k+1}) = (\tilde{d}^k)^t(x^* - x^k) - \lambda_k \|\tilde{d}^k\|^2 = 0.$ ∘

Lemma 2.8 *In each iteration k, $x^k \neq x^*$, the new direction $\tilde{d}^k$ is well defined and satisfies*

i) $(\tilde{d}^k)^t(x^* - x^k) = (d^k)^t(x^* - x^k) = \sum_{i=1}^q w_i \|d_i^k\|^2$.

ii) $\|\tilde{d}^k\| = \|Q_k(d^k)\| < \|d^k\|$, *if* $d^k = \sum_{i=1}^q w_i d_i^k$.

Proof Since $(\tilde{d}^{k-1})^t(x^* - x^k) = 0$ by Lemma 2.7, and considering $(d^k)^t(x^* - x^k) = \sum_{i=1}^q w_i \|d_i^k\|^2 > 0$, it follows the projection $\|Q_k(d^k)\|$ is at least equal to the norm the projection of d^k onto $x^* - x^k$, which is positive. As a consequence of the previous result $\tilde{d}^k$ is well defined.
Furthermore, since $(x^* - x^k)$ is orthogonal to $\tilde{d}^{k-1}$, it follows that $(\tilde{d}^k)^t(x^* - x^k) = (Q_k(d^k))^t(x^* - x^k) = (d^k)^t(x^* - x^k)$. Thus, (i) follows.
As a consequence of x^k satisfies $(\tilde{d}^{k-1})^t(x^* - x^k) = 0$ by the previous Lemma, we get $d^{k-1} - ((d^{k-1})^t \tilde{d}^{k-2})\tilde{d}^{k-2}/\|\tilde{d}^{k-2}\|^2$ is orthogonal to $(x^* - x^k)$. Thus, considering $\tilde{d}^{k-2}$ is orthogonal to both $x^* - x^{k-1}$ and $\tilde{d}^{k-1}$, we also obtain that $(d^{k-1})^t(x^* - x^k) = 0$.
Since d^k has constant weights w_i, it satisfies $(d^k)^t(x^* - x^{k-1}) = (d^{k-1})^t(x^* - x^k)$. Therefore $(d^k)^t(x^* - x^{k-1}) = 0$.
Considering that $(d^k)^t(x^* - x^{k-1}) = (d^k)^t(x^* - x^k) + (d^k)^t(x^k - x^{k-1}) = 0$, and $(d^k)^t(x^* - x^k) > 0$ we obtain $(d^k)^t(x^k - x^{k-1}) < 0$. Therefore, as a consequence of $(d^k)^t \tilde{d}^{k-1} < 0$, we obtain $\|\tilde{d}^k\| = \|Q_k(d^k)\| < \|d^k\|$. ∘

In algorithm ACIMM , the main difference with the basic algorithm PPAM in [8] is the new direction $\tilde{d}^k$, which is the projection of the combination $\sum_{i=1}^q w_i d_i^k$ onto the orthogonal subspace to $\tilde{d}^{k-1}$. From this definition and Lemma 2.8, we get the next result.

Lemma 2.9 *The sequence $\{x^k\}$ generated by ACIMM satisfies*
$\|x^{k+1} - x^*\|^2 = \|x^k - x^*\|^2 - \tilde{\alpha}_k$, *where*

$$\tilde{\alpha}_k = (\lambda_k)^2 \|\tilde{d}^k\|^2 = \frac{(\sum_{i=1}^q w_i \|d_i^k\|^2)^2}{\|\tilde{d}^k\|^2}, \quad \text{with } \tilde{\alpha}_k > \alpha_k. \tag{2.13}$$

Where α_k is the value given by (2.5) for the weights w_i.

Proof This result follows from Lemma 2.8 and comparing $\|\tilde{d}^k\|^2$ with $\|d^k\|^2$. ∘

Remark 2.3 *In particular, when each block is composed by a row of the matrix A, the method* ACIMM *accelerates the convergence of the classical Cimmino algorithm.*

Recently Y. Censor et al. have published a new iterative parallel algorithm(CAV) in [5] which uses oblique projections used for defining a generalized convex combination. They show that its practical convergence rate, compared to Cimmino's method, approaches the one of ART (Kaczmarz's row-action algorithm [4]). They considered a modification of Cimmino in which the factor $wi = 1/m$ is replaced by a factor that depends only of the nonzero elements in each column of A. For each $j = 1, \ldots, n$, they denote s_j the number of nonzero elements of column j.

Their iterative step can be briefly described, using the notation a_{ij} for the j-th component of the i-th row of A, as follows

Initialization:$x^0 \in \Re^n$ *arbitrary.*

Iterative step:*Given* x^k, *compute* x^{k+1} *by using, for* $j = 1, \ldots, n$, *the formula:*

$$x_j^{k+1} = x_j^k + \lambda_k \sum_{i=1}^{m} \frac{b_i - a_i^t x^k}{\sum_{l=1}^{n} s_l (a_{il})^2} \cdot a_{ij}$$

where $\{\lambda_k\}_{k \geq 0}$ *are relaxation parameters and* $\{s_l\}_{l=1}^{n}$ *are as defined above* . ◇

The CAV algorithm, with $\lambda_k = 1$ for all $k \geq 0$, generates a sequence $\{x^k\}$ which converges ([5]), regardless of the initial point x^0 and independently of the consistency of the system $Ax = b$.

We have considered a modified version of ACIMM, called ACCAV, where each block is composed by a row of the matrix, with weigths w_i equal to those of the CAV algorithm. This particular scheme, is described by:

Algorithm (ACCAV):
Initialization :*Compute* s_l *for* $j = 1, \ldots, n$
Do for $i = 1, \ldots, m$ *in parallel*
compute $w_i = \frac{1}{\sum_{l=1}^{n} s_l (a_{il})^2}$
End do
Iterative Step: *Given* x^k, *and* Q_k *defined as in ACIMM*
Do for $i = 1, \ldots, m$ *in parallel*
Compute $r_i^k = b_i - a_i^t(x^k)$
Define $d_i^k = r_i^k a_i$.
Compute $\tilde{d}_i^k = Q_k(d_i^k)$.
End do.
Define $\tilde{d}^k = \sum_{i=1}^{m} w_i \tilde{d}_i^k$,
$x^{k+1} = x^k + \lambda_k \tilde{d}^k$, λ_k *as given by (2.2)* . ◇

Remark 2.4 *The new weights* w_i *are constant with respect to* k *as they were those considered in the hypotheses of Lemmas 2.7 and 2.8. Thus, all steps needed for proving Lemma 2.8, can be repeated using the new weights. Hence, we can prove a result similar to Lemma 2.9 showing that the sequence generated by the ACCAV speeds up the convergence*

of the algorithm CAV, if the system $Ax = b$ is consistent. Therefore, the sequence $\{x^k\}$ generated by ACCAV also satisfies
$||x^{k+1} - x^*||^2 = ||x^k - x^*||^2 - \tilde{\alpha}_k^s$, *where*

$$\tilde{\alpha}_k^s = (\lambda_k)^2 ||\tilde{d}^k||^2 = \frac{(\sum_{i=1}^m w_i ||d_i^k||^2)^2}{||\tilde{d}^k||^2}, \text{ with } \tilde{\alpha}_k^s > \alpha_k. \tag{2.14}$$

Where α_k is the value given by (2.5) for the weights w_i of the CAV algorithm.

2.3. Convergence

For studying the convergence of our methods we used a theory developed by Gubin et al., [9].

We shall use the notation $\mathcal{P} = \{1, \ldots, q\}$, $L = \bigcap_{i \in \mathcal{P}} L_i$ where
$L_i = \{x \in \Re^n \ : \ A_i x = b_i, b_i \in \Re^{m_i}\}$.

Denote by $d(x, L_i)$ the Euclidean distance between a point $x \in \Re^n$ and a set L_i, and define $\Phi(x) = \max_{i \in \mathcal{P}} \{d(x, L_i)\}$.

Definition. A sequence $\{x^k\}_0^\infty$ is called Fejèr-monotone with respect to the set L, if for $x^* \in L$, and for all $k \geq 0$, $||x^{k+1} - x^*|| \leq ||x^k - x^*||$.

It is easy to check that every Fejèr-monotone sequence is bounded.

The fundamental theorem of Gubin et al. [9], is:

Theorem 2.3 *Let $L_i \subset \Re^n$, be a closed convex set for each $i \in \mathcal{P}$, $L = \bigcap_{i \in \mathcal{P}} L_i$, $L \neq \emptyset$. If the sequence $\{x^k\}_0^\infty$ satisfies the properties :*
i) $\{x^k\}_0^\infty$ is Fejèr-monotone with regard to L, and
ii) $lim_{k \to \infty} \Phi(x^k) = 0$,
then $\{x^k\}_0^\infty$ converges to x^*, $x^* \in L$.

Proof. It follows from Lemma 5 and Lemma 6 of Gubin et al. in [9]. ∘
We proved in [14] that any sequence $\{x^k\}_0^\infty$ generated by ACPAM satisfies (i) and (ii) of Theorem 2.3, provided $x^k \notin L$, for all $k \geq 0$.

Lemma 2.10 *Any sequence $\{x^k\}_0^\infty$ generated by ACIMM and ACCAV satisfies (i) and (ii) of Theorem 2.3, provided $x^k \notin L$, for all $k \geq 0$.*

Proof. The proof of (i) follows immediately from Lemma 2.9 and Remark 2.4. Moreover, both $\tilde{\alpha}_k$ in (2.13) and $\tilde{\alpha}_k^s$ in (2.14) tend to 0 when $k \to \infty$, because the corresponding sequences $\{||x^k - x^*||\}$ converge since they are bounded and monotonically decreasing.
First, considering the expression of $\tilde{\alpha}_k$, we have $\tilde{\alpha}_k = \frac{(\sum_{i=1}^q w_i ||d_i^k||^2)^2}{||\tilde{d}^k||^2} > \frac{(\sum_{i=1}^q w_i ||d_i^k||^2)^2}{||d^k||^2}$.
Since $||d^k|| \leq \sum_{i=1}^q w_i ||d_i^k|| \leq \max_i ||d_i^k||$, then $||d^k||^2 \leq \max_i ||d_i^k||^2$. On the other hand, $(\sum_{i=1}^q w_i ||d_i^k||^2)^2 \geq (w_{i_{max}} \max_i ||d_i^k||^2)^2$, being i_{max} the index corresponding to $\max_i ||d_i^k||^2$. Hence, $\tilde{\alpha}_k > (w_{i_{max}})^2 \max_{i \in \mathcal{P}} ||d_i^k||^2$. Thus, we get $lim_{k \to \infty} \Phi(x^k) = 0$ if $\{x^k\}_0^\infty$ is generated by ACIMM.
Now, we consider $\tilde{\alpha}_k^s$ in (2.14), where its $w_i = \frac{1}{\sum_{j=1}^n s_j (a_{ij})^2}$.
As a consequence of $1 \leq s_j \leq m$ and $||a_i|| = 1$ from our hypotheses, we obtain $1/m \leq w_i \leq 1$. Since $\tilde{\alpha}_k^s = \frac{(\sum_{i=1}^m w_i ||d_i^k||^2)^2}{||\tilde{d}^k||^2} > \frac{(\sum_{i=1}^m w_i ||d_i^k||^2)^2}{||d^k||^2}$, using $||d^k|| \leq \sum_{i=1}^m w_i ||d_i^k||$, it follows $\tilde{\alpha}_k^s > \frac{\max_{i \in \mathcal{P}} ||d_i^k||^2}{m^3}$. Therefore, we also get $lim_{k \to \infty} \Phi(x^k) = 0$ if $\{x^k\}_0^\infty$ is generated by ACCAV. ∘

3. NUMERICAL EXPERIENCES

In [14] we presented a comparison of the numerical results obtained with ACPAM(called Alg2 in that paper) in regard to other row projection methods and Krylov subspace based algorithms([11]). Here we include a couple of those experiences for the sake of completeness, plus some new results concerning the comparison with CAV in order to validate the theory.

The first purpose of our experiments was to compare the behavior of the *ACPAM* with two versions of the parallel block method described in [8]. In order to carry out the comparisons we wrote an experimental code for each algorithm.

The numerical experiences were made with $A \in \Re^{n\times n}$ and run on a PC Pentium III, 408MHz, with 256 Mb Ram and 128 Mb Swap using FORTRAN 77 for Linux.

In what follows we will present a brief description of the implemented parallel block algorithms. In all of them the intermediate directions are defined by the projection of x^k onto each one of the blocks. They differ in the ways in which weights are chosen.

If d_i^k is the direction given by the projection of x_k onto $L_i = \{x \in \Re^n : A_i x = b_i, b_i \in \Re^{m_i}\}$ in the *kth* iteration, the different algorithms compared in the following experiences can be described as follows:

PACI1 (Projected Aggregate Cimmino with equal weights): From an iterate x^k, we define the direction $d^k = \sum_{i=1}^{q} w_i^k d_i^k$ where $w_i^k = 1/q$. The new iterate is $x^{k+1} = x^k + \lambda_k d^k$, where λ_k is defined in (2.3).

PACI2 (Projected Aggregate Cimmino with weights defined by the residuals): From an iterate x^k, we define the direction $d^k = \sum_{i=1}^{q} w_i^k d_i^k$ where $w_i^k = \|A_i x^k - b_i\| / \sum_{j=1}^{q} \|A_j x^k - b_j\|$. The new iterate is $x^{k+1} = x^k + \lambda_k d^k$, where λ_k is defined in (2.3).

ACPAM : Define the direction $d^k = \sum_{i=1}^{q} w_i^k \hat{d}_i^k$ where $\hat{d}_i^k = P_{v^\perp}(d_i^k)$ with $v = d^{k-1}$, and

$$(w_i^k)_{i=1}^{q_k} = ((\hat{D}^k)^t \hat{D}^k)^{-1}(\|\hat{d}_1^k\|^2, \ldots, \|\hat{d}_{q_k}^k\|^2)^t$$

is the solution of the quadratic problem (2.10), and $\hat{D}^k = [\hat{d}_1^k, \hat{d}_2^k, \ldots, \hat{d}_{q_k}^k]$, and q_k is the number of linearly independent directions. The new iterate is $x^{k+1} = x^k + d^k$.

The splitting method used for partitioning the matrix A into blocks and for obtaining the Cholesky decomposition([2]) of the matrices $(A_i {A_i}^t)^{-1}$ used for computing projections has been described in Section 3 of [14]. The intermediate directions used in these algorithms are calculated by means of the projectors obtained from the preprocessing.

For the block splitting procedure the used upper bound for the condition number([2]) was $\kappa = 10^5$. The maximum number of rows μ allowed for each block has been chosen as a function of the problem dimension. The stopping conditions are either the residual $\|Ax^k - b\|^2$ is less than 10^{-9} or when more than ITMAX iterations have been performed.

In *ACPAM*, the quadratic problem (2.10) is solved by means of the Cholesky decomposition of the involved matrices. In order to guarantee the numerical stability of *ACPAM*, the Cholesky decomposition is computed recursively adding up only intermediate directions such that the estimates of the condition numbers of $(\hat{D}^k)^t \hat{D}^k$ do not exceed the upper bound $\kappa = 10^8$.

Test problems. The first set of problems consisted of solving linear systems $Ax = b$ with 500 equalities and 500 unknowns, where A in each case is a matrix whose entries were randomly generated between [-5,5] and $b = Ae$ with $e = (1, \ldots, 1)$ to ensure the consistency of $Ax = b$. The starting point was a random vector with all of its components belonging to [-1,1]. The maximum number of rows μ allowed for each block was 50.

The time required by the block splitting algorithm was 2.6 seconds. In all cases matrices were splitted into 10 blocks; in other words at most 10 directions had been combined in each iteration.

Figure 1 shows the total average time (preprocessing included), in minutes and seconds required to achieve convergence in 10 test problems.

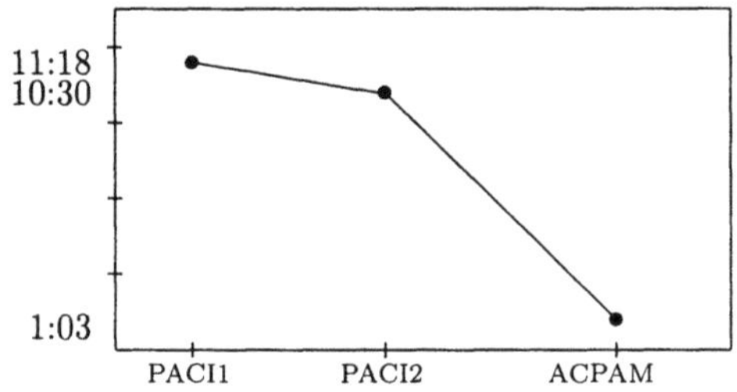

Figure 1. Average time in minutes and seconds for solving 10 random problems with each algorithm.

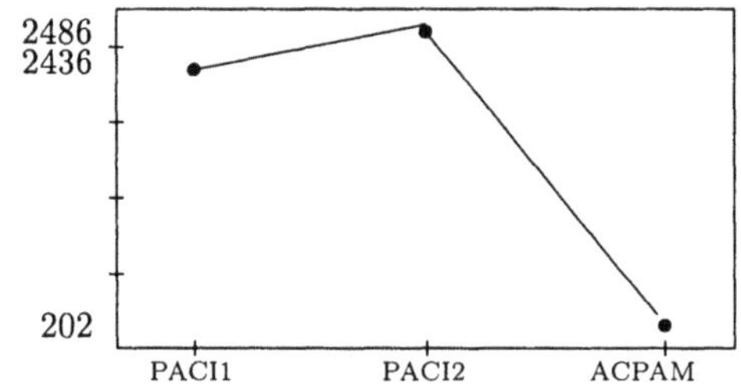

Figure 2. Average number of iterations used by each algorithm for the test problems.

For testing the $ACIMM$ and $ACCAV$ algorithms, the problems $P1-P6$ proposed in [3], were run. Those problems arise from the discretization, using central differences, of elliptic partial differential equations of the form $au_{xx} + bu_{yy} + cu_{zz} + du_x + eu_y + fu_z + gu = F$, where $a - g$ are functions of (x, y, z) and the domain is the unit cube $[0, 1] \times [0, 1] \times [0, 1]$.

The Dirichlet boundary conditions were imposed in order to have a known solution against which errors can be calculated. When the discretization is performed using n_1 points in each direction, the resulting non-symmetric system is of order $n = n_1^3$, and therefore the dimension grows rapidly with n_1 when the grid is refined. If a grid of size $n_1 = 24$ is used, it leads to a problem of dimension $n = m = 13824$.

P1 : $\Delta u + 1000u_x = F$ with solution $u(x, y, z) = xyz(1-x)(1-y)(1-z)$.

P2 : $\Delta u + 10^3 e^{xyz}(u_x + u_y - u_z) = F$ with solution $u(x, y, z) = x + y + z$.

P3 : $\Delta u + 100xu_x - yu_y + zu_z + 100(x + y + z)(u/xyz) = F$ with solution $u(x, y, z) = e^{xyz}\sin(\pi x)\sin(\pi y)\sin(\pi z)$.

P4 : $\Delta u - 10^5 x^2(u_x + u_y + u_z) = F$ with solution idem *P3*.

P5 : $\Delta u - 10^3(1+x^2)u_x + 100(u_y + u_z) = F$ with solution idem *P3*.
P6 : $\Delta u - 10^3((1-2x)u_x + (1-2y)u_y + (1-2z)u_z) = F$ with solution idem *P3*.
We present in the Table 1-3 the obtained results for Problems P1-P6 with $n_1 = 24$, $n = 13824$, comparing *ACIMM*(using single row blocks), *ACCAV*, *PAMCAV*(the PPAM algorithm using CAV weights, and $x^{k+1} = x^k + \lambda_k d^k$ defined by (2.3)) and the *CAV algorithm*.

The starting point was $x^0(i) = 0$ for $i = 1,..,n$, using the following notation:
Iter: number of iterations
Error: $\|x_s - x^*\|_\infty$
Rsn: $\|Ax_s - b\|_2$
CPU: time measured in seconds.
It was considered *the stopping criteria*: if $\|Rsn\|^2 < 10^{-9}$, or $Iter = ITMAX$, with $ITMAX = 5000$.

Table 1
P1-P2: $n = 13824$, block size $m_i = 1$.

	Problem P1				Problem P2			
Method	Iter	Error	Rsn	CPU	Iter	Error	Rsn	CPU
ACIMM	46	3.3d-6	2.8d-5	2.4	334	4.5d-6	3.1d-5	17.2
ACCAV	64	5.6d-6	2.9d-5	3.0	331	5.1d-6	3.1d-5	15.2
PAMCAV	1766	1.6d-5	3.2d-5	75.0	5000	1.5d-1	5.7d-2	232.7
CAV	5000	3.9d-4	8.2d-4	193.9	5000	1.7d0	6.4d-1	216.7

Table 2
P3-P4: $n = 13824$, block size $m_i = 1$.

	Problem P3				Problem P4			
Method	Iter	Error	Rsn	CPU	Iter	Error	Rsn	CPU
ACIMM	1920	6.5d-4	3.1d-5	99.1	1164	1.0d-5	3.2d-5	59.9
ACCAV	1908	6.6d-4	3.1d-5	87.9	1168	1.3d-5	3.1d-5	53.5
PAMCAV	5000	1.2d-3	1.4d-4	238.0	5000	3.9d-1	1.5d-1	231.7
CAV	5000	1.7d-3	2.5d-4	225.0	5000	8.8d-1	2.6d-1	220.8

With the aim of comparing the accelerated behaviour of the *ACIMM* scheme with *PACI1*, when using blocks, we have run problems P3-P4. For that purpose we have considered a splitting into q blocks with q=18, obtaining in such a way a block size of $m_i = 768$.
The final experience was to compare the iterative scheme of both algorithms leaving aside both the sort of blocks and the algorithm for computing the projections onto them. For that purpose we have formed blocks composed of orthogonal rows for computing projections trivially. The results were:

Table 3
P5-P6: $n = 13824$, block size $m_i = 1$.

Method	Problem P5				Problem P6			
	Iter	Error	Rsn	CPU	Iter	Error	Rsn	CPU
ACIMM	82	5.8d-6	2.7d-5	4.2	75	1.9d-6	2.9d-5	3.8
ACCAV	111	8.4d-6	2.9d-5	5.1	78	1.8d-6	2.7d-5	3.7
PAMCAV	2810	1.3d-5	3.2d-5	120.4	756	9.3d-6	3.1d-5	32.1
CAV	5000	2.5d-2	5.1d-2	195.1	1917	1.3d-5	3.2d-5	74.5

Table 4
P3-P4: $n = 13824$, block size $m_i = 768$.

Method	Problem P3				Problem P4			
	Iter	Error	Rsn	CPU	Iter	Error	Rsn	CPU
ACIMM	1924	6.5d-4	3.0d-5	116.0	1161	1.1d-5	3.1d-5	70.1
PACI1	5000	1.1d-3	8.8d-5	282.7	5000	4.1d-1	1.6d-1	282.1

Conclusion: the general acceleration schemes presented in this paper turned out to be efficient when applied to different projection algorithms. Since they are easily parallelizable, it seems that they can be applied to a variety of problems.

Acknowledgments. To Y. Censor for many enlightening discussions during his visit to Argentina, and also to D. Butnariu and S. Reich for having invited us to the outstanding workshop held in Haifa.

REFERENCES

1. R. Aharoni and Y. Censor, Block-iterative projection methods for parallel computation of solutions to convex feasibility problems, *Linear Algebra Appl.* **120** (1989) 165–175.
2. A. Björck, *Numerical Methods for Least Squares Problems* (SIAM, Philadelphia, 1996).
3. R. Bramley and A. Sameh, Row projection methods for large nonsymmetric linear systems, *SIAM J. Sci. Statist. Comput.* **13** (1992) 168–193.
4. Y. Censor and S. Zenios, *Parallel Optimization: Theory and Applications* (Oxford University Press, New York, 1997).
5. Y. Censor, D. Gordon, R. Gordon, *Component Averaging: An Efficient Iterative Parallel Algorithm for Large and Sparce Unstructured Problems* (Technical Report, Department of Mathematics, University of Haifa, Israel, November 1998) (accepted for publication in Parallel Computing).
6. G. Cimmino, Calcolo approssimato per le soluzioni dei sistemi di equazioni lineari, *Ric. Sci.* **16** (1938) 326–333.
7. U. M. García-Palomares, Projected aggregation methods for solving a linear system of equalities and inequalities, in: *Mathematical Research, Parametric Programming*

and Related Topics II (Akademie-Verlag 62, Berlin, 1991) 61–75.
8. U. M. García-Palomares, Parallel projected aggregation methods for solving the convex feasibility problem, *SIAM J. Optim.* **3** (1993) 882–900.
9. L. G. Gubin, B. T. Polyak, and E. V. Raik, The method of projections for finding the common point of convex sets, *USSR Comput. Math. and Math.Phys.* **7** (1967) 1–24.
10. S. Kaczmarz, Angenäherte Auflösung von Systemen linearer Gleichungen, *Bull. Intern. Acad. Polonaise Sci. Lett.* **35** (1937) 355–357.
11. Y. Saad and M. Schultz, Conjugate gradient-like algorithms for solving nonsymmetric linear systems, *Math. Co.* **44** (1985) 417–424.
12. H. D. Scolnik, New Algorithms for Solving Large Sparse Systems of Linear Equations and their Application to Nonlinear Optimization, *Investigación Operativa* **7** (1997) 103–116.
13. H. D. Scolnik, N. Echebest, M. T. Guardarucci, M. C. Vacchino, A New Method for Solving Large Sparse Systems of Linear Equations using row Projections, in: *Proceedings of IMACS International Multiconference Congress Computational Engineering in Systems Applications* (Nabeul-Hammamet, Tunisia, 1998) 26–30.
14. H. D. Scolnik, N. Echebest, M. T. Guardarucci, M. C. Vacchino, A class of optimized row projection methods for solving large non-symmetric linear systems , *Report Notas de Matemática-74, Department of Mathematics, University of La Plata, AR, 2000* (submmited to Applied Numerical Mathematics).

Inherently Parallel Algorithms in Feasibility and Optimization and their Applications
D. Butnariu, Y. Censor and S. Reich (Editors)

THE HYBRID STEEPEST DESCENT METHOD FOR THE VARIATIONAL INEQUALITY PROBLEM OVER THE INTERSECTION OF FIXED POINT SETS OF NONEXPANSIVE MAPPINGS

Isao Yamada[a]

[a]Department of Communications and Integrated Systems, Tokyo Institute of Technology, Ookayama, Meguro-ku, Tokyo 152-8552 JAPAN

This paper presents a simple algorithmic solution to the variational inequality problem defined over the nonempty intersection of multiple fixed point sets of nonexpansive mappings in a real Hilbert space. The algorithmic solution is named the *hybrid steepest descent method*, because it is constructed by blending important ideas in the *steepest descent method* and in the fixed point theory, and generates a sequence converging strongly to the solution of the problem.

The remarkable applicability of this method to the *convexly constrained generalized pseudoinverse problem* as well as to the *convex feasibility problem* is demonstrated by constructing nonexpansive mappings whose fixed point sets are the feasible sets of the problems.

1. INTRODUCTION

The *Variational Inequality Problem* [6,52,118,119] has been and will continue to be one of the central problems in nonlinear analysis and is defined as follows: given monotone operator $\mathcal{F}:\mathcal{H}\to\mathcal{H}$ and closed convex set $C\subset\mathcal{H}$, where $\mathcal{H}$ is a real Hilbert space with inner product $\langle\cdot,\cdot\rangle$ and induced norm $\|\cdot\|$, find $x^*\in C$ such that $\langle x-x^*,\mathcal{F}(x^*)\rangle\geq 0$ for all $x\in C$. This condition is the optimality condition of the convex optimization problem: min Θ over C when $\mathcal{F}=\Theta'$. The simplest iterative procedure for the variational inequality problem (VIP) may be the well-known *projected gradient method* [119] : $x_{n+1}=P_C(x_n-\mu\mathcal{F}(x_n))$ $(n=0,1,2,\ldots)$ where $P_C:\mathcal{H}\to C$ is the convex projection onto C and μ is a positive real number (This method is an example of so called the *gradient projection method* [57,74]. In the following, the method specified by this formula is called the *projected gradient method* to distinguish it from *Rosen's* gradient projection method [90,91]). Indeed, when $\mathcal{F}$ is strongly monotone and Lipschitzian, the *projected gradient method*, with any $x_0\in C$ and certain $\mu>0$, generates a sequence $(x_n)_{n\geq 0}$ converging strongly to the unique solution of the VIP. The use of the *projected gradient method* requires the closed form expression of P_C, which unfortunately is not always known.

Motivated by the tremendous progresses in the fixed point theory of nonexpansive mappings (see for example [60,48,75,104,106,39,5,8,9,16–18,87,88,101,102,55,56] and references therein), the *hybrid steepest descent method* [108,109,47,110,111,80,112] has been developed as a steepest descent type algorithm minimizing certain convex functions over the intersection of fixed point sets of nonexpansive mappings. This method is essentially an algorithmic solution to the above VIP that does not require the closed form expression of P_C but instead requires a closed form expression of a nonexpansive mapping T, whose fixed point set is C.

The generalization made by the *hybrid steepest descent method*, of the direct use of the convex projection P_C to that of a nonexpansive mapping, is important in many practical situations where no closed form expression of P_C is known but where the closed form expression of a nonexpansive mapping whose fixed point set is C can be based on fundamentals of fixed point theory and on algorithms for convex feasibility problems [39,8,10,32,37,100]. A notable advantage of the *hybrid steepest descent method*, over the methods using the closed form expression of P_C, is that it is applicable in the frequent cases in which the set C is not simple enough to have a closed form expression of P_C but is defined as the intersection of multiple closed convex sets C_i ($i \in J \subset \mathbb{Z}$) each of which is simple enough to have a closed form expression of P_{C_i}.

The first objective of the paper is to present in a simple way the ideas underlying the *hybrid steepest descent method* [108,109,47]. The second objective is to demonstrate the applications of its simple formula to the *convexly constrained generalized pseudoinverse problem* [51,93,110] as well as to the *convex feasibility problem* [37,32,8,10,100] of broad interdisciplinary interest in various areas of mathematical science and engineering.

The rest of this paper is divided into four sections. The next section contains preliminaries on fixed points, nonexpansive mappings, and convex projections, as well as brief introductions to the variational inequality problem and to the weak topology in a real Hilbert space. Other necessary mathematical facts are also collected there. The third section contains the main theorems of the *hybrid steepest descent method*, where it will be shown that the variational inequality problem, defined over the fixed point set of nonexpansive mappings, under certain condition, can be solved algorithmically by the surprisingly simple formulae. The fourth section, after discussing applications of the *hybrid steepest descent method* to the convex feasibility problems, introduces important fixed point characterization of the *generalized convex feasible set* in [108,109] together with its another fixed point characterization presented independently in [41]. Based on these fixed point characterizations, we demonstrate how the *hybrid steepest descent method* can be applied, in mathematically sound way, to the *convexly constrained generalized pseudoinverse problem*. Lastly in the final section, we conclude the paper with some remarks on our recent partial generalization showing that the *hybrid steepest descent method* is also suitable to the variational inequality problems under more general conditions, which includes the case where the problem may have multiple solutions over the *generalized convex feasible set*.

2. PRELIMINARIES

A. Fixed points, Nonexpansive mappings, and Convex projections

A fixed point of a mapping $T : \mathcal{H} \to \mathcal{H}$ is a point $x \in \mathcal{H}$ such that $T(x) = x$. $Fix(T) := \{x \in \mathcal{H} \mid T(x) = x\}$ denotes the set of all fixed points of T. A mapping

$T : \mathcal{H} \to \mathcal{H}$ is called κ-Lipschitzian (or κ-Lipschitz continuous) over $S \subset \mathcal{H}$ if there exists $\kappa > 0$ such that

$$\|T(x) - T(y)\| \leq \kappa \|x - y\| \text{ for all } x, y \in S. \tag{1}$$

In particular, a mapping $T : \mathcal{H} \to \mathcal{H}$ is called (i) *strictly contractive* if $\|T(x) - T(y)\| \leq \kappa\|x - y\|$ for some $\kappa \in (0,1)$ and all $x, y \in \mathcal{H}$ [The *Banach-Picard fixed point theorem* guarantees the unique existence of the fixed point, say $x^* \in Fix(T)$, of T and the strong convergence of $(T^n(x_0))_{n\geq 0}$ to x^* for any $x_0 \in \mathcal{H}$.] ; (ii) *nonexpansive* if $\|T(x) - T(y)\| \leq \|x - y\|$ for all $x, y \in \mathcal{H}$; (iii) *firmly nonexpansive* if $\|T(x) - T(y)\|^2 \leq \langle x - y, T(x) - T(y)\rangle$ for all $x, y \in \mathcal{H}$; (iv) *averaged* (or *α-averaged*) if there exists $\alpha \in [0,1)$ and a nonexpansive mapping $\mathcal{N} : \mathcal{H} \to \mathcal{H}$ such that $T := (1-\alpha)I + \alpha\mathcal{N}$ where I is identity. (Note: This definition is slightly different from the one in the sense of [5] where $\alpha = 0$ is excluded to ensure $Fix(T) = Fix(\mathcal{N})$. A firmly nonexpansive mapping is alternatively characterized as $\frac{1}{2}$-averaged mapping [see Fact 2.1(d)].); and (v) *attracting nonexpansive* if T is nonexpansive with $Fix(T) \neq \emptyset$ and $\|T(x) - f\| < \|x - f\|$ for all $f \in Fix(T)$ and all $x \notin Fix(T)$ (We regard the identity I as a special attracting nonexpansive mapping due to $Fix(I) = \mathcal{H}$). If $T : \mathcal{H} \to \mathcal{H}$ is averaged with $Fix(T) \neq \emptyset$, then T is attracting (see for example [8, Lemma 2.4]).

Recall that a nonempty set $C \subset \mathcal{H}$ is called *convex* if $x, y \in C$ and $\lambda \in [0,1]$ imply $\lambda x + (1-\lambda)y \in C$. Given a nonempty closed convex set C in $\mathcal{H}$, the mapping that assigns every point in $\mathcal{H}$ to its unique nearest point in C is called the *metric projection* or *convex projection* onto C and is denoted by P_C; i.e., $\|x - P_C(x)\| = d(x, C)$, where $d(x, C) := \inf_{y \in C} \|x - y\|$. The metric projection P_C is characterized by the relation:

$$x^* \in C \text{ satisfies } x^* = P_C(x) \Leftrightarrow x^* \in C \text{ satisfies } \langle x - x^*, y - x^*\rangle \leq 0 \text{ for all } y \in C, \tag{2}$$

and therefore P_C is firmly nonexpansive with $Fix(P_C) = C$. In particular, if $\mathcal{M}$ is a closed subspace in $\mathcal{H}$, $P_{\mathcal{M}}$ is linear and $x - P_{\mathcal{M}}(x) \in \mathcal{M}^\perp$ for all $x \in \mathcal{H}$.

Some closed convex set C is simple in the sense that the closed form expression of P_C is known, which implies that P_C can be computed within a finite number of arithmetic operations. This will be the case, for example, when C is certain linear variety, closed ball, closed cone or closed polytope [107] etc.

The following proposition collects part of remarkable properties of nonexpansive mappings. For the details, see for example [116,55,56,88,23,8,10,100,109,39] and references therein.

Fact 2.1 (Selected properties of nonexpansive mappings)

(a) *If a nonexpansive mapping $T : \mathcal{H} \to \mathcal{H}$ has at least one fixed point, $Fix(T) \subset \mathcal{H}$ is closed convex and expressed as*

$$Fix(T) = \bigcap_{y \in \mathcal{H}} \{x \in \mathcal{H} \mid 2\langle y - T(y), x\rangle \leq \|y\|^2 - \|T(y)\|^2\}.$$

(b) *For nonexpansive mappings $T_i : \mathcal{H} \to \mathcal{H}$ $(i = 1, 2, \ldots, m)$, both $T_m T_{m-1} \cdots T_1$ and $\sum_{i=1}^m w_i T_i$ are also nonexpansive, where $(w_i)_{i=1}^m \subset [0,1]$ and $\sum_{i=1}^m w_i = 1$.*

(c) *For nonexpansive mappings* $T_i : \mathcal{H} \to \mathcal{H}$ $(i = 1, 2, \ldots, m)$ *satisfying* $\bigcap_{i=1}^m Fix(T_i) \neq \emptyset$, *it follows* $Fix(\sum_{i=1}^m w_i T_i) = \bigcap_{i=1}^m Fix(T_i)$ *where* $(w_i)_{i=1}^m \subset (0,1]$ *and* $\sum_{i=1}^m w_i = 1$.

(d) $T : \mathcal{H} \to \mathcal{H}$ *is firmly nonexpansive if and only if* $2T - I$ *is nonexpansive. This fact implies that* $I - T : \mathcal{H} \to \mathcal{H}$ *is firmly nonexpansive if* $T : \mathcal{H} \to \mathcal{H}$ *is firmly nonexpansive [because* $2(I - T) - I = I - 2T$ *is nonexpansive]. Moreover, for given firmly nonexpansive mappings* $T_i : \mathcal{H} \to \mathcal{H}$ *and* $w_i \geq 0$ $(i = 1, 2, \ldots, m)$ *satisfying* $\sum_{k=1}^m w_k = 1$, $\sum_{i=1}^m w_i T_i$ *is firmly nonexpansive because* $2\left(\sum_{i=1}^m w_i T_i\right) - I = \sum_{i=1}^m w_i (2T_i - I)$ *is nonexpansive.*

(e) *For a firmly nonexpansive mapping* $T : \mathcal{H} \to \mathcal{H}$ *with* $Fix(T) \neq \emptyset$, $(1-\alpha)I + \alpha T$ *is attracting nonexpansive for all* $\alpha \in [0, 2)$.

(f) *For attracting nonexpansive mappings* $T_i : \mathcal{H} \to \mathcal{H}$ $(i = 1, 2, \ldots, m)$ *satisfying* $\bigcap_{i=1}^m Fix(T_i) \neq \emptyset$, $T_m T_{m-1} \cdots T_1$ *is attracting and* $Fix(T_m T_{m-1} \cdots T_1) = \bigcap_{i=1}^m Fix(T_i)$.

(g) *Suppose that* $T_i : \mathcal{H} \to \mathcal{H} (i \in J \subset \mathbb{Z})$ *is a countable family of firmly nonexpansive mappings and* $F_\infty := \bigcap_{i \in J} Fix(T_i) \neq \emptyset$. *Let* $T := \sum_{i \in J} w_i T_i$, *where* $(w_i)_{i \in J} \subset (0, 1]$ *and* $\sum_{i \in J} w_i = 1$. *Then* T *is nonexpansive and* $Fix(T) = F_\infty$.

(h) *For any pair of firmly nonexpansive mappings* T_1 *and* T_2 *that do not necessarily have a common fixed point,* $(1-\alpha)I + \alpha T_1 T_2$ *is nonexpansive for all* $\alpha \in [0, 3/2]$. *Conversely, for every* $\alpha \notin [0, 3/2]$, *there exists a pair of firmly nonexpansive mappings* T_1 *and* T_2 *such that* $(1-\alpha)I + \alpha T_1 T_2$ *is not nonexpansive (see [109]).*

B. Variational Inequality Problem (VIP)

For a given set $S \subset \mathcal{H}$, a mapping $\mathcal{F} : \mathcal{H} \to \mathcal{H}$ is said to be (i) *monotone* over S if $\langle \mathcal{F}(x) - \mathcal{F}(y), x - y \rangle \geq 0$ for all $x, y \in S$. In particular, a mapping $\mathcal{F} : \mathcal{H} \to \mathcal{H}$, which is monotone over S, is called (ii) *paramonotone* over S if $\langle \mathcal{F}(x) - \mathcal{F}(y), x - y \rangle = 0 \Leftrightarrow \mathcal{F}(x) = \mathcal{F}(y)$ holds for all $x, y \in S$; (iii) *strictly monotone* over S if $\langle \mathcal{F}(x) - \mathcal{F}(y), x - y \rangle = 0 \Leftrightarrow x = y$ holds for all $x, y \in S$; (iv) η*-strongly monotone* over S if there exists $\eta > 0$ such that $\langle \mathcal{F}(x) - \mathcal{F}(y), x - y \rangle \geq \eta \|x - y\|^2$ for all $x, y \in S$.

Problem 2.2 (Variational Inequality Problem: $VIP(\mathcal{F}, C)$ [6,52,118,119]) *Given* $\mathcal{F} : \mathcal{H} \to \mathcal{H}$ *which is monotone over a nonempty closed convex set* $C \subset \mathcal{H}$, *the Variational Inequality Problem (VIP), denoted by* $VIP(\mathcal{F}, C)$, *is the problem:*

Find $u^* \in C$ *such that* $\langle v - u^*, \mathcal{F}(u^*) \rangle \geq 0$ *for all* $v \in C$.

(Note: For a general discussion on the existence of the solution, see for example [6, Theorem II.2.6], [52, Prop.II.3.1] or [119, Theorem 54.A(b)].)

Proposition 2.3 (Solution set of $VIP(\mathcal{F}, C)$ [33]) *Let* $\mathcal{F} : \mathcal{H} \to \mathcal{H}$ *be monotone and continuous over a nonempty closed convex set* $C \subset \mathcal{H}$. *Then,*

(a) $u^* \in C$ *is a solution of* $VIP(\mathcal{F}, C)$ *if and only if*

$\langle v - u^*, \mathcal{F}(v) \rangle \geq 0$ *for all* $v \in C$.

(b) *if $\mathcal{F}$ is paramonotone over C, u^* is a solution of $VIP(\mathcal{F}, C)$, and $u \in C$ satisfies $\langle u - u^*, \mathcal{F}(u)\rangle = 0$, then u is also a solution of $VIP(\mathcal{F}, C)$.*

Let U be an open subset of $\mathcal{H}$. Then a function $\varphi : \mathcal{H} \to \mathbb{R} \cup \{\infty\}$ is called *Gâteaux* differentiable over U if for each $u \in U$ there exists $a(u) \in \mathcal{H}$ such that

$$\lim_{\delta \to 0} \frac{\varphi(u + \delta h) - \varphi(u)}{\delta} = \langle a(u), h\rangle \text{ for all } h \in \mathcal{H}.$$

In this case, $\varphi' : U \to \mathcal{H}$ defined by $\varphi'(u) := a(u)$ is called *Gâteaux* derivative of φ over U.

On the other hand, a function $\varphi : \mathcal{H} \to \mathbb{R} \cup \{\infty\}$ is called *Fréchet* differentiable over U if for each $u \in U$ there exists $a(u) \in \mathcal{H}$ such that

$$\varphi(u + h) = \varphi(u) + \langle a(u), h\rangle + o(\|h\|) \text{ for all } h \in \mathcal{H},$$

where $r(h) = o(\|h\|)$ means $\lim_{h \to 0} r(h)/\|h\| = 0$. In this case, $\varphi' : U \to \mathcal{H}$ defined by $\varphi'(u) = a(u)$ is called *Fréchet* derivative of φ over U. If φ is *Fréchet* differentiable over U, φ is also *Gâteaux* differentiable over U and both derivatives coincide. Moreover, if φ is *Gâteaux* differentiable with continuous derivative φ' over U, then φ is also *Fréchet* differentiable over U.

For details including the notion of higher differentials and higher derivatives, see for example [117,118].

Recall that a function $\varphi : \mathcal{H} \to \mathbb{R} \cup \{\infty\}$ is called *convex* over a convex set $C \subset \mathcal{H}$ if

$$\varphi(\lambda x + (1 - \lambda)y) \leq \lambda\varphi(x) + (1 - \lambda)\varphi(y) \text{ for all } \lambda \in [0, 1] \text{ and all } x, y \in C.$$

The next fact shows that the convex optimization problem is reduced to a variational inequality problem, which obviously shows an invaluable role of $VIP(\mathcal{F}, C)$ in real world applications.

Fact 2.4 (Convex optimization as a variational inequality problem. For the details, see for example [52, Prop.I.5.5, Prop.II.2.1 and their proofs], [118, Prop.25.10] and [119, Theorem 46.A])

(a) (Monotonicity and convexity) *Let $C \subset \mathcal{H}$ be a nonempty closed convex set. Suppose that $\varphi : \mathcal{H} \to \mathbb{R} \cup \{\infty\}$ is Gâteaux differentiable over an open set $U \supset C$. Then φ is convex over C if and only if $\varphi' : U \to \mathcal{H}$ is monotone (indeed more precisely 'paramonotone') over C.* [Note: The paramonotonicity of φ' is shown for example in [33, Lemma 12] when $\mathcal{H}$ is a finite dimensional real Hilbert space (Euclidean space) $\mathbb{R}^N$, but essentially the same proof works for a general real Hilbert space $\mathcal{H}$.]

(b) (Characterization of the convex optimization problem) *Let $C \subset \mathcal{H}$ be a nonempty closed convex set. Suppose that $\varphi : \mathcal{H} \to \mathbb{R} \cup \{\infty\}$ is convex over C and Gâteaux differentiable with derivative φ' over an open set $U \supset C$. Then, for $x^* \in C$, $\varphi(x^*) = \inf \varphi(C)$ if and only if $\langle x - x^*, \varphi'(x^*)\rangle \geq 0$ for all $x \in C$.*

The characterization in (2) of the convex projection P_C yields at once an alternative interpretation of the VIP as a fixed point problem.

Proposition 2.5 (VIP as a fixed point problem) *Given $\mathcal{F} : \mathcal{H} \to \mathcal{H}$ which is monotone over a nonempty closed convex set C, the following four statements are equivalent.*

(a) *$u^* \in C$ is a solution of VIP($\mathcal{F}, C$); i.e.,*

$$\langle v - u^*, \mathcal{F}(u^*)\rangle \geq 0 \text{ for all } v \in C.$$

(b) *For an arbitrarily fixed $\mu > 0$, $u^* \in C$ satisfies*

$$\langle v - u^*, (u^* - \mu\mathcal{F}(u^*)) - u^*\rangle \leq 0 \text{ for all } v \in C.$$

(c) *For an arbitrarily fixed $\mu > 0$,*

$$u^* \in Fix\,(P_C\,(I - \mu\mathcal{F}))\,. \tag{3}$$

When some additional assumptions are imposed on the mapping $\mathcal{F} : \mathcal{H} \to \mathcal{H}$ and the closed convex set C in $VIP(\mathcal{F}, C)$, the mapping $P_C\,(I - \mu\mathcal{F})\,P_C : \mathcal{H} \to \mathcal{H}$ can be strictly contractive [or equivalently $P_C\,(I - \mu\mathcal{F}) : \mathcal{H} \to \mathcal{H}$ can be strictly contractive over C] for certain $\mu > 0$ as follows.

Lemma 2.6 *Suppose that $\mathcal{F} : \mathcal{H} \to \mathcal{H}$ is κ-Lipschitzian and η-strongly monotone over a nonempty closed convex set $C \subset \mathcal{H}$. Then we have*

$$\|P_C\,(I - \mu\mathcal{F})\,(u) - P_C\,(I - \mu\mathcal{F})\,(v)\|^2 \leq \{1 - \mu(2\eta - \mu\kappa^2)\}\|u - v\|^2 \text{ for all } u, v \in C, \tag{4}$$

which ensures that $P_C\,(I - \mu\mathcal{F}) : \mathcal{H} \to \mathcal{H}$ is strictly contractive over C for $\mu \in \left(0, \frac{2\eta}{\kappa^2}\right)$.

Proof: This fact is well-known (see for example [119, Theorem 46.C]) and almost obvious by applying the conditions on $\mathcal{F}$ to the simple inequality:

$$\begin{aligned} & \|P_C\,(I - \mu\mathcal{F})\,(u) - P_C\,(I - \mu\mathcal{F})\,(v)\|^2 \\ \leq\ & \|u - v - \mu\,(\mathcal{F}(u) - \mathcal{F}(v))\|^2 \\ =\ & \|u - v\|^2 - 2\mu\langle u - v, \mathcal{F}(u) - \mathcal{F}(v)\rangle + \mu^2\,\|\mathcal{F}(u) - \mathcal{F}(v)\|^2 \text{ for all } u, v \in C. \end{aligned}$$

(Q.E.D.)

The interpretation shown in (3) of Proposition 2.5, Lemma 2.6 and *Banach-Picard fixed point theorem* yield at once the following proposition.

Proposition 2.7 (Projected gradient method [57,74,119]) *Suppose that $\mathcal{F} : \mathcal{H} \to \mathcal{H}$ is κ-Lipschitzian and η-strongly monotone over a closed convex set $C \neq \emptyset$. Then,*

(a) *$VIP(\mathcal{F}, C)$ has its unique solution $u^* \in C$,*

(b) *for any $u_0 \in C$ and $\mu \in \left(0, \frac{2\eta}{\kappa^2}\right)$, $(u_n)_{n\geq 0}$ generated by the formula*

$$u_{n+1} = P_C\,(I - \mu\mathcal{F})\,(u_n) \text{ for } n = 0, 1, 2, \ldots \tag{5}$$

converges strongly to the unique solution $u^* \in C$ of $VIP(\mathcal{F}, C)$. The iteration (5) is so called the projected gradient method.

Remark 2.8 *For minimizing a convex function Θ over a closed convex set C, a more general formula : $u_{n+1} = P_C\left(I - \mu_n\Theta'\right)(u_n)$ $(n = 0, 1, 2, \ldots)$ is given in [57,74], where the conditions on Θ are discussed to ensure the convergence to the solution.*

The next lemma and example introduce typical convex functions whose derivatives are κ-Lipschitzian as well as η-strongly monotone over a convex set $C \neq \emptyset$. These facts suggest the great applicability of the *projected gradient method.*

Lemma 2.9 [47] *Denote the set of all bounded linear operators from $\mathcal{H}$ to $\mathcal{H}$ by $B(\mathcal{H})$. Let $C \subset \mathcal{H}$ be a nonempty convex set and $\Theta : \mathcal{H} \to \mathbb{R}\cup\{\infty\}$ be twice Fréchet differentiable on some open set $U \supset C$. Then $\Theta'' : U \to B(\mathcal{H})$ is said to be uniformly strongly positive and uniformly bounded over C if $\Theta''(x)$ is self-adjoint for all $x \in C$, and there exist scalars $\kappa_{max} \geq \kappa_{min} > 0$ such that*

$$\kappa_{min}\|v\|^2 \leq \langle \Theta''(x)v, v\rangle \leq \kappa_{max}\|v\|^2 \qquad \text{for all } x \in C \text{ and } v \in \mathcal{H}. \tag{6}$$

In this case, $\Theta' : U \to \mathcal{H}$ is κ_{max}-Lipschitzian and κ_{min}-strongly monotone over C.

Proof: Fix any distinct points $x, y \in C$ and define $f : [0,1] \to \mathbb{R}$ by

$$f(t) := \left\langle \Theta'\left(x + t(y-x)\right), y - x\right\rangle.$$

Then, by the continuity of the inner product and *Fréchet* differentiability (hence *Gâteaux* differentiability) of Θ', it follows that

$$\begin{aligned} f'(t) &= \lim_{\delta\to 0}\frac{1}{\delta}\left[\left\langle \Theta'\left(x + (t+\delta)(y-x)\right), y-x\right\rangle - \left\langle \Theta'\left(x + t(y-x)\right), y-x\right\rangle\right] \\ &= \left\langle \lim_{\delta\to 0}\frac{1}{\delta}\left[\Theta'\left(x + (t+\delta)(y-x)\right) - \Theta'\left(x + t(y-x)\right)\right], y - x\right\rangle \\ &= \left\langle \Theta''\left(x + t(y-x)\right)(y-x), y-x\right\rangle \text{ for all } t \in (0,1). \end{aligned}$$

Then, by (6), the standard *mean value theorem* [2] ensures the existence of some $t_* \in (0,1)$ satisfying

$$\begin{aligned} f'(t_*) &= f(1) - f(0) = \langle x - y, \Theta'(x) - \Theta'(y)\rangle \\ &= \left\langle \Theta''\left(x + t_*(y-x)\right)(y-x), y-x\right\rangle \\ &\geq \kappa_{min}\|x-y\|^2, \end{aligned}$$

which implies that Θ' is κ_{min}-strongly monotone over C.

On the other hand, applying a well known counterpart [76] of the *mean value theorem* for the *Fréchet* derivative to $\Theta' : U \to \mathcal{H}$, we obtain

$$\begin{aligned} \|\Theta'(x) - \Theta'(y)\| &\leq \|x-y\| \sup_{0<\alpha<1}\|\Theta''(x + \alpha(y-x))\| \\ &\leq \kappa_{max}\|x-y\| \quad \text{for all } x, y \in C, \end{aligned}$$

where we used the fact: $\|A\| = \sup_{x\neq 0}|\langle Ax, x\rangle|\|x\|^{-2}$ for any self-adjoint operator $A \in B(\mathcal{H})$ (see for example [73, Sec.9.2]).

This implies that Θ' is κ_{max}-Lipschitzian over C. (Q.E.D.)

Example 2.10 [109] *Suppose that $r \in \mathcal{H}$ and $Q : \mathcal{H} \to \mathcal{H}$ is a self-adjoint bounded linear operator and is strongly positive; i.e., $\langle Qx, x\rangle \geq \alpha \|x\|^2$ for some $\alpha > 0$ and all $x \in \mathcal{H}$ (such a Q is nothing but a positive definite matrix when $\mathcal{H}$ is finite dimensional). Define a quadratic function $\Theta : \mathcal{H} \to \mathbb{R}$ by*

$$\Theta(x) := \frac{1}{2}\langle Qx, x\rangle - \langle r, x\rangle \text{ for all } x \in \mathcal{H}.$$

Then $\Theta''(x) = Q$ for all $x \in \mathcal{H}$ and $\alpha\|v\|^2 \leq \langle Qv, v\rangle \leq \|Q\|\|v\|^2$ for all $v \in \mathcal{H}$, which implies that $\Theta'' : \mathcal{H} \to B(\mathcal{H})$ is uniformly strongly positive and uniformly bounded over $\mathcal{H}$. Indeed, $\Theta'(x) = Qx$ is $\|Q\|$-Lipschitzian and α-strongly monotone over $\mathcal{H}$.

The *projected gradient method* in Proposition 2.7 (see also Remark 2.8) has been widely used to solve many practical variational inequality problems $VIP(\mathcal{F}, C)$, mainly because of its simple structure and fast convergence. Two major practical limitations, however, have been pointed out. One is that the additional assumptions on $\mathcal{F} : \mathcal{H} \to \mathcal{H}$, i.e.; κ-Lipschitz continuity and η-strongly monotonicity of $\mathcal{F} : \mathcal{H} \to \mathcal{H}$, are rather strong and seem to be restrictive in some applications. The second is that the use of formula (5) is based on the assumption that the closed form expression of $P_C : \mathcal{H} \to C$ is known, whereas in many situations it is not.

A great deal of effort has gone into relaxing the first limitation, in particular when $\mathcal{H}$ is a finite dimensional real Hilbert space (Euclidean space) $\mathbb{R}^N$. For example, Korpolevich removed the requirement that $\mathcal{F} : \mathbb{R}^N \to \mathbb{R}^N$ be strongly monotone by splitting the formula (5) into two similar stages [72]. Iusem later proposed a *modified Korpolevich method* where the requirement for κ-Lipschitz continuity is removed by introducing a search for an appropriate constant sequence during the iteration [68]. The *modified Korpolevich method* requires only monotonicity and continuity for $\mathcal{F} : \mathbb{R}^N \to \mathbb{R}^N$. When $\mathcal{F} : \mathbb{R}^N \to \mathbb{R}^N$ is paramonotone and continuous, an interior point method for $VIP(\mathcal{F}, C)$ was also developed by introducing pseudo-translation related to a Bregman function whose zone is the interior of C [33]. Indeed the method in [33] has also some potential advantage to relax the above second limitation by which the *projected gradient method* is faced (see [33, Sec.6]).

In this paper, mainly for simplicity, we present the *hybrid steepest descent method* by concentrating our discussion on the relaxation of the above second limitation, while a partial but notable relaxation, of the first limitation; i.e., the strong monotonicity, of $\mathcal{F}$ was made recently for the *hybrid steepest descent method* as well. For the evolution in the latter direction, see the remarks in Section 5 of this paper and [80,112]. It will be shown in Section 3 that by using a closed form expression of any nonexpansive mapping $T : \mathcal{H} \to \mathcal{H}$, not necessarily $T = P_C$, with $Fix(T) = C$, the proposed *hybrid steepest descent method* generates a sequence converging strongly to the uniquely existing solution of $VIP(C, \mathcal{F})$ when $\mathcal{F}$ is κ-Lipschitzian and η-strongly monotone over $T(\mathcal{H})$. This relaxation provides us with a method, based on sound mathematical foundations, that can be used to solve a significantly wider class of variational inequality problems.

C. Weak Topology and Convexity

Recall that a sequence $(x_n)_{n\geq 0} \subset \mathcal{H}$ is said to converge *strongly* to a point $x \in \mathcal{H}$ if $(\|x_n - x\|)_{n\geq 0}$ converges to 0, and to converge *weakly* to $x \in \mathcal{H}$ if $(\langle x_n - x, y\rangle)_{n\geq 0}$ converges to 0 for every $y \in \mathcal{H}$. If $(x_n)_{n\geq 0}$ converges strongly to x, then $(x_n)_{n\geq 0}$ converges weakly to x. The converse is true if $\mathcal{H}$ is finite dimensional.

A set $S \subset \mathcal{H}$ is said to be *weakly sequentially compact* if every sequence $(x_n)_{n\geq 0} \subset S$ contains a subsequence that converges weakly to a point in $\mathcal{H}$. In the following fact, (a) presents a necessary and sufficient condition for $S \subset \mathcal{H}$ to be weakly sequentially compact. (b) ensures that the weak limit, of a sequence $(x_n)_{n\geq 0}$ in a closed convex set $C \subset \mathcal{H}$, belongs to C.

Fact 2.11 (Weakly sequentially compact set and weak limit of a sequence [49,113]):

(a) *A set $S \subset \mathcal{H}$ is weakly sequentially compact if and only if S is bounded.*

(b) *Let a sequence $(x_n)_{n\geq 0} \subset \mathcal{H}$ converge weakly to some point $x_\infty \in \mathcal{H}$. Then there exists, for any $\varepsilon > 0$, a convex combination $\sum_{j=1}^N \alpha_j x_j$ $\left(\alpha_j \geq 0, \sum_{j=1}^N \alpha_j = 1\right)$ such that $\|x_\infty - \sum_{j=1}^N \alpha_j x_j\| \leq \varepsilon$.*

The following is a key fact in the proofs of the convergence of the *hybrid steepest descent method.*

Fact 2.12 (Opial's demiclosedness principle [82])*: Suppose $T : \mathcal{H} \to \mathcal{H}$ is a nonexpansive mapping. If a sequence $(x_n)_{n\geq 0} \subset \mathcal{H}$ converges weakly to $x \in \mathcal{H}$ and $(x_n - T(x_n))_{n\geq 0}$ converges strongly to $0 \in \mathcal{H}$, then x is a fixed point of T.*

D. Simple Properties of a Sequence

The hybrid steepest descent method presented in this paper uses a sequence $(\lambda_k)_{k\geq 1} \subset [0,1]$ to generate $(u_n)_{n\geq 0} \subset \mathcal{H}$ converging strongly to the solution of the variational inequality problem. Loosely speaking, when the monotone mapping is defined as the derivative of a convex function Θ, u_{n+1} is generated by the use of the steepest descent direction $\Theta'(T(u_n))$, of which effect is controlled by λ_{n+1}. The following fact will be used to design $(\lambda_k)_{k\geq 1}$.

Fact 2.13 *(Relation between a series and a product)*

(a) *For any $(\lambda_k)_{k\geq 1} \subset [0,1]$ and $n \geq m$,*

$$\sum_{i=m}^{n} \left\{ \lambda_i \prod_{j=i+1}^{n} (1-\lambda_j) \right\} \leq 1.$$

(b) *Suppose $(\lambda_k)_{k\geq 1} \subset [0,1)$ is a sequence which converges to 0. Then*

$$\sum_{k=1}^{\infty} \lambda_k = +\infty \Leftrightarrow \prod_{k=1}^{\infty}(1-\lambda_k) := \lim_{n\to\infty} \prod_{k=1}^{n}(1-\lambda_k) = 0.$$

Proof: (a) follows by induction, while (b) is well known (see for example [92, Theorem 15.5]).

(Q.E.D.)

3. HYBRID STEEPEST DESCENT METHOD

The *hybrid steepest descent method* for minimization of certain convex functions over the set of fixed points of nonexpansive mappings [108,109,47,111,112,80] has been developed by generalizing the results for approximation of the fixed point of nonexpansive mappings. To demonstrate simply the underlying ideas of the *hybrid steepest descent method*, we present it as algorithmic solutions to the variational inequality problem (VIP) defined over the fixed point set of nonexpansive mappings in a real Hilbert space.

The next lemma plays a central role in analyzing the convergence of the proposed algorithms. (d) is a generalization of a pioneering theorem by Browder [17].

Lemma 3.1 *Let $T : \mathcal{H} \to \mathcal{H}$ be a nonexpansive mapping with $Fix\,(T) \neq \emptyset$. Suppose that $\mathcal{F} : \mathcal{H} \to \mathcal{H}$ is κ-Lipschitzian and η-strongly monotone over $T(\mathcal{H})$. By using arbitrarily fixed $\mu \in (0, \frac{2\eta}{\kappa^2})$, define $T^{(\lambda)} : \mathcal{H} \to \mathcal{H}$ by*

$$T^{(\lambda)}(x) := T(x) - \lambda\mu\mathcal{F}\,(T(x)) \text{ for all } \lambda \in [0, 1]. \tag{7}$$

Then :

(a) *$\mathcal{G} := \mu\mathcal{F} - I$ satisfies*

$$\|\mathcal{G}(x) - \mathcal{G}(y)\|^2 \leq \{1 - \mu(2\eta - \mu\kappa^2)\}\|x - y\|^2 \text{ for all } x, y \in T(\mathcal{H}), \tag{8}$$

which implies that $\mathcal{G}$ is strictly contractive over $T(\mathcal{H})$. Moreover the obvious relation

$$0 < \tau := 1 - \sqrt{1 - \mu(2\eta - \mu\kappa^2)} \leq 1 \tag{9}$$

ensures that the closed ball

$$C_f := \left\{x \in \mathcal{H} \mid \|x - f\| \leq \frac{\|\mu\mathcal{F}(f)\|}{\tau}\right\} \tag{10}$$

is well defined for all $f \in Fix\,(T)$.

(b) *$T^{(\lambda)} : \mathcal{H} \to \mathcal{H}$ satisfies*

$$T^{(\lambda)}(C_f) \subset C_f \text{ for all } f \in Fix\,(T) \text{ and } \lambda \in [0, 1]. \tag{11}$$

In particular, $T(C_f) \subset C_f$ for all $f \in Fix\,(T)$.

(c) *$T^{(\lambda)} : \mathcal{H} \to \mathcal{H}$ (for all $\lambda \in (0, 1]$) is a strictly contractive mapping having its unique fixed point $\xi_\lambda \in \bigcap_{f \in Fix\,(T)} C_f$.*

(d) *Suppose that the sequence of parameters $(\lambda_n)_{n \geq 1} \subset (0, 1]$ satisfies $\lim_{n\to\infty} \lambda_n = 0$. Let ξ_n be the unique fixed point of $T_{<n>} := T^{(\lambda_n)}$; i.e., $\xi_n := \xi_{\lambda_n} \in Fix(T_{<n>})$ for all n. Then the sequence (ξ_n) converges strongly to the unique solution $u^* \in Fix(T)$ of $VIP(\mathcal{F}, Fix(T))$.*

Proof: The existence and uniqueness of the solution u^* of $VIP(\mathcal{F}, Fix(T))$ is guaranteed by Fact.2.1 (a) and Proposition 2.7 (a).
(a): By applying the κ-Lipschitz continuity and η-strong monotonicity of $\mathcal{F}$ over $T(\mathcal{H})$ to $\mathcal{G} := \mu\mathcal{F} - I$, we obtain

$$\begin{aligned} & \|\mathcal{G}(x) - \mathcal{G}(y)\|^2 \\ = \ & \mu^2\|\mathcal{F}(x) - \mathcal{F}(y)\|^2 - 2\mu\langle x - y, \mathcal{F}(x) - \mathcal{F}(y)\rangle + \|x - y\|^2 \\ \leq \ & \mu^2\kappa^2\|x - y\|^2 - 2\mu\eta\|x - y\|^2 + \|x - y\|^2 \\ = \ & \{1 - \mu(2\eta - \mu\kappa^2)\}\|x - y\|^2 \text{ for all } x, y \in T(\mathcal{H}). \end{aligned}$$

The remaining is obvious from $\mu \in (0, \frac{2\eta}{\kappa^2})$
(b): By the inequality (8) and the nonexpansiveness of T, it follows

$$\begin{aligned} & \|T^{(\lambda)}(x) - T^{(\lambda)}(y)\| \\ = \ & \|\{(1 - \lambda)\{T(x) - T(y)\} - \lambda\{\mathcal{G}(T(x)) - \mathcal{G}(T(y))\}\| \\ \leq \ & (1 - \lambda)\|T(x) - T(y)\| + \lambda\|\mathcal{G}(T(x)) - \mathcal{G}(T(y))\| \\ \leq \ & (1 - \lambda)\|T(x) - T(y)\| + \lambda\sqrt{1 - \mu(2\eta - \mu\kappa^2)}\|T(x) - T(y)\| \\ \leq \ & (1 - \lambda\tau)\|x - y\| \text{ for all } x, y \in \mathcal{H} \end{aligned} \tag{12}$$

The relation (12) yields

$$\begin{aligned} & \|T^{(\lambda)}(x) - f\| \leq \|T^{(\lambda)}(x) - T^{(\lambda)}(f)\| + \|T^{(\lambda)}(f) - f\| \\ \leq \ & (1 - \lambda\tau)\|x - f\| + \lambda\|\mu\mathcal{F}(f)\| \\ = \ & (1 - \lambda\tau)\|x - f\| + \lambda\tau\frac{\|\mu\mathcal{F}(f)\|}{\tau} \qquad \text{for all } x, y \in \mathcal{H} \text{ and } f \in Fix(T), \end{aligned}$$

which implies (11).
(c): By (12), $T^{(\lambda)} : \mathcal{H} \to \mathcal{H}$ and $T^{(\lambda)}P_{C_f} : \mathcal{H} \to \mathcal{H}$ are strictly contractive for all $\lambda \in (0, 1]$. Therefore, *Banach-Picard fixed point theorem* ensures that both $T^{(\lambda)}$ and $T^{(\lambda)}P_{C_f}$ have their own unique fixed points. Since $Fix(P_{C_f}) = C_f$ and $Fix(T^{(\lambda)}P_{C_f}) \subset C_f$ from (b), it follows that the unique fixed point of $T^{(\lambda)}P_{C_f}$ is nothing but the unique fixed point ξ_λ of $T^{(\lambda)}$, which implies $\xi_\lambda \in C_f$ for all $f \in Fix(T)$.
(d): It suffices to show that every subsequence $(\xi_{m(j)})_{j\geq 1}$ of $(\xi_n)_{n\geq 1}$ has its subsequence which converges strongly to u^*.

Since C_f, for any $f \in Fix(T)$, is closed bounded and convex, it is *weakly sequentially compact* (see Fact 2.11). We may assume (by passing to further subsequence if necessary) that $(\xi_{m(j)})$ converges weakly to a point $v \in C_f$. To simplify, we use the notation $v_j := \xi_{m(j)}$.

Claim 1:

$$T(v) = v. \tag{13}$$

For an arbitrarily fixed $f \in Fix(T)$ and every $x \in C_f$, it follows that

$$\|\mathcal{F}(T(x)) - \mathcal{F}(f)\| \leq \kappa\|T(x) - f\| \leq \kappa\|x - f\| \leq \kappa\frac{\|\mu\mathcal{F}(f)\|}{\tau}, \tag{14}$$

which implies the boundedness of $(\mathcal{F}T(x))_{x\in C_f}$ and hence the boundedness of $(\mathcal{F}T(v_j))_{j\geq 1}$ by $v_j \in C_f$ [from (c)]. Obviously, by $T_{<m(j)>}(v_j) = v_j$ and $\lim_{n\to\infty}\lambda_n = 0$, we obtain

$$\|T(v_j)-v_j\| = \|T(v_j)-T(v_j)+\lambda_{m(j)}\mu\mathcal{F}(T(v_j))\| = \lambda_{m(j)}\mu\|\mathcal{F}(T(v_j))\| \to 0 \text{ as } j\to\infty. \quad (15)$$

Therefore, by Fact 2.12 and (15), we conclude $Tv = v$ as promised.

Again, by the weak convergence of (v_j) to $v \in Fix(T)$, to prove $\lim_{j\to\infty} v_j = u^*$, and hence to complete the proof of the theorem, it suffices to show

Claim 2:

$$\|v_j - u^*\|^2 \leq \frac{\mu}{\tau}\langle\mathcal{F}(u^*), v - v_j\rangle. \quad (16)$$

By $T_{<m(j)>}v_j = v_j$ and $Tu^* = u^*$, simple observations yield that

$$(1-\lambda_{m(j)})(I-T)(v_j) + \lambda_{m(j)}(I+\mathcal{G}T)(v_j) = 0 \quad (17)$$

and

$$(1-\lambda_{m(j)})(I-T)(u^*) + \lambda_{m(j)}(I+\mathcal{G}T)(u^*) = \lambda_{m(j)}(I+\mathcal{G}T)(u^*). \quad (18)$$

Subtracting (18) from (17) and taking the inner product with $v_j - u^*$, we obtain

$$\begin{aligned} & (1-\lambda_{m(j)})\langle v_j - u^* - \{T(v_j) - T(u^*)\}, v_j - u^*\rangle \\ & +\lambda_{m(j)}\langle v_j - u^* + \{\mathcal{G}T(v_j) - \mathcal{G}T(u^*)\}, v_j - u^*\rangle \\ = \;& -\lambda_{m(j)}\langle u^* + \mathcal{G}T(u^*), v_j - u^*\rangle. \end{aligned} \quad (19)$$

By Schwarz inequality, the first term on the left hand side of (19) is bounded as

$$\begin{aligned} \langle v_j - u^* - \{T(v_j) - T(u^*)\}, v_j - u^*\rangle &\geq \|v_j - u^*\|^2 - \|T(v_j) - T(u^*)\|\|v_j - u^*\| \\ &\geq \|v_j - u^*\|^2 - \|v_j - u^*\|^2 = 0. \end{aligned} \quad (20)$$

Again, by Schwarz inequality and (8), the second term is also bounded as

$$\begin{aligned} \langle (v_j - u^*) + \{\mathcal{G}T(v_j) - \mathcal{G}T(u^*)\}, v_j - u^*\rangle &\geq \|v_j - u^*\|^2 - \|\mathcal{G}T(v_j) - \mathcal{G}T(u^*)\|\|v_j - u^*\| \\ &\geq \tau\|v_j - u^*\|^2. \end{aligned} \quad (21)$$

Moreover, by noting (13) and the variational inequality satisfied by u^*, for the right hand side of (19), we obtain

$$\begin{aligned} \langle u^* + \mathcal{G}T(u^*), v_j - u^*\rangle &= \langle u^* + \mathcal{G}(u^*), v_j - u^*\rangle \\ &= \mu\langle\mathcal{F}(u^*), v_j - u^*\rangle \\ &= \mu\left(\langle\mathcal{F}(u^*), v_j - v\rangle + \langle\mathcal{F}(u^*), v - u^*\rangle\right) \\ &\geq \mu\langle\mathcal{F}(u^*), v_j - v\rangle. \end{aligned} \quad (22)$$

Finally, applying (20)–(22) to (19), we get (16). (Q.E.D.)

Now, by using a parameter sequence $(\lambda_n)_{n\geq 1}$ introduced by P.L Lions [75], we can prove the first main theorem on the hybrid steepest descent method. This is a generalization of the fixed point theorems for $\mathcal{F}(x) := x$ [60] and for $\mathcal{F}(x) := x - a$ [75].

Theorem 3.2 (Hybrid steepest descent method for $VIP(\mathcal{F}, Fix(T)))$: *Let $T : \mathcal{H} \to \mathcal{H}$ be a nonexpansive mapping with $Fix(T) \neq \emptyset$. Suppose that a mapping $\mathcal{F} : \mathcal{H} \to \mathcal{H}$ is κ-Lipschitzian and η-strongly monotone over $T(\mathcal{H})$. Then with any $u_0 \in \mathcal{H}$, any $\mu \in (0, \frac{2\eta}{\kappa^2})$ and any sequence $(\lambda_n)_{n\geq 1} \subset (0, 1]$ satisfying (L1) $\lim_{n\to+\infty} \lambda_n = 0$, (L2) $\sum_{n\geq 1} \lambda_n = +\infty$, (L3) $\lim_{n\to+\infty}(\lambda_n - \lambda_{n+1})\lambda_{n+1}^{-2} = 0$, the sequence $(u_n)_{n\geq 0}$ generated by*

$$u_{n+1} := T^{(\lambda_{n+1})}(u_n) := T(u_n) - \lambda_{n+1}\mu\mathcal{F}(T(u_n)) \tag{23}$$

converges strongly to the uniquely existing solution of the $VIP(\mathcal{F}, Fix(T))$: find $u^* \in Fix(T)$ such that $\langle v - u^*, \mathcal{F}(u^*)\rangle \geq 0$ for all $v \in Fix(T)$. (Note: An example of sequence $(\lambda_n)_{n\geq 1}$ satisfying (L1)–(L3) is $\lambda_n = 1/n^\varrho$ for $0 < \varrho < 1$.)

Proof: By (d) of Lemma 3.1, to prove the theorem, it suffices to show that

$$\lim_{n\to\infty} \|u_n - \xi_n\| = 0, \tag{24}$$

where ξ_n is the uniquely existing fixed point of $T_{<n>} := T^{(\lambda_n)}$ for $n = 1, 2, \ldots$.

By (12), we get

$$\begin{aligned} \|u_n - \xi_n\| &= \|T_{<n>}(u_{n-1}) - T_{<n>}(\xi_n)\| \leq (1 - \lambda_n\tau)\|u_{n-1} - \xi_n\| \\ &\leq (1 - \lambda_n\tau)(\|u_{n-1} - \xi_{n-1}\| + \|\xi_{n-1} - \xi_n\|) \\ &\leq (1 - \lambda_n\tau)\|u_{n-1} - \xi_{n-1}\| + \|\xi_{n-1} - \xi_n\|. \end{aligned} \tag{25}$$

Similarly, by $T_{<m>}(\xi_m) = \xi_m$ for all $m \geq 1$ and simple observations in terms of $\mathcal{G} := \mu\mathcal{F} - I$, it follows that

$$\begin{aligned} &\|\xi_l - \xi_{l-1}\| \\ = \ &\|T_{<l>}(\xi_l) - T_{<l-1>}(\xi_{l-1})\| \\ = \ &\|(1 - \lambda_l)T(\xi_l) - \lambda_l\mathcal{G}T(\xi_l) - (1 - \lambda_{l-1})T(\xi_{l-1}) + \lambda_{l-1}\mathcal{G}T(\xi_{l-1})\| \\ = \ &\|(1 - \lambda_l)\{T(\xi_l) - T(\xi_{l-1})\} \\ &-\lambda_l\{\mathcal{G}T(\xi_l) - \mathcal{G}T(\xi_{l-1})\} + (\lambda_{l-1} - \lambda_l)\{T(\xi_{l-1}) + \mathcal{G}T(\xi_{l-1})\}\| \\ \leq \ &(1 - \lambda_l)\|T(\xi_l) - T(\xi_{l-1})\| + \lambda_l\|\mathcal{G}T(\xi_l) - \mathcal{G}T(\xi_{l-1})\| + |\lambda_{l-1} - \lambda_l|\|\mu\mathcal{F}T(\xi_{l-1})\| \\ \leq \ &(1 - \lambda_l)\|\xi_l - \xi_{l-1}\| + \lambda_l\sqrt{1 - \mu(2\eta - \mu\kappa^2)}\|\xi_l - \xi_{l-1}\| + |\lambda_{l-1} - \lambda_l|\|\mu\mathcal{F}T(\xi_{l-1})\|, \end{aligned}$$

for all $l \geq 2$, which implies

$$\lambda_l\tau\|\xi_l - \xi_{l-1}\| \leq |\lambda_{l-1} - \lambda_l|\|\mu\mathcal{F}T(\xi_{l-1})\| \text{ for all } l \geq 2.$$

By this inequality and the boundedness of $(\mu\mathcal{F}T(\xi_{l-1}))_{l\geq 2}$ [from Lemma 3.1(c) and (14)], there exists some $c > 0$ such that for all $l \geq 2$

$$\frac{\|\xi_l - \xi_{l-1}\|}{\lambda_l} \leq \frac{\|\mu\mathcal{F}T(\xi_{l-1})\|}{\tau}\frac{|\lambda_{l-1} - \lambda_l|}{\lambda_l^2} \leq c\frac{|\lambda_l - \lambda_{l-1}|}{{\lambda_l}^2}. \tag{26}$$

Now let

$$M_p := \frac{c}{\tau}\sup_{l\geq p}\frac{|\lambda_l - \lambda_{l-1}|}{{\lambda_l}^2}.$$

Then, by (L3) and (26), it follows that

$$M_p \to 0 \text{ as } p \to \infty \tag{27}$$

and

$$\|\xi_n - \xi_{n-1}\| \leq \lambda_n \tau M_p \text{ for all } n \geq p. \tag{28}$$

Applying (28) to (25), we obtain

$$\|u_n - \xi_n\| \leq (1 - \lambda_n \tau)\|u_{n-1} - \xi_{n-1}\| + \lambda_n \tau M_p$$

and thus, by induction, for all $n \geq p+1$

$$\|u_n - \xi_n\| \leq \|u_p - \xi_p\| \prod_{i=p+1}^{n} (1 - \lambda_i \tau) + M_p \left[\sum_{i=p+1}^{n} \left\{ \lambda_i \tau \prod_{j=i+1}^{n} (1 - \lambda_j \tau) \right\} \right].$$

Moreover, by Fact 2.13(a), it follows that

$$\|u_n - \xi_n\| \leq \|u_p - \xi_p\| \prod_{i=p+1}^{n} (1 - \lambda_i \tau) + M_p. \tag{29}$$

Furthermore, applying $\prod_{i=p+1}^{\infty}(1 - \lambda_i \tau) = 0$ (by (L2) and Fact 2.13 (b)) to (29), we obtain

$$\limsup_{n\to\infty} \|u_n - \xi_n\| \leq M_p \text{ for all } p.$$

Finally, by letting $p \to \infty$, (27) yields $\limsup_{n\to\infty} \|u_n - \xi_n\| = 0$, which completes the proof.

(Q.E.D.)

The next theorem is a generalization of fixed point theorems that were developed in order to minimize $\Theta(x) = \|x - a\|^2$ over $\bigcap_{i=1}^{m} Fix(T_i) \neq \emptyset$ for $N = 1$ by Wittmann [106] and for general finite N by Bauschke [9].

Theorem 3.3 *Let $T_i : \mathcal{H} \to \mathcal{H}$ $(i = 1, \ldots, N)$ be nonexpansive mappings with $F := \bigcap_{i=1}^{N} Fix\,(T_i) \neq \emptyset$ and*

$$F = Fix\,(T_N \cdots T_1) = Fix\,(T_1 T_N \cdots T_3 T_2) = \cdots = Fix\,(T_{N-1} T_{N-2} \cdots T_1 T_N). \tag{30}$$

Suppose that a mapping $\mathcal{F} : \mathcal{H} \to \mathcal{H}$ is κ-Lipschitzian and η-strongly monotone over $\Delta := \bigcup_{i=1}^{N} T_i(\mathcal{H})$. Then with any $u_0 \in \mathcal{H}$, any $\mu \in (0, \frac{2\eta}{\kappa^2})$ and any sequence $(\lambda_n)_{n\geq 1} \subset [0,1]$ satisfying (B1) $\lim_{n\to+\infty} \lambda_n = 0$, (B2) $\sum_{n\geq 1} \lambda_n = +\infty$, (B3) $\sum_{n\geq 1} |\lambda_n - \lambda_{n+N}| < +\infty$, the sequence $(u_n)_{n\geq 0}$ generated by

$$u_{n+1} := T_{[n+1]}^{(\lambda_{n+1})}(u_n) := T_{[n+1]}(u_n) - \lambda_{n+1}\mu \mathcal{F}\left(T_{[n+1]}(u_n)\right) \tag{31}$$

converges strongly to the uniquely existing solution of the $VIP(\mathcal{F}, F)$: find $u^* \in F$ such that $\langle v - u^*, \mathcal{F}(u^*)\rangle \geq 0$ for all $v \in F$, where $[\cdot]$ is the modulo N function defined by

$$[i] := [i]_N := \{i - kN \mid k = 0, 1, 2, \ldots\} \cap \{1, 2, \ldots, N\}.$$

(Note: An example of sequence $(\lambda_n)_{n\geq 1}$ satisfying (B1)–(B3) is $\lambda_n = 1/n$.)

Proof: Note that the existence and uniqueness of the solution u^* of $VIP(\mathcal{F}, F)$ is guaranteed by Fact.2.1 (a) and Proposition 2.7 (a). Let us first assume that

$$u_0 \in C_{u^*} := \left\{ x \in \mathcal{H} \mid \|x - u^*\| \le \frac{\|\mu\mathcal{F}(u^*)\|}{\tau} \right\},$$

where $0 < \tau := 1 - \sqrt{1 - \mu(2\eta - \mu\kappa^2)} \le 1$ (see Lemma 3.1(a)). We will later show that the general case when $u_0 \notin C_{u^*}$ can be reduced to this case (see also Bauschke [9] for a similar reduction).

In this case, Lemma 3.1(b) implies that

$$(u_n)_{n\ge0} \subset C_{u^*} \text{ and } \left(T_{[n+1]}(u_n)\right)_{n\ge0} \subset C_{u^*} \tag{32}$$

(since $\lambda = 0$ implies $T^{(0)}_{[n+1]} = T_{[n+1]}$) for all $n \ge 0$. Hence both sequences (u_n) and $\left(T_{[n+1]}(u_n)\right)$ are bounded.

By (8) of Lemma 3.1 and the nonexpansiveness of $T_{[n+1]}$, it follows that

$$\begin{aligned} \left\|\mathcal{G}T_{[n+1]}(u_n) - \mathcal{G}(u^*)\right\| &\le (1-\tau)\left\|T_{[n+1]}(u_n) - u^*\right\| \le (1-\tau)\|u_n - u^*\| \\ &\le \frac{1-\tau}{\tau}\|\mu\mathcal{F}(u^*)\| \text{ for all } n, \end{aligned}$$

where (32) was used for the last inequality. This and (14) respectively imply the boundedness of $\left(\mathcal{G}T_{[n+1]}(u_n)\right)\big)_{n\ge0}$ and $\left(\mathcal{F}T_{[n+1]}(u_n)\right)\big)_{n\ge0}$.

By (31), (B1) and the boundedness of $\left(\mathcal{F}T_{[n+1]}(u_n)\right)\big)_{n\ge0}$, it is easy to verify that

$$u_{n+1} - T_{[n+1]}(u_n) \to 0 \text{ as } n \to \infty. \tag{33}$$

We need to prove the following three claims.

Claim 1. $u_{n+N} - u_n \to 0$ as $n \to \infty$. (34)

By (31) and the fact that $T_{[n+N]} = T_{[n]}$, it follows that

$$\begin{aligned} u_{n+N} - u_n &= T^{(\lambda_{n+N})}_{[n+N]}(u_{n+N-1}) - T^{(\lambda_n)}_{[n]}(u_{n-1}) \\ &= T^{(\lambda_{n+N})}_{[n+N]}(u_{n+N-1}) - T^{(\lambda_{n+N})}_{[n+N]}(u_{n-1}) - \left(T^{(\lambda_n)}_{[n]}(u_{n-1}) - T^{(\lambda_{n+N})}_{[n+N]}(u_{n-1})\right) \\ &= T^{(\lambda_{n+N})}_{[n+N]}(u_{n+N-1}) - T^{(\lambda_{n+N})}_{[n+N]}(u_{n-1}) + (\lambda_n - \lambda_{n+N})\mu\mathcal{F}T_{[n]}(u_{n-1}). \end{aligned} \tag{35}$$

Again, by the boundedness of $(u_n)_{n\ge0}$ and $\left(\mu\mathcal{F}(T_{[n+1]}(u_n))\right)_{n\ge0}$, there exists some $c > 0$ satisfying

$$\left.\begin{aligned} &\|u_{n+N} - u_n\| \le c\tau \\ &\left\|\mu\mathcal{F}T_{[n]}(u_{n-1})\right\| \le c\tau \end{aligned}\right\} \text{ for all } n \ge 1. \tag{36}$$

Applying (12) and (36) to (35), we obtain

$$\|u_{n+N} - u_n\| \le c\tau|\lambda_{n+N} - \lambda_n| + (1 - \lambda_{n+N}\tau)\|u_{n+N-1} - u_{n-1}\|$$

and thus, by induction,

$$\begin{aligned}\|u_{n+N}-u_n\| &\le c\sum_{k=m+1}^{n}\tau|\lambda_{k+N}-\lambda_k| \\ &\quad +\|u_{m+N}-u_m\|\prod_{k=m+1}^{n}(1-\lambda_{k+N}\tau) \text{ for all } n>m\ge 0. \end{aligned} \tag{37}$$

Moreover, since the assumption (B2) and Fact 2.13(b) ensure $\prod_{k=m+1}^{\infty}(1-\lambda_{k+N}\tau)=0$, it follows by (37) and (36) that

$$\begin{aligned}\limsup_{n\to\infty}\|u_{n+N}-u_n\| &\le c\tau\sum_{k=m+1}^{\infty}|\lambda_{k+N}-\lambda_k| + c\tau\prod_{k=m+1}^{\infty}(1-\lambda_{k+N}\tau) \\ &= c\tau\sum_{k=m+1}^{\infty}|\lambda_{k+N}-\lambda_k|. \end{aligned} \tag{38}$$

Finally, by letting $m\to+\infty$, (38) and (B3) yield (34).

Claim 2. $u_n - T_{[n+N]}\cdots T_{[n+1]}(u_n)\to 0$ as $n\to\infty$. (39)

By (33),

$$u_{n+N-k}-T_{[n+N-k]}(u_{n+N-k-1})\to 0 \text{ as } n\to\infty \text{ for } k=0,\ldots,N-1. \tag{40}$$

Applying the nonexpansive mapping $T_{[n+N]}T_{[n+N-1]}\cdots T_{[n+N-k+1]}$ for $k=1,\ldots,N-1$, we get

$$T_{[n+N]}T_{[n+N-1]}\cdots T_{[n+N-k+1]}(u_{n+N-k})-T_{[n+N]}T_{[n+N-1]}\cdots T_{[n+N-k]}(u_{n+N-k-1})\to 0$$

as $n\to\infty$ for $k=1,\ldots,N-1$. Finally, adding these sequences (for $k=1,\ldots,N-1$) and the sequence of (40) (for $k=0$) yields

$$u_{n+N}-T_{[n+N]}\cdots T_{[n+1]}(u_n)\to 0 \text{ as } n\to\infty.$$

Claim 2 now follows by (34).

Claim 3. $\limsup_{n\to\infty}\langle T_{[n+1]}(u_n)-u^*,-\mathcal{F}(u^*)\rangle\le 0.$ (41)

Let (u_{n_j}) be a subsequence of (u_n) such that

$$\lim_{j\to\infty}\langle T_{[n_j+1]}(u_{n_j})-u^*,-\mathcal{F}(u^*)\rangle=\limsup_{n\to\infty}\langle T_{[n+1]}(u_n)-u^*,-\mathcal{F}(u^*)\rangle \tag{42}$$

and $[n_j+1]=i$ for some fixed $i\in\{1,2,\ldots,N\}$ and all $j\ge 0$. The existence of such a subsequence $(u_{n_j})_{j\ge 0}$ follows by $N<\infty$. Moreover, since C_{u^*} is closed bounded and convex, it is weakly sequentially compact (due to Fact 2.11). We may thus assume (by passing to a further subsequence if necessary) that

$$\left.\begin{array}{l}(u_{n_j+1})_{j\ge 0} \text{ converges weakly to some } \hat{u}\in C_{u^*} \text{ and}\\ [n_j+1]=i \text{ for some } i\in\{1,2,\ldots,N\} \text{ and all } j\ge 0.\end{array}\right\} \tag{43}$$

By (39), it follows that

$$u_{n_j+1} - T_{[i+N]} \cdots T_{[i+1]}(u_{n_j+1}) = u_{n_j+1} - T_{[n_j+1+N]} \cdots T_{[n_j+2]}(u_{n_j+1}) \to 0 \tag{44}$$

as $j \to \infty$. Applying Fact 2.12 to (43) and (44), we obtain that

$$\hat{u} \in Fix\,(T_{[i+N]} \cdots T_{[i+1]}) = F. \tag{45}$$

Therefore, by (42), (33), (43), (45) and the variational inequality satisfied by u^*, we deduce that

$$\begin{aligned} & \limsup_{n\to\infty} \langle T_{[n+1]}(u_n) - u^*, -\mathcal{F}(u^*)\rangle \\ = \; & \lim_{j\to\infty} \langle T_{[n_j+1]}(u_{n_j}) - u_{n_j+1}, -\mathcal{F}(u^*)\rangle + \lim_{j\to\infty} \langle u_{n_j+1} - u^*, -\mathcal{F}(u^*)\rangle \\ = \; & \langle \hat{u} - u^*, -\mathcal{F}(u^*)\rangle \le 0. \end{aligned}$$

This verifies Claim 3.

The final step of the proof proceeds as follows. By (31), it follows that

$$\begin{aligned} \|u_{n+1} - u^*\|^2 \;=\; & \left\| T^{(\lambda_{n+1})}_{[n+1]}(u_n) - T^{(\lambda_{n+1})}_{[n+1]}(u^*) \right\|^2 + 2\lambda_{n+1}\mu \left\langle T_{[n+1]}(u_n) - u^*, -\mathcal{F}(u^*)\right\rangle \\ & + \lambda_{n+1}^2 \mu^2 \left[\|\mathcal{F}(u^*)\|^2 + 2\left\langle \mathcal{F}\left(T_{[n+1]}(u_n)\right) - \mathcal{F}(u^*), \mathcal{F}(u^*)\right\rangle \right]. \end{aligned} \tag{46}$$

By (32), the boundedness of $\left(\mathcal{F}T_{[n+1]}(u_n)\right)_{n\ge 0}$, and Claim 3, for an arbitrary $\varepsilon > 0$, there exists some $n_\varepsilon > 0$ such that for all $n \ge n_\varepsilon$,

$$2\mu \left\langle T_{[n+1]}(u_n) - u^*, -\mathcal{F}(u^*)\right\rangle < \frac{\varepsilon}{2}\tau$$

and

$$\lambda_{n+1}\mu^2 \left[\|\mathcal{F}(u^*)\|^2 + 2\left\langle \mathcal{F}\left(T_{[n+1]}(u_n)\right) - \mathcal{F}(u^*), \mathcal{F}(u^*)\right\rangle \right] < \frac{\varepsilon}{2}\tau.$$

Applying these and (12) to (46), we obtain

$$\begin{aligned} \|u_{n+1} - u^*\|^2 \;\le\; & \lambda_{n+1}\tau\varepsilon + (1 - \lambda_{n+1}\tau)^2 \|u_n - u^*\|^2 \\ \le\; & \lambda_{n+1}\tau\varepsilon + (1 - \lambda_{n+1}\tau) \|u_n - u^*\|^2. \end{aligned}$$

Using induction and Fact 2.13(a), it follows that for all $n \ge n_\varepsilon$,

$$\begin{aligned} \|u_{n+1} - u^*\|^2 \;\le\; & \varepsilon \sum_{i=n_\varepsilon+1}^{n+1} \left\{ \lambda_i\tau \prod_{j=i+1}^{n+1} (1 - \lambda_j\tau) \right\} + \|u_{n_\varepsilon} - u^*\|^2 \prod_{k=n_\varepsilon+1}^{n+1} (1 - \lambda_k\tau) \\ \le\; & \varepsilon + \|u_{n_\varepsilon} - u^*\|^2 \prod_{k=n_\varepsilon+1}^{n+1} (1 - \lambda_k\tau). \end{aligned}$$

Finally, using (B2) and Fact 2.13(b), we get $\limsup_{n\to\infty} \|u_n - u^*\|^2 \le \varepsilon$. This completes the proof of the theorem for the special case when $u_0 \in C_{u^*}$.

For the general case, assume that $u_0 \in \mathcal{H}$ is arbitrarily given. Let (s_n) be another sequence generated by (31) for some $s_0 \in C_{u^*}$. Since the proof for the special case guarantees that $s_n \to u^*$ as $n \to +\infty$, it suffices to show $\lim_{n\to\infty} \|u_n - s_n\| = 0$. By (31) (or (12)), it follows that

$$\|u_{n+1} - s_{n+1}\| \leq (1 - \lambda_{n+1}\tau)\|u_n - s_n\|$$

and hence, by induction, for all $n \geq 0$,

$$\|u_n - s_n\| \leq \|u_0 - s_0\| \prod_{k=1}^{n}(1 - \lambda_k \tau).$$

Finally, by (B2) and Fact 2.13(b), we obtain $\lim_{n\to\infty} \|u_n - s_n\| = 0$, which completes the proof. (Q.E.D.)

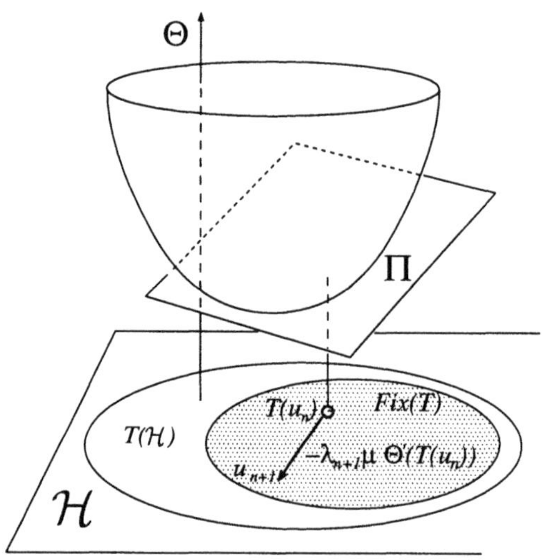

Figure 1. Sequence generation by the hybrid steepest descent method

Remark 3.4

(a) *Suppose that we need to minimize a convex function Θ over $Fix(T)$, where $T : \mathcal{H} \to \mathcal{H}$ is nonexpansive and the derivative $\Theta' : \mathcal{H} \to \mathcal{H}$ is κ-Lipschitzian and η-strongly monotone over $T(\mathcal{H})$. Then the hybrid steepest descent method (23) or (31) [for $N = 1$], for $\mathcal{F} = \Theta'$, generates a sequence $(u_n)_{n\geq 0}$ converging strongly to the unique minimizer of Θ over $Fix(T)$. In the formulae, a new point u_{n+1} is generated by combining the operation of nonexpansive mapping T and a small change in the steepest descent direction as illustrated in Fig.1, where Π is the tangent plane of* *the surface* $\{(x, \Theta(x)) \in \mathcal{H} \times \mathbb{R} \mid x \in \mathcal{H}\}$ *at* $(T(u_n), \Theta(T(u_n))) \in \mathcal{H} \times \mathbb{R}$. *Intuitively, the conditions [(L1) and (L2)], or [(B1) and (B2)] on* $(\lambda_n)_{n \geq 1}$ *ensure that the effect of the steepest descent direction in the formulae declines but influences forever.*

(b) *In Theorems 3.2 and 3.3, any choice of the starting point* $u_0 \in \mathcal{H}$ *does not violate the strong convergence to the solution. Moreover even if the nonnegative sequence* $(\lambda_n)_{n \geq 1}$ *is not restricted in* $[0,1]$, *the sequence* $(\lambda_n \mu)_{n \geq N_0}$ *for sufficiently large* $N_0 > 0$ *satisfies essentially the same conditions as [(L1)–(L3)], or [(B1)–(B3)] . These facts yield slightly simpler expressions, without explicitly using* μ, *of the hybrid steepest descent method as*

$$u_{n+1} := T(u_n) - \lambda_{n+1} \mathcal{F}(T(u_n)) \text{ or } u_{n+1} := T_{[n+1]}(u_n) - \lambda_{n+1} \mathcal{F}\left(T_{[n+1]}(u_n)\right).$$

As seen from Fact 2.1, there exist many nonexpansive mappings having the same fixed point set. The key for the successful applications of Theorem 3.2 or Theorem 3.3 to the variational inequality problem over a nonempty closed convex set $C \subset \mathcal{H}$ is the construction of a simple nonexpansive mapping T or a set of simple nonexpansive mappings $\{T_i\}_{i \in J}$ for some $J \subset \mathbb{Z}$ satisfying the conditions in the theorems as well as $C = Fix(T)$ or $C = \bigcap_{i \in J} Fix(T_i)$.

By applying Fact 2.1 to Theorem 3.2 or Theorem 3.3 (for $N = 1$), the following generalizations of [75,9,39] are immediately deduced.

Corollary 3.5 : *Let* $T_i : \mathcal{H} \to \mathcal{H}$ $(i = 1, \ldots, N)$ *be nonexpansive mappings with* $F := \bigcap_{i=1}^{m} Fix(T_i) \neq \emptyset$ *and*

$$F = Fix(T_m T_{m-1} \cdots T_1) = Fix(T_1 T_m \cdots T_3 T_2) = \cdots = Fix(T_{m-1} T_{m-2} \cdots T_1 T_m).$$

Let $T := T_m T_{m-1} \cdots T_1$. *Suppose that a mapping* $\mathcal{F} : \mathcal{H} \to \mathcal{H}$ *is* κ*-Lipschitzian and* η*-strongly monotone over* $T(\mathcal{H})$. *Then with any* $u_0 \in \mathcal{H}$, *any* $\mu \in (0, \frac{2\eta}{\kappa^2})$ *and any sequence* $(\lambda_n)_{n \geq 1}$ *satisfying (L1)–(L3) in* $(0,1]$ *or (B1)–(B3) in* $[0,1]$ *(for* $N = 1$*), the sequence* $(u_n)_{n \geq 0}$ *generated by*

$$u_{n+1} := T(u_n) - \lambda_{n+1} \mu \mathcal{F}(T(u_n)) \tag{47}$$

converges strongly to the uniquely existing solution of the $VIP(\mathcal{F}, F)$: *find* $u^* \in F$ *such that* $\langle v - u^*, \mathcal{F}(u^*) \rangle \geq 0$ *for all* $v \in F$.

Corollary 3.6 : *Let* $(T_i)_{i \in J}$ *be a countable family of firmly nonexpansive mappings on* $\mathcal{H}$, *and* $F_\infty := \bigcap_{i \in J} Fix(T_i) \neq \emptyset$. *Define* $T := \sum_{i \in J} w_i T_i$ *for any sequence* $(w_i)_{i \in J} \subset (0,1]$ *satisfying* $\sum_{i \in J} w_i = 1$. *Suppose that a mapping* $\mathcal{F} : \mathcal{H} \to \mathcal{H}$ *is* κ*-Lipschitzian and* η*-strongly monotone over* $T(\mathcal{H})$. *Then with any* $u_0 \in \mathcal{H}$, *any* $\mu \in (0, \frac{2\eta}{\kappa^2})$ *and any sequence* $(\lambda_n)_{n \geq 1}$ *satisfying (L1)–(L3) in* $(0,1]$ *or (B1)–(B3) in* $[0,1]$ *(for* $N = 1$*), the sequence* $(u_n)_{n \geq 0}$ *generated by*

$$u_{n+1} := T(u_n) - \lambda_{n+1} \mu \mathcal{F}(T(u_n)) \tag{48}$$

converges strongly to the uniquely existing solution of the $VIP(\mathcal{F}, F_\infty)$: *find* $u^* \in F_\infty$ *such that* $\langle v - u^*, \mathcal{F}(u^*) \rangle \geq 0$ *for all* $v \in F_\infty$.

4. VARIATIONAL INEQUALITY PROBLEM OVER GENERALIZED CONVEX FEASIBLE SET

A. Projection algorithms for best approximation and convex feasibility problems

The best approximation problem of finding the projection of a given point in a Hilbert space onto the (nonempty) intersection $C = \bigcap_{i=1}^{m} C_i$ of a finite number of closed convex sets C_i $(i = 1, 2\dots, m)$ arises in many branches of applied mathematics, the physical and computer sciences, and engineering. One frequently employed approach to solving this problem is algorithmic. The idea of this approach is to generate a sequence of points which converges to the solution of the problem by using projections onto the *individual* sets C_i. Each individual set C_i is usually assumed to be 'simple' in the sense that the projection onto C_i can be computed easily.

One of the earliest successful iterative algorithms by this approach was the method of *alternating projections* due to von Neumann [79] for two closed subspaces. Dykstra proposed a modified version of von Neumann's *alternating projections algorithm* that solves the problem for closed convex cones C_i in a Euclidean space [50], and Boyle and Dykstra [12] later showed that Dykstra's algorithm solves the problem for general closed convex sets C_i in any Hilbert space. This algorithm was also rediscovered later, from the viewpoint of duality, by Han [61]. The rate of convergence of Dykstra's algorithm has been investigated by Deutsch and Hundal [45,46], who also presented two extended versions of the algorithm. One of these handles the intersection of a *countable* family of convex sets and the other handles a *random ordering* of projections onto the individual sets [65]. A novel important direction on the extensions of the Dykstra's algorithm is found in [31,11, 15] where *Dykstra's algorithm with Bregman projections* is developed to find the Bregman projection [14,28,32] of a point onto the nonempty intersection of finitely many closed convex sets. A close relationship between the algorithm and the Dual Block Coordinate Ascent (DBCA) methods [105] was also discovered [11]. On the other hand, Iusem and De Pierro [67] applied Pierra's *product space formulation* [84,85] to Dykstra's algorithm and proposed a parallel scheme for the quadratic programming problem in Euclidean space, where an inner product corresponding to the quadratic objective was employed. This algorithm was generalized to infinite-dimensional Hilbert space by Bauschke and Borwein [7]. Crombez [44] proposed a parallel algorithm that allows variable weights and relaxation parameters. In [38] a parallel algorithm was proposed by applying Pierra's idea to P.L. Lions' formula [75].

A remarkable property of these parallel algorithms is that, under certain conditions, they generate sequences converging to the projection onto

$$\mathcal{H}_\Phi := \{u \in \mathcal{H} \mid \Phi(u) = \inf \Phi(\mathcal{H})\}, \tag{49}$$

for $\Phi(x) := \sum_{i=1}^{m} w_i d(x, C_i)^2$ with $w_i > 0$ for $i = 1, \dots, m$, provided that $\mathcal{H}_\Phi$ is a nonempty closed convex set. Since the function $\Phi : \mathcal{H} \to \mathbb{R}$ is a continuous convex function, $\mathcal{H}_\Phi$ is nonempty, closed, convex and bounded if at least one C_i is bounded (see [38, Prop.7] or [119, Prop.38.15]).

In estimation or design problems, since each closed convex set usually represents *a priori* knowledge or a design specification, it is often sufficient to solve the problem of finding a point in the intersection $C = \bigcap_{i=1}^{m} C_i$, not necessarily the projection onto C. This general problem, including the above best approximation problem as its special important example, is referred to as the *convex feasibility problem* [37,8,32,100] and many algorithmic solutions have been developed based on the use of convex projections P_{C_i} or on certain other key operations, for example *Bregman projections* or *subgradient projections* (see for example [13,59,115,114,37,8,10,24,31,32,100,21,22,40] and references therein). In practical situations, some collection of convex sets C_i $(i = 1, 2, \ldots, m)$ may often become inconsistent, (i.e., $C := \bigcap_{i=1}^{m} C_i = \emptyset$) because of the low reliabilities due to noisy measurements or because of the tightness of design specifications [37,38,41]. However since the well-known classical theory of the pseudoinverse problems covers inconsistent convex feasibility problem only when the closed convex sets are all given as linear varieties, a new strategy for solving general, possibly inconsistent, convex feasibility problem has been desired for many years. In the last decade, the use of the set $\mathcal{H}_\Phi$ or its generalized version has become one of the most promising strategies because $\mathcal{H}_\Phi$ reflects all closed convex sets C_i's through the weights $w_i's$ and is reduced to $C := \bigcap_{i=1}^{m} C_i$ when $C \neq \emptyset$. Indeed, many parallel algorithmic solutions to the convex feasibility problems have attracted attention not only because of their fast convergence but also because of their convergence to a point in such sets, even in the inconsistent case $C = \emptyset$. (Note: For the other methods to the convex feasibility problem including linear problems, consult the extensive surveys [37,8,10,22,32,100], other papers for example [1,19,20,27,29,30,25,26,89,103,53,38,40,66,70] and references therein.)

B. Fixed Point Characterization of Generalized Convex Feasible Set

Strongly motivated by the positive use of the set $\mathcal{H}_\Phi$, instead of the set $C := \bigcap_{i=1}^{m} C_i$, in the above many parallel algorithmic solutions to convex feasibility problems, the *generalized convex feasible set* K_Φ in Definition 4.1 was firstly introduced in [108,109]. It will be shown that the *generalized convex feasible set* K_Φ can play flexible roles in many applications.

Definition 4.1 (Generalized convex feasible set)
For given nonempty closed convex sets $C_i (\subset \mathcal{H})$ $(i = 1, 2, \ldots, m)$ and $K \subset \mathcal{H}$, define a proximity function $\Phi : \mathcal{H} \to \mathbb{R}$ by

$$\Phi(x) := \frac{1}{2} \sum_{i=1}^{m} w_i d(x, C_i)^2, \tag{50}$$

where $(w_i)_{i=1}^{m} \subset (0, 1]$ and $\sum_{i=1}^{m} w_i = 1$. Then the generalized convex feasible set (or a specially important example of the sets used for the hard-constrained inconsistent signal feasibility problems in [41]): $K_\Phi \subset K$ is defined by

$$K_\Phi := \{u \in K \mid \Phi(u) = \inf \Phi(K)\}.$$

We call the set K the absolute constraint because every point in K_Φ has at least the guarantee to belong to K.

Obviously, $K_\Phi = K \cap (\bigcap_{i=1}^{m} C_i)$ if $K \cap (\bigcap_{i=1}^{m} C_i) \neq \emptyset$. Even if $K \cap (\bigcap_{i=1}^{m} C_i) = \emptyset$, the subset K_Φ, of arbitrarily selected closed convex set K, is well defined as the set of all minimizers

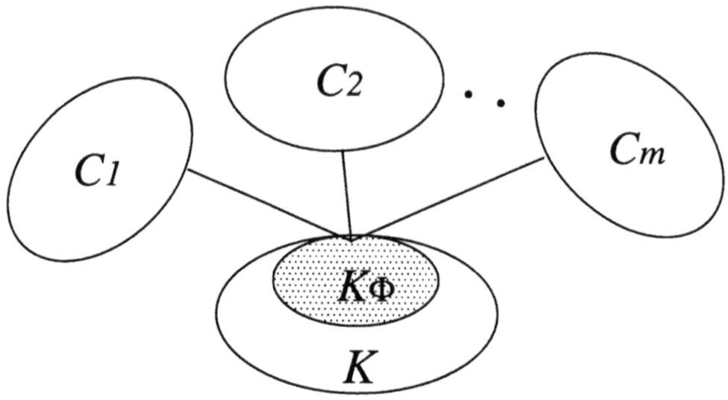

Figure 2. The absolute constraint K and the generalized convex feasible set K_Φ

of Φ over K (see Fig.2), while any point in the set $\mathcal{H}_\Phi$ of (49) has no guarantee to belong to any set among C_i's or K.

Next we introduce two constructions of nonexpansive mapping T satisfying $Fix(T) = K_\Phi$, which makes Theorem 3.2 or Theorem 3.3 (for $N = 1$) applicable to the variational inequality problem over K_Φ. Proposition 4.2(a), the fixed point characterization of K_Φ, was originally shown in [108,109] by applying Pierra's *product space formulation* [84,85] and Fact 2.1(h) to the well-known fact $Fix(P_{C_1}P_{C_2}) = \{x \in C_1 \mid d(x, C_2) = d(C_1, C_2)\}$ for a pair of nonempty closed convex sets C_1 and C_2 [34]. In this paper, for simplicity, we derive this fixed point characterization by combining Fact 2.1(h) and a proof in [41].

Proposition 4.2 (Fixed point characterization of K_Φ in Definition 4.1)

(a) *For any* $\alpha \neq 0$, *it follows*

$$K_\Phi = Fix\left((1-\alpha)I + \alpha P_K \sum_{i=1}^{m} w_i P_{C_i}\right).$$

In particular, $T := (1-\alpha)I + \alpha P_K \sum_{i=1}^{m} w_i P_{C_i}$ *becomes nonexpansive for any* $0 < \alpha \leq 3/2$ [109].

(b) *For any* $\beta > 0$, *it follows*

$$K_\Phi = Fix\left(P_K\left[(1-\beta)I + \beta\sum_{i=1}^{m} w_i P_{C_i}\right]\right).$$

In particular, $T := P_K\left[(1-\beta)I + \beta\sum_{i=1}^{m} w_i P_{C_i}\right]$ *becomes nonexpansive for any* $0 < \beta \leq 2$ [41].

Proof: It is well known that the functional $\frac{1}{2}d(\cdot, C_i)^2 : \mathcal{H} \to \mathbb{R}$ is *Gâteaux* differentiable over $\mathcal{H}$ with derivative $I - P_{C_i}$ (due to [3, Theorem 5.2]), which implies

$$\Phi' = \sum_{i=1}^{m} w_i (I - P_{C_i}) = I - \sum_{i=1}^{m} w_i P_{C_i}$$

is firmly nonexpansive [by Fact 2.1(d)] and monotone over $\mathcal{H}$. By applying Fact 2.4(b) and Proposition 2.5 to $\mathcal{F} := \Phi'$, it follows that

$$K_\Phi = Fix\left(P_K\left[I - \mu\Phi'\right]\right) \text{ for arbitrarily fixed } \mu > 0. \tag{51}$$

The firmly nonexpansiveness of $\Phi' : \mathcal{H} \to \mathcal{H}$ and Fact 2.1(b),(d) ensure that $P_K\left[I - \mu\Phi'\right]$ is nonexpansive for any $0 < \mu \leq 2$. Moreover, application of Fact 2.1(h) to a pair of firmly nonexpansive mappings P_K and $I - \Phi'$ ensures the nonexpansiveness of $(1-\alpha)I + \alpha P_K(I - \Phi')$ for any $0 < \alpha \leq 3/2$, and

$$K_\Phi = Fix\left((1-\alpha)I + \alpha P_K(I - \Phi')\right) \text{ for arbitrarily fixed } \alpha \neq 0.$$

(Q.E.D.)

Remark 4.3

(a) *The condition $K_\Phi \neq \emptyset$ holds if at least one of K and C_i $(i = 1, \ldots, m)$ is bounded [109].*

(b) *Although both nonexpansive mappings in Proposition 4.2 contains commonly the mapping $P_K(\sum_{i=1}^{m} w_i P_{C_i})$, it is not simple to compare these due to the nonlinearity of P_K. On the other hand, it is easy to construct a wider class of nonexpansive mappings whose fixed point set is K_Φ. For example the convex combination of the nonexpansive mappings in Proposition 4.2 and other nonexpansive mapping T with $Fix(T) \supset K_\Phi$ as well becomes such a nonexpansive mapping.*

(c) *Suppose that $\mathcal{F} : \mathcal{H} \to \mathcal{H}$ be κ-Lipschitzian and η-strongly monotone over K. In this case any nonexpansive mapping T satisfying $T(\mathcal{H}) \subset K$ and $Fix(T) = K_\Phi$ is applicable as the mapping T in Theorem 3.3. For example, if we have an attracting nonexpansive mapping T with $Fix(T) = K_\Phi$, we can use $P_K T$ as the nonexpansive mapping in Theorem 3.2 or Theorem 3.3 (for $N = 1$) because of $P_K T(\mathcal{H}) \subset K$ and $Fix(P_K T) = K_\Phi$.*

C. Application to convexly constrained generalized pseudoinverse problem

Convexly constrained linear inverse problems have attracted much attention due to their broad applicability to *numerical linear algebra* (see for example [58, Chap.12],[32, Chap.6]), to *approximation theory* [77,78,35,36] and to *signal processing* [71,86,98,115,114, 64,94–97,99,4,81,100,62].

In particular, it is well known that the following *convexly constrained generalized pseudoinverse problem* unifies a wide class of signal processing problems such as: (i) time or band-limited extrapolation with subspace constraints [71,86,54,83], (ii) image reconstruction with positivity constraint [98,115,114,96,97,99,62], and (iii) signal recovery (or design) [86,64,94,95,4,81] accompanied by constraints on amplitude, support, energy etc.

In this section, we remark that certain class of this important inverse problem can be eventually elucidated by directly applying a theorem of the *Hybrid steepest descent method.*

The *convexly constrained generalized pseudoinverse problem* is formulated as follows.

Problem 4.4 *(K-constrained generalized pseudoinverse problem) Let K be a nonempty closed convex set in $\mathcal{H}$. Denote the standard norm defined in Euclid space $\mathbb{R}^m$ by $\|\cdot\|_{(m)}$. Suppose that*

$$\mathcal{S} := \arg\inf_{x\in K} \|Ax - \boldsymbol{b}\|_{(m)} \neq \emptyset. \tag{52}$$

for a given bounded linear (experiment) operator $A : \mathcal{H} \to \mathbb{R}^m$ and a possibly perturbed measurement $\boldsymbol{b} = (b_1, \ldots, b_m)^T \in \mathbb{R}^m$. Then the problem is

Find a point $\hat{x} \in \arg\inf_{x\in\mathcal{S}} \Theta(x)$,

where $\Theta : \mathcal{H} \to \mathbb{R}$ is a convex function.

Remark 4.5

The condition (52) holds if and only if $\mathcal{P}_{\overline{A(K)}}(\boldsymbol{b}) \in A(K)$, where $\mathcal{P}_{\mathcal{C}} : \mathbb{R}^m \to \mathcal{C}$ is the metric projection, assigns every point in $\mathbb{R}^m$ to its unique nearest point in a nonempty closed convex set $\mathcal{C} \subset \mathbb{R}^m$ [93].

In [51,93], Problem 4.4 is studied for a special cost function $\Theta(x) = \frac{1}{2}\|x\|^2$. In this case, since the uniqueness of the solution of Problem 4.4 is automatically guaranteed under condition (52), by defining $\mathcal{D}(A_K^\dagger) := \{\boldsymbol{b} \in \mathbb{R}^m \mid \mathcal{P}_{\overline{A(K)}}(\boldsymbol{b}) \in A(K)\}$, a mapping $A_K^\dagger : \mathcal{D}(A_K^\dagger) \to K$ named K-constrained pseudoinverse is well defined by

$$A_K^\dagger(\boldsymbol{b}) := \arg\inf_{x\in\mathcal{S}} \|x\|.$$

$\mathcal{D}(A_K^\dagger) = \mathbb{R}^m$ holds if K is bounded. In case $K = \mathcal{H}$, $A_K^\dagger$ reduces to the well-known Moore-Penrose pseudoinverse [76]. In [51,93], it is shown that projected Landweber iteration: $u_{n+1} := P_K(\lambda A^*\boldsymbol{b} + \beta_n(I - \lambda A^*A)u_n)$ (for $\lambda \in (0, 2\|A\|^{-2})$ and $\beta_n := (1 + n^{-c})^{-1}$ with $\beta_0 = 0$ and $0 < c < 1$) generates a sequence (u_n) that converges strongly to $A_K^\dagger(\boldsymbol{b})$ for $\boldsymbol{b} \in \mathcal{D}(A_K^\dagger)$.

By interpreting Problem 4.4 as a simple example of a variational inequality problem over the generalized convex feasible set K_Φ of Definition 4.1, we demonstrate that the *hybrid steepest descent method*, presented in Theorems 3.2 and 3.3, for a nonexpansive mapping T with special weights $(w_i)_{i=1}^m$ in Proposition 4.2, produces a sequence converging strongly to the unique solution of Problem 4.4 when the derivative Θ' is κ-Lipschitzian and η-strongly monotone over $T(\mathcal{H})$.

The following well-known facts bridge the gap between Problem 4.4 and the convex optimization over the generalized convex feasible set K_Φ.

Lemma 4.6

(a) (F.Riesz's representation theorem[113,73]) *For any bounded linear functional $f \in \mathcal{H}^*$ (i.e., bounded linear mapping $f : \mathcal{H} \to \mathbb{R}$), there exists unique $a \in \mathcal{H}$ satisfying*

$$f(x) = \langle a, x \rangle \text{ for all } x \in \mathcal{H}.$$

(b) *For a given hyperplane expressed as $\Pi := \{x \in \mathcal{H} \mid \langle a, x \rangle = b\}$ with $a \in \mathcal{H} \setminus \{0\}$ and $b \in \mathbb{R}$, the distance from every point x^* in $\mathcal{H}$ to Π is given by*

$$d(x^*, \Pi) = \frac{|\langle a, x^* \rangle - b|}{\|a\|}.$$

By using the hyperplanes in Lemma 4.6, we can interpret Problem 4.4 geometrically, which leads to the following algorithmic solution.

Theorem 4.7 *[110] Suppose $a_i \in \mathcal{H}$ $(i = 1, \ldots, m)$ are determined to follow $Ax = (\langle a_1, x \rangle, \ldots, \langle a_m, x \rangle)^T$ (by Lemma 4.6). Let $C_i := \{x \in \mathcal{H} \mid \langle a_i, x \rangle = b_i\}$, and T a non-expansive mapping in Prop. 4.2 with special weights $w_i := \frac{\|a_i\|^2}{\sum_{j=1}^m \|a_j\|^2}$ for $i = 1, 2, \ldots, m$. Suppose that $\Theta : \mathcal{H} \to \mathbb{R}$ is convex and differentiable with derivative Θ' κ-Lipschitzian and η-strongly monotone over $T(\mathcal{H})$. Then for any $\mu \in (0, \frac{2\eta}{\kappa^2})$ and any sequence $(\lambda_n)_{n\geq 1}$ satisfying (L1)-(L3) in $(0,1]$ or (B1)-(B3) in $[0,1]$ (for $N = 1$), the sequence $(u_n)_{n\geq 0}$ generated, with arbitrary $u_0 \in \mathcal{H}$, by*

$$u_{n+1} := T(u_n) - \lambda_{n+1}\mu\Theta'(T(u_n)) \tag{53}$$

converges strongly to the uniquely existing solution of Problem 4.4.

Proof: By defining

$$\Phi(x) := \frac{1}{2}\|Ax - \mathbf{b}\|_{(m)}^2 = \frac{1}{2}\sum_{i=1}^{m}(\langle a_i, x \rangle - b_i)^2 = \frac{1}{2}\sum_{i=1}^{m}\|a_i\|^2 d(x, C_i)^2 \text{ for every } x \in \mathcal{H},$$

it follows $\mathcal{S} = K_\Phi$ (in Definition 4.1).

Application of Theorem 3.2 or Theorem 3.3 (for $N = 1$) to the nonexpansive mapping T in Proposition 4.2 immediately yields the statement of the theorem. (Q.E.D.)

Remark 4.8 *The method in Theorem 4.7 can be applied to Problem 4.4 if a closed form expression $P_K : \mathcal{H} \to K$ is known. As seen in the discussion on the projected gradient method (see Section 2.B), the closed form expression of P_K may not be known in many situations. On the other hand, many convexly constraint inverse problems do not necessarily require any optimality for points in $\arg\inf_{x\in K} \|Ax - \mathbf{b}\|_{(m)}$ (in Problem 4.4). Indeed, for such problems, [42] presented a scheme based on a fixed point theorem [106] with a special inner product corresponding to the problem as well as a scheme based on outer approximation method [43]. In the next section, we will add some remarks on the application of the hybrid steepest descent method to the non-strictly convex minimization over the fixed point set of certain nonexpansive mapping. The result shows that the problem: find a point in $\arg\inf_{x\in K} \|Ax - \mathbf{b}\|_{(m)}$, can also be resolved at least when $\mathcal{H}$ is finite dimensional, by the hybrid steepest descent method if K is characterized as $K = Fix(T)$ with such a nonexpansive mapping T.*

5. CONCLUDING REMARKS

In this paper, we have presented in a simple way the underlying ideas of the *hybrid steepest descent method* as an algorithmic solution to a VIP defined over the fixed point set of a nonexpansive mapping in a real Hilbert space. As seen from the discussion in section 4, the proposed *hybrid steepest descent method* plays notable roles in various inverse problems. Some straightforward applications of the *hybrid steepest descent method* to certain signal estimation or design problems have been shown [63,69].

Our last remark concerns the relaxation of the monotone operator $\mathcal{F}$ in $VIP(\mathcal{F}, Fix(T))$.

Let $T : \mathcal{H} \to \mathcal{H}$ be an attracting nonexpansive mapping satisfying

$$\sup_{\|x\| \geq R} \frac{\|T(x)\|}{\|x\|} < 1 \text{ for some } R > 0. \tag{54}$$

Suppose that $\mathcal{F} : \mathcal{H} \to \mathcal{H}$ is κ-Lipschitzian over $T(\mathcal{H})$ and paramonotone [but not necessarily strictly monotone] over $Fix(T) \neq \emptyset$. Now let's consider the variational inequality problem $VIP(\mathcal{F}, Fix(T))$. Recently we have shown in [80] that

1. *If a nonexpansive mapping $T : \mathcal{H} \to \mathcal{H}$ satisfies the condition (54), then $Fix(T)$ is nonempty closed bounded. Moreover the nonexpansive mapping $T := P_K [\sum_{i=1}^m w_i P_{C_i}]$ in Prop. 4.2 is attracting nonexpansive as well as satisfies the condition (54) if at least one of K and C_i $(i = 1, \ldots, m)$ is bounded (see also Remark 4.3(a)).*

2. *With any $u_0 \in \mathcal{H}$ and any nonnegative sequence $(\lambda_n)_{n\geq 1} \in l^2 \cap (l^1)^C$, the sequence $(u_n)_{n\geq 0}$ generated by the hybrid steepest descent method:*

   $$u_{n+1} := T(u_n) - \lambda_{n+1}\mathcal{F}(T(u_n)) \text{ (see Remark 3.4)} \tag{55}$$

   satisfies $\lim_{n\to\infty} d(u_n, \Gamma) = 0$, where $\Gamma := \{u \in Fix(T) \mid \langle v-u, \mathcal{F}(u)\rangle \geq 0$ for all $v \in Fix(T)\} \neq \emptyset$ (at least when $\mathcal{H}$ is finite dimensional). [Note: The requirement for $(\lambda_n)_{n\geq 1}$ is different from one employed in Theorem 3.3 although the simplest example $\lambda_n = \frac{1}{n}$ $(n = 1, 2, \ldots)$ satisfies both of them. $\Gamma \neq \emptyset$ is guaranteed due to [6, Theorem II.2.6], [52, Prop.II.3.1] or [119, Theorem 54.A(b)]. The closedness and convexity of Γ is also immediately verified by Proposition 2.3.]

Intuitively the above result shows that the *hybrid steepest descent method* in (55) is straightforwardly applicable to the minimization of a non-strictly convex function $\Theta : \mathbb{R}^N \to \mathbb{R}$ [for example $\Theta(x) := \|Ax - \boldsymbol{b}\|^2_{(m)}$ (see Problem 4.4 and Remark 4.8)] as well over the generalized convex feasible set $Fix(T) = K_\Phi$ in Prop. 4.2 when at least one of K and C_i $(i = 1, \ldots, m)$ is bounded. This partial relaxation clearly makes the *hybrid steepest descent method* applicable to significantly wider real world problems including more general convexly constrained inverse problems [80,112].

Acknowledgments

It is my great honor to thank Frank Deutsch, Patrick L. Combettes, Heinz H. Bauschke, Jonathan M. Borwein, Boris T. Polyak, Charles L. Byrne, Paul Tseng, M. Zuhair Nashed, K. Tanabe and U. M. García-Palomares for their encouraging advice at the Haifa workshop (March 2000). I also wish to thank Dan Butnariu, Yair Censor and Simeon Reich for giving me this great opportunity and their helpful comments.

REFERENCES

1. R. Aharoni and Y. Censor: Block-iterative projection methods for parallel computation of solutions to convex feasibility problems, *Linear Algebra an Its Applications* **120** (1989) 165–175.
2. T.M. Apostol: *Mathematical Analysis (2nd ed.)*, (Addison-Wesley, 1974).
3. J.-P. Aubin: *Optima and Equilibra — An Introduction to Nonlinear Analysis*, (Springer-Verlag, 1993).
4. C. Sánchez-Avila, An adaptive regularized method for deconvolution of signal with edges by convex projections, *IEEE Trans. Signal Processing* **42** (1994) 1849–1851.
5. J.-B. Baillon, R. E. Bruck and S. Reich, On the asymptotic behavior of nonexpansive mappings and semigroups in Banach spaces, *Houston J. Math.* **4** (1978) 1–9.
6. V. Barbu and Th. Precupanu, *Convexity and Optimization in Banach Spaces, 3rd ed.*, (D. Reidel Publishing Company, 1986).
7. H.H. Bauschke and J.M. Borwein, Dykstra's alternating projection algorithm for two sets, *J. Approx. Theory* **79** (1994) 418–443.
8. H.H. Bauschke and J.M. Borwein, On projection algorithms for solving convex feasibility problems, *SIAM Review* **38** (1996) 367–426.
9. H.H. Bauschke, The approximation of fixed points of compositions of nonexpansive mappings in Hilbert space, *J. Math. Anal. Appl.* **202** (1996) 150–159.
10. H.H. Bauschke, J.M. Borwein and A.S. Lewis, The method of cyclic projections for closed convex sets in Hilbert space, *Contemp. Math.* **204** (1997) 1–38.
11. H.H. Bauschke and A.S. Lewis, Dykstra's algorithm with Bregman projections: a convergence proof, *Optimization* **48** (2000) 409–427.
12. J.P. Boyle and R.L. Dykstra, A method for finding projections onto the intersection of convex sets in Hilbert spaces, *Advances in Order Restricted Statistical Inference", Lecture Notes in Statistics* (Springer-Verlag, 1985) 28–47.
13. L.M. Bregman, The method of successive projection for finding a common point of convex sets, *Soviet Math. Dokl.* **6** (1965) 688–692.
14. L.M. Bregman, The relaxation method of finding the common point of convex sets and its application to the solution of problems in convex programming, *USSR Computational Mathematics and Mathematical Physics* **7** (1967) 200–217.
15. L. M. Bregman, Y. Censor and S. Reich, Dykstra's algorithm as the nonlinear extension of Bregman's optimization method, *J. Convex Analysis* **6** (1999) 319–333.
16. F.E. Browder, Nonexpansive nonlinear operators in Banach space, *Proc. Nat. Acad. Sci. USA* **54** (1965) 1041–1044.
17. F.E. Browder, Convergence of approximants to fixed points of nonexpansive nonlinear mappings in Banach spaces, *Arch. Rat. Mech. Anal.* **24** (1967) 82–90.
18. F.E. Browder and W.V. Petryshyn, Construction of fixed points of nonlinear mappings in Hilbert space, *J. Math. Anal. Appl.* **20** (1967) 197–228.
19. D. Butnariu and Y. Censor, On the behavior of a block-iterative projection method for solving convex feasibility problems, *International Journal of Computer Mathematics* **34** (1990) 79–94.
20. D. Butnariu and Y. Censor, Strong convergence of almost simultaneous block-iterative projection methods in Hilbert spaces, *Journal of Computational and Applied Mathe-*

matics **53** (1994) 33–42.
21. D. Butnariu, Y. Censor and S. Reich, Iterative averaging of entropic projections for solving stochastic convex feasibility, *Computational Optimization and Applications* **8** (1997) 21–39.
22. D. Butnariu and A.N. Iusem, *Totally Convex Functions for fixed point computation and infinite dimensional optimization* (Kluwer Academic Publishers, 2000).
23. R. E. Bruck and S. Reich, Nonexpansive projections and resolvents of accretive operators in Banach spaces, *Houston J. Math.* **3** (1977) 459–470.
24. C.L. Byrne, Iterative projection onto convex sets using multiple Bregman distances, *Inverse Problems* **15** (1999) 1295–1313.
25. C.L. Byrne and Y. Censor, Proximity function minimization for separable jointly convex Bregman distances, with applications, Technical Report (1998).
26. C.L. Byrne and Y. Censor, Proximity function minimization using multiple Bregman projections, with applications to split feasibility and Kullback-Leibler distance minimization, Technical Report (2000).
27. Y. Censor, Row-action methods for huge and sparse systems and their applications, *SIAM Review* **23** (1981) 444–464.
28. Y. Censor and A. Lent, An iterative row-action method for interval convex programming, *Journal of Optimization Theory and Applications* **34** (1981) 321–353.
29. Y. Censor, Parallel application of block-iterative methods in medical imaging and radiation therapy, *Math. Programming* **42** (1988) 307–325.
30. Y. Censor and T. Elfving, A multiprojection algorithm using Bregman projections in a product space, *Numerical Algorithms* **8** (1994) 221–239.
31. Y. Censor and S. Reich, The Dykstra algorithm with Bregman projections, *Communications in Applied Analysis* **2** (1998) 407–419.
32. Y. Censor and S.A Zenios, *Parallel Optimization: Theory, Algorithms, and Applications* (Oxford University Press, New York, 1997).
33. Y. Censor, A.N. Iusem and S.A. Zenios, An interior point method with Bregman functions for the variational inequality problem with paramonotone operators, *Math. Programming* **81** (1998) 373–400.
34. W. Cheney and A.A. Goldstein, Proximity maps for convex sets, *Proc. Amer. Math.Soc.* **10** (1959) 448–450.
35. C.K. Chui, F. Deutsch and J.W. Ward, Constrained best approximation in Hilbert space, *Constr. Approx.* **6** (1990) 35–64.
36. C.K. Chui, F. Deutsch, and J.W. Ward, Constrained best approximation in Hilbert space II, *J. Approx. Theory*, **71** (1992) 213–238.
37. P.L. Combettes, Foundation of set theoretic estimation, *Proc. IEEE* **81** (1993) 182–208.
38. P.L. Combettes, Inconsistent signal feasibility problems: least squares solutions in a product space, *IEEE Trans. on Signal Processing* **42** (1994) 2955–2966.
39. P.L. Combettes, Construction d'un point fixe commun à une famille de contractions fermes, *C.R. Acad. Sci. Paris Sèr. I Math.* **320** (1995) 1385–1390.
40. P.L. Combettes, Convex set theoretic image recovery by extrapolated iterations of parallel subgradient projections, *IEEE Trans. Image Processing* **6** (1997) 493–506.
41. P.L. Combettes and P. Bondon, Hard-constrained inconsistent signal feasibility prob-

lem, *IEEE Trans. Signal Processing* **47** (1999) 2460–2468.
42. P.L Combettes, A parallel constraint disintegration and approximation scheme for quadratic signal recovery, *Proc. of the 2000 IEEE International Conference on Acoustics, Speech and Signal Processing* (2000) 165–168.
43. P.L. Combettes, Strong convergence of block-iterative outer approximation methods for convex optimization, *SIAM J.Control Optim.* **38** (2000) 538–565.
44. G. Crombez, Finding projections onto the intersection of convex sets in Hilbert spaces, *Numer. Funct. Anal. Optim.* **15** (1996) 637–652.
45. F. Deutsch and H. Hundal, The rate of convergence of Dykstra's cyclic projections algorithms: the polyhedral case, *Numer. Funct. Anal. Optim.* **15** (1994) 537–565.
46. F. Deutsch and H. Hundal, The rate of convergence for the method of alternating projections II, *J. Math. Anal. Appl.* **205** (1997) 381–405.
47. F. Deutsch and I. Yamada, Minimizing certain convex functions over the intersection of the fixed point sets of nonexpansive mappings, *Numer. Funct. Anal. Optim.* **19** (1998) 33–56.
48. W.G. Dotson, On the Mann iterative process, *Trans. Amer. Math. Soc.* **149** (1970) 65–73.
49. N. Dunford and J.T Schwartz, *Linear Operators Part I: General Theory*, Wiley Classics Library Edition, (John Wiley & Sons, 1988).
50. R.L. Dykstra, An algorithm for restricted least squares regression, *J. Amer. Statist. Assoc.* **78** (1983) 837–842.
51. B. Eicke, Iteration methods for convexly constrained ill-posed problems in Hilbert space, *Numer. Funct. Anal. Optim.* **13** (1992) 413–429.
52. I. Ekeland and R. Themam, *Convex Analysis and Variational Problems*, Classics in Applied Mathematics 28 (SIAM, Philadelphia, PA, USA, 1999).
53. U.M. García-Palomares, Parallel projected aggregation methods for solving the convex feasibility problem, *SIAM J.Optim.* **3** (1993) 882–900.
54. R.W Gerchberg, Super-restoration through error energy reduction, *Optica Acta* **21** (1974) 709–720.
55. K. Goebel, and W.A. Kirk, *Topics in Metric Fixed Point Theory* (Cambridge Univ. Press, 1990).
56. K. Goebel and S. Reich, *Uniform Convexity, Hyperbolic Geometry, and Nonexpansive Mappings* (Dekker, New York and Basel, 1984).
57. A.A. Goldstein, Convex programming in Hilbert space, *Bull. Amer. Math. Soc.* **70** (1964) 709–710.
58. G.H. Golub and C.F.Van Loan, *Matrix computations 3rd ed.* (The Johns Hopkins University Press, 1996).
59. L.G. Gubin, B.T. Polyak, and E.V. Raik, The method of projections for finding the common point of convex sets, *USSR Computational Mathematics and Mathematical Physics* **7** (1967) 1–24.
60. B. Halpern, Fixed points of nonexpanding maps, *Bull. Amer. Math. Soc.* **73** (1967) 957–961.
61. S.-P. Han, A successive projection method, *Math. Programming* **40** (1988) 1–14.
62. J.C. Harsanyu and C.I. Chang, Hyperspectral image classification and dimensionality reduction: an orthogonal subspace projection approach, *IEEE Trans. Geosci. Remote*

Sensing **28** (1994) 779–785.

63. H. Hasegawa, I. Yamada and K. Sakaniwa, A design of near perfect reconstruction linear phase QMF banks based on Hybrid steepest descent method, *IEICE Trans. Fundamentals* **E83-A** (2000), to appear.
64. M.H. Hayes, M.H. Lim, and A.V. Oppenheim, Signal reconstruction from phase or magnitude, *IEEE Trans. Acoust., Speech, Signal Processing* **28** (1980) 672–680.
65. H. Hundal and F. Deutsch, Two generalizations of Dykstra's cyclic projections algorithm, *Math. Programming* **77** (1997) 335–355.
66. A.N. Iusem and A.R. De Pierro, Convergence results for an accelerated nonlinear Cimmino algorithm, *Numerische Mathematik* **49** (1986) 367–378.
67. A.N. Iusem and A.R. De Pierro, On the convergence of Han's method for convex programming with quadratic objective, *Math. Programming* **52** (1991) 265–284.
68. A.N. Iusem, An iterative algorithm for the variational inequality problem, *Comp. Appl. Math.* **13** (1994) 103–114.
69. M. Kato, I. Yamada and K. Sakaniwa, A set theoretic blind image deconvolution based on Hybrid steepest descent method, *IEICE Trans. Fundamentals* **E82-A** (1999) 1443–1449.
70. K.C. Kiwiel, Free-steering relaxation methods for problems with strictly convex costs and linear constraints, *Mathematics of Operations Research* **22** (1983) 326–349.
71. D.P. Kolba and T.W. Parks, Optimal estimation for band-limited signals including time domain consideration, *IEEE Trans. Acoust., Speech, Signal Processing* **31** (1983) 113–122.
72. G.M. Korpolevich, The extragradient method for finding saddle points and other problems, *Ekonomika i matematicheskie metody* **12** (1976) 747–756.
73. E. Kreyszig *Introductory Functional Analysis with Applications*, Wiley Classics Library Edition, (John Wiley & Sons, 1989).
74. E.S. Levitin and B.T. Polyak, Constrained Minimization Method, *USSR Computational Mathematics and Mathematical Physics* **6** (1966) 1–50.
75. P.L. Lions, Approximation de points fixes de contractions, *C. R. Acad. Sci. Paris Sèrie A-B* **284** (1977) 1357–1359.
76. D.G. Luenberger, *Optimization by Vector Space Methods* (Wiley, 1968).
77. C.A. Micchelli, P.W. Smith, J. Swetits, and J.W. Ward, Constrained L_p Approximation, *Constr. Approx.* **1** (1985) 93–102.
78. C.A. Micchelli and F. Utreras, Smoothing and interpolation in a convex set in a Hilbert space, *SIAM J.Sci. Statist. Comput.* **9** (1988) 728–746.
79. J. von Neumann, Functional operators, Vol. II. The Geometry of Orthogonal Spaces, *Annals of Math. Studies* **22** (Princeton University Press, 1950) [Reprint of mimeographed lecture notes first distributed in 1933].
80. N. Ogura and I. Yamada, Non-strictly convex minimization over the fixed point set of the asymptotically shrinking nonexpansive mapping, *submitted for publication* (2001).
81. S. Oh, R.J. Marks II, and L.E. Atlas, Kernel synthesis for generalized time frequency distribution using the method of alternating projections onto convex sets, *IEEE Trans. Signal Processing* **42** (1994) 1653–1661.
82. Z. Opial, Weak convergence of the sequence of successive approximations for nonexpansive mapping, *Bull. Amer. Math. Soc.* **73** (1967) 591–597.

83. A Papoulis, A new algorithm in spectral analysis and band limited extrapolation, *IEEE Trans. Circuits and Syst.* **22** (1975) 735–742.
84. G. Pierra, Eclatement de contraintes en parallèle pour la minimisaition d'une forme quadratique, *Lecture Notes Computer Sci.* **40** (1976) 200–218.
85. G. Pierra, Decomposition through formalization in a product space, *Math. Programming* **28** (1984) 96–115.
86. L.C. Potter and K.S. Arun, Energy concentration in band-limited extrapolation, *IEEE Trans. Acoust., Speech, Signal Processing* **37** (1989) 1027–1041.
87. S. Reich, Weak convergence theorems for nonexpansive mappings in Banach spaces, *J. Math. Anal. Appl.* **67** (1979) 274–292.
88. S. Reich, Some problems and results in fixed point theory, *Contemp. Math.* **21** (1983) 179–187.
89. S. Reich, A limit theorem for projections, *Linear and Multilinear Algebra* **13** (1983) 281–290.
90. J.B. Rosen, The gradient projection method for non-linear programming, Part I. Linear Constraints, *SIAM J. Applied Mathematics* **8** (1960) 181–217.
91. J.B. Rosen, The gradient projection method for non-linear programming, Part II. Nonlinear Constraints, *SIAM J. Applied Mathematics* **9** (1961) 514–532.
92. W. Rudin, *Real and complex analysis*, 3rd Edition, (McGraw-Hill, 1987).
93. A. Sabharwal and L.C. Potter, Convexly Constrained Linear Inverse Problems: Iterative Least-Squares and Regularization, *IEEE Trans. on Signal Processing* **46** (1998) 2345–2352.
94. J.L.C. Sanz and T.S. Huang, Continuation techniques for a certain class of analytic functions, *SIAM J. Appl. Math.* **44** (1984) 819–838.
95. J.L.C. Sanz and T.S. Huang, A unified approach to noniterative linear signal restoration, *IEEE Trans. Acoust., Speech, Signal Processing* **32** (1984) 403–409.
96. J.J. Settle and N.A. Drake, Linear mixing and the estimation of ground cover proportions, *Int. J. Remote Sensing* **14** (1993) 1159–1177.
97. J.J. Settle and N. Campbell, On the error of two estimators of sub-pixel fractional cover when mixing is linear, *IEEE Trans. Geosci. Remote Sensing* **36** (1998) 163–170.
98. Y.S. Shin and Z.H. Cho, SVD pseudoinversion image reconstruction, *IEEE Trans. Acoust., Speech, Signal Processing* **29** (1989) 904–909.
99. B. Sirkrci, D. Brady and J. Burman, Restricted total least squares solutions for hyperspectral imagery, *Proceedings of 2000 IEEE International Conference on Acoustics, Speech, and Signal Processing* **1** (2000) 624–627.
100. H. Stark and Y. Yang, *Vector Space Projections: A Numerical Approach to Signal and Image Processing, Neural Nets, and Optics* (John Wiley & Sons Inc, 1998).
101. W. Takahashi, Fixed point theorems and nonlinear ergodic theorem for a nonexpansive semigroups and their applications, *Nonlinear Analysis* **30** (1997) 1283–1293.
102. W. Takahashi, *Nonlinear Functional Analysis: Fixed Point Theory and its Applications* (Yokohama Publishers, 2000).
103. K. Tanabe, Projection method for solving a singular system of linear equations and its applications, *Numerische Mathematik* **17** (1971) 203–214.
104. P. Tseng, On the convergence of the products of firmly nonexpansive mappings, *SIAM J. Optim.* **2** (1992) 425–434.

105.P. Tseng, Dual coordinate ascent methods for non-strictly convex minimization, *Math. Programming* **59** (1993) 231–247.
106.R. Wittmann, Approximation of fixed points of nonexpansive mappings, *Arch. Math.* **58** (1992) 486–491.
107.P. Wolfe, Finding the nearest point in a polytope, *Math. Programming* **11** (1976) 128–149.
108.I. Yamada, N. Ogura, Y. Yamashita and K. Sakaniwa, An extension of optimal fixed point theorem for nonexpansive operator and its application to set theoretic signal estimation, *Technical Report of IEICE* **DSP96-106** (1996) 63–70.
109.I. Yamada, N. Ogura, Y. Yamashita and K. Sakaniwa, Quadratic optimization of fixed points of nonexpansive mappings in Hilbert space, *Numer. Funct. Anal. Optim.* **19** (1998) 165–190.
110.I. Yamada, Approximation of convexly constrained pseudoinverse by Hybrid Steepest Descent Method, *Proceedings of 1999 IEEE International Symposium on Circuits and Systems*, **5** (1999) 37–40.
111.I. Yamada, Convex projection algorithm from POCS to Hybrid steepest descent method (in Japanese), *Journal of the IEICE* **83** (2000).
112.I. Yamada, Hybrid steepest descent method and its applications to convexly constrained inverse problems, The annual meeting of the American Mathematical Society, New Orleans, January 2001.
113.K. Yosida, *Functional analysis*, 4th Edition, (Springer-Verlag, 1974).
114.D.C. Youla, Generalized image restoration by the method of alternating orthogonal projections, *IEEE Trans. Circuits and Syst.* **25** (1978) 694–702.
115.D.C Youla and H. Webb, Image restoration by the method of convex projections: Part 1–Theory, *IEEE Trans. Medical Imaging* **1** (1982) 81–94.
116.E.H Zarantonello, ed., *Contributions to Nonlinear Functional Analysis* (Academic Press, 1971).
117.E. Zeidler, *Nonlinear Functional Analysis and its Applications, I - Fixed-Point Theorems* (Springer-Verlag, 1986).
118.E. Zeidler, *Nonlinear Functional Analysis and its Applications, II/B - Nonlinear Monotone Operator* (Springer-Verlag, 1990).
119.E. Zeidler, *Nonlinear Functional Analysis and Its Applications, III - Variational Methods and Optimization* (Springer-Verlag, 1985).

www.ingramcontent.com/pod-product-compliance
Ingram Content Group UK Ltd.
Pitfield, Milton Keynes, MK11 3LW, UK
UKHW050949290926
15470UKWH00005B/1726

9 780444 505958